Advances in Underwater Technology, Ocean Science and Offshore Engineering

Volume 28

Offshore Site Investigation and Foundation Behaviour

ADVANCES IN UNDERWATER TECHNOLOGY, OCEAN SCIENCE AND OFFSHORE ENGINEERING

Vol. 1. Developments in Diving Technology
Vol. 2. Design and Installation of Subsea Systems
Vol. 3. Offshore Site Investigation
Vol. 4. Evaluation, Comparison and Calibration of Oceanographic Instruments
Vol. 5. Submersible Technology
Vol. 6. Oceanology
Vol. 7. Subsea Control and Data Acquisition
Vol. 8. Exclusive Economic Zones
Vol. 9. Stationing and Stability of Semi-submersibles
Vol. 10. Modular Subsea Production Systems
Vol. 11. Underwater Construction: Development and Potential
Vol. 12. Modelling the Offshore Environment
Vol. 13. Economics of Floating Production Systems
Vol. 14. Submersible Technology: Adapting to Change
Vol. 15. Technology Common to Aero and Marine Engineering
Vol. 16. Oceanology '88
Vol. 17. Energy for Islands
Vol. 18. Disposal of Radioactive Waste in Subsea Sediments
Vol. 19. Diverless and Deepwater Technology
Vol. 20. Subsea International '89: Second Generation Subsea Production Systems
Vol. 21. NDT: Advances in Underwater Inspection Methods
Vol. 22. Subsea Control and Data Acquisition: Technology and Experience
Vol. 23. Subtech '89. Fitness for Purpose
Vol. 24. Advances in Subsea Pipeline Engineering and Technology
Vol. 25. Safety in Offshore Drilling. The Role of Shallow Gas Surveys
Vol. 26. Environmental Forces on Offshore Structures and their Prediction
Vol. 27. Subtech '91. Back to the Future
Vol. 28. Offshore Site Investigation and Foundation Behaviour
Vol. 29. Wave Kinematics and Environmental Forces
Vol. 30. Subsea International '93: Low Cost Subsea Production Systems

CONFERENCE PLANNING COMMITTEE

D.A. Ardus, *British Geological Survey*; D. Clare, *Ove Arup & Partners*; A. Hill, *BP Exploration*; R. Hobbs, *Lloyds Register*; R. Jardine, *Imperial College*; J. Pritchard, *Society for Underwater Technology*; J. Squire, *BP Exploration.*

Advances in Underwater Technology, Ocean Science and Offshore Engineering

Volume 28

Offshore Site Investigation and Foundation Behaviour

Papers presented at a conference
organized by the Society for Underwater Technology
and held in London, UK, September 22–24, 1992.

edited by
D. A. Ardus
British Geological Survey, Edinburgh
D. Clare
Ove Arup & Partners, London
A. Hill
BP Exploration, Aberdeen
R. Hobbs
Lloyds Register, London
R.J. Jardine
Imperial College of Science, Technology & Medicine, London
J.M. Squire
BP Exploration, Aberdeen

KLUWER ACADEMIC PUBLISHERS
DORDRECHT / BOSTON / LONDON

Library of Congress Cataloging-in-Publication Data

Offshore site investigation and foundation behaviour / edited by
D.A. Ardus, D. Clare, A. Hill, R. Hobbs, R.J. Jardine and
J.M. Squire.
p. cm. -- (Advances in underwater technology, ocean science,
and offshore engineering ; v. 28)
Includes bibliographical references.
ISBN 0-7923-2363-7 (alk. paper)
1. Ocean engineering--Congresses. 2. Foundations--Congresses.
I. D.A. Ardus (et al.). II. Series.
TC1505.0336 1993
620'.4162--dc20 93-8336

ISBN 0-7923-2363-7

Published by Kluwer Academic Publishers,
P.O. Box 17, 3300 AA Dordrecht, The Netherlands.

Kluwer Academic Publishers incorporates
the publishing programmes of
D. Reidel, Martinus Nijhoff, Dr W. Junk and MTP Press.

Sold and distributed in the U.S.A. and Canada
by Kluwer Academic Publishers,
101 Philip Drive, Norwell, MA 02061, U.S.A.

In all other countries, sold and distributed
by Kluwer Academic Publishers Group,
P.O. Box 322, 3300 AH Dordrecht, The Netherlands.

Printed on acid-free paper

Printed in the Netherlands

Society for Underwater Technology

The Society was founded in 1966 to promote the further understanding of the underwater environment. It is a multi-disciplinary body with a worldwide membership of scientists and engineers who are active or have a common interest in underwater technology, ocean science and offshore engineering.

Committees

The Society has a number of Committees to study such topics as:

Diving and Submersibles
Offshore Site Investigation and Geotechnics
Environmental Forces and Physical Oceanography
Ocean Resources
Subsea Engineering and Operations
Education and Training

Conference and Seminars

An extensive programme is organized to cater for the diverse interests and needs of the membership. An annual programme usually comprises four conferences and a much greater number of one-day seminars plus evening meetings and an occasional visit to a place of technical interest. The Society has organized over 100 seminars in London, Aberdeen and other appropriate centres during the past decade. Attendance at these events is available at significantly reduced levels of registration fees for Members or staff of Corporate Members.

Publications

Proceedings of the more recent conferences have been published in this series of *Advances in Underwater Technology, Ocean Science and Offshore Engineering.* These and other publications produced separately by the Society are available through the Society to members at a reduced cost. A careers pack 'Oceans of Opportunity' has been produced by the Society in response to the growing demand by students schools and colleges for up-to-date information.

Journal

The Society's quarterly journal *Underwater Technology* caters for the whole spectrum of the inter-disciplinary interests and professional involvement of its readership. It includes papers from authoritative international sources on such subjects as:

Diving Technology and Physiology
Civil Engineering
Submersible Design and Operation
Geology and Geophysics

Subsea Systems
Naval Architecture
Marine Biology and Pollution
Oceanography
Petroleum Exploration and Production
Environmental Data

An Editorial Board has responsibility for ensuring that a high standard of quality and presentation of papers reflects a coherent and balanced coverage of the Society's diverse subject interests; through the Editorial Board, a procedure for assessment of papers is conducted.

Endowment fund

A separate fund has been established to provide tangible incentives to students to acquire knowledge and skills in underwater technology or related aspects of ocean science and offshore engineering. Postgraduate students have been sponsored to study to MSc level and subject to the growth of the fund it is hoped to extend this activity.

Awards

An annual President's Award is presented for a major achievement in underwater technology. In addition there is a series of sponsored annual awards by some Corporate Members for the best contribution to diving operations and oceanography, and for the best technical paper in the Journal

FURTHER INFORMATION

If you would like to receive further details, please contact
Society for Underwater Technology, The Memorial Building, 76 Mark Lane,
London EC3R 7JN.
Telephone: 071-481 0750; Telex: 886481 I Mar E G; Fax: 071-481 4001.

TABLE OF CONTENTS

SESSION 3: ADVANCED INTERPRETATION TECHNIQUES

SESSION 4: INTEGRATED INTERPRETATIONS

SESSION 5: GRAVITY FOUNDATIONS

SESSION 6: FOUNDATIONS PERFORMANCE MONITORING

SESSION 7: PILING RESEARCH

SESSION 8: DESIGN CRITERIA

SESSION 1

INTRODUCTION

OPENING ADDRESS

PROFESSOR T. D. PATTEN
Chairman, Marine Technology Directorate Ltd., 19 Buckingham St., London WC2N 6EF, U.K.

Mr. Chairman, Ladies and Gentlemen, it is an honour for me to open this SUT International Conference on "Offshore Site Investigation and Foundation Behaviour" and a privilege for me to welcome you here on behalf of the Planning Committee and of the Organisers. A meeting such as this with its range of underlying science and engineering is an ideal opportunity for exchange between professionals of the different disciplines involved.

The conference has been convened to consider developments in two related areas of activity. The first is concerned with the need for an integrated approach to the use of geological, geophysical and geotechnical data in the determination of site conditions. The second is to review foundation behaviour in the light of field experience, tests, monitoring and research.

Contributions concerning geotechnical sampling and testing include appraisal of new developments, improvements in understanding the effects of sampling on soil properties and conditions in frontier areas.

Derivation of geotechnical information from high resolution seismic data, advanced geophysical interpretation techniques and case studies on integrated interpretations are illustrated.

The evolution and performance of a number of novel forms of gravity structure are appraised and developments in piling research are presented.

A considerable amount of research has been done by industry and by universities into the behaviour of offshore foundations including offshore monitoring at full scale, in large scale trials onshore, by means of intensive research in small scale experiments in the field and laboratory, and through theoretical work.

One of the aims of the conference is to draw all this information together, much of it previously confidential, and to encourage open debate in the hope of promoting more efficient interpretation, design and development.

In all, the Planning Committee selected 33 papers, twenty from industrial companies, five from academia and the remaining eight are the result of collaboration between industrial and academic authors. Together you authors and other participants are a powerful source of expertise, potentially capable of new and profitable ideas, given the right interactive environment.

Although I confess to being completely out of my depth in matters geological and geophysical, papers by Paul *et al* in Session 2 and Haynes *et al* in Session 3

Volume 28: Offshore Site Investigation and Foundation Behaviour, 3–5, 1993.

highlight the relationship between the geotechnical and the seismic characteristics of sediments but, as Stoker *et al* point out in Session 4, the interpretation of seismic facies in terms of specific lithological and geotechnical characteristics can be problematic.

However, when the subject gets on to research of foreseeable application to the exploitation of offshore oil and gas, I feel marginally more at ease, and certainly more at home in the case of SERC/MTD funded research for application to practice.

For example, in Session 6 Stock *et al* use the results of a long term monitoring programme, developed from SERC/MTD funded research, on the Hutton tension leg platform. In the same session Sharp and Kenley summarise a joint industry research programme concerned with the monitoring of the Magnus foundation which included SERC/MTD funded elements, while in Session 7 papers by Bond and Jardine consider studies of piles in clay and sand, respectively. Recent axial load tests on piles in clay, which have improved the understanding of the physical processes involved, have enabled the validity of established design procedures to be reviewed.

While small scale tests in sand have also been carried out, recent changes in API RP2A recommendations, which have not been accepted in North Sea practice, emphasise the need for large-scale tests in sand similar to those carried out in the BP large diameter pile tests in clay, described by Clark and Lambson in Session 7, which more clearly replicate offshore conditions.

In Session 8 Hobbs reviews design and certification of offshore piles, and the evolution of pile design codes is considered by Toolan and Horsnell. Changes may come with the introduction of Eurocodes and it is important that experienced practitioners have full involvement in the drafting process to ensure acceptability.

Session 5 considers aspects of gravity based structures which have been used to support topside facilities in the central and northern North Sea for many years, particularly in the Norwegian sector. Recently Hamilton Brothers Oil & Gas reversed this trend with the first gravity based structure in the UK southern North Sea in the Ravenspurn Field.

O'Riordan and Seaman describe a key feature of the design, the optimisation of drainage systems to limit build up of water pressure in the foundation sand, with consequent economy in the required self-weight and ballast of gravity structures.

Christophersen describes a novel application of gravity foundations to provide anchorages for the Snorre tension leg platform in soft soils in 300m water depth, which is noteworthy for its simplicity and cost effectiveness.

Increasing use of gravity foundations in shallower water requires a better understanding of the scouring of the near surface foundation sediments. Model tests are often used to study the problem but cannot adequately resolve the poor understanding of the scouring process. This is an area where joint industry research funding could be appropriate.

New foundation concepts assessed or proposed, including gravity based tension leg platforms, lightweight gravity structures, skirt piled gravity structures and suction caissons, leave much scope for new research.

The Marine Technology Directorate Ltd., of which I am Chairman, is the channel to University investigators for research funds in the multi-disciplinary area of marine technology. In the case of managed programmes of work, investigators often secure additional funds and support from companies or Government Agencies/Departments with oil and gas involvements. Since 1985, MTD has funded 18 projects related to Offshore Site Investigations through SERC, of which two are still running. This represents a total of £ 835,000 of SERC money and £ 315,000 of other sponsorship.

The two current MTD projects are being conducted by researchers at Imperial College and Glasgow University. At Imperial College, the behaviour of offshore piles is being examined with a view to filling vital gaps in existing knowledge, concerning piles in dense sand and low plasticity clays. The Glasgow project is looking at the biological strengthening of marine sediments.

It is MTD policy to encourage quality research proposals, but we can only consider applications for research into Offshore Site Investigation where there is an explicit connection with Marine Technology preferably related to offshore oil and gas, otherwise the application has to be dealt with by another committee of the Science and Engineering Research Council.

Mr. Chairman, I congratulate you and your Committee on your planning and for providing the basis for this potentially exciting conference, and to all the delegates I wish you success in accepting the challenge it offers.

KEYNOTE ADDRESS: OFFSHORE FOUNDATION SAFETY

M. BIRKINSHAW
Offshore Safety Division, Health and Safety Executive

The title of this keynote address is 'Offshore Foundation Safety'. This is a wide subject and could be controversial, however nothing I am going to say is particularly new rather I am going to present old information with a new emphasis. I will be concentrating on the UK offshore sector but the principles that I will be elaborating on are the same worldwide.

Like all good systems the offshore regulatory regime is one that is constantly evolving particularly following the major change of emphasis following the Cullen Report (Reference 1). This change involves as one of its facets a far more explicit risk evaluation approach to safety issues than previously adopted and is a challenge to engineers that are involved with the more traditional structural and geotechnical disciplines to demonstrate and document how hazards are identified and assessed and effectively communicate this in what is called a *safety case*.

The objective of this conference as stated in the programme are:

- to address the need for an integrated approach to the use of geological, geophysical and geotechnical data in determining site conditions; and
- to review foundation behaviour in light of field experience tests, monitoring and research.

To meet these broad objectives the organising committee have put together a three day programme containing papers on many aspects of the above. This is the third such international conference to be held with these two objectives in mind the others being in 1979 and 1985. I will be drawing extensively from addresses to these latter two conferences. The objectives are laudable and broad and it is to be hoped that the first objective – that of integration – is becoming a reality, as it seems to dominate any conversation in this field, with all agreeing the necessity but having more difficulty in achieving the reality!

In view of the objectives the question may be asked as to what relevance, other than my insatiable appetite for knowledge, is this conference to me in my new found role of risk assessor? This leads on to the questions:

Volume 28: Offshore Site Investigation and Foundation Behaviour, 7–14, 1993.

- How does technology fit in with safety?
- How do geotechnical/geological/geophysical specialists assess and *communicate the risks* associated with their technology?
- In other words how safe are we?

I would like to emphasis that the thoughts, view and asides I make are my own. To answer these questions in the short time available to me I would like to present an overview of what we have learnt from the past conferences, and then suggest a few topics that may influence the future. I will not dwell on the many minor miracles of technology that the fraternity have performed in the offshore sector as I cannot improve on the recent paper given at BOSS'92 by STATOIL (Reference 2). This makes impressive reading of just how much advancement has been made in the last thirty years and those responsible for these technical achievements are to be congratulated. Perhaps to their current disadvantages the non technologist has come to expect a continuation of these achievements for little if any effort! With this in mind there is a need to explain and communicate the safety and risk message and I will concentrate on the themes from the past and possible themes for the future in the specific area of *formal safety assessment*.

First and foremost I would like to emphasis that I am not aware of anything to suggest that offshore foundations are unsafe. After all nothing has suffered significant failure on the UK Continental Shelf through foundation failure (with the notable exception of the special case of the first Christchurch Bay Tower (Reference 3). However this statement on the lack of failure is in itself not helpful as I am reminded of what I have named 'the Sevenoak Tree syndrome'. As all analogies, it is not perfect but I think it is sufficient to make the point. Before the extreme storm of October 1987 there were seven large oaks on the green in Sevenoaks, Kent but after the storm there was only one! In other words since we have not had the design storm we cannot make substantive statements on safety in the extreme design conditions let alone for the total range of hazards that may affect foundation safety. There is nothing new in this statement as it was also given by the Chairman of the Planning Committee for this conference – Mr. Ardus – in an article on the 1979 conference (Reference 4). Of course the situation is not as unknown as perhaps I am making out and full scale data obtained from offshore, to which HSE has played its part and continues to play its part, gives some reassurance and confidence in existing safety levels for extreme weather design events (Reference 5 and 6).

However, how do we communicate the case for safety between ourselves and to other, perhaps less technical orientated parties? This communication is the very essence of Safety Case assessment in the proposed new legislation (Reference 7). It involves an understanding of safety assessment. As in all disciplines safety assessment is seen at its best when the methods etc. involved are recognised ones. This is called *formal* safety assessment. The fundamentals of safety assessment involve:

– hazard identification
– hazard assessment

There are many tools available for hazard identification most of which have their origins in the chemical industry but, with a little ingenuity, can be adapted for offshore structures and foundation aspects (Reference 8).

Hazard Identification

In the review of the 1979 conference of Dr. Burland (Reference 9) remarked that one had to be careful in using the term hazard and obviously there was much debate on trying to take away the then doomwatch association with the word. Various alternatives were suggested (feature, problem, anomaly). Today I do not see the reluctance to talk about hazards. All things are hazardous to some extent and whether they are trivial or have the potential to cause a major accident they are still hazards. Thus I dispense with the apparent sensitivity that surrounded the term in 1979.

I am not here to give a lecture on hazard identification but a basic rule is that formalisation of the process will reduce the risk of omitting hazards. Dr. Burland's 1979 review states a most important feature in support of formal hazard identification: 'the cause of failures are often not those things one has remembered and got wrong but those that one has forgotten about',

Checklists are a good start in hazard identification. The Guidance to the proposed new Safety Case Regulations (Reference 10) has such a list with foundation failure being amongst the hazards. Obviously, a subsidiary checklist for foundation aspects is required. This may depend on structure and foundation type. As a little exercise I have compiled a list from the past conferences.

1. Muir-Wood (1979) (Reference 11)
   - design uncertainties
   - dynamic behaviour
   - computational dexterity compensating
   - relative paucity of reliable information

2. Burland (1979) (Reference 9)
   - shallow gas pockets
   - seismicity (earthquakes)
   - submarine slope instability
   - design uncertainties
     = soil strength sampling
     = soil strength interpretation
     = awareness of model limitations

3. Ardus (1979) (Reference 4)
   - pock marks
   - buried channels

4. Marsland (1985) (Reference 12)
   - design uncertainties
   - non homogeneity of soil
   - interpretation

I am sure that a more thorough review of the papers presented would unearth other hazards and that we all have others that we would add to the list. Other hazards that I have found in the literature are:

1. Preslan/Merrill (1983) (Reference 13)
   - mudslides

2. Campbell (1991) (Reference 14)
   - landslides
   - active faults
   - gas hydrates
   - rocky topography
   - sea floor erosion
   - unusual soils

I am relieved to say that the last two references were dealing specifically with the deep water Gulf of Mexico but may become important on the western margins of the UK continental shelf.

Some hazards that I would definitely like to include are:

- scour
- overload behaviour (settlement/pullout)
- punch through
- foundation/foundation interaction
- specific design uncertainties
  = preload effectiveness
  = fixity
  = cyclic effects

And of course probably one of the main hazards is lack of knowledge of soil conditions at the site of interest for whatever reason.

Hazard Assessment

Having arrived at our list of hazards the job is now one of assessment. As you will see there is a need to fully understand the impact of the hazard on the chosen design. This, of course will call upon the skills of geologists, geophysicists and geotechnical engineers if the hazard is to be fully described and the potential effectiveness of various means of mitigation evaluated. I am not saying anything new here. It was said by Professor Muir-Wood, Dr. Woodland and Dr. Burland in 1979 and just in case you missed it then, it was repeated by Wroth (Reference 15) and Green *et al* (Reference 16) in 1985.

Hazard can be assessed in many ways and many tools brought to bear on the assessment. It is beyond the scope of this address to be exhaustive but let me try and give a flavour of what I see as involved.

Inherent in assessment is the fact that sufficient information is available. In most cases this will require *quantitative and qualitative* knowledge and it should be self explanatory that it should include *site investigation* somewhere along the line. Not just any old investigation but a planned one orientated towards the specifics of the engineering tasks in hand and broad enough to allow for changes to this plan. This, once again, is not new and was emphasised by Dr. Woodland (Reference 17) in his closing address to the 1979 Conference where he made a plea for *standardisation of techniques* and looked to Committees such as the SUT Offshore Site Investigation and Geotechnics Committee to take the lead. This obviously had some effect as Professor Wroth was able to report in 1985 on 'a steady and marked development since 1979 in reliability of equipment, maturity of industry and the growing confidence in results'. HSE is playing its role as our own guidance document is currently being updated.

I cannot emphasise this point enough. One cannot hope to effectively assess a hazard on the basis of insufficient information.

Having got our data we must now be aware of the warning given by Muir-Wood and Burland at the 1979 Conference – *over sophistication* of models compared to the data. Once again nothing new here. I found a useful reference on this topic in the 1973 Rankine Lecture (Reference 18). I have taken the illustration from Lambe's paper (Figure 1). I suggest that the model is true for all engineering disciplines and occasionally we should shake the dust from the illustration and by plotting some disciplines on it (e.g. fluid loading; fatigue calculations; foundation design techniques) we may get some insights for the direction of future research effort. I emphasis this area as, in the absence of failure experience, it is the risk associated with *how our models represent reality that we are assessing*.

A further way of assessing some hazards is, of course, by the use of recognised codes and standards. There is a need to establish harmonisation of methodologies and techniques in this area.

Whichever way one chooses to assess the hazards, two things are important – *being methodical and the ability to communicate*. I read with some interest in

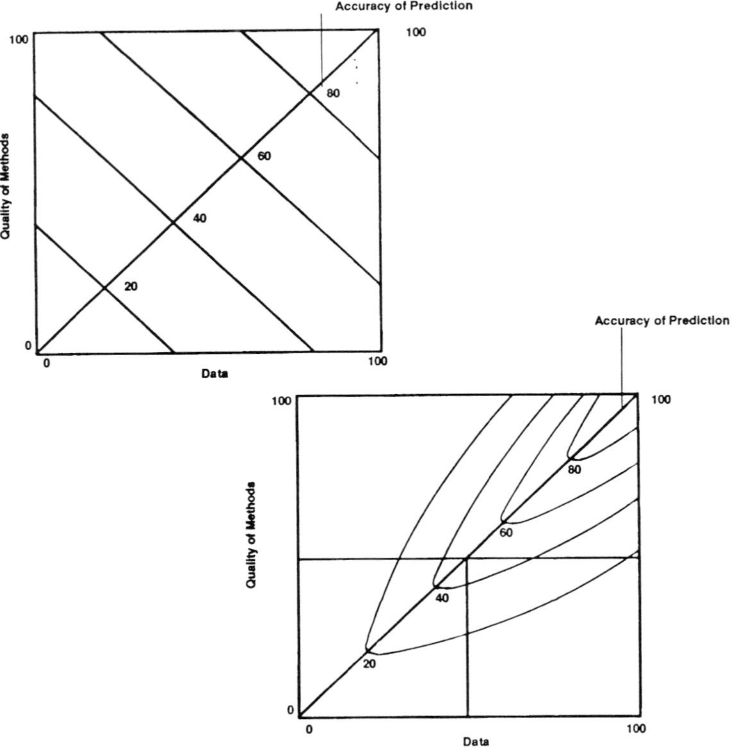

Consistency in the sophistication of:

- **Method of prediction**

  and

- **Quality of data employed**

Fig. 1. Accuracy of prediction (Lambe, 1973).

the Offshore Engineer (Reference 19) that at the recent Conference on Piling in Clays it was pronounced that design techniques for offshore platforms are ‘valid, if slightly conservative’. (Note that the quotes are those of the magazine and the quotation is not attributed to any particular body or individual). This somewhat puzzled me as there appears to be a wide range of existing design techniques with significant differences between our American colleagues through API and UK practice even for the extreme storm condition let alone for all the other hazards that require assessment. This shows that a balance needs to be struck between the over simple statement and the lengthy statement that communicates nothing more than confusion.

Where does it all lead. There is no doubt that the post Cullen regulatory regime with emphasis on Safety Cases will require an improvement of hazard identification and assessment techniques as applied to foundation issues. There is still some way to go in this area. There is a need for a joint geological/geophysical/geotechnical approach to this issue and perhaps this is what has long been awaited to bring about the integration of these closely related disciplines.

Other topics that will probably affect the future and in which HSE is actively involved are in two area

- **Codes and Standards**
- **Research**

I do not need to tell any practising engineer of the movement into limit state and reliability based foundation codes on the European scene (Reference 20) and of the load and resistance type codes being developed on the international scene (Reference 21). There is a need for much communication and harmonisation in these areas if the implementation of these types of codes and standards is to be successful.

In the research areas, we are particularly concerned with developing our understanding of jack up foundation behaviour and the modes of failure. Other areas of activity include seismic assessment of structures. If you require any further information on our research programme, I am sure Martin Thompson will be happy to oblige.

We are looking forward to closer dialogue with the Industry on research and technology issues, particularly with SUT's Offshore Site Investigation and Geotechnics Committee and looking forward to seeing issues in their 'statement of research needs' published in 1987 (Reference 21) coming through to working practice. Perhaps it may be opportune to review this document in the light of the new regulatory emphasis that I have described here. I see no room for complacency on the research front as this has enabled so many of the innovations achieved in the last thirty years to take place and I see no lessening of the innovative pressure.

I started my address by posing the question 'How safe are we'. As you will have gathered I have not given an answer but I hope I have provided a means whereby the question can be answered. I see that all the major ingredients of formal risk assessment as applied to foundations are available. What is required next is to put them in place!

I wish you well in the next three days.

References

1. Cullen, The Hon Lord (1990), 'The Public Inquiry into the Piper Alpha Disaster', HMSO.
2. Tjelta, T. I. (1992), 'Historical overview of geotechnical design in the North Sea', *BOSS '92*, London.

3. Burnett, S. J. (1979), 'OSFLAG project: Christchurch Bay Tower', *Offshore Research Focus* **14**, August 1979.
4. Ardus, D. A. (1979), 'Rare Foundation Failures No Cause for Complacency', Offshore Services, August 1979.
5. (1992) 'Magnus Foundation Monitoring Project', *Conf. on Recent Large Scale Fully Instrumented Pile Tests in Clay*, Session 2, Institution of Civil Engineers, London.
6. Brekke, J. N., Campbell, R. B., Lamb, W. C., and Murff, J. D. (1990), 'Calibration of a jack up structural analysis procedure using measurement from a North Sea jack up', *Proc. Offshore Technology Conference*, Houston, OTC 6465.
7. Caldwell, S., (1992), 'Safety case: Setting the agenda for the 1990s', *Proc. I Mar E/RINA Conference on Offshore Safety*.
8. McIntosh, A. R., Birkinshaw, M. (1992), 'The offshore safety case: Structural considerations', *Proc. ERA Technology Conference*.
9. Burland, J. (1979), 'Conference review and appraisal', *Proc. of Conference on Offshore Site Investigations*, SUT, London.
10. Health and Safety Executive (1992), 'Draft Offshore Installations (Safety Case) Regulations 199-', HMSO.
11. Muir-Wood, A. M., (1979), 'Opening address', *Proc. of Conference on Offshore Site Investigations*, SUT, London.
12. Marsland, A. (1985), 'The influence of geological processes and test procedures of measured and evaluated parameters', *Proc. of International Conference on Offshore Site Investigations*, SUT, London.
13. Preslan, W. L. and Merrill, K. S. (1983), 'Design must deal with mudslide problems', *Offshore* **43**(3).
14. Campbell, K. J. (1981), 'Deepwater geohazards: An engineering challenge', *Offshore*, October 1981.
15. Wroth, C. P. (1985), 'Summary', *Proc. of International Conference on Offshore Site Investigations*, SUT, London.
16. Green, C. D., Heijna, B., and Walker, P. (1985), 'An integrated approach to the investigation of new development areas', *Proc. of International Conference on Offshore Site Investigations*, SUT, London.
17. Woodland, A. W. (1979), 'Closing address', *Proc. of Conference on Offshore Site Investigations*, SUT, London.
18. Lambe, T. W. (1973), 'Predictions in soil engineering', The Rankine Lecture, *Gétechnique* **XXIII**(2).
19. The Offshore Engineer, July 1992.
20. Institution of Civil Engineers and The British Geotechnical Society (1992), 'Limit State Design in Geotechnics – Will EC7 Work?', Notes for Half Day Meeting, London.
21. Thomas, G. A. N. and Snell, R. O. (1992), 'Application of API RP2A LRFD to a North Sea platform structure', *Proc. Offshore Technology Conference*, Houston, OTC 6932.
22. Offshore Site Investigation and Geotechnics Committee, SUT (1987), 'Statement of Research Needs in Offshore Foundation Design', *SUT Underwater Technology* **13**(2).

THE USE OF EXPLORATION GEO-SCIENCE DATA IN THE PLANNING AND EXECUTION OF SITE INVESTIGATIONS FOR OFFSHORE DEVELOPMENT FACILITIES

M. R. COOK
Hydrosearch Associates Limited, Chandler House, Anchor Hill, Knaphill, Woking, Surrey GU21 2NL

J. M. SQUIRE
BP Exploration, 301 St. Vincent Street, Glasgow G2 5DD

and

A. W. HILL
BP Exploration, Farburn Industrial Estate, Dyce, Aberdeen AB2 0PD

Abstract. Prior to development of an offshore hydrocarbon field large sums are spent by Operating Companies on the acquisition and processing of exploration geo-science data. These data, augmented by other publicly available data, can be cost-effectively integrated and reviewed to aid field development conceptual design, selection of suitable sites for development facilities and planning of detailed, site-specific, geotechnical and geophysical site investigations.

This paper illustrates the range of geo-science data that normally exists and the uses to which such data can be put. A case history of two field development sites in the southern North Sea is presented to emphasise the benefits of such a data review and the pitfalls that can occur if such a review is not undertaken.

The benefits of multi-use of existing data are summarised. Recommendations are made for future geo-science data acquisition to provide further information for subsequent field development site investigations.

1. Introduction

Before a decision is made to develop an offshore hydrocarbon field Operating Companies invest large sums on exploration and field appraisal in the acquisition and processing of geo-science data. The costs incurred are, obviously, highly variable and dependent on the nature of the field being appraised. Figure 1 provides a simplified estimate of the magnitude of costs incurred.

Much of the top-hole (upper 1000 metres of geological section) information that is derived from these activities can be re-used for site evaluation purposes in the conceptual design phase of a field development. The data can also be used to aid cost effective planning and execution of detailed geotechnical and geophysical site investigations.

Volume 28: Offshore Site Investigation and Foundation Behaviour, 15–36, 1993.

	£'s (pounds sterling)
2D SEISMIC SURVEY - ACQUISITION & PROCESSING	500,000
3D SEISMIC SURVEY - ACQUISITION & PROCESSING	2,500,000
DRILLING 5 WELLS @ £3,500k PER WELL (incl. logging/site surveys etc)	17,500,000
TOTAL COST PER FIELD	**20,500,000**

Fig. 1. Typical costs of acquiring geo-science data during the exploration and appraisal phases of a North Sea field.

This paper illustrates how existing exploration and high-resolution geo-hazard seismic data, augmented by top-hole well information and other available geo-science data, can be effectively integrated to provide a database for planning and selection of suitable sites for development facilities.

The paper is split into two parts. Part 1 illustrates the range of data that is available, and the information that can be produced from integrating such data. Part 2 comprises a case history of two southern North Sea gas fields.

In conclusion, the benefits of front-end site investigation data integration studies are summarised and minimum predicted cost savings resulting from such studies are presented. In addition, recommendations are made for future geo-science data acquisition to aid future field developments from a site investigation standpoint.

2. Existing Geo-Science Data Sources

By the time an Operating Company decides to develop a hydrocarbon field, a wealth of geo-science data will have been collected over or adjacent to the area during exploration and appraisal. These data will include:

- exploration seismic (2D and 3D);
- high resolution seismic for drilling rig site surveys;
- debris clearance surveys;
- well drilling reports;
- top-hole well logs (wireline and mud logs);
- soil borings (undertaken for jack-up rigs or other purposes);
- peripheral field data (possibly acquired by and available from other Operators);

– internal reports on specific aspects of exploration or drilling.

Additional data may have been acquired, or compiled, by other organisations such as:

– Research Institutes (British Geological Survey (BGS), Deacon Laboratories);
– Universities;
– Government Departments (Ministry of Defence – Hydrographic Department);
– UKOOA (e.g. pipelines database);
– Fishing Organisations;
– Local Authorities.

Review and integration of these data can produce a database of information to aid front-end conceptual design of development facilities and detailed site investigation planning.

3. Results of Geo-Science Data Integration

BP Exploration (BPX) has recognised the cost and planning benefits of multi-use of such geo-science data for the planning and execution of detailed site investigations for offshore development facilities. Since 1986 several data integration desk studies have been performed in advance of detailed site investigations (Figure 2).

The information used for site investigation planning is wide-ranging and can be compiled/interpreted from the diverse array of geo-science data sources indicated earlier.

Most data integration reports compiled to date have included the following maps and geological cross-sections:

– existing data coverage
– bathymetry and seabed morphology
– seabed sediments and features
– relevant soils isopachs
– soil provinces
– geo-hazards
– 'shallow' and 'deep' geological cross-sections.

In addition, seismic processing velocities and downhole checkshot data have been used to compile representative and calibrated top-hole seismic time to depth conversion curves for the field areas.

The main sources of information used to compile the maps and sections are summarised and briefly described below:

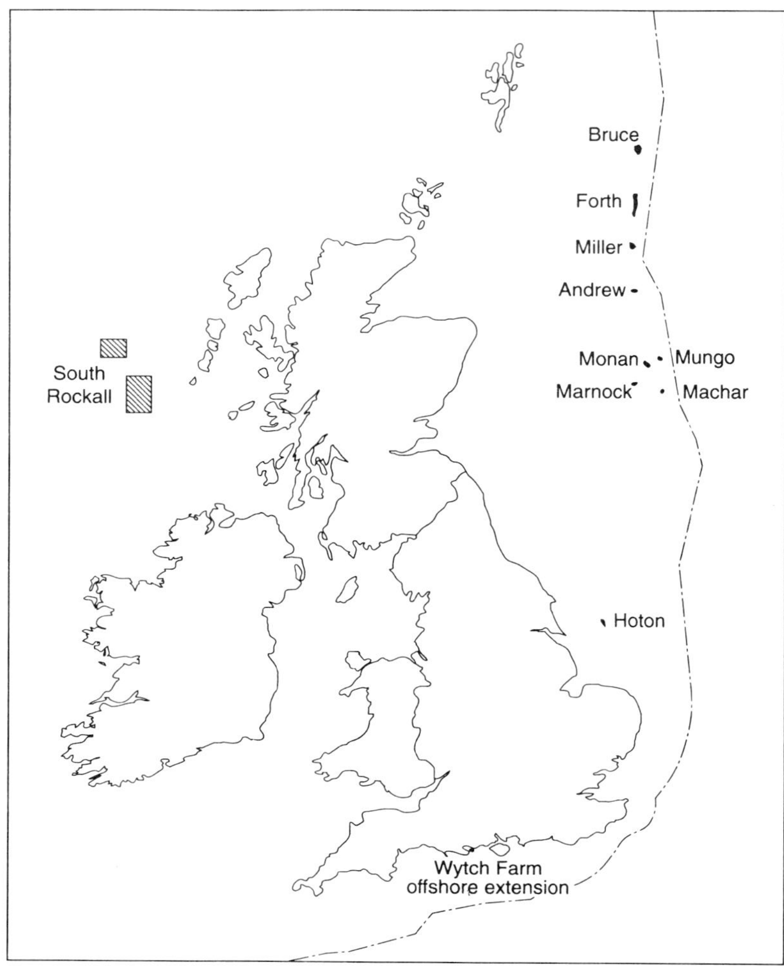

Fig. 2. Location of site investigation data review studies performed by BP Exploration 1986–1992.

Bathymetric and seabed morphology data are mainly derived from drilling rig site surveys. Bathymetry is mapped on all such surveys without exception. This is supplemented in areas where no site survey data coverage is available by water depths recorded on exploration seismic sections and data taken from Admiralty Charts. Water depths given on exploration seismic sections are notoriously unreliable and are used with caution. Whilst water depths are routinely recorded on all exploration surveys it is not unusual for little attention to be paid to such information.

Seabed sediments and seabed features data can be obtained from a variety of sources. Seabed sediments are usually sampled on drilling rig site surveys. Large quantities of information have also been gathered by research institutes and government bodies; particularly BGS. In addition, experience in anchoring semi-submersible drilling rigs or spud-can penetrations for jack-up rigs, is frequently recorded, and can be used for broad assessment of the geotechnical properties of the seabed soils where actual soils data are not available.

Data on seabed features are available from a host of sources including site surveys, debris clearance surveys, the 'Admiralty Wrecks Database', British Telecom International and Fishing Organisations. However, some of these data can be unreliable in terms of feature origin and positioning and should therefore be treated with caution. In addition, seabed features data are 'out-of-date' within a short period of being acquired. Nevertheless, it is obviously important from a field development standpoint to know if a ship wreck, telephone cable etc. is located in the general area of interest. Such a discovery some time into the field development can have undesirable consequences.

Foundation soils (i.e. upper 100 metres of geological section) information is primarily gleaned from soil borings (if any exist), analogue and digital seismic site survey data and data acquired by research institutes and government departments; particularly BGS. In certain instances top-hole well log data is available from shallow depths sub-seabed. However, this is relatively rare for wells drilled on the U.K.C.S., and the reliability of such data is often suspect. Shallow soils data can be integrated to produce isopachs of relevant lithological or geotechnical units.

Where data allows, it is often possible to construct a predictive Foundation Soils Model which is often presented in plan form as a Soil Province Map. A Soil Province is defined as "an area within which soil conditions to some specified depth are generally the same or within a relatively narrow range" (Campbell, 1984). An example of such a map prepared for a northern North Sea field is presented as Figure 3. Four main Soil Provinces were identified over this field. Often such provinces are sub-divided into sub-provinces; a sub-province being an area where soil conditions are generally similar to those within the province, but differ by some identifiable, and mappable aspect. Schematic vertical soils profiles are prepared for each province and sub-province, and are shown on the Soil Province Maps. Prior to such studies preliminary geotechnical investigations may have been performed and it is possible to use the soil boring data from these to calibrate the Foundation Soils Model. Nevertheless, due to the complex nature of the soils that are often encountered, many of the predicted provinces and sub-provinces cannot be calibrated. Therefore, an element of interpretation is used to predict the likely lithologies and geotechnical properties of the soils in such areas.

The vertical relationships of foundation soils can be presented in the form of shallow geological cross-sections compiled from interpreting and integrating the available geophysical data. These are calibrated, where possible, by any existing geotechnical or geological soil boring data.

In all cases, in the UK sector of the North Sea, Quaternary sediments identified

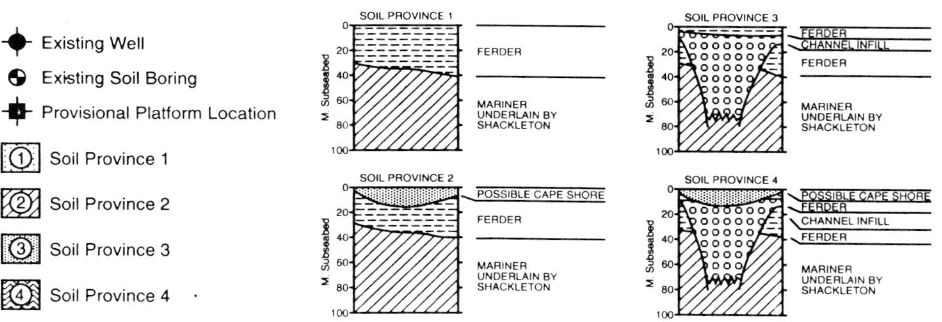

Fig. 3. Simplified soil province map, northern North Sea.

on seismic records are correlated with seismo-stratigraphic Quaternary formations recognised by the BGS. Where possible, the likely lithologies and geotechnical properties common to these 'type' formations are used in the construction of the Foundation Soils Model. In this respect, extensive use is made of the 1:250,000 offshore geology sheets (Quaternary geology, Solid geology and Seabed Sediments) published by the BGS.

Where data exists and geology allows it is sometimes possible to predict Foundation Soil lithologies using exploration seismic and well log data, as will be illustrated in the Case History that follows.

Geo-hazards. An essential consideration for platform site selection and detailed site investigation planning, is the identification of 'geo-hazards' in relation to platform foundations and top-hole drilling. Geo-hazards include the following:

- faulting,
- shallow gas,
- buried channelling,
- boulder beds.

It is usually possible to provide good information on potential geo-hazards from an integration of existing high resolution seismic, exploration seismic, well log data and drilling reports for previously drilled wells. An example of a geo-hazard chart is illustrated on Figure 4. Identification of geo-hazards not only highlights areas that should be avoided in platform site selection but also provides invaluable information for planning detailed site investigations. In the case of one central North Sea Field no high resolution seismic data had previously been acquired over the preferred platform site area. However, an extensive exploration 2D data-set had been acquired. These data were reviewed and evidence of high amplitude seismic anomalies possibly indicating shallow gas at approximately 600ms TWT were apparent (Figure 5).

On the basis of this, a provisional drilling casing design programme was devised for the field with the 20" casing shoe being set above the potential gas level. This enabled a site investigation work scope to be specified which comprised a stand-alone Ultra-high Resolution (UHR) digital seismic survey (0.5ms sample interval, 4×20 cu. in sleeve gun, shallow towed source and streamer), and allowed a conventional high resolution seismic survey to be waived at a cost saving of approximately £50,000 in 1992 terms. As can be seen from the UHR data example (Figure 6) there is a marked increase in resolution of the amplitude anomaly providing very detailed information on the nature and extent of the potential gas hazard. In addition, the UHR data provided high quality information on the distribution of the foundation soils; in particular, in the 'twilight' zone between conventional very resolute analogue seismic data penetration and less resolute conventional high resolution seismic data (i.e. 30–100 metres below seabed).

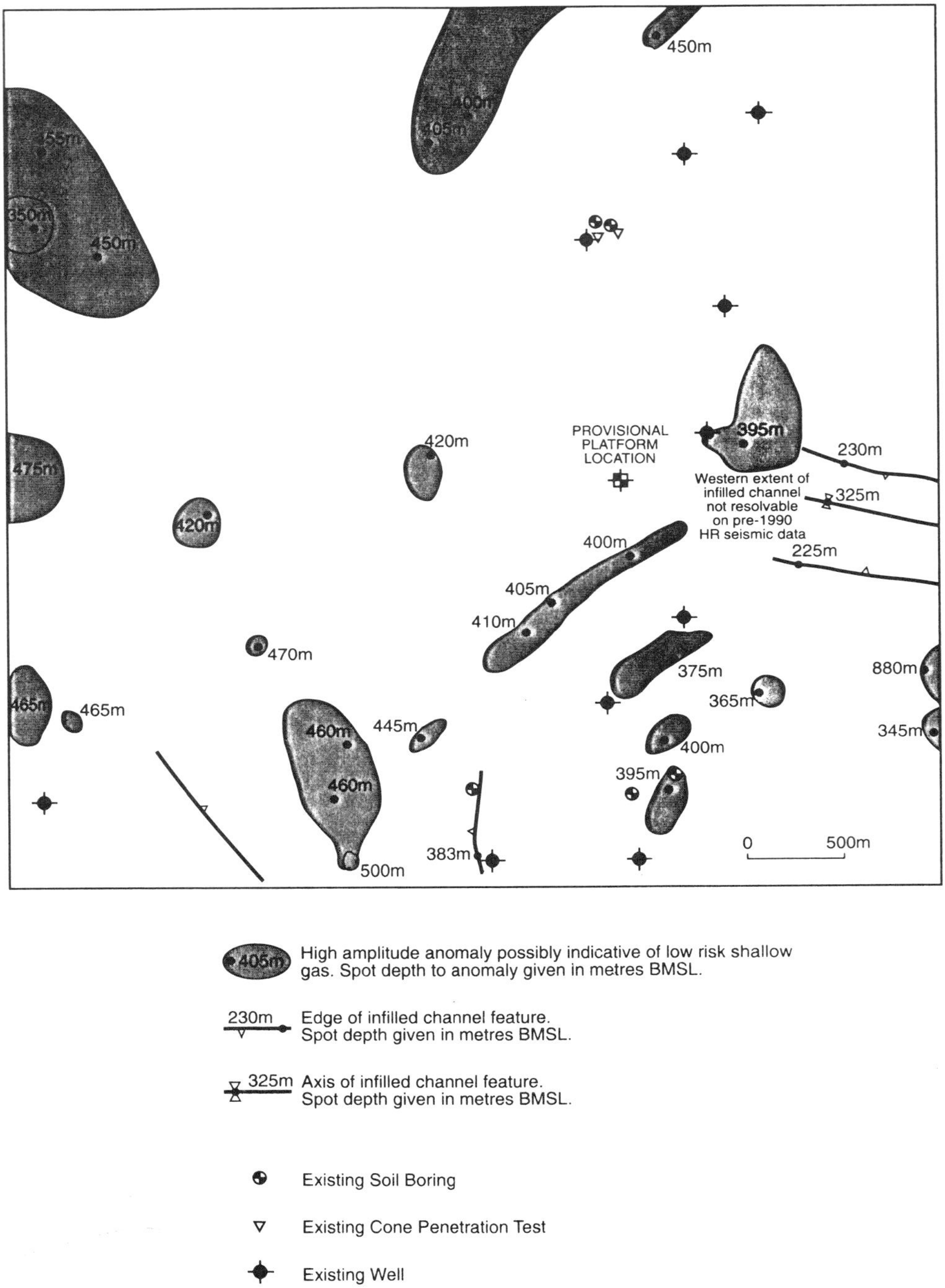

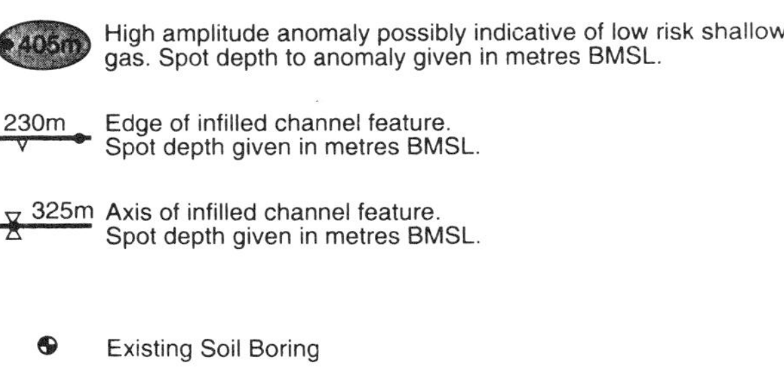

Fig. 4. Seismic anomalies and "deep" infilled channels, northern North Sea

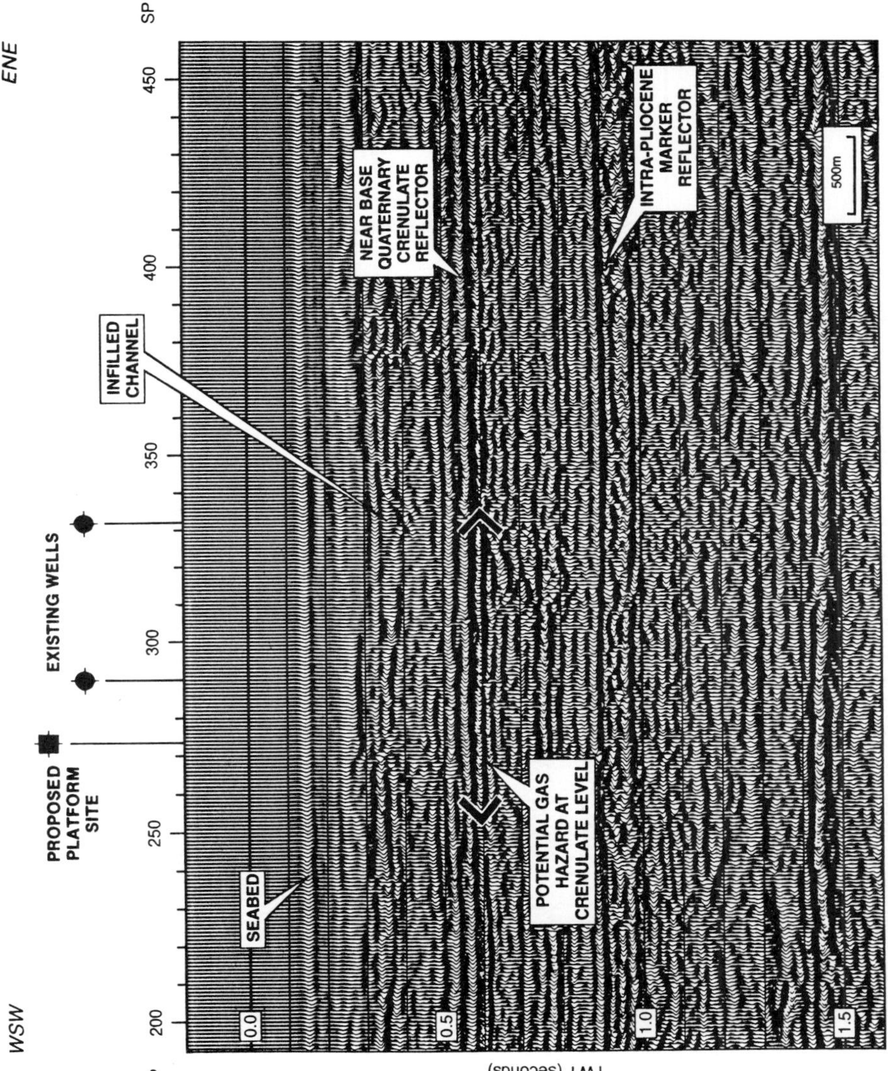

Fig. 5. 2D exploration seismic line through potential platform site, central North Sea, 60-fold migrated stack.

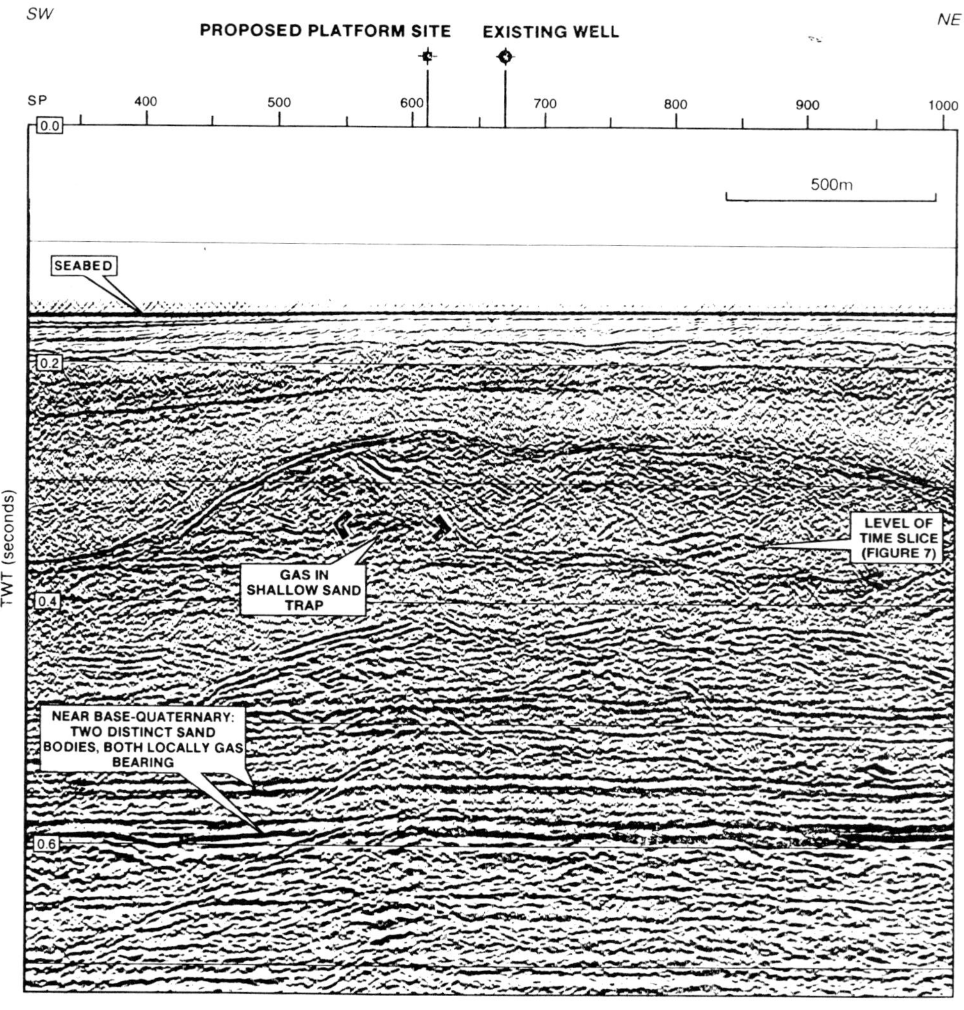

Fig. 6. Ultra-high resolution seismic line through proposed platform site, central North Sea, 48-fold migrated stack.

It is now common practise for a 3D exploration seismic survey to be acquired over a hydrocarbon field prior to development. As these data-sets become more widely available they will become increasingly useful in pre-development site investigation data review studies for identification of potential geo-hazards. Figure 7 shows buried channels imaged on 3D seismic survey data and is an example of the sort of information that can be extracted from existing exploration data.

Data coverage. One of the most important maps to result from a geo-science data review is that showing existing data coverage. It is common to find that different departments within an Operating Company are unaware of (a) the extent of coverage and type of geo-science data that exists over 'their' Field and (b) the potential usefulness of these data from a site investigation standpoint.

Data reviews undertaken to date have made extensive use of computer mapping and electronic production of charts. This facilitates amendment and update of information for use by the 'project team' as detailed site investigation results are acquired. By collating the results of the data integration into a formal report, a comprehensive database of site investigation information is produced. This generally has a wide circulation within an Operating Company and the uses to which the results are put have been found to be many and varied.

4. Case History: Two Southern North Sea Gas Fields

Some of the benefits of conducting a data review study and some of the potential pitfalls in *not* conducting such a study, are well illustrated in the following Case History.

4.1. BACKGROUND

The Case History describes two, as yet undeveloped, southern North Sea gas fields. Final site investigations have been performed for *Field A*. The development scenario being proposed at the time of the site investigation required results to be delivered within a very short time frame. The schedule did not permit a detailed data review study to be performed in advance of the field work. Adverse ground conditions for the driven pile foundation scheme proposed were encountered at the preferred platform location, and a series of additional borings had to be drilled, first to locate a suitable site, and subsequently to investigate it in detail.

Final platform site investigations have not yet been performed for *Field B*. However, the shallow geology of the field is expected to be very similar to that of Field A, and a site investigation data review study has been performed. A primary objective of the study was to ensure that the problems encountered in selecting a suitable platform location at Field A were not experienced at Field B. An integral part of the study was the preparation of a 'Platform Installation Constraints' chart

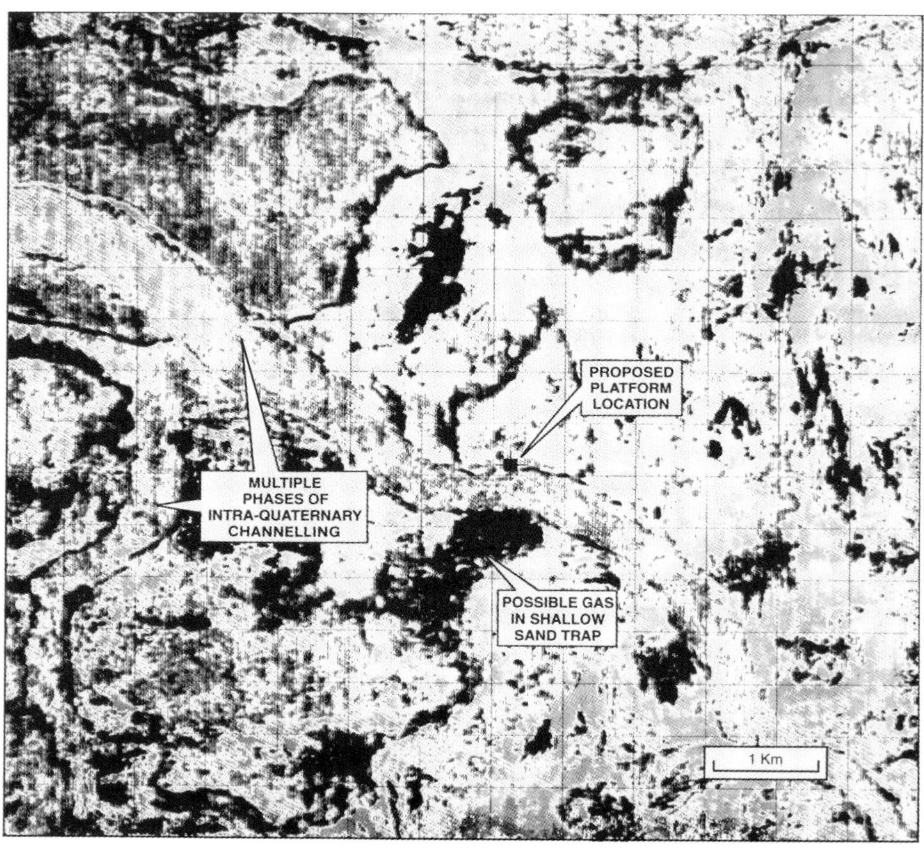

Fig. 7. 3D seismic timeslice at 350ms TWT, central North Sea.

highlighting those parts of the fields most, and least, suited to a piled foundation scheme.

4.2. Field A

Field A lies on the northwestern limb of a north-northwesterly plunging anticline. Prior to commencement of the site investigations in December 1989, it was expected that soil conditions would comprise 10–15m of Quaternary sediments, overlying Mesozoic bedrock. The Quaternary sequence was expected to consist predominantly of stiff to hard glacial clays, while underlying Mesozoic strata were forecast to comprise very hard clays with limestone interbeds of Middle Jurassic age. This very general appraisal was based on the results of previous shallow geotechnical borings drilled in the field and geological boreholes drilled in the general vicinity of the site by the BGS. No attempt was made to map the distribution of Quaternary or pre-Quaternary formations in any detail by integrating the results of previous site surveys with borehole and top-hole well log data.

Instructions to proceed with detailed geophysical and geotechnical site investigations were received, at short notice, in late-November 1988. The position of the proposed platform location to be investigated was supplied by the Asset Team. In selecting this location, no reference was made to likely seabed or shallow geological conditions. The unavailability of a site survey vessel meant that field work for the geophysical investigation could not commence until mid-December. Bad weather meant that the full survey work scope had still not been completed by mid-January.

With the imminent arrival onsite of the geotechnical drilling vessel, the outstanding part of the survey, the UHR digital seismic survey, was cancelled. Less than twenty-four hours after the departure of the survey vessel from location, the geotechnical investigation was commenced.

The first boring at the preferred platform site encountered well cemented sandstone, at 11m below seabed (BSB). These ground conditions were unsuitable for a driven pile foundation scheme. Attempts were made to select an alternative location following inspections of sub-tow boomer and mini-airgun data. A second boring encountered moderately strong limestone at 7m BSB, conditions that were, again deemed unsuitable. The drillship moved onto a third borehole location. Results here indicated that the site was marginal, but just acceptable, for driven piles, and this site was adopted as the platform location. Three further borings revealed variable, but broadly similar ground conditions.

The extra costs of this extended geotechnical programme are estimated as 2 days of drillship time with all associated ancillary costs (circa £80,000 at 1992 prices). Weather conditions during the period of the investigation were reasonably favourable, which helped to keep the supplementary costs to a minimum. The Authors believe that a detailed evaluation of the site in advance of mobilising any survey vessels would have significantly reduced costs.

Following the field work, subsequent integration of the geophysical and geotechnical data has enabled a detailed Subcrop Chart, showing the variation in the Middle Jurassic sediments that underlie the Quaternary, to be prepared. The site is underlain by rocks belonging to the West Sole Group, with sediments of the overlying Oxford Clay Formation probably subcropping just to the west of the survey area. Lithologies of the West Sole Group sediments range from well-cemented sandstones and limestones, in the centre and west of the site, to very hard clays and weak to moderately strong mudstones, with thin sandstone/limestone intercalations, which occur within a 500–600m strip across the eastern side of the site. The former lithologies are unsuitable for the installation of driven piles, while the latter are considered more acceptable foundation materials. On this basis, an 'Installation Constraints Chart' (Figure 8) was prepared to illustrate the parts of the survey area in which the most, and least, favourable conditions for pile driving might be found. The original platform location, and the first alternative site both lie within the 'unsuitable' area. The platform site finally adopted, is located on the eastern margin of an area considered to be of marginal suitability for pile installation. With the information now available, the Authors believe that significantly better foundation conditions would have been encountered at a site located approximately 300m west of the final platform location.

It is the Authors' contention that the absence of a detailed data review study in advance of final site investigations at Field A resulted in: (a) a geotechnical site investigation costing a minimum of £80,000 (in 1992 terms) more than it need otherwise have done and (b) adoption of a less-than-ideal platform site, for which a compromise foundation scheme has had to be utilised.

4.3. Field B

Field B lies about 15km east of Field A. When development of Field B was first considered, it was immediately apparent that broadly similar shallow geological conditions to those encountered at Field A could be present, as the two fields lie at similar stratigraphic levels on opposing flanks of the anticline referred to above (Figure 9). A proposal was therefore put to the Project Team to review all existing shallow seismic, borehole and top-hole well information in the vicinity of the field in order to highlight any shallow geological or geotechnical conditions that might adversely affect the siting, or installation, of development facilities. The total cost of the study was about £20,000 (in 1992 terms).

The review study considered an area 8km by 4km, within which six wells had previously been drilled. Data utilised for the study included:

- top-hole well data;
- seabed and shallow seismic survey data from six rig site surveys;

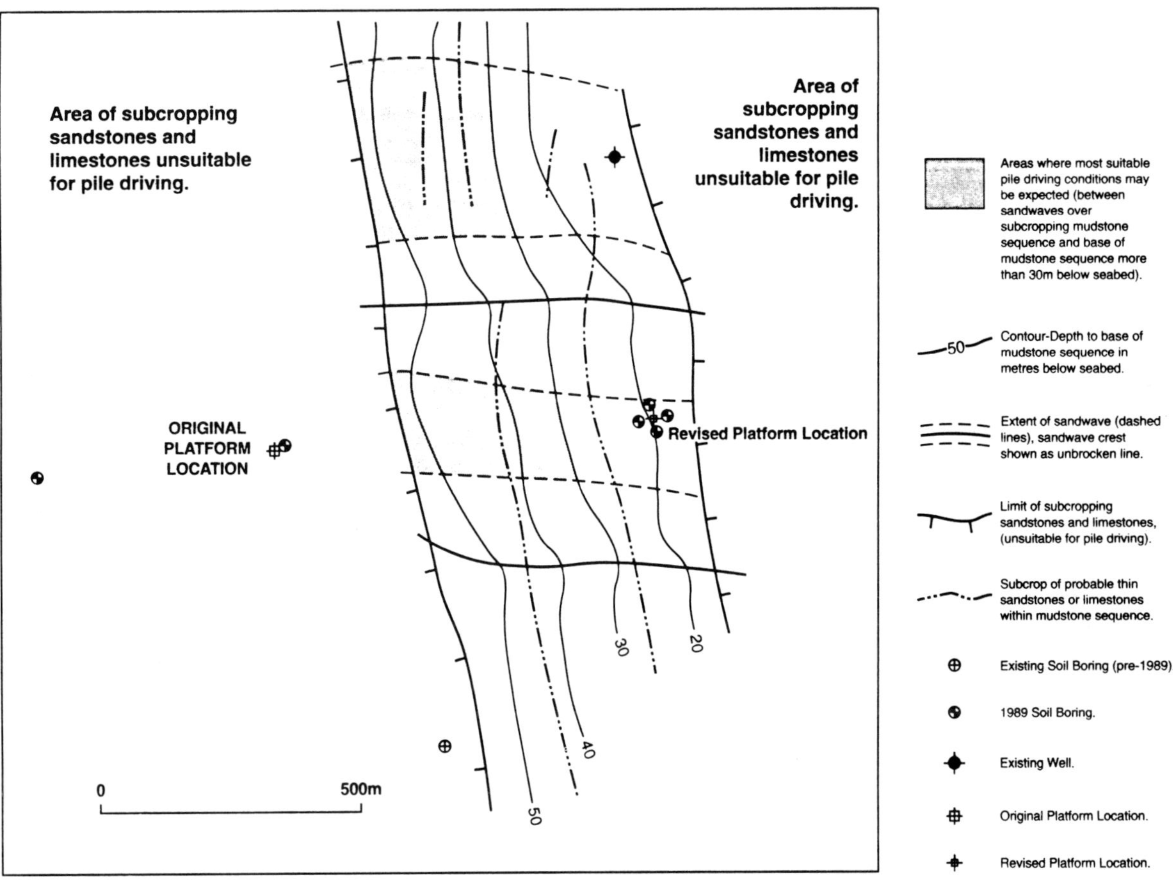

Fig. 8. Installation constraints map – Field A (post-detailed site investigation).

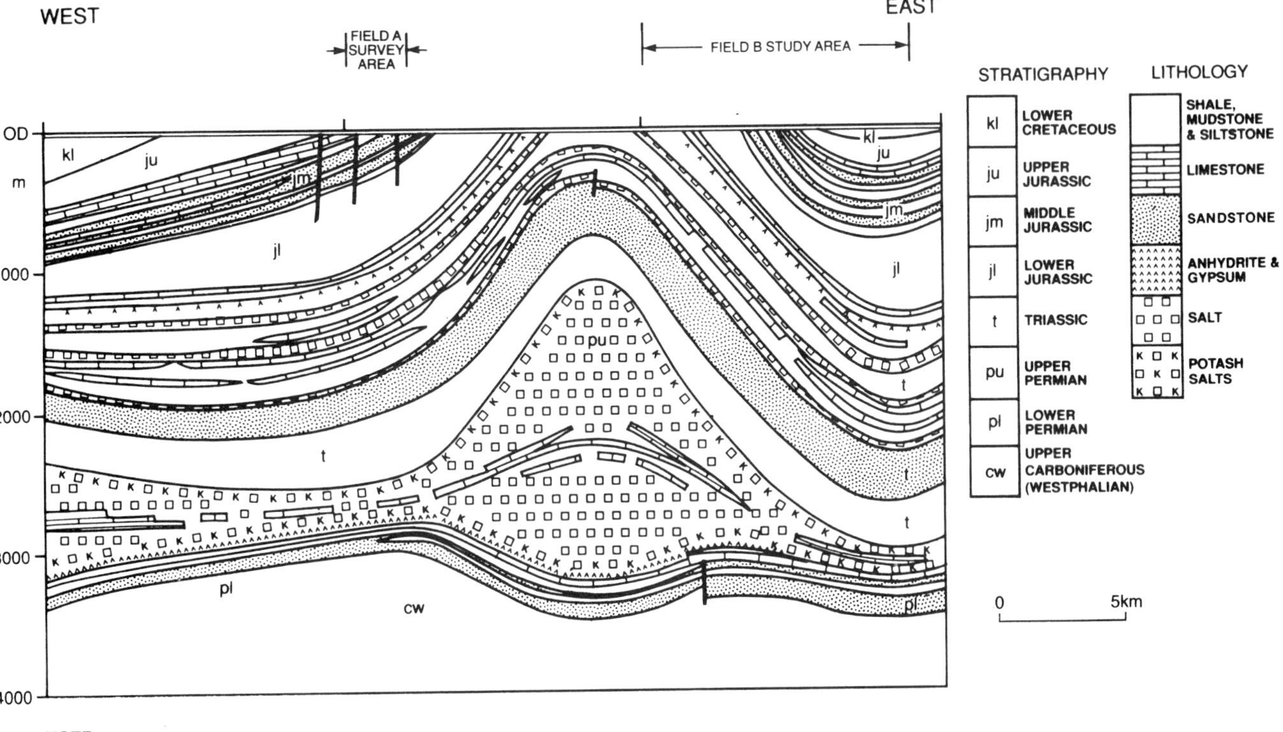

Fig. 9. Generalised geological cross section showing structural and stratigraphic relationships between Field A and Field B.

- exploration seismic data over the field;
- BGS regional seismic data;
- published BGS maps and charts;
- geological and geotechnical borehole data from offset sites.

Of these, the first two provided the bulk of the information for the study.

Deliverables from the study included charts showing Survey Data Coverage, Bathymetry and Seabed Features, Quaternary Isopach, Mesozoic Bedrock Subcrop and Platform Installation Constraints. Interpreted 'shallow' and 'deep' geological profiles were also prepared. The following significant conclusions were reached:

1. The thickness of Quaternary sediments lies in the range 15 to 30m over most of the study area. However, a significant channel feature, within which Base Quaternary is deeper than 30m BSB, crosses the western part of the study area with a northwest-southeast orientation. The position of this channel, and the thickness and composition of Quaternary sediments infilling it, could influence the choice of the type and position of a production platform.

2. Several distinct Mesozoic Formations subcrop at Base Quaternary level in the study area. These formations, which range from Speeton Clay to the Lias Group in age, are expected to have significantly different geotechnical characteristics. The areal extent of these formations was mapped using HR digital seismic data correlated with picked formation boundaries identified on top-hole well logs.

3. The location and type of production facility, and in particular the type of foundation scheme, selected to develop Field B may in part be determined by the findings of this data review study. Options might include a platform founded on either driven piles, or drilled and grouted piles. Alternatively, a gravity based structure may be considered. The feasibility of using driven piles at Field B is dependent on the nature of the sediments within the foundation zone. This will be determined by both the thickness of the Quaternary overburden, and the geotechnical properties of the underlying Mesozoic bedrock.

Previous experience in the area indicates that piles can easily be driven into Lias Formation bedrock (Clarke *et al*, 1985). The geotechnical investigations at Field A indicate that the West Sole Group deposits are not generally suitable for driven piles. Top-hole well logs from Field B confirm that similar lithologies are likely to be present in areas where sediments of the West Sole Group, Oxford Clay and Corallian Limestone subcrop the Quaternary. By considering both of these factors, installation constraints can be identified (Figure 10). The most suitable parts of the field, from a pile driving standpoint, are those areas in which the Quaternary exceeds 30m in thickness, and the underlying bedrock comprises predominantly mudstone/claystone (i.e. in certain parts of the western side of the field). Much of the eastern part of the study area is forecast to be underlain by mudstones of the Kimmeridge Clay Formation, which might also prove to be suitable for pile driving.

4.4. Implications

Although it cannot be proved that a data review study would have saved the supplementary site investigation costs at Field A, the Authors suggest that a review study would have more clearly identified some of the potential platform site and installation constraints at an early stage in project planning, thus improving the chances of selecting an optimum site before commissioning. It should be noted that the absence of any elapsed time between the geophysical and geotechnical surveys precluded detailed interpretation and evaluation of the shallow seismic data to allow site selection to be made on engineering geological grounds.

The results of the data review study of Field B will aid selection of a suitable platform location, type and foundation scheme for the shallow geological conditions disclosed by the study. Results of the study will also help to design appropriate and cost-effective final site investigations. The study is considered by the Authors to add significant value to the project, and reduce the likelihood of encountering unexpected, and potentially unsuitable sub-surface conditions during the site investigation phase.

5. Conclusions and Recommendations

5.1. Conclusions

Large investments are made by Operating Companies in acquiring geo- science data in the exploration and appraisal phases of an offshore field. This paper has illustrated the benefits that a front-end review of such data, from a site evaluation perspective, can provide. At relatively little cost to the Operator such data integration studies will, as a minimum, reduce the chances of selecting unsuitable sites for development facilities, and help to ensure that cost-effective final site investigations are performed.

Maps showing existing geo-science data coverage and integrated interpretations of these data are considered to be an essential pre-requisite for planning detailed site investigations. Knowledge of existing data coverage, and quality, reduces the potential for data acquisition duplication with consequent cost and time savings. Charts of bathymetry and seabed features aid platform site selection, as do a predicted Foundation Soils Model and maps showing potential geo-hazards.

5.2. Recommendations

The value added by front-end data review studies could be further enhanced if Operating Companies were to implement some relatively small changes to the manner

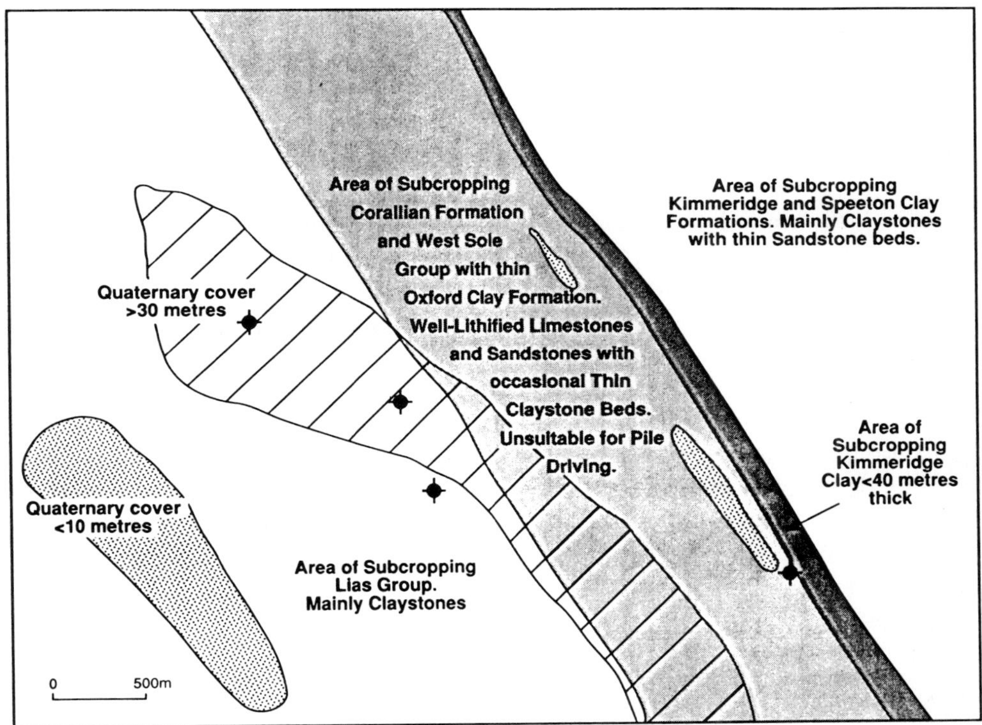

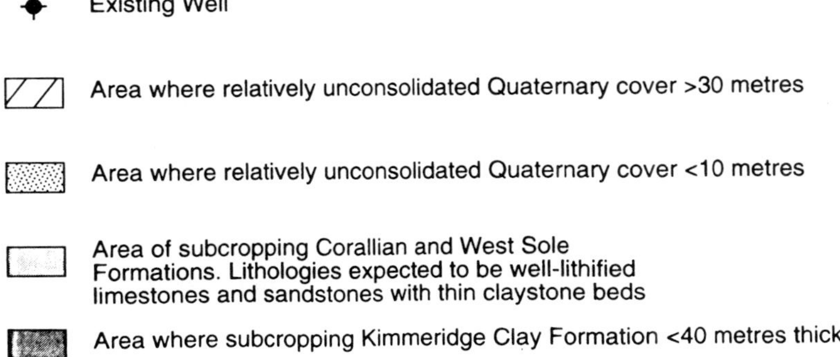

Fig. 10. Installation constraints map – Field B (pre-detailed site investigation).

in which they acquire geo-science data. The most significant of these are as follows:

1. To pay more attention to echo sounding data that are acquired during exploration seismic surveys. Depths should be recorded 'accurately' using calibrated equipment and should be stored digitally (if not already done so) for subsequent processing.

2. To consider recording single trace mini-airgun (for instance) data during exploration seismic data acquisition. Such data would provide an informative overview of the disposition of the foundation soils over a potential field and would not interfere with seismic exploration operations.

3. To log the top-hole section of wells where practical and extend downhole check-shot surveys as close to seabed as noise conditions will permit. Where well casing design precludes logging of the shallow section consideration should be given to acquiring MWD logs while drilling the top-hole sections of wells in fields in which shallow geology may influence the development scheme.

4. To record conventional analogue seismic data collected on drilled rig site surveys in digital format for subsequent loading to computer workstations and detailed analysis and integration with soils data.

The majority of these recommendations would be relatively inexpensive to implement but would certainly enhance the potential value of the data for site investigation purposes.

Acknowledgements

In preparing this paper, the Authors have benefited from discussion with colleagues at both BP Exploration and Hydrosearch Associates Ltd. The permission of BP Exploration to publish this paper is gratefully acknowledged.

References

1. Campbell, K. J. (1984), 'Predicting offshore soil conditions', *Proc. Offshore Technology Conference*, Paper OTC 4692, pp. 391–396.
2. Clarke, J., Rigden, W. J., and Senner, D. W. F. (1985), 'Re- interpretation of the West Sole Platform 'WC' pile load tests', *Géotechnique* **35**(4), 393–412.

Discussion

Question from John Arthur of J. Arthur and Associated, Twickenham, Middlesex, UK: It has been proposed that a mini-airgun should be run with exploration data acquisition and we have been requested to reprocess 3D exploration data to assess gas risk: in that the site survey industry has been be-devilled by short time planning for rig site surveys, does the speaker consider that we are about to experience the demise of the site-specific survey?

Authors' response: The subject raised in the second half of your question is one that we have been considering within BP over the last year.

Certainly there would be a cost advantage to an oil company if their 3D data could be used in this way. Over the last few years within BP we have seen a dramatic improvement in the quality of results the explorers have been attaining in the shallow section (the first one second of record). This has been the result of a realisation that there was a benefit to the explorer in this added effort. The results in some cases have been startling following on from the use of workstation technology and image enhancement techniques and can even be applied to tasks such as pipeline routing!

However, there remain two fundamental resolution considerations in the possible use of these data. Firstly, the frequencies present in the data and thus, regardless of processing effort, the vertical resolution seen is not as good as purpose shot high resolution seismic. On the other hand we have a line spacing of, say, 25m and a spatial understanding, not necessarily to say resolution, that exceeds what we might see with a grid of 2D high resolution seismic. In this latter area the work of Statoil at Haltenbank, Norway in mapping buried, gas charged, sand prone ice-berg scours stands out and would be thought to be conclusive. The problem is whether the improvement in the lateral picture more than covers the loss in the vertical resolution.

At BP we are yet to be convinced that the balance weighs in favour of dropping high resolution seismic altogether.

We understand that this is a stance being followed in Norway by the Norwegian Petroleum Directorate. In the meantime BP will use 3D, when available, to provide a planning basis for future work. By making use of the 3D data combined with our regional understanding of the habitat of Shallow Gas a prospect would be reviewed and risked for the presence of shallow gas.

On the basis of this work a cost-effective high resolution seismic grid would be acquired, which might entail as few as only 2–3 lines. The data would be shot to assess the key risks anticipated at any one site, and would be interpreted together with the 3D data, the results of which could then be fed directly into a field review study.

In summary, therefore, we at BP are not yet convinced that even in a mature exploration area such as the North Sea that exploration 3D data can entirely, and safely, displace the purpose shot high resolution seismic survey. However, we do believe it offers opportunities for improved planning of these, and other, surveys and in so doing will probably reduce the capital outlay of operators on this type of work in the medium term.

Question from D. Long, British Geological Survey, Edinburgh, UK: Although you indicate that in the 'field A' case an additional £20K was spent at the site investigation stage of development due to the absence of an integrated data study, would there not also have been savings in installation design, construction and emplacement if such a study had been undertaken, as in the case of 'field B' where you were able to select an optimum site? Would such savings have been of a magnitude to more than justify the cost of such an initial integrated study?

Authors' response: We firmly believe that had a review study been performed in the case of Field A, the savings would have brought, in terms of reduced site investigation and installation costs, would have far exceeded the initial cost of the review study described.

Furthermore, we would not now proceed to survey any development location without first having performed such a study, due to the risk of finding that that location was unsuitable due to foundation conditions, shallow gas or other such problems, and that, as a result, the work required extension at extra cost.

SESSION 2

GEOTECHNICAL SAMPLING AND TESTING

SUCCESSFUL CABLE BURIAL – ITS DEPENDENCE ON THE CORRECT USE OF PLOUGH ASSESSMENT AND GEOPHYSICAL SURVEYS

JON NOAD
BT (Marine) Limited, Berth 203, Western Docks, Southampton SO1 0HH

Abstract. Burial of subsea cables began in the early 1980s when excessive cable faults due to increasing and changing fishing activity, was resulting in a significant loss of income to the cable owners. Several methods are now used to bury the cables (usually to a depth of 0.6m), the primary method being the use of a subsea plough pulled and controlled from a host ship. However, water jetting tools mounted on tracked or swimming ROV systems are also used.

The BT (Marine) Limited (BT(M)) cable plough is towed behind a vessel, using a share to cut a trapezoidal prism of sediment from the seabed. This is lifted up by an inclined ramp, while the cable is laid underneath it. Before undertaking this type of cable burial, it is very important to assess whether burial depth can be achieved. To this end a geophysical route survey and a plough assessment survey are carried out, and the correct analysis, correlation and interpretation of these determine the success of the burial operation.

This paper assesses the present state of these surveys, and considers new methods that could be used to improve the results obtained. The use of high resolution sonar scans, coupled with an improved plough assessment tool, could give an excellent picture of the conditions extant at the sea floor. This would also allow an accurate prediction of the way the plough behaves in layered sediments, helping to control burial depth. The new processes will be used to develop a new strategy to the benefit of cable owners.

1. Introduction

No fewer than 16 optical fibre telecommunication cables form a network extending from the shores of the United Kingdom (see Figure 1), a figure that is planned to double within the next decade. These cables have a very high call capacity, and because of this it is extremely important that they are fault free. Considerable customer inconvenience can be caused should a cable be damaged necessitating a switch to a back up satellite system.

Excluding faults due to production or cable lay related errors, the most common causes of cable downtime are damage due to fishing or anchors. Two methods of protection against these threats to the cable have been considered:

Volume 28: Offshore Site Investigation and Foundation Behaviour, 39–56, 1993.

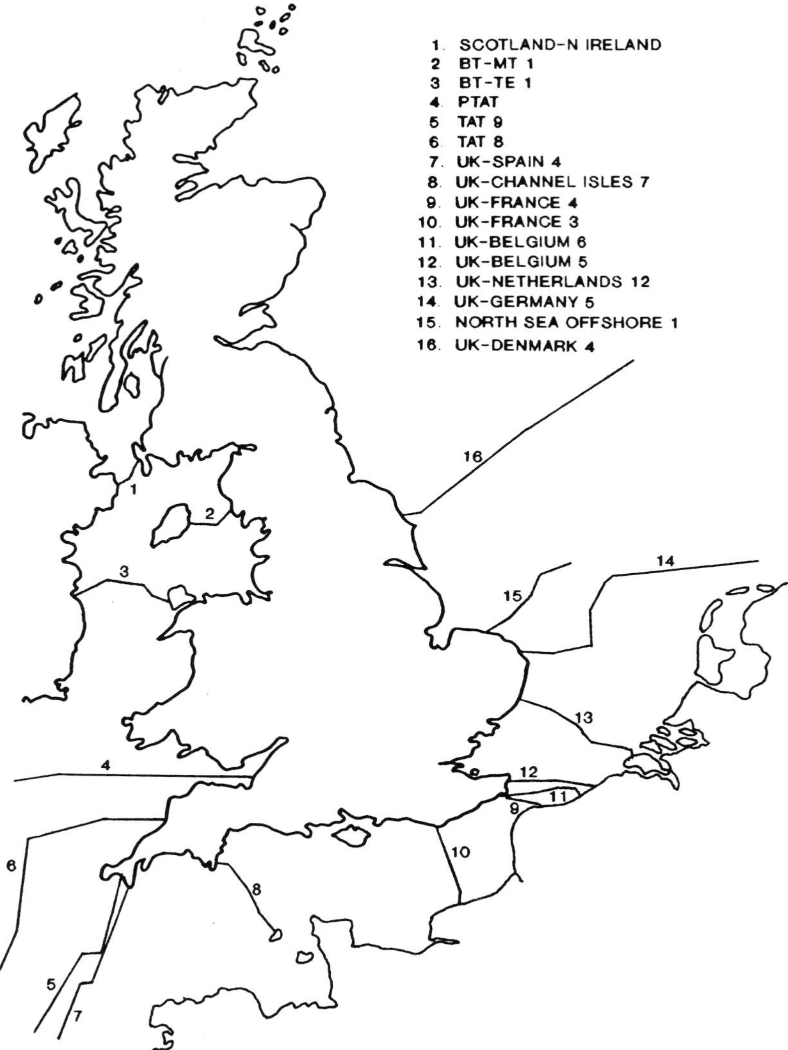

Fig. 1. Optical fibre cables extending from the United Kingdom.

i Heavy duty cables, strong enough to withstand the impact of a beam trawl. These are very heavy, making them impractical for long routes, and are very expensive to make and install.

ii Cable burial, either by the use of simultaneous lay and burial using a sea plough, or post lay burial using a jetting device. Burial is less expensive than heavy duty cables, and is generally preferred within the industry.

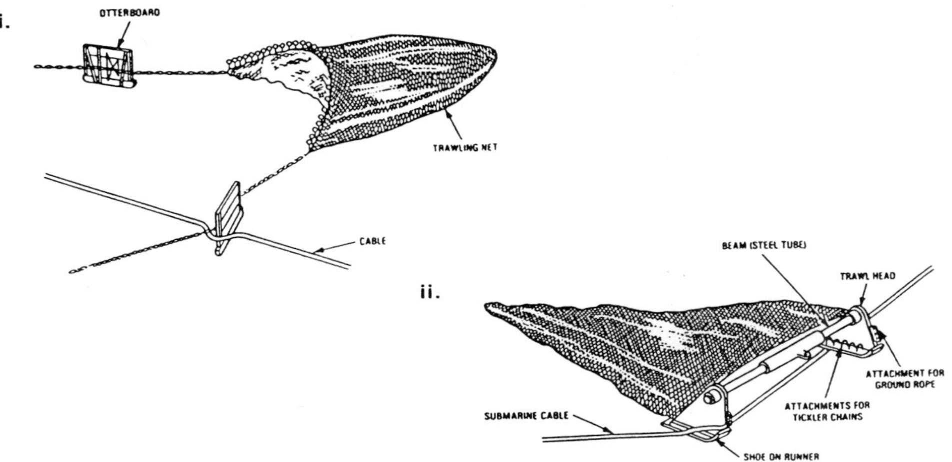

Fig. 2. The (i) Otterboard and (ii) beam trawl methods of fishing, showing the potential for snagging cable.

2. Threats to As Laid Cables

Before considering cable burial, the nature of the potential hazards to the cable itself should be considered. As already stated, there are two main threats and the burial strategy used to protect the cable has been planned after considering these in detail.

2.1. FISHING

Several methods of fishing are carried out off the UK coast, and even more in other parts of the world. Generally, the beam type trawl has the greater potential for cable damage. These are usually 8m–10m in width, and as Figure 2 shows, they are so designed as to allow submarine cables to snag on them. This can be exacerbated by the welding of additional 'runners', usually steel bars, to the base of the shoe. Typically the beam trawl shoes penetrate the sediment to perhaps 10cm, but this may increase to at least 40cm in very soft sediments.

Subsidiary risks are provided by clam dredgers, which operate similarly to the beam trawl but with a diameter of about 2m, and also the doors used to hold nets open. The doors are designed to run along the sea floor, and penetrate slightly to raise a cloud of sediment which helps to herd fish into the net itself. They may

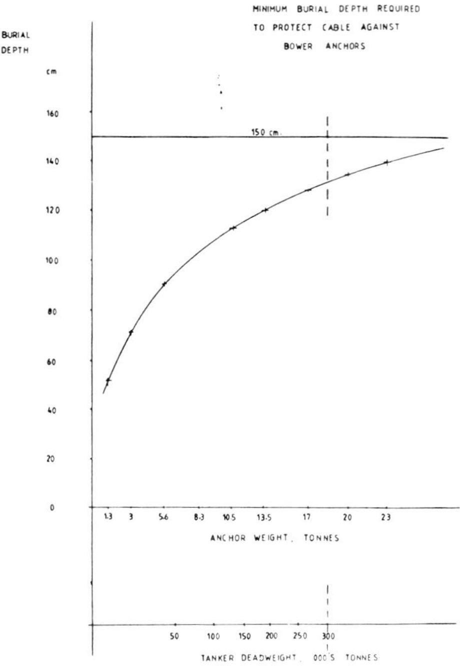

Fig. 3. Graphs showing the range of anchor and fishing gear penetrations into the seabed.

also have steel runners attached to the base, but are unlikely to penetrate more than 20cm into the sediment, even if the door falls over pushing the towing bridle into the sea floor.

2.2. Anchors

There are many types of anchors in use throughout the world, but of these the biggest concern to the cable industry is the bower type. High holding power anchors, as in those used in the offshore petroleum industry, can penetrate to great depths (in some cases to as much as 5m), but do so over a very short distance, after which they will be left in that position, for several years in some cases. This presents a small threat to cables.

However bower anchors, which usually have a holding to weight ratio of between 3 and 10:1, are preferred and generally considered efficient in most ships. Except in very soft sediments they tend not to penetrate fully, but rather to drag, or penetrate and then eventually ‘flip’ over as tensions increase. The penetration of such anchors depends on anchor weight, tripping angle, anchor weight distribution, crown thickness and most importantly sediment type. Figure 3 demonstrates the range of anchor penetration for ships of varying deadweight tonnage.

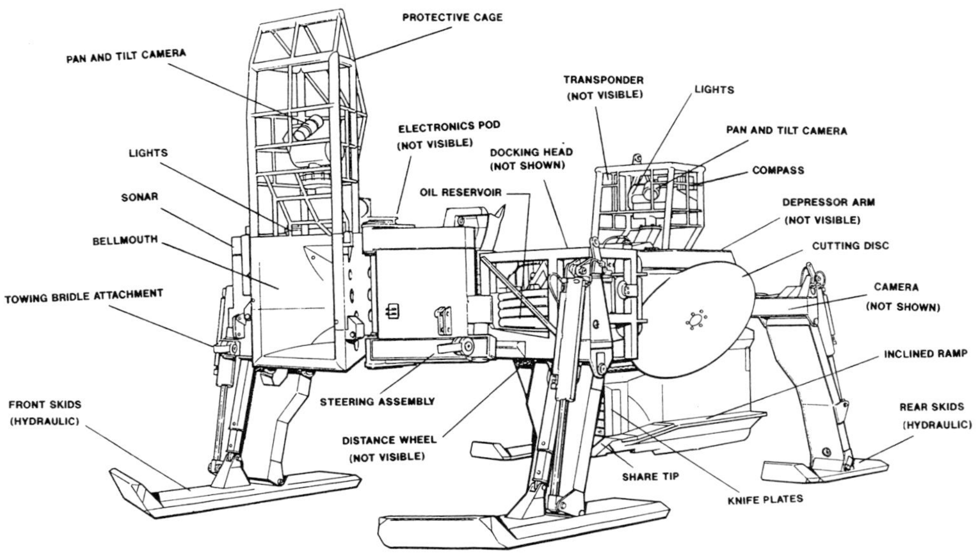

Fig. 4. Submersible plough (BT(M)).

It is considered that burial to depths of greater than 1.5m to guard against damage by even the heaviest bower anchors would make the submarine cables almost impossible to recover using present day technology. Accordingly a compromise has been reached whereby burial of the cables takes place to a cover depth of 60cm, thus providing limited protection from small anchors and a factor of safety of 20cm from damage by beam trawlers.

3. Methods of Burial

To assess the ploughability of a route it is obviously important to understand those factors relevant to the burial of the cable. This depends partially on the method of burial and the two main techniques used are:

3.1. Simultaneous Lay and Burial Using a Submarine Plough (see Figure 4)

The cable passes through the plough which is deployed from the back of the cableship. As the cableship also stores the cable prior to laying this is a highly

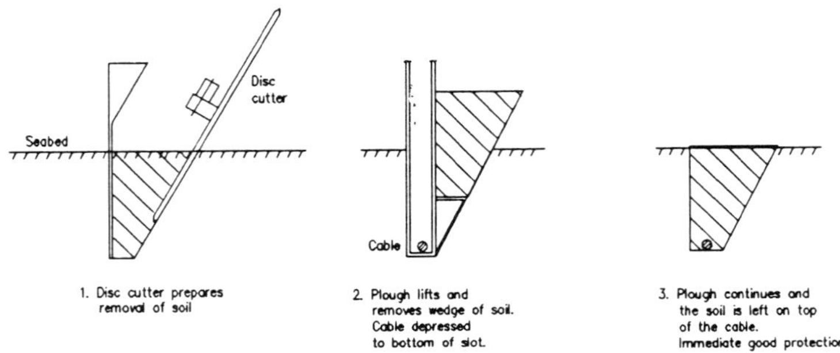

Fig. 5. Method whereby the plough lifts a prism of sediment from the seabed and lays the cable beneath.

efficient system. The plough is lowered to the sea floor and is towed behind the ship. A trapezoidal prism of soil is cut by the free rotating disc cutter, inclined at 35° from the vertical, and the sharp vertical knife. The prism is gently lifted from below by a ramp, rising along the sloping surface cut by the disc. The cable is placed underneath it and the wedge is then returned into place (see Figure 5). Backfill is thus automatic and the seabed is left unscarred.

3.2. Post Lay Burial Using a 'Trenching' Device

Another method of burying cable is to first lay it on the seabed, and then to use water jets to fluidise the sediment. The slurry is then dredged and sucked out, leaving a trench into which the cable then sinks. Many tools of widely varying appearance and specifications are used, of which BT(M) Trencher (see Figure 6) is the one best known to the author. Although the cable is left in an open trench, until naturally or artificially backfilled, this provides enough protection to guard against damage from trawling vessels. Jetting remains one of the best ways of burying cable already on the sea floor, for example after a repair has taken place, as there is no risk of cable damage from the water jets.

To successfully bury submarine cables the following data is essential:

i The nature of the seabed sediments along the route, and the positions where transitions from one type to another may occur. This information can include data on sediment types, densities and shear strengths. Details on bearing capacity may also be important.

ii The thickness of the sediment cover over underlying bedrock. If the bedrock lies close to the surface its characteristics should be tested in the same way as the sediments (in case it might be ploughable, although it is rare).

iii An estimate of the tensions the plough will experience as it cuts through the sediments while burying cable.

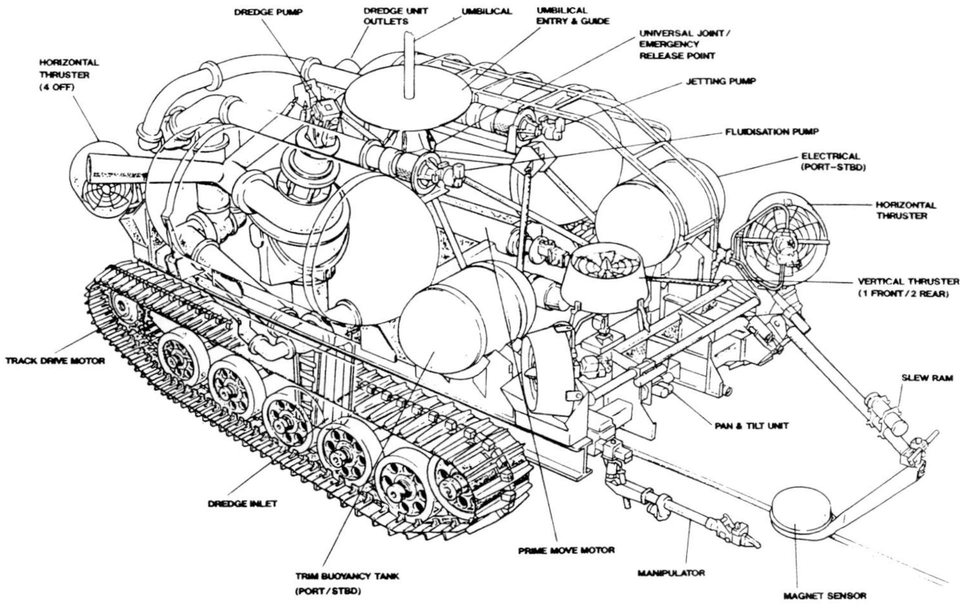

Fig. 6. Submersible trencher (BT(M)).

iv The precise position where the above data was collected, to as high a navigational accuracy as possible.

v An accurate bathymetric picture of the seabed, to ensure that the slope gradients are not too steep for the deployment of Plough or Trencher. This should include data on large boulders and other obstacles.

It is obvious that for cable burial to be carried out successfully a considerable amount of data is required. Failure to assess the planned route can result in damage to both the cable and burial equipment. It may also leave cable unburied and unprotected on the seabed, which may place the burial company in breach of a performance based contract.

4. Data Collection

The collection of data prior to the laying of the cable falls into two parts. The first is the 'Electronic Survey', carried out by a subcontracted survey company, which generally comprises bathymetric profiles, and sidescan and sub-bottom surveys. Following this the 'Plough Assessment Survey' (PAS) is run along the route, usually by towing a grapnel or plough-like device behind the ship and measuring penetration, tensions and attitude. The two types of survey are discussed below in more detail.

4.1. Electronic Survey

As stated above this falls into three separate categories:

a. Bathymetric Survey
b. Sidescan Sonar Survey
c. Sub-bottom Profiles

a. *Bathymetric Survey*

Until recently a picture of the seabed contours was built up by running several survey lines along the route, parallel to one another, using a Precision Depth Recorder (or echo sounder). This will be augmented, especially in shallower water, by extra lines run at right angles to the planned route. If necessary a picture of the contours can be built up from the resulting bathymetric data.

This is very important in picking out those sections of the route where the gradients are too steep to permit operation of the Plough or Trencher. Thin troughs or holes may also appear on the profiles. The spacing of the lines is generally dependent on depth, and is closer in shallower water. This means that the coverage is broader but the resolution less at depth, which could lead to 'missing' potential hazards that could damage plough or cable.

A relatively new bathymetric survey system is swathe bathymetry, where a line of transducers collect information to either side of the towed fish. Thus a grid of surveyed points can be built up and then contoured. This is a much better system with regard to picking up potential hazards along a planned cable route. The only drawbacks would be that the coverage is still on a fairly broad scale at greater depths, and that the computer undertakes the data interpolation. In this way it would be possible to miss a relatively small hole or trough.

b. *Sidescan Sonar Survey*

This type of survey is used to image the surface of the seabed itself. It works as follows. A transducer assembly is towed on a steady course and at a constant depth through the water (the assembly may be hull mounted or in a tow fish). Sound pulses are emitted at regulated intervals, and the system receives the returning echoes from the water column and sea floor immediately afterwards.

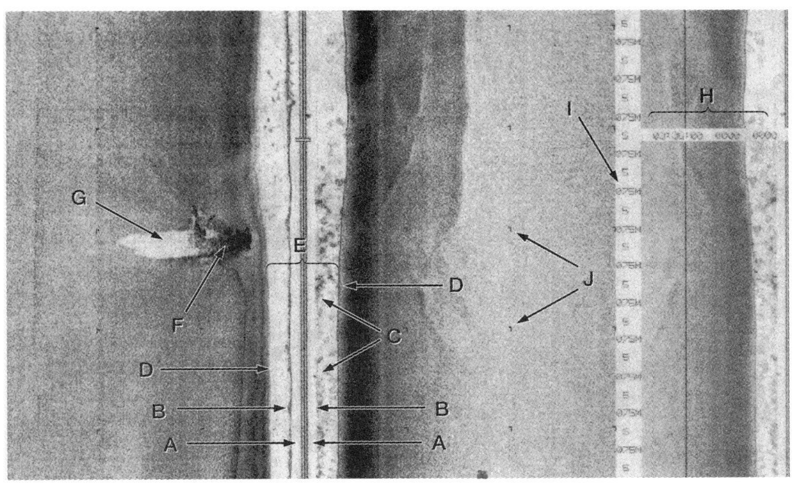

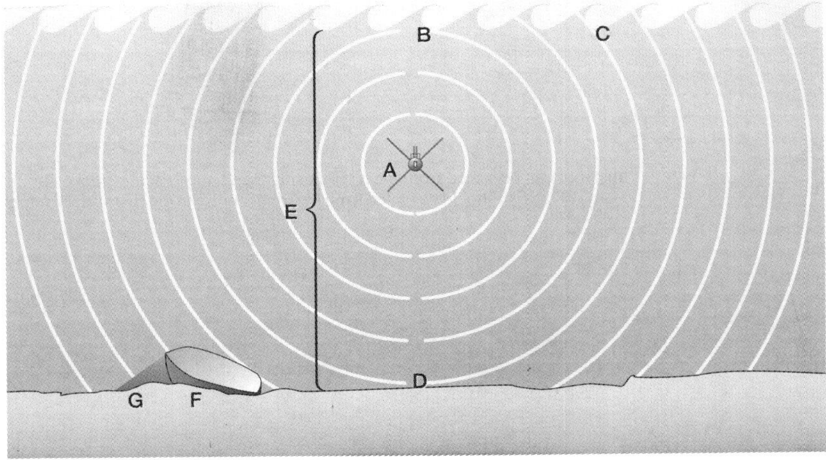

Fig. 7. The portions of this record are as follows.

This continues until the next pulse is transmitted. The echoes from one pulse are displayed on the recorder as a single line, the darker portions of the line showing stronger echoes. As the lines juxtapose, a picture of the sea floor appears (see Figure 7).

Many variables affect the sonar data, including wind, waves and temperature gradients. However the most important variables are seabed topography and sediment type. It is for the delineation and interpretation of the variations in these features that the sidescan sonar survey is undertaken.

With regard to cable burial, a great deal of data can be collected from sidescan survey results. This falls into several categories:

Hazards to submarine plant and cable:

i Boulders, with which ploughs or jetting tools could collide. Even with steering it may be difficult to avoid large 'house sized' clasts.

ii Gullies, valleys, etc., into which plant may fall. These are particularly common close to the Continental Shelf edge, but may occur elsewhere, for example close to large river deltas.

iii Pockmarks may be indicative of areas where the sediment is extremely gas rich. It is possible that plant could tip over as a result.

iv Bedrock, especially if very rugged. This category includes sea mounts and dissected rock outcrops. Plant may become trapped under overhangs or cable can be left in suspension.

v Sandwaves and other smaller scale current induced bedforms tend to lead to cable suspensions either as a result of the initial lay, or possibly due to seasonal sandwave mobility (Smith 1988).

vi An abrupt change in sediment type from very soft sediments, such as silts, to very stiff sediments (perhaps boulder clays), while travelling at speeds of over a knot, could lead to potential damage to the tow rope.

vii Although unlikely, it is possible that old cables might already be present on the seabed. If these show suspensions it could be due to high sediment mobility in that area. This could suggest a change of route to avoid such an area.

viii Wrecks and other manmade debris, with which submarine plant could become entangled. Were the cable to be laid over such an obstacle it would result in suspensions.

ix The type of sediment can normally be differentiated, which could help to estimate the tensions likely to be encountered during the PAS. This data is not always reliable, however, and should where possible be backed up by ground truthing.

Several of the above hazards are extremely unlikely to be picked up by any other type of survey, so it is vital that the importance of this type of survey is not underestimated. Picking up details of seabed topography on a small scale can make the difference between a smooth lay or damage to plant and cable.

c. *Sub-bottom Profiling*

In this type of survey data is collected from below the sea floor. A sound wave is generated and pulsed down towards the seabed. Some of the signal will penetrate into the seabed (dependent on the frequency), and is reflected off sub-bottom reflectors back towards the surface. These are picked up by transducers and converted into electrical signals, which can then be used to plot up a profile of sub- bottom conditions. A narrow beam is used to avoid reflections from other parts of the seabed.

Generally, the higher the frequency employed, the less the penetration, but the better the resolution. Some systems use a sparker to produce high energy sonar pulses, where a large electrical voltage is built up and then discharged. This is used in deep seismic applications. Other systems use airguns, with a compressed air discharge. For cable burial surveys the usual methods are to deploy hull or fish mounted transducers to generate pulses, generally between 5 and 15kHz. These can achieve a penetration of up to 50m, depending on the nature of the sub- bottom sediments encountered by the pulse.

It should be noted that there are certain difficulties in the interpretation of sub-bottom data. Estimates of the sound velocity through the sub-bottom sediments have to be made to draw meaningful conclusions, and identifying sediments can be something of an educated guess. However, the use of vibrocores and other sampling techniques aid the process greatly by ground truthing the data collected.

The most important information that sub-bottom profiling can provide is whether bedrock or very rocky sediments occur close to the sea floor. These might make cable burial impossible. As has been mentioned earlier, the cable layer is generally only interested in the top 1m of sediment, but with many systems it is extremely difficult to resolve this part of the sub-bottom succession. This is because a great deal of the pulse energy is reflected from the sea/sea floor boundary, and it tends to obscure any meaningful returns from the seabed.

In response to this, systems have been developed that give a specified resolution of less than 10cm from the seabed downwards. Simrad's Chirp system (Figure 8) is one of these, although it has yet to be tested on cable laying operations. Obviously a system successful in this area would prove very popular in the cable burial industry.

Generally a cable plough can cut through almost any type of unconsolidated sediment. The degree of consolidation is indicated from sub-bottom returns, but the interpretation is subjective. Definitive work in this area is still required.

Jetting tools are much more limited in their application, and as such require accurate sub-bottom information. They are usually capable of jetting sands, whatever their density, as the sands do not have cohesive strength. Clays are another matter, as they may have a very high cohesive shear strength. In this case it is not enough to know that clay layers occur close to the seabed. The only way to assess the shear strength accurately is to take core samples or to

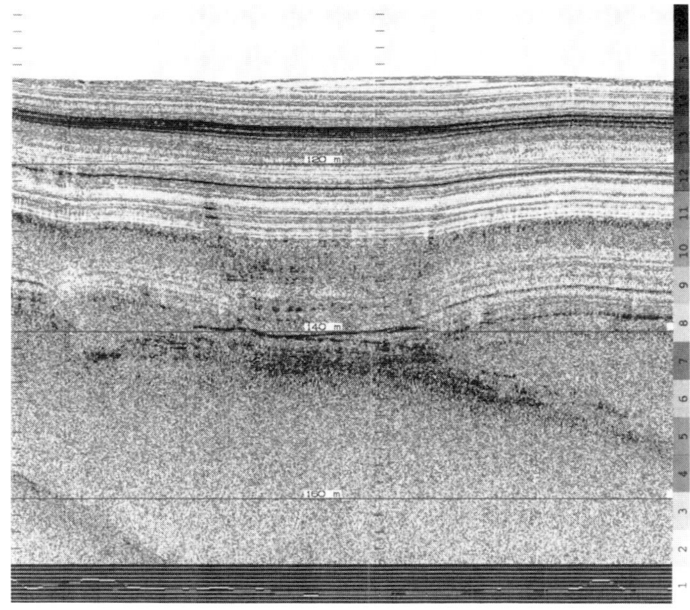

Fig. 8. An example of a chirp profile recorded off Harwich.

use cone penetrometer tests. As this type of sampling is carried out routinely to ground truth the sub-bottom profiles, it provides an excellent opportunity to collect more data to assist in the accurate assessment of cable burial potential.

It should also be mentioned that a great deal can be gained from sub-bottom profiles by examining the relationship between the different sedimentary layers. It can often be estimated where a certain lithofacies (one individual sedimentary layer) will outcrop at the surface. Even more importantly, it is occasionally possible to pick up sediments consisting of folding and shallow dipping shear planes indicative of 'ice push' deformation, characterised by shear strengths over 200 kPa. Very few jetting tools are able to trench in clays of this nature.

5. Plough Assessment Surveys

Plough Assessment Surveys (PAS) are extremly important with respect to the burial of cable in that they are the only type of survey to continuously penetrate the sea floor. Electronic surveys provide a great deal of information, but collect data through remote sensing which is then processed through subjective interpretation.

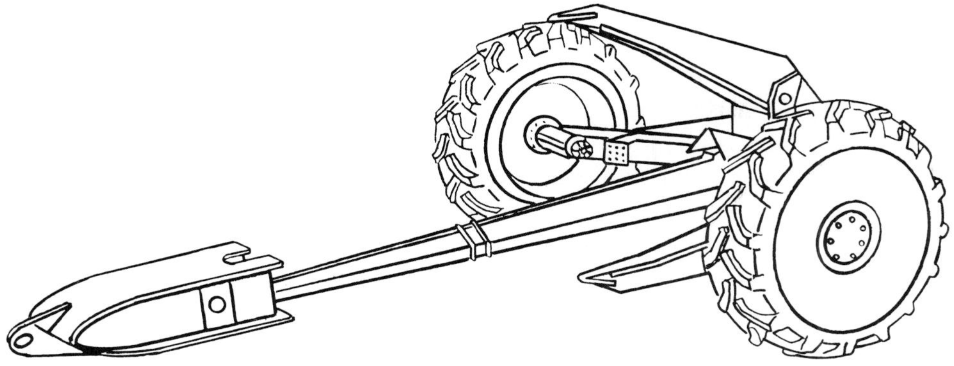

Fig. 9. The Mark II wheeled detrenching grapnel (BT(M)).

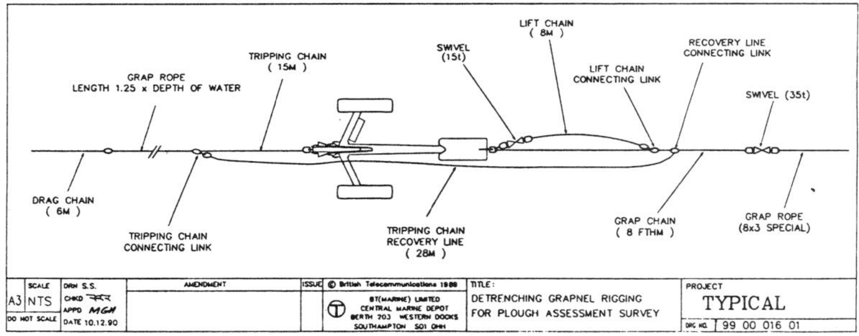

Fig. 10. The rigging of the Mark II wheeled grapnel.

The PAS data is real in that measurements are collected of penetration into the seabed and the resultant tensions on the tow rope.

Several PAS tools are in use by companies worldwide, most of which are model ploughs. These are generally similar to the ploughs that will subsequently bury the cable, although usually considerably lighter. The tool used by BT(M) is the detrenching grapnel (see Figure 9). Originally designed to recover buried cables, it consists of a long shank, weighted by chain at the front to hold it horizontal, to which an axle and wheels are mounted at the rear. Between those wheels a fluke projects from the shank, and can penetrate up to 0.8m into the seabed. A shear pin will release the fluke should it become trapped in rock or other obstructions. Figure 10 shows the rigging of the detrenching grapnel for PAS. There is also a cable-caught sensor located at the hinge point between shank and fluke.

The grapnel collects several types of information about the seabed, which fall into three categories:

5.1. INFORMATION RELAYED DIRECTLY TO THE SHIP USING ACOUSTIC TELEMETRY

a. *Pitch and Roll*

These are measured using a clinometer, and allow for the indirect calculation of the fluke penetration. They also indicate areas where the slopes are too great for the plough to operate safely.

b. *Penetration*

The fluke is attached to the shank with a scissor type of arrangement, so that a change in penetration is reflected by a corresponding change in the angle of the shank of the detrenching grapnel. This will also be measured by a clinometer, and a computer program automatically corrects the data relative to the pitch and roll measurements.

5.2. INFORMATION COLLECTED DURING PAS AT THE SHIP

a. *Tensions experienced on the Grapnel Tow Rope*

This is one of the most important sets of data and using pre-established equations can give an accurate indication of the potential tensions that will be seen on the plough tow rope during cable burial.

b. *Samples*

When the detrenching grapnel is retrieved at the end of a run, sediment will often have been caught in the region of the cable-caught sensor. This can be removed and analysed, although it should be stressed that no accurate location for such samples can be assumed.

c. *Fluke Share Weight*

Whenever the grapnel is recovered at the end of a run the share is weighed. The wear and thus gradual lightening in weight of the share can be used to estimate share wear during cable burial itself.

5.3. INFORMATION INFERRED FROM PAS DATA

a. *Tension Trace Signatures*

A great deal of information on the nature of the sea floor can be inferred by examining the trace of the tensions on the grapnel tow rope during the PAS. By running the ship's bow accelerometer trace simultaneously, the effect of the ship's heave can be removed and a clear picture of the tension trace gained. Figure 11 shows some typical examples.

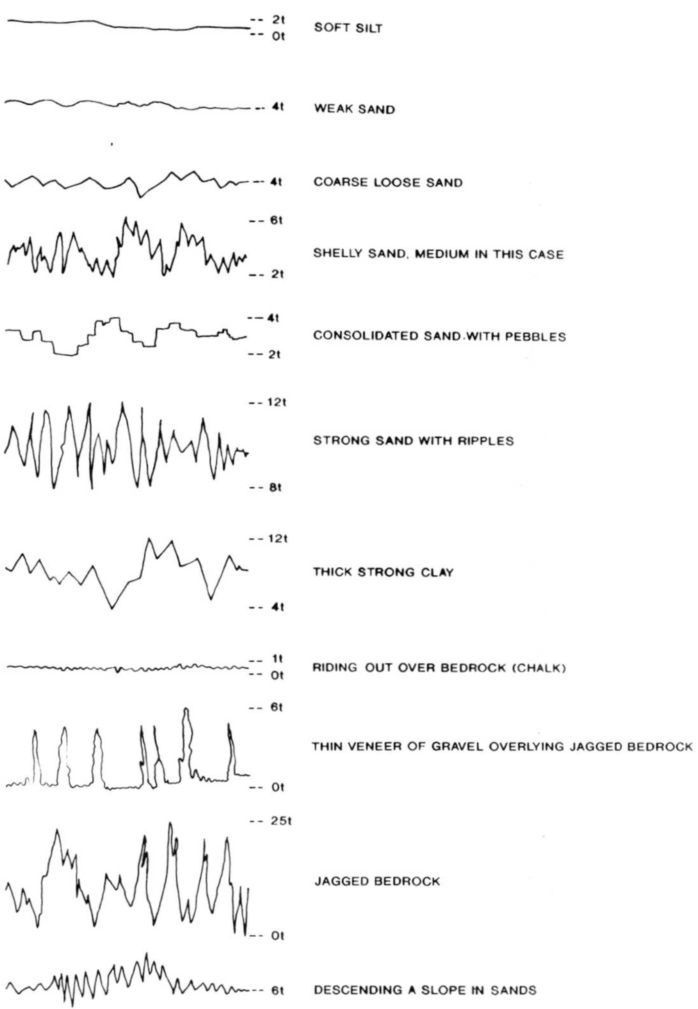

Fig. 11. Examples of 'tension trace signatures' for various types of sediment.

b. *Cable Retrieval Potential*

By running the detrenching grapnel along the planned burial route, a precise record of the penetration of the fluke tip is made. This confirms that the grapnel will be able to recover cables buried to the planned depth on this route, should a fault occur.

6. PAS Results

From the data collected, a summary is made of the changes in sediment type and tensions experienced along the route. The results of the electronic survey are also invoked to make this as accurate as possible. From this data every portion of the route is classified according to the degree of burial expected.

This data is then used to choose:

i The type of cable to be used for each section of the route.

ii The depth of burial to be attempted. This is not as straightforward as it might seem, and is related to the risk of damage to the cable both at the time of burial and in the future. As much information as possible should be included in the PAS report to make this decision easier.

iii The number and position of any plough share changes along the route.

iv Those portions of the route where burial should not be attempted. This could be due to topography, bedrock or other reasons.

7. Developments in PAS Technology

7.1. Detrenching Grapnel

One of the first tools to be used for Plough Assessment Surveys was the detrenching grapnel. Initially the depth of fluke penetration was relayed to the ship using a pinger system, although this has now been updated to acoustic telemetry linked to the signal from an LVDT. The grapnel is a relatively primitive tool but has many advantages in PAS. A summary of pros and cons is shown in Table 1.

7.2. Plough-Like Tool

The next step forward from the detrenching grapnel is the use of a plough-like tool. By its very nature, such a tool would present a clearer picture of the likely behaviour of the plough itself. As it would run on skids, while using a share to cut through sediments, scaling up the recorded tensions to predict plough performance would be straightforward. The large subframe could also be used to mount cameras and a sonar device to give extra information about the seabed.

Disadvantages would lie in the weight and size of the tool, making it difficult to deploy, especially in poor weather. The presence of an umbilical, necessary for the tool's operation, would add to these problems, as well as making the plough-like tool rather more vulnerable on the seabed. These more complex sensors would also add to its cost. Lastly, unless a full scale plough was used for the assessment (without cable), the lesser weight of the tool would tend to allow it to ride out. This is because it is the plough's weight that engages the share, rather than the action of

TABLE 1.
Detrenching grapnel.

ADVANTAGES	DISADVANTAGES
Very lightweight (1.9 tonnes) so it can be deployed from small low cost vessels.	The grapnel's wide fluke means that it drags through sediments. This is unlike the plough whose share cuts through them.
Relatively robust and the fluke has a shear pin (25 tonnes) should it become jammed.	The grapnel's wheels encounter minimal friction, thus behaving differently to the plough skids.
Uncomplicated design means that little can go wrong, and grapnel is inexpensive.	All data collected by the grapnel must be carefully interpreted in order to extrapolate it to the plough.
Soil forces tend to dig the grapnel fluke in deeper, ensuring penetration wherever possible.	The grapnel operates close to the seabed, so it would be very risky to use cameras or sonar, particularly as it may land upside down.
No umbilical so no umbilical winch required.	

soil forces on the share tip. In some ways this would be an advantage, in that the tool would under-predict the plough burial depth.

8. Future Developments

Considerable advances have been made in the field of electronic surveys. Recent ones include the development of 'Roxanne', which analyses the sediment type from the first sonar return. 'Chirp' is another system which calculates the sediment density from the strength of the return. 'Gloria' is configured to achieve the maximum ranges permitted by propagation conditions in the deep ocean, and has sidescan sonar mapped more than 2% of the world's ocean floor.

Another area of rapid improvement is in swathe bathymetry, as more and more companies invest in both EM12 and EM1000, the deep and shallow water versions. In navigation too, DGPS (Differential Global Positioning System) has allowed new levels of positional accuracy, with an accuracy to within 5m over hundreds of kilometres. This will help to promote the accuracy of all types of survey undertaken prior to and including the cable lay itself. For all the above technological innovations, the general computing powers have enhanced the processing speed and volume of data collected, thus allowing greater survey accuracy.

As survey techniques continue to develop, it is felt that the need for PAS may be reduced to those areas where changes in sediment type occur, or where the possibility of burial is marginal. In other areas the survey and sampling data should prove sufficient to make a determination on burial potential. It is worth bearing in mind that, however accurate the navigational equipment, the PAS run and the cable lay will not follow precisely the same route. This is in contrast to the survey data which would cover a swathe across the seabed and thus far more than just one line.

As navigational accuracy increases, and as the demand for low cost, low risk cable burial grows, it will be interesting to see how route surveys develop in years to come.

Acknowledgements

I would like to thank my colleagues at BT(M) who have contributed to the development of plough assessment over a number of years, in particular Mr. J. Butler of BTL.

References

1. Bonnon, I. D. (1988), *A Comprehensive Study of the Methods Available for the Repair of the Buried Subsea Fibre Optic Cable.*
2. Buller, A. and McManus, J. (1981), 'Sediment sampling and analysis', from *Estuarine Hydrology and Sedimentation*, K. Dyer (ed.), EBSA Publishers.
3. De Boer, P. (1988), 'Bypassing of sand over sandwaves and through a sandwave field in the central region of the Southern North Sea', from *Tide Influenced Sedimentary Environments and Facies*, D. Reidel Publishing Company.
4. Danish Hydraulic Institute (1990), *Modelling of Coastal Sediment Transport.*
5. Fish, J. and Carr, H. (1990), *Sound Underwater Images*, Lower Cape Publishing, Orleans, MA.
6. Kenyon, N. and Hunter, P. (1985), 'A Long Range Sidescan Sonar Survey of the Meriadzek Terrace, Bay of Biscay', Report No. 210, Institute of Oceanographic Sciences.
7. Meigh, A. (1988), *Cone Penetrometer Testing*, Butterworths.
8. Peuch, A. (1984), *The Use of Anchors in Offshore Petroleum Operations*, Editions Technip, Paris.
9. Stride, A. (1982), *Offshore Tidal Sands*, Chapman and Hall.
10. Struzyna, R. J. (1989), 'Subsea Fibre Optic Cables: Modern Methods of Protection, Installation and Detection', *Common Defense '89.*
11. Vincent-Brown, P. (1983), 'Protection of Submarine Cables', *British Telecommunications Engineering* **2**, October 1983.

A MEASURABLE CLASSIFICATION SYSTEM FOR NON-CALCAREOUS MARINE SOILS

U. F. KARIM* and M. R. DE RUIJTER
Fugro-McClelland Engineers B.V., P.O. Box 250, 2260 AG Leidschendam, The Netherlands

Abstract. In this paper a practical engineering classification system for non- calcareous marine soils is proposed as an alternative to BS-5930 and ASTM-D2487. The system is based upon measurable data furnished by Atterberg limits and particle size distributions. This system is applied by means of simple flow charts for naming the main soil type (MST), the secondary soil type (SST) and the soil group symbol. These flow charts are easily adaptable in computer software and can be extended to include more categories. A total of 46 soil classes are possible in the new system. This paper also addresses differences between BS, ASTM and the new system and suggests consistency description criteria for fine and coarse grained soils. The system proposed in this paper is currently adopted by Fugro-McClelland Engineers (FME), The Netherlands in offshore soil investigations.

1. Introduction

The continuous upgrading of existing classification systems and introduction of new ones forces companies to evaluate their classification practice on a regular basis. ASTM-D2487 (1991) and BS- 5930 (1981) are the two most widely used systems in offshore soil classification for engineering purposes. Casagrande (1948), Liu (1976) and Carter and Bentley (1991) among others have reported differences between several classification standards. Criticism of BS-5930, particularly in the classification of fine grained soils, have surfaced (Norbury *et al.*, 1986).

BS-5930 and ASTM-D2487 differ in their starting point in defining the boundaries between fine and coarse grained soils (see Al-Hussaini, 1977). In BS-5930, a soil with grains smaller than 60 mm is considered fine grained when the percentage by dry weight passing the 0.06 mm sieve is greater than 35%. Instead, ASTM-D2487 employs 75 mm, 0.075 mm and 50% boundaries respectively. Particle sizes over 2 mm are considered gravel by BS but may be coarse sand according to ASTM.

Distinctions between silt, clay and organic soils are based on Atterberg Limits criteria applied to the part of the soil smaller than 425 μm. The plasticity charts defining these criteria in ASTM-D2487 and BS-5930 are also different. ASTM-D2487 defines silty clays on the plasticity chart in a zone bounded by Plasticity Indices of 4 and 7 percent, and Casagrande's A-line (Casagrande, 1948). In BS-

* Current address: Vrijhof Ankers B.V., P.O. Box 105, 2920 AC Krimpen aan den IJssel, The Netherlands

Volume 28: Offshore Site Investigation and Foundation Behaviour, 57–76, 1993.

clay, and the border between the two is formed by a plasticity index of 6 and Casagrande's A-line.

Many differences in the use of group symbols, main and secondary soil type names, terminologies, test procedures, and order of presentation in a sample description statement are also evident when comparing BS-5930 and ASTM-D2487.

For internationally operating companies, such as Fugro-McClelland, the application of different classification systems for different clients and areas is cumbersome and may create confusion. This led various companies to adopt their own practice for coping with the soil types encountered offshore. These practices have remained, by and large, inaccessible to the soil investigation community of practitioners and researchers.

This paper makes accessible a new system now in use by FME for the engineering classification of non-calcareous marine soils. The system is proposed as an alternative to BS-5930 and ASTM-D2487, and provides a consistent basis from which further extensions to other soil types can be included. The main criteria are given in detail after discussing key aspects of the system. This paper also suggests consistency description criteria and addresses differences between BS-5930, ASTM-D2487 and the new system by means of triangles based on particle size distributions. It will be shown that it is largely similar to BS-5930, but is more concise and provides other details on the composition of fine grained soils.

2. The Measurable Classification System

2.1. Description and Classification

Classification systems are used to identify the general properties of samples, thereby ordering the samples in an easily accessible system that permits correlation and interpretation. As such, classification forms part of the sample description, which describes general and unique features of samples.

Description of soil samples for engineering purposes can be made on the basis of relatively simple visual and manual procedures. Laboratory particle size distribution and Atterberg Limits tests serve to confirm the results of these procedures and can add confidence in estimation of engineering behaviour.

A sample description is expressed in a logical order via a sample description statement. Several sample description statement orders are shown in Table 1. Item 2 to 4 of the FME system cover the classification as discussed in this paper. Item 1 and 5 allow for future extensions. For sands the grading range is included in the main soil type name (see Figures 3 and 8). Other items mentioned by ASTM and BS are either included as additional information or covered elsewhere in a soil investigation report. This paper is concerned with items 2 to 5 of the FME system.

TABLE 1.
Sample description statement order.

Classification systems		
BS-5930	ASTM-D2488	FME System
1. Consistency (in capitals)	1. Group name	1. Particle type
2. Discontinuities	2. Group symbol	2. Main soil type (in capitals)
3. Colour	3. Percent of cobbles or boulders, or both (by volume)	3. Secondary soil type
4. Particle shape (gravel only)	4. Percent of gravel, sand or fines, or all three (by dry mass)	4. Group symbol (in brackets, optional)
5. Secondary soil types	5. Particle size range: Gravel – fine, coarse; Sand – fine, medium coarse	5. Consistency/cementation/ lithification
6. Grain size (sand or gravel)	6. Particle shape	6. Discontinuities
7. Primary soil type (in capitals)	7. Maximum particle size or dimension	7. Colour/colour code (in brackets)
8. Inclusion	8. Hardnesss of coarse sand and larger particles	8. Additional constituents
9. Geological formation (between brackets)	9. Plasticity of fines	9. Structure
	10. Dry strength	10. Additional information
	11. Dilatancy	
	12. Toughness	
	13. Colour	
	14. Odour	
	15. Moisture	
	16. Reaction with HCl	
	For intact samples:	
	17. Consistency (fine-grained soils only)	
	18. Structure	
	19. Cementation	
	20. Local name	
	21. Geological interpretation	
	22. Additional comments	

2.2. Key Aspects of the System

When developing the new system, the following key objectives were set out:

- It must embody and merge criteria from existing engineering classification systems that are found relevant in offshore practice.

- The system must be based on measurable data and translates into only one possible classification per sample. Particle size distribution and Atterberg Limits provide the measurable data.

- The system should be practical and simple to users. It should also be adaptable and extendable to include all the categories of soils encountered offshore in database and knowledge systems.

Categories to be covered in a geotechnical classification system are:

1. Coarse grained soils (gravels and sands)
2. Fine grained soils (silts and clays)
3. Organic soils
4. Cemented soils and rocks
5. Bioclastic soils.

This classification system addresses items 1 to 3. Of the organic soils, only organic clays and organic silts are included in this system. Future extensions will provide for peats, cemented soils, rocks and bioclastic soils.

Items 4 and 5 include the calcareous (and carbonate) soils. According to Clark and Walker (1977) calcareous soils are identified by a carbonate content of 10% or more. This definition would exclude a number of soils from the proposed classification system, which we believe should be included. Non-calcareous soils are therefore redefined as soils which offer no visible or mechanical clues to suspect a high carbonate content.

Main soil type names (MST), secondary soil type names (SST), group symbols and consistency descriptions are obtained from the following steps:

- Use flow chart in Figure 1 to determine MST name and identity number. Six MST categories and six grading ranges for sands are possible.

- Depending on MST, enter flow charts in Figures 2, 3 or 4 to determine SST name and identity number. A total of 46 SST combinations are possible.

- From the MST and SST identity numbers, determine the group symbol using Table 2. The grading criteria for sands and gravels are shown below. These criteria determine if the soil is poorly (P) or well (W) graded.

- Define the consistency using Table 3 or 4.

Following above steps will lead to only one possible classification per sample.

TABLE 2.
Soil group symbols.

		MAIN SOIL TYPE (MST)					
		GRAVEL	SAND	SILT	CLAY	ORGANIC SILT	ORGANIC CLAY
	G1–G3	(GW)*, (GP)*					
	G4–G9	(GM)					
	G10–G15	(GC)					
Secondary	S1–S3		(SW)**, (SP)**				
Soil	S4–S9		(SM)				
Type	S10–S15		(SC)				
(SST)	L1–L8			(ML)	(CL)	(OL)	(OL)
	H1–H8			(MH)	(CH)	(OH)	(OH)

* GW: $c_u > 4;\ 1 \leq C_c \leq 3$
GP: not meeting all gradation requirements for GW

** SW: $c_u > 6;\ 1 \leq C_c \leq 3$
SP: not meeting all gradation requirements for SW

where

C_u: coefficient of uniformity = D_{60}/D_{10}
C_c: coefficient of curvature = $(D_{30})^2/D_{10} * D_{60}$
D_{10}, D_{30}, D_{60}: average diameters corresponding to 10, 30, and 60% passing by dry weight respectively.

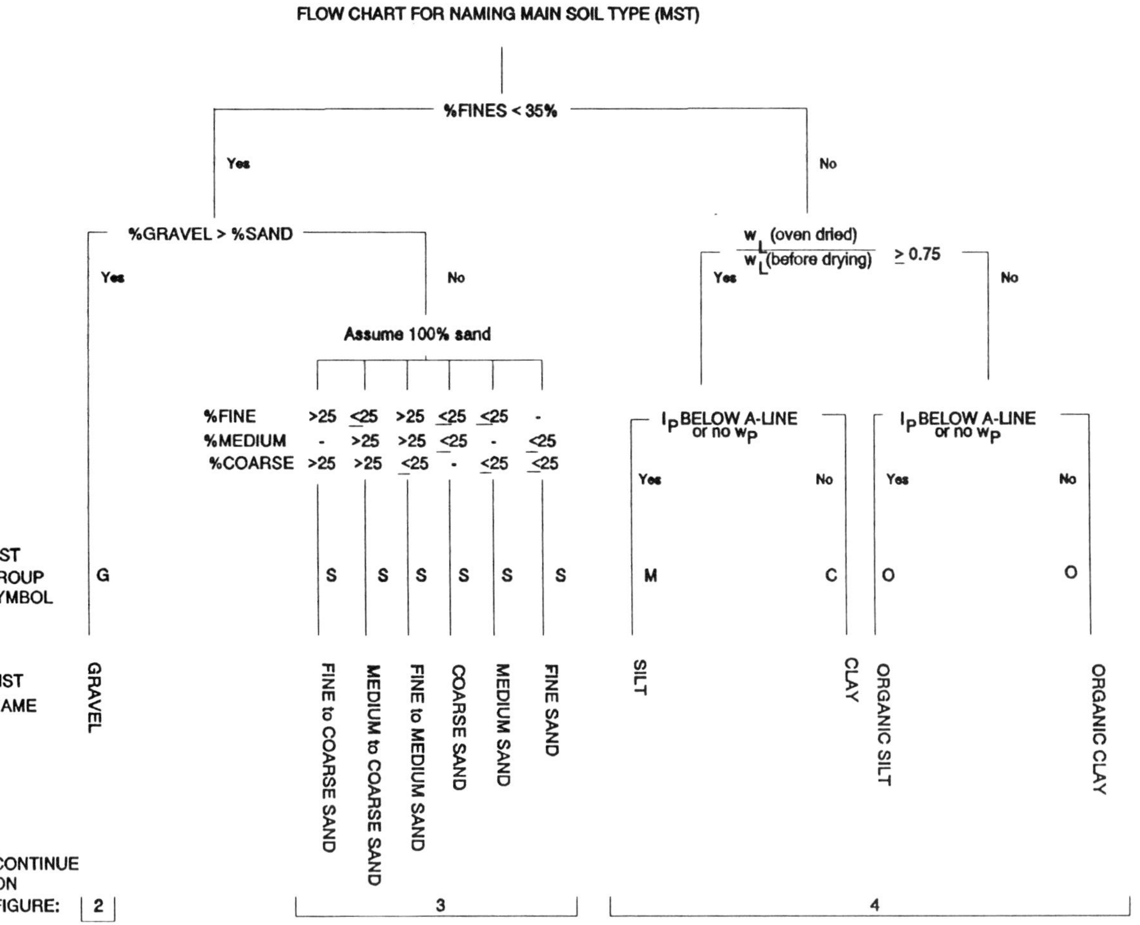

Fig. 1. Flow chart for naming main soil type (MST).

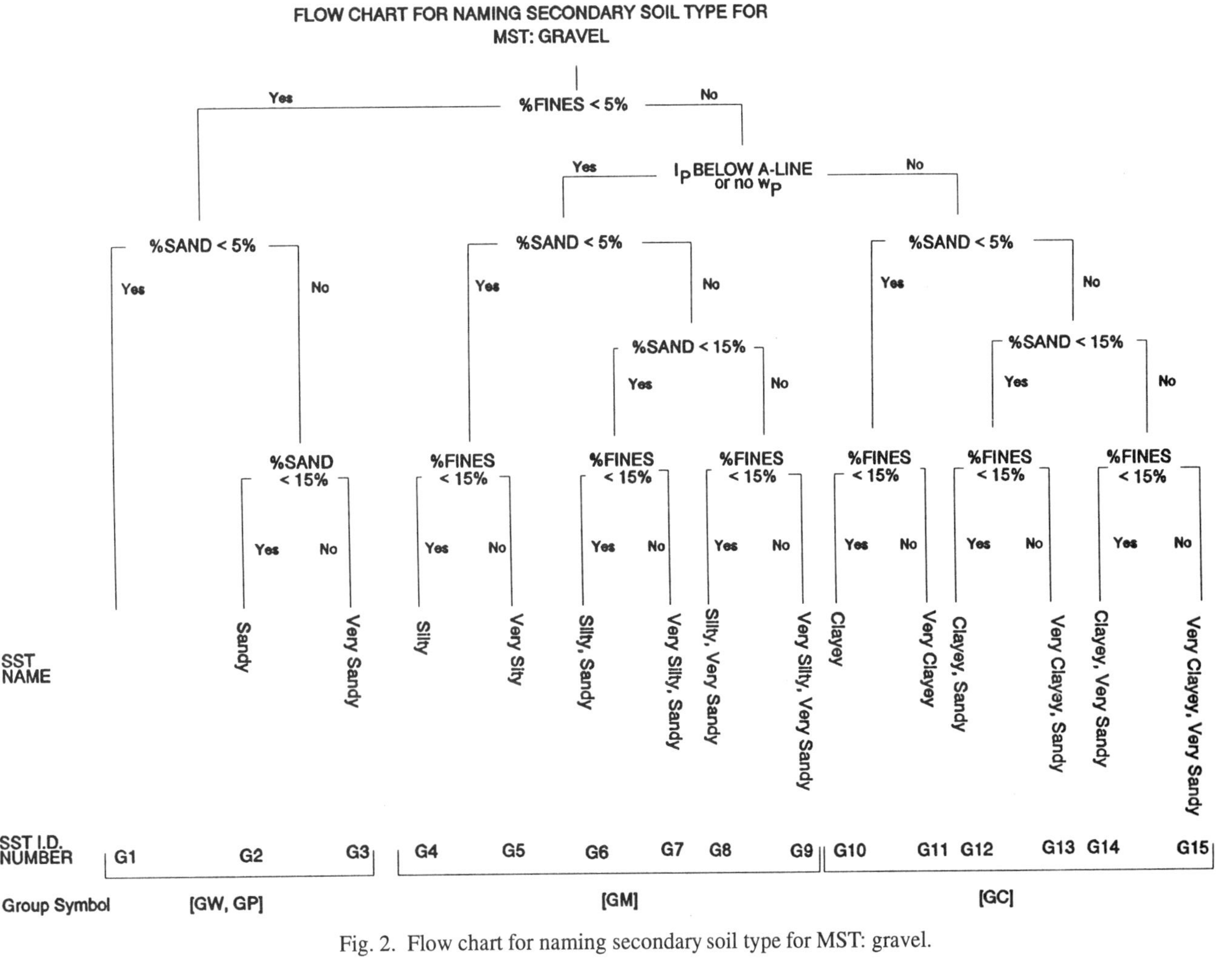

Fig. 2. Flow chart for naming secondary soil type for MST: gravel.

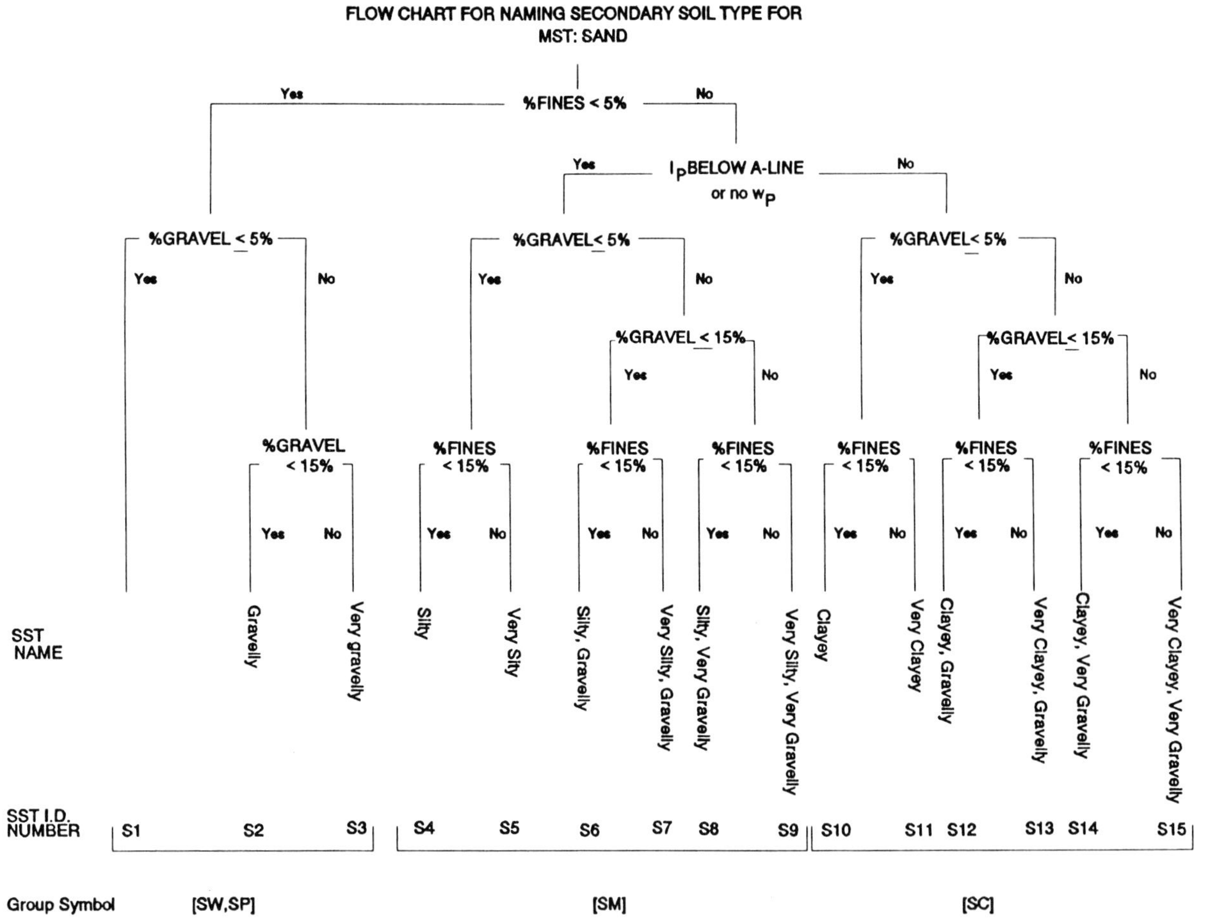

Fig. 3. Flow chart for naming secondary soil type for MST: sand.

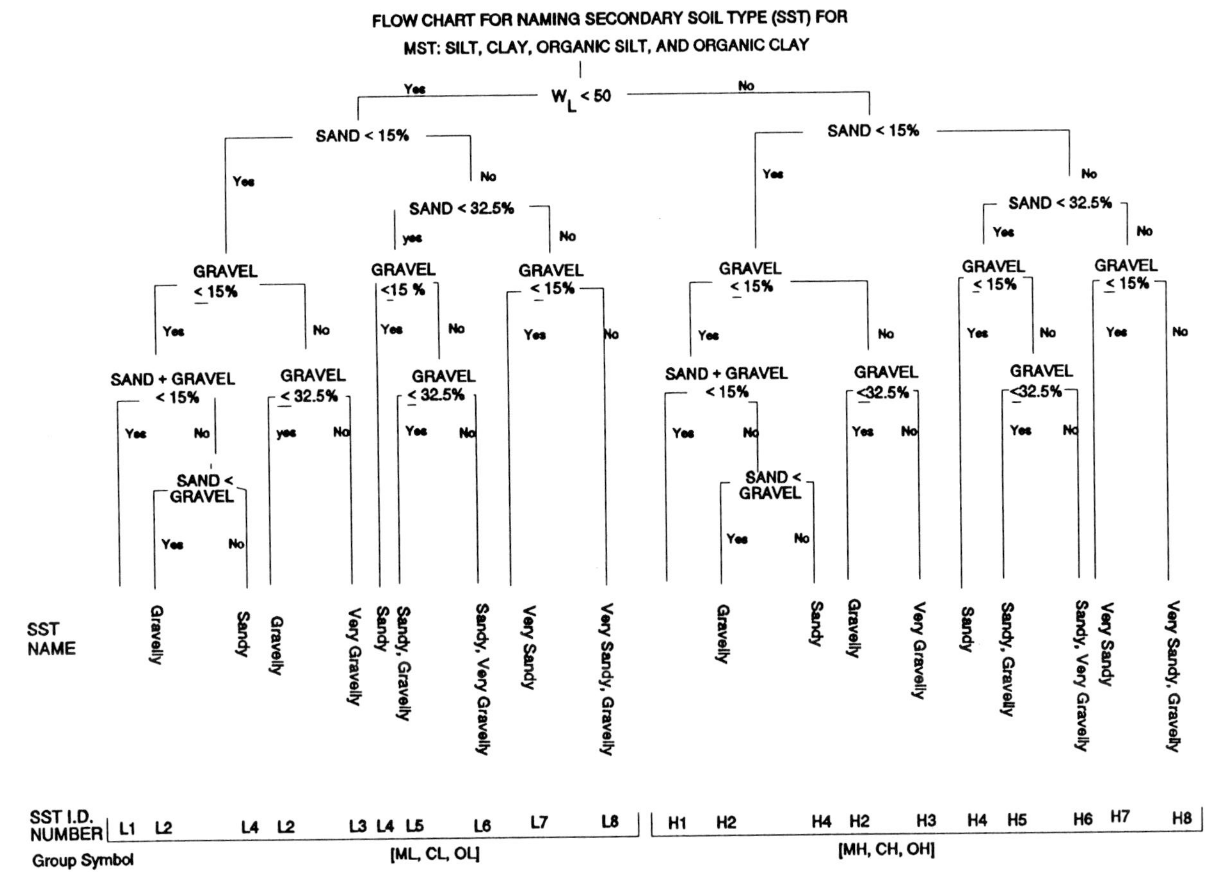

Fig. 4. Flow chart for naming secondary soil type (SST) for MST: silt, clay, organic silt, and organic clay.

2.3. Main Criteria

The main criteria of the new system are addressed in the following clauses:

1. The main soil type is clay or silt when the percentage fines (particles smaller than 0.06 mm) is 35% or more, as in BS-5930. Though the ASTM adopts a border of 50% (for particles smaller than 0.075 mm), in our experience more weight should be given to the percentage fines. Norbury *et al.* (1986) note that fine-grained and coarse-grained soils should be separated by Atterberg limits alone, which is by itself correct but would be impractical and could lead to breakdown of the "measurability" of the classification system.

2. Gravel is defined as the soil fraction ranging from 2 mm up to and including 60 mm, as in British Standards. For other soil types refer to Figure 5. This is a standard particle size distribution chart with soil subdivisions also based on BS-5930.

3. As a consequence of clauses (1) and (2), the used sieve set must include the 63 μm and 2 mm sieves. These sizes are specified for example by BS-1377 (1990), NEN 2560 and ISO 565. Since 60 mm sieves are not available, a standard 63 mm sieve is used.

4. Six grain size ranges for MST = SAND are used. The descriptions FINE, MEDIUM, COARSE, FINE to MEDIUM, FINE to COARSE and MEDIUM to COARSE are given to indicate a range for grain sizes. Ranges with a proportion of less than 25% of the sand fraction are ignored when describing the grain size range. The system is complemented by assigning the grading symbols.

5. Atterberg limit tests are carried out on materials smaller than 425 μm as in BS-5930 and ASTM-D2487. Drying and sieving are only performed when a significant coarse fraction cannot be removed by hand, since drying may considerably influence the Liquid Limit behaviour (Rao *et al.*, 1989).

6. No hydrometer test is required to distinguish between silts and clays. Instead, the distinction is based on the plasticity chart shown in Figure 6. In this chart, the A-line is defined by the equations $I_p = 0.73(w_L - 20)$ and $I_p = 6$ as shown. I_p and w_L are the Plasticity Index and Liquid Limit respectively. Casagrande's A-line (Casagrande, 1948) is used to separate silt and organic silt below from clay and organic clay on or above this line. The Liquid Limit boundary of 50%, separating low ($w_L < 50\%$) from high ($w_L \geq 50\%$) plasticity soils (taken from ASTM-D2487), is considered adequate for engineering classification purposes. Note that ASTM and BS specify a different Casagrande apparatus, which may result in differences of 3 to 4% in Liquid Limit values (Norman, 1958).

7. There is no silty clay or clayey silt classification either as primary or secondary soil constituents (as in BS-5930). Boundaries for such subdivisions are yet not well established.

8. Classify a sample with $\geq$ 35% fines as silt when the Plastic Limit test is not possible. This is based on practical observations. Soils containing a fair proportion of fine to medium sands and silt may give rise to difficulties in carrying out the Plastic Limit test.

9. The criteria for organic silts and organic clays are based on the Liquid Limits before and after oven drying as in ASTM-D2487 and BS-5930. The Liquid Limit of organic soils after oven drying is reduced by 25% or more.

10. SST names are qualified as given below:

    a for MST = SAND (or GRAVEL):
    - ignore associated SST name when fines < 5%, gravel $\leq$ 5% (or sand < 5%)
    - use gravelly (and/or sandy, silty, clayey) when SST is in the range 5% to 15%
    - use very gravelly (and/or very sandy, silty, clayey) for fractions of 15% or more
    - the SST name for the fine fraction (silt *or* clay) is defined in accordance with plasticity criteria from clauses (6) through (8)

    b for MST = CLAY, SILT, ORGANIC CLAY and ORGANIC SILT:
    - ignore fractions less than 15%
    - use gravelly (and/or sandy) when SST is in the range 15% to 32.5%
    - use very sandy (and/or very gravelly) when sand $\geq$ 32.5% (and/or gravel > 32.5%)
    - when sand and gravel fractions are less than 15% but the total of these fractions is 15% or more, use sandy (or gravelly) depending on largest fraction (as in ASTM).

11. In the case when the particle size distribution curve does not extend to D_{10} (in the case of 10% passing), extrapolate to obtain D_{10}. This is in accordance with ASTM-D2487.

12. Group symbols are kept simple. There are 14 group symbols in this system as shown in Table 2. Use the grading quality coefficients of uniformity C_u and curvature C_c, as defined below Table 2, to assign P or W for sands and gravels containing less than 5% fines. The criteria shown below the table are from ASTM-D2487.

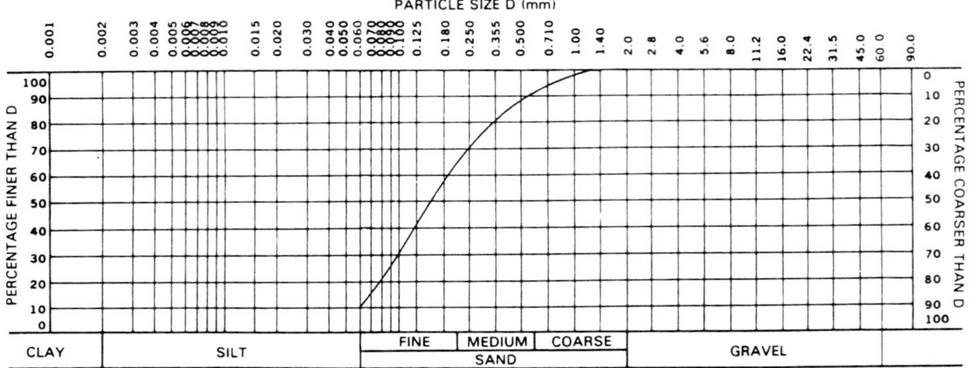

Fig. 5. Particle size distribution chart.

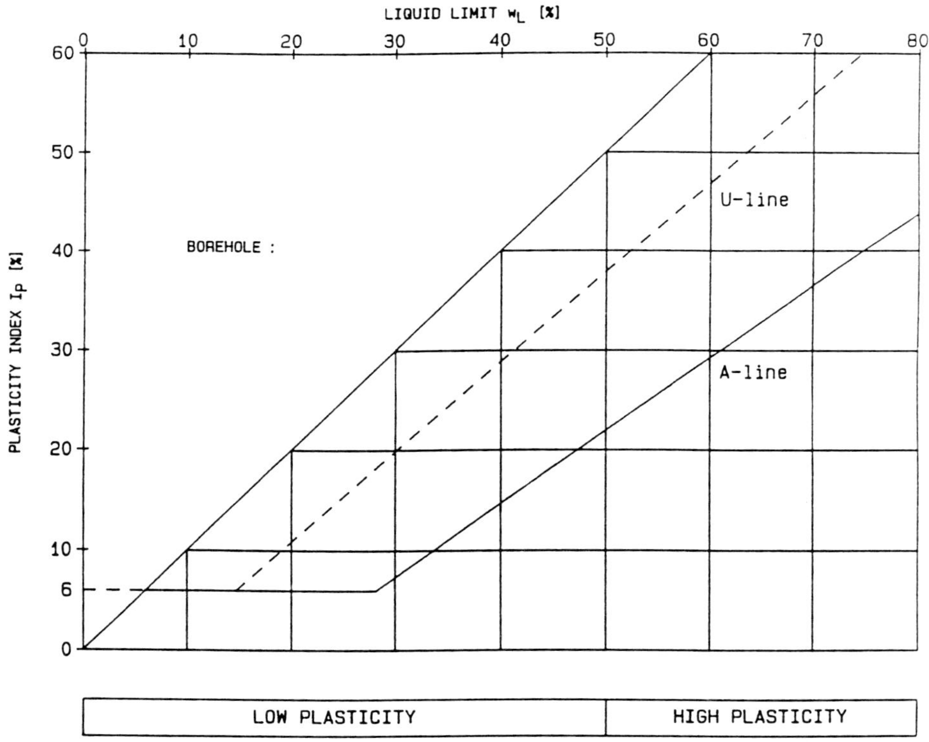

Fig. 6. Plasticity chart.

TABLE 3.
Consistency description of fine grained soils.

CLASSIFICATION SYSTEM					
BS-5930		ASTM–D2488*		FME SYSTEM**	
Term	c_u[kPa]	Term	c_u[kPa]	Term	c_u[kPa]
Very soft	< 20	Very soft		Very soft	0–12.5
Soft	20–40	Soft	not defined	Soft	12.5–25
Firm	40–75	Firm		Firm	25–50
Stiff	75–150	Hard		Stiff	50–100
Very stiff or hard	> 150	Very hard		Very stiff	100–200
				Hard	200–400
(also)				Very hard	> 400
Soft to Firm	40–50				
Firm	50–75				
Firm to Stiff	75–100				
Stiff	100–150				

* – gives manual procedure only
** – taken from the Canadian Foundation Engineering Manual (1985), Table 3.3, p. 31.

TABLE 4.
Density description for granular soils.

CLASSIFICATION SYSTEM			
Term	BS-5930*	ASTM	FME (Relative Density)
Very loose			0–0.2
Loose			0.2–0.4
Medium dense	Based	Not	0.4–0.5
Dense	on	required/	0.6–0.8
Very dense	SPT	not defined	0.8–1.0

* – manual description only gives loose or dense in Table 6 (BS-5930) for sands and gravels.

2.4. Consistency Descriptions

Consistency descriptions are given in Tables 3 and 4. The undrained shear strength classes for fine grained soils (Table 3), and relative density descriptions for granular soils (Table 4), are based on the sources quoted below these tables. These descriptions vary between different institutes as shown. The use of generally applied definitions is particularly important in marine soils for correlation purposes. References to the density descriptions applied by FME may be found in e.g. Sanglerat (1972) and Fang (1991). For cohesive soils FME has adopted the definitions from the Canadian Geotechnical Society, because they are unambiguous. These definitions also provide the much-needed detail for geotechnical analysis. The very hard description ($c_u > 400$ kPa) has been added because of the high shear strengths often encountered offshore.

Estimation of density conditions of coarse grained soil samples is usually difficult without complementary techniques such as Cone Penetration Tests. The reason for this is the inevitable substantial sample disturbance occurring during conventional sampling operations. Where practicable, an estimate of density condition is recorded as part of a sample description statement, so as to allow some comparison between laboratory and *in-situ* techniques.

A similar comment on recording the consistency of fine grained soils also applies. In case of high quality undisturbed samples, estimates of sample consistency can be reliable, but complementary laboratory and *in-situ* testing are recommended. Note that when dealing with inhomogeneous cohesive soils, manual procedures may be the only way to determine consistency.

3. Comparisons

The criteria shown in Figures 1 to 4 and discussed in Section 2 are compared in Figures 6 to 8. Figure 6 is the familiar plasticity chart, presenting the Atterberg limits for distinguishing clays and silts. Other criteria based on grain size distribution are represented by the triangles of Figures 7 and 8. All borders in these triangles are marked with two lines. The dotted line indicates that the border is included in the soil group shown on the same side of the drawn line. These triangles express precisely the boundaries separating soil types and hence facilitate comparison with other systems.

Similar triangles as shown in Figure 7 are drawn for the BS and ASTM criteria (Plates 9 and 10). Figure 9 clearly shows that a number of boundaries are undefined by BS-5930, i.e. samples possessing these properties may be classified in several ways. While the effects of this are only marginal given the inaccuracies in determining grain size distributions, undefined boundaries are undesirable for measurable classification systems (by definition).

The drawn triangles also show that the classification of coarse grained soils is roughly the same using all three systems. The fundamental difference is the separation between fine and coarse grained soils. As discussed previously, we have selected the boundary %fines < 35%, which corresponds to BS.

Figure 9 indicates that the British Standards show little detail in describing the composition of fine grained soils. In our experience it is possible to provide more differentiation by both manual procedures and laboratory testing. We have therefore subdivided the fine grained soils more than BS does, yielding a system

FME

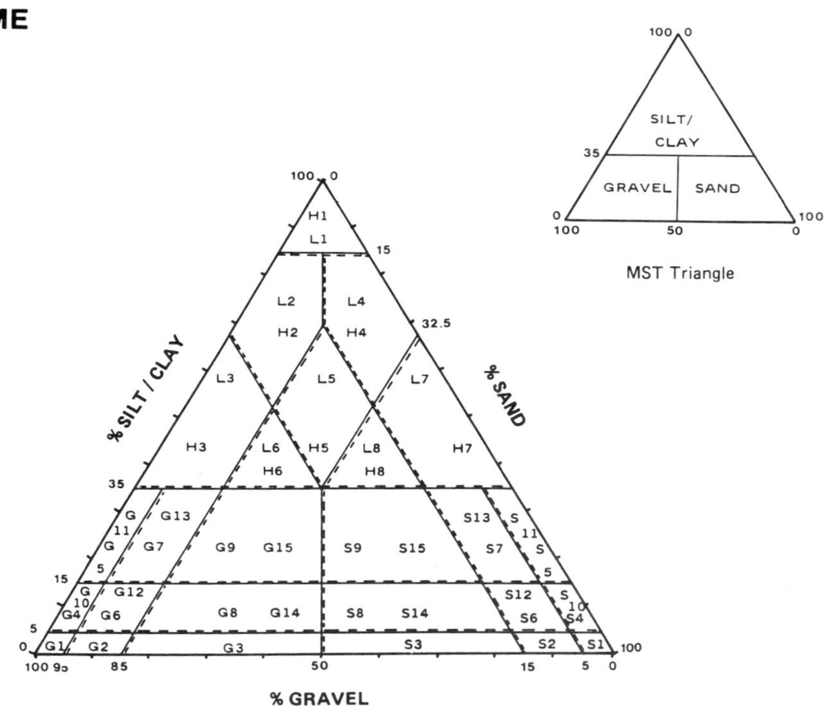

Fig. 7. SST triangle.

similar to the ASTM (Figure 10). Because of the adoption of the boundary %fines $< 35\%$, the ASTM could not be applied fully for fine grained soils. The borders %sands $\geq 32.5\%$ and %gravel $> 32.5\%$ are introduced which removes the necessity of a boundary of %fines $> 70\%$.

The new system thus shows general agreement with BS-5930 main soil names and grouping. Differences are mainly in the names of secondary soils and the plasticity symbol. BS-5930 employs more divisions on Liquid Limits by assigning L, I, H, V and E in combination with silt (M) and clay (C). The elaborate division by BS-5930 for the Liquid Limit is found cumbersome and unnecessary in engineering practice and is replaced in the new system by the simpler divisions from ASTM-D2487.

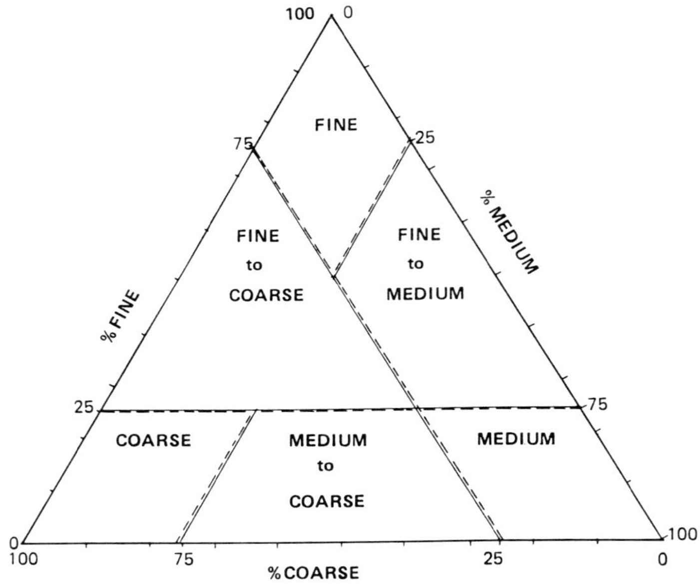

Fig. 8. FME grading triangle for sand.

4. Conclusions and Recommendations

A measurable soil classification system now used by FME has been described in this paper. It is proposed as an alternative to BS-5930 and ASTM-D2487 for non-calcareous marine soils. Use of the new system results in one possible classification per soil sample. This is achieved by means of simple flow charts and a table of soil group symbols. The measurable criteria are furnished by Atterberg Limits and particle size distribution test results. The system can thus be translated into computer code quite easily.

Combinations of grading names for sands, main and secondary soil names, group symbols and a consistency description can be concisely determined from the new system. The system gives a total of 46 soil names. Boundaries are determined by grain size, grading, percent of material type and plasticity, separating the various soil classes.

No silty clay or clayey silt classification is used in the new system since appropriate boundaries for these soils did not emerge from this study. These boundaries need to be defined by fundamental research before they are incorporated in any system.

Generally, the new system was found to be more in agreement with BS-5930 than with ASTM-D2487. All these systems, however, differ in some respects as outlined in the paper.

This system can be recommended for future extensions to include peat, cemented soils, rocks and bioclastic soils.

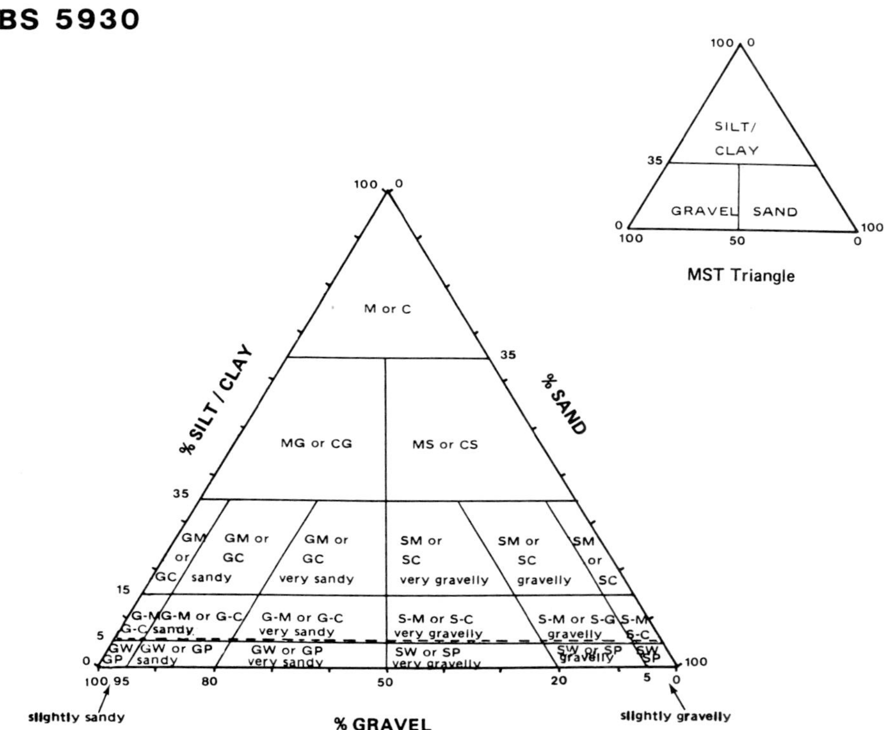

Fig. 9. SST triangle.

Acknowledgements

The authors wish to acknowledge the useful remarks and discussions with Mr. J. Peuchen, Senior Engineer during the development of the system, and further comments on the paper from the senior engineering staff. The quality typing by Miss E. van Noort, and drafting by Mr. M. Goodett are gratefully acknowledged. Permission to publish has been granted by the Directors of Fugro-McClelland Engineers B.V., The Netherlands.

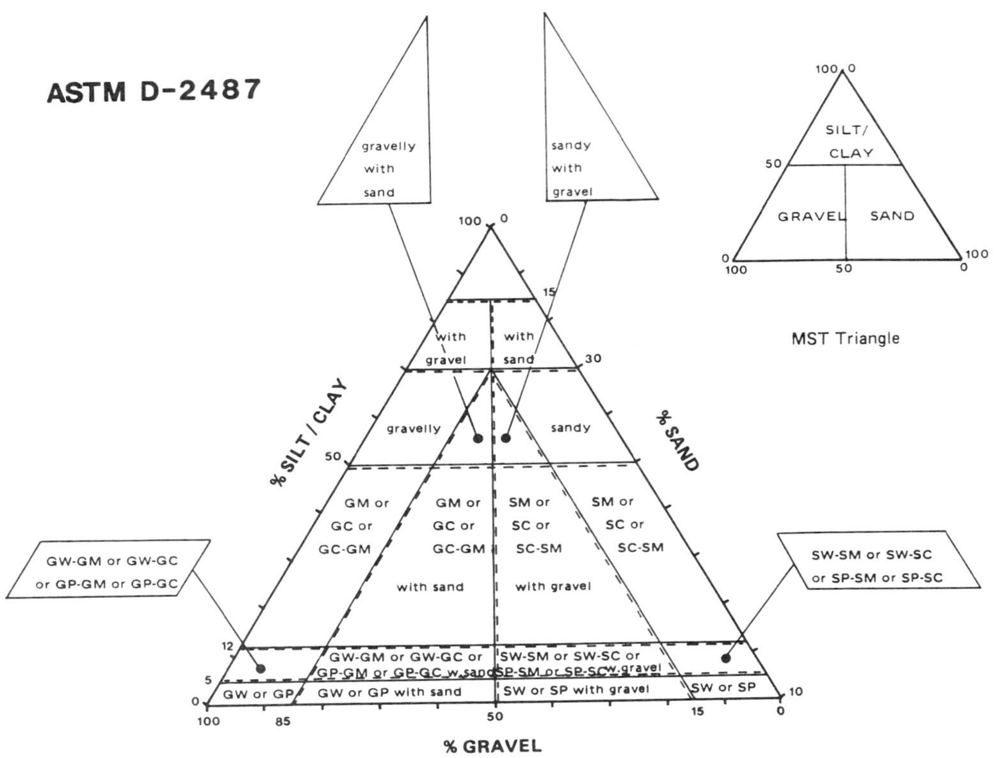

Fig. 10. SST triangle.

List of Symbols

C_c	Coefficients of Curvature
C_u	Coefficient of Uniformity
c_u	Undrained Shear Strength
D_n	Average diameter corresponding to $n\%$ passing by weight
I_p	Plasticity Index
w_L	Liquid Limit
w_P	Plastic Limit

References

1. Al-Hussaini, M. M. (1977), 'Contribution to the Engineering Soil Classification of Cohesionless Soils', Final Report Papers S-77-21, U.S. Army Eng. Waterways Experiment Station, Vicksburg, Mississippi.
2. ASTM-D2487 (1991), 'Standard Test Method for Classification of Soils for Engineering Purposes', American Society for Testing and Materials, U.S.A.
3. ASTM-D2488 (1991), 'Standard Practice for Description and Identification of Soils – Visual/Manual Procedure', American Society for Testing and Materials, U.S.A.
4. BS-1377 (1990), 'British Standard Methods of Test for Soils for Civil Engineering Purposes', Part 2, British Standards Institution, London.
5. BS-5930 (1985), 'Code of Practice for Site Investigations', Section 8, British Standards Institution, London.
6. Canadian Geotechnical Society (1985), 'Canadian Foundation Engineering Manual', 2nd edition, BiTech Publishers Ltd., Vancouver, Canada.
7. Carter, M. and Bentley, M. (1991), *Correlation of Soil Properties*, Pentech Press Ltd., London.
8. Casagrande, A. (1948), 'Classification and identification of soils', *ASCE Proc.* **73**(6), 783–810.
9. Clark, A. R. and Walker, B. F. (1977), 'A proposed scheme for the classification and nomenclature for use in the engineering description of Middle Eastern sedimentary rocks', *Géotechnique* **17**(1), 93–99.
10. Fang (1991), *Foundation Engineering Handbook*, 2nd edition, Von Nostrand Reinhold, New York.
11. Fookes, P. G. and Higgenbotham, I. E. (1975), 'The classification and description of near-shore carbonate sediments for engineering purposes', *Géotechnique* **25**(2), 406–411.
12. Liu, T. K. (1967), 'A review of engineering soil classification systems', *Highway Research Board Records* **156**, 1–22.
13. NEN 2560 (1980), 'Controlezeven – Draadzeven en Plaatzeven met Ronde en Vierkante Gaten', Dutch Standardisation Institute, Delft.
14. Norbury, D. R., Child, G. H., and Spink, T. W. (1986), 'A critical review of Section 8 (BS-5930) – Soil and rock description', *Geological Society Special Publication* **2**, 331-342.
15. Norman, L. E. J. (1985), 'A comparison of values of Liquid Limit determined with apparatus having bases of different hardness', *Géotechnique* **8**(2), 79–83.
16. Poulos, H. G. (1988), *Marine Geotechnics*, Unwin Hyman Ltd., London.
17. Rao, S. M., Sridharan, A., and Chandrakaran, S. (1989), 'Influence of drying on the Liquid Limit behaviour of a marine clay', *Géotechnique* **39**(4), 715–719.
18. Sanglerat, G. (1972), *Developments in Geotechnical Engineering*, Elsevier Publishing Company, Amsterdam, The Netherlands.

Discussion

Question from C. H. Price, Kent, UK: While I agree with the general principle of not using the term silty clay or clayey silt, I would suggest that care is taken not to put the field engineering geologist into a straight jacket which forces him to make errors. Descriptions often have to be made without laboratory test results as back-up, using judgement and experience. With most soils a small mis-estimate will result in a slightly different description. However, in the case of clay/silt soils it is very easy to misjudge a silty clay as a clayey silt and *vice versa*. Using these terms, we have a very good idea of the type of material concerned. However, without the intermediate silty/clayey option the hand descriptions could easily become badly incorrect. What is more, there will be no warning that the particular soil should be more carefully tested for re-categorisation as there is at present when soil is called a silty clay or clayey silt.

Authors' response: The proposed system is *measurable*, as implied by the paper title, and *not* a manual/visual system. Therefore, the final verdict is always based on data furnished by engineering tests (Atterberg and grading tests).

In the field, a clay/silt description (not classification) or even clayey silt and silty clay description should give sufficient warning to carry out engineering classification tests and invoke the proposed system.

Furthermore, field engineers must be warned of the difference between visual and laboratory-based systems regarding this issue. Assigning SILT or CLAY to a sample on site must only be done when experience and manual procedures suffice.

GEOTECHNICAL PROPERTIES OF SEDIMENTS FROM THE CONTINENTAL SLOPE NORTHWEST OF THE BRITISH ISLES

M. A. PAUL and L. A. TALBOT
Department of Civil and Offshore Engineering, Heriot-Watt University, Edinburgh EH14 4AS, Scotland

and

M. S. STOKER
British Geological Survey, Murchison House, West Mains Road, Edinburgh EH9 3LA, Scotland

Abstract. To the west of the Shetland Islands the continental margin is constructed from a thick sequence of sediments of Pliocene to Holocene age. In this paper the sedimentology and geotechnical properties of five vibrocores are described. The cores selected for study can be divided into two groups on the basis of their sedimentology, which relate to the acoustic signatures of the horizons from which they were collected. The first group consists of clay-silts of probable mass flow origin. These sediments are acoustically transparent and the cores are lithologically very uniform throughout their depth. The second group is drawn from sediments which are acoustically well-layered at the sample sites, and the cores show major facies variations between muds, muddy silts and sandy silts. Consistent differences are seen between the two groups over a wide range of engineering properties, including water content, grading, plasticity limits, undrained shear strength and compressibility. These differences arise from the sedimentological character of the cores, in particular the mud content and the presence of depositional facies variations, which exert a strong control on the engineering behaviour. Those facies which are rich in mud have high plasticity, high water contents, low strength and high compressibility, whereas those facies which are low in mud content are of correspondingly low plasticity, high density and low compressibility.

1. Introduction

This paper reports the results of geotechnical investigations that have been carried out on sediments collected from the northern UK continental margin during the BGS regional mapping programme. This geotechnical study was conducted in order to capitalise on the availability of soft sediment cores collected in deep water from an area of potential scientific and commercial interest. Its primary objective was to collect geotechnical data within a well studied regional geological framework in order to relate the geotechnical properties of the samples to both their sedimentology and seismic setting, and hence to the depositional processes which formed them. In so doing a better understanding has been gained of the possible complexities in both geotechnical and seismic facies models for this area.

Volume 28: Offshore Site Investigation and Foundation Behaviour, 77–106, 1993.

2. Geological Setting

The study area is located at the southwest end of the Faeroe-Shetland Channel, a narrow, northeasterly-trending, deep-water basin separating the West Shetland and Faeroe shelves (Figure 1a). To the northeast, the channel slopes down into the Norwegian Sea; to the southwest, the channel turns northwestwards and continues as the Faeroe Bank Channel which spills out into the Iceland Basin. The Wyville-Thomson Ridge separates the Faeroe-Shetland Channel from the Rockall Trough to the south. Water depths in the channel increase from about 100m in the southwest to about 1700m in the northeast. The width of the channel (at the 100m isobath) increases to the northeast, from 15 to 100km.

Although a major basinal area has probably existed in this region since the late Palaeozoic (Haszeldine *et al.*, 1987), the present bathymetric configuration largely developed during the Tertiary and Quaternary periods. Following the early Tertiary opening of the northeast Atlantic Ocean, the Faeroe-Shetland area was dominated by tectonic quiescence and regional thermal subsidence, with the deposition of a Neogene to Quaternary clastic wedge on the southeast margin of the Faeroe-Shetland Channel (Stoker, 1990). This clastic wedge varies in thickness from 400m on the outer shelf and upper slope to less than 60m in the channel. The relatively restricted accumulation of sediment in the channel is probably a reflection of intense bottom current activity which has prevailed since the mid-Tertiary (Miller and Tucholke, 1983; Edholm, 1990).

The Pliocene to Holocene section of the clastic wedge has been subdivided into a number of seismostratigraphical sequences (Figures 1b and 1c: cf. Stoker, 1990), distinguished primarily by their external geometry and internal acoustic signature. The sediments described in this paper belong predominantly to the Morrison and Faeroe-Shetland Channel sequences. The former is a major slope-front succession which records the Plio-Pleistocene progradation of the West Shetland margin. The Faeroe-Shetland Channel sequence is the deep-water basinal equivalent of the Morrison sequence (Figure 1d).

On seismic profiles, the Morrison sequence is characterised by numerous transparent and structureless to chaotic lensoid units, forming packages up to 50m thick, interbedded with acoustically well-layered sediments. The lensoid bodies are interpreted as mass flow deposits which, in the upper part of the sequence, are predominantly glacigenic in origin (Stoker *et al.*, 1991). The accumulation of these deposits was episodic and related to specific rapid phases of downslope resedimentation, most probably concomitant with ice-marginal deposition on the West Shetland Slope, fed by ice-sheets located at the shelf edge (Stoker and Holmes, 1991; Stoker *et al.*, in press). The acoustically well layered sediments may represent the more pervasive 'background' glaciomarine and marine sedimentation, often diluted in areas of thick debris flow accumulation. On parts of the West Shetland Slope the layered signature is locally disturbed and shows a crumpled appearance, possibly the result of small-scale downslope movement. A thin, Late to Postglacial

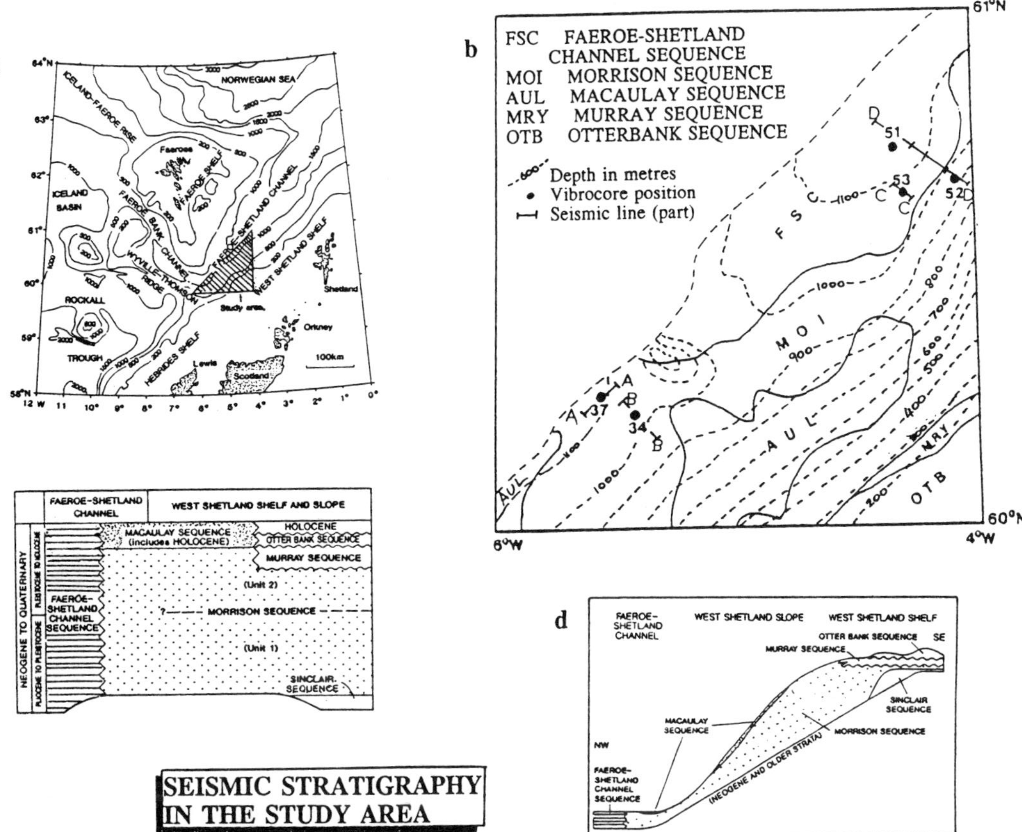

SEISMIC STRATIGRAPHY IN THE STUDY AREA

Fig. 1. Location maps and seismic sequences in the study area. (a) The location of the study area north of Scotland. (b) Locations of seismic cross sections and vibrocores in the study area in relation to the outcrop of the seismo-stratigraphical sequences. (c) Schematic age relationships of the seismo-stratigraphical sequences. (d) Schematic geological section across the study area.

veneer (< 10m thick), termed the MacAulay sequence, has been locally identified on the slope and channel floor overlying the uppermost debris flow units of the Morrison sequence.

The Faeroe-Shetland Channel sequence displays a uniform, well layered reflection character lacking any seismically resolvable lensoid bodies. Basal downlap and onlap, lateral accretion, thinning over palaeotopographic highs and internal convergence of reflectors all suggest that bottom current activity has greatly influenced the development of the sequence. Evidence of biological activity (Akhurst, 1991, and below) suggests that sedimentation in the basin was intermittent and thus the sediments were able to be reworked by bottom currents. This includes the reworking of material derived through rain-out from surface plumes and iceberg rafting, together with sediment carried into the channel by downslope processes.

The boundary between the Morrison and Faeroe-Shetland Channel sequences is taken at the seismically resolvable limit of the slope-front debris flow units. Although the sequence boundary is depicted as a line on the geological map (Figure 1b) it is most probably an interdigitating contact over several kilometres (Stoker, 1990). With the exception of the thick debris flow units, there is probably considerable overlap in the nature and style of sedimentation between the two sequences, particularly in the area of the lower slope and channel floor.

3. Sediment Description

Figure 1b shows the locations of five vibrocores (each 5–6m in length) which were selected for study. These cores were collected from water depths between 900 and 1200m on the southeast flank of the Faeroe-Shetland Channel and penetrated sediments from the upper part of the Morrison and Faeroe-Shetland Channel seismic sequences (Figures 2a to d). Both laminated and structureless seismic facies occur in this area and the cores were chosen to illustrate examples of both seismic signatures. Their lithology is described below: they fall into two distinct groups based on their sedimentology, a grouping which also in this case matches their seismic signature (but cf. Stoker *et al.* (this volume)).

3.1. Group One Cores

This group contains cores 60-06/34 and 60-06/37, both of which are from acoustically structureless horizons within the Morrison sequence, and are composed of uniform sandy muddy silts which vary little down their length (Figure 3). At the top of both cores there is a loosely consolidated layer of dark greyish brown (2.5Y 4/2) muddy sand or sandy silt, which is coarser than the remainder of the cores. This seabed veneer, which is believed to be of Holocene age, may be equivalent to the MacAulay sequence. Below this the cores are composed of very dark grey (5Y 3/1) sandy silty clay. All samples in each core have a very similar grain size distribution (Figure 4) with only slight differences between the two cores. Core 34

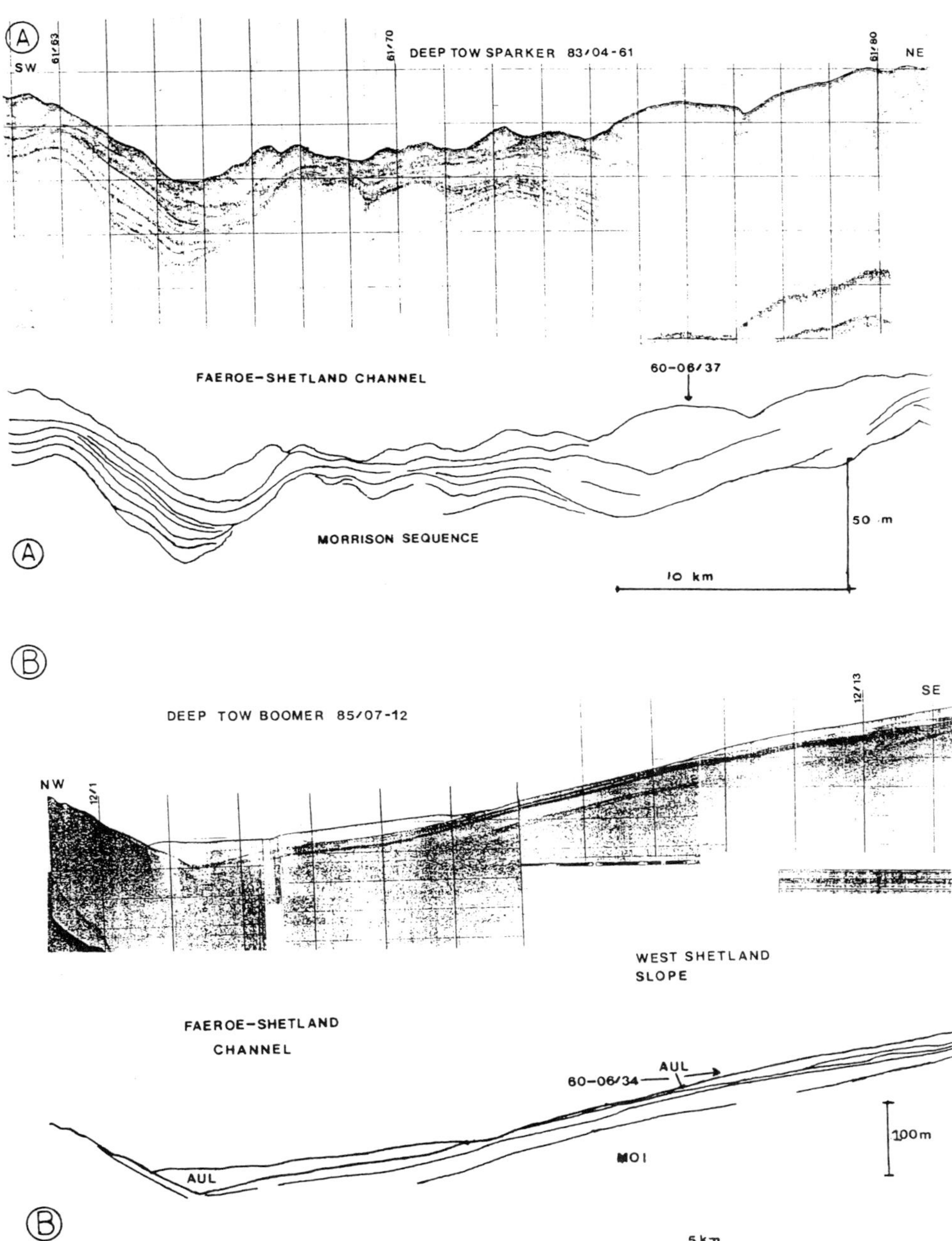

Fig. 2a-b. Seismic cross sections at the locations of the five vibrocores discussed in the text (locations of sections shown in Figure 1(b). (a) Section AA: core 37; (b) Section BB: core 34.

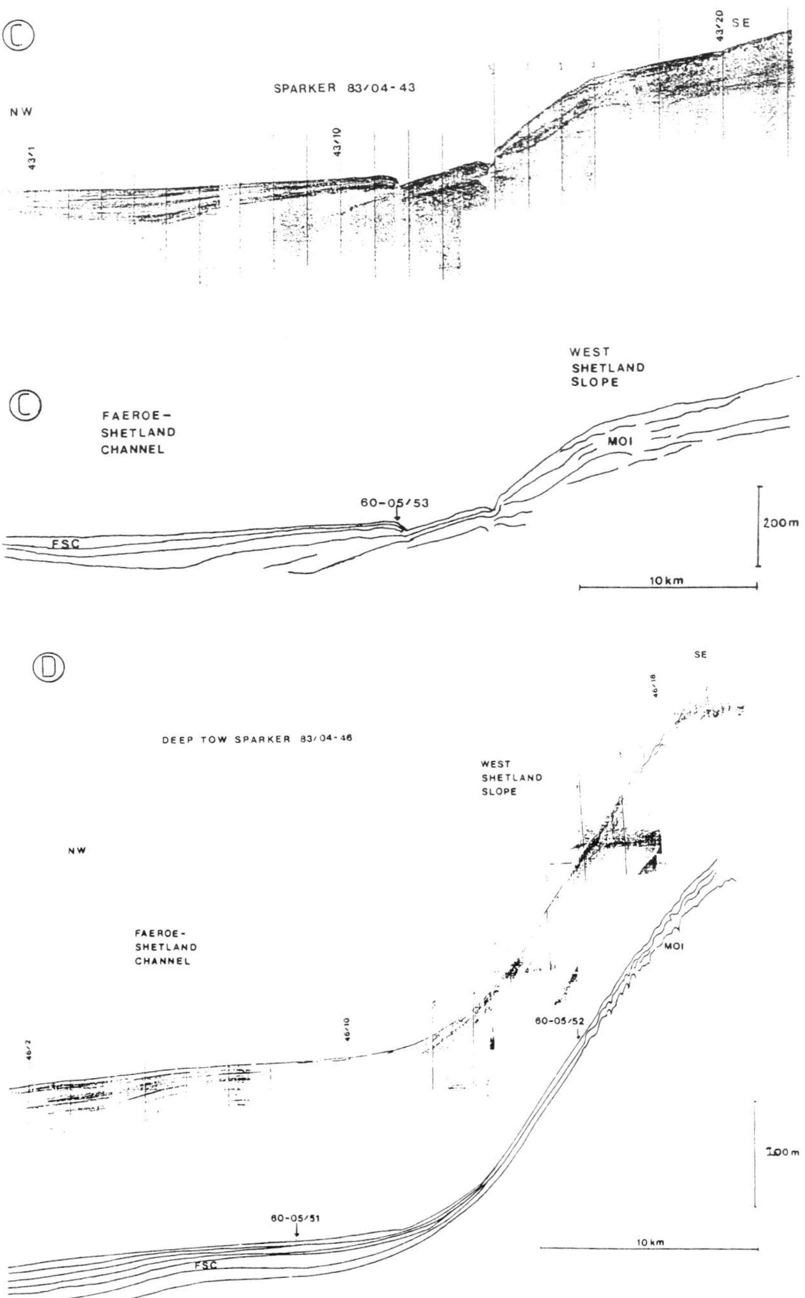

Fig. 2c-d. (c) Section CC: core 53; (d) Section DD: cores 51 and 52.

contains approximately 60% sand (> 4μm) with a 1:1 split between silt (4–8μm) and clay (< 8μm); core 37 contains approximately 50% sand, approximately 30% silt and approximately 20% clay. The only exception is the sandy layer at 0.10m in core 37 with 40% coarser than 3μm (fine sand) and only 12% clay (9μm). The variation of the clay percentage with depth is shown in Figure 5. This consistency leads to similar consistency in other geotechnical properties, as discussed below.

Gravel sized particles (> 2mm) comprise less than two weight percent. Clasts up to 50mm diameter occur infrequently and may have originated as glacial dropstones, indicating derivation from a glaciomarine source, and have been rafted into final position by downslope flow. A few broken shell fragments are present but bioturbation is limited, the burrows being irregular in shape and usually mud filled. Monosulphide knots and streaks occur at intervals.

3.2. Group Two Cores

This group comprises cores 60-05/51, 60-05/52 and 60-05/53 which are from acoustically well layered horizons within both the Morrison and Faeroe-Shetland Channel sequences. The sediments are lithologically variable (from sandy silt to mud: Figure 6) and contain major biogenic features.

The uppermost layer of cores 51 and 53, both from the Faeroe- Shetland Channel sequence, is formed from Holocene sand; in core 52, from the Morrison sequence higher up the continental slope, the uppermost layer is a slightly finer muddy sand. These layers may be laterally equivalent to the MacAulay sequence.

The remaining sediments in these cores can be divided into four lithofacies, three of which occur in all three cores. These lithofacies are very similar to those reported by Akhurst (1991) from the floor of the Faeroe-Shetland Channel northeast of our study area. The vertical transitions between facies in these cores leads to considerable changes in the clay percentage with depth (Figure 7), unlike the cores in Group One. This has important implications for the geotechnical properties of these sediments, as discussed below.

The vertical distribution of the lithofacies is shown in Figure 3, and the sedimentological details are as follows:

Mud lithofacies (comprises about 40% of the core material). The sediment in this facies is dark grey (5Y 4/1) and contains 30–50% clay size. The remainder is 10–30% fine silt, up to 40% coarse silt and almost no sand size or larger. The two dominant populations in all samples of the mud lithofacies are at approximately 5μm (coarse silt) and 9μm (clay), with minor amounts of fine silt in all samples. The sorting is generally poor. Bioturbation is present in roughly 20% of the facies: the burrows are usually isolated and of *Zoophycos* type. Small shell fragments are also found associated with the burrowing. Black monosulphide patches and streaks are occasionally present. The beds are between 15cm and 2m thick, and tend to be gradational with the underlying and overlying sandy mud. Any original bedding surfaces have been completely destroyed by bioturbation.

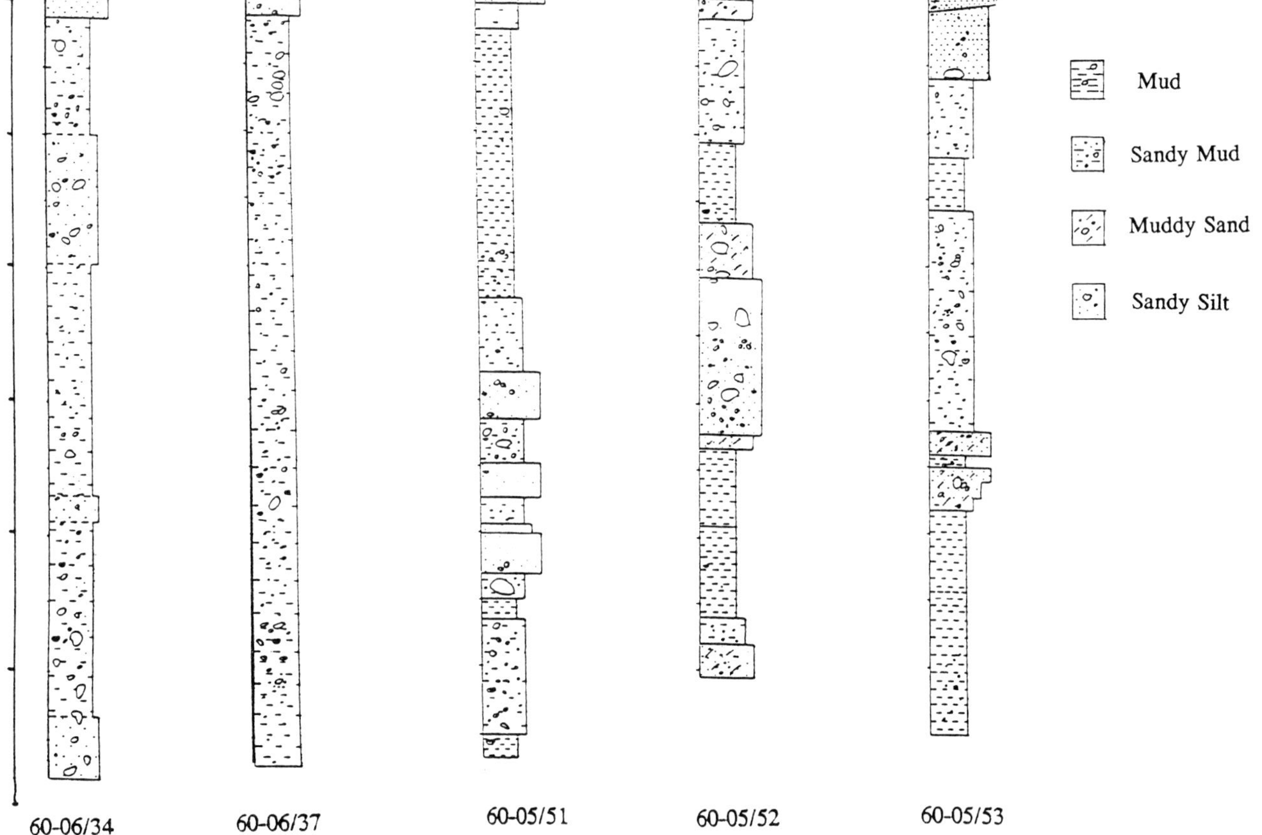

Fig. 3. Lithological logs of the vibrocores described in the text. Sample locations are shown in Figure 1(b).

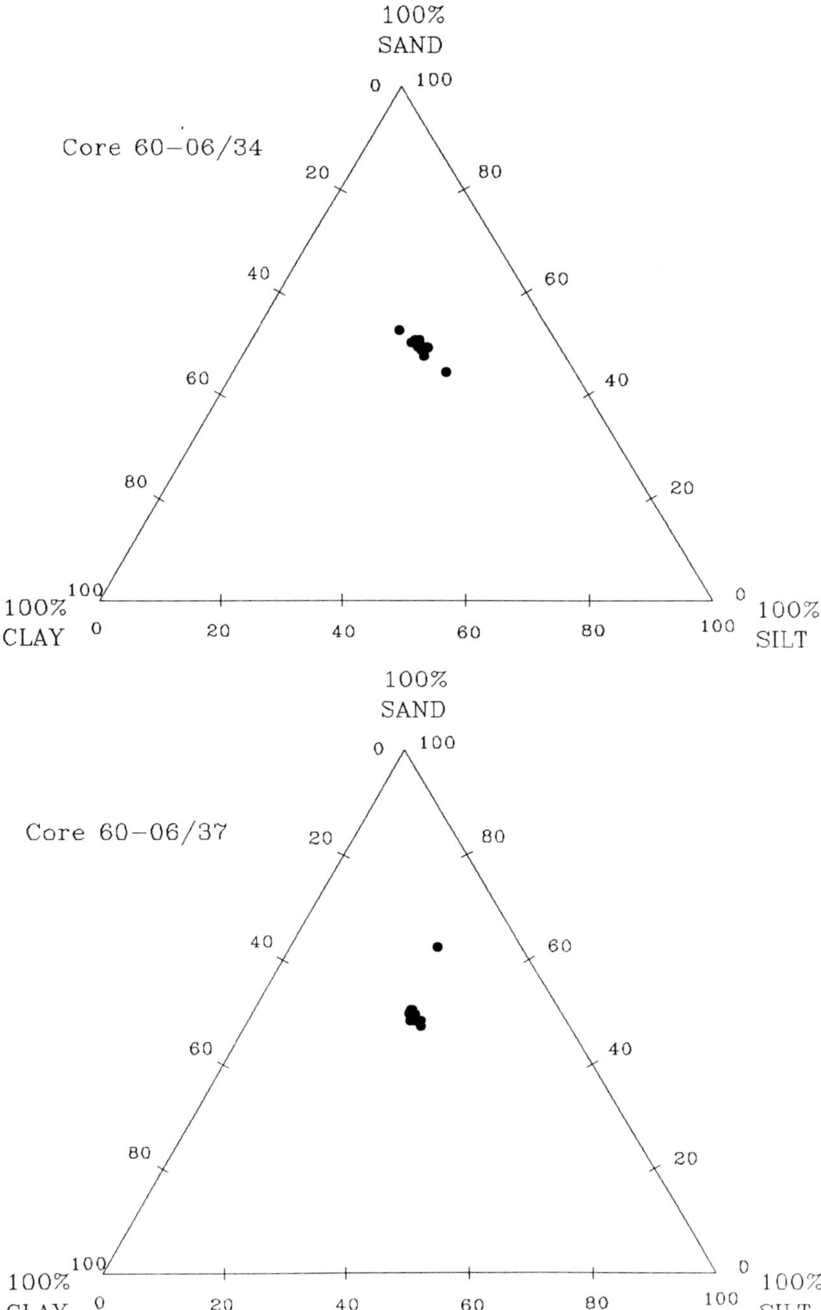

Fig. 4. Particle size composition of Group One cores.

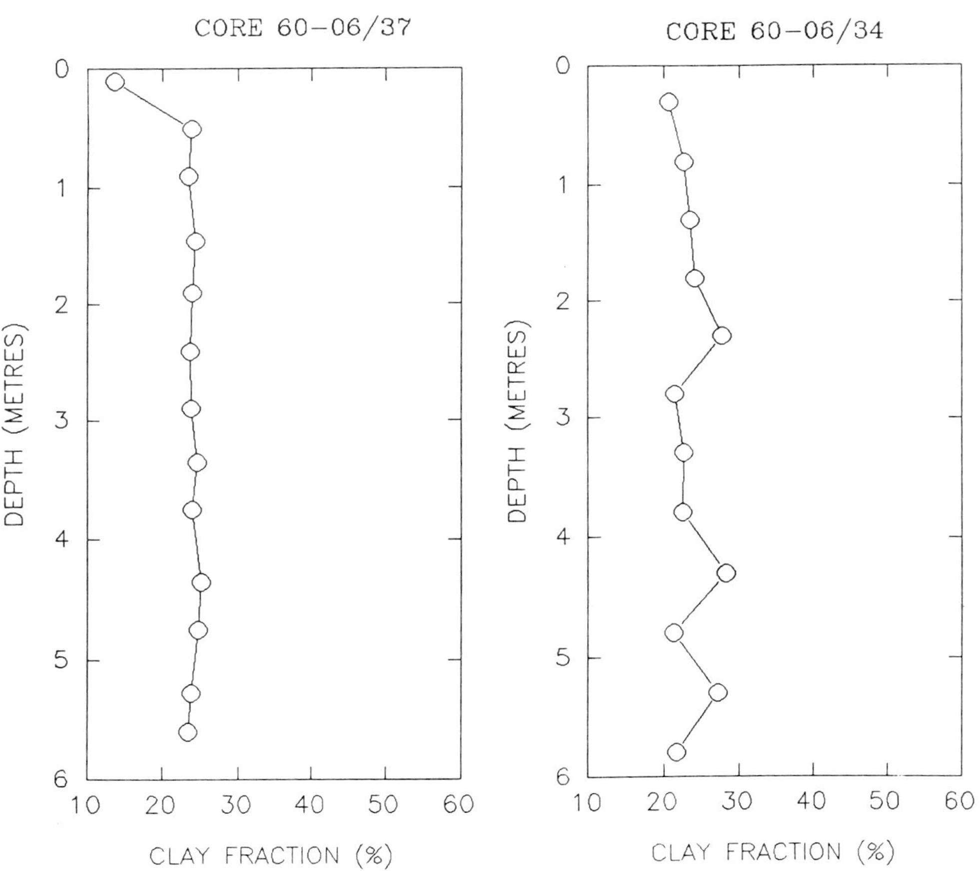

Fig. 5. Depth profiles of clay percentage in Group One cores.

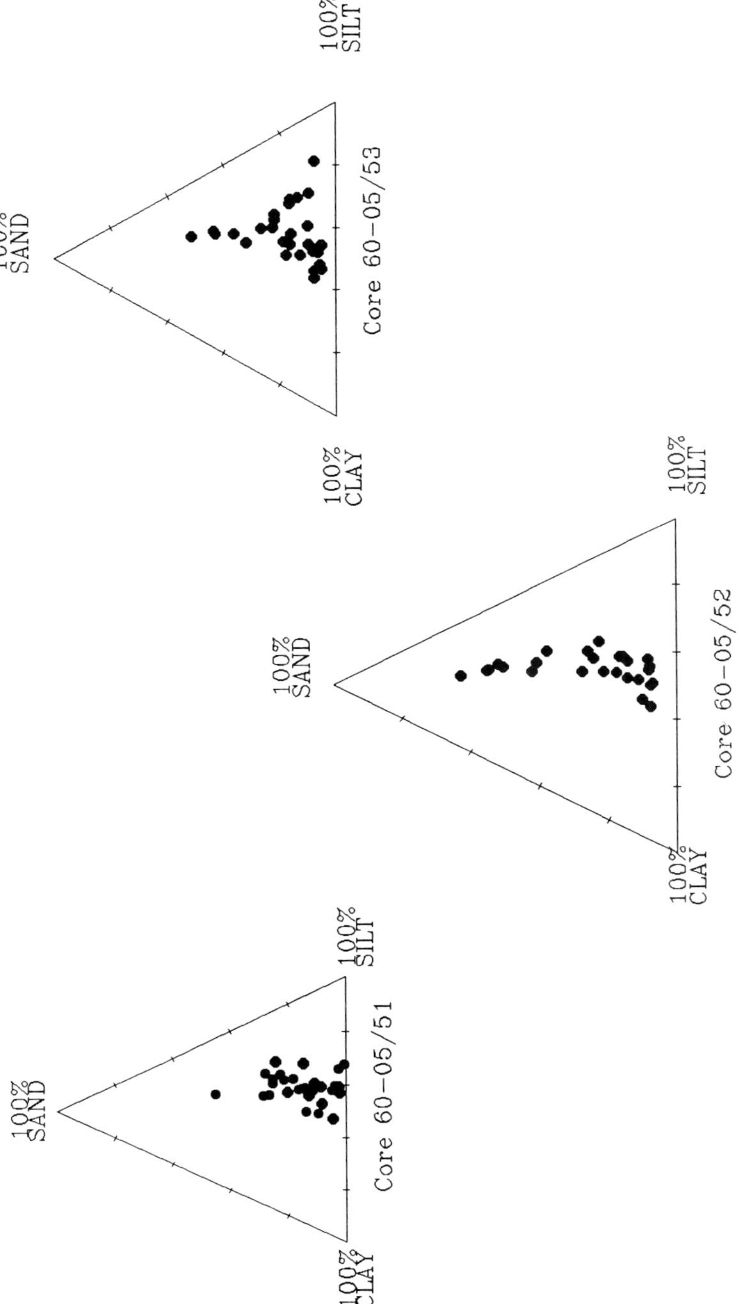

Fig. 6. Particle size composition of Group Two cores.

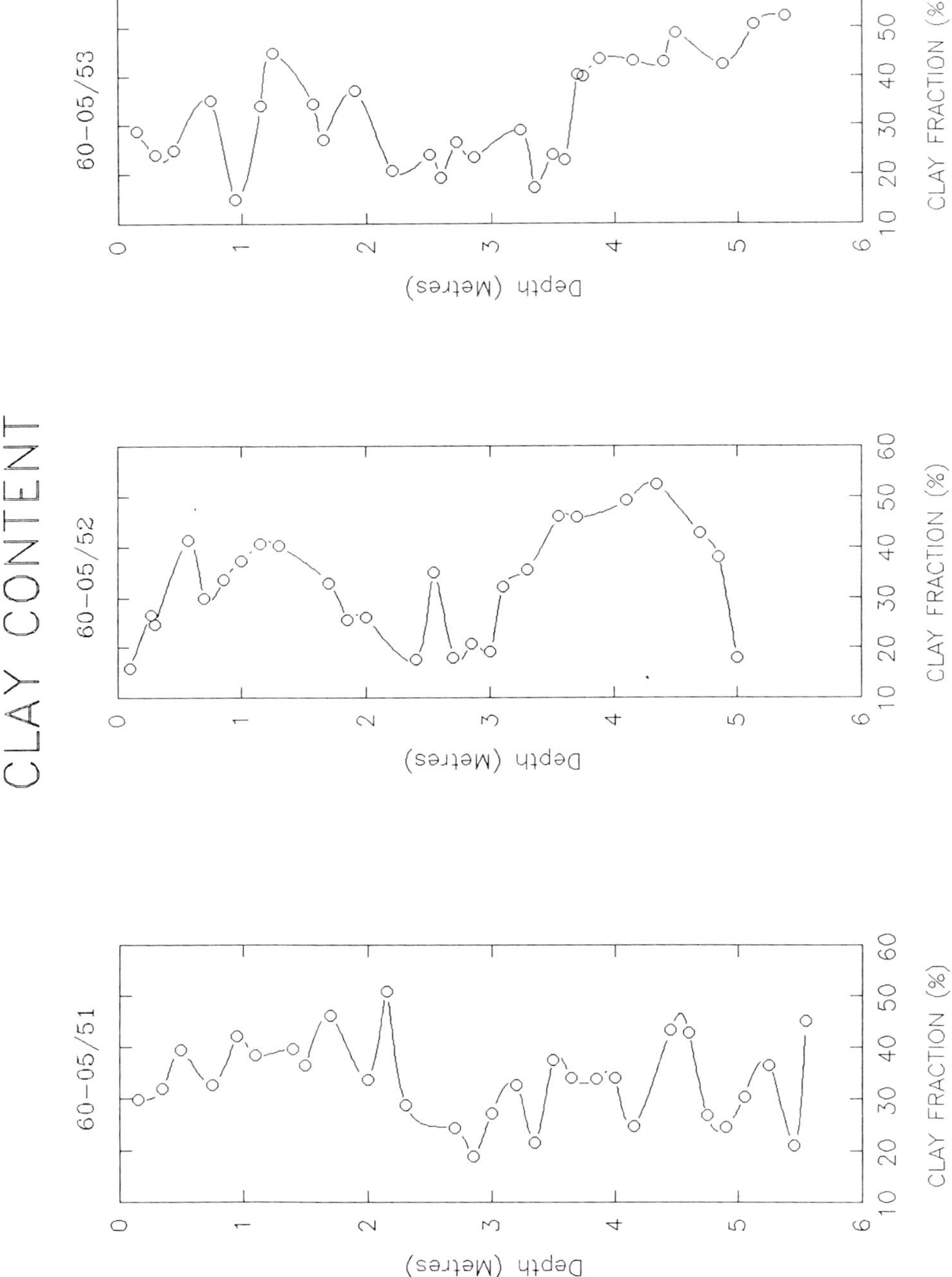

Fig. 7. Depth profiles of clay percentage in Group Two cores.

Sandy mud lithofacies (comprises between 20–35% of core material). The distinguishing feature of this facies is that up to 10% sand size fraction is present. The sediment is generally poorly sorted and contains 20–40% clay sized material, together with 10–15% fine silt and 30–40% coarse silt. The sediment is very dark grey (5Y 3/1) with silty lenses either as horizontal layers (as at the top of core 51), within *Planolites* burrows or as undefinable patches. Small pebbles occur throughout the facies, with larger clasts present in cores 51 and 53. Bioturbation is present in about 20% of the facies, the burrow being isolated and usually silty in composition. No shells are found in core 52, whereas a few are present in core 51 and they are abundant in core 53. In this last core 53 a hard clay concretion is present at one horizon.

Sandy silt lithofacies (comprises up to 20% of the cores). In general, the sediment in this facies is very poorly sorted and contains less than 20% clay size. Coarse silt is more abundant (30–40%) than fine silt (< 10%) and a high proportion (10–40%) of sand is also present. Bioturbation is common with both silt-filled *Planolites* and vertical mud filled burrows present. Shell fragments are found within cores 51 and 52, though none were found in core 53. Both small and large pebbles are also found. Hard clay concretions similar to those of the sandy mud lithofacies, seen in core 53, are found in this facies in both core 53 and core 52. These concretions are dark grey (5Y 3/1) in colour; in core 53 they are about 10cm in length and 6cm in diameter, whereas in core 52 they are slightly smaller.

Muddy sand lithofacies. This is only found as a distinct facies in core 52 where it forms less than 10% of the core. In cores 51 and 53, the muddy sand is gradational with the underlying and overlying layers of sandy mud and sandy silt, and never more than a few centimetres thick. In core 52 the facies is dark grey (5Y 4/1) containing many dropstones both small and large. The sediment comprises over 50% sand; the clay content is reduced to no more than 30% and the fine silt content to approximately 10%. There is a corresponding increase in the coarse silt fraction to 25–45% and in the sand fraction to 10–30%. The facies is visually structureless: no bioturbation is present and no shells were found.

3.3. Mineralogy

The mineralogy of the sediments exerts a fundamental control on their plasticity and related geotechnical properties. It was studied on air dried samples taken from all of the cores at 300–500mm intervals using X-ray powder diffraction (XRD) between diffraction angles of 5°–50° (Cu_α radiation: d-spacing 1.54– 44.13 Å).

The suite of minerals making up the silt fraction was found to be fairly uniform throughout the five cores. Quartz is the dominant mineral: calcite, muscovite, chlorite, corundum and dolomite are also seen. Halite is present in all samples probably as a precipitate from saline pore water. Plagioclase feldspar is always present: this is usually albite, although in cores 51 and 53 a more calcic variety appears to be present also. The ferro-magnesian minerals include augite, diopside, olivine and one or more amphiboles. They are mainly absent from cores 34 and 37 apart from large quantities of amphibole in core 37 and traces at 2.30m in core 34. The iron minerals present in all cores are haematite and magnetite though difficult to identify on many traces due to a high abundance of minerals with a similar d-spacing.

Due to the high proportion of silt-sized material in the samples, the only clay minerals to show above background level are chlorite (or kaolinite) and illite. Both are present in all the cores. Four samples from core 53 were taken by sedimentation from settling columns to produce samples of near clay-size ($< 8\mu m$). These samples showed the same suites of minerals as found in the coarser samples, apart from no amphibole, olivine or anorthite being found in any of the samples. In addition, however, this technique revealed in all the samples the possible presence of smectite in very limited amounts.

A visual examination of the minerals found by XRD was carried out by binocular microscope, on two of the samples, one from core 52 and one from core 53. The grain size was between 38–63μm diameter. Quartz was the main mineral present, plus a large number of foram tests which may account for the calcite content. Also identifiable were muscovite, biotite (which did not show up on the traces due to peak overlap), chlorite, plagioclase, olivine, haematite, magnetite and a few grains of a ferro-magnesian silicate, either pyroxene or an amphibole.

Scanning electron microscopy of the clay particle morphology, together with energy dispersive X-ray spectroscopy (EDX) analysis, supports the XRD identification of illite as the main clay mineral. Limited kaolinite was found as a probable alteration product of felspar. EDX analysis of various silt grains supports the XRD identification of quartz, felspar and ferromagnesian minerals. This mixed suite would be expected from a source area on the adjacent Scottish mainland and Islands.

4. Geotechnical Properties

Measurements have been made of a range of geotechnical properties, including water content, Atterberg limits, undrained shear strength and compressibility. Details of these are given below. It has been found in general that, given the constant mineralogical composition of the sediment, the near-surface geotechnical properties are largely controlled by two simple variables: the percentage of clay sized particles and the water content. The former directly determines the Atterberg limits and indirectly influences the compressibility; the latter controls the liquidity index (given the Atterberg limits) and hence the undrained shear strength. The water content is itself controlled by the stress history of the sediment: this, and the clay percentage, are the product of the sedimentary processes by which the material formed.

TABLE 1.
Index properties of sediments from the study area.

Group 1			
Core	Liquid Limit	Plastic Limit	Plasticity Index
34	26–27%	12–13%	12–14%
37	25–26%	c.14%	11–12%
Group 2			
Facies	Liquid Limit	Plastic Limit	Plasticity Index
Mud	40–55%	20–25%	20–25%
Sandy Mud	25–40%	15–20%	15–20%
Muddy Sand	c.30%	c.15%	c.15%
Sandy Silt	25–35%	15–20%	10–20%

4.1. Water Content and Atterberg Limits

Figures 8 and 9 show the depth profiles of water content and Atterberg limits from the five vibrocores studied. The Group One cores (34 and 37: Figure 8) show similar profiles of water content which are almost constant with depth. In core 34 the water content varies from 21% at 1m depth to 15% at 6m depth, and in core 37 from 21% at 1m depth to 17% at 6m depth. Above this the less compact nature of the sample allows more water to be retained and a sharp increase in water content, up to 40% in core 34, is noted. The decrease in water content with depth is to be expected due to consolidation by increasing selfweight.

The liquid and plastic limits are almost constant with depth (about 14% and 27% respectively) and are very similar in value in both cores. The results are as expected for sandy muds with a low clay content (Table 1). The remarkable lack of variation in the values is consistent with the uniform lithology of the cores. The liquidity index is however somewhat different in the two cores: in 34 it is close to zero (the plastic limit), whereas in 37 it is around 0.5. This has implications for both the undrained shear strength and compression history which are explored below.

The results for the Group Two cores are shown in Figure 9. In core 51 the water content ranges from 20% to 65%. There appears to be an overall decrease with depth, but variations can be more closely related to differences in sedimentology. The highest values are found in the mud facies (65% to 53%) and decrease with depth. The sandy mud facies contains 45–50% at 2–4m depth reducing to 30–40% at 5–6m. The sandy silt facies contains about 35–40%. In core 52 water content values are between 20–50% and decrease with depth is masked by facies variation. The mud lithofacies contains 40–50% water, the sandy mud and muddy sand 30–40% and the sandy silt varies from 20–40%. Core 53 has a smaller range of 30–50% only, again with no obvious decrease with depth. Taken as a whole, the

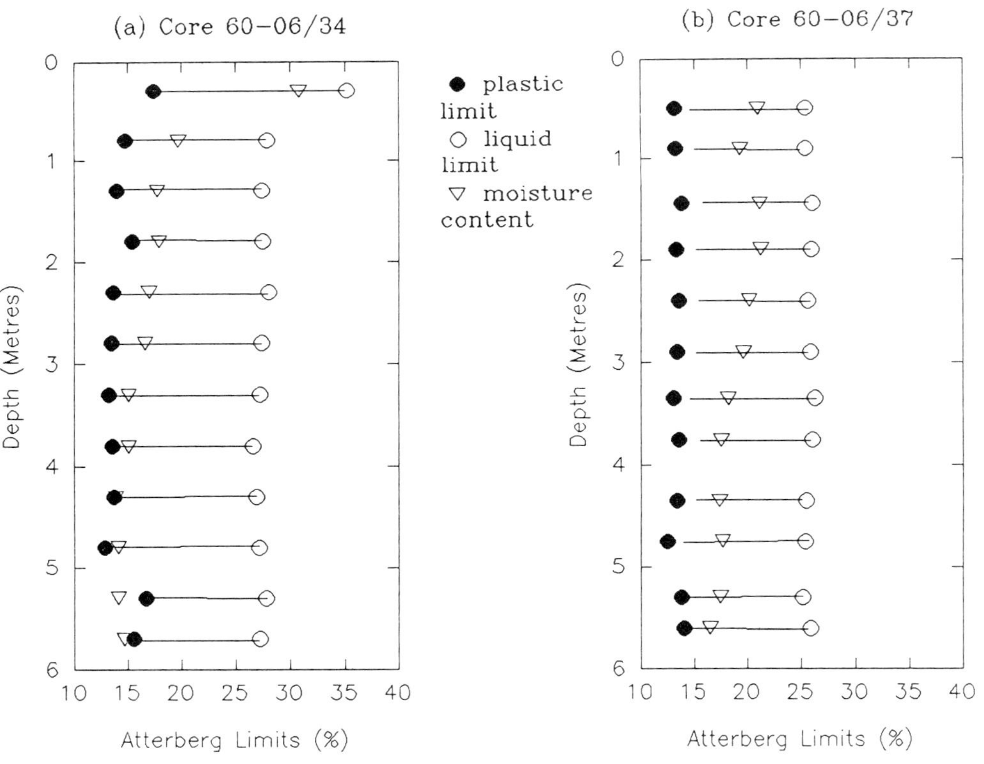

Fig. 8. Depth profiles of water content and Atterberg limits in Group One cores.

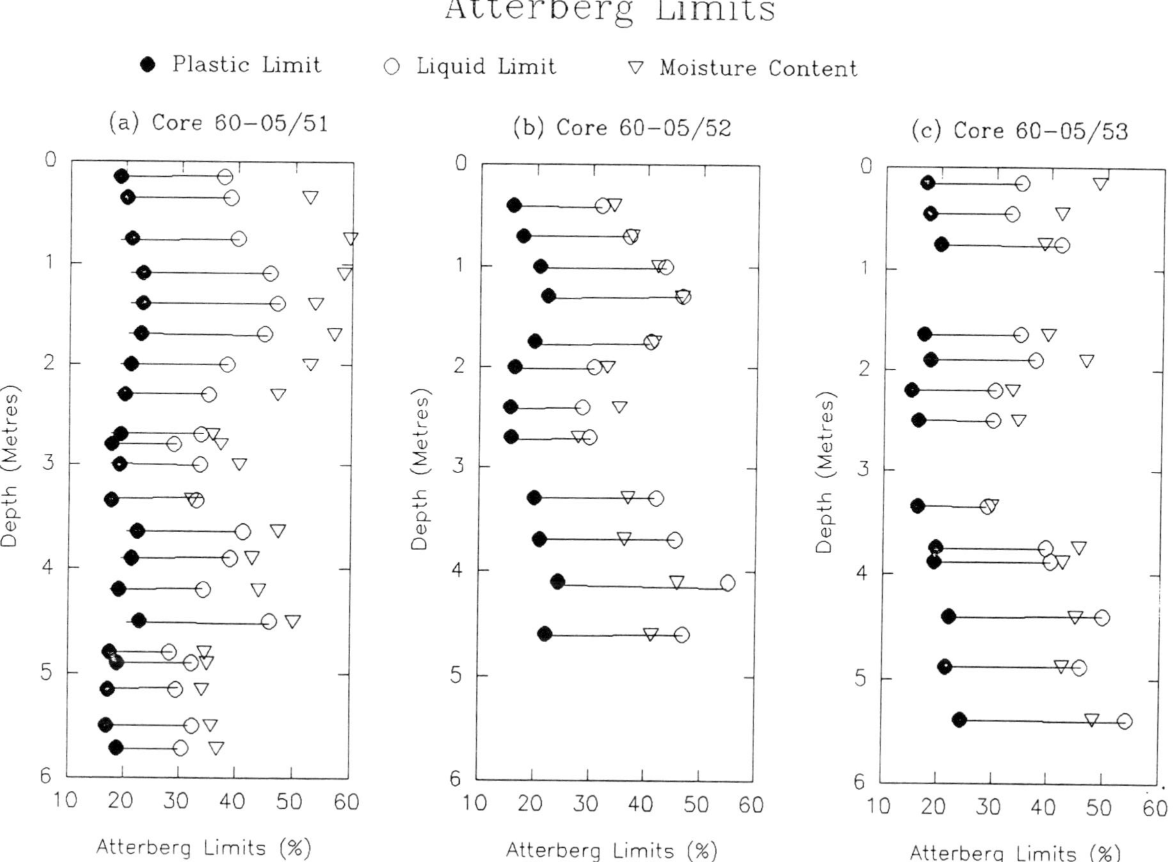

Fig. 9. Depth profiles of water content and Atterberg limits in Group Two cores.

results indicate that the water content is controlled by the local mud content, which are superimposed on and may mask the general selfweight trend.

The Atterberg limit profiles are also shown in Figure 9. The variation down the core is a direct reflection of the local clay percentage and hence of the facies (Table 1). The clay content has a greater influence on the liquid limit than on the plastic limit. The liquidity index in the mud and sandy mud facies is high, from 1.0 to nearly 2.0 in the upper parts of core 51.

4.2. Soil Plasticity and Activity

Figures 10 and 11 show respectively the plasticity charts for samples from the cores in Groups One and Two. For comparison they also show the marine clay line (Skempton, 1970) and the till line (Boulton and Paul, 1976). In Figure 10 (cores 34 and 37) all the samples plot close to the till line, as might be expected for samples of probable mass flow origin from a glacial source. Such an origin has resulted in a sediment which is similar to a generic till and accounts for the position of the samples on the chart. The restricted scatter of points reflects the lack of grading changes within the cores. Figure 11 shows the results from cores 51, 52 and 53. These data plot above the marine clay line showing the influence of coarse fraction on the Atterberg limits, and suggesting a likely glacigenic origin for the sediment in the first instance. The left to right trend of the data points results from the changing clay content of the samples, which increases from about 20% (sandy silt facies) to about 50% (mud lithofacies).

The relationship between clay content and plasticity index for all the cores is shown in Figure 12. The ratio of plasticity index/clay fraction (the activity of the clay fraction) lies between 0.5 and 0.7 in all the cores. This ratio is a fundamental property of the clay mineralogy; the sediments in all the cores studied have a relatively low activity. This is expected since their clay fraction is dominated by inactive clay minerals such as kaolinite and illite and by inert quartz and feldspar flour, as shown by the XRD results. The proportions of chlorite and illite have some effect on the activity of the sample, and those samples with a slightly higher illite content have a marginally higher activity. Detailed XRD analysis of the clay fraction revealed possible smectites in these cores (in very small quantities), and it may be presumed that these will also have raised the activity.

4.3. Undrained Shear Strength

Figure 13 profiles the undrained shear strength down the cores in Group One. The intact strength in core 37 increases with depth from 10 to 40 kPa, as would be expected due to selfweight consolidation. The intact strength in core 34 increases from 10 to 120 kPa, with a noticeable increase below a depth of 3m. These higher shear strengths are probably due to the lower water content, which is very close to or below the plastic limit. The reasons for the relative densification of core

PLASTICITY CHARTS

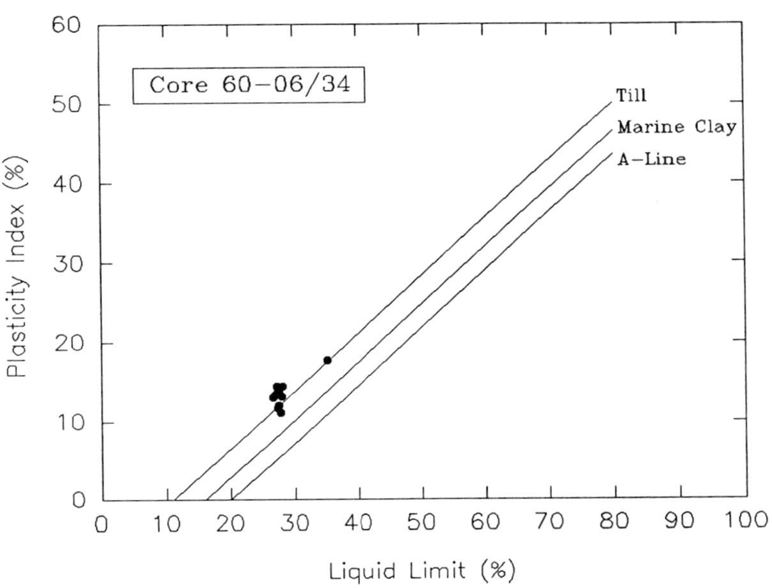

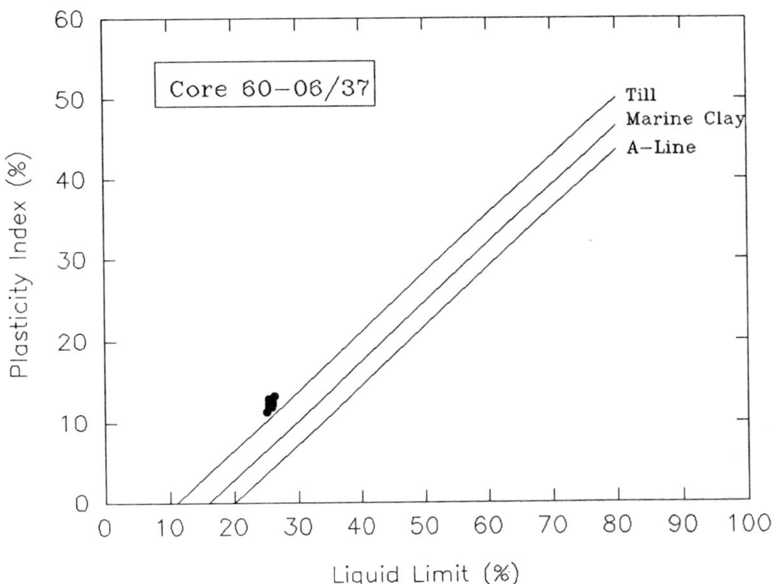

Fig. 10. Plasticity charts for Group One cores.

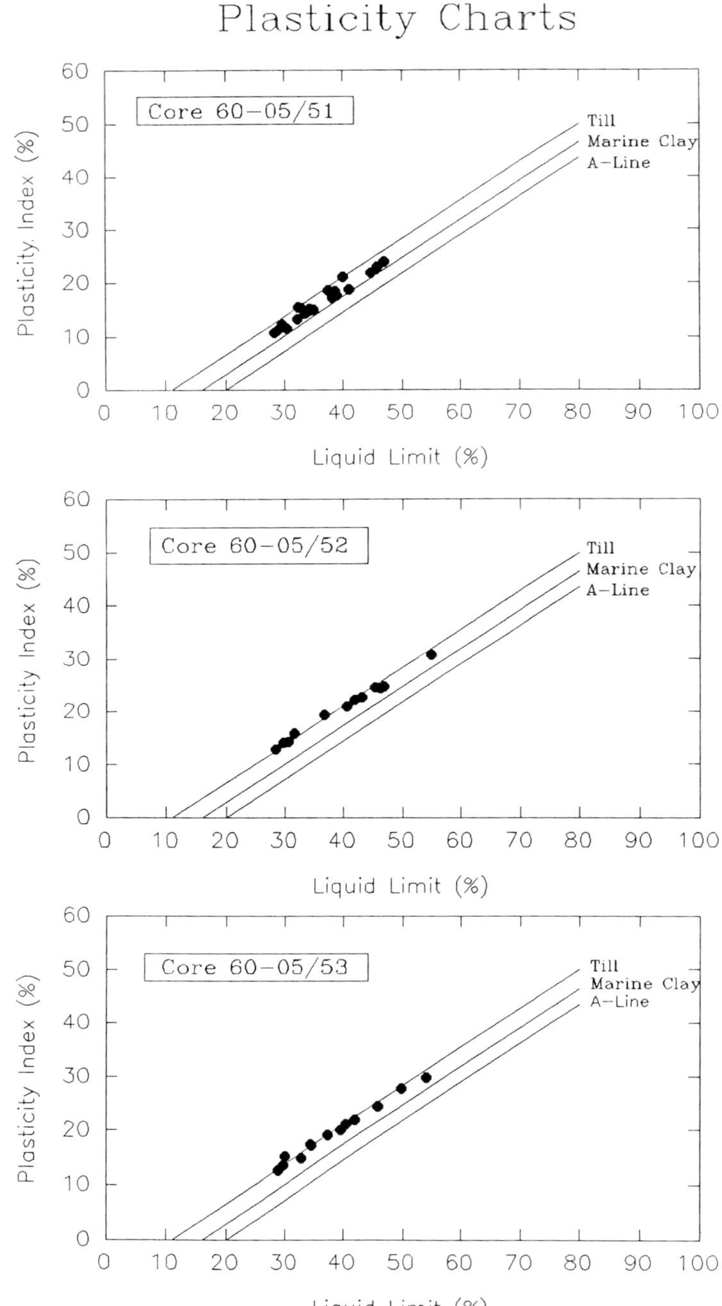

Fig. 11. Plasticity charts for Group Two cores.

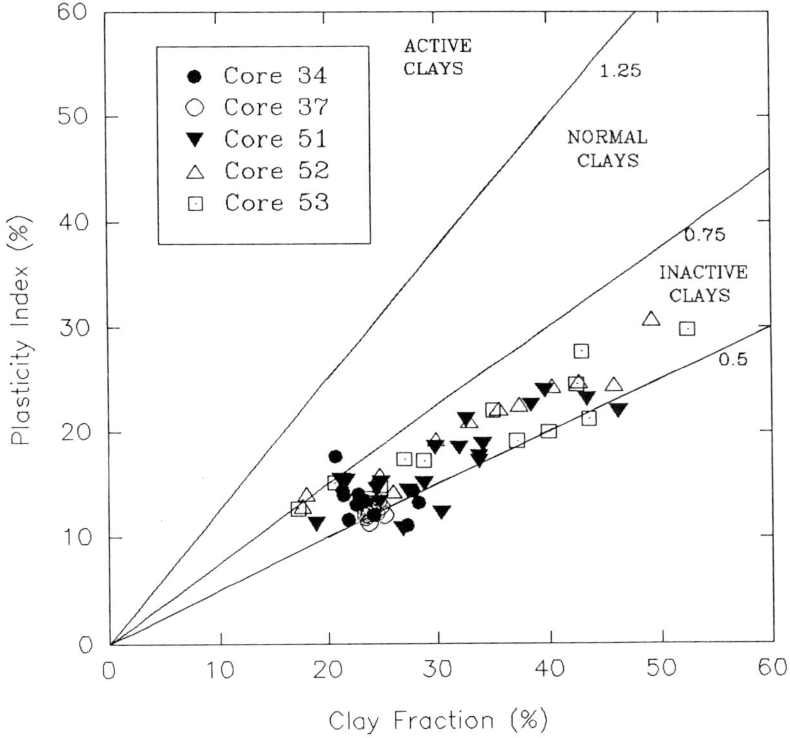

Fig. 12. Activity chart: all cores.

34 are unclear: depositional processes or post-depositional disturbance (including sampling) are both plausible and are discussed below.

In the case of the Group Two cores (Figure 14) the increased fines content of most of the facies produces strength profiles typical of clayey soils, albeit with evidence of overconsolidation in some cases. Changes in grain size and composition of the cores have small effects on changes in shear strength but in general these changes are hidden by the stronger control of decreasing void ratio with depth. In detail, in core 51 the intact shear strength varies from 3 kPa at seabed to 15 kPa at 6m depth, with a very slight reduction in strength in the muddy layers relative to the more silty/sandy layers. In core 52 the intact shear strength is generally higher than at the equivalent depth in core 51, increasing from 5 kPa at seabed level to about 30 kPa at 5m depth. This suggests a modest degree of overconsolidation in

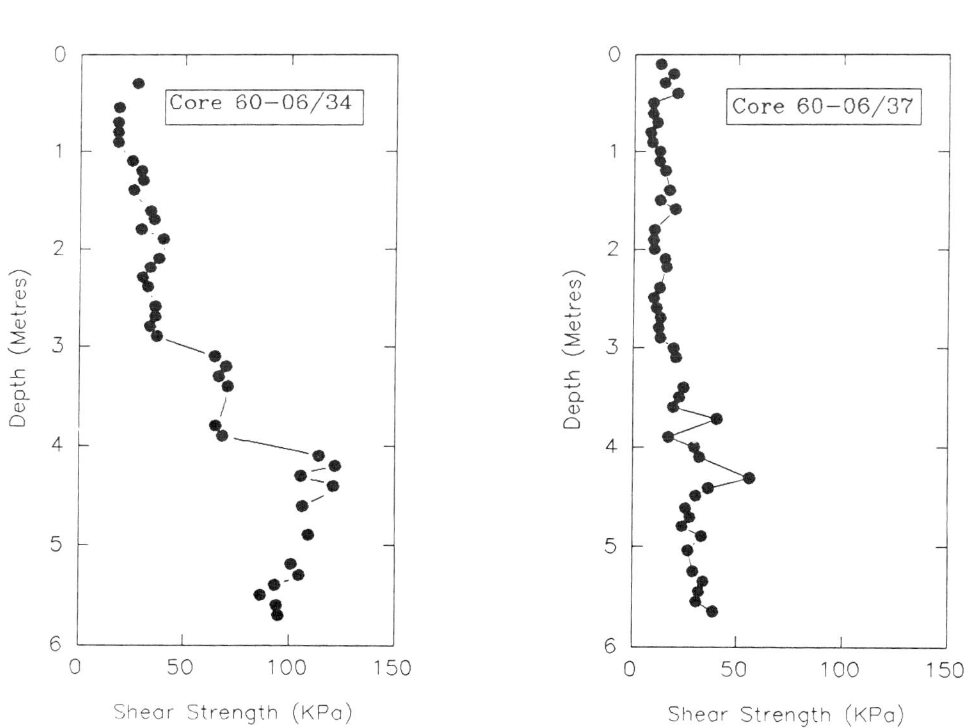

Fig. 13. Depth profile of undrained shear strength for Group One cores.

keeping with the compression results (below). Finally, in core 53 the intact shear strength increases from 10 kPa near seabed level to 20 kPa at 4m depth, which again suggest a modest degree of overconsolidation.

4.4. Compression and Preconsolidation

A series of 45 one dimensional compression tests has been carried out in a standard front loading oedometer using applied loads of up to 2500 kPa. Figures 15 and 16 show some typical results from this large data set for each group of cores. The compression curves were analysed using the method of Casagrande to find the maximum past pressure to which the soil had been subjected. These values are compared in Figures 17 and 18 with the vertical effective stress at the depth in question, using an effective stress gradient based on the measured water content. It should be noted that since the construction is best suited to clay soils its application to the silty sediments from cores 34 and 37 must be regarded as notional, however the results from these cores are included for comparison.

Consider first the results from the Group Two cores (Figures 16 and 18). These suggest that core 51 is normally consolidated, whereas both cores 52 and 53 show some degree of apparent overconsolidation. In core 52 the past pressure of 85 kPa at 5m depth would imply that around five metres of sediment have been eroded since deposition. This is a reasonable suggestion for a slope face deposition. This is a reasonable suggestion for a slope face deposit. Likewise, in core 53 an extra six metres of sediment would be necessary to produce the measured past pressure of 95 kPa at the present depth of 6m.

The results from the Group One cores (Figures 15 and 17) present a different picture. The samples are of relatively cohesionless soil, whose *in situ* densities are relatively high. This almost certainly reflects the initial densification rather than stress history, and thus flexure in the compression curve cannot be taken as evidence of past pressure. In addition, the curvature in compression curve of these samples, like many such soils, is relatively gradual and a definite flexure is therefore difficult to pinpoint. For these reasons, the ostensible past pressures of 200 kPa in core 34, which suggest erosion of around 20m of sediment, are treated with caution.

5. Discussion

Individual aspects of the geotechnical behaviour have been discussed above in their context. Core 37 is a sandy, clayey silt whose plasticity is consistent with its sedimentology and mineralogy, and whose water content, shear strength and compression behaviour is consistent with a lack of precompression. Core 34 exhibits an apparent precompression consistent with a lowered water content and raised shear strength: this is probably the result of densification either during deposition (as a mass flow) or during sampling. Taken together, these Group One cores appear characteristic of a generally poorly consolidated mass flow which has lost much

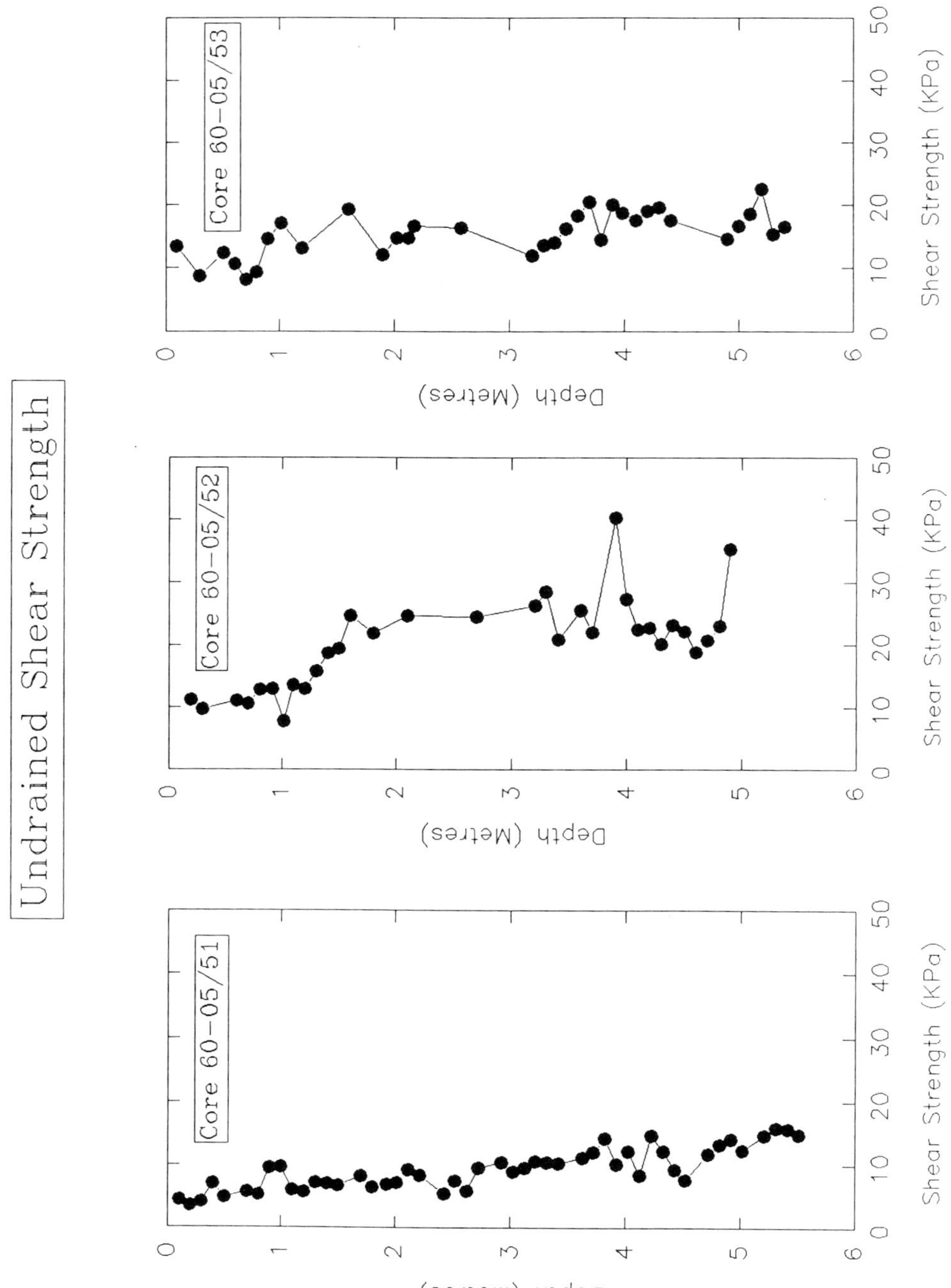

Fig. 14. Depth profile of undrained shear strength for Group Two cores.

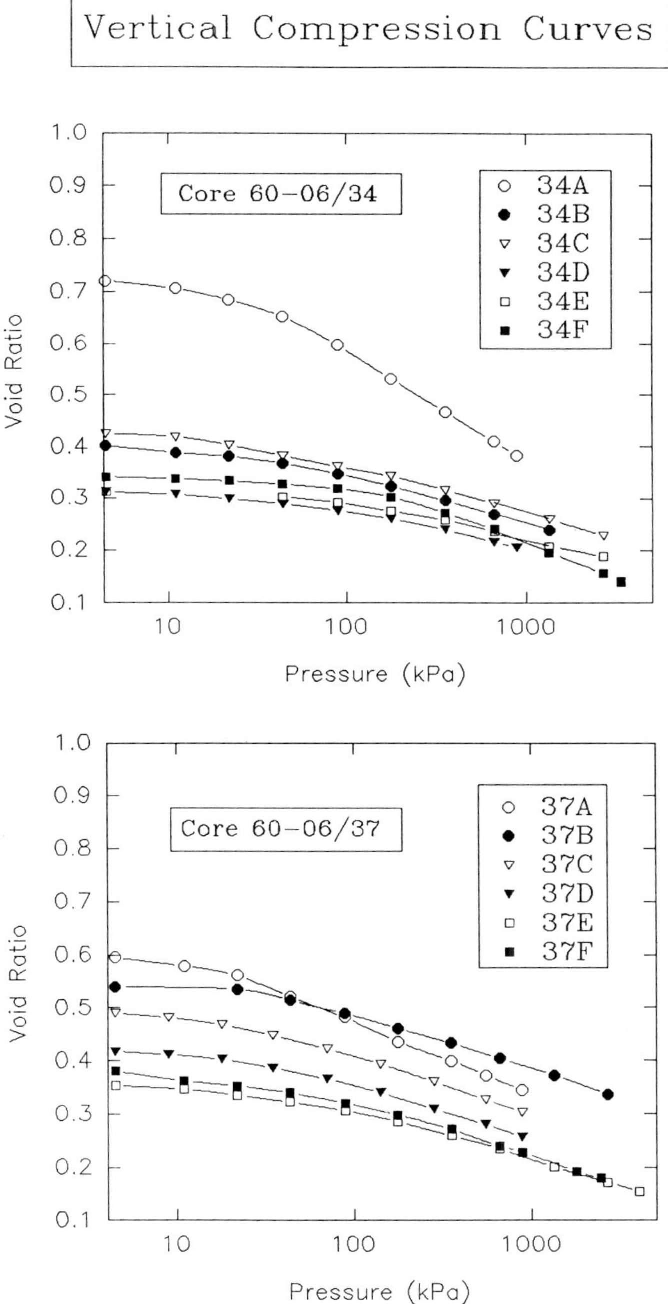

Fig. 15. One dimensional compression (oedometer) results for Group One cores.

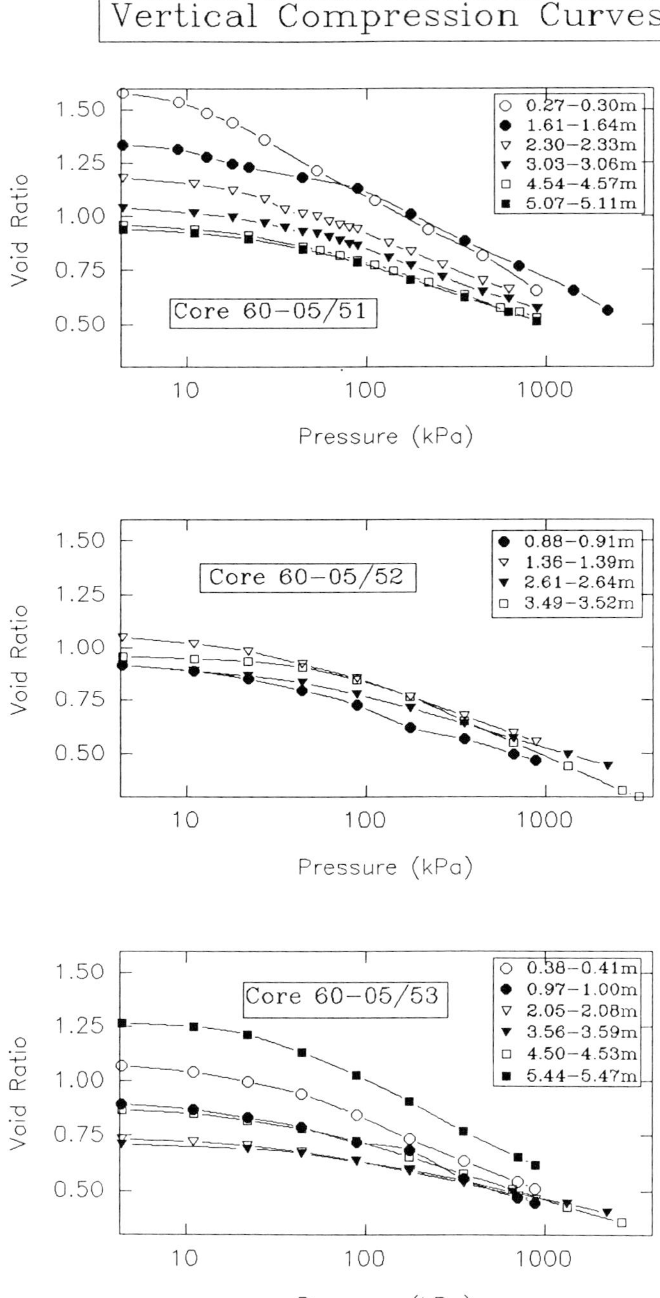

Fig. 16. One dimensional compression (oedometer) results for Group Two cores.

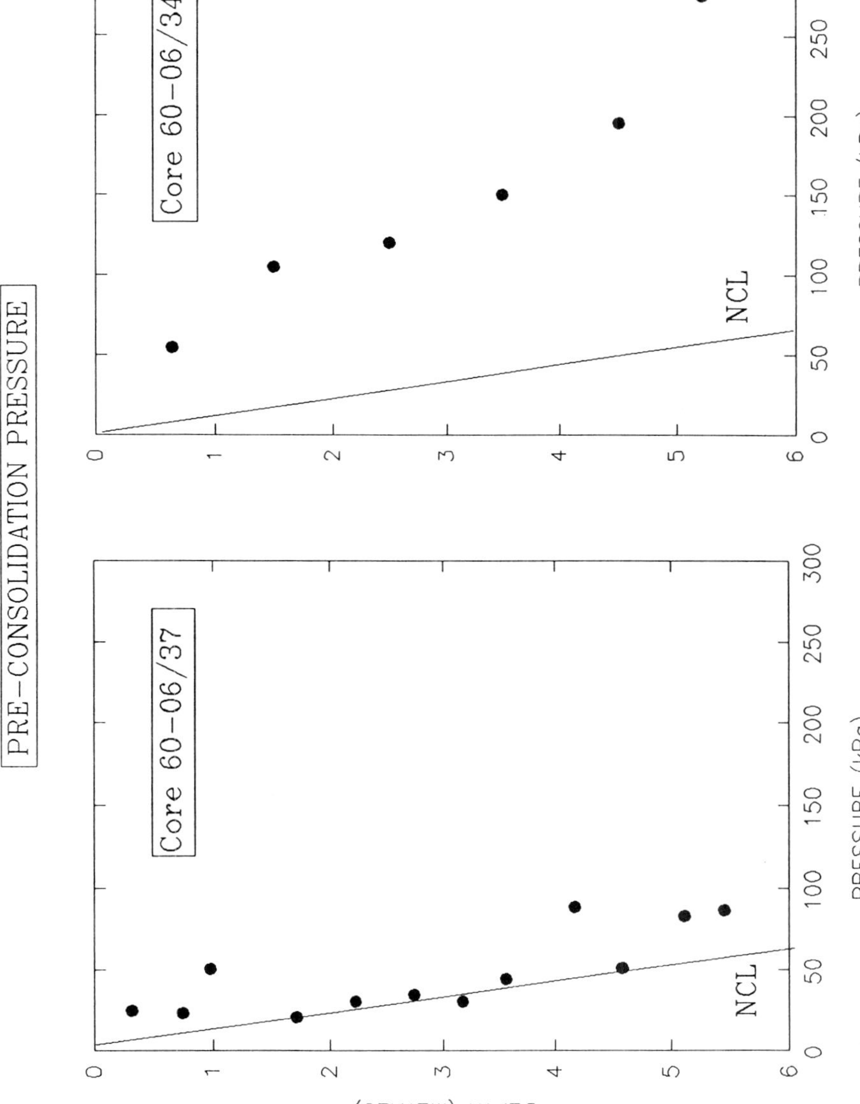

Fig. 17. Depth profiles of *in situ* vertical stress and preconsolidation pressure for Group One cores.

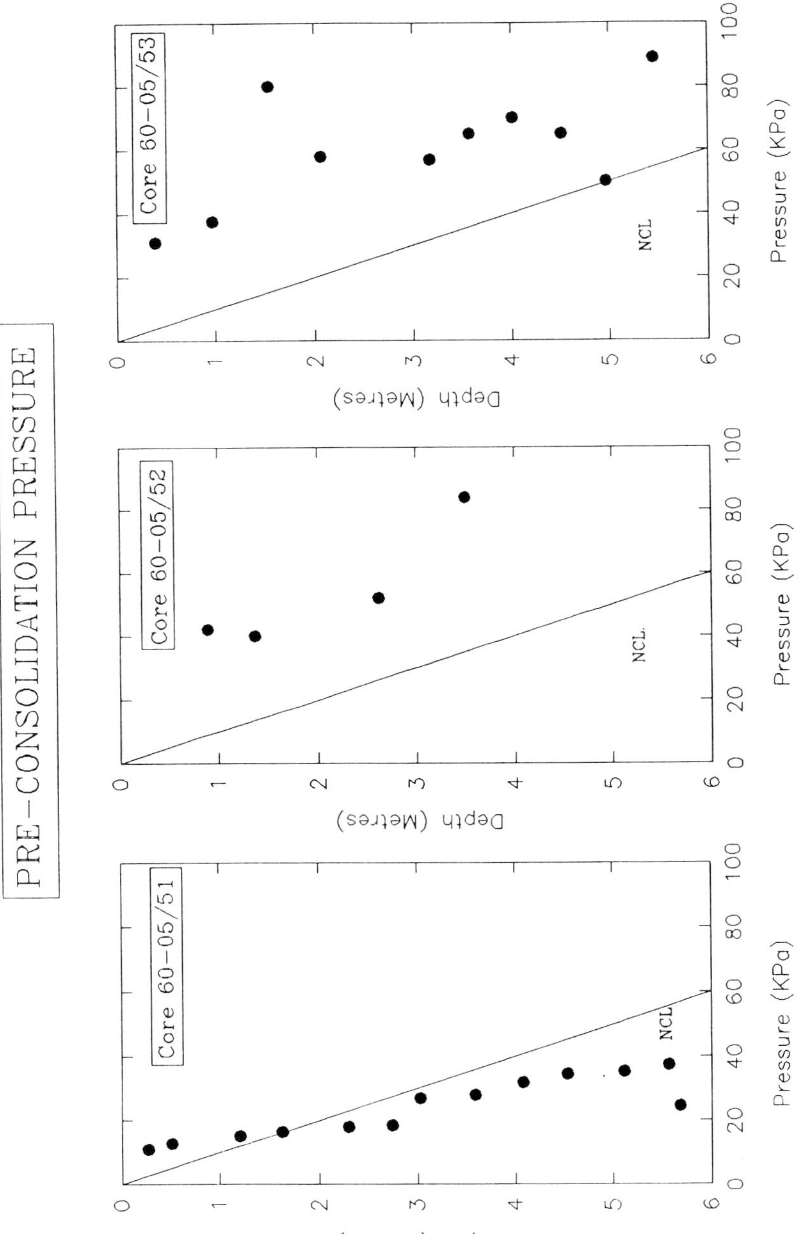

Fig. 18. Depth profiles of *in situ* vertical stress and preconsolidation pressure for Group Two cores.

of its fine component. The seismic evidence suggests that this flow (or flows) now forms an area of hummocky, acoustically structureless sediment at the foot of the slope in the southwest of the study area. We note that Akhurst (1991) has observed a mud drape facies on the floor of the Faeroe-Shetland Channel, and we speculate that this may represent the laterally equivalent fine grained component missing from the flows in our area.

Cores 51, 52 and 53 have a more complex sedimentology, in which several facies are present. When due account is taken of this complexity, we find that their plasticity is consistent with their being derived from the finer fraction of a glacigenic diamiction, probably transported by downslope flow. The sediment in core 51 is normally consolidated, and in 52 and 53 is lightly overconsolidated: this is consistent with their probable erosional history on the lower slope, and accounts for their present shear strengths.

These cores have been recovered from the acoustically layered sequences from which the body of the lower slope is constructed. This acoustic layering may arise from the lithofacies variations described in this paper. Despite their differences in stress history, they form an homogenous group both sedimentologically and geotechnically. This suggests that in this area the seismic character of the sediments can be used as an indicator of their likely geotechnical character, at least at the reconnaissance level. We note, however, that there are limits to which such an approach can be taken (cf. Stoker *et al.*, this volume) and stress the need for an adequate geological framework, based on both processes and chronology, within which these indicators should be applied.

6. Conclusions

The cores selected for study can be divided into two groups on the basis of their sedimentology and the acoustic signatures of the horizons from which they were collected. This work has shown that consistent differences exist between these two groups over a wide range of engineering properties and suggests that these differences arise from the sedimentological character of the cores, in particular the mud content. These differences are linked to facies variations which are believed to have resulted from the operation of differing geological processes during the deposition of the sediments. Thus in this area past patterns of sedimentation exert a strong control on the present engineering behaviour: the reconstruction of these patterns from seismic and sedimentological data provides a useful framework within which to understand the geotechnical properties of the seabed sediments.

Acknowledgements

We thank Prof. G. S. Boulton and Mr. G. Angell (Edinburgh University) for the use of the XRD and Mr. C. Cameron (St. Andrews University) for the use of the SediGraph. Mr. H. Barras and Mrs. B. F. Barras gave valuable advice and help with many aspects of laboratory testing. Several colleagues at Heriot-Watt University and at the British Geological Survey gave generous advice and constructive criticism. The financial support of MTD Ltd. for the geotechnical testing programme is gratefully acknowledged (Grant GR/E/81289). The contribution of M. S. Stoker is published with the permission of the Director, British Geological Survey, NERC.

References

1. Akhurst, M. C. (1991), 'Aspects of Late Quaternary Sedimentation in the Faeroe-Shetland Channel, Northwest U.K. Continental Margin', British Geological Survey Technical Report, WB/91/2.
2. Boulton, G. S. and Paul, M. A. (1976), 'The influence of genetic processes on some geotechnical properties of glacial tills', *Quarterly Journal of Engineering Geology* **9**, 159–193.
3. Eldholm, O. (1990), 'Paleogene North Atlantic magmatic-tectonic events: Environmental implications', *Memoir Geological Society Italy* **44**, 13–28.
4. Haszeltine, R. S., Ritchie, J. D., and Hitchen, K. (1987), 'Seismic and well evidence for the early development of the Faeroe-Shetland Basin', *Scottish Journal of Geology* **23**, 283–300.
5. Miller, K. G. and Tucholke, B. E. (1983), 'Development of Cenozoic abyssal circulation south of the Greenland-Scotland Ridge', in *Structure and Development of the Greenland-Scotland Ridge*, M. H. P. Bott, S. Saxov, M. Talwani, and J. Theide (eds.), Plenum Press, New York, 549–589.
6. Scrutton, R. A. (1986), 'The geology, crustal structure and evolution of the Rockall Trough and the Faeroe-Shetland Channel', *Proceedings of the Royal Society*, 88B, 7–26.
7. Skempton, A. W. (1970), 'The consolidation of clays by gravitational compaction', *Quarterly Journal of the Geological Society of London* **125**, 373–412.
8. Skempton, A. W. and Northey, R. D. (1952), 'The sensitivity of clays', *Géotechnique* **3**, 30–53.
9. Stoker, M. S., Harland, R., Morton, A. C., Graham, D. K. (1989), 'Late Quaternary stratigraphy of the northern Rockall Trough and the Faeroe- Shetland Channel, northeast Atlantic Ocean', *Journal of Quaternary Science* **4(3)**, 211–222.
10. Stoker, M. S. (1990), 'Judd, 60°N–06°W, Quaternary Geology', British Geological Survey 1:250 000 map series.
11. Stokes, M. S. and Holmes, R. (1991), 'Submarine end-moraines as indicators of Pleistocene ice-limits off northwest Britain', *Journal of the Geological Society of London* **148**, 431–434.
12. Stoker, M. S., Harland, R., and Graham, D. K. (1991), 'Glacially influenced basin plain sedimentation in the southern Faeroe-Shetland Channel, northwest United Kingdom continental margin', *Marine Geology* **100**, 189–199.
13. Stoker, M. S., Stewart, F. S., Paul, M. A., and Long, D. (1992), 'Problems associated with seismic facies analysis of Quaternary sediments on the northern UK continental margin', this volume.
14. Stoker, M. S., Hitchen, K., and Graham, C. C. (in press), 'The geology of the Hebrides and West Shetland shelves and adjacent deep-water areas', HMSO for the British Geological Survey, London.

GEOSIS PROJECT:INTEGRATION OF GEOTECHNICAL AND GEOPHYSICAL DATA

J. F. NAUROY
Institut Français du Pétrole, BP 311, 92506 Rueil Malmaison, France

and

J. MEUNIER
IFREMER, BP 70, 29263 Plouzané, France

Abstract. The GEOSIS project aims to integrate seismic and geotechnical data for improving the extrapolation of soil data around geotechnical boreholes. The project includes the development and the testing of two techniques:

– vertical seismic profiling in geotechnical boreholes,
– multi-channel very high resolution seismic surveying.

The paper describes the state of development of these techniques and comments on the first field test results.

1. Introduction

The installation of offshore facilities needs good spatial understanding of the seabed. A typical site investigation programme comprises two principal components: the geophysical survey and the geotechnical investigation. These two phases are usually carried out with different vessels and at different times. The information obtained remain relatively unconnected. There is poor correlation between the seismic records and the geotechnical data, and it is often difficult to relate the observed seismic profiles to the cone penetrometer or the borehole soil log. One of the reasons for this is that the seismic section is obtained and processed in terms of time, and conversion from time to depth requires knowledge of the seismic velocity of each layer of the profile.

In this context, a group of research organizations, oil companies and offshore contractors, including, IFP, IFREMER, ELF Aquitaine Production, TOTAL, BEICIP and GEODIA, decided in 1990 to conduct a Joint Research and Development programme called GEOSIS. The GEOSIS project aims to integrate seismic and geotechnical data for improving the extrapolation of soil data around geotechnical boreholes.

This paper outlines the main objectives of the GEOSIS project and describes the first results.

Volume 28: Offshore Site Investigation and Foundation Behaviour, 107–113, 1993.

2. Current Practice in Offshore Site Investigation

A typical site investigation programme comprises two principal components: the seismic survey and the geotechnical investigation.

In current practice, the seismic reflection survey for offshore civil engineering is carried out according to two approaches:

- single-channel seismic, analog for the most part, with sources and receivers of very high frequency (0.5 to 3 kHz),
- multi-channel seismic with sources and receivers of lower frequency (< 500 Hz).

The digital multi-channel seismic survey concerns the first 500 meters under the seabed and is essentially devoted to the detection of drilling hazards.

Several digital very high resolution seismic data acquisition systems have been developed recently (SINUP by IFREMER, DELPH 1 by ELICS, etc.). The greatest advantage of digital recording lies in the ability to apply powerful processing (Girault and Mathevon, 1990, Lericolais *et al*, 1991).

Geotechnical investigation campaigns are carried out with vessels equipped with a drilling rig and a motion compensator. Two techniques are used according to the strength of the soil:

- standard drilling with the 5 inch diameter API drill pipe and sampling and *in situ* testing with wireline probes,
- drilling in piggyback mode with wireline sampling.

In situ probes are essentially the penetrometer (CPT) and the vane probe, more seldom a pressuremeter or dilatometer.

The incorporation of geophones in a cone penetrometer offer the possibility of combining standard geotechnical testing with seismic testing in the same operation. The seismic cone penetrometer has been used in offshore site investigation since 1987 for measuring *in situ* shear wave velocity (De Lange, 1991).

3. Geotechnical Seismic Correlations

In practice, a very high resolution seismic section cannot be directly compared to a geotechnical profile. Seismic sections are obtained and processed in terms of time. Geotechnical data are given in terms of depth. Conversion from time to depth requires knowing of the P-wave velocity of each soil layer. This knowledge is indispensable to engineers for extrapolating soil data around a borehole.

To date, the methods which may be used by geophysicists for evaluating Vp are:

- processing of multi-channel seismic data,

- velocity logging,
- laboratory tests on samples (resonant column or traveltime measurements),
- vertical seismic profiling.

Vertical Seismic Profiling (VSP) consists in recording seismic data using a surface energy source and a receiver lowered by a wireline into a borehole. VSP surveys are used like checkshot surveys to convert surface seismic sections to depth (Justice *et al*, 1984).

But as previously said, very high resolution seismic data recorded to date is single-channel and yet analog for the most part. Data processing is not possible, and so velocity cannot be obtained by this way. Moreover velocity logging is seldom carried out in shallow formations because it requires an uncased borehole, and the stability of hole walls are not ensured. It is evident that this situation must be changed. Only a comprehensive approach will enable seismic and geotechnics to be integrated. This is why the objective of the GEOSIS project includes the development and the testing of two techniques:

- vertical seismic profiling in geotechnical boreholes,
- multi-channel very high resolution seismic surveying.

4. Vertical Seismic Profiling in Geotechnical Borehole

Two probes can be used to perform a VSP:

- wireline seismic cone penetrometer,
- standard wireline VSP unit (only used in an open hole).

Both probes are equipped with triaxial sets of geophones (one for vertical motion, two for horizontal motion). The seismic cone penetrometer has been developed to provide a rapid means of obtaining the shear wave velocity and has already been used offshore (De Lange, 1991) and can be easily adapted to P-waves.

A lot of VSP wireline units can be used in an open hole.

Tests were performed in January 1992 on an onshore site at Calais in the north of France. P and S wave sources (hammers) were located at the soil surface and signals obtained in three boreholes down to 40 m were recorded and processed as standard VSP.

Figure 1 shows an example of VSP obtained with a seismic cone penetrometer. Values of Vs and Vp, whatever technique was used.

In its fullest application the technique requires the recording of not just the wave transmitted directly from the surface, but also the downgoing and upgoing wave reflected from seismic interfaces located above and below the position of the probe, therefore the recording time must be long enough.

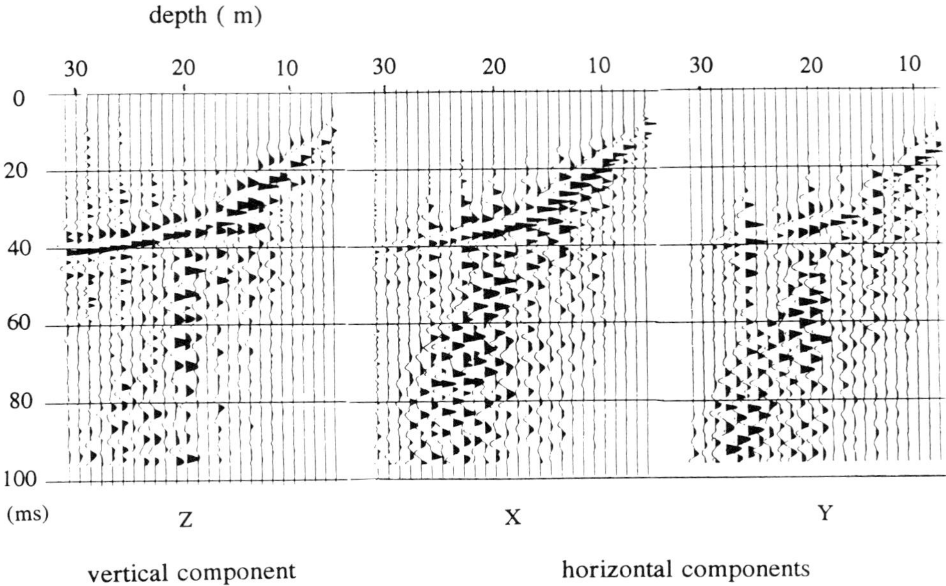

Fig. 1. Example of VSP obtained with a seismic cone penetrometer on Calais site (P-wave source – stack 4).

These techniques will be used at the end of 1992 on a site in Mediterranean Sea in 50 m of water.

5. Multi-Channel Very High Resolution Seismic Survey

No system for multi-channel very high resolution seismic survey was available in 1990. A special streamer was designed and built early in 1991 (two lengths with 24 channels and 50 m long). The streamer must be towed near the surface of the sea to highlight a central frequency of 1000 Hz.

As there was no digital and recording system available corresponding to the specifications of GEOSIS, IFREMER joined ELICS to develop a new system. The main characteristics of DELPH 24 are:

- 24 channels
- sampling frequency = 12 kHz
- maximum shooting rate = 250 ms
- 70 dB adjustable gain
- 16 bits A/D convertor.

The nominal version of DELPH 24 is now available. Two offshore test programmes were performed, in Brest and Monaco, with the following objectives:

- testing the multi-channel streamer
- testing several sources: sparker, water gun SODERA S15, boomer SEAOTTER
- recording multi-channel data to extract P wave velocity.

In conclusion of these campaigns, the complete set of multi-channel VHR seismic has to be considered as operational and to fit the requirements of the GEOSIS project.

Figure 2 compares sections of single-channel and multi-channel seismic on the same site. The multi-channel seismic improves the signal to noise ratio. The surface of erosion clearly appears in the middle of the profile. The migrated section expressed as a function of depth can be directly compared to a geotechnical log. Moreover P-wave velocity obtained through processing may be compared to values measured by other methods.

6. Conclusions

The GEOSIS project aims to integrate seismic and geotechnical data for improving the extrapolation of soil data around geotechnical boreholes. The project includes the development and the testing of two techniques:

- vertical seismic profiling in geotechnical boreholes,
- multi-channel very high resolution seismic surveying.

Techniques for performing P and S wave VSP in a geotechnical borehole have been tested on an onshore site. The seismic cone penetrometer and VSP unit can be used to calibrate velocities for high resolution seismic surveying.

A complete multi-channel very high resolution system has been developed and tested under offshore conditions. Results are already encouraging.

At the end of 1992 both geotechnical and seismic experimental campaigns will be carried out on a Mediterranean site.

Acknowledgements

The GEOSIS project is sponsored by CLAROM (Club for Research Activities on Offshore Structures). We would like to thank IFP, IFREMER, ELF Aquitaine Production, TOTAL, GEODIA and BEICIP for their permission to publish this paper. We wish to express our sincere gratitude to our colleagues of these companies for their invaluable help in various discussions during the course of this study. We also thank GEOCEAN for their cooperation during the onshore field tests.

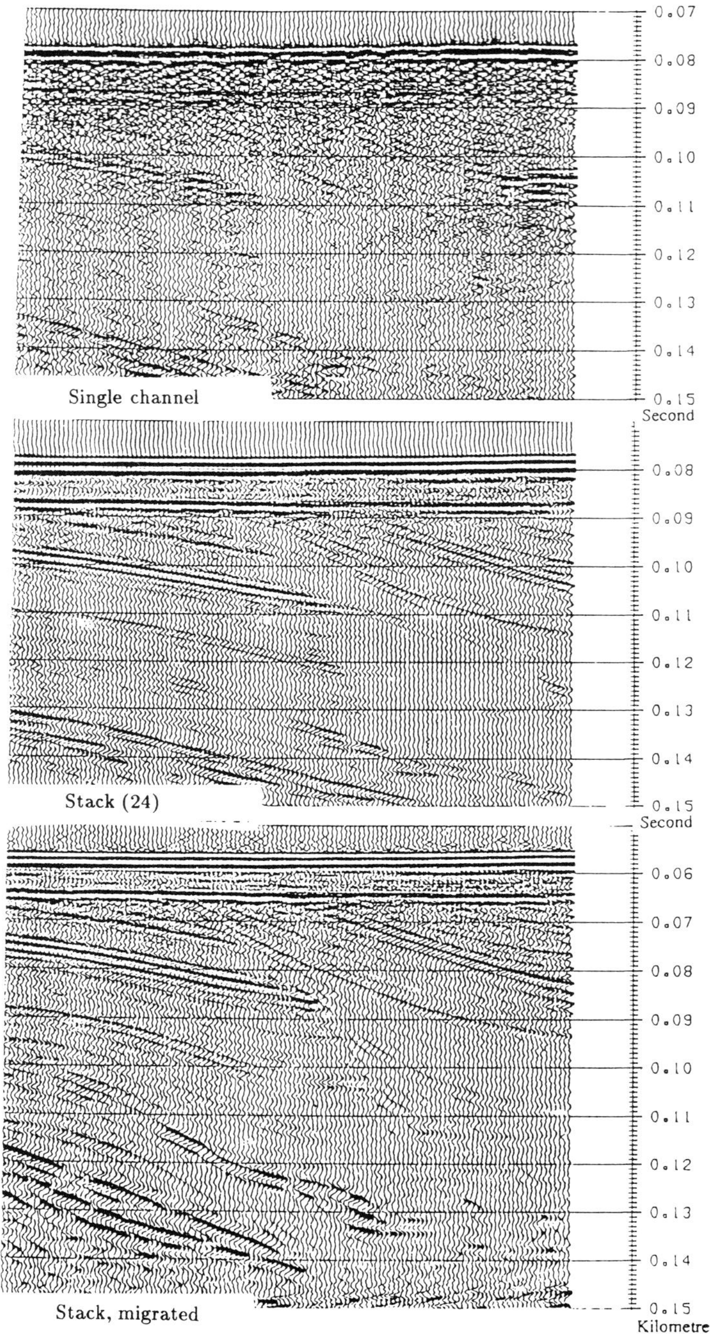

Fig. 2. Comparison of single-channel and multi-channel seismic sections on the Monaco site.

References

1. Campbell, K. J., Quiros, G. W., and Young, A. G. (1988), 'The importance of integrated studies to deep water site investigation', *Proc. 20th Offshore Technology Conference*, Houston, Paper OTC 5757.
2. De Lange, G., Rawlings, C. G., and Willet, N. (1990), 'Comparison of shear moduli from offshore seismic cone tests and resonant column and piezoceramic bender element laboratory tests', Oceanology '90.
3. De Lange, G. (1991), 'Experience with the seismic cone penetrometer in offshore site investigation', in *Shear Waves in Marine Sediments*, J. M. Hovem, M. D. Richardson and R. Stoll (eds.), Kluwer Academic Publishers, Dordrecht, pp. 275–282.
4. Girault, R. and Mathevon, G. (1990), 'Real time digital signal processing for high resolution seismic survey', *Proc. 22nd Offshore Technology Conference*, Houston, Paper OTC 6341.
5. Justice, J. M., Hinds, R., and Stirbys, A. F. (1984), 'The use of vertical seismic profiling in geotechnical site investigation', *Proc. 16th Offshore Technology Conference*, Houston, Paper OTC 4756.
6. Lericolais, G., Girault, R., Tofani, R., and Olagnon, M. (1991), 'Recent advances in shallow seismic reflection processing', *Proc. 23rd Offshore Technology Conference*, Houston, Paper OTC 6556.

Discussion

Question from John Arthur, of J. Arthur Associates, Twickenham, Middlesex, UK: Were any problems experienced in utilizing a sparker for the seismic cone p/s wave experiment in view of the length of the pulse?

Authors' response: In general a sparker is an excellent source so long as it is used in good condition. The length of the pulse did not entail problems, a deconvolution was used.

A REVIEW OF SAMPLING EFFECTS IN CLAYS AND SANDS

D. W. HIGHT
Geotechnical Consulting Group, 1a Queensberry Place, London SW7 2DL

Abstract. Predictions of the strains caused by tube sampling are combined with a simple framework for soil behaviour to examine both the response of soils to tube sampling and the effect of sampling on the soils' subsequent behaviour. Factors shown to be important are the soil's plasticity, stress history and structure, and the geometry of the sampler. Attention is paid to comparing field and laboratory measurements of dynamic shear modulus as a means of evaluating sample disturbance in clays and sands.

1. Introduction

Current sampling practice offshore involves the use of tube samplers for the full range of soil types and strengths, from loose to dense sands and from soft to hard clays. Rotary coring is generally used only for identification purposes and the developing onshore practice of rotary coring to obtain test quality 100mm diameter samples of stiff and hard clays (Hight *et al*, 1992a) and of sands (Scarrow and Gosling, 1986) has not yet been adopted.

The availability of almost unlimited reaction force means that tube samplers are generally pushed and thin wall tubes are used successfully in soils having undrained strengths up to 500 or 600 kPa. Lower strength limits apply to pushed sampling onshore where limited reaction forces result in sample damage from unsteady penetration and rebound. Pushing rates offshore are fast, typically 2cm/sec. At these rates tube sampling is likely to be an undrained process in soils having a permeability less than 1×10^{-7}m/sec and to be drained when permeability exceeds 1×10^{-4}m/sec.

Thicker walled sampling tubes are driven offshore to recover dense sands and hard clays, when the pushing force exceeds the limit of the hydraulics. Piston samplers are used in soft and firm clays when sample retention would otherwise be poor.

Strains accompanying penetration of the sampling tube into the ground represent an inevitable source of disturbance, therefore, in offshore sampling. Attention is focused in this paper on the response of soil to tube sampling strains and on the effect of these on the soils' subsequent behaviour. The effects of other sources of disturbance, for example, by borehole instability and during sample withdrawal, transport, extrusion, storage and specimen preparation, all of which

Volume 28: Offshore Site Investigation and Foundation Behaviour, 115–146, 1993.

can be controlled to a large extent and, indeed, are probably better controlled in offshore than onshore operations, have been considered by Hight and Burland (1988).

2. Tube Sampling Strains

Probably the most important advances in understanding the effects of tube sampling have been made possible by the application of the Strain Path Method – SPM (Baligh, 1985). Using this method, Baligh, Azzouz and Chin (1987) have predicted the deformation pattern and soil strains caused by a simple open ended tube, the Simple Sampler, of outer diameter, B, and wall thickness, t, when pushed steadily under undrained conditions into a saturated clay having no shear resistance. The predictions from the SPM have drawn attention to the types of strain involved, the sequence in which they are applied and their likely magnitude.

The geometry of the Simple Sampler is shown in Figures 1 and 2. It is characterised by the aspect ratio B/t. (Note that B/t and area ratio are directly related: for B/T of 40, the area ratio is 11%.) It has rounded tips to the walls and there is a slight increase in inner diameter over a distance of $2B$ from the tip. Inside clearance is, therefore, crudely modelled, but there is no allowance for a knife edge at the tips.

The detailed strain pattern for a Simple Sampler with B/t of 40 and inside clearance ratio of 1% is presented in Figure 1 in terms of contours of the following four components of strain: ε_{rr}, radial strain, $\varepsilon_{\theta\theta}$, tangential strain, ε_{rz}, meridional shear strain, ε_{zz}, vertical strain. Each set of strain contours is shown for half a sample.

The following should be noted:

- strains and, therefore, pore pressures for undrained penetration, vary across the diameter;
- disturbance, as expressed by level of shear distortions, decreases towards the centre;
- near the walls, there are steep gradients of strain and pore pressure; shear strains are large and the zone affected has been shown to increase as B/t reduces; the thickness of the highly distorted zone is controlled by t rather than by B;
- strains, and, therefore, pore pressures, vary along the length of the sample; the greatest variations occur within $B/2$ of the tip;
- the soil entering the sampler experiences a complex sequence of strains which involve unloading; above the tip, the strains to which the sample is finally subjected are relatively small;
- over the inner half of the specimen, strains are relatively uniform; ε_{zz} is the dominant strain, ε_{rr} equals $\varepsilon_{\theta\theta}$ and ε_{rz} is approximately zero; conditions approach those of triaxial compression and extension;

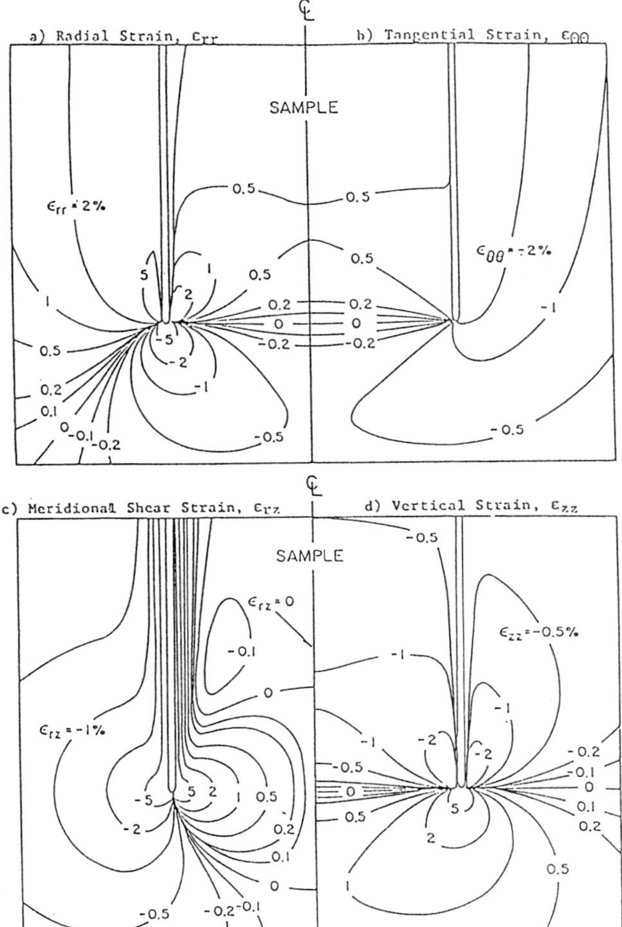

Fig. 1. Strain contours during undrained Simple Sampler penetration in saturated clays (from Baligh, Azzouz and Chin, 1987).

- strains are a minimum on the centre-line and $\varepsilon_{rr} = \varepsilon_{\theta\theta} = -1/2\varepsilon_{zz}$ and $\varepsilon_{rz} = 0$; therefore, conditions of triaxial compression and extension apply.

It is a convenient simplification to consider the tube sampling strains as comprising two different patterns:

- one applying over the central portion of the specimen, for which the minimum strains are represented by the triaxial compression and extension strains at the centre-line – centre-line sampling strains;
- one applying in the peripheral zone, where boundary shear dominates.

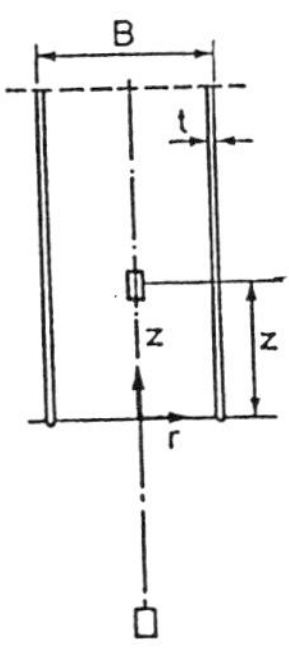

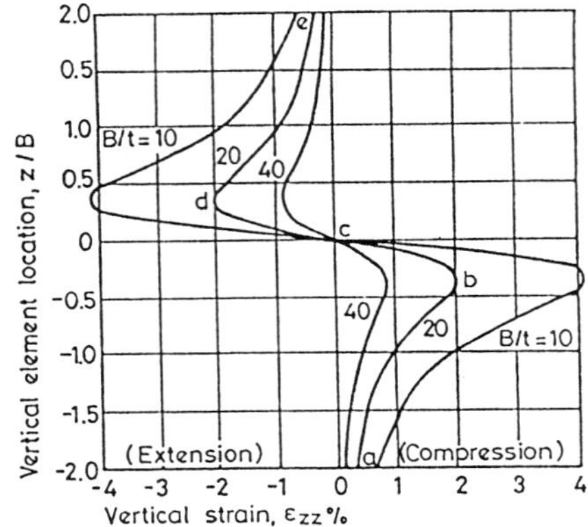

Fig. 2. Strain history of element on centre-line of penetrating tube sampler (from Baligh, 1985).

The centre-line strain history is particularly valuable since the response of soil to triaxial compression and extension is reasonably well understood, and can be investigated in laboratory triaxial tests. The effect of these strains on subsequent soil behaviour provides a lower bound to the effects of sampling. The centre-line strain history for an element of soil being sampled and its dependence on sampler geometry is shown in Figure 2. It involves triaxial compression ahead of the sampler, with the maximum compression strain developing when the element is at $0.35B$ below the tip; there is triaxial extension as the element enters the tube; after entry into the tube, the strain history depends on the internal geometry, i.e. inside clearance ratio – for the Simple Sampler there is a second compression cycle beyond z of $0.35B$, associated with the modelling of inside clearance and the sample being able to expand, and this restores ε_{zz} towards zero.

3. Framework for Soil Behaviour

To appreciate both the response of soil to centre-line tube sampling strains and the effect of these strains on its subsequent behaviour, the following simple framework for soil behaviour is introduced. In triaxial stress space, two kinematic sub-yield surfaces Y1 and Y2 in Figure 3, exist, surrounding the current stress point O. Y1 is the boundary to linear elastic behaviour and Y2 is the boundary to non-linear elastic behaviour. As the stress point moves in stress space, surfaces Y1 and Y2 are engaged and dragged with the stress point. Irrecoverable strains start to develop after reaching Y2 and increase progressively as the initial bounding surface (BS) is approached. The BS marks the onset of gross fabric distortion. For simplicity of presentation, the surface shown in Figure 3 is a section of constant water content, w_o, taken through the full bounding surface. It is a typical section for K_o consolidated clay and it represents the surface beyond which effective stress

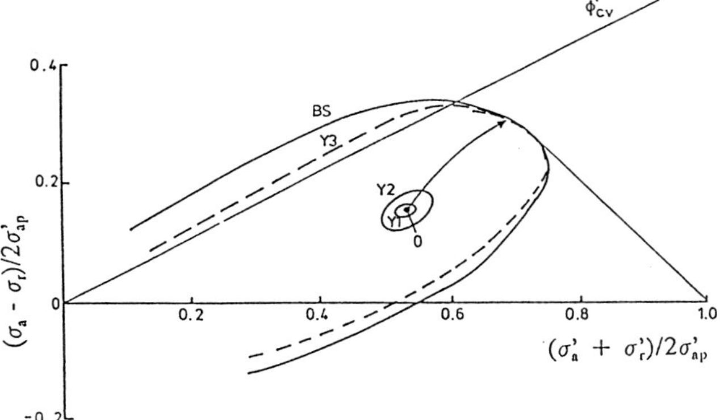

Fig. 3. A simple framework for undrained soil behaviour (adapted from Jardine, St. John, Hight, and Potts, 1991).

paths in undrained shear of the soil at w_o cannot stray. This section of the BS will be sufficient for examining the effects of undrained tube sampling (σ'_{ap} is the axial preconsolidation stress).

For soil sheared undrained from a lightly overconsolidated state the BS forms the main yield surface (Y3 in the terminology of Jardine, St. John, Hight and Potts, 1991). For soil in a heavily overconsolidated state a yield surface inside the BS can often be detected and this will be identified as a continuation of the Y3 surface.

Of particular relevance to understanding the response to tube sampling are:

- the shape of the BS, i.e. the section shown in Figure 3;
- the strains associated with reaching the boundaries to Y1, Y2, Y3 and the BS (ε_{Y1}, ε_{Y2}, ε_{Y3}, ε_{bs});
- the effects of passing through the Y2 and Y3 boundaries and reaching and travelling along the BS.

3.1. Shape of Bounding Surface

In reconstituted soils, the shape of the bounding surface can be defined using the effective stress paths from undrained triaxial compression and extension tests on the K_o normally consolidated soil. The shapes of bounding surface defined in this way for a range of reconstituted soils of different plasticity are shown in Figure 4, where the data has been normalised by the axial consolidation stress, σ'_{ac}.

It is immediately apparent that the shape of the bounding surface is strongly influenced by the plasticity of the soil, particularly in triaxial extension. As the plasticity of the soil reduces, the surface swings towards the origin in stress space on the extension side, reflecting the increasing anisotropy of strength with reducing plasticity. In triaxial compression, reducing plasticity is associated with a rise in

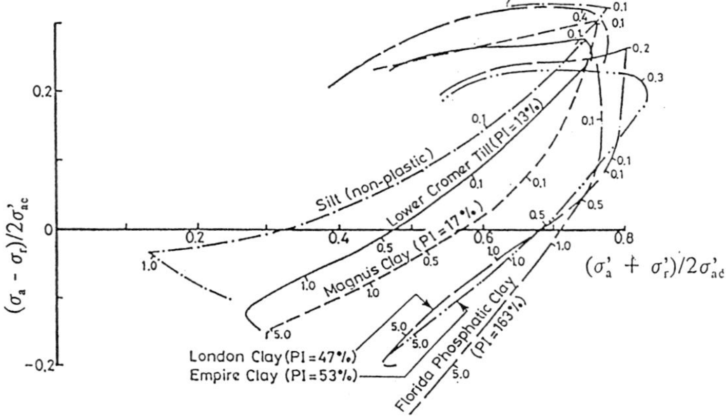

Fig. 4. A constant water content section through the BS of a range of reconstituted soils.

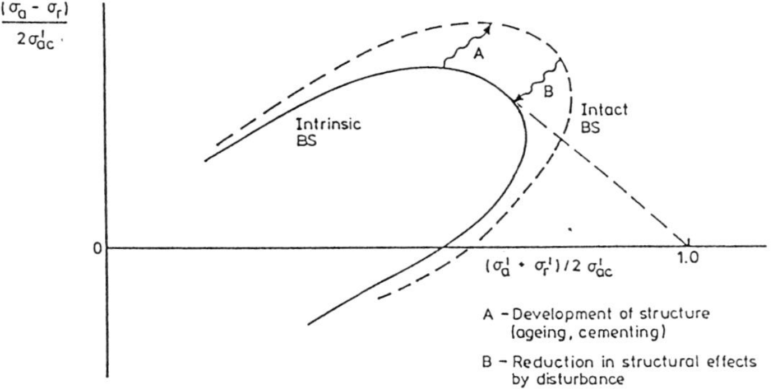

Fig. 5. Effects of post-depositional processes on the shape of the BS.

the location of the surface and with reducing strains to the peak of the surface.

A bounding surface defined for reconstituted soil (the intrinsic bounding surface, after Burland, 1990) provides a useful lower bound to the location of the surface for the intact structured soil. The effects of post-depositional processes such as ageing and cementing are to expand the surface as illustrated in Figure 5[1]. Even in intact natural clays, the effect of soil plasticity on the shape of the bounding surface remains. This is illustrated in Figure 6 where the bounding surfaces for a range of natural and carefully sampled clays are compared in terms of ϕ' (which, of course, reflects plasticity).

The effects of ageing and cementing will also expand the Y2 surface and in strongly cemented soils, the Y2, Y3 and BS may coincide.

[1] Slow rates of sedimentation will also result in the BS being located outside that for the reconstituted soil.

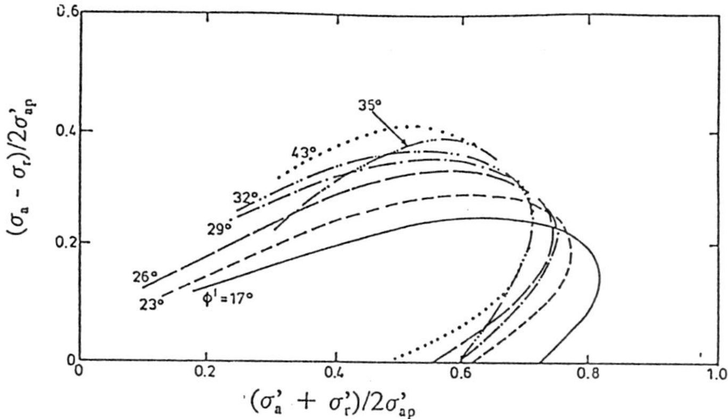

Fig. 6. A constant water content section through the BS of a range of natural soft clays (from Diaz-Rodriguez, Leroueil, and Aleman, 1992).

3.2. Critical Strain Levels

Strain limits to linear elastic behaviour, i.e. to reach the Y1 surface, are extremely small in both uncemented soils ($\varepsilon_{Y1} < 0.001\%$ in compression) and cemented soils and weak rocks ($\varepsilon_{Y1} < 0.01\%$ in compression). Strain limits to recoverable behaviour (i.e. non-linear elastic behaviour) are also relatively small; typical values are presented by Jardine *et al*(1991), and these show that ε_{Y2} tends to increase with PI but does not appear to exceed 0.04% in clays.

For lightly overconsolidated soils (OCR $\leq$ 4, ε_{Y3} is effectively the same as ε_{bs}. In most soils, ε_{bs} for triaxial compression is similar to the strain required to mobilise peak compression strength, ε_{ap}. In triaxial compression, the axial strains required to mobilise peak strength, ε_{ap}, appear to be determined principally by:

- soil composition and fabric (reflected in plasticity);
- stress history (OCR) and current stress state.

Figure 7 illustrates the effect of both stress history and soil composition on ε_{ap}, using data from a range of reconstituted soils which have been subject to a simple one-dimensional loading-unloading stress history. The following observations may be made:

- strains increase with OCR;
- at any OCR, strains increase with soil plasticity.

There is evidence that the processes which introduce structure into the soil, i.e. which expand the soil's bounding surface (Figure 5) do not necessarily change the levels of critical strain marking the boundaries to the zones under consideration.

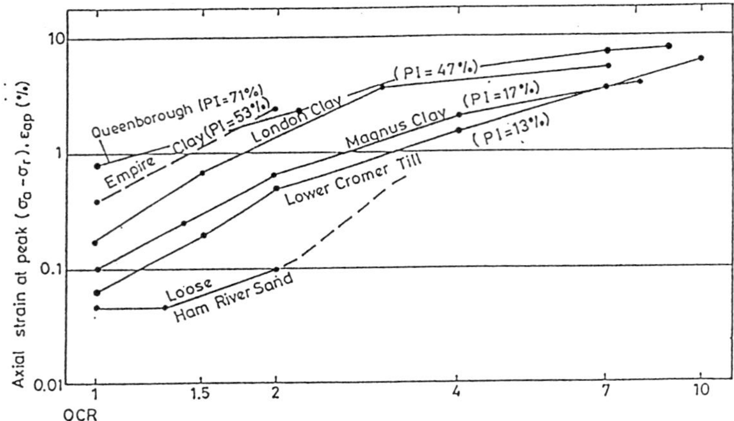

Fig. 7. Axial strain to peak in triaxial compression versus OCR.

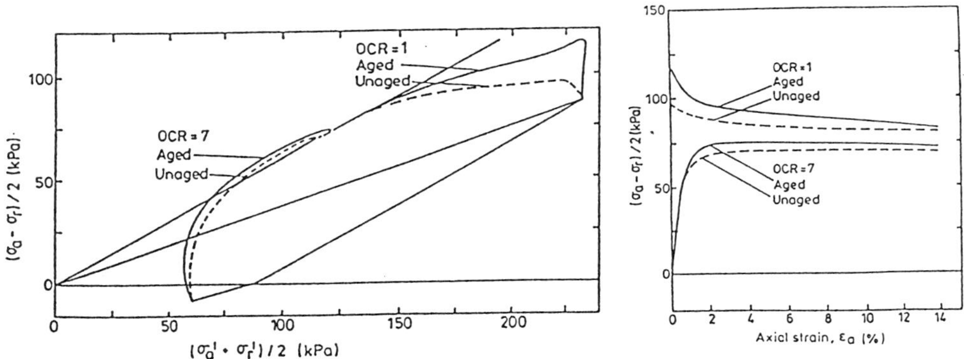

Fig. 8. Effects of ageing on the behaviour of a low plasticity clay.

Figure 8 compares the behaviour of young low plasticity clay at OCRs of 1 and 7 with that for clay aged at the same OCRs. Similar strains at yield and on reaching the BS are observed. Hanzawa (1983) has demonstrated the same point in natural soft clays, by comparing values of ε_{ap} in both aged normally consolidated clay and mechanically overconsolidated clay. Despite high apparent OCRs in some of the aged clays, they retain the same ε_{ap} as the normally consolidated soil. Even in cemented sands the strains to yield appear to be controlled to some extent by the initial uncemented fabric, although cementing tends to reduce the strains (Clough *et al*, 1981). Evidence can be found, see for example Hight, Jardine and Gens (1987) that the strain levels shown in Figure 7 are also largely independent of rate of shearing and of drainage.

TABLE 1.
Yield-strains, ε_{Y3}, in triaxial compression tests on stiff overconsolidated clays.

CLAY	ϵ_{Y3} (%)	REFERENCE
London Clay, Ashford Common	0.7-1.0	Ward et al (1965)
London Clay, Broad Oak	0.5-1.0	Sandroni (1977)
London Clay Basement Beds, Waterloo	≈0.5	GCG Files
Woolwich and Reading Beds Mottled Clay, Islington	0.55-0.75	GCG Files
Todi Clay	2.1-2.3	Georgiannou (pers. comm.)
Vallerica Clay	0.9	Georgiannou (pers. comm.)
Pietrafitta Clay	0.6-1.0	Georgiannou (pers. comm.)
Corinth Marl	$1.0\text{-}0.7^{+}$	Georgiannou (pers. comm.)

$^{+}$ reducing with increasing stress level

The strains shown in Figure 7 are the strains required to mobilise peak strength, ε_{ap}, in triaxial compression. For heavily overconsolidated soils (OCR > 4), it is possible to identify a yield point prior to the mobilisation of peak strength in both the stress- strain curve and effective stress path. A typical example is presented in Figure 9, which shows the stress-strain and pore pressure-strain curves observed in an unconsolidated undrained triaxial compression test on London Clay. The axial compression strain at yield, ε_{Y3}, is approximately 0.7%.

The results of a brief review of data on ε_{Y3} strains in triaxial compression of stiff and hard plastic or cemented clays are summarised in Table 1. This data has been taken from tests on block or rotary cored samples in which axial displacement measurements have been made locally on the specimen. With occasional exceptions, ε_{Y3} lies between 0.7 and 1.0%. It is of interest to note that these strains are similar to ε_{Y3} strains in the aged normally consolidated soil and may well indicate the initiation of the breakdown of a structure developed while ageing under normal

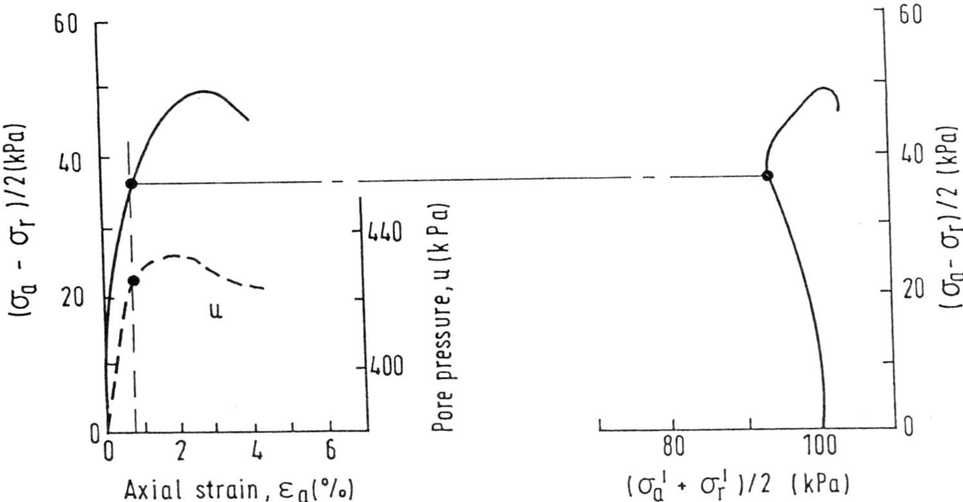

Fig. 9. Yielding in an unconsolidated undrained triaxial compression test on intact London Clay (from Sandroni, 1977).

consolidation. This requires the initial structure to be retained during overconsolidation and to continue to dominate response to strain. (A limit may well be reached when swelling to low effective stresses, if volumetric strains are sufficiently large to disrupt the initial fabric – this will influence the location of the Y3 surface.) Soils showing significantly higher apparent yield strains than 1% in Table 1 may well owe their structural effects to some other form of bonding or to bonding while in an overconsolidated state.

Observations reported by Ward *et al* (1965) and shown in Figure 10 suggests that in fissured plastic clays the ε_{Y3} strains may well mark the initiation of movement on existing fissures. Certainly movement on fissures well before reaching the BS is consistent with observations by Sandroni (1977).

4. Predicted Response During Tube Sampling

By combining the tube sampling strains predicted by the SPM (Figures 1 and 2) with this simple framework for soil behaviour and its critical strain levels it is possible to speculate on both:

- the response of the soil during tube sampling, and
- the effects on subsequent behaviour of soil which has been subjected to tube sampling strains.

Hight, Gens and Jardine (1985) used this approach to illustrate the different response of normally and overconsolidated soil to tube sampling strains. For sim-

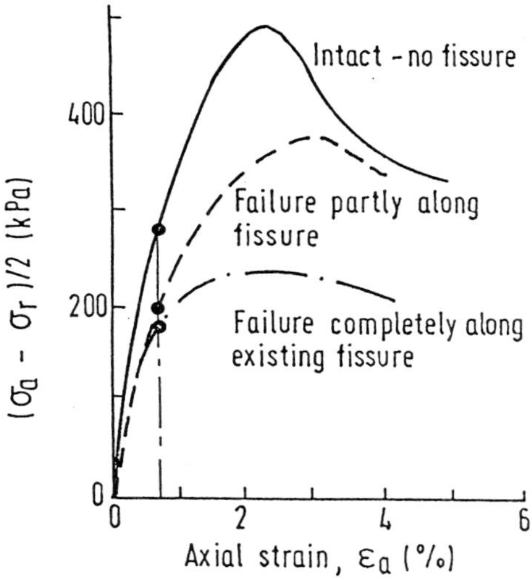

Fig. 10. Effects of fissures on laboratory stress-strain characteristics of London Clay (from Ward, Marsland and Samuels, 1965).

plicity they considered separately the effects of the strains on elements in the path of the tube centre-line and those at the tube's periphery. It was shown that in lightly overconsolidated soil there is a significant reduction in mean effective stress, p', (Figure 11a) while in heavily overconsolidated soils p' is affected much less (Figure 11b): in fact, p' may increase, largely as a result of intense shearing of the dilatant soil at the periphery.

Since 1985, laboratory experiments, summarised in Table 2, have been carried out to examine the effect of the strains at the sample tube centre-line – triaxial compression and extension – on the subsequent behaviour of reconstituted and intact clays. The findings from these experiments confirm the assessments made by Hight *et al* and allow their approach to be extended with some confidence to examine the response to tube sampling further, taking into account, for example in this paper, the effects of:

- soil plasticity, through its influence on the shape of the bounding surface,
- sampler geometry (or quality), and
- soil structure.

4.1. Comparison of Tube Sampling Strains and Critical Strain Levels

As discussed by Clayton, Hight and Hopper (1992), it appears that the strain levels predicted by the SPM are a reasonable approximation to those that occur due to tube sampling. Assuming this to be the case and taking typical levels of maximum compression-extension strain ($\varepsilon_{zz\,\max}$) to be in excess of 0.7%, it follows from

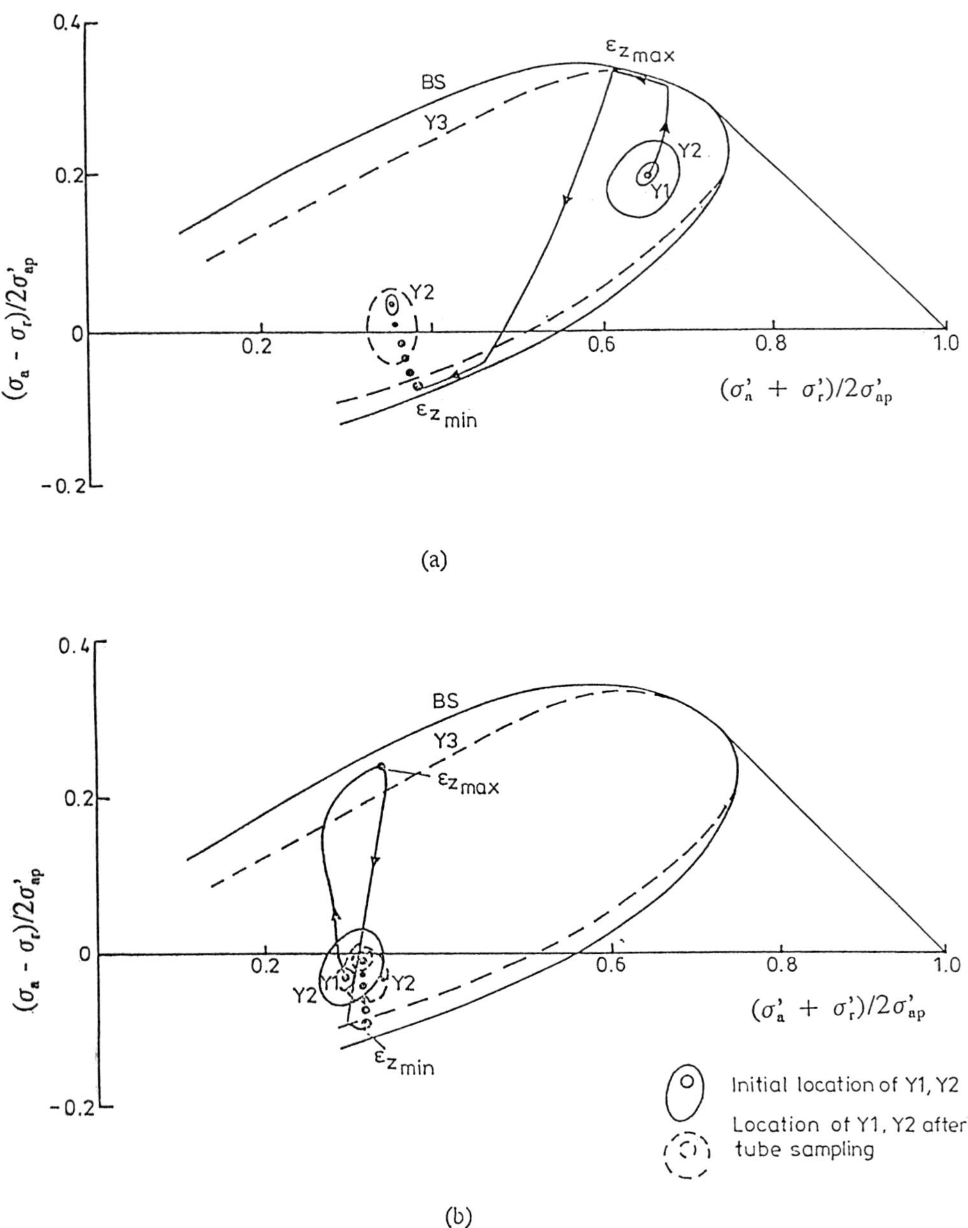

Fig. 11. Response to tube sampling strains of (a) lightly overconsolidated soil, and (b) heavily overconsolidated soil.

TABLE 2.
Laboratory investigations of the effects of centre-line tube sampling strains.

REFERENCE	SOIL TYPE	STRESS HISTORY
Baligh, Azzouz and Chin (1987)	Reconstituted Boston Blue Clay	Normally consolidated
Lacasse and Berre (1988)	Destructured Drammen Clay	Normally and lightly overconsolidated (OCR = 2.5)
Siddique (1990)	Reconstituted London Clay	Normally consolidated
Hajj (1990)	Kaolin	Normally consolidated and overconsolidated (OCR = 4)
Clayton, Hight and Hopper (1992)	Intact Bothkennar Clay	Lightly overconsolidated
Hopper (1992)	Reconstitued London Clay	Overconsolidated (OCR = 3.7)

Figure 7 and Section 3.2. that:

- the linear elastic strain limit, ε_{Y1}, is exceeded by the imposed tube sampling strains;
- the recoverable strain limit, ε_{Y2}, is also exceeded resulting in irrecoverable strains, and in Y1 and Y2 surfaces being carried around in stress space and relocated as a result of tube sampling – as described previously by Hight, Gens and Jardine (1985), and shown in Figure 11;
- the gross yield strain limit, ε_{Y3}, and the strain to reach the BS, ε_{bs}, may or may not be exceeded, depending on the geometry of the sampling tube and on the soil's plasticity and stress history.

In soils of low plasticity and low OCR, $\varepsilon_{zz\,max}$ will exceed $\varepsilon_{Y3}/\varepsilon_{bs}$, while in higher plasticity clay, particularly at high OCR, $\varepsilon_{zz\,max}$ will be less than ε_{bs} but may exceed ε_{Y3}. Soils of high plasticity and OCR will better survive the rigours of tube sampling.

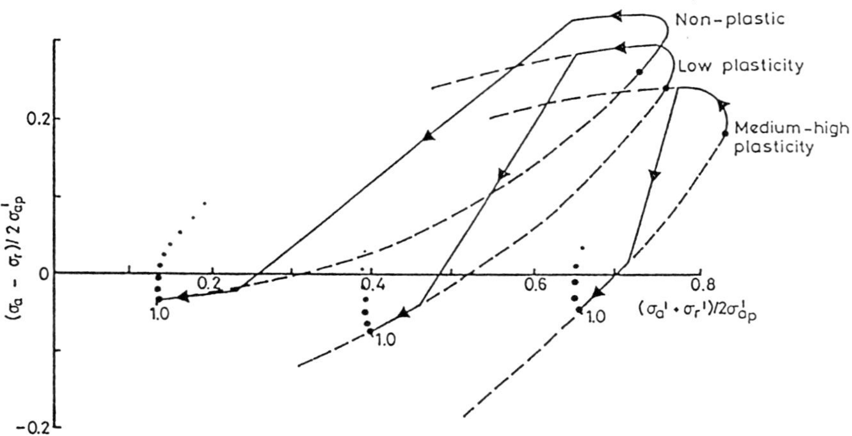

Fig. 12. The effect of soil plasticity on response to tube sampling strains in lightly overconsolidated soil.

4.2. Effect of Soil Plasticity

The effect of soil plasticity on response to tube sampling of lightly overconsolidated soils is indicated in Figure 12. The range in BS shapes shown in Figure 4 has been simplified to the three shown in Figure 12 for non-plastic, low plasticity and medium-high plasticity soil. Superimposed on each are the stress paths that would be followed by centre-line elements during tube penetration; these take into account both the differences in shape of the BS and the increasing strains to reach and travel along the BS as plasticity increases. It can be seen that both damage, related to reaching the BS and the distance travelled along it, and change in mean effective stress become less as the plasticity of the soil increases.

The effect of reducing sampler quality, in this case through reduction in the ratio B/t, i.e. increase in area ratio, is illustrated in Figure 13. As the magnitude of the tube sampling strains increases, the stress path travels further round the BS on both the compression and extension sides, causing increasing reductions in mean effective stress and further damage to structural components of resistance, i.e. further shrinking of the BS, as described below.

4.3. Modifications to Bounding Surface

We need to consider now the effects of exceeding ε_{Y2} and ε_{Y3} and of reaching and travelling on the bounding surface for intact natural soils.

For natural soils, it has been suggested that the effects of ageing and bonding[2] are to cause an expansion of the BS (Figure 5). Bearing in mind the fabric of natural soils it is reasonable to consider at least two levels of bonding:

[2] Bonding includes the effects of cementing by chemical deposition and the effects of cold welding.

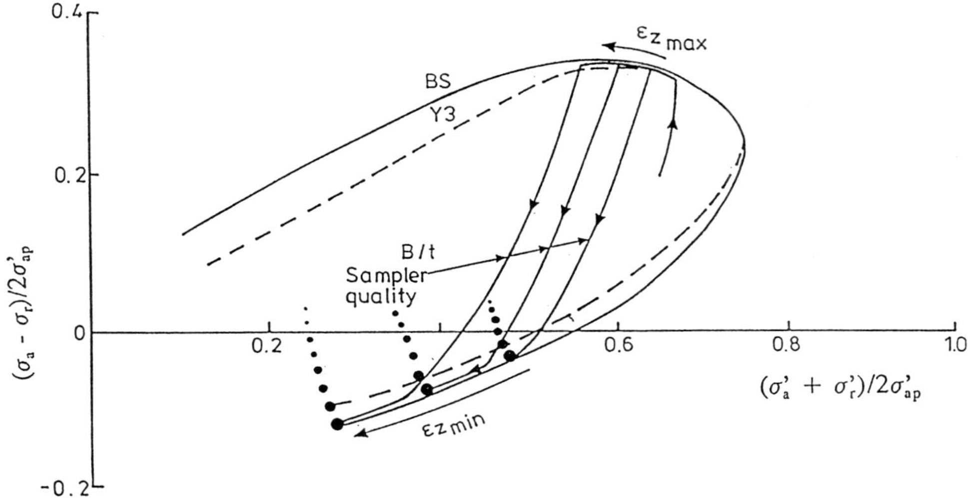

Fig. 13. The effect of sampler geometry on the response to tube sampling strains in lightly overconsolidated soil.

- intra-aggregate – i.e. between particles forming the aggregate;
- inter-aggregate – or inter-particle.

The strains which will cause damage to these bonds will depend on the compliance of the soil structure. In a clast dominated structure, such as sand or silt, the scope for compliance between contacts is limited and damage will occur to the inter-particle bonds at relatively small strains. In an aggregated clay, compliance within the aggregate will enable the soil to sustain increased levels of overall strain before damage occurs to the inter-aggregate bonds. In both cases, damage will occur over a range of strains, because of the variability in particle or aggregate size and arrangement (or in inter-aggregate pore size). The process of breakdown of bonding will be progressive and will increase with strain.

It has already been stated that irrecoverable strains which develop beyond Y2 increase as the BS is approached and that reaching the BS marks the onset of gross fabric distortion. It seems reasonable, therefore, to distinguish between two stages in any breakdown process that may be associated with irrecoverable strains caused by tube sampling:

Stage 1 $- \varepsilon_{Y2} < \varepsilon_{zz\,\max} < \varepsilon_{Y3}$

Stage 2 $- \varepsilon_{Y2} < \varepsilon_{Y3} < \varepsilon_{zz\,\max}$.

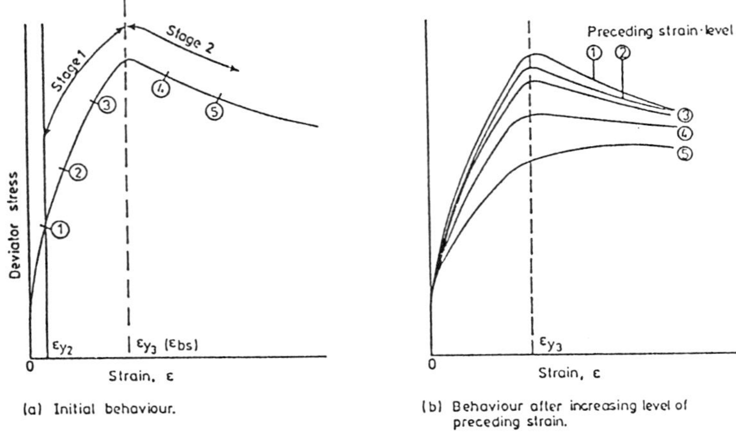

Fig. 14. Stages in the breakdown of soil structure.

The differences between these two stages are indicated in Figure 14, using behaviour in only triaxial compression for clarity.

For strains between ε_{Y2} and ε_{Y3}, it is suggested that on subsequent loading there is a small reduction in peak strength, the size of the reduction increasing with the strain amplitude and indicating a progressive reduction in the component of resistance imparted by structural effects. For strains greater than ε_{Y3} (ε_{bs}) the breakdown accelerates; for $\varepsilon_{zz\,max}$ only just greater than ε_{bs}, the disruption to the fabric is sufficiently limited for ε_{Y3} still to be discernible but less obvious in subsequent shear.

The behaviour illustrated in Figure 14 is symptomatic of a shrinking of the soil's BS as irrecoverable strains develop. In other words, the expansion of the BS by ageing and cementing is progressively reversed, as indicated in Figure 15.

It follows from Figure 10 that, in overconsolidated fissured plastic clays, the effect of $\varepsilon_{zz\,max}$ exceeding ε_{Y3} is to cause displacement on fissures.

5. Evidence for the Response to Tube Sampling Strains

5.1. Modifications to Bounding Surface

Evidence for shrinking of the soil's BS as a result of disturbance by tube sampling in soft clays has been presented by Tavenas and Leroueil (1987). Figure 15 compares the bounding surfaces defined using samples taken with the 200m diameter Laval sampler (La Rochelle *et al*, 1981) and with a 50mm diameter piston sampler. The BS defined on samples taken with the 50mm piston sampler falls well inside that based on the Laval samples, in which disturbance was probably negligible, as shown by the agreement with block samples.

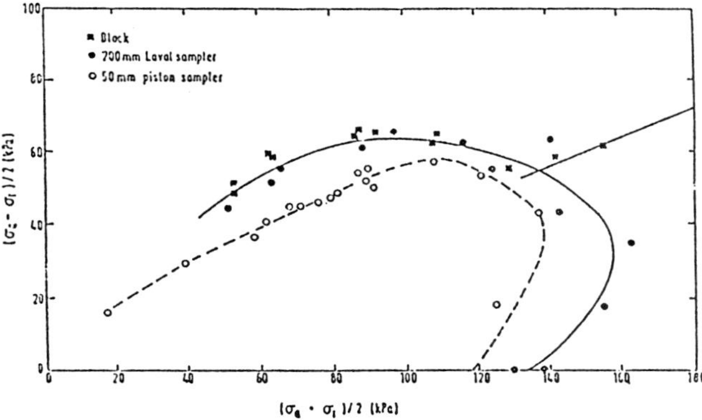

Fig. 15. Shrinking of the BS for Saint Louis Clay by disturbance during sampling (from Tavenas and Leroueil, 1987).

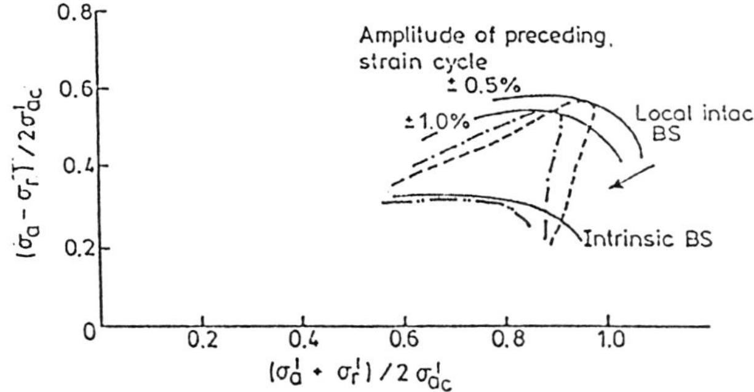

Fig. 16. Progressive destructuring of the Bothkennar clay by increasing tube sampling strains.

The fact that the shrinking of the BS takes place progressively with increasing strain was demonstrated in a set of triaxial strain path tests on structured Bothkennar clay, described by Clayton, Hight and Hopper (1992). Figure 16 compares the behaviour in undrained triaxial compression of samples which had previously been subject to increasing levels of triaxial compression-extension strains, to simulate tube sampling with different quality samplers. The location of the BS local to each stress path is shown tentatively and compared to that of the intrinsic soil. Increasing levels of tube sampling strain cause increased shrinking of the BS towards the intrinsic BS.

An indication of damage to stiff overconsolidated clays by tube sampling, revealed as shrinking of the BS, is presented in Figure 17. Effective stress paths observed in UU and CIU tests on rotary cored and pushed thin wall samples of London Clay are compared. Atypical data and data from tests failing prematurely on fissures has been omitted. It can be seen that the thin wall samples form a lower

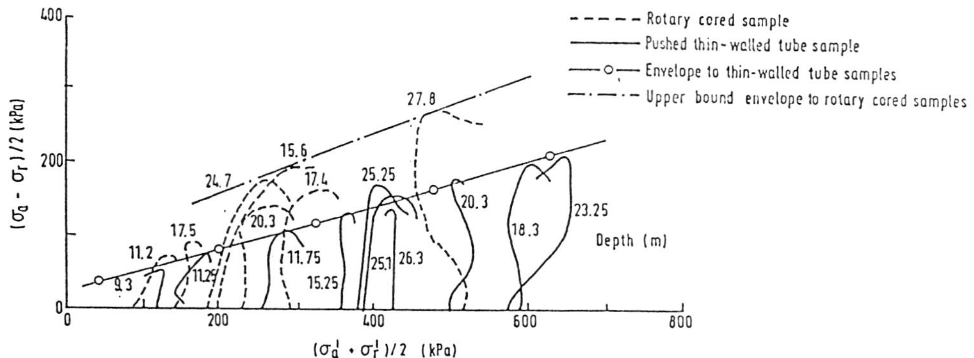

Fig. 17. Shrinking of the BS in London Clay: comparison of UU and CIU tests on rotary cored and thin wall tube samples.

envelope. Effective stress paths for rotary cored samples consistently rise above this lower envelope by varying amounts. In the London Clay, tube sampling strains exceed ε_{Y3} ($\approx 0.7\%$) and caused the damage, probably by displacement on fissures; in the rotary cored samples, although the tube sampling strains were avoided, it seems likely that smaller but varying amounts of damage may have occurred.

5.2. Changes in Mean Effective Stress

Evidence to support the predictions for the effects of stress history, soil plasticity and sampler geometry has been assembled in the form of changes in mean effective stress. The following figures compare the measurements of initial effective stress, p_i', in triaxial specimens prepared from tube samples, with the best estimates of mean effective stress *in situ*, p_o', based on pressuremeter lift- off, spade cell measurements and geological history.

- Figure 18(a) shows data for the Bothkennar soft clay. In this lightly overconsolidated and aged clay, p_i' values measured in 100mm fixed piston samples show more than a 50% reduction from p_o'. The clay has a PI of 18–22%, after removal of its organic content, and shows ε_{Y3} strains in compression of 0.5–0.6% when sheared from a lightly overconsolidated state – reductions in p' of this order are to be expected, therefore, and damage to its structure will also have occurred (Hight *et al*, 1992b). Reductions in p_i' in the 200mm diameter Laval samples are much less. This is the result partly of the better sampler geometry, both in terms of the higher B/t ratio, sharper cutting edge, and the lower relative volume of the highly distorted peripheral zone.

- Figure 18(b) shows data from an offshore site where thin wall sampling tubes were pushed into lightly overconsolidated clayey silts having a high calcium carbonate content. In these soils there was almost complete loss of effective stress, despite the samples remaining saturated. This is consistent with the very low PI of the silts (< 15%) and low ε_{Y3} strains (0.2% in triaxial compression). The higher values of p'_i were always measured in the slightly more plastic soils.

- Figure 18(c) shows data for the heavily overconsolidated plastic London Clay. In this material, the mean effective stress is not reduced, but because of dilatancy in the peripheral zone, p'_i may actually exceed p'_o. The increase of p'_i above p'_o is even more marked in the driven thick walled samples (U100) – here the thicker wall leads to a thicker distorted peripheral zone where negative excess pore pressures are generated. Figure 18(c) illustrates a potential problem with thin wall sampling tubes pushed into very stiff or hard clays – in onshore investigations distortion of the tube cross section sometimes occurs leading to additional disturbance and increase in p'.

- Figure 18(d) presents data for the heavily overconsolidated low plasticity Cowden Till (Powell, personal communication). Here again, p'_i remains close to p'_o after tube sampling, and this is consistent with its stress history.

6. Effects of Sampling in Clays

The effects of tube sampling on the subsequent behaviour of soil can be appreciated once it is recognised that tube sampling causes:

- a change in mean effective stress;
- relocation of the sub-yield surfaces Y1 and Y2, and a change in their shape;
- modification to the shape of the bounding surface, in particular the progressive removal of structural effects imparted by ageing and cementing.

The extent of these changes is determined by the plasticity of the soil, the initial level of structure, the quality of the tube sampler, in particular its geometry, and other factors referred to in Section 1.

The effects of tube sampling will be manifest in different ways depending on the type of laboratory test that is undertaken and the reconsolidation path that is followed. This is illustrated below for selected laboratory tests on soft and stiff clays.

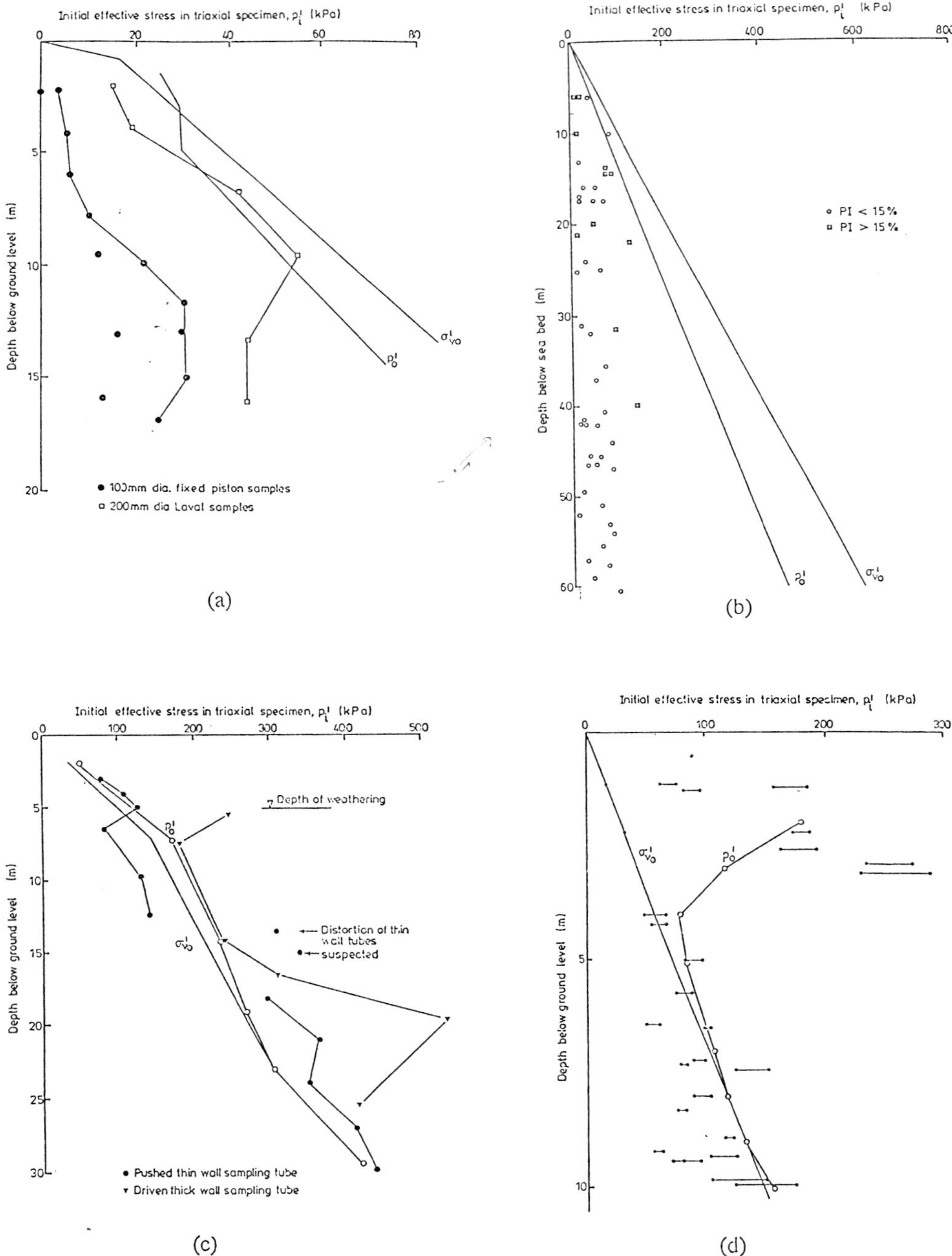

Fig. 18. Comparison of mean effective stresses *in situ* and after sampling. (a) Bothkennar Clay (PI = 18–22%); (b) Offshore clayey silts (PI = 5–15%); (c) London Clay, and (d) Cowden Till (Powell personal communication).

6.1. Soft Clays

In Figure 19, the framework for soil behaviour is used to anticipate the behaviour of soft clay after tube sampling in unconsolidated undrained triaxial tests (Figure 19a), in CK_oU triaxial tests (Figure 19b), and in oedometer tests (Figure 19c). Two qualities of sample are shown, the poorer quality being associated with the larger reduction in p' and greater shrinking of the BS. Behaviour of the soil in situ is also shown.

Differences between *in situ* and sampled soil behaviour are greatest in UU triaxial compression tests. In all cases the poorer quality sample exhibits lower strength, stiffness and preconsolidation stress.

Data supporting the hypotheses in Figure 19 is presented in Figure 20, taking as an example results of tests on the Bothkennar clay. In these comparisons, the piston sampler represents the poorer quality sampler; the Sherbrooke sampler (Lefebvre and Poulin, 1979) is a down-hole block sampler and is associated with the minimum disturbance.

6.2. Stiff Clays

We will restrict the discussion to the behaviour of overconsolidated medium to high plasticity clays, which are usually fissured. In these clays, the peak strength that is mobilised is determined by the initial effective stress in the sample, p'_i, and not by water content. The peak strength is dependent on a cohesive component of strength and is dependent on whether or not displacement occurs on fissures during loading to peak.

In these soils, tube sampling strains are likely to exceed ε_{Y3} (Section 4.1), causing displacement on existing fissures and some reduction in the cohesive component of strength – shrinking of the BS (Figure 17); mean effective stresses may increase as a result of tube sampling (Figure 18c). The effect of these changes in the location of the BS and in p' are illustrated in Figure 21, for the case of UU triaxial compression tests on different types of sample. Four different locations of the BS are shown, each descending with increasing level of disturbance, from block to rotary-cored to thin wall tube to thick wall tube. Each type of sample has a different value of initial effective stress; the thick wall and thin wall tubes have $p'_i > p'_o$ while the rotary cores have $p'_i < p'_o$ as a result of exposure to drilling muds. It can be seen that opposing effects of changes in p' and in BS location can result in similar strengths for rotary-cored, thin wall and thick wall tube samples in UU triaxial compression.

Some support of this hypothesis is provided by data in Figure 22 which compares the behaviour in undrained triaxial compression of block, thin wall and thick wall tube samples on London Clay. The much lower stiffness in the tube samples, in particular that from the thick wall tube, may reflect the effect of displacement on fissures caused by the tube sampling strains.

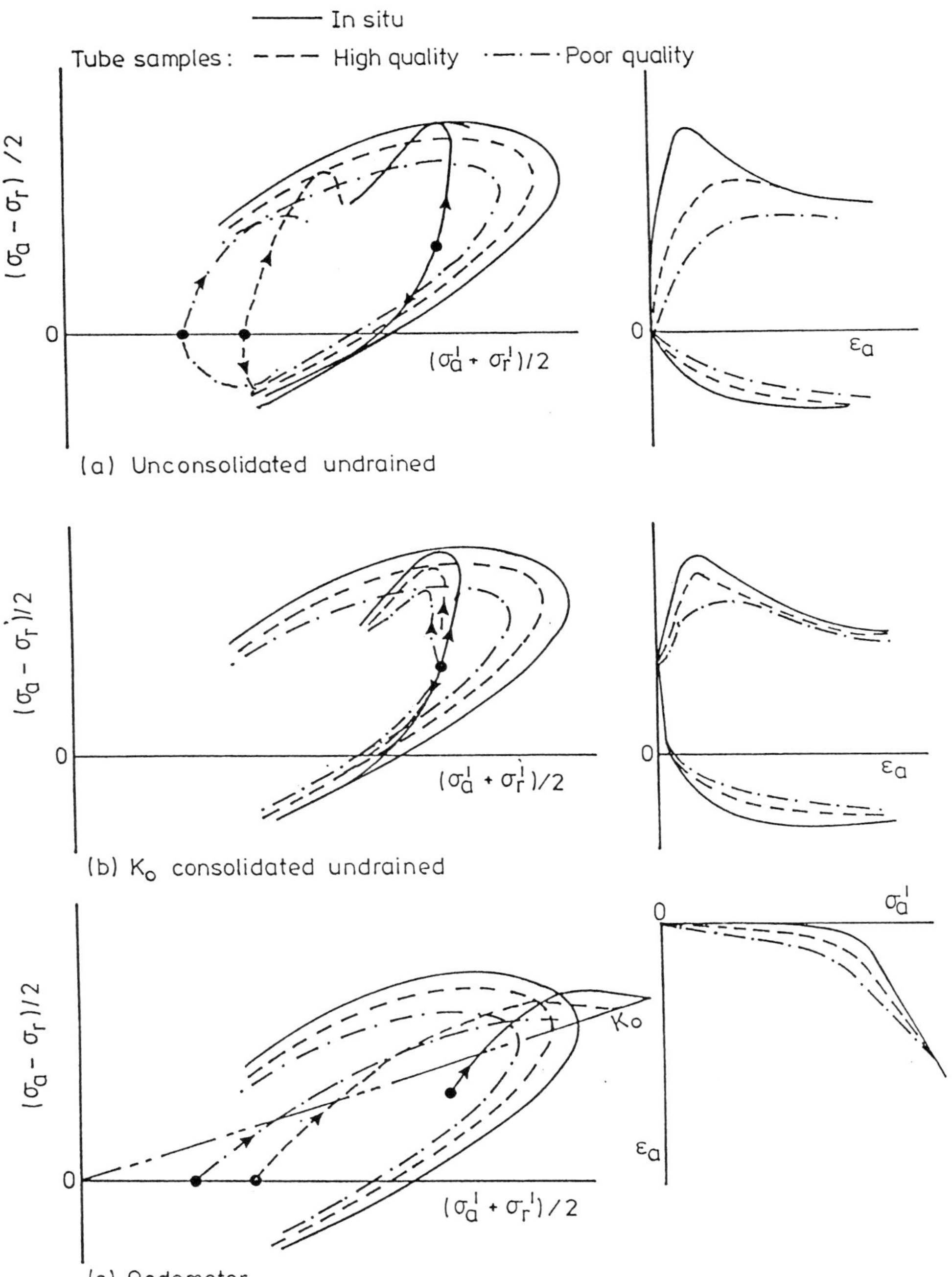

Fig. 19. Predicted effects of sampling of lightly overconsolidated clay in (a) UU triaxial tests, (b) CK_oU triaxial tests, and (c) Oedometer tests.

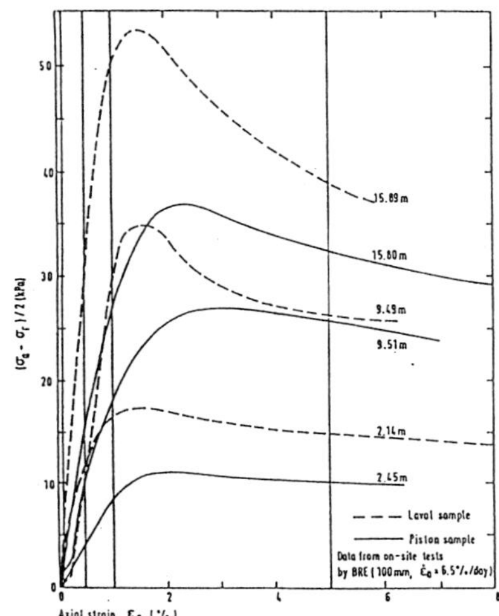

(a) UU triaxial compression

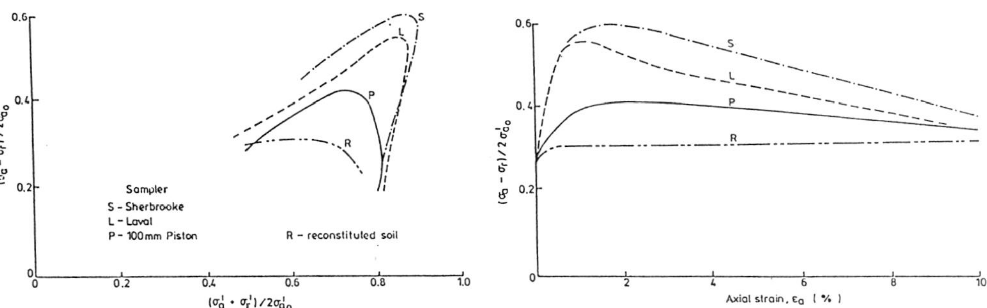

(b) CK_oU triaxial compression

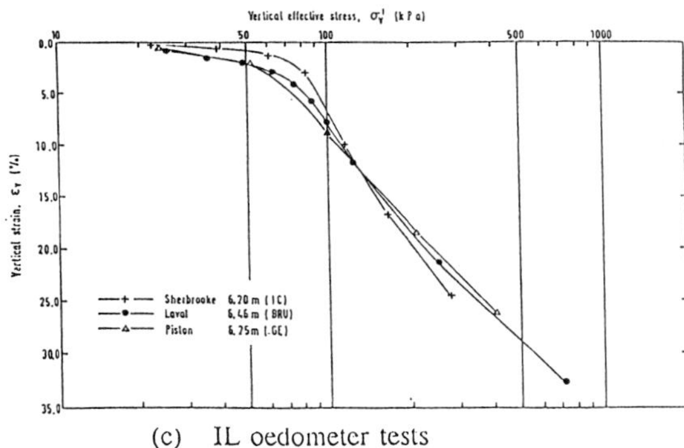

(c) IL oedometer tests

Fig. 20. Observed effects of sampling of Bothkennar clay.

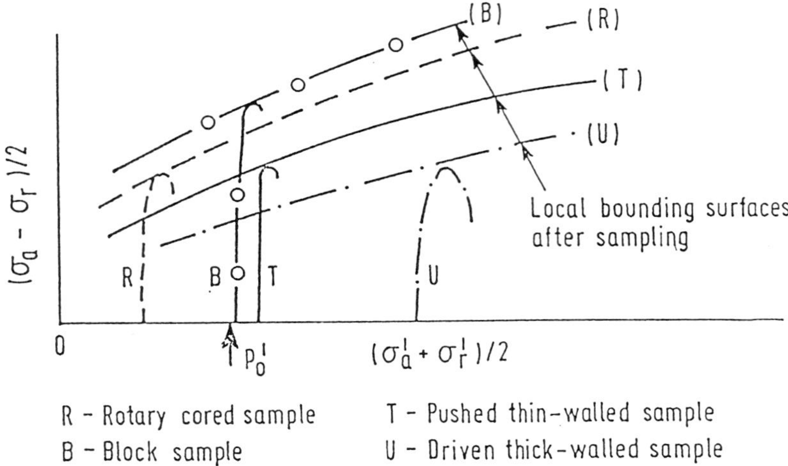

Fig. 21. Predicted effects of sampling of stiff plastic clays in UU triaxial compression.

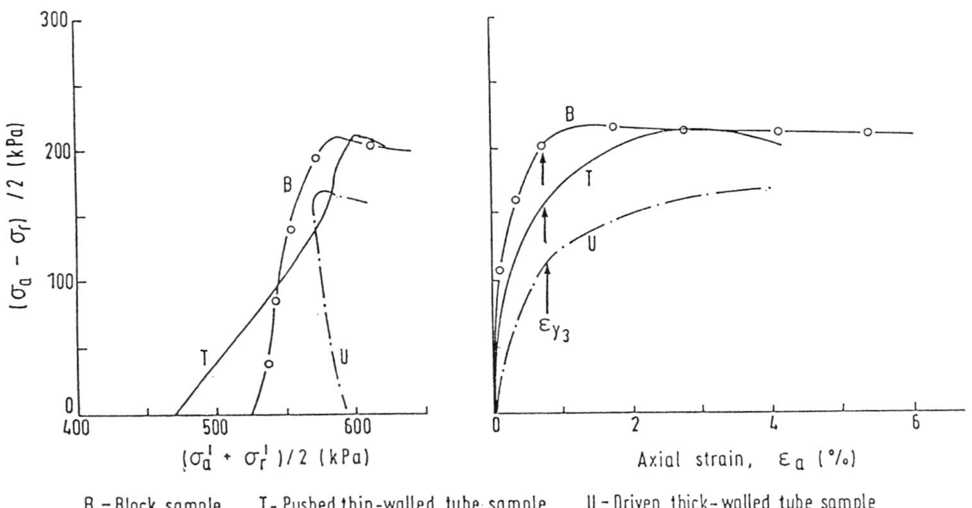

Fig. 22. Observed effects of sampling of London Clay in triaxial compression.

7. Effects of Sampling in Sands

In sands the tube sampling process is drained so that both shear and volumetric strains occur. Although predictions of tube sampling strains for drained or partially drained penetration are not yet available, it is reasonable to expect axial strains of a similar order to those shown in Figure 2; on the centre-line triaxial compression

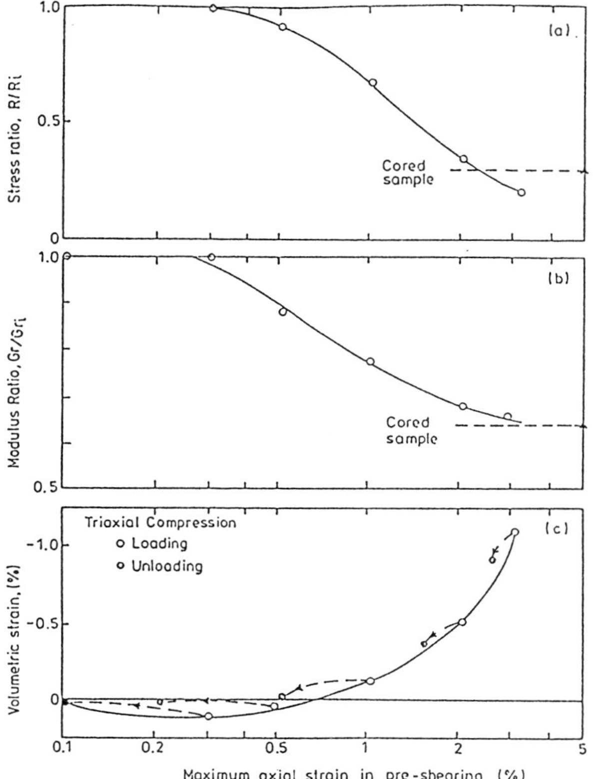

Fig. 23. Disturbance related to preceding strain level in Niigata Sand ($\sigma_c' = 98$ kPa, $D_r = 88\%$) (from Tokimatsu and Hosaka, 1986).

and extension strains will again apply because of symmetry. Because of the very low values of ε_{ap} in triaxial compression of K_o consolidated sands (Figure 7), and because these values may be reduced further by cementing, it follows that $\varepsilon_{zz\,\max}$ will exceed ε_{ap} (ε_{Y3}) so that the initial fabric will be disrupted and damage will occur to bonding.

The importance of small shear strains on the cyclic behaviour of sand is illustrated in Figure 23. Samples of structured sands were subject to increasing levels of triaxial compression strain, which simulate part of the tube sampling strain cycle, prior to undrained cyclic triaxial compression-extension loading. The disturbance increases with increasing level of pre-shearing strain; this is shown by the reduction in the stress ratio, R, to cause a 5% double amplitude axial strain in 20 cycles, and in the reference shear modulus, G_r. Disturbance begins at axial strains of 0.3% – significant volumetric strains do not occur until axial strains are in excess of 1%.

The volume changes which occur as a result of tube sampling depend on the initial density and grading. Loose sands will compact and dense sands will expand, as a result of dilation. Evidence to support this is presented by Marcuson and Franklin (1979) although there is much scatter in the data. In cemented sands, it is probable that volumetric strains are suppressed until bonds are broken. It follows that in loose cemented sands, the shear and volumetric strains caused by sampling may have opposing effects, i.e. the damage to cementing resulting in a reduction in small strain stiffness and strength while compaction causes an increase. In dense cemented sands the disturbances caused by shear and volumetric strains are additive.

Even rotary coring of clean sands leads to volume changes which depend on initial density. This can be seen by comparing relative densities quoted by Yoshimi *et al* (1989) for samples taken by rotary coring, both with and without *in situ* freezing – coring after in situ freezing is thought to provide a reliable measure of *in situ* density.

Relative Density (%)	
In Situ	**After Sampling**
55	78
78	83
87	72

The disturbance caused by even the most careful sampling of sands may be subtle as illustrated in Figure 24. Differences between cored samples of Niigata sand taken with and without in situ freezing are relatively small in triaxial compression; in triaxial extension the differences are more apparent but only markedly so in volume change characteristics, which will have a large influence on cyclic response.

8. Evaluating Sample Disturbance

Various methods have been proposed to assess the likely level of disturbance to samples. The value of some of these methods can be judged using the framework for soil behaviour and the tube sampling strain histories that have been described.

8.1. Fabric Inspection

Although of major importance for assessing some of the potential effects of sampling, in particular the risk of there having been water content redistribution between adjacent sand and clay layers, and for interpreting results of in situ and laboratory tests, fabric inspection is not sufficient to determine the likely level of disturbance. Only gross distortion, for example in the distorted peripheral zone, can be seen, whereas only relatively small strains cause yield and damage to bonded structure. In addition, strain histories in the central zone of the sample involve unloading (see Section 2) so that the maximum imposed strains cannot be deduced.

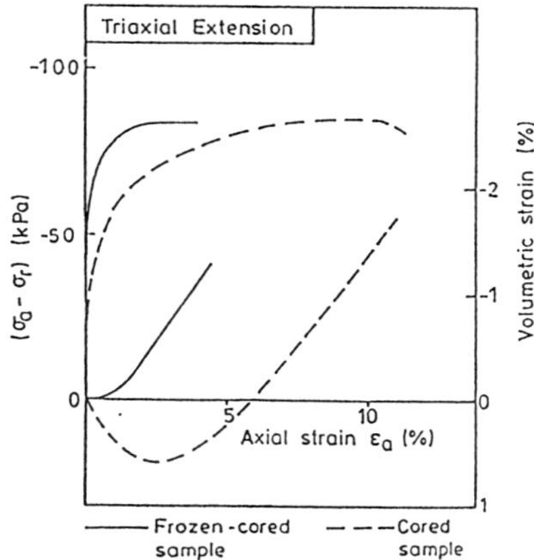

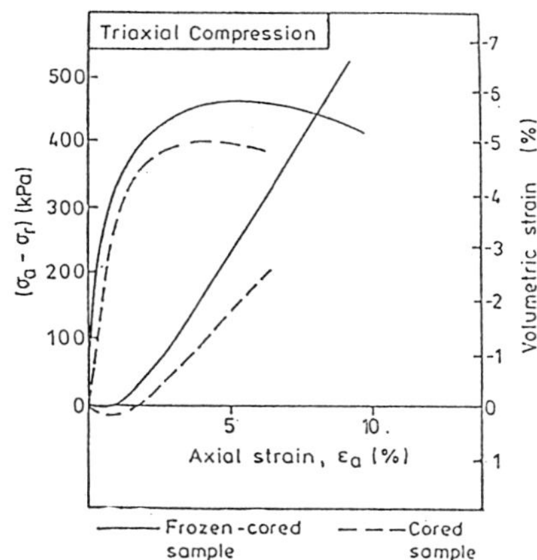

Fig. 24. Effects of sampling disturbance on the behaviour of Niigata Sand ($\sigma_c' = 98$ kPa) (from Tokimatsu and Hosaka, 1986).

8.2. Measurement of Initial Effective Stress

Measurement of initial effective stress, p_i', in laboratory specimens has been advocated for many years, see, for example Ladd and Lambe (1963), yet these measurements are not routinely made in even high quality investigations. The value of p_i' measurements has been illustrated by the dependence of changes in p' on soil type, stress history and sampler quality. The measurement is not sufficient, however, as it cannot indicate the amount of destructuring – shrinking of the BS – that has occurred.

8.3. Measurement of Strains During Reconsolidation

In reconsolidating samples, for example, to in situ stresses, the strains will depend on the loss in effective stress that has occurred and on the amount of destructuring. In this respect, they are a useful comparative measure of quality between samples. The absolute value of the strains will depend, however, on the stress path followed, the specimen size and the local soil fabric, as discussed by Hight *et al* (1992b).

8.4. Comparison of Tube Sampling Strains and Yield Strains

It is suggested here that an indication of whether or not destructuring is likely to have occurred can be obtained by comparing the maximum centre-line strain imposed by a particular sampler with the critical strains in the soil, ε_{Y3} and ε_{bs}. This will require a classification of tube samplers on the basis of the strains they impose. Critical strain levels for the soil can be established from carefully controlled laboratory triaxial tests on samples reconsolidated to in situ stresses, following the soil's recent stress history.

8.5. Comparison of Field and Laboratory Measurements of Dynamic Shear Modulus

Comparisons between the results of a laboratory test on a retrieved sample and the results of an in situ test on the same material are generally not valid for assessing sample quality, because of differences in stress or strain path, rate of shearing and drainage conditions. An exception appears to be the measurement of dynamic shear modulus *in situ* and in the laboratory, on samples brought to the *in situ* stress state.

Cementing leads to a significant increase in dynamic shear modulus, G_{max}, (Lovelady and Picornell, 1989); correspondingly, removal of bonding reduces G_{max}, as demonstrated, for example, by Nishi, Ishiguro and Kudo (1989) when artificially weathering soft rocks. Damage to any cementing by sampling strains might be expected to show a reduction in G_{max} from the field to the laboratory. The following case histories are of interest in this respect.

- In Figure 25(a) comparisons are made for the Bothkennar clay of in situ measurements of shear wave velocity, from the seismic cone, and laboratory measurements on piston samples, from resonant column and bender element tests. In this structured clayey silt, comparisons between piston, Laval and Sherbrooke samples in CK_oU triaxial compression tests (Figure 20b) demonstrate that damage occurs in the piston samples. This damage also appears to show as reductions in G_{max} in the laboratory.

- Figure 25(b) presents a similar comparison between field and laboratory measurements of G_{max}. In these clayey silts of very low plasticity, the high carbonate content was found by fabric examination to form a cement only below 35m. Above 35m, there was good agreement between the field and laboratory measurements, despite the major loss in p' that occurred in the samples (Figure 18b) and the low values of ε_{Y3}. This would suggest that the upper materials are not structured. Below 35m there are some large discrepancies between field and laboratory measurements, consistent with the cementing having been damaged by the tube sampling strains.

- Figure 25(c) compares field and laboratory measurements of dynamic shear modulus of Fucino Clay; *in situ* measurements are from cross-hole and down-hole tests and laboratory measurements are from resonant column tests on samples taken with a modified Osterberg piston sampler. The Fucino Clay is described by Burghignoli *et al* (1991) as a 'soft, homogeneous highly structured $CaCO_3$ cemented lacustrine clay'. It is an organic clay of high plasticity, having a plasticity index generally between 50% and 70% and a liquid limit between 90% and 120%. Results of CK_oU tests, from which ε_{Y3} and ε_{ap} could be assessed, are not reported. However, it is reasonable on the basis of Figure 7 and the soil's high plasticity, to anticipate relative large values of

ε_{Y3} and ε_{ap}, larger than $\varepsilon_{zz\,\max}$ associated with the high quality sampler used, so that the samples survive the tube sampling operation. This would then explain the agreement between the in situ and laboratory measurements of $G_{\max}$.

- Figure 25(d) presents a comparison of field and laboratory measurements of $G_{\max}$ in two types of sand taken by conventional means and in a sand cored after freezing in situ. The ratio of field to laboratory $G_{\max}$ reduces with increasing $(G_{\max})_{\text{insitu}}$ in the conventional samples. Increasing $(G_{\max})_{\text{insitu}}$ is an indicator of increasing relative density. At low relative densities, the reclaimed sands compact during sampling and $G_{\max}$ is higher in the laboratory. (In the reclaimed sands the effects of ageing on $(G_{\max})_{\text{insitu}}$ will have been small.) At high relative densities the samples expand and $G_{\max}$ reduces – damage to cementing in the Pleistocene sands also plays its part, as it can be seen that in these aged sands $G_{\max}$ is always lower in the laboratory. Samples taken after *in situ* freezing appear to retain their structure as $G_{\max}$ is unchanged.

These case records suggest that shear wave measurements in the field and laboratory provide a potential method for assessing whether destructuring has occurred. However, the Author is aware of measurements in soft clays (Skomedal, personal communication) and stiff clays (Butcher, personal communication) which show significantly higher values of $G_{\max}$ in the laboratory. Further investigation of these differences is required.

9. Conclusions

Typical levels of tube sampling strain exceed yield strains in sands and in lightly overconsolidated low plasticity clays. In structured soils these strains cause a reduction in the component of resistance imparted by structural effects. This may be viewed as progressive shrinking of the soil's bounding surface. In clays of high plasticity, samples may survive the tube sampling strains. Measurements of dynamic shear modulus *in situ* and in the laboratory may indicate whether destructuring has occurred.

In soft clays, mean effective stress, p', is reduced, by an amount which reduces with increasing soil plasticity and increasing sampler quality. In stiff clays, there is an increase in p', which is related to the wall thickness of the sampling tube. In fissured stiff clays, tube sampling strains may cause displacement on existing fissures.

The effects of sampling can be related to the changes in p', modifications to the bounding surface, relocation of the Y1 and Y2 sub-yield surfaces, and the stress path followed in a particular test.

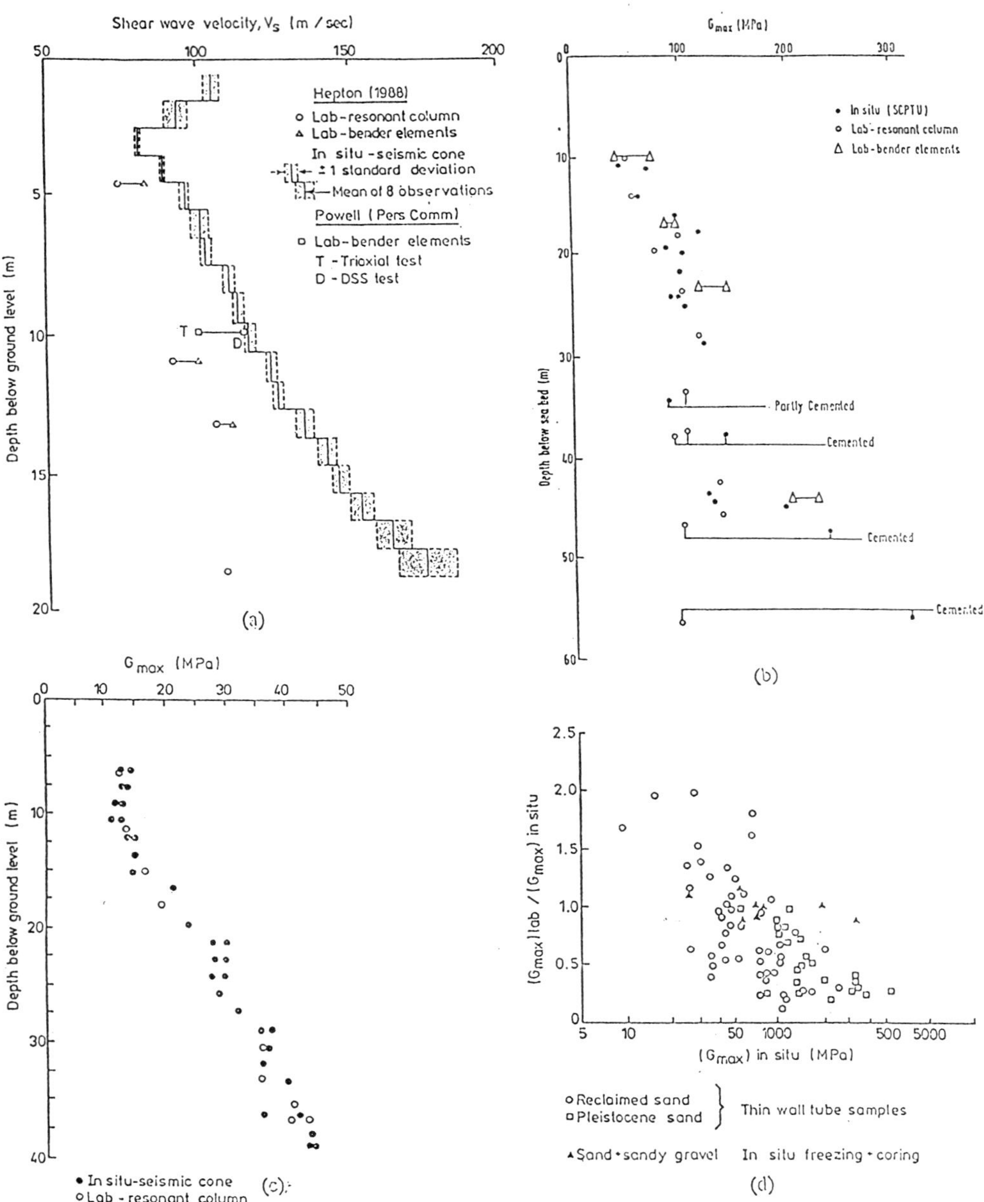

Fig. 25. Comparisons of *in situ* and laboratory measurements of dynamic shear modulus or shear wave velocity. (a) Bothkennar Clay (PI = 18–22% (from Hepton, 1988)); (b) Offshore clayey silts (PI = 5–15%); (c) Fucino Clay (PI = 50–70%) (based on Burghignoli *et al*, 1991); and (d) Sands (from Tatsuoka and Shibuya, 1992, based on Yasuda *et al*, 1980, Tokimatsu and Oh-hara, 1990).

Acknowledgements

The Author wishes to thank John Powell and Tony Butcher of BRE for making data available, and Pat Power and Steve Hoare of Fugro-McClelland for discussing offshore sampling techniques. Tom Henderson of GCG was also most helpful in this respect. The Author has benefitted from discussions with Richard Jardine, Serge Leroueil, Peter Vaughan and Mohsen Baligh.

References

1. Baligh, M. M. (1985), 'Strain path method', *J. Geotech. Engng. Div., ASCE* **111**(GT9), 1108–1136.
2. Baligh, M. M., Azzouz, A. S., and Chin, C. T. (1987), 'Disturbance due to ideal tube sampling', *J. Geotech. Engng. Div., ASCE* **113**(GT7), 739–757.
3. Burghignoli, A., Cavalera, L., Chieppa, V., Jamiolkowski, M., Marchetti, S., Pane, V., Paolini, P., Silvestri, F., Vinale, F., and Vittori, E (1991), 'Geotechnical characterisation of Fucino Clay', *Proc. 10th European Conf. on Soil Mech. and Fdn. Eng.*, Florence, Vol. 1, pp. 27–40.
4. Burland, J. B. (1990), 'On the compressibility and shear strength of natural clays', *Géotechnique* **40**(3), 329–378.
5. Clayton, C. R. I., Hight, D. W., and Hopper, R. J. (1992), 'Progressive destructuring of Bothkennar clay: Implications for sampling and reconsolidation procedures', *Géotechnique* **42**(2), 219–239.
6. Clough, G. W., Sitar, N., Bachus, R. C., and Rad, N. S. (1981), 'Cemented sands under static loading', *J. Geotech. Engng. Div., ASCE*, **107**(GT6), 799–817.
7. Diaz-Rodriguez, J. A., Leroueil, S., and Aleman, J. D. (1991), 'On yielding of Mexico City Clay and other natural clays', *J. Geotech. Engng. Div., ASCE* **118**(GT7), 981–995.
8. Hajj, A. R. (1990), 'The Simulation of Sampling Disturbance and Its Effects on the Deformation Behaviour of Clays', Ph.D. Thesis, University of Sheffield.
9. Hanzawa, H. (1983), 'Undrained strength characteristics of normally consolidated aged clay', *Soils and Foundations* **23**(3), 39–49.
10. Hepton, P. (1989), 'Shear Wave Velocity Measurements During Penetration Tests', Ph.D. Thesis, University of Wales.
11. Hight, D. W., Gens, A., and Jardine, R. J. (1985), 'Evaluation of geotechnical parameters from triaxial tests on offshore clay', *Proc. Int. Conf. on Offshore Site Investigation*, SUT, London, pp. 253–268.
12. Hight, D. W., Jardine, R. J., and Gens, A. (1987), 'The behaviour of soft clays, *Embankments on Soft Clays*, Public Works Research Centre, Athens, Ch. 2, pp. 33–158.
13. Hight, D. W. and Burland, J. B. (1988), 'Review of Soil Sampling and Laboratory Testing', Science and Engineering Research Council Report.
14. Hight, D. W., Pickles, A. R., DeMoor, E. K., Higgins, K. G., Jardine, R. J., Potts, D. M., and Nyirenda, Z. M. (1992a), 'Predicted and measured tunnel distortions associated with construction of Waterloo International Terminal', *Proc. Wroth Memorial Symposium.*
15. Hight, D. W., Boese, R., Butcher, A. P., Clayton, C. R. I., and Smith, P. R. (1992b), 'Disturbance of the Bothkennar clay prior to laboratory testing', *Géotechnique* **42**(2), 199–217.
16. Hopper, R. J. (1992), 'The Effects and Implications of Sampling Clay Soils', Ph.D. Thesis, University of Surrey.
17. Jardine, R. J., St. John, H. D., Hight, D. W., and Potts, D. M. (1991), 'Some practical applications of a non-linear ground model', *Proc. 10th European Conf. on Soil Mech. and Fdn. Eng.*, Florence, Vol. 1, pp. 223–228.
18. La Rochelle, P., Sarrailh, J., Tavenas, F., Roy, M., and Leroueil, S. (1981), 'Causes of sampling disturbance and design of a new sampler for sensitive soils', *Can. Geotech. J.* **18**(1), 52–66.
19. Lacasse, S. and Berre, T. (1988), 'Triaxial testing methods for soils: State-of-the-art report', *Advanced Triaxial Testing of Soil and Rock*, ASTM STP 977, pp. 264–289.

20. Ladd, C. C. and Lambe, T. W. (1963), 'The strength of undisturbed clay determined from undrained tests', *Symp. on Laboratory Shear Testing of Soils*, ASTM STP 361, pp. 342–371.
21. Lefebvre, G. and Poulin, C. (1979), 'A new method of sampling in sensitive clay', *Can. Geotech. J.* **16**(1), 226–233.
22. Lovelady, P. L. and Picornell, M. (1989), 'Sample coupling in resonant column testing on cemented soils', *Dynamic Elastic Modulus, Measurement in Material*, ASTM STP 1045, p-p. 180–194.
23. Marcuson, W. F. and Franklin, A. G. (1979), 'State-of-the-art of undisturbed sampling of cohesionless soils', *Proc. Int. Symp. on Soil Sampling*, Singapore, pp. 57–72.
24. Nishi, K., Ishiguro, T., and Kudo, K. (1989), 'Dynamic properties of weathered sedimentary soft rocks', *Soils and Foundations* **29**(3), 67–82.
25. Powell, J. J. M. and Butcher, A. P. (1991), 'Assessment of ground stiffness from field and laboratory tests', *Proc. 10th European Conf. on Soil Mech. and Fdn. Eng.*, Florence, Vol. 1, pp. 153–156.
26. Sandroni, S. S. (1977), 'The Strength of London Clay in Total and Effective Stress Terms', Ph.D. Thesis, University of London.
27. Scarrow, J. A. and Gosling, R. C. (1986), 'An Example of Rotary Core Drilling in Soils', Geol. Soc. Eng. Geol. Special Publication No. 2, Site Investigation Practice Asssessing BS 5930, 357–363.
28. Siddique, A. (1990), 'A Numerical and Experimental Study of Sampling Disturbance', Ph.D. Thesis, University of Surrey.
29. Tatsuoka, F. and Shibuya, S. (1992), 'Deformation Characteristics of Soils and Rocks from Field and Laboratory Tests', Report of the Institute of Industrial Science, The University of Tokyo, Vol. 37, No. 1, 136 pp.
30. Tavenas, F. and Leroueil, S. (1987), 'Laboratory and *in situ* stress- strain-time behaviour in soft clays: A state-of-the-art', *Proc. International Symposium on Geotechnical Engineering of Soft Soils*, Mexico City, Vol. 2, pp. 1–46.
31. Tokimatsu, K. and Hosaka, Y. (1986), 'Effects of sample disturbance on dynamic properties of sand', *Soils and Foundations* **26**(1), 53–64.
32. Tokimatsu, K. and Oh-Hara, J. (1990), '*in situ* freezing. Tsuchi-to- Kiso', *Proc. JSSMFE*, 38-11, pp. 61–68 (in Japanese).
33. Ward, W. H., Marsland, A., and Samuels, S. G. (1965), 'Properties of the London Clay at the Ashford Common Shaft: *in situ* and undrained strength tests', *Géotechnique* **15**(4), 321–344.
34. Yasuda, S., Sakajo, S., and Kaizu, N. (1980), 'Comparison of dynamic shear modulus by P- and S-wave logging and dynamic triaxial tests', *Proc. 15th Japan National Conf. on SMFE, JSSMFE*, pp. 545–548 (in Japanese).
35. Yoshimi, Y., Tokimatsu, K., and Hosaka, Y. (1989), 'Evaluation of liquefaction resistance of clean sands based on high-quality undisturbed samples', *Soils and Foundations* **29**(1), 93–104.

RECENT DEVELOPMENTS IN *IN SITU* TESTING IN OFFSHORE SOIL INVESTIGATIONS

T. LUNNE
Norwegian Geotechnical Institute, P.O. Box 3930 Ullevaal Hageby, N-0806 Oslo, Norway

and

J. J. M. POWELL
Building Research Establishment, Garston, Watford WD2 7JR, England

Abstract. This paper gives a review of developments in offshore *in situ* testing since about 1985. Deployment of the devices is now so well developed that it is difficult to imagine that more efficient operations can be achieved. Advances in methods of measurement and data acquisitions have facilitated multisensor devices. This has resulted in very useful tools like the triple element piezocone and the seismic piezocone which can now be carried out on a routine basis. Other useful tools like the dilatometer and BAT probe are also available.

After a period of intense equipment development it is thought that the main emphasis should now be on consolidating present measurement techniques and concentrating on interpretation in terms of soil design parameters. A lot of research work has also been done in this field but there is still much room for further development.

A recent trend has been to carry out large scale *in situ* model tests to check out the feasibility of new foundation concepts especially for complex and difficult soil conditions. Some examples are given.

1. Introduction

The scope of this paper is to present recent developments in *in situ* testing in connection with offshore soil investigations. The main emphasis will be on the developments since the last SUT conference on Offshore Site Investigations (OSSI) organized in London in 1985.

Developments in *in situ* testing tool design, sensors, procedures and data acquisition will be covered. Interpretation in terms of soil parameters will not be dealt with in detail but a summary of the present status will be given. There is an increasing trend for site and structure specific large scale model tests to be carried out for new foundation concepts for difficult soil conditions. Some case histories will be covered.

Some thoughts on possible future developments will be included as well as recommendations as to when special *in situ* tests are most relevant. Seismic reflection or refraction profiling will *not* be included in this paper.

Volume 28: Offshore Site Investigation and Foundation Behaviour, 147–180, 1993.

2. Deployment of *in situ* Testing Tools

As outlined by Zuidberg *et al* (1986), there are 2 modes in which *in situ* tests can be deployed during offshore soil investigations:

a) *Down-the-hole mode* where the *in situ* probe (e.g. cone penetrometer) is pushed into the soil below the drill bit. Drill string control is mainly achieved through the ship's heave compensator and a seabed jack, that can grip the drill string when the *in situ* test is to be carried out at the required depth. In soft soil additional drill string control is achieved by using a "hard tie" system where the drill string compensator is connected to the sea floor jacking unit compensator (Amundsen *et al*, 1985).

   In the North Sea the *in situ* tool is pushed into the soil below the drill bit using an hydraulic cylinder powered by an umbilical while real time results can be inspected on deck. A pushing force of 10- 12 tonnes can normally be applied and the stroke length is limited to 3 m. The system used by Fugro-McClelland is called the WISON while Geocean has named their system MASCOT.

   In deep waters Fugro-McClelland operates an alternative "DOLPHIN" system which does not use any umbilical cable. The tool (e.g. piezocone or vane) is dropped down the drill string and pushed into the soil in the bottom of the borehole using mud pressure. Dolphin stores all data in its memory module. The tool is retrieved with a wireline overshot and the data are plotted on board the drilling ship (see Peterson and Johnson, 1985).

   The tool has recently been used in deep waters offshore Brazil where tests were carried out to 100 m below seabed in 1150 m water depth (Geise, 1992).

b) *Seabed mode* where the *in situ* probe is pushed into the soil from the seabed. Up to 1984 the two major companies operating in the North Sea, Fugro and McClelland, used rigs with discontinuous penetration. The sounding rods were pushed into the seabed by hydraulic cylinders in strokes of up to 0.9 m. In 1983 the Dutch company A. P. v.d. Berg made an important improvement by developing the ROSON with continuous penetration of the sounding rods through a steel wheel system as shown in Figure 1.

   Since 1985 Fugro and from 1987 Fugro-McClelland has operated a similar system called the Wheeldrive Seacalf offering very efficient operation with penetration capability of 45–50 m in soft to medium stiff soils. A lighter version called the SEASPRITE is available for pipeline investigations. This type of equipment, when used from modern dynamic positioning vessels, offers extremely efficient operations. For pipeline studies a daily production of 25–30 testing stations (3–5 m penetration) several km apart is frequently obtained. The Danish Geotechnical Institute has recently developed a seabottom CPT rig (SCOPE) which has the particular feature that it automatically ensures vertical penetration even if the seabed slopes by as much as 10° to the

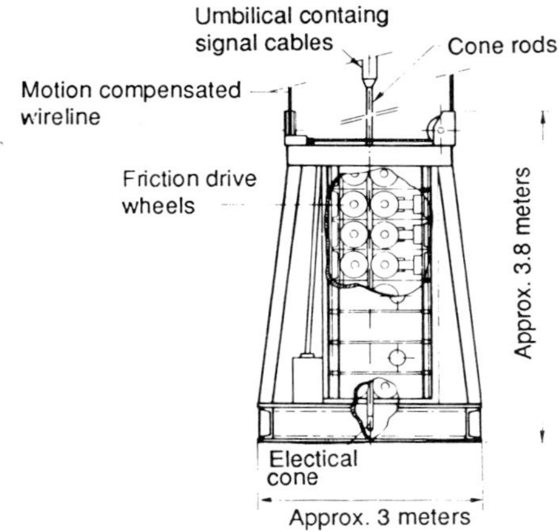

Fig. 1. The Roson seabed frame.

horizontal (Denver and Riis, 1992). A special rig for *in situ* testing in deep water is presently under development in Norway by Seabed Exploration A/S (a subsidiary of Fugro-McClelland, Houston, and Rapp Marine, Bodø). One particular feature of this rig is that the CPT rods are rolled up with a diameter of about 1.8 m (see Offshore Norge, 1990). A similar system has been used for onshore purposes for sometime by Fugro-McClelland in the USA.

3. "Standard" Size in *in situ* Tests

3.1. PIEZOCONE

Since 1985 the piezocone has consolidated its position as *the* most important *in situ* tool for offshore use. It is beyond doubt the most efficient tool for soil profiling and the data can also be used to assess a variety of soil parameters in various soil types. Before 1985 most offshore piezocone tests were performed with the filter element on the cone face (e.g. Lunne *et al*, 1985a). The tendency nowadays is to have the filter behind the cone (or shoulder position), as this facilitates a more realistic correction of cone resistance for pore pressure effects (e.g. Lunne *et al*, 1986). The shoulder position is also the preferred position according to the International Society of Soil Mechanics and Foundation Engineering (ISSMFE, 1988).

There are, however, advantages of measuring the pore pressure on the cone face a more sensitive pore pressure response can generally be achieved and thereby the potential for the detection of thin layers is probably the most significant advantage.

In order to correct sleeve friction for the pore water pressure effects it is necessary to also measure pore pressure at the upper end of the sleeve; according to Wroth (1988) this is also a good position, theoretically, for the pore pressure dissipation test. Hence, ideally pore pressures should be measured at two or preferably

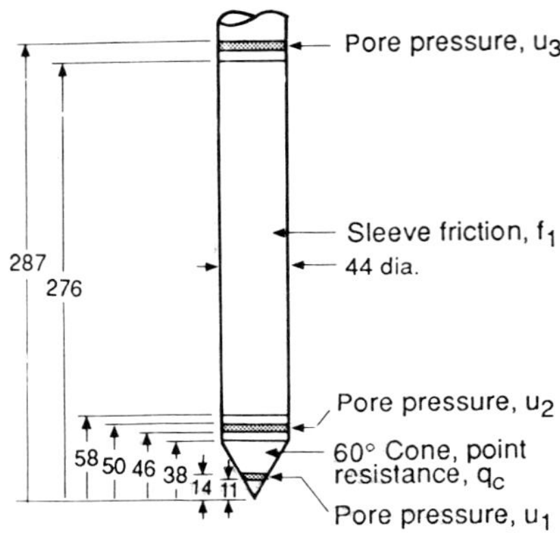

Fig. 2. Triple element piezocone (after Bayne and Tjelta, 1987).

three locations. For the above reasons this has led to the development of dual or triple element piezocones with pore pressures measured as shown in Figure 2 and denoted u_1, u_2 or u_3. Bayne and Tjelta (1987) describe the development of a triple element piezocone.

The importance of the pore water pressure corrections to cone resistance and friction sleeve results is now generally accepted and for triple element piezocones the corrected cone resistance, q_t, and sleeve friction, f_t, are given by:

$$\begin{aligned} q_t &= q_c + (1 - a)u_2 \\ f_t &= fs - [u_2 \cdot A_{sb} - u_3 \cdot A_{st}]/A_s \end{aligned}$$

where

$$\begin{aligned} q_c &= \text{measured cone resistance} \\ f_s &= \text{measured sleeve friction} \\ a &= \text{area of ratio of cone tip } = d_s^2/d_1^2 \\ A_{sb} &= \text{end area of sleeve at base} \\ A_{st} &= \text{end area of sleeve at top.} \end{aligned}$$

See Figure 3 for definitions.

Figure 4 shows examples of results of triple element cone tests at the Gullfaks "C" site in the North Sea. The importance of correcting for pore pressure effects is included in this figure (see Skomedal and Bayne (1988) for more examples). The measurement of pore pressure at two or three locations thus facilitates important corrections and allows higher quality data to be obtained from the tests.

The use of corrected data also leads to more reliable interpretation of soil parameters since the results from piezocone tests from one organization and cone type should then be closer to those from another organization, or cone type.

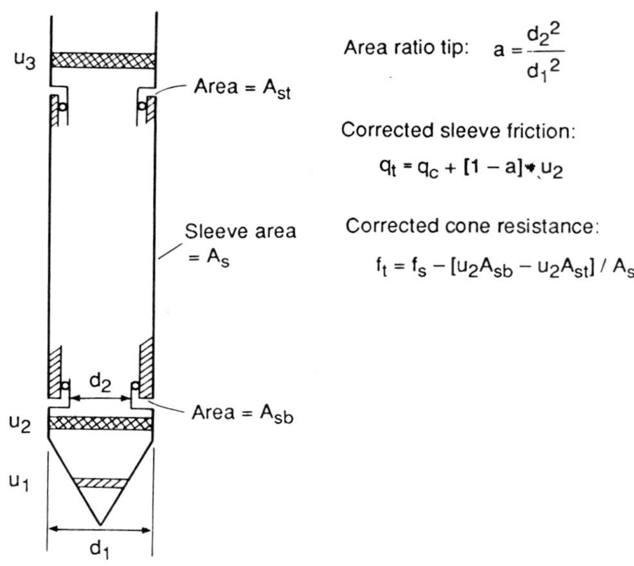

Fig. 3. Definition of pore pressure correction factors.

In addition with triple element cones the excess pore pressure ratio $(u_2 - u_0)/(u_1 - u_0)$ or the pore pressure difference (PPD) $(u_1 - u_2)/u_0$ can be used for interpretation purposes. For example Sully *et al* (1988, 1991) have presented promising correlations between PPD and the important soil parameters overconsolidation ratio (OCR) and lateral stress coefficient (K_o).

Recent studies on the interpretation of the coefficient of consolidation, c_h, from piezocone dissipation tests from the various filter positions look promising. A study by Robertson *et al* (1992) gives a simplified interpretation procedure.

3.2. Lateral Stress Cone

Several of the theories used to interpret strength parameters from the CPT or piezocone test require the input of *in situ* lateral stress. This parameter is difficult to assess and cannot presently be reliably determined from the CPT/PCPT. In order to improve this situation, several research workers have instrumented the friction sleeve so that lateral stresses can be determined during a test (e.g. Huntsman *et al*, 1986; Jefferies *et al*, 1987; Masood, 1990). One example of a lateral stress cone design is shown in Figure 5 (after Jefferies *et al*, 1987). Tests have been done offshore in the Beaufort Sea (Huntsman, *et al*, 1986) and at Gullfaks “C” (Bayne and Tjelta, 1987).

However, there are still problems with the instrumentation of this type of device in order to maintain a robust cone and at the same time to obtain the required sensitivity in the readings. Further, no reliable interpretation method is available

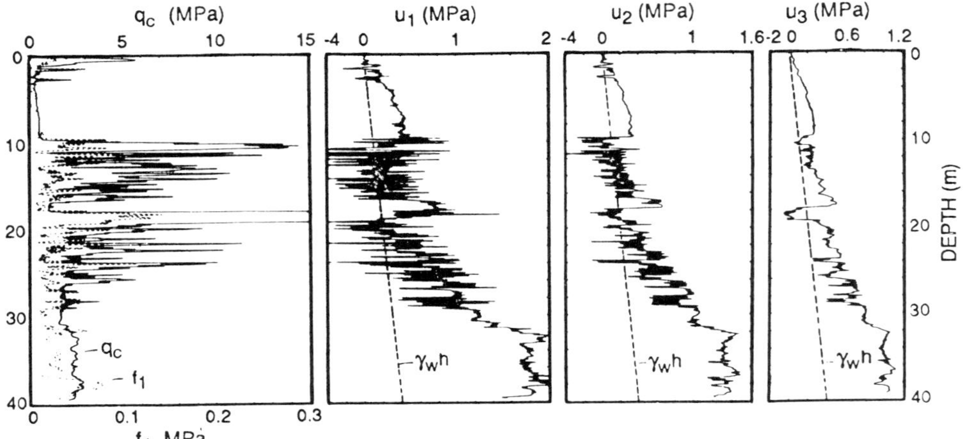

Piezocone type A – Cone point resistance, sleeve friction, and pore pressure

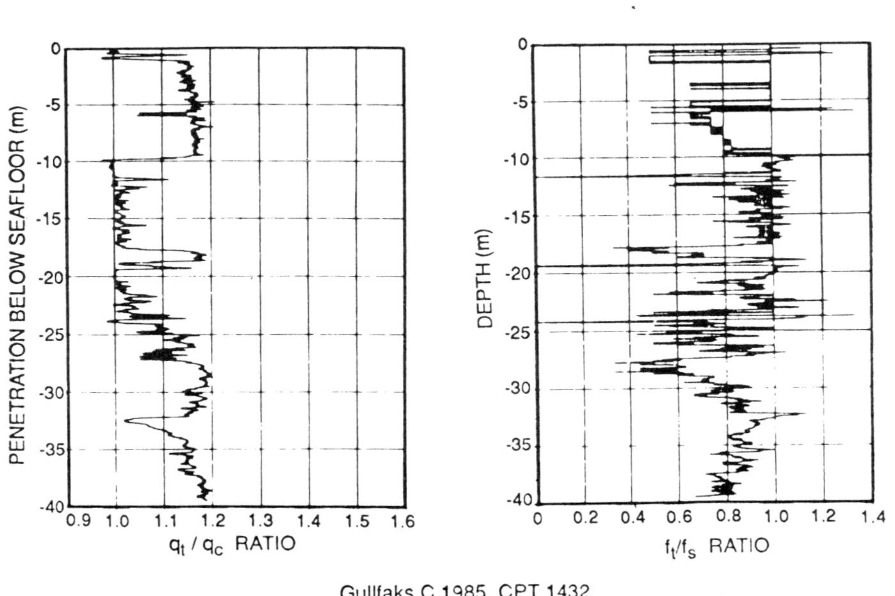

Gullfaks C 1985, CPT 1432

Fig. 4. Results from test with a triple element piezocone at Gullfaks "C" (from Skomedal and Bayne, 1988).

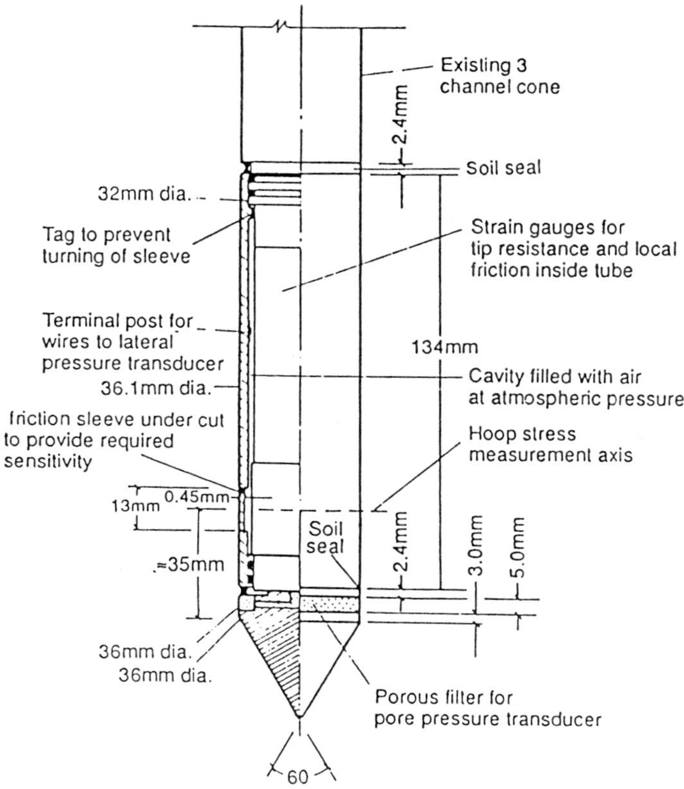

Fig. 5. Lateral stress cone (after Jefferies *et al*, 1987).

(e.g. Jamiolkowski *et al*, 1988). However, future improvements in these two aspects may possibly lead to the lateral stress cone becoming a useful tool.

3.3. Seismic Cone

The dynamic (or initial or maximum) shear modulus, G_{max}, is an important soil parameter for offshore structures, and especially for deep water platforms where dynamic behaviour may be the critical issue.

Onshore G_{max} has traditionally been measured *in situ* with the cross-hole method which requires two or more boreholes. Recently the seismic cone has facilitated more efficient measurements.

The modern version of the seismic cone was developed at the University of British Columbia (UBC) initially for onshore use (Campanella *et al*, 1986).

The first near offshore tests were done by Campanella *et al* (1987) in the MacKenzie Delta near the Beaufort Sea. The operations were performed through ice as illustrated in Figure 6 and revealed the potential for this test offshore.

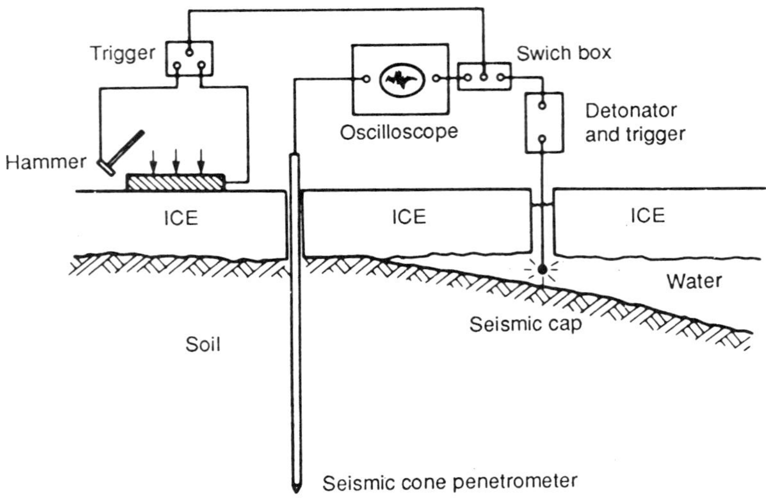

Fig. 6. Nearshore seismic cone tests in Beaufort Sea (after Campanella *et al*, 1987).

Subsequently, seismic piezocone testing has become a practical test offshore, notably in the North Sea, and tests may be carried out both in the seabed and the downhole modes.

Figure 7 shows the set up used by Fugro-McClelland for downhole testing (Lange *et al*, 1990). The seismic source is now an Hydraulic Underwater Shear Wave Box (HUSHBOX). As this system is integrated in the underwater hydraulic power system, an unlimited number of blows can be generated for signal stacking. According to Lange *et al* one blow of the HUSHBOX produces sufficiently strong shear wave signals to a depth of 40–50 m. Below this depth several blows can be stacked to improve signal to noise ratio and testing has successfully been achieved down to 100 m.

According to Butcher and Powell (1992) onshore studies have shown that the use of two sets of receivers either 0.5 or 1 m apart can greatly improve the quality of the data produced by eliminating problems related to triggering times. The data is also specific to that particular depth.

When coupled with piezocone testing the seismic cone is a very powerful tool. Also while doing the seismic testing at a particular depth a dissipation test can be performed simultaneously.

Normally travel times are recorded at different depth intervals so that average shear wave velocities (V_s) can be computed in each depth interval. G_{max} can then be computed in these intervals using elastic theory:

$$G_{max} = V_s^2 \cdot \rho$$

where ρ is soil density.

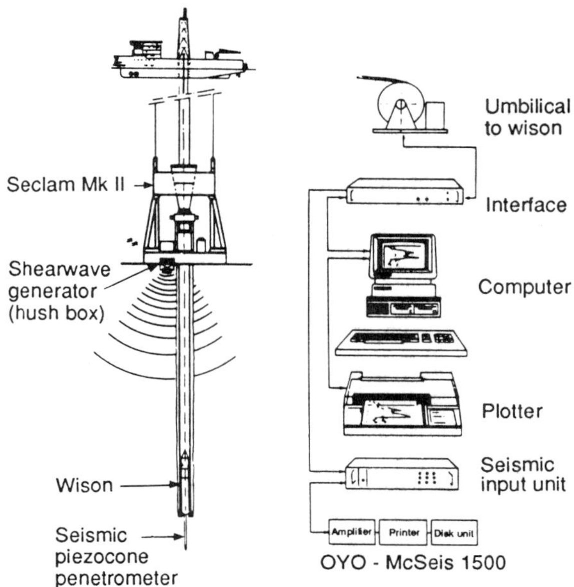

Fig. 7. Set up for seismic cone testing in down hole mode (after Lange *et al*, 1990).

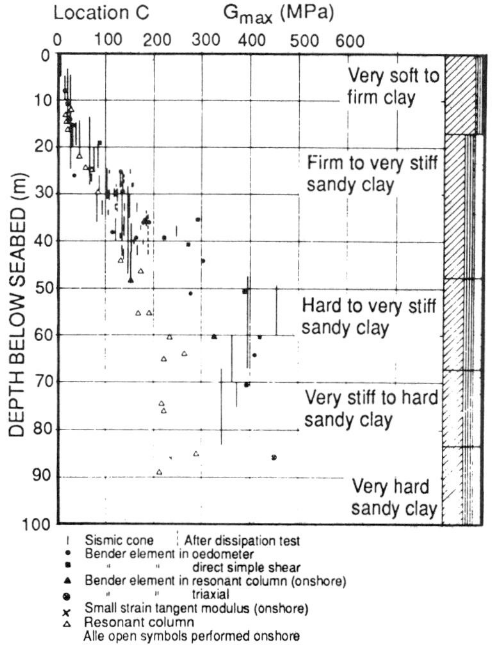

Fig. 8. Results of G_{max} measurements at an offshore site (after Lange *et al*, 1990).

Figure 8 shows an example from an offshore site where G_{max} computed in this way is compared to results of laboratory measurements (from Lange *et al*, 1990).

Agreement between laboratory and *in situ* testing for G_{max} results is not always to be expected even in the best quality tests. The anisotropy of the deposit can significantly affect the results depending on the orientation of the shear waves. This will be particularly true in heavily overconsolidated clays.

In connection with dynamic analysis the damping parameter D is frequently needed in addition to G_{max}. A limitation with seismic cone testing up to now has been that only G_{max} has been obtained, while D has had to be determined from laboratory tests.

Recent work at UBC has shown that damping parameters may also be obtained from seismic cone testing using special analysis techniques (Stewart and Campanella, 1991).

3.4. Pressuremeters

During the period 1980 to 1985 the push in pressuremeter (PIP) developed by Building Research Establishment (BRE) and operated by Stressprobe Ltd., was used in a number of offshore soil investigations (e.g. Fyffe *et al*, 1986). Since 1985 the use of the PIP has been very limited, one of the main reasons being that it was found in practice that the results yielded little in addition to what was already being obtained during the "standard" part of an offshore soil investigation consisting of laboratory testing on recovered samples and cone penetration/piezocone testing. Test repeatability was also frequently a problem with PIP, although this may be less true in land based work.

One of the main incentives for carrying out pressuremeter tests offshore is to determine the *in situ* horizontal stress. Several onshore evaluation programs including those reported by Lacasse *et al* (1990) and Powell (1990) concluded that this is not possible with the PIP test. However, the selfboring pressuremeter test if properly performed may be used for this purpose.

A large selfboring pressuremeter rig (PAM) was developed by IFP (e.g. Fäy *et al*, 1985), but due to the complexity of operation and high cost of deployment it has never found much practical use. A new wireline selfboring pressuremeter (WSBP) has been developed by IFP and is shown in Figure 9 (from Fäy and Le Tirant, 1990). This device is a further development of the onshore PAF (Pressiomètre Auto-Foureur) and the PAM and can be used in downhole mode interchangeably with sampling and other *in situ* testing devices (like the piezocone). The WSBP is now operated by Geocean. So far limited offshore experience exists but onshore calibration tests (including cyclic loading) have been performed.

Several organizations are presently working on developing equipment, test procedures and interpretation methodology for the cone pressuremeter. The tool consists of a standard size cone penetrometer or piezocone with a pressuremeter module located behind it (see Figure 10). Important advantages of this test are

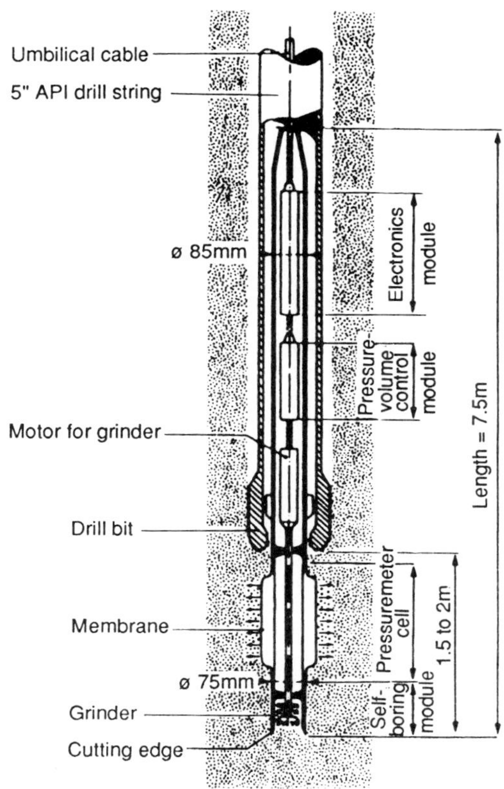

Fig. 9. The wire-line shelf boring pressuremeter (from Fäy and le Tirant, 1990).

the potential for high repeatability, good cost efficiency and, by continuously monitoring piezocone data during penetration, optimal test depths can be found for the pressuremeter part of the test.

Recent theoretical and experimental work at Oxford (Schnaid and Houlsby, 1990) has shown the potential for using both the CPT part and PMT parts of the test to obtain important parameters in sand like σ'_h, D_r and elastic shear modulus in sand. Similar work has been done in clay (Houlsby and Withers, 1988) although the usefulness of the theory for assessing σ'_h in clays is still questionable (Powell, 1990). Although the cone pressuremeter is still only used onshore it is expected that it will be available for offshore use in the near future (Houlsby and Nutt, 1992; Kolk, 1992).

3.5. Dilatometer

The dilatometer was developed in Italy for onshore use by Marchetti (1980) in the seventies. The dilatometer test has the great advantage of being a simple and rapid

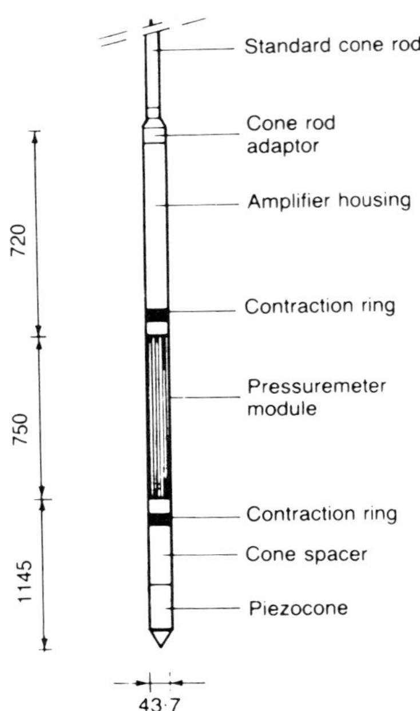

Fig. 10. Fugro cone pressuremeter (after Houlsby and Withers, 1988).

test with good repeatability. The dilatometer has become a valuable instrument that is now widely used onshore in many parts of the world – particularly in Europe and North America. In 1985, NGI modified the onshore equipment for application offshore. The offshore dilatometer is smaller than the onshore device because of the inner diameter of the drill pipe through which the dilatometer is lowered when the test is run in the downhole mode (see Figure 11). The offshore dilatometer is equipped with a filter located on the opposite side of the centre of the membrane so that pore pressure can be measured continuously. This opens up new possibilities for the interpretation of the results relative to the original Marchetti device.

The penetration of the blade is normally halted every 0.2 m and readings of the membrane expansion pressure as a function of the deflection of the centre of the membrane are taken continuously; in particular the contact pressure (p_o) and the 1 mm expansion pressure (p_1) are recorded. Pore pressure is recorded continuously during penetration of the blade and during membrane inflation/deflation.

A comparative laboratory and field study of the standard Marchetti and the offshore dilatometers confirmed that there are no significant differences in the results obtained with the two devices (Lunne *et al*, 1987). Hence correlations obtained with the Marchetti dilatometer (e.g. Lacasse and Lunne, 1988, Powell and Uglow, 1988) are applicable for the offshore dilatometer.

In the authors opinion it is in the assessment of the coefficient of earth pressure at rest, K_o, that the dilatometer is particularly useful, since K_o cannot be reliably

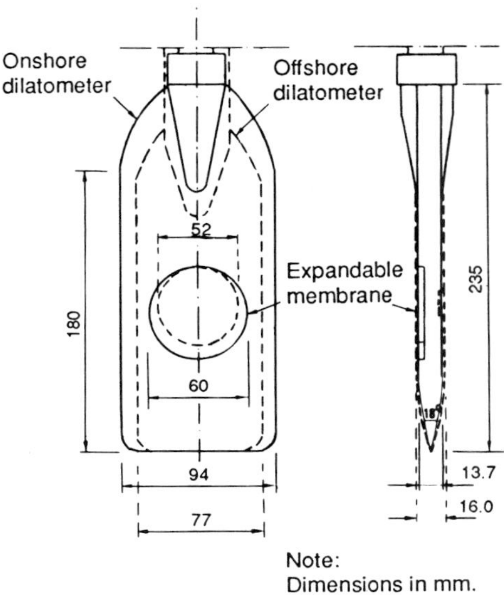

Fig. 11. Comparison of dimensions of the onshore Marchetti dilatometer and the offshore dilatometer.

determined from the piezocone or other presently available offshore *in situ* tests.

Figure 12a shows the results of dilatometer profiling at Gullfaks "C" where the measured parameters p_o and p_1 are included as well as the derived dilatometer parameters I_D and K_D (definitions given in the Figure) and the penetration pore pressure.

Figure 12b shows the interpreted soil parameters K_o, OCR, s_u and M using correlations after Lacasse and Lunne (1988).

The most recent correlations for K_o have been presented by Lunne *et al* (1990). A recent research programme has also demonstrated the potential of the dilatometer for the evaluation of axial and lateral pile behaviour in clay (Gabr *et al*, 1991). Figure 13 shows, as an example, the predicted (using DMT results) and measured (on field model pile) lateral pile capacities in the Norwegian Haga clay, good agreement is seen. Similarly good agreement has been found in several other British and Norwegian clays (Gabr *et al*, 1991).

3.6. BAT PROBE

The BAT Ground Water Monitoring system was first developed in Sweden for onshore use (Torstensson, 1984). The offshore version of BAT was developed at NGI with the close cooperation of Dr. Torstensson and can be deployed in similar fashion to other *in situ* tools (e.g. the cone penetrometer).

The offshore applications of BAT are mainly to measure the *in situ* permeability or to detect and quantify the amount of gas, either in free or dissolved form.

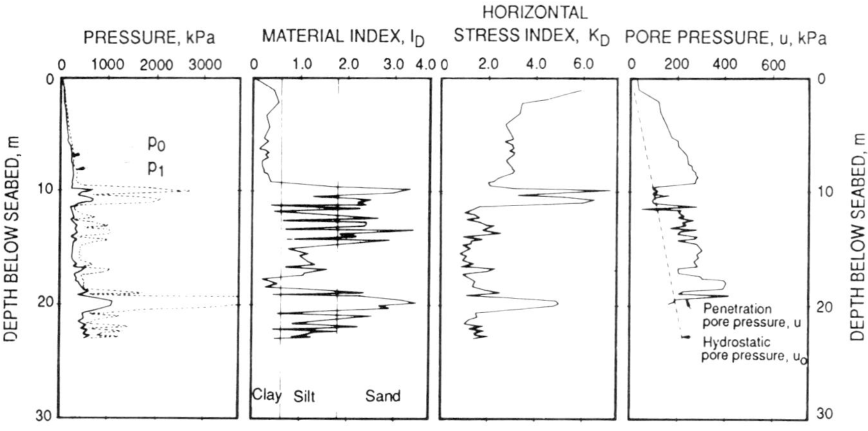

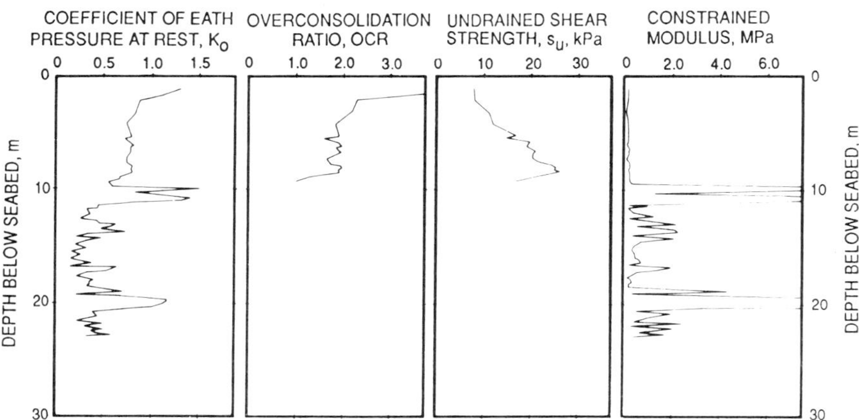

Fig. 12. Dilatometer test results and derived soil parameters from Gullfaks "C".

Figure 14 shows the offshore BAT which is described in detail by Rad *et al* (1988) and Rad and Lunne (1991, 1992). Once penetrated to the desired test level the motor is activated to connect the container to both the transmitter and the filter through the spring and double-ended hypodermic needle units. Normally the pressure in the container is much lower than the ambient pore water pressure and water will start to flow into the container. The change in pressure is read continuously with time and this information can be used to compute permeability of the soil in the vicinity of the filter. Figure 15 shows a typical result from an offshore test. Soil disturbance of the BAT-tip causes some uncertainties in the results but nevertheless comparative tests onshore have shown that the BAT system

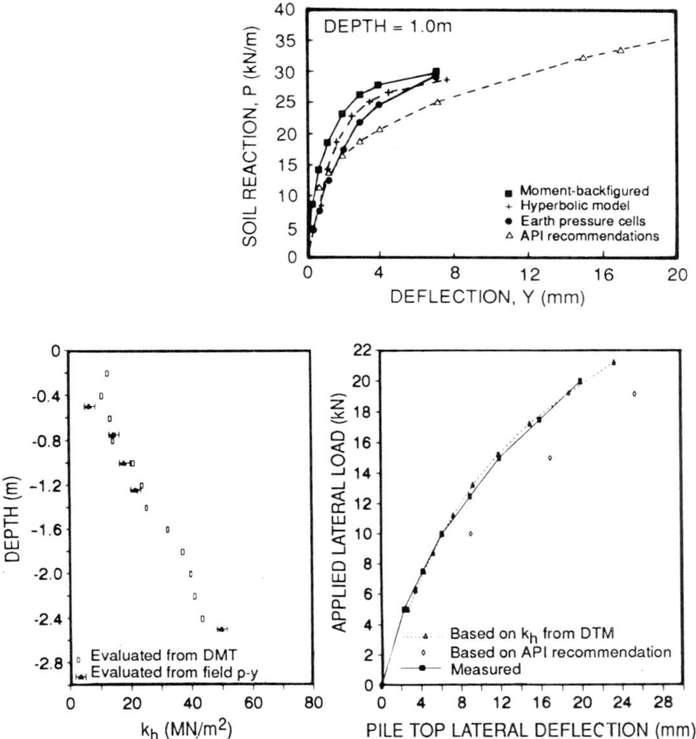

Fig. 13. Lateral pile capacity predicted from dilatometer test (after Gabr *et al*, 1991).

provides estimates of soil permeabilities which are similar to those obtained from oedometer tests on recovered samples (Rad *et al*, 1988).

However, the main use of the offshore BAT has been gas detection. Pore water with any gas in it is collected in the same manner as for permeability testing. When sufficient fluid has been collected, the motor is reversed and the container is disconnected from the filter and the transmitter. The precompressed rubber disc closes again and the pore fluid is thus kept sealed in the container. The BAT equipment is then retrieved to deck level and the container is removed and weighed.

The *in situ* gas collected in the container is analysed on board the vessel using a gas chromatograph to determine the gas composition and the concentration of each gas. Following the analysis (see Rad and Lunne, 1991) the water-gas saturation or degree of water-gas saturation, η, can be computed. η is defined as the amount of gas dissolved in the pore water, expressed as a percentage of the maximum amount of gas that can be dissolved in the pore water at the relevant pore pressure.

An η-value of 100% indicates that the *in situ* pore water is fully gas saturated and free gas may be present at the test location.

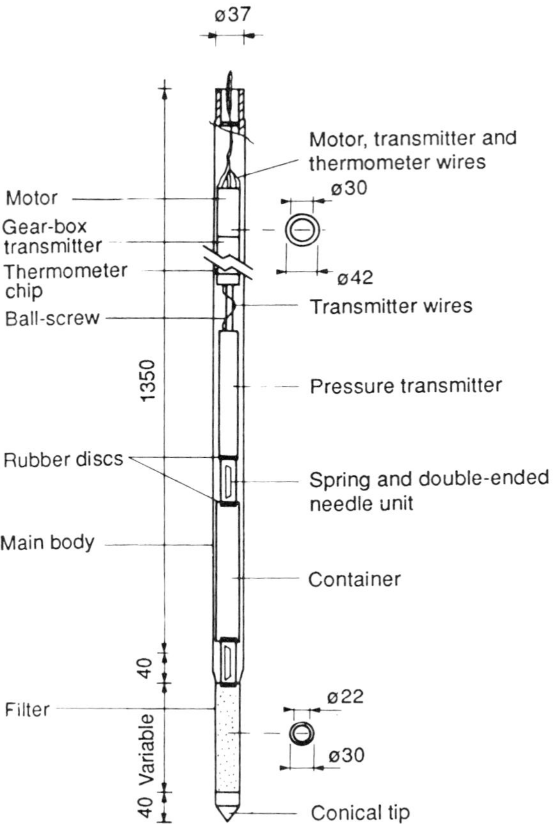

Fig. 14. The offshore BAT probe.

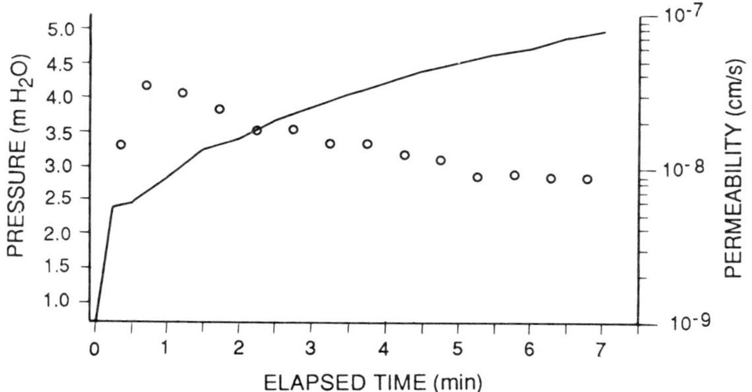

Fig. 15. *In situ* permeability measured with BAT probe.

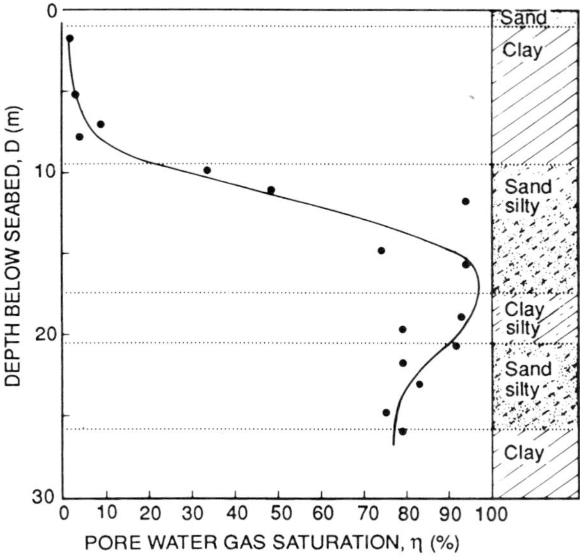

Fig. 16. *In situ* water-gas saturation, η, profile from BAT tests, Gullfaks "C".

The trend of dissolved gas content with depth may be used to predict gas pockets ahead of the drill bit during soil investigation or drilling operations. In many cases it has been observed that sediments overlying a gas charged permeable zone, e.g. a sand layer, will have a gas content that increases with depth and reaches an η-value close to or equivalent to the gas charged layer. Thus possible blowouts may be prevented.

Also soil samples can be taken and, in the onshore laboratory, be restored to the *in situ* gas content and *in situ* stress conditions, before tests are run to obtain soil parameters under the correct *in situ* conditions.

In the Gullfaks area in the North Sea a large number of pockmarks are found, some of them are still being formed. The results from a high resolution seismic survey had indicated that free gas might be present within the area of interest. However, since gas was not encountered during previous soil investigations at the proposed location, the possibilities of a gas blowout had been ruled out. Nonetheless, the *in situ* gas content can have a significant effect on a deep skirted structure like Gullfaks "C" and Statoil requested BAT tests to be carried out as a part of the final Gullfaks "C" soil investigation to establish the η-profile. Figure 16 shows the results of a series of 19 BAT tests carried out at the site. The η-values were used in an advanced laboratory testing programme for re-evaluating the previously suggested design parameters, to take account of the possible effect of gas on the soil behaviour. The tests were carried out in 217 m of water and covered a depth interval from 2 to 25 m beneath the seabed.

The test results shown in Figure 16 revealed that the gas content of the pore water increased with depth to a maximum η-value of 90–100% at a depth of 16 to 17 m beneath seabed. This indicated that minor amounts of occluded gas bubbles might exist at the 16 to 17 m depth interval, and that these could have resulted in seismic turbidity without showing any signs during the previous soil investigations. The *in situ* η-profile also indicated that with time gas could possibly seep upwards along weakness zones, like the conductors, to seabed and be collected inside the skirts beneath the drill shaft area of the Gullfaks "C" gravity base structure. A gas evacuation system was provided for in these areas, partly based on the results of the BAT investigations (Tjelta, 1992).

The BAT probe has been used for η-profiling in several soil investigations in the North Sea, offshore Congo and in Hong Kong (Rad and Lunne, 1992).

3.7. Vane Test

The *in situ* vane test is primarily used in soft to medium clay with undrained shear strengths less than 200 kPa. It can be used in the downhole or seabed mode. Amundsen *et al* (1985) gave some good examples of vane testing in the Norwegian Trench.

Over the last 5–10 years the test equipment and procedures have remained more or less the same.

Geise *et al* (1988) gave a comprehensive review of vane design and presented results from a large number of investigations. As an example they showed how drilling with the hard tie system caused less disturbance in the soil below the drill bit resulting in higher shear strength values in the soil within 1 m below the drill bit (see Figure 17).

Regarding interpretation of vane test results, Chandler (1988) has presented a comprehensive summary. Aas *et al* (1986) presented a new method for finding K_o and also updated correlations for correcting results before using them in design.

3.8. Thermal Conductivity

Fugro-McClelland has developed a heatflow probe as shown in Figure 18. This combines the accurate measurement of temperature with the measurement of thermal conductivity *in situ* (Zielinski *et al*, 1986). The probe has been used on the Norwegian continental shelf as part of a survey of hydrocarbon potential. It can also be used in pipeline investigations and in studies concerning the burial of heat generating materials, such as nuclear waste (Geise, 1992).

The measurement of thermal conductivity is done by heating a wire in the thin steel tube shown in Figure 18. The temperature rise in the middle of the tube is then monitored and plotted on a log time scale (Figure 19). The straight portion of the temperature rise curve can be used to calculate thermal conductivity.

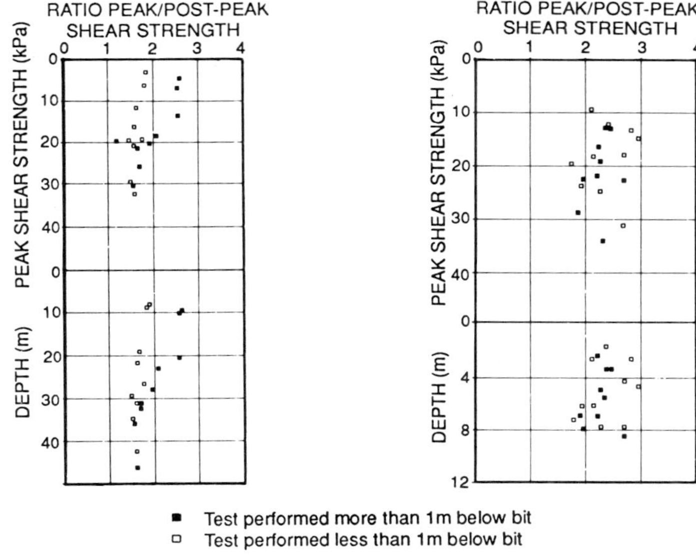

Fig. 17. Vane results for analysis of disturbance caused by drilling (after Geise *et al*, 1988).

3.9. Electrical Conductivity Probe

As reported by Lunne *et al* (1985b) the electrical resistivity probe has been used on some offshore investigations for the purpose of determining *in situ* density or porosity.

The bulk soil resistivity (ρ_s), the resistivity of pore water (ρ_w), and porosity measurements on obtained samples, are needed in order to interpret the results and the testing is quite time consuming. Hence this approach has not been used much in practice for this purpose during recent years.

For *onshore* purposes the electrical conductivity probe is being increasingly used for mapping of contaminated sites (e.g. Campanella and Weemees, 1990). *Offshore* the problem of contamination may become important in connection with pipeline and land fall studies. Further the results of offshore electrical conductivity tests can be useful for the evaluation of corrosivity in the upper soil layers.

3.10. Hydraulic Fracture Tests

Since about 1980 hydraulic fracture tests (HFT) have been carried out during offshore soil investigations in order to determine conductor setting depths.

The conductor is the first casing string to be installed when drilling an oil or gas well. The purpose of the conductor is to seal off the surface formations to a

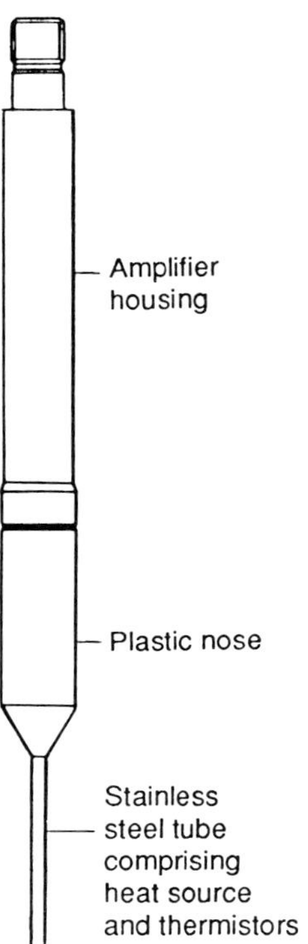

Fig. 18. Heat flow probe.

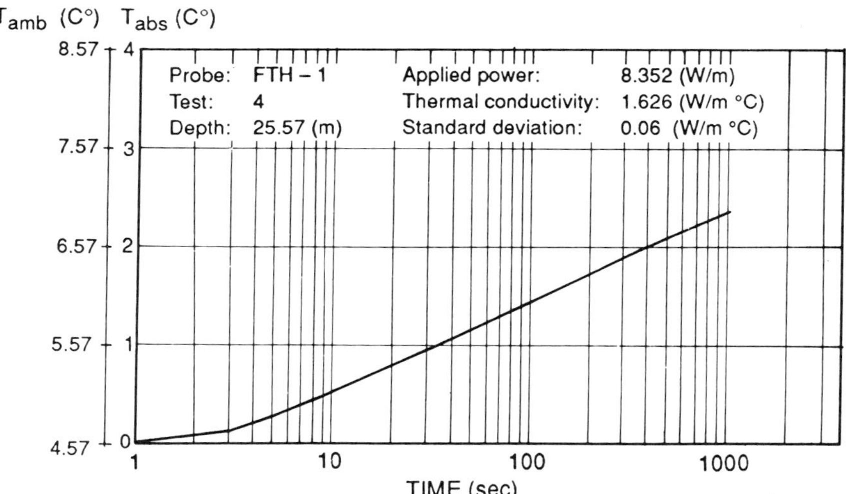

Fig. 19. Result of heat flow tests.

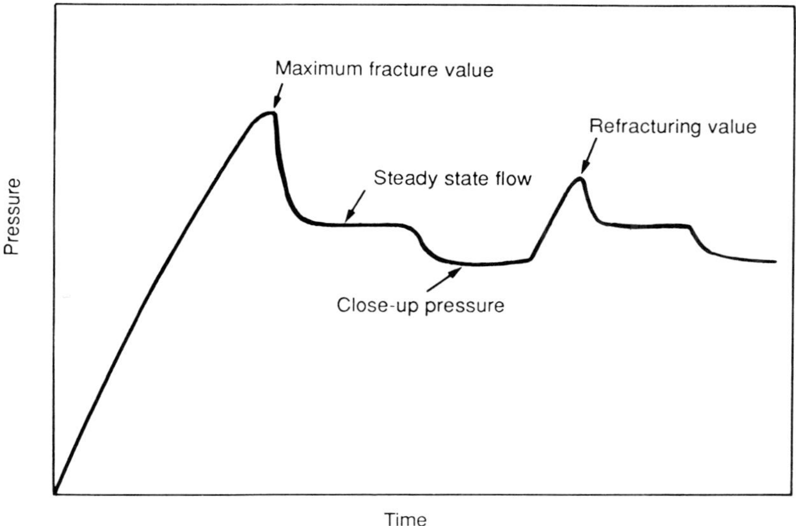

Fig. 20. Result of an ideal hydraulic fracture test.

predetermined depth below mudline. The conductors also provide a passage way between the mudline and the platform through which the smaller casings and drill strings are run. When drilling out below the conductor drilling mud with cuttings is frequently returned to deck level, thus causing high excess mud pressure in the borehole. Further, high mud pressures may be necessary to maintain control of high gas pressures/gas pockets when drilling out for the next casing. It is therefore necessary to know the maximum allowable mud pressure before the occurrence of hydraulic fracture, which cause loss of circulation and other negative effects (e.g. Parkin and Lunne, 1986).

HFTs are performed to establish a profile of maximum allowable mud pressure vs. depth which is essential input to determining the conductor setting depth.

Since 1985 several attempts have been made to find the best procedures for both carrying out offshore HFTs and interpreting the results in a rational manner (e.g. Parkin and Lunne, 1986; Overy and Dean, 1986; Aldridge and Håland, 1991 and Wright and Tan, 1991; NGI, 1992).

The present conclusions regarding equipment and test performance are that the tested volume should be as large as possible and that a steady rate of flow test should be performed with the closing pressure and refracture pressure being recorded in addition to the first fracture pressure. The procedure best resembles the conditions during conductor installation and subsequent drilling for the next casing. The results of an ideal test are shown in Figure 20.

The Fugro-McClelland rough hole packer, testing the full diameter borehole produced by drilling with the standard 4 1/2 API drill string, is the equipment most

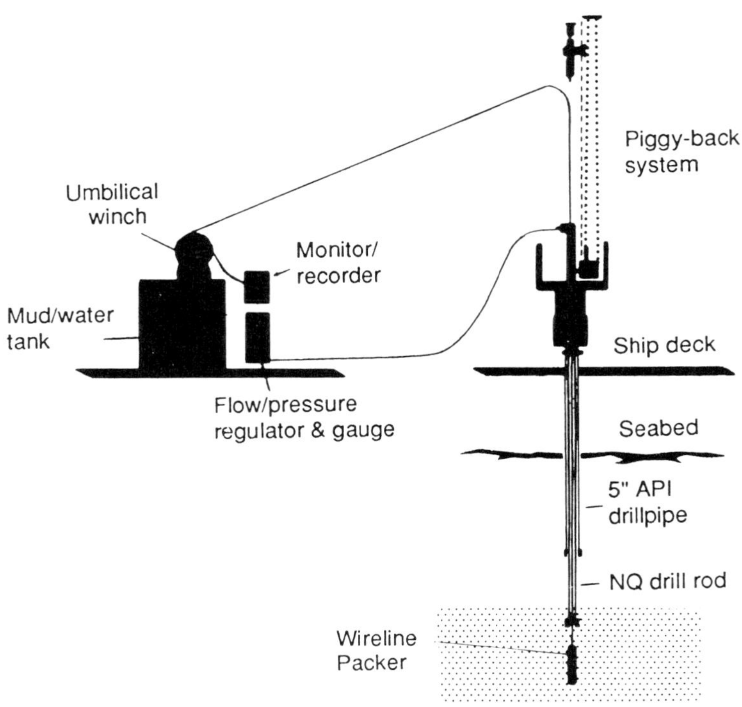

Fig. 21. Hydraulic fracture test in piggy-back mode (from Wright and Tan, 1991).

frequently used.

Piggy-back drilling with standard NQ diameter (2.87") and associated packer system has also been used (e.g. Wright and Tan, 1991), as illustrated in Figure 21.

Figure 22 shows in principle how results from offshore HFTs have been used to determine conductor setting depth.

Regarding interpretation of the results in terms of soil parameters, recent work at NGI (1992) has shown that all of the existing theories (elasticity theory, cavity expansion theory or initial yielding) have significant shortcomings mainly because they do not take the following two factors into account:

1. Non-linearity of the stress-strain properties of the soil.
2. Pore pressure changes induced by changes in mean total stress and shearing of the soil.

A new theory has been developed at NGI incorporating these two factors and it has been found to fit measured *in situ* HF pressures better than existing theories. More details will be given in a future publication (NGI, 1992).

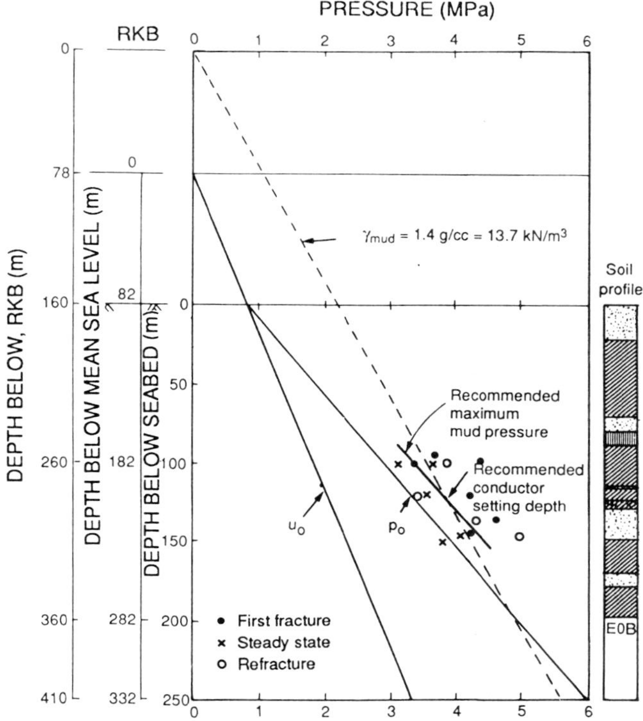

Fig. 22. Use of HFT results to determine conductor setting depth.

4. Summary of Interpretation Status

Some aspects of the interpretation of *in situ* test results for engineering purposes have been mentioned in the preceding section.

Theoretical and experimental work is going on in many parts of the world to refine existing interpretation methods and to extend the use of the *in situ* tests to unusual soil types and to new foundation problems. Notably much work has been and is being done related to the piezocone test. It is beyond the scope of this paper to describe all the various methods and theories available at this stage. A summary of the state of the art in 1989 was given by Lunne *et al* (1989) for a large number of *in situ* test devices including most of the ones described above. Table 1 intends to summarize the present potential of each device for deriving soil parameters and also indicates the offshore potential. In short it is felt the piezocone is the most versatile device and should be included in most important investigations.

The other tools should be considered for special purposes, as can be evaluated from Table 1, not instead of the piezocone but to complement it.

5. Field Model Tests

5.1. General

The *in situ* testing tools described above can be used to derive soil parameters for foundation analysis, but they generally only test small volumes of soil.

In some cases it is necessary to test larger volumes of soil and also to load the soils in a manner that closely simulates the full scale structure that is going to be installed. The results from such model tests can then be used to check *both* the soil design parameters evaluated from the soil investigation and the foundation analysis procedure.

Model tests in connection with offshore structures can be divided into four categories:

1. 1 g model tests in the laboratory;
2. centrifuge tests;
3. 1 g model tests at an onshore site with soil conditions reasonably similar to those at the actual platform location; and
4. 1 g model tests at the actual platform location.

In most cases offshore model tests in the last category are the most complicated and expensive. However, in some cases the soil conditions are so complicated or other boundary conditions, like the total pressure, are so dominating that site specific tests are necessary. In the following, some examples are given of actual offshore model tests.

5.2. Gullfaks "C" Penetration Test

The Gullfaks "C" Condeep Gravity Base structure was ordered by Statoil in 1985 and was installed in 1989 (Tjelta *et al*, 1990). The water depth is 220 m and the platform is the largest gravity base structure that has so far been installed. The soil conditions are complex comprising normally consolidated clays, relatively loose clayey and silty sands, and medium dense sands in the upper 45 m (see Figure 23).

The platform was designed with 1200 running metres of 22 m high concrete skirts. During the foundation design it became evident that the penetration of these 22 m high platform skirts, which were required for stability reasons, could become difficult. Predicted skirt penetration resistance exceeded the available driving force from the structural submerged weight and ballast water, and it was realized that underbase suction would most probably be necessary to achieve the required driving force. Even with this additional force the platform had a rather low margin against penetration refusal.

Due to these uncertainties Statoil decided to perform a large scale field test at the Gullfaks "C" site by penetrating a segment of the skirt wall into the seabed.

TABLE 1.

Evaluation of engineering soil parameters from *in situ* tests and offshore applicability (modified from Lunne *et al*, 1989).

TEST EQUIPMENT		INTERPRETATION																	
Section	Soil Type	Initial State Parameter					Strength Parameters		Deformation Charact.			Flow Charact.		Direct Application					
		γ,D_r	ψ	K_o	OCR	S_t	s_u	φ'a	E,G	M	G_{max}	K	C_h	APC	LPC	LP	SPR	SF	SG
Cone Penetrometer[a]	Clay	3-4		4-5	3	2-3	2-3	3-4	5	5	5	2-4	2-3	1-2	P		1-2		
	Sand	2-3	2	4-5				2	2-4	2-4	2-3			1-2	P	2-3	1-2	2-3	
Dilatometer	Clay	3		2-3	3		3	P	2-4	2-4	4	2-4	2-4	3	2				
	Sand	2-3	3	3	4			3	2-3	2-3	3					3		2-3	
Field Vane	Clay			2	2-3	1	1-2	4	5										
	Sand																		
Pressuremeter	Clay			1-3[b]			3	3	1-2[c]		P	2-4	2-4	4	2			3-4	
	Sand			2-3[b]				1-3	1-2[c]		P			4	2-4	P		2-3	
Seismic Cone[a]	Clay	3-4		4-5	3	2-3	2-3	3-4	5	5	1	2-4	2-3	1-2	P		1-2		
	Sand	2-3		4-5				2	2-4	2-4	1			1-2	P	2-3	1-2	2-3	
Density Probes	Clay	2																	
	Sand	2																	
BAT Probe	Clay			2								1-2	1-3						1-2
	Sand											1-2	2-3						1-2

RATING

1 - High reliability
2 - High to moderate reliability
3 - Moderate reliability
4 - Moderate to low reliability
5 - Low reliability

LEGEND

a) Including piezocone
b) Self-boring pressuremeter only
c) Unload-reload modulus (elastic response)
P Device has potential

ABBREVIATIONS

APC - Axial Pile Capacity
LPC - Lateral Pile Capacity
LP - Liquefaction Potential
SG - Shallow gas detection
SPR - Skirt Penetration Resistance
SF - Shallow Foundation

SOIL PARAMETERS

γ = Soil unit weight
D_r = Relative density
K_o = Coefficient of earth pressure at rest
ψ = State parameter
OCR = Overconsolidation ratio
s_t = Sensitivity
s_u = Undrained shear strength
ϕ',a = Effective stress shear strength parameters
E = Young's modulus
G = Shear modulus
M = Constrained modulus
G_{max} = Initial or small strain shear modulus
K = Coefficient of permeability
C_h = Coefficient of consolidation

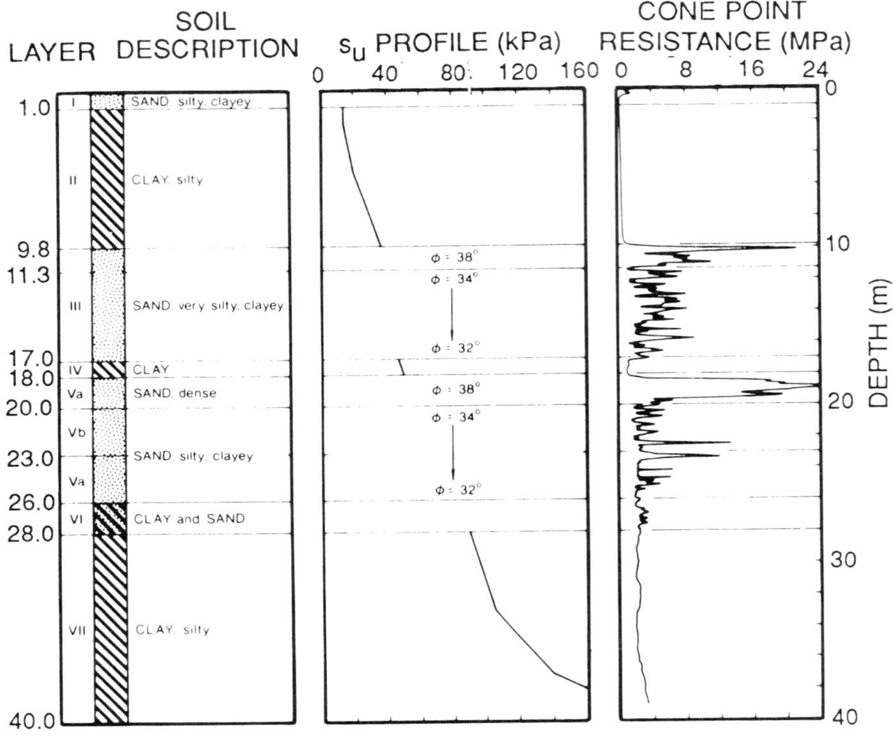

Fig. 23. Soil profile at Gullfaks "C" (from Tjelta *et al*, 1986).

This was done to improve the confidence in the predicted soil response and to clarify uncertain aspects with respect to penetrating the concrete skirts (Tjelta *et al*, 1986).

The model test structure is seen in Figure 24 which also shows the extensive instrumentation. On the concrete skirt element, earth pressure, pore pressure and soil friction measurement gauges were installed. Altogether 70 data measuring points were included in order to monitor and control structural behaviour during testing and to collect geotechnical information.

The test structure was deployed from McDermott's 2000 tonnes capacity heavy lift vessel DB101. As described in detail by Tjelta *et al* (1986) two tests were carried out including cyclic loading. Herein only some limited test results will be included.

The tests demonstrated that suction was a very efficient way of achieving penetration and also, excess pressure was found to be useful to jack the structure out of the soil (see Figure 25).

Very importantly the tests also showed that penetration resistance was much less than predicted (Figure 25) and increased Statoil's confidence in the platform

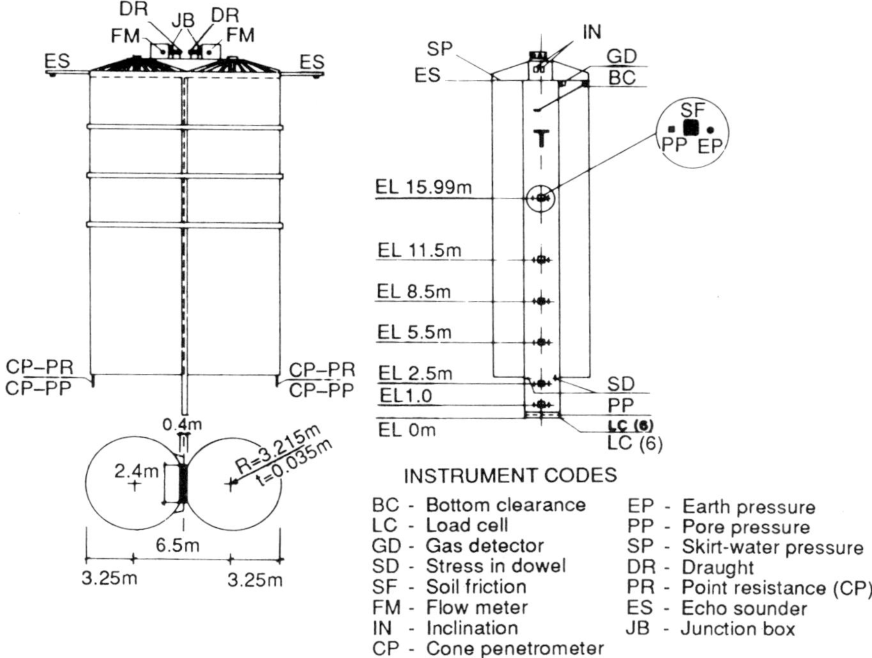

Fig. 24. Gullfaks "C" skirt penetration model test structure (from Tjelta *et al*, 1986).

design. Although the costs associated with this test amounted to some 20 mill NOK (~ £2 mill) the test results showed that a jetting system, incorporated into the skirt tip to reduce penetration resistance, could in fact be eliminated. The total cost savings considerably exceeded the costs of the large scale penetration test (Tjelta, 1992).

5.3. Statoil Suction Test

In the search for more cost effective foundation solutions suction anchors have recently been given much attention. One recent successful example for a clay foundation is the Snorre Tension Leg Platform (Christophersen, 1993).

The beneficial effects of suction has also been utilised with the introduction of a skirt piled steel jacket as advocated by Statoil (Baerheim and Tjelta, 1992). To prove the feasibility of this foundation method in sand, Statoil, in late 1991 commissioned NGI and Fugro- McClelland to carry out an offshore model test at the Sleipner Field. The main purpose was to check the penetration resistance of confined steel skirts in very dense sand, using the effect of underbase suction and also to check the static and cyclic tension capacities of steel skirt caissons. Fugro-McClelland's seabed jack was modified to be able to penetrate a 1.5 m dia,

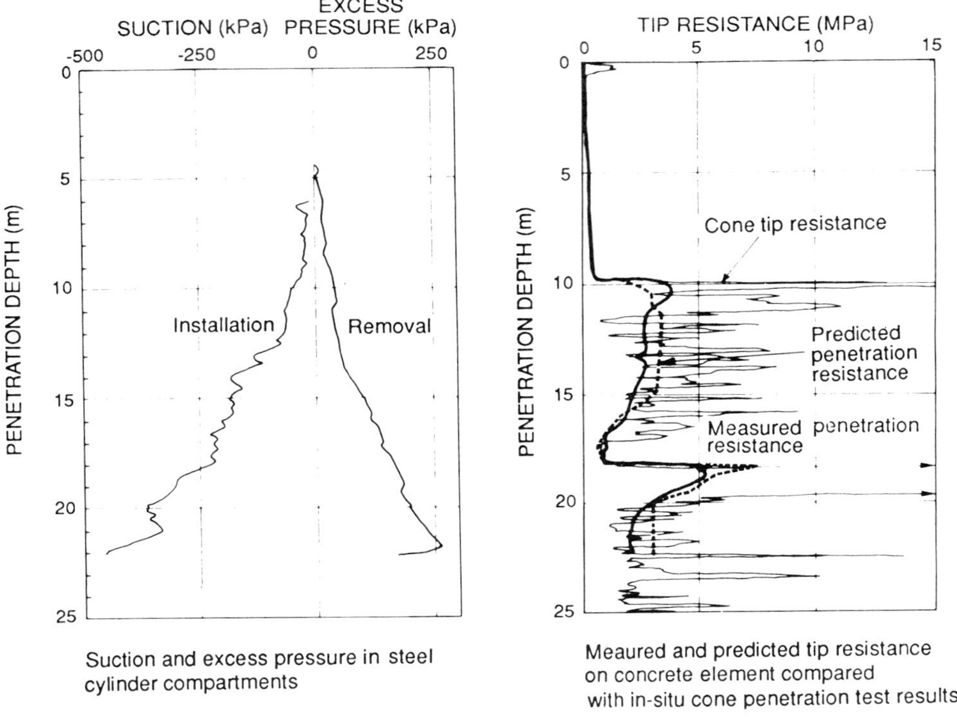

Fig. 25. Results from Gullfaks "C" penetration test (from Tjelta *et al*, 1986).

1.7 m deep skirt into the soil and to perform static and cyclic pull-out tests. A sketch of the test set up is included in Figure 26. A number of tests were performed in 100 m of water in dense sand and the results indeed showed the great potential of this foundation concept.

More details are given by Baerheim and Tjelta (1992) and in another paper to this conference (Tjelta *et al*, 1993).

5.4. Instrumented Plough

Over the last few years soil investigations for large pipeline projects (e.g. Zeepipe and Europipe) have become a substantial part of the offshore soil investigation market.

The usual procedure for geotechnical pipeline surveys is to perform sampling, using vibrocoring or gravity coring, and to perform CPTs at intervals generally varying between 1 and 5 km. There is a need to have more continuous information in the upper 3.5 m, but presently geophysical investigation methods do not have sufficient resolution to do this.

Fig. 26. Statoil suction test set up (after Baerheim and Tjelta, 1992).

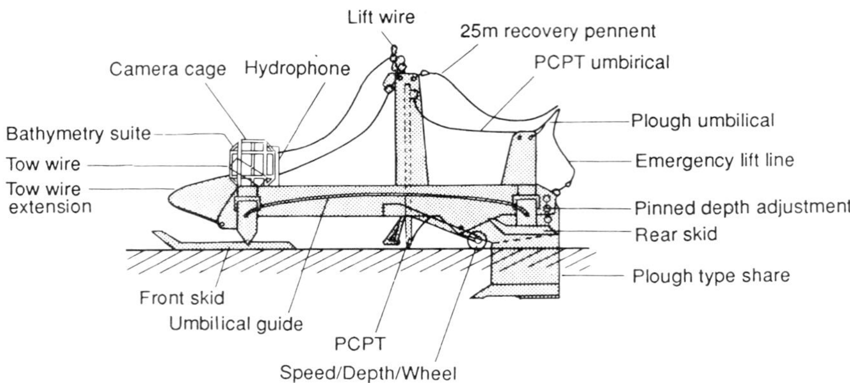

Fig. 27. Instrumented plough surveyor (after NOS brochure).

One approach that has been used to overcome this problem is to instrument a plough and to pull this along the route measuring pull speed and pull force. Northern Ocean Services Ltd. offers such a plough shown in Figure 27. This has a trenching capacity of 0.9 m. In addition the plough can be equipped with a small CPT rig. If the plough is stopped at a chosen location a CPT can be performed to 1.5 m penetration. Wright (1991) summarized the main assets of an instrumented plough to be:

1. Where final cover is required to prevent upheaval buckling, information along the pipeline route is essential to determine achievability of required trench depth and likely required quantities of rock dumping (if any).

2. Continuous soils information may be useful in assessment of axial friction acting on the pipe.

3. Continuous information may be useful where variable soil conditions are expected so that uneven trench depths or founding conditions are known before pipeline laying or burial has started.

4. Buried obstructions along the pipeline route – not detected by other means – can be found.

Statoil has in the summer of 1992 carried out a trial using Fugro- McClelland's newly developed GEOSLEDGE for the Europipe project (Tjelta, 1992). The GEOSLEDGE has a penetration capability of 0.5 m.

6. Summary, Conclusions and Recommendations

The deployment of *in situ* testing devices can now be done very efficiently, often even more efficiently than for onshore operations. The developments in measurement and data acquisition techniques have led to an increase in the reliability and quality of the data obtained, and the ability to both take more frequent readings and to include an increasing number of sensors in the penetration tool.

The piezocone has manifested itself as *the* most versatile and efficient tool, being unsurpassed for soil profiling, and giving data that lends itself to interpretation in terms of a wide range of soil parameters and soil types, especially when the pore pressure is recorded at two or three locations. Considerable amounts of research have been carried out in recent years increasing the reliability of the soil parameters interpreted from the piezocone test. However, there is still room for further improvement.

The use of the seismic cone for offshore soil investigations has been an important development, facilitating reliable *in situ* measurements of dynamic soil properties. The offshore version of BAT has effectively covered the specific need for shallow gas detection, and the dilatometer is an important supplement to piezocone testing with a potential for use in pile design.

Other *in situ* tests designed for special purposes such as assessing heat and electrical conductivity are also available.

Generally the measurements obtained with all the new probes developed are of high quality. However, it is important to maintain and improve calibration procedures, documentation and quality control schemes.

After a period of intense probe development it is now an expressed wish in the industry to consolidate the methods already available and to concentrate on improving the reliability of parameters interpreted from the test results. A lot of work has already been done and more is underway in this respect regarding specifications and standards.

Finally there is also a trend for carrying out large scale *in situ* model tests to check out the feasibility of new foundation concepts and/or complex and difficult soil conditions. These appear to be most valuable.

Acknowledgement

The authors would like to acknowledge the very useful comments by J. Geise and H. Kolk of Fugro-McClelland and T. I. Tjelta of Statoil.

References

1. Aas, G., Lacasse, S., Lunne, T., and Høeg, K. (1986), 'Use of *in situ* tests for foundation design on clay', *ASCE Spec. Conf. In Situ '86. Use of In Situ Tests in Geotechnical Engineering*, Blacksburg, Virginia, USA, pp. 1–30.
2. Aldridge, T. R. and Håland, G. (1991), 'Assessment of conductor setting depth', *Proc. Offshore Technology Conference*, Houston, Paper No. 6713.

3. Amundsen, T., Lunne, T., Christophersen, H. P., Bayne, J. M., and Barnwell, C. L. (1985), 'Advanced deep-water investigation at the Troll East Field', *Proc. of an International Conference 'Offshore Site Investigation', London 1985*, Advances in Underwater Technology and Offshore Engineering, London, Vol. 3, pp. 165–168.
4. Bayne, J. M. and Tjelta, T. I. (1987), 'Advanced cone penetrometer development for *in situ* testing at Gullfaks C', *Proc. Offshore Technology Conference*, Houston, Paper No. 5420.
5. Baerheim, M. and Tjelta, T. I. (1992), 'Skirt–plate foundations for offshore jackets', *International Symposium on Offshore and Polar Engineering, ISOPE*, San Francisco, June 1992.
6. Butcher, A. P. and Powell, J. J. M. (1992), 'Seismic cone testing in clays', BRE Report No. G/GP/9218.
7. Campanella, R. G., Robertson, P. K., and Gillespie, D. (1986), 'A seismic cone penetrometer for offshore applications', *Oceanology International '86, Proc. of an International Conference*, Brighton, UK, Advances in Underwater Technology, Ocean Science and Offshore Engineering, Vol. 6, Chapter 51.
8. Campanella, R. G., Robertson, P. K., Gillespie, D., Laing, N., and Kurfurst, P. J. (1987), 'Seismic cone penetration testing in the near offshore of the MacKenzie Delta', *Canadian Geotechnical Journal* **24**(1), 154–159.
9. Chandler, R. J. (1988), 'The *in situ* measurement of undrained shear strength using the field vane', *ASTM STP1014, Int. Symp. on Laboratory and Field Vane Strength Testing*, Tampa, Florida, USA, pp. 13– 44.
10. Christophersen, H. P. (1993), 'The non-piled foundation systems for the Snorre Field', *Proc. Conf. on Offshore Site Investigations and Foundation Behaviour*, SUT, London, September 1992.
11. Denver, H. and Riis, H. (1992), 'CPT offshore rig', *Proc. 11th Nordic Geotechnical Meeting*, Aalborg, May 1992, Vol. 2, pp. 261–266.
12. Fäy, B. and Le Tirant, P. (1990), 'Offshore wireline self-boring pressuremeter', *Proc. 3rd International Symposium on Pressuremeters*, Oxford, April 1990, pp. 55–64.
13. Fäy, J. B., Montarges, R., Le Tirant, P., and Brucy, F. (1985), 'Use of the PAM self-boring pressuremeter and the STACOR large-sized fixed- piston corer for deep seabed surveying', *Proc. International Conference on Offshore Site Investigation*, SUT, London, 1985, Vol. 3, pp. 187–200.
14. Fyffe, S., Reid, W. M., and Summers, J. B. (1986), 'The push-in pressuremeter: 5 years of offshore experience', *Proc. Second International Symposium on The Pressuremeter and Its Marine Applications*, ASTM STP50, pp. 22–37.
15. Gabr, M. A., Lunne, T., Mokkelbost, K. H., and Powell, J. J. M. (1991), 'Dilatometer soil parameters for analysis of piles in clay', *Proc. X European Conference Soil Mechanics and Foundation Engineering*, Florence, May 1991, Vol. 1, pp. 403–406.
16. Geise, J. M., ten Hoope, J., and May, R. E. (1988), 'Design and offshore experience with an *in situ* vane', *Proc. International Symposium on Laboratory and Field Vane Shear Strength Testing*, ASTM STP1014, January 1987, pp. 318–338.
17. Geise, J. M. (1992), Personal Communication.
18. Houlsby, G. T. and Withers, N. J. (1988), 'Analysis of the cone pressuremeter test in clay', *Géotechnique* **38**(4), 575–587.
19. Houlsby, G. and Nutt, N. R. F. (1992), 'Development of the cone pressuremeter', *Proc. Wroth Memorial Symposium*, Oxford.
20. Huntsmann, S. R., Mitchell, J. K., Klejbuk, L. W., and Shinde, S. B. (1986), Lateral stress measurements during cone penetration', *Proc. ASCE Spec. Conf. In Situ '86*, pp. 617–634.
21. Jefferies, M. G., Jønsson, L, and Been, K. (1987), 'Experience with measurement of horizontal geostatic stress in sand during cone penetration test profiling', *Géotechnique* **37**(4), 484–498.
22. Kolk, H. (1992), 'A recent development in pressuremeter testing: The cone pressuremeter', presented at Int. Sem. in Lisboa, May 1992.
23. Lacasse, S., d'Orazio, T. B., and Bandis, C. (1990), 'Interpretation of self-boring and push-in pressuremeter tests', *Proc. 3rd International Symposium on Pressuremeters*, Oxford, April 1990, pp. 273– 286.
24. Lacasse, S. and Lunne, T. (1988), 'Calibration of dilatometer correlations', *Proc. of First Int.*

Symp. on Penetration Testing, ISOPT-1, Florida, March 1988, Vol. 1, pp. 539–548.

25. Lange, G. de, Rawlings, C. G., and Willet, N. (1990), 'Comparison of shear moduli from offshore seismic cone tests and resonant column and piezoceramic bender element laboratory tests', Society of Underwater Technology, Vol. 16, No. 3, pp. 13–20.
26. Lunne, T., Powell, J. J. M., Hauge, E. A., Mokkelbost, K. H., and Uglow, I. M. (1990), 'Correlation of dilatometer readings to lateral stress', Specialty Session on Measurement of Lateral Stress, *Proc. 69th Annual Meeting of the Transportation Research Board*, Washington, DC, USA, Transportation Research Board Record No. 1278, pp. 183–193.
27. Lunne, T., Lacasse, S., and Rad, N. S. (1989), 'SPT, CPT, pressuremeter testing and recent developments on *in situ* testing of soils', Proc. International Conference on Soil Mechanics and Foundation Engineering, Rio de Janeiro, General Report, Vol. 4. Also NGI Publication No. 179.
28. Lunne, T., Jonsrud, R., Eidsmoen, T., and Lacasse, S. (1987), 'The offshore dilatometer', *Proc. International Symposium of Offshore Engineering, Brasil '87*, Rio de Janeiro, pp. 256–266.
29. Lunne, T., Eidsmoen, T., Gillespie, D., and Howland, J. D. (1986), 'Laboratory and field evaluation of cone penetrometers', *Proc. of In Situ '86*, GT. Div. ASCE, June 1986, pp. 714–729.
30. Lunne, T., Christophersen, H. P., and Tjelta, T. I. (1985a), 'Engineering use of piezocone data in North Sea clays', *Proc. 11th International Conference on Soil Mechanics and Foundation Engineering*, San Francisco, CA, Vol. 2, pp. 907–912.
31. Lunne, T., Lacasse, S., Aas, G., and Madshus, C. (1985b), 'Design parameters for offshore sands, use of *in situ* tests', *Proc. International Conference on Offshore Site Investigations*, Society of Underwater Technology, London, Vol. 3, pp. 269–293.
32. Marchetti, S. (1980), '*In situ* tests by flat dilatometer', *JGED, ASCE* **106**(GT3), 299–321.
33. Masood, T. (1990), 'Comparison of In Situ Methods to Determine Lateral Earth Pressure at Rest in Soils', PhD Thesis, Department of Civil Engineering, University of California, Berkeley, USA.
34. Norwegian Geotechnical Institute (1992), 'Theory and Backcalculations of Laboratory and Offshore Tests in Clay', Joint Industry Project on Hydraulic Fracture and Conductor Installation, NGI Rep. No. 521620-4, April 1992.
35. Offshore Norge (1990), 'New Geotechnical Tool for Deep Waters', (in Norwegian), p. 50.
36. Overy, R. F. and Dean, A. R. (1986), 'Hydraulic fracture testing of cohesive soil', *Proc. Offshore Technology Conference*, Houston, Paper No. 5226.
37. Parkin, A. K. and Lunne, L. (1986), 'Hydraulic fracture testing offshore', *Proc. Speciality Geomechanics Symposium; Interpretation of Field Tests for Design Parameters*, Institute of Engineering, Adelaide, Australia, pp. 123–127.
38. Peterson, L. M. and Johnson, G. W. (1985), 'Deep water site investigations', *Proc. 3rd Offshore Technology Conference*, Sorrento, Italy.
39. Powell, J. J. M. (1990), 'A comparison of four different pressuremeters and their methods of interpretation in stiff heavily overconsolidated clays', *Proc.3rd Int. Symp. on Pressuremeters*, Oxford, pp. 287–298.
40. Powell, J. J. M. and Uglow, I. M. (1988), 'The interpretation of the Marchetti dilatometer test in U.K. clays', *Proc. Geotechnology Conference on Penetration Testing in the UK*, Birmingham, pp. 269–273.
41. Rad, N. S. and Lunne, T. (1992), 'Gas in soils: Detection and η- profiling', *Journal of Geotechnical Engineering, ASCE*, accepted for publication.
42. Rad, N. S. and Lunne, T. (1991), 'Use of BAT probe for shallow gas detection', Society of Underwater Technology, Vol. 17, No. 2, pp. 16–20.
43. Rad, N. S., Sollie, S., Lunne, T., Torstensson, B. A. (1988), 'A new offshore soil investigation tool for measuring the *in situ* coefficient of permeability and sampling pore water and gas', *Proc. 5th International Conference on the Behaviour of Offshore Structures, BOSS'88*, Trondheim, Norway, Vol. 1, pp. 409–417.
44. Robertson, P. K., Sully, J. P., Woeller, D. J., Lunne, T., Powell, J. J. M., and Gillespie, D. (1992), 'Estimating coefficient of consolidation from piezocone test', *Canadian Geotechnical Journal* **29**(4), 539–550.
45. Schnaid, F. and Houlsby, G. T. (1990), 'Calibration chamber tests on the cone pressuremeter,

Proc. 3rd International Symposium on Pressuremeters, Oxford, pp. 263–272.
46. Skomedal, E. and Bayne, J. M. (1988), 'Interpretation of pore pressure measurements from advanced cone penetration testing', *Proc. Geotechnology Conference on Penetration Testing in the UK*, Birmingham, pp. 131–135.
47. Stewart, W. P. and Campanella, R. G. (1991), '*In situ* measurement of damping of soils', *Proc. 2nd International Conference on Recent Advances in Geotechnical Engineering and Soil Dynamics*, No. 1. 33, Vol. 1, pp. 83–93.
48. Sully, J. P. and Campanella, R. G. (1991), 'Effect of lateral stress on CPT penetration pore pressure', *Journal of Geotechnical Engineering, ASCE* **117**(7), 1082–1088.
49. Sully, J. P., Campanella, R. G., and Robertson, P. K. (1988), 'Interpretation of penetration pore pressures to evaluate stress history in clay', *Proc. International Symposium on Penetration Testing, ISOPT-1*, Orlando, Florida, USA, Vol. 2, pp. 993–999.
50. Tjelta, T. I. (1992), Personal Communication.
51. Tjelta, T. I., Baerheim, M., and Håland, G. (1993), 'Novel foundation concepts for jackets finding its place', *Proc. Int. Conf. Offshore Site Investigations and Foundation Behaviour*, SUT, London.
52. Tjelta, T. I., Aas, P. M., Hermstad, J., and Andenæs, E. (1990), 'The skirt piled Gullfaks C platform installation', *Proc. Offshore Technology Conference*, Houston, Paper No. 6473.
53. Tjelta, T. I., Guttormsen, T. R., and Hermstad, J. (1986), 'Large scale penetration test at a deep-water site', *Proc. Offshore Technology Conference*, Houston, Paper No. 5103.
54. Torstensson, B. A. (1984), 'A new system for ground water monitoring', *Ground Water Monitoring Review* **4**(4), 131–138.
55. Wright, A. (1991), 'Experience with an instrumented plough', Presentation at Soil Investigation Forum, Aberdeen, 1991.
56. Wright, N. D. and Tan, M. (1991), 'Hydraulic fracture tests in heavily overconsolidated clays', *Proc. 1st International Symposium on Offshore and Polar Engineering*, Vol. 1, pp. 198–206.
57. Wroth, C. P. (1988), 'Penetration testing – More rigorous approach to interpretation', *Proc. International Symposium on Penetration Testing, ISOPT-1*, Orlando, Florida, Vol. 1, pp. 303–311.
58. Zielinski, G. W., Gunleiksrud, T., Sættem, J., Zuidberg, H. M., and Geise, J. M. (1986), 'Deep heatflow measurements in quaternary sediments on the Norwegian continental shelf', *Proc. Offshore Technology Conference*, Paper No. 5183.
59. Zuidberg, H. M., Hoope, J. T., and Geise, J. M. (1988), 'Advances in *in situ* measurements', *Proc. 2nd International Symposium on Field Measurements in Geomechanics*, Kobe, Japan, pp. 279–291.
60. Zuidberg, H. M., Richards, A. F., and Geise, J. M. (1986), 'Soil exploration offshore', *Proc. 4th International Geotechnical Seminar on Field Instrumentation and In Situ Measurements*, Nanyang Technical Institute, Singapore, pp. 3–11.

THE INFLUENCE OF STONE AND BOULDER INCLUSIONS ON OFFSHORE SITE INVESTIGATION AND FOUNDATION BEHAVIOUR

M. R. COOPER
Gifford and Partners, Carlton House, Ringwood Road, Woodlands, Southampton SO4 2HT, and Department of Civil Engineering, University of Southampton, Southampton SO9 5NH

and

T. LUNNE and T. BY
Norwegian Geotechnical Institute, Postboks 3930, Ullevål Hageby, N-0806 Oslo

Abstract. This paper examines the origin of stony and bouldery soils offshore and describes the techniques and equipment available for their investigation. The experiences of the Norwegian Geotechnical Institute and British Geological Survey are reviewed in presenting a detailed discussion of direct drilling techniques, and the usefulness of indirect methods is also discussed. The applicability of statistical methods is briefly covered, and the paper concludes with a consideration of the foundation engineering implications of these geotechnically difficult materials.

1. Introduction

Soil investigations carried out since 1986 in the Haltenbanken area of the Norwegian Sea have encountered clay strata containing an appreciable content of stones (i.e. cobbles, 60–200mm, and boulders > 200mm). The presence of the stone and boulder inclusions has caused problems for both the conduct and interpretation of investigation work, whilst the need to make allowance for possible boulder encounters presents additional design problems. This paper reviews current knowledge and experience available at the Norwegian Geotechnical Institute (NGI) for investigating, interpreting and designing for the boulder content of predominantly finer matrixed soils.

2. Geological Context

The origin of the extremely bi-modal gap-graded soils considered in this paper can generally be attributed to glacial erosion and transport. Similar materials occur elsewhere in quaternary geological environments similar to those of the Norwegian Sea.

Volume 28: Offshore Site Investigation and Foundation Behaviour, 181–193, 1993.

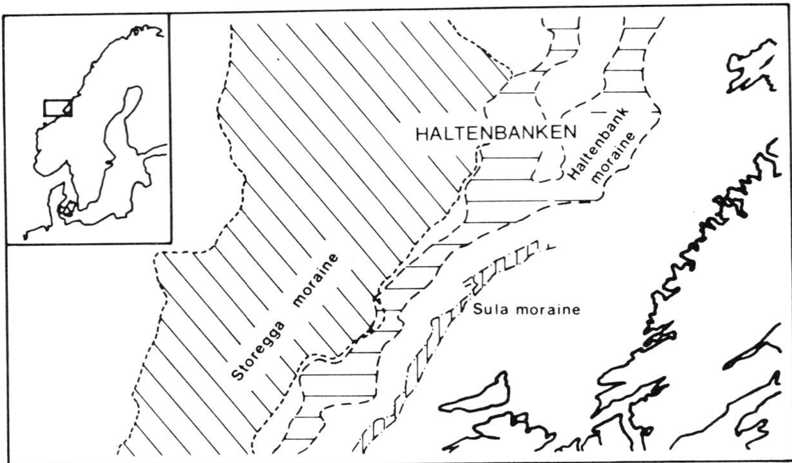

Fig. 1. Location of Haltenbanken field and distribution of moraine deposits off the central Norway coast (after Bugge, 1980).

Brown (1986) for example reports till material on the Labrador Shelf as being too consolidated or bouldery for piston cores. The transportation and deposition may be by one of two mechanisms; either direct ice-transport or ice-rafting. Sættem (1987) points out that in both cases the material origins would be similar and similar grain size distributions would result. The capacity of the ice-rafting mechanism, either below icebergs or a floating ice-shelf, is uncertain and may have been insufficient to produce widespread boulder rich deposits of appreciable thickness. Cooper (1986, reporting Forsberg) suggests that the large stones and boulders held at the ice base would quickly melt out and so not be widely distributed. Isolated boulder patches could be the result of "boulder dumping" during iceberg rotation (Lewis *et al*, 1986). However, the stony and bouldery soils of the Norwegian Sea are believed to be generally associated with morainic structures (see Figure 1). Unusual concentrations of cobbles or boulders at a particular level could be expected immediately above an ice-erosion junction in such an environment, in which the boulder rich layer would lie on or directly above a marked increase in matrix soil stiffness.

The supposed nature of the transport mechanisms thus indicates two contrasting modes of occurrence for cobbles and boulders; either evenly distributed through morainic or ice-rafted deposits, or as concentrated layers on ice-erosion surfaces or the current seabed. This is confirmed by recent NGI experience.

3. Indirect Boulder Detection

Since boulders represent inclusions of much more rigid material within the more compressible soil matrix there is considerable potential for the application of acoustic techniques for boulder detection. Existing offshore surface seismic methods are able to detect the distinct seismic reflector resulting from a boulder layer, which may also be aligned along a change of matrix stiffness as described above. They are not able to distinguish the boulder intensity except in the broadest terms as it affects the apparent rigidity of the reflecting surface.

Standard seismic methods are much less successful for the detection of individual boulders. The minimum size of anomaly which can be detected is fixed by the wavelength of the excitation. It is not possible to detect any feature which is smaller than the order of one wavelength. This implies a minimum detectable size of 6 to 8 metres for standard surface methods. Using special equipment and procedures the detection limit may be reduced to 1 to 2 metres.

The use of diffraction interpretation and closer spacial sampling, as discussed by Fulton and Hsiao (1983) can be expected to reveal anomalies less than a wavelength in size.

Perhaps a more significant shortcoming of conventional surface techniques is their inability to give a precise location for any anomaly positively, or tenuously, identified. Fine adjustment of foundation, conductor or borehole location to either avoid, or locate and examine a suspected anomaly is therefore not possible.

The Acoustic Core approach of the IMAP method (Guigné *et al*, 1991) developed by the Centre for Cold Ocean Resources Engineering uses a stationary C-core "sparker" source (English *et al*, 1991) operating close to seabed level and operating at frequencies up to 30kZ. This method promises the capability of boulder detection down to possibly 0.5m diameter at much greater depths than surface methods, but at its present development stage it relies on diver support for positioning, and re-positioning, the beam mounted source and hydrophones, see Figure 2. It is understood that further development work to produce a system capable of fully automated remote operation in deep water is planned.

The IMAP method produces a vertical record of a stationary test "column" about 10 metres in diameter and is thus well suited to the investigation of localised test sites such as foundation or conductor positions. A collaborative proposal between NGI, GEODIA (France) and ISMES (Italy) is currently being submitted to develop a high resolution seismic seabed imager with the capability of continuous scanning to a depth of up to 20 metres along proposed pipeline routes. This will be achieved through improved processing methods applied in conjunction with diffraction tomography techniques.

Cross-hole seismic methods have the potential to give good anomaly images in the inter-borehole plane, but are unavoidably restricted in terms of cost and coverage by their associated borehole requirements. The practical application of cross-hole techniques is restricted to shallow water sites.

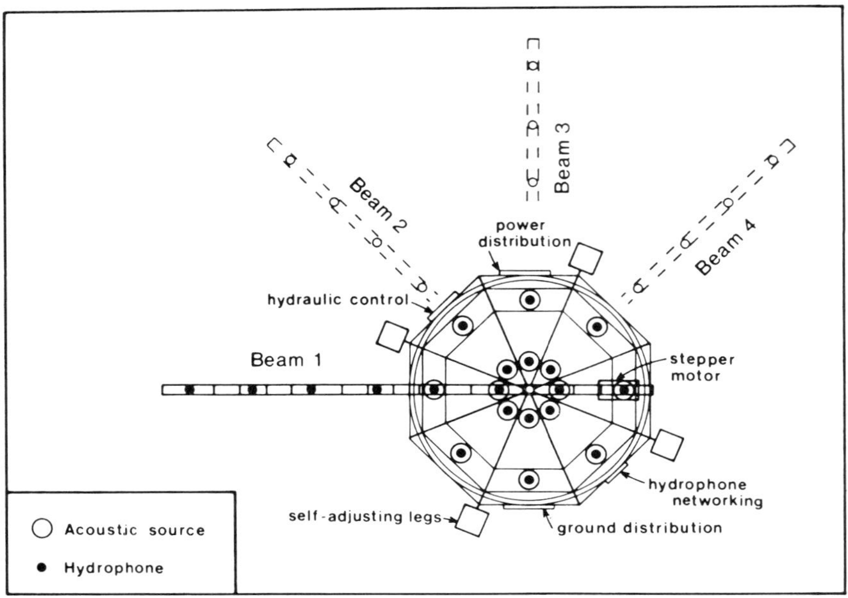

Fig. 2. General plan arrangement of the C-core IMAP acoustic core (modified from Guigné *et al*, 1991).

4. Direct Detection

The role of direct detection in respect of stony or bouldery deposits is one of conflict between largely irreconcilable requirements and constraints. The ideal requirements of the perfect direct detection method might be seen as follows.

i. Accurate positioning of each and every boulder strike in order to provide the frequency data for any proposed statistical analysis.

ii. Recording of boulder size, which may also be an important input into statistical analysis, and is also useful as simple unrefined data.

iii. Recovery of matrix material in the boulder rich zone, since the strength and compressibility of the soils surrounding the boulders will generally remain the prime geotechnical parameters.

iv. Penetration of the boulders and boulder zone; preferably in such a way that normal drilling, sampling or probing can continue below the bouldery zone.

No single technique is currently available which will satisfy all four requirements, but recent NGI experience on several investigations in the Haltenbanken area of the Norwegian Sea, Figure 1, allows some valuable observations to be made in respect of drilling, sampling and in-situ testing. The material encountered in nearly all cases had a distributed content of cobbles and boulders. The inclusion frequency was less in the upper layers and local concentrations of cobbles and boulders at specific depths were common.

The distribution of stones, boulders and cobbles and the consistency of the soil matrix vary from site to site. The optimum combination of drilling, sampling and in-situ testing equipment must be mobilised for each investigation.

4.1. Drilling

The first soil investigations on the Haltenbanken area encountered the problem of penetrating stones and boulders, and very frequently target depth was not reached as the open-throat drill bits simply wore out or were unsuitable for the formation, thus preventing further penetration (see Figure 3). The situation has improved over the last five to seven years, mainly as a result of drill bit modification.

The current approach is to start with an open-throat drag bit with cutting faces coated with tungsten carbide chips. If this meets refusal then the API drill string is normally pulled up and the drill bit changed to an open-throat four-cone roller bit, Figure 4. Experience has shown that this is able to drill down and penetrate small boulders and stones, but an insert bit has to be used to prevent stones being stuck inside the drill bit and blocking the drill string. This will to some extent slow down the drilling operations. The use of a dragbit may also require an insert bit for the same reason but it is preferred wherever possible because it is cheaper and penetrates rapidly in all but the most stony of formations, even though this may mean tripping the pipe when larger cobbles or boulders are encountered.

If a lot of drilling is to be done in bouldery soil then it may be beneficial to use "piggy-back" coring as described, for instance, by Sættem (1987). In this technique a small diameter string is run inside the standard 4-inch I.D. API string, and a 54.5mm core is cut by a diamond bit (see Figure 5). However, piggy-back equipment is expensive to mobilise, the system is best suited to non-tidal conditions, and normal soil investigation sampling and in-situ testing equipment cannot be used with it. There is also a danger of string twist-off unless the formation has a hard and stable matrix which will suspend cored material without jamming or running a casing string.

Skinner (1987) pointed out the need for a carefully designed drilling mud programme when drilling in stony soils. Logging or observation of drilling parameters, notably mud pressure and rate of penetration, can be very useful in evaluating layering.

4.2. Sampling

The two aims of sampling are:

a. to get as good coverage of the soil profile as possible, and

Fig. 3. Examples of severe (B) and very severe (C) wear of drill bits (original condition (A)) due to drilling in stony soils.

Fig. 4. Four-cone roller bit (with throat blocked by a cobble).

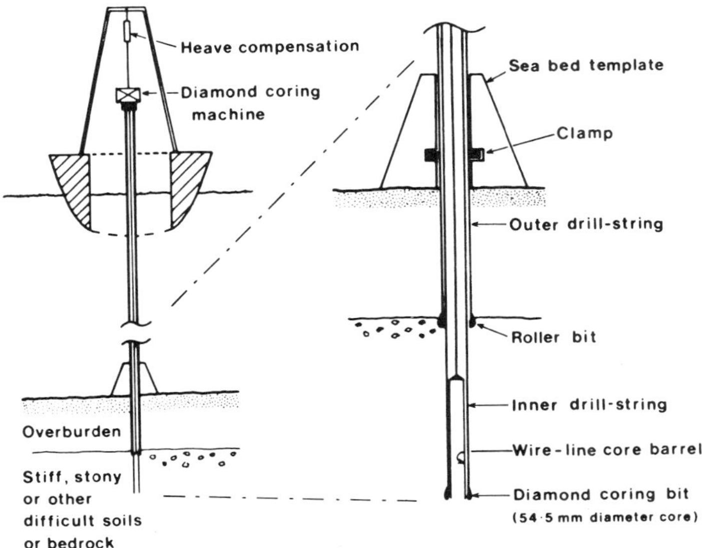

Fig. 5. General principles of "piggy-back" coring (modified from Sættem, 1987 and Rise and Sættem, in preparation).

b. to obtain undisturbed samples for subsequent laboratory testing to give the strength and deformation parameters needed for foundation analysis.

The sampling programme usually evolves through trial and error, assisted by information from previous investigations and drilling observations.

Thin walled 75mm diameter push sampling is the preferred method, with a change to thick walled push samplers if tube damage is too great. If sufficient penetration cannot be achieved with push sampling then hammer sampling must be tried.

The main sampling problem is satisfactory recovery of stone or pebble rich horizons. One potential method is to use wireline rotary coring with the 5-inch API drill string (4-inch I.D. and core barrel), but this method is not expected to work well when the stone content is not firmly bound by strong matrix soils. Sættem (1987) reports a successful use of piggy-back coring for such horizons.

Rygg and Andresen (1990) consider the applicability of percussion samplers. They describe laboratory tests on three percussion samplers and conclude that high frequency percussion is well suited for sampling the particular material they tested (gravelly sand with some cobbles), but its applicability to isolated cobble or boulder inclusions is not yet proven. A Vibra-Percussive Corer (VPC), able to operate by wireline methods within a 5-inch API string (4-inch I.D.), has been developed by ODP/BGS and a prototype has been field tested (Pheasant and Skinner, 1992, personal communication).

5. Statistical Methods

Methods exist (e.g. Tang and Quek, 1986) which enable the probability density function of boulder diameters, and hence the probability of boulder strikes (Ang and Tang, 1965), to be estimated – provided the input data can be obtained during the investigations. Unfortunately, the Tang and Quek method takes as its base data the observed lengths of the parts of boulders intercepted along the line of boring and, more significantly, assumes negligible error in these measurements. This implies either continuous core drilling, core drilling of every intercepted inclusion (of above the minimum detectable size), or through drilling with control and logging of a standard not currently practised or indeed attainable (in order to determine entry to and exit from each inclusion). It also requires a matrix of sufficient stiffness to restrain the inclusion firmly enough for the above procedures to be carried out, and would necessitate drilling standards currently unachievable offshore.

Tang and Quek demonstrate the method with an onshore case history based on 500 measured chord lengths in the size range < 0.2 to 2.5 metres. It is believed that their data was obtained from continuous small diameter core drilling, and that the deposit studied was sufficiently dense, indurated or cemented to keep all stones in place during coring.

At a more fundamental level statistical interpretation is applied by NGI (1988) in attempting to develop strategies for specific site investigation aims. One such strategy is the number of investigation locations required to prove the presence of (but not to delineate) an anomaly. Using NGI's results gives a good insight into the problem of using direct boring or probing methods to prove the absence of boulders. This is the most basic function of a boulder investigation since once the presence of a single potentially damaging boulder is confirmed within the investigated area, it will almost certainly be necessary to allow in the design for the presence of boulders. Assuming a potentially damaging boulder to be 2 metres in diameter, and that the acceptance criterion for proof of absence is that 2 metres will correspond to the mean undetectable boulder size plus one standard deviation, the acceptably undetectable anomaly represents as little as 0.1% of a base only 20 metres across. Extrapolation of NGI's results suggests that perhaps as many as 50 probes or borings, accurately positioned on a regular grid, would be required to prove statistically that no damaging boulder was present. Real life "noise" of probes stopped by smaller but non-critical boulders and positioning inaccuracies causing a randomness in the supposedly regular grid, could increase further the required number of borings.

At present the main function of the statistical approach in the investigation of isolated hard inclusions is to make clear the difficulty of defining the problem at any location by direct investigation methods; and perhaps to suggest that ultimately the answer to the successful delineation of the risks posed by boulders must be through the more regionally effective geophysical methods provided the resolution can be made high enough, coupled with an understanding of the geological processes which have caused the boulders to be present.

6. Engineering Works

The effects of boulders on the engineering works are principally concerned with their influence on pile driving, skirt penetration and conductor installation. These effects can be considered in respect of three main factors,

- design features necessitated by prior knowledge or suspicion of boulder presence;
- damage caused by boulders but safely detected and repaired;
- damage caused by boulders and remaining undetected.

The effects will be dependent not only on the size and frequency of boulders but also on the matrix strength which determines the rigidity with which the potentially damaging boulders are held in place.

Prior knowledge of the need to design for the effects of boulders will follow either from a fully successful site investigation where the risk has been quantified or, more likely at present, from an investigation which has encountered some boulders and has led to a judgement based decision that a "design boulder" of a certain size and frequency must be allowed for.

The design decision for piles will be heavily dependent on the reliability of the estimate of boulder size and frequency. In the worst case the frequency and size of boulders may simply be too great within the required pile length to allow a piled solution in which case an alternative must be sought. At low intensities of even quite large boulders the design strategy may merely allow for a certain proportion of piles either to be relocated to spare slots provided as a contingency measure, or else accepted at a reduced capacity. If the boulders are both small and infrequent it may be decided that the driving energy will be sufficient to displace them without damage to the pile, perhaps with the use of strengthened or specially profiled shoes. Further advances in pile shoe design may allow breaking up or displacement of larger boulders.

For skirts the presence of boulders poses a combination of problems. The boulder, or a number of boulders, may present sufficient additional resistance to limit penetration. For gravity bases this may not be disastrous, provided that the possibility has been foreseen and that the skirt is sufficiently reinforced to withstand the worst case load concentrations which would result, again a decision which is highly dependent on the quality of the site investigation data. For suction anchor skirts the situation will be more serious as limited penetration has a more significant effect on the intended pull-out capacity, and encounters with larger boulders could seriously reduce the effectiveness of the suction seal.

The problems for conductor installation through bouldery soils will be very similar to the general drilling problems described earlier, and again the most efficient approach will be to use the same equipment to penetrate both boulders and matrix. Over-sized holes will be required through the bouldery materials both to allow free passage of the conductor through large boulders and to prevent damage to the conductor from displaced boulders or boulder remnants in the hole wall.

Damage caused to piles and skirts during installation is most dangerous if undetected, and is more likely to occur in the dynamic piling process. The recent advances in dynamic pile analysis give at least the promise of detection of major boulder strikes, and more usefully of detecting any significant pile damage from the altered dynamic response which would result. The current consensus among operators of dynamic analysis equipment is that current techniques could detect pile damage only by comparison with the response of undamaged piles, and that further development of sacrificial transducers would be necessary for economic application in deep water offshore environments.

Skirt penetration is a static process and the loads required to push down or push aside the "design boulder" can be estimated and allowed for more easily in the design. The main danger of unforeseen skirt damage therefore arises when boulders are unknowingly encountered when their presence has not been predicted from the site investigation, and no design precautions have been taken.

The problem of undetected boulder strikes is unlikely to arise in conductor installation provided the necessary degree of over-sizing has been allowed.

7. Bulk Soil Properties

The presence of cobbles and boulders in an otherwise fine grained material will affect the bulk density, shear strength and compressibility of the composite soil, though in soils so far encountered by NGI the effect has not been significant.

The effect on bulk density is direct and obvious as denser boulders replace less dense matrix.

Work such as that by Lupini *et al* (1981), and previously Fedorev and Sergovina (1973) and Vasileva *et al* (1979), suggests that the shear strength is little altered by strong inclusions up to a certain "threshold concentration", below which matrix strength properties can be used directly. In all cases the "threshold concentration" was reported as being greater than 20%, often considerably so, and this value would seem an appropriate guideline at this stage.

The compressibility will be directly altered by the presence of rigid inclusions. The soil will contain a certain volume of compressible matrix, V_m, and so the total volume change $\Delta V = f(\Delta p' \cdot V_m)$ only, the inclusions being rigid. A less predictable effect concerns the modification of stress distribution due to the larger incompressible elements. This reduces mean $\Delta p'$ in the matrix and hence further reduces the compressibility. Parametric studies carried out by Cooper (1986) indicate that the effect can be appreciable. Figure 6 shows the mean axial strain against inclusion content for an unconfined soil block analysed in plane strain under a uniform vertical pressure. This can be seen to be appreciably less than the strain estimated simply on the basis of the reduced volume of compressible material.

8. Conclusions

The most promising area of development for defining the boulder content of clay/cobble/boulder mixtures offshore is in remote high resolution seabed geophysical methods such as the IMAP method, which is presently limited to shallow water applications.

Drilling and sampling procedures are available which allow successful penetration of cobble/boulder rich zones with continued push sampling of the matrix. Where the content of hard inclusions is too great for acceptable push sample recovery recourse must be had to percussion sampling or piggy-back rotary coring.

With the presently available investigation methods it is unlikely that sufficiently accurate or extensive databases can be obtained to realise fully the benefits of statistical analysis, indeed such analyses can help to indicate the limitations of current data intensity.

Bouldery soils are still a comparatively infrequent deposit in offshore Geotechnical engineering. The problem is therefore unlikely to attract large amounts of development money for special "boulder investigating" equipment. Site investigation strategy will therefore continue to rely on careful application of, and modification to, existing standard methods.

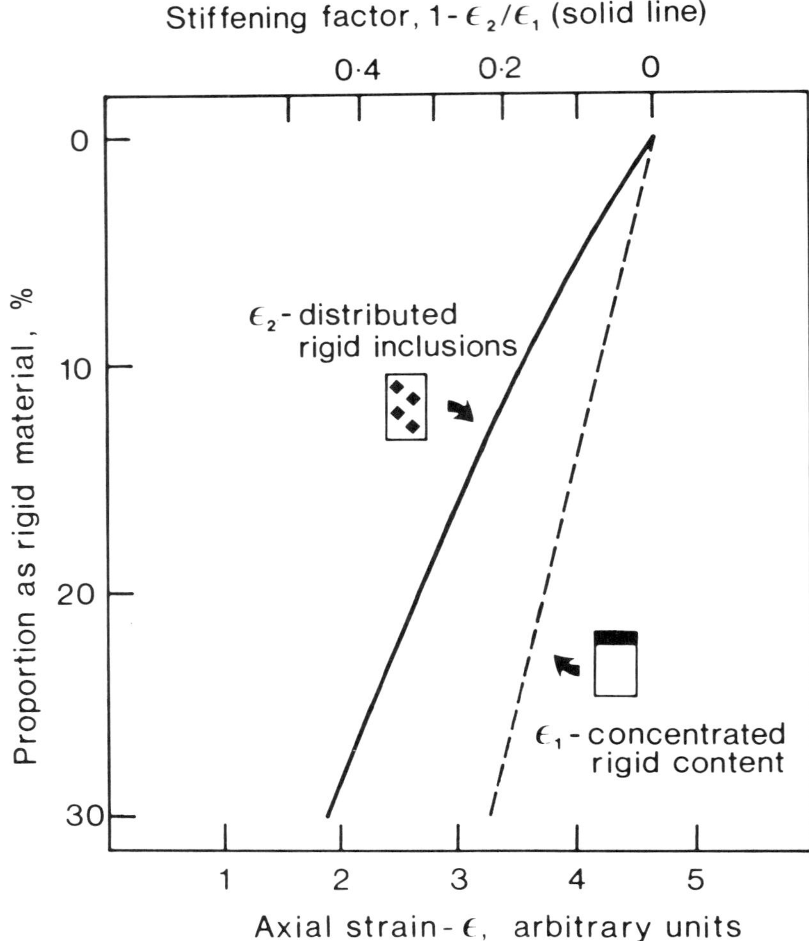

Fig. 6. Results of numerical parametric study of the stiffening effect of rigid inclusions.

Acknowledgements

The authors would like to thank the following people for their assistance during the preparation of the paper: S. Lacasse, A. Andresen, T. L. By and R. Lauritzen, all of the Norwegian Geotechnical Institute and A. Skinner of the British Geological Survey.

References

1. Ang, A. H. S. and Tang, W. (1975), *Probability Concepts in Engineering Planning and Design. Volume 1, Basic Principles*, John Wiley and Sons.
2. Brown, J. D. (1986), 'Geotechnical engineering offshore eastern Canada',*Proc. 3rd Canadian Conference on Marine Geotechnical Engineering*, Vol. 3, pp. 852–878.

3. Bugge, T. (1990), 'Shallow Geology on the Continental Shelf off Møre and Trøndelag, Norway', Continental Shelf Institute (IKU), Trondheim, Publication No. 104.
4. Cooper, M. R. (1986), 'Geotechnical Problems Related to Boulders', NGI Report No. 59092-1 (dated 12 September 1986).
5. English, J., Inkpen, S. T., and Guigné, J. Y. (1991), 'A new high frequency, broadband seismic source', *Proc. Offshore Technology Conference 1991*, Copyright Offshore Technology Conference, 1991.
6. Fedorev, V. and Sergovina, V. (1973), 'Effect of the clay fraction on the strength parameter of clayey gravels', *Osnov. Fund. Mech. Grunt.* **15**(6), 13–15 (in Russian).
7. Fulton, T. K. and Hsiao, R. T. (1983), 'Diffractions reveal drilling hazards', *Proc. 15th Annual Offshore Technology Conference*, Houston, pp. 533–538.
8. Guigné, J. Y., Pike, C. J., and Inkpen, S. T. (1991), 'IMAP – Interactive Marine Acoustic Probe, Acoustic Site Evaluation, Terrenceville, Newfoundland, C-Core Report No. 91-31 (dated 10 October 1991).
9. Guttormsen, T. (1988), 'Optimum Site Investigation and Laboratory Testing Strategy', NGI Report No. 41411-9 (dated 15 March 1988).
10. Lewis, C. F. M., Parrott, D. R., Josenhans, H. W., Barrie, J. V., and Gaskell, H. S. (1986), 'Iceberg scouring: Hazard for seabed development', *Proc. 3rd Canadian Conference on Marine Geotechnical Engineering*, Discussion, Vol. 3, pp. 1031–1021.
11. Lupini, J. F., Skinner, A. E., and Vaughan, P. R. (1981), 'The drained residual strength of cohesive soils', *Géotechnique* **31**(2).
12. Rise, L. and Sættem, J. 'Shallow coring in bedrock offshore Norway', in preparation.
13. Rygg, N. O. and Andresen, A. (1990), 'Site investigation in moraine, gravelly and stony soils for strait crossings', *Proc. Strait Crossing Symposium*, Trondheim.
14. Sættem, J. (1987), 'Coarse fragments in clays offshore Norway in relation to drilling and sampling, and suggested improvements of methods', *International Course on Drilling and Sampling in Stony Soils*, Discussion Contribution, NGI, Oslo.
15. Skinner, A. (1987), 'Drilling and downhole sampling', *International Course on Drilling and Sampling in Stony Soils*, NGI, Oslo.
16. Tang, W. and Quek, S. T. (1986), 'Statistical model of boulder size and fraction', *ASCE Journal of Geotechnical Engineering* **11.2**(1), 79–90.
17. Vasileva, A., Tkachenko, G., and Lebedëv, V. (1979), 'Strength properties of gravels containing clay', *Osnov. Fund. Mech. Grunt.* **21**(4), 16–17 (in Russian).

SESSION 3

ADVANCED INTERPRETATION TECHNIQUES

IMPROVEMENT OF GEOPHYSICAL INTERPRETATION BY USE OF DELPH1 PROCESSED DATA

T. DES VALLIERES and T. L. ARMSTRONG
TOTAL, Tour TOTAL, Cédex 47, 92069 Paris La Défense, France

and

R. GIRAULT
ELICS, 53 rue de la Jarry, 94300 Vincennes, France

Abstract. The paper presents several pinger, boomer and sleeve gun data acquired for drilling site investigation. The data were digitally recorded and processed by DELPH1. Different processing modules (noise control, swell filtering, true amplitude recovery, deconvolution, time varying filtering, scaling, screen interpretation, ...) show the enhancement of the seismic data leading to an improved interpretation.

1. Introduction

DELPH1 is a digital *seismic* recording and processing system designed to replace conventional single channel analogue equipment normally used in engineering geophysics and geohazard surveys.

The sources used for these surveys range from pinger, boomer and sparker to other deeper investigation sources (G.I. guns, sleeve gun, ...).

DELPH1 is actually designed to cover the full range of the above sources. The digital recording achieved by the system allows any form of seismic processing to be performed. This is a great advantage over an analogue recording mode.

2. Description of the DELPH1 System

2.1. General Features

Processing with DELPH1 is done in real time. The processed data are displayed on a computer screen and on a thermal plotter. The raw data are stored on disk giving the opportunity to undertake post-processing. The post-processing can be performed using either DELPH1 or data can be exported to another workstation using a SEGY format.

Volume 28: Offshore Site Investigation and Foundation Behaviour, 197–214, 1993.

2.2. Hardware

The DELPHI system is based on an IBM-PC computer running under DOS. A dedicated board is plugged into the PC for acquisition and processing.

The analogue stage of this board features:

- An analogue amplifier with a software-driven gain, allowing input signal in the range of 1mV up to 3 Volts.

- An antialias filter of Chebychev type with nine coefficients, having a 160 dB/octave slope and -80 dB of rejection.

- A 16-bit analogue to digital converter, giving a dynamic range of 96 dB.

- A sampling frequency, user selectable within the range from 1 kHz to 32 kHz.

The digital stage of the board features a digital signal processor (DSP) of Texas Instruments.

The recorded and processed data are displayed on a high resolution monitor.

As part of the hardware the DELPH1 system offers triggering capabilities so that the system can be stand-alone for recording, processing, display and shot triggering.

A link with a navigation unit or an annotator is provided by an RS232 serial port.

2.3. Software

The software offers the following capabilities:

- digital acquisition,
- real-time quality control
- real-time processing;
- interpretation facilities.

In real-time mode, upon reception of an order from the PC, the DSP starts the acquisition of the data and stores them in memory using a direct memory access channel, whilst the DSP processes the data of the previous shot stored in the DSP memory.

During acquisition, data are compressed for storage and the following user-selectable processing modules can be applied:

- band pass filters;
- automatic gain control;

- swell filter,
- time-varying filter.

On completion of processing, the DSP informs the PS by sending a message. The PC can then store the raw data onto disk and send the processed data to the graphic processor for display. In the meantime the DSP is ready to acquire another shot.

In post-processing, in addition to the previous modules, horizontal stacking, multiple attenuation based on a predictive deconvolution and a spiking filter can be applied.

For the *interpretation*, a module allows interactive horizon picking directly on the screen. This information is stored as the two- way time of the reflectors and the location or fix number. The resulting file can then be transferred to a cartography software package for automatic contouring.

3. The System in Use

3.1. Real-Time Quality Control

In seismic work, the primary objective is obviously to record the best possible data.

Therefore, the DELPH1 system offers a built-in facility to observe the data on a shot by shot basis. This gives a good idea of what is really recorded and enables recording parameters to be adjusted until the optimum configuration is reached.

This facility is called the Control Mode.

3.1.1. Analogue amplifier gain control

The first step when setting-up the system is to adjust the gain of the amplifier in order to use the full dynamic range of the analogue to digital converter, whatever the signal input level.

Figure 1 shows a screen display. The top of the figure shows the seismic signal. The horizontal scale is time and the vertical scale is amplitude in mV. The full vertical scale of this window represents the maximum input voltage in accordance with the selected amplifier gain.

The lower left window displays the energy spectrum of the signal. The OdB reference is defined by a sine curve with a maximum amplitude coded on 16 bits.

The lower right window displays the user-accessible keys to select and change parameters.

The central window displays the parameter settings.

It can be seen on Figure 1 that the selected gain was too small, leading to a loss of dynamic range.

Figure 2 shows a selected gain which was too high, leading to a loss of information due to clipping.

Figure 3 shows a good setting of the amplifier gain.

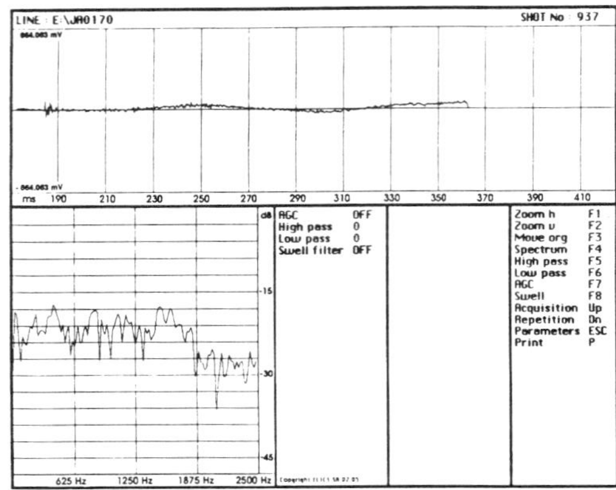

Fig. 1. Screen display (selected gain too small).

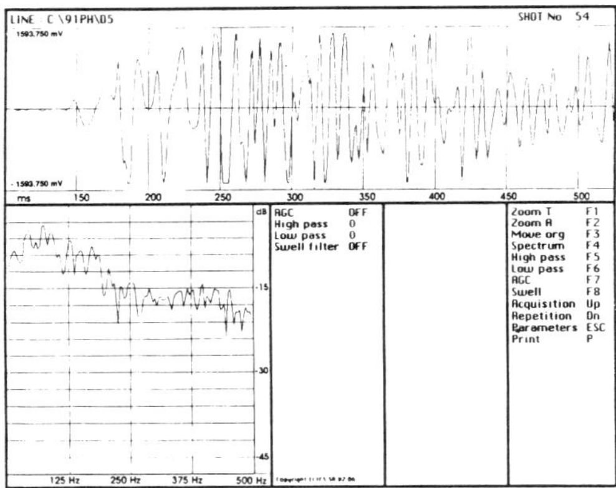

Fig. 2. Screen display (selected gain too high).

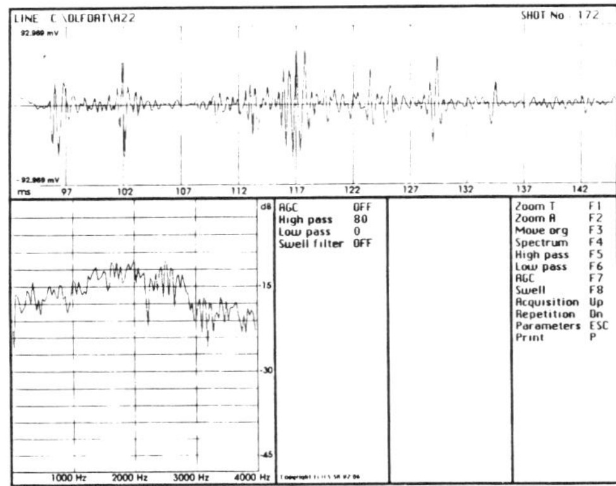

Fig. 3. Screen display (selected gain correct).

This Control Mode acts as a sophisticated oscilloscope and energy spectrum analyzer.

3.1.2. Sampling frequency

Within Control Mode, the energy spectrum of the signal can be displayed. This information can be used to select a suitable sampling frequency. The user has to adjust the sampling frequency according to the source used and can directly check on the screen the best setting.

If the sampling frequency is too small, the digitized data will not contain high frequencies. Too high a sampling frequency may record more noise as the bandwidth of the signal is wider.

Figure 2 shows an example of over-sampled data.

3.1.3. Noise control

Noise can easily be detected on the record as shown on Figure 4 where a 50 Hz signal is present on the data.

This noise should be removed before digitisation in order to avoid loss of information due to clipping of the data due to high amplitude noise.

However, the data can be digitally filtered to reduce this noise (Figure 5) provided that the dynamic range of the A/D converter has not been exceeded.

The energy spectrum gives a good indication of the signal to noise ratio. For example, Figure 6 displays the energy spectrum of the noise in the water. The energy spectrum is computed on a 10 ms temporal window starting at 80 ms. Figure 7 displays the energy spectrum of the sea bed reflected signal of the same shot. A comparison of the two spectra gives the signal to noise ratio. This feature can be used to check the influence of sea state on the receiver sensors as well as other types of noise.

Figure 8 shows a near trace section of a high resolution multichannel streamer with a sleeve gun array source. It clearly reveals bad shots.

3.2. Real-Time Processing

3.2.1. Swell filter

Sea-swell is part of life on a vessel and it is very troublesome for high resolution seismic data recording. DELPH1 efficiently removes swell effects as shown on Figures 9 and 10.

The swell filter is not a heave compensator. Unless the sea floor is affected by regular ripples in phase with the swell, which is an exceptional situation, the swell filter operates satisfactorily.

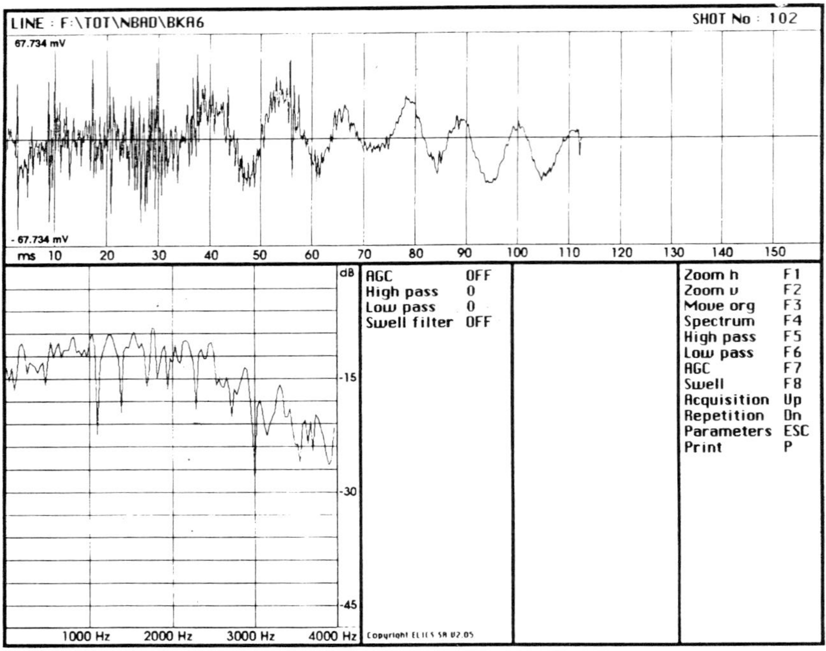

Fig. 4. Noise control (50 Hz interference).

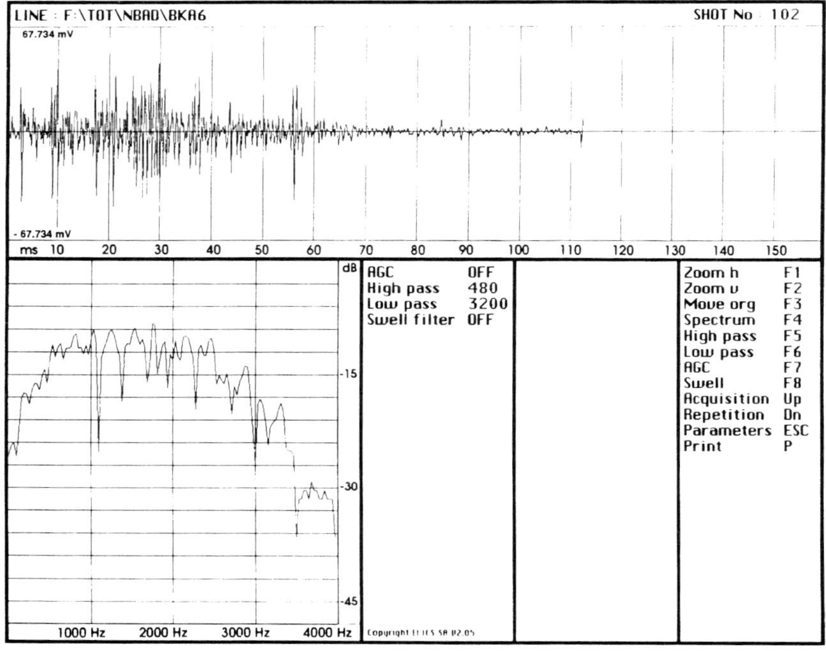

Fig. 5. Noise control (50 Hz filtered).

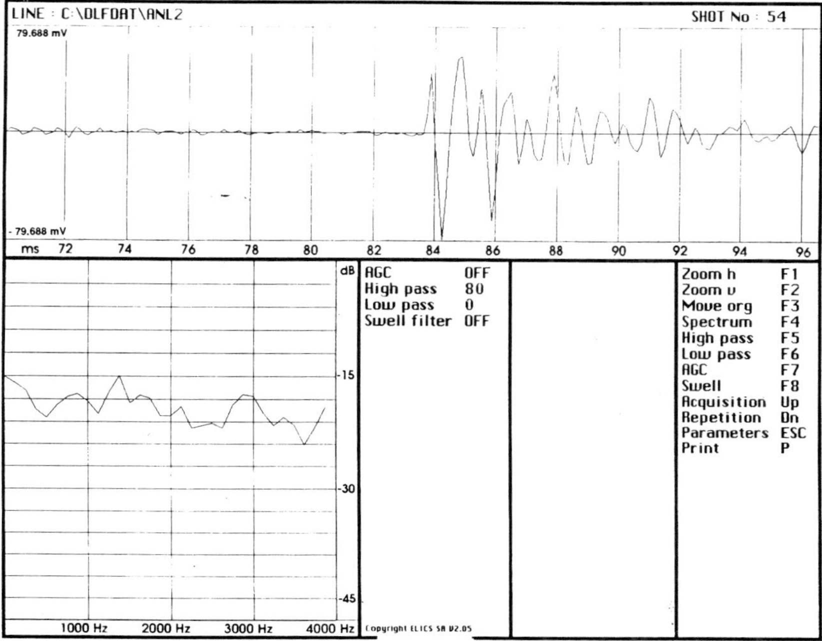

Fig. 6. Noise control (sea water noise).

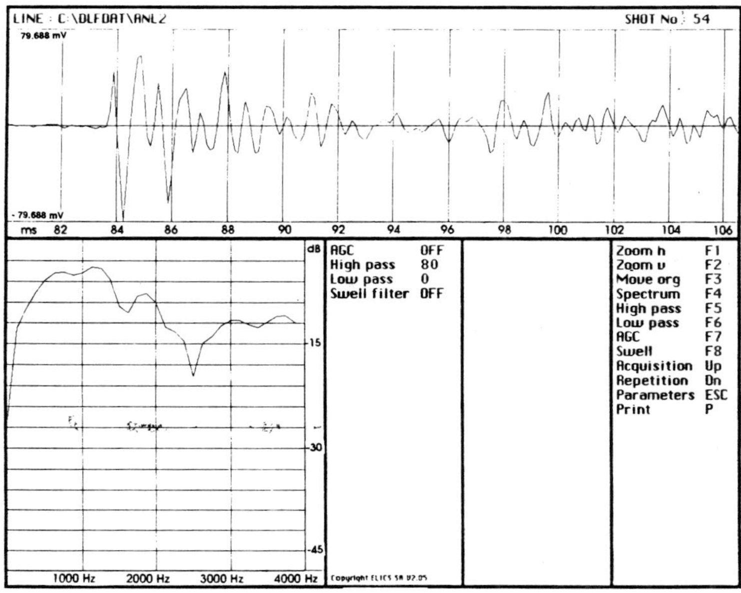

Fig. 7. Noise control (signal to noise ratio).

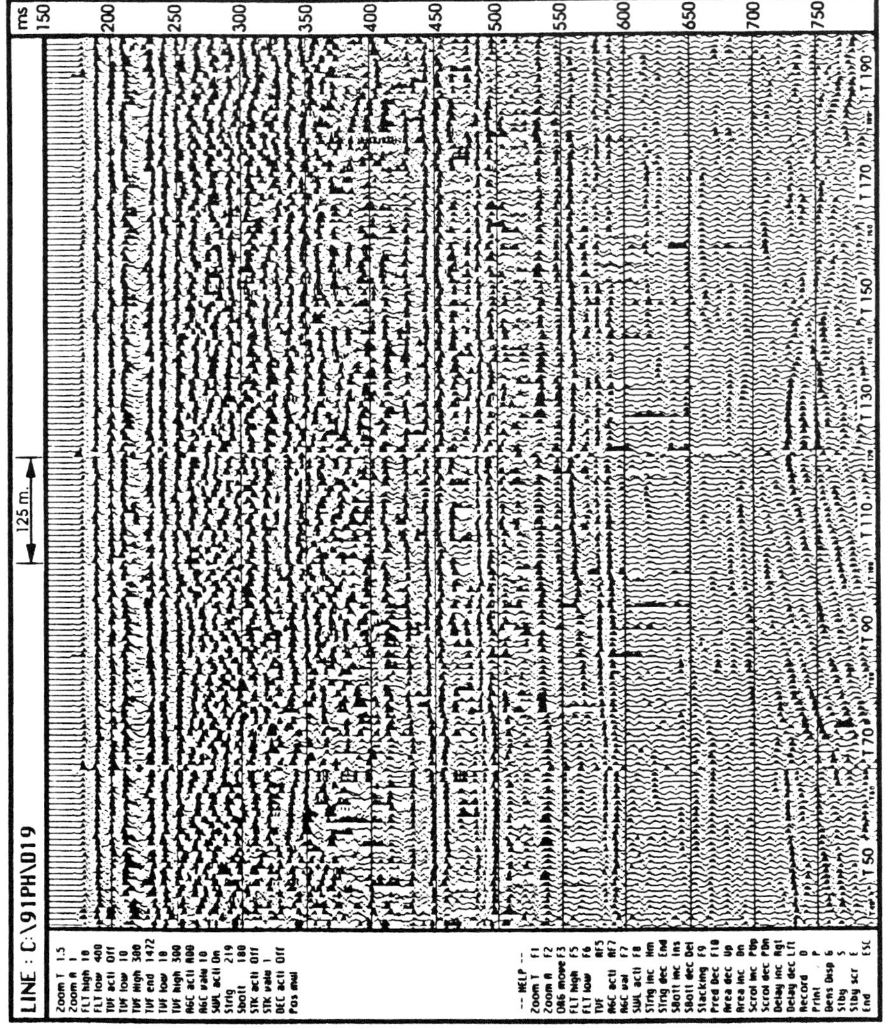

Fig. 8. Multichannel near trace section (sleeve gun).

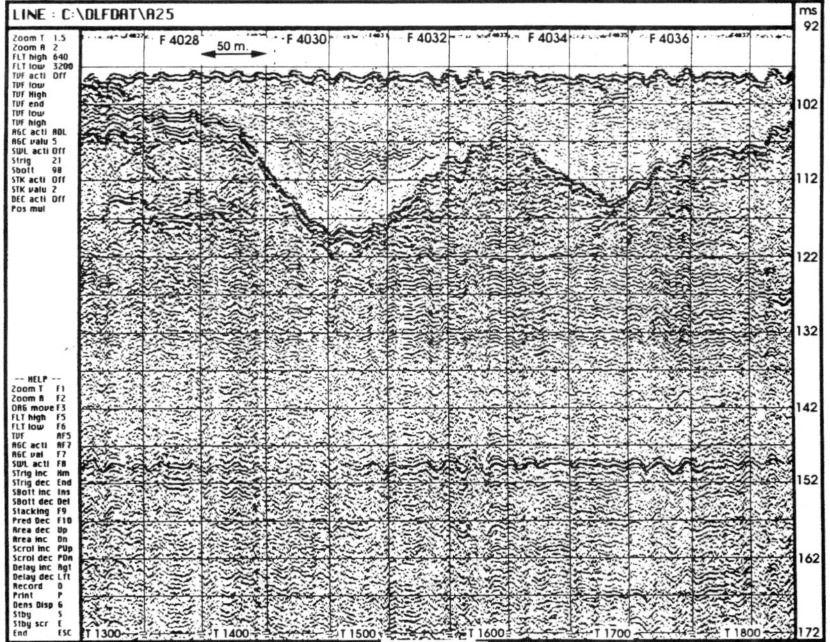

Fig. 9. Swell filtering off (boomer section).

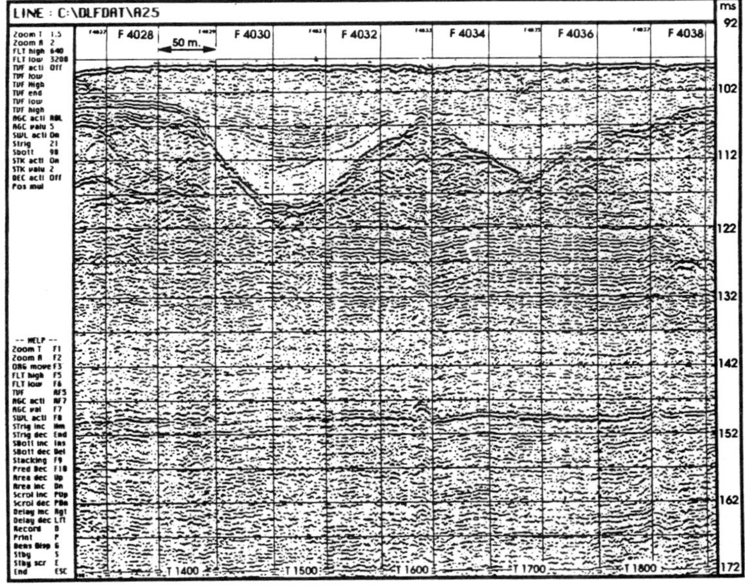

Fig. 10. Swell filtering on (boomer section).

For high resolution data, we consider that the swell filter should be the first item to be applied in the processing sequence. In addition, this program smooths any possible jitter induced by delays in shot triggering.

In all the following examples, dedicated swell filtering is systematically applied.

3.2.2. Band pass filters

The aim of a band pass filter is to eliminate noise and to shape the spectrum of the signal. This is a classical straight forward step in seismic digital processing. The effect on the applied filters is seen clearly on the spectra of Figures 4 and 5.

3.2.3. Gain recovery

DELPH1 offers different gain recovery options:

- *Time variable gain* (TVG); the signal is multiplied by a variable function which increases with respect to time (depth).

- *Decreasing adaptative gain*; this option assumes that the reflected amplitudes decrease with time (i.e. the deeper reflectors produce lower signal amplitudes). This assumption holds true except in the presence of multiple or surface e-choes due to a deep-tow source.

- *Exponential or linearly-adaptative gain*; with this option, a time varying gain is applied to equal length windows along the seismic traces. The size of the windows is user selectable. For each window, a value proportional to the in-verse of the absolute maximum amplitude is calculated and these values along the trace are then smoothed according to an exponential or linear variation.

The selection of the option to be applied depends on the geology and on the intended interpretation.

A TVG will be selected, for example, to observe amplitude anomalies of re-flectors, as shown in Figure 11. The energy absorption due to shallow gas is such that data are highly attenuated under the gas- affected area. A linearly adaptative gain may be chosen in order to improve the continuity of deeper reflectors (see Figure 12).

3.2.4. Time varying filter

Our objective is to get as close as possible to the earth response. To do so, a TVF can be applied. The TVF is used to remove noise outside the signal bandwidth. This improves both resolution and effective penetration by increasing the signal to noise ratio.

Figures 13 and 14 show the same line with and without a TVF applied.

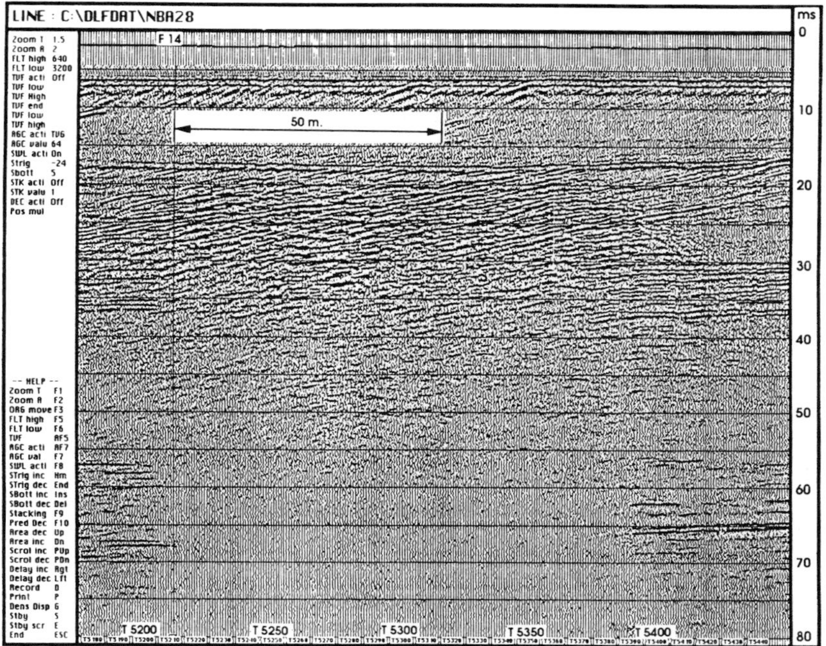

Fig. 11. Gain recovery: TVG.

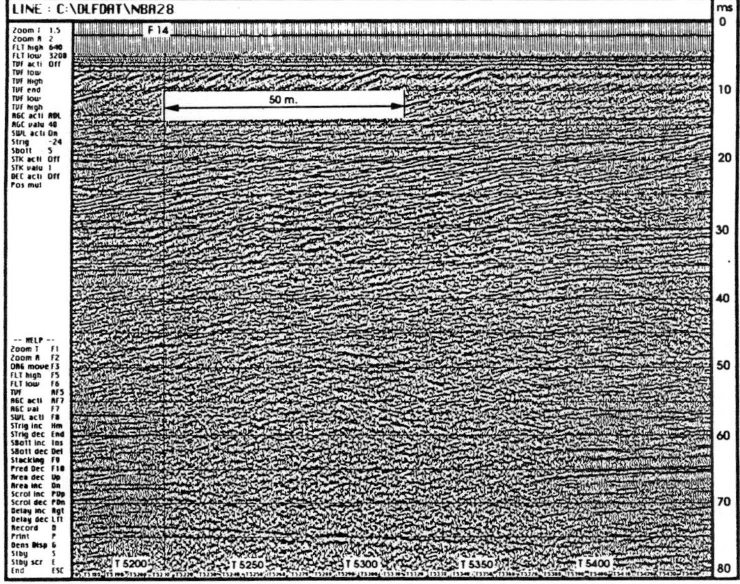

Fig. 12. Gain recovery: linear adaptative.

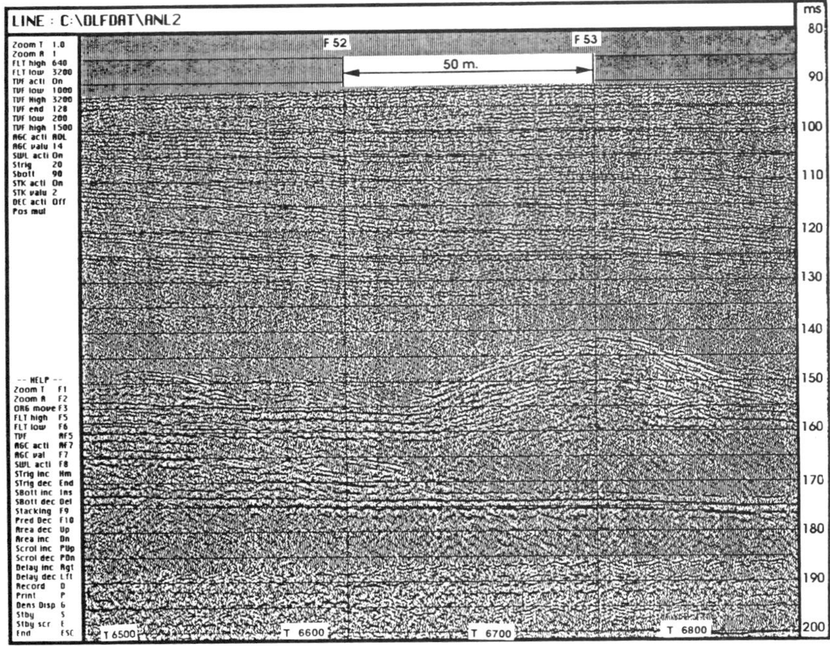

Fig. 13. Time varying filter on.

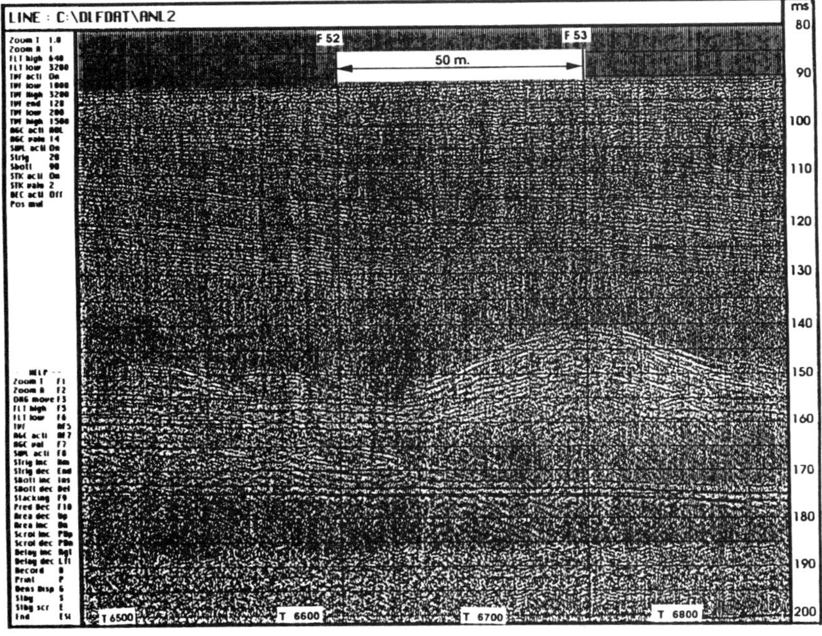

Fig. 14. Time varying filter off.

3.3. Post-Processing

The real-time processing sequence is stored in memory, but does not affect the raw field data which are stored. In other words, the real-time processing sequence can be, if necessary, totally reworked. At this stage three additional processing modules can be applied:

- Horizontal stacking
- Multiple removal
- Spiking deconvolution.

3.3.1. Horizontal stacking

The aim of horizontal stacking is to improve the signal to noise ratio and the continuity of reflectors by summing consecutive shots. The DELPH1 stacking software allows the summation of up to five consecutive shots. This is a tool to be used by the interpreter but care must be taken not to introduce artefacts or a loss of resolution when summing too many consecutive shots.

Figures 15 and 16 show an improvement of signal to random noise when summing two consecutive shots.

3.3.2. Multiple removal

This is achieved by a deterministic deconvolution using the signal measured from the sea-bottom return.

Examples in Figures 17 and 18 clearly show multiple removal and permit discrimination between the real reflector and a multiple.

3.3.3. Spiking deconvolution

Spiking deconvolution compresses the signal and in this application produces a zero phase signal which improves resolution.

3.4. Interpretation

DELPH1 also permits graphic enhancement, horizon picking and data transfer to a workstation if required by the interpreter.

3.4.1. Graphic enhancement

Any possible *scaling* can be performed by a software zoom (which maintains resolution) and the seismic section can be enlarged or reduced in both time and distance (independently of each other, if required). In addition, amplitudes can be adjusted as required and data can be visualized in both variable area and colour-coded variable density modes. The latter represents the amplitude strength of a reflector.

The section (Figures 19 and 20) has been recorded with a 3.5kHz pinger and is presented with two different time scales, allowing a more detailed interpretation.

3.4.2. Screen interpretation/Horizon picking

A reflector can be interactively digitized on the screen using a mouse. The digitized events (time and position) are stored in an ASCII file and can be used by any cartography package.

Although computerized interpretation is common in today's exploration seismic, it is presently a new product for single channel seismic. Another application of the screen interpretation (Figure 21) is to output an interpreted and labelled example quickly in order to illustrate on board interpretation as well as final reporting.

3.4.3. Workstation

DELPH1 recorded data can be played back on a workstation. Hence, all available options of such computers, ranging from trace attributes to horizon picking and cartography can be used.

Figure 22 is an example of data transferred on to a SIERRA workstation.

4. Conclusion

Better control of data acquisition, the ability to replay data with different processing sequences and the choice of different graphical enhancements as required by the interpreter represent important progress for high resolution engineering geophysics surveys.

The possibilities of interactive interpretation and data transfer on to a workstation, as offered by DELPH1, give a new dimension to so-called analogue high resolution seismic.

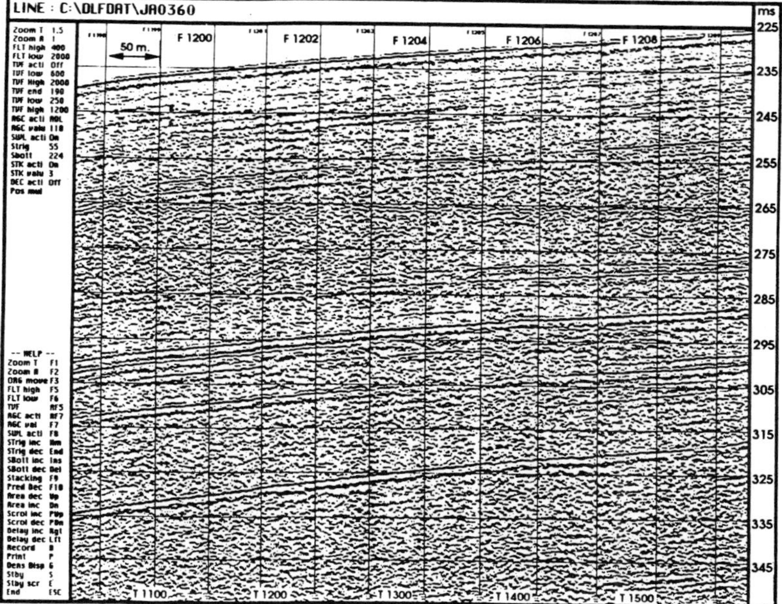

Fig. 15. Horizontal stacking on.

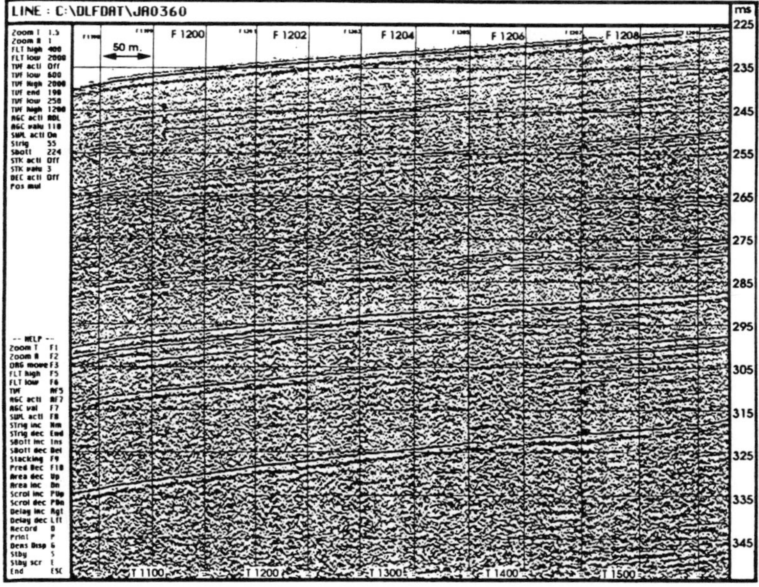

Fig. 16. Horizontal stacking off.

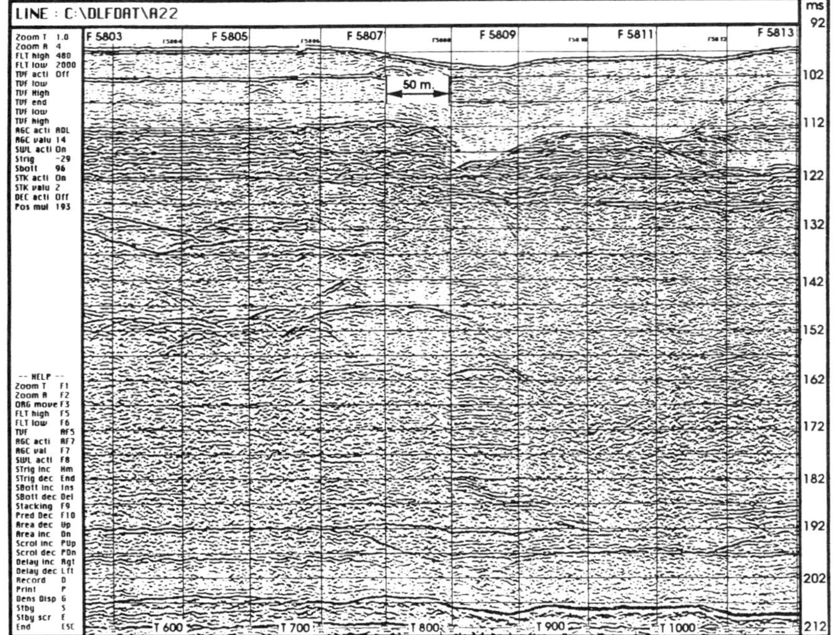

Fig. 17. Multiple removal off.

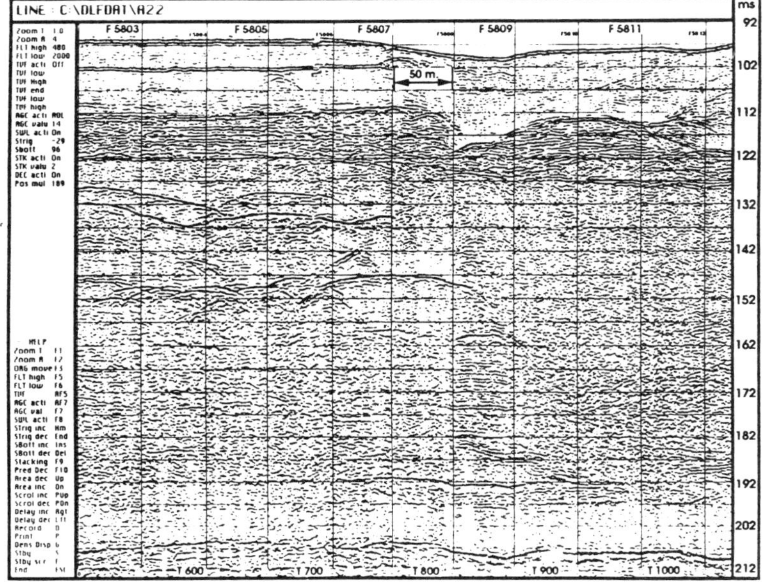

Fig. 18. Multiple removal on.

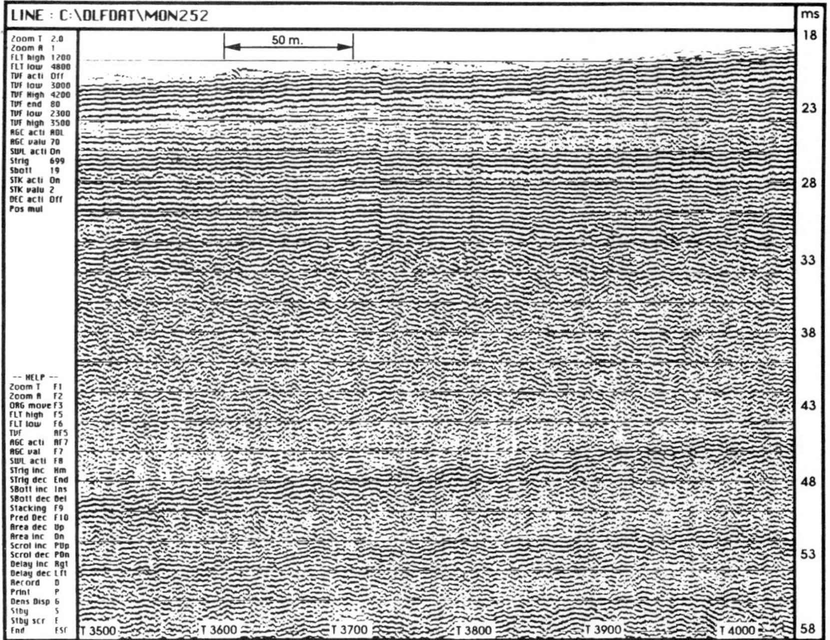

Fig. 19. Pinger record.

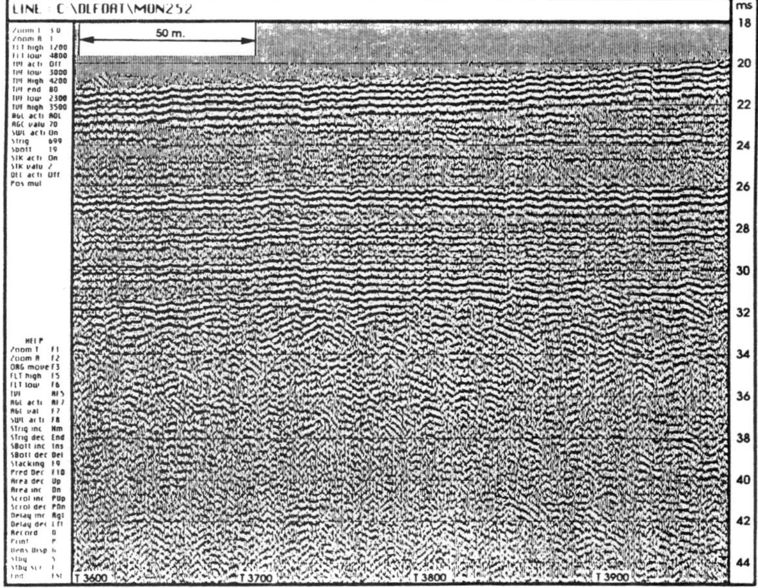

Fig. 20. Same pinger record zoomed.

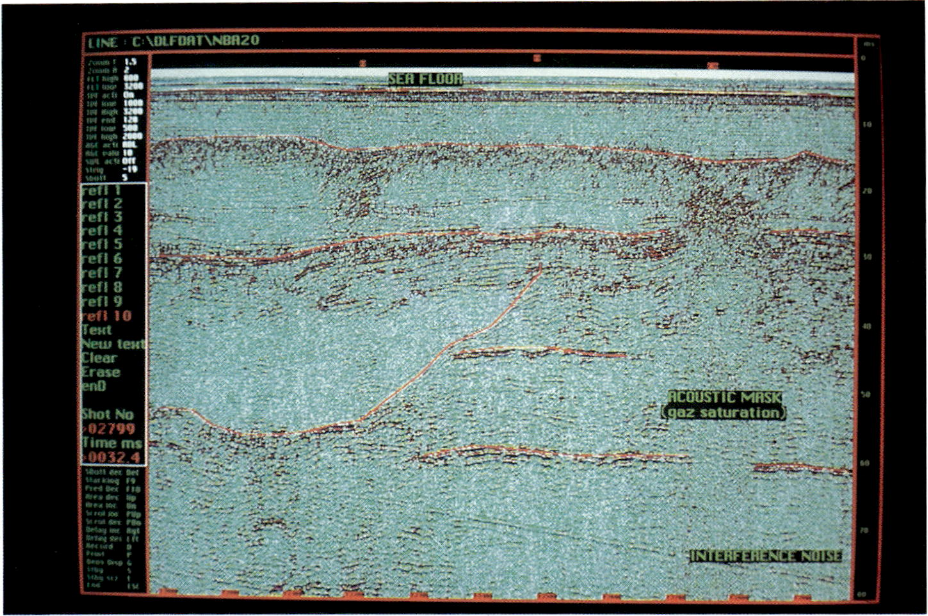

Fig. 21. Screen interpretation.

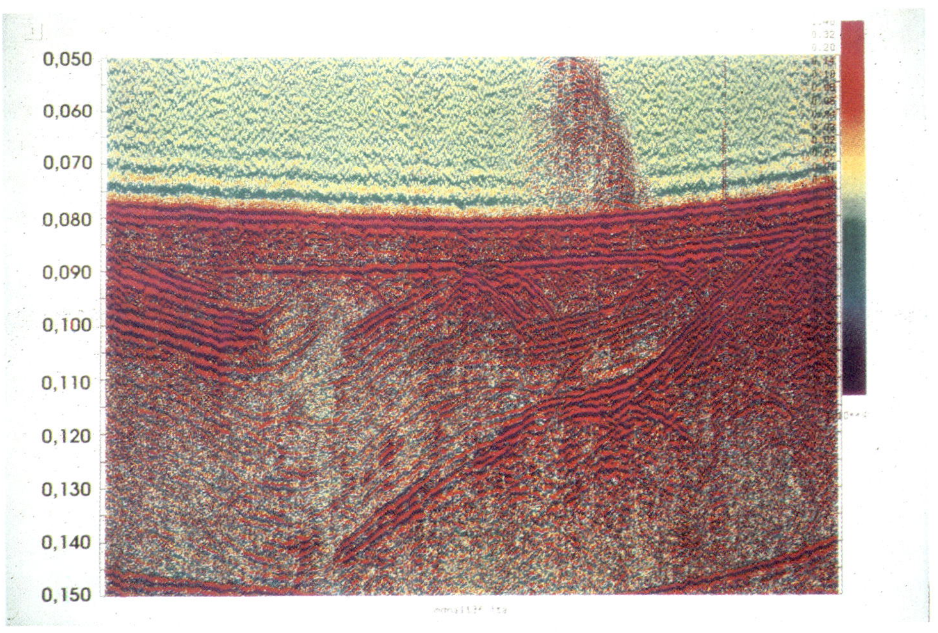

Fig. 22. Workstation display.

THE EXTRACTION OF GEOTECHNICAL INFORMATION FROM HIGH-RESOLUTION SEISMIC REFLECTION DATA

R. HAYNES
School of Ocean Sciences, University College of North Wales, Menai Bridge, Gwynedd, North Wales LL59 5EY and Applied Geology (NW) Ltd., Techbase 3, Newtech Square, Deeside, Clwyd, North Wales CH5 2NT

A. M. DAVIS
School of Ocean Sciences, University College of North Wales, Menai Bridge, Gwynedd, North Wales LL59 5EY

and

J. M. REYNOLDS and D. I. TAYLOR
Applied Geology (NW) Ltd., Techbase 3, Newtech Square, Deeside, Clwyd, North Wales CH5 2NT

Abstract. There are strong empirical and theoretical correlations between the geotechnical and seismic properties of a marine sediment. With the development of modern digital acquisition techniques, and the advent of PC/workstation-based data processing methods it is becoming possible to extract geotechnical information from seismic data in a cost-effective manner.

In a collaborative project between U.C.N.W., Bangor and Applied Geology (NW) Ltd., attempts are being made to develop improved capabilities in high-resolution seismic data acquisition and processing for geotechnical site investigation purposes. The ultimate objective is the creation of a rapid and cost-effective method by which the physical properties of seafloor materials can be determined. The potential users of such a method might include offshore contractors working within the hydrocarbon industry (for rig-site surveys, investigation of pipeline routes, etc.), coastal engineers, dredging companies, and river and harbour authorities.

It is envisaged that via the development of robust computer programmes designed to extract a number of different seismic parameters (velocity, acoustic impedance, attenuation) it will be possible to produce depth profiles (to 20–30 metres below seabed surface) of a marine soil's bulk properties (e.g. density, void ratio, moisture content and grain size). Given this information, it is then theoretically possible to calculate order-of-magnitude estimates of parameters such as shear strength and permeability. Ultimately, with a limited amount of borehole control, it should be possible to improve upon these initial estimates and provide information on the spatial variability of the *in situ* geotechnical properties.

1. Introduction

A joint project between UCNW, Bangor and Applied Geology (NW) Ltd., Deeside, Clwyd, has been established as part of a SERC Teaching Company Scheme. The objective of the project is to develop methods to extract geotechnical information from high-resolution seismo-acoustic data obtained from underway sub-bottom

Volume 28: Offshore Site Investigation and Foundation Behaviour, 215–228, 1993.

profiling systems (for example, boomer, sparker, pinger systems). Such data are collected routinely, usually in the form of analogue paper records but increasingly in the form of digitally-stored records, during offshore site investigation studies.

The ultimate goal of the project is to create a rapid and moderately inexpensive methodology for investigating the physical properties of marine sediments. It is presently envisaged that such a method will be capable of investigating physical properties to depths of the order of 30–50 m and have a resolution of layer thickness to better than 0.5 m. The development of such a methodology offers large potential cost and technical benefits to the offshore industry.

To date, the major emphasis of the project has been on evaluating what is theoretically possible and what is presently technically feasible. This paper is a brief precis of what the authors believe can be expected realistically as regards the extraction of geotechnical information from seismo-acoustic data. The theoretical relationships between the geotechnical and acoustic properties of a marine sediment, and the strong empirical relationships between the two are discussed briefly. The important role of acoustic models is mentioned, before the implications of recent technological advances are discussed.

Recent developments in the processing of seismic data for the hydrocarbon industry and in the processing of basic sonar data have also led to methodologies which can be applied to the extraction of geotechnical information from acoustic data. However, one of the most important technological improvements has been the advent of moderately inexpensive but very powerful computers. This makes it possible (probably for the first time) to process the large amounts for acoustic data required to obtain useful geotechnical information quickly, cheaply and efficiently.

2. Geotechnical Relationships

The most important parameters that control the acoustic response of a marine sediment are given by Stoll (1989) as:

i. porosity,

ii. density,

iii. overburden stress,

iv. degree and type of lithification,

v. grain size and distribution,

vi. dynamic strain amplitude,

vii. material property of grains,

viii. sediment structure.

TABLE 1.
Problems associated with the acquisition of sub-bottom profiler data.

INSTRUMENTAL EFFECTS			
(i)	source		- repeatability
			- directionality
			- bandwidth
(ii)	receiver		- array directionaility
(iii)	acquisition system		- dynamic range
			- sampling rate
ENVIRONMENTAL PROBLEMS			
(i)	noise	- self-induced	- ship noise
			- electrical noise
			- hydrophone turbulence
		- ambient	- cultural
			- natural
(ii)	waves		
GEOLOGICAL EFFECTS			
(i)	multiples		- water-bottom multiples
(ii)	diffractions		
(iii)	apparent attenuation		- peg-leg multiples

From the above list it can be seen that many of the properties that affect the acoustic response are of direct interest to geotechnical engineers (e.g. porosity, density, grain size and distribution). It is this inter-dependence between the acoustic and geotechnical properties of a marine sediment that theoretically allows the extraction of geotechnical information from sub-bottom profiler data.

The problem of deriving geotechnical information from seismo- acoustic data has been taxing geophysicists and engineers for at least the last thirty years and as yet there is no commercially available system that can extract meaningful results consistently. The measurements of seismic properties (e.g. velocity, acoustic impedance and attenuation) is not always straightforward and a number of intrinsic problems need to be addressed and overcome before any meaningful geotechnical parameters can be derived. These intrinsic problems can be split into three categories (Table 1), the first relating to instrumental effects, the second to environmental problems, and the third to geological effects.

When using analogue systems these problems cannot often be overcome. This results in a serious degradation of the quality of the output which renders the data difficult, if not impossible, to interpret adequately. In the light of recent technological advances, particularly in the digital recording of data, many of these problems can be eliminated, or substantially reduced, by careful design of the data acquisition and by subsequent data processing.

Within the literature, a large number of empirical relations exist between the acoustic impedance, velocity and attenuation of acoustic waves and the geotechnical properties of the sediment (e.g. Buchan *et al*, 1972, Hamilton, 1980). These relationships between the three acoustic properties and geotechnical parameters are schematically summarised in Figure 1. These empirical relationships have led to the development of specific hardware e.g. a number of 'sediment classifiers', which are usually modified echo-sounders, which attempt to determine the nature of the seafloor sediment type using acoustic methods.

Looking at one example of the relationships between the acoustic and geotechnical properties of a surficial marine sediment, that between the acoustic impedance and the porosity, there is an excellent linear relationship between the two with little scatter of results (Figure 2). Therefore if the acoustic impedance of a sediment is precisely known the porosity of that sediment can be predicted to within a few percent. If the velocity of the sediment can also be accurately determined the bulk density of the material can be derived from the simple relationship that acoustic impedance is the product of density and velocity. This clearly demonstrates therefore, that several of the bulk properties of the sediment can be readily determined from seismo-acoustic data if the acoustic results can be derived accurately.

The degree of correlation of the relationship between the acoustic and geotechnical parameters can vary widely. For example, in the previous case of the acoustic impedance-porosity relationship there is an excellent degree of correlation. However, in the case of many relationships the degree of correlation may be poor. This is to be expected as the range and variety of marine sediments is large with different chemical and physical processes being dominant in different sediments. There still remains a large amount of research to be undertaken on determining the precise nature of the geo-acoustic relationships. In particular the standard deviations of the relationships needs to be determined so that appropriate errors can be assessed. This will become increasingly important as commercial systems are developed.

In a recent review paper by Anstey (1991) it was intuitively argued that the empirical relationship between porosity and velocity for rocks, extensively used by hydrocarbon geophysicists, was an incidental, rather than a causal effect. These incidental inter-relationships are also likely to be quite common for marine sediments. In the case of the velocity-porosity relationship of rocks it is the fact that the number and length of grain contacts varies that is more important than the change of porosity *per se*. However, the fact that the relationship is incidental does not make the empirical relationship invalid, but such factors should be borne in mind when interpreting any results derived using such formulae. It also explains the large scatter observed for many relationships. For example, the number and length of grain contacts can vary due to a number of interdependent physical and chemical properties.

As previously mentioned, the relationships between the acoustic properties of a surficial sediment and its geotechnical parameters are relatively well known.

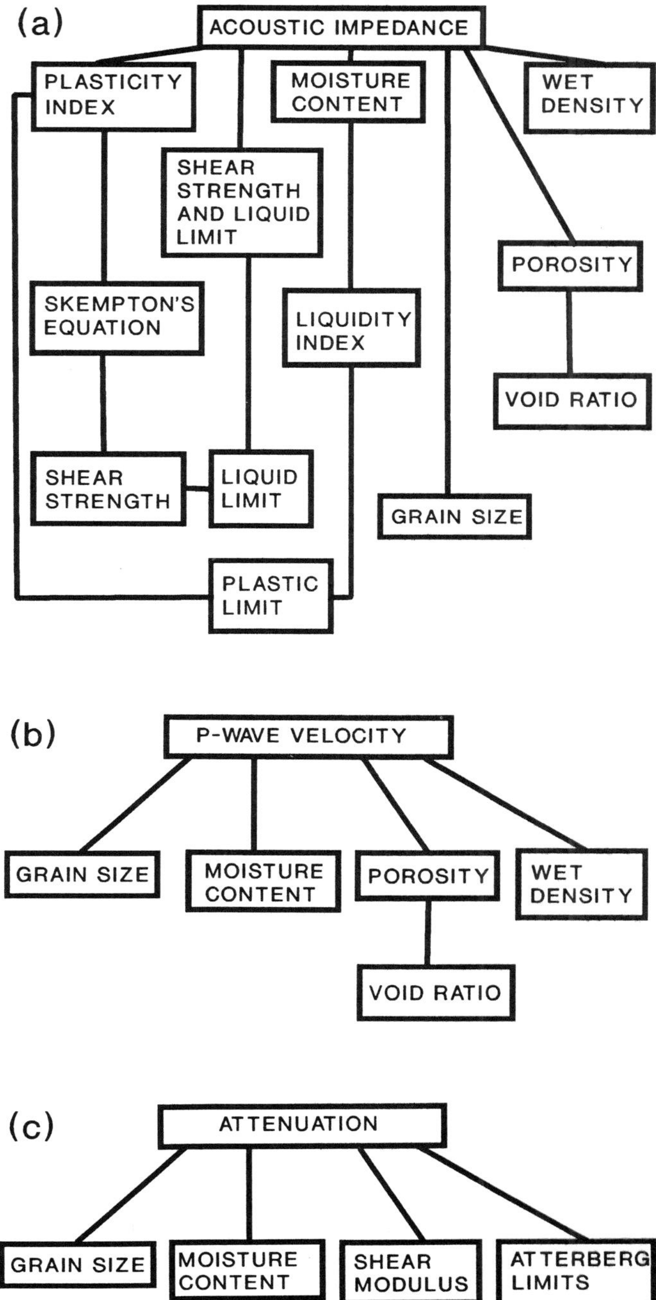

Fig. 1. Schematic diagrams showing the empirical inter-relationships between the geotechnical properties of a marine sediment and its seismo-acoustic properties ((a) acoustic impedance, (b) P-wave velocity, and (c) attentuation).

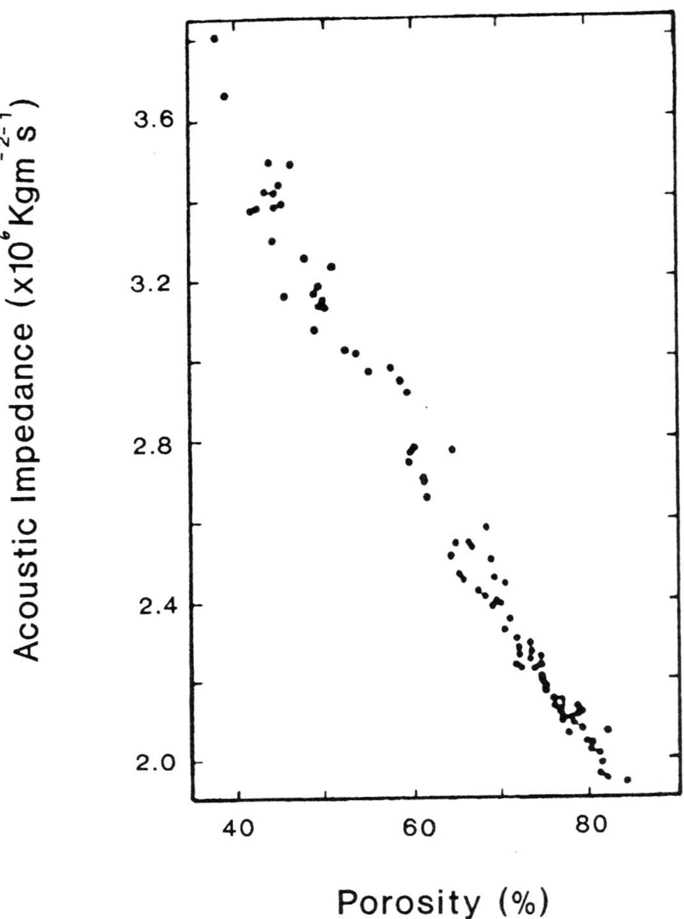

Fig. 2. Relationship between acoustic impedance and porosity (after Hamilton, 1970).

Variations of acoustic properties with depth (or effective stress), however, are not nearly so well understood. It is well established, for instance, that there is an increase in velocity with depth. This is primarily as a result of the increase in sediment frame stiffness (related to the number and length of the grain contacts); such an increase could be due to many factors, e.g. effective stress, chemical bonding, and maybe geological time itself (Taylor Smith, 1986). However, relations do exist (Taylor Smith, 1986) between the porosity (void ratio), velocity and effective stress (depth) such that it is theoretically possibly to estimate the porosity and sediment type given the velocity and effective stress (depth) (Figure 3).

Many similar relations exist between other geotechnical and acoustic parameters. Given recent improvements in the processing of acoustic profiler data, it should be possible to extract estimates of the bulk geotechnical properties of ma-

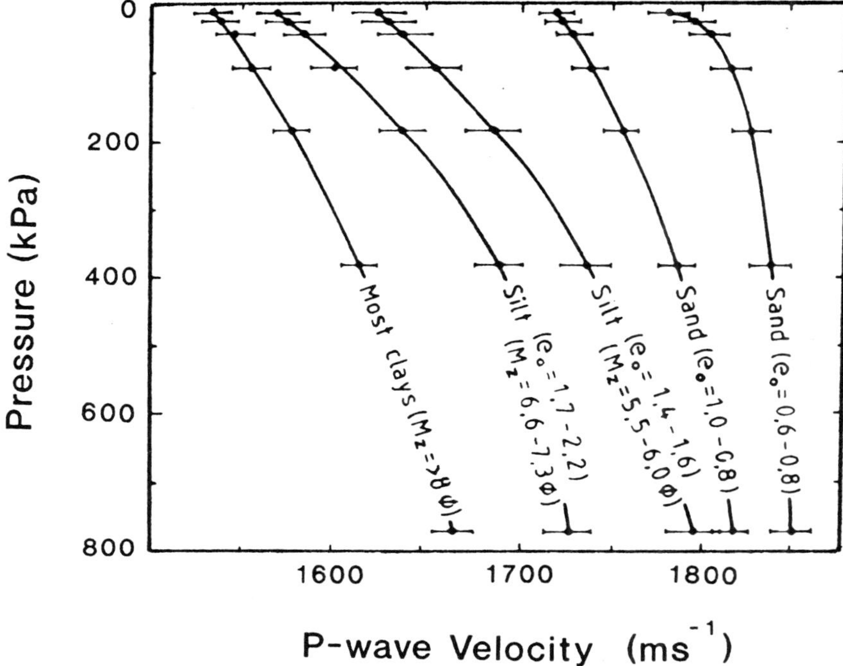

Fig. 3. Compressional wave velocity variations with effective stress and gain size (after Taylor Smith 1986).

rine sediments using data from underway continuous profiling systems.

3. Acoustic Models

The formulation and application of acoustic models of marine sediments, in particular the Biot-Stoll model (Biot, 1962, Stoll, 1989) has led to a much greater theoretical understanding of the inter-relationships between the acoustic and geotechnical parameters. The Biot-Stoll model enables the prediction of the acoustic properties of a sediment to be achieved almost entirely from geotechnical quantities. The model allows the geo-acoustic behaviour of a sediment to be unravelled. In particular, it predicts the velocity and attenuation of both compressional and shear waves.

The use of geo-acoustic models allows the relationships between the velocity and attenuation versus frequency of acoustic waves to be examined. At present there is a paucity of observational data relating to frequency dependence of velocity and attenuation. Potentially this represents one of the greatest sources of error in deriving geotechnical parameters from acoustic data. These errors may arise as many of the empirical relationships connecting geotechnical and acoustic properties are laboratory-derived using frequencies in the 100's kHz – 1 MHz range. It is a matter of debate (Stoll, 1989) whether these relationships, particularly those involving attenuation, can be applied to the frequencies of interest for *in situ* geotechnical site investigations (1–10 kHz). Clearly, more studies need to be carried out to resolve this question and determine the precise frequency-dependent relations between the geotechnical and acoustic properties of marine sediments.

The use of models, such as the Biot-Stoll, represents one potential method to study these relations. A number of authors, notably Stoll (1989) and Ogushwitz (1985a, b, c), have started to address these issues. Their work is leading to a much better theoretical understanding of the acoustic response of marine sediments.

It is not possible at present, however, to use such models to derive the geotechnical parameters as many of the same parameters are required to solve the equations involved to form the model in the first instance. This means that whilst models allow us to study the effects of varying sediment properties on the acoustic response, the user must also supply meaningful geotechnical input to the model to obtain reasonable results. In other words there is a classic "Catch 22" situation if the Biot Stoll model is used as a predictive tool.

4. Technological Advances

Advances in the last few decades in the electronics field have had an immeasurable effect on many aspects of human life in the developed world; the field of geophysics is no exception. It is these technological advances that make it feasible, possibly for the first time, to attain the goal of determining the geotechnical parameters of a marine sediment from continuous acoustic sub-bottom profiling data.

Probably the two most important advances are (i) the development of fast A/D converters which can operate at the rates required for use with sub-bottom profiling systems (i.e. sampling rates well in excess of 10 kHz) (McGee, 1990), and (ii) the advent of relatively inexpensive, powerful computers, which enable the data acquisition, and more importantly, the data processing, to be fast, efficient and cost- effective (Hatton *et al*, 1986).

It should be noted that in nearly every aspect of shallow high- resolution marine seismo-acoustics there have been recent improvements in technology, from source and receiver design, digital data acquisition systems, to output (both hardcopy and digitally-recorded). The cumulative effect of all these advances is to enable high-quality digitally-recorded seismo-acoustic data to be more readily available. This availability of digitally-recorded data then enables subsequent post-processing to be undertaken thus facilitating the derivation of appropriate geotechnical parameters.

5. Data Processing

It is the application of data processing techniques which are already routinely employed in the processing of seismic data in the hydrocarbon exploration industry that will allow the geophysicist operating in the marine site investigation industry to produce acoustic data in a form suitable for the subsequent derivation of geotechnical parameters. The three data processing procedures of most interest from this point of view are (i) velocity analysis, (ii) seismic inversion to determine acoustic impedance, and (iii) estimation/determination of attenuation.

All three data processing procedures are becoming increasingly sophisticated as the power of computers increases. In the case of velocity analysis, recent high-resolution techniques have emerged from the field of multi-channel sonar array processing which can be readily applied to seismic reflection data (Key and Smithson, 1990). These techniques provide results superior to the widely used semblance measure (Hatton *et al*, 1986) and offer a degree of resolution not previously attainable. This improved degree of resolution will be required if meaningful estimates of geotechnical parameters are to be made from the velocity determinations.

The recovery of acoustic impedance data from high-resolution reflection records is potentially the most rewarding and productive avenue of approach in determining the geotechnical parameters of a marine sediment. This requires that the 'seismic inverse problem' is solved. A large number of methods have and are being developed to solve this 'inverse' problem (Oldenburg *et al*, 1983; Berteussen and Ursin, 1983; Walker and Ulrych, 1983; Caulfield and Yim 1983) and it is presently an active area of research in the hydrocarbon exploration industry due to its potential value in reservoir studies.

The problem of determining attenuation from seismo-acoustic data is probably a much more exacting problem than the other two and it is likely that a number of methodologies will need to be developed to ascertain the most reliable techniques. Indeed it may prove to be the case that accurate determination of attenuation will be difficult to achieve *in situ* and only relative estimates will be able to made (Janssen *et al*, 1985; Tarif and Bourbie, 1987; Tonn, 1991). However, even given estimates it should be possible to gain some useful information on the grain size of marine sediments (Hamilton, 1972).

6. Practical Aspects

The approach that the authors are pursuing towards the ultimate goal of this project, namely the extraction of useful geotechnical information from seismo-acoustic data obtained during continuous seismic profiling site investigation surveys, can be summarised as follows.

6.1. Data Acquisition

The acquisition of high-quality digitally-recorded data is the basis on which the whole success, or failure, of the project will depend. Therefore, a lot of effort has been placed in these early stages of the project on designing optimum acquisition procedures. The source should have the broadest frequency bandwidth that the required depth of penetration to the investigation target will allow. If the depth of interest is less than 30–50 m experiments undertaken in Lake Windermere, in conjunction with the Universities of Utrecht (Holland) and Århus (Denmark) and the British Geological Survey, indicate that a 'boomer'-type system is presently the best commercially available source type (McGee, 1992). The hydrophone streamer should be suitable for high-resolution work with a group length of the order of 1 m. There are presently few commercially available streamers that are suitable. Similarly, and probably most importantly, there are very few commercially available digital data acquisition systems that are suitable for very high-resolution work. Ideally such a system should have 10–12 channels capable of recording at sampling rates in excess of 25 kHz (40 μs) with a dynamic range of at least 16 bit. At present such systems exist in research institutes only but commercial systems approaching this are appearing (e.g. GeoAcoustics Sonar Enhancement System and Elics – Delph 1). The system presently used at UCNW, Bangor, is a Carrack SAQ-V which is capable of sampling up to twelve channels with a total throughput of 48 kHz (i.e. 3 channels at 16 kHz, 12 channels at 4 kHz, etc.). This is a PC-based system which can be installed on small boats making it suitable for very shallow water work. An example of the quality of the data that can be acquired using this system is given in Figure 4, this shows that in ideal conditions it is possible to resolve reflectors to better than 20 cm down to sediment depths in excess of 15 m.

6.2. Data Processing

There are two avenues of approach to the processing of seismo-acoustic data for the extraction of geotechnical information. In the first, the data are digitally recorded and then post-processed at a later date. In the second, the data are processed in real-time using the acquisition computer. At UCNW the approach presently favoured is the first, the data being digitally recorded on magnetic tape and then processed using a commercial data processing package. The processing package used is the Sierra Geophysics ISX/SierraSeis system which is run on a Sun IPX workstation. The Sierra ISX system permits rapid processing of the acoustic data into a form suitable for the derivation of the geotechnical information. The type of processing that can be done includes: deconvolution, filtering, multiple suppression, velocity analysis, and pseudo-acoustic impedance inversion. Computer modules are presently being written that will derive geotechnical parameters from the processed seismo-acoustic data. It is possible that if some optimum processing parameters can be determined then some processing can be undertaken in real time, especially if multi-tasking computers are employed as acquisition systems.

7. Future Developments

From the above discussions it should be apparent that in the future there are likely to be:

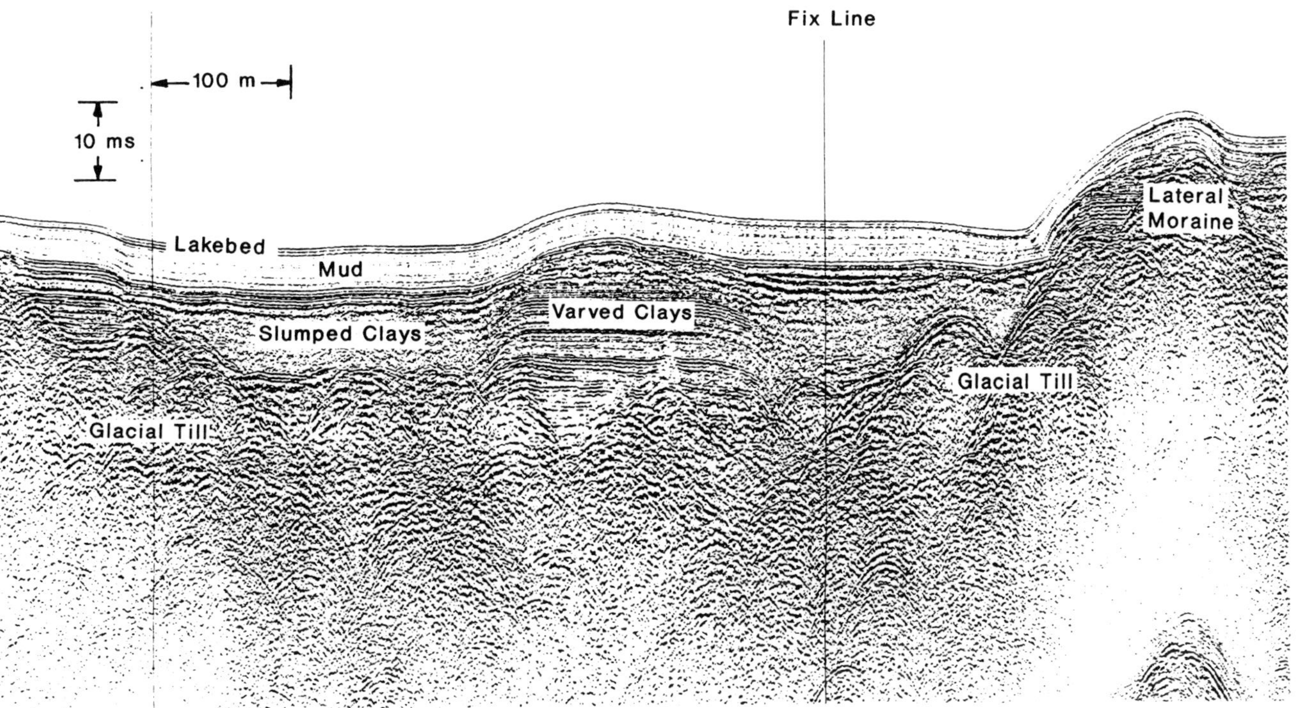

Fig. 4. An example of high-resolution seismic data obtained from Lake Windermere using a 'boomer'-type source. The actual source being an O.R.E. "Geopulse" Model 5810B, operated at 280 J with a trigger cycle length of 0.5 s, and a sweep length of 150 ms. The signals were received by an E.G.&G. Model 265 hydrophone streamer and recorded using a Dowty Waverley Model 3700 thermal linescan recorder.

TABLE 2.
Potential Benefits for offshore industries.

1. RECONNAISSANCE USE
- rapid surveying of large areas
- gross interpretation of bulk properties
- improved data quality leads to identification of difficult areas ⇒ better placement of boreholes
2. DETAILED SURVEY
- interpolation between boreholes leads to better geotechnical design
- identification of areas of rapid lateral variation
LEADS TO:
(i) cost savings due to a possible reduction in the number of boreholes required and/or increased cost-effectiveness that a given number of boreholes are located with maximum benefit
(ii) increased confidence in geological/geotechnical interpretations
(iii) cost savings due to more appropriate and efficient geotechnical designs being produced

a. Continued technical improvements especially in the fields of data processing techniques and data acquisition hardware. (Particularly if the defence industry under the changed political climate decides, or is forced, to make available for general use some of its acoustic expertise.)

b. An improved theoretical understanding of the acoustic behaviour of marine sediments and the inter-relationships with the geotechnical properties. This should come about from laboratory experiments and by the use of acoustic models.

c. The integration into intelligent computer systems of a number of different seismo-acoustic processing techniques (e.g. velocity determination, acoustic impedance analysis, and attenuation determination) to determine the geotechnical properties of a marine sediment. The combination of a number of techniques should enable much better estimates of the geotechnical parameters to be made.

These technological improvements should enable the derivation of geotechnical properties of marine sediments remotely using sub-bottom profiler data. Such a methodology offers large potential benefits for the offshore site investigation industry (Table 2).

8. Conclusions

It can be seen that the extraction of geotechnical parameters from seismo-acoustic data is possible theoretically, and with recent technological advances is, or is becoming, practically feasible. These recent improvements now make it possible to digitally record sub-bottom profiler data with enough signal fidelity and at the required sampling rates to enable digital seismic data processing techniques to be subsequently applied. The data processing of the seismo-acoustic data should then produce data of sufficiently high quality to enable the many empirical relationships that exits between geotechnical and acoustic properties to be used to extract geotechnical parameters. It is envisaged that it will be possible to produce depth profiles (to 20–30 metres below seabed) of a marine sediment's bulk properties (e.g. density, void ratio, moisture content and grain size). Given this information it is then theoretically possible to calculate order-of- magnitude estimates of parameters such as shear strength and permeability. Ultimately, with a limited amount of borehole control, it should be possible to improve upon these initial estimates and provide information on the spatial variability of the *in situ* geotechnical properties.

With the advent of powerful, but moderately inexpensive, computers it should be possible to extract this geotechnical information from seismo-acoustic data in a cost-effective manner. The derivation of geotechnical properties remotely and rapidly from sub-bottom profiler data offers large potential technical and cost benefits to the offshore site investigation industry.

Acknowledgements

This project is jointly funded by the SERC/DTI Teaching Company Scheme and Applied Geology (NW) Ltd. D. Taylor Smith and J. Bennell provided useful comments on various aspects of this work.

References

1. Anstey, N. A. (1991), 'Velocity in thin section', *First Break* **9**, 449–457.
2. Berteussen, K. A. and Ursin, B. (1983), 'Approximate computation of the acoustic impedance from seismic data', *Geophysics* **48**, 1351–1358.
3. Biot, M. A. (1962), 'Mechanics of deformation and acoustic propagation in porous dissipative media', *Journal of Applied Physics* **33**, 1482–1498.
4. Buchan, S., McCann, D. M., and Taylor Smith, D. (1972), 'Relations between the acoustic and geotechnical properties of marine sediments', *Quarterly Journal of Engineering Geology* **5**, 265–284.
5. Caulfield, D. D. and Yim, Y.-C. (1983), 'Prediction of shallow sub- bottom sediment acoustic impedance while estimating absorption and other losses', *Journal of the Canadian Society for Exploration Geophysicists* **19**, 44–50.
6. Hamilton, E. L. (1972), 'Compressional-wave attenuation in marine sediments', *Geophysics* **37**, 620–646.
7. Hamilton, E. L. (1980), 'Geoacoustic modelling of the sea floor', *Journal of the Acoustic Society of America* **68**, 1313–1340.
8. Hatton, L., Worthington, M. H., and Makin, J. (1986), *Seismic Data Processing*, Blackwell Scientific Publications, Oxford.

9. Janssen, D., Voss, J., and Theilen, F. (1985), 'Comparison of methods to determine Q in shallow marine sediments from vertical reflection seismograms', *Geophysical Prospecting* **33**, 479–497.
10. Key, S. C. and Smithson, S. B. (1990), 'New approach to seismic-reflection event detection and velocity determination', *Geophysics* **55**, 1057–1069.
11. McGee, T. M. (1990), 'The use of marine seismic profiling for environmental assessment', *Geophysical Prospecting* **38**, 861– 880.
12. McGee, T., Davis, A., Anderson, H., and Verbeek, N. (1992), 'High-resolution marine seismic source signatures', in *European Conference on Underwater Acoustics*, M. Weydert (ed.), Elsevier, pp. 639–643.
13. Ogushwitz (1985a, b, c), 'Applicability of the Biot Theory. I. Low-porosity materials, II. Suspensions, III. Wave speeds versus depth in marine sediments', *Journal of the Acoustical Society of America* **77**, 429–464.
14. Oldenburg, D. W., Scheuer, T., and Levy, S. (1983), 'Recovery of the acoustic impedance from reflection seismograms', *Geophysics* **48**, 1318–1337.
15. Stoll, R. D. (1989), *Sediment Acoustics* **26**, Lecture Notes in Earth Sciences, Springer-Verlag, Berlin, pp. 155.
16. Tarif, R. and Bourbie, T. (1987), 'Experimental comparison between spectral rise and rise time techniques for attenuation measurements', *Geophysical Prospecting* **35**, 668–680.
17. Taylor Smith, D. (1986), 'Geotechnical characteristics of the sea bed related to seismo-acoustics', in *Ocean Seismo-Acoustics*, T. Akal and J. M. Berkson (eds.), Plenum Press, New York, pp. 483–500.
18. Tonn, R. (1991), 'The determination of the seismic quality factor Q from VSP data: A comparison of different computational methods', *Geophysical Prospecting* **39**, 1–27.
19. Walker, C. and Ulrych, T. J. (1983), 'Autoregressive recovery of the acoustic impedance', *Geophysics* **48**, 1338–1350.

WORKSTATION DATA INTEGRATION TECHNIQUES FOR OFFSHORE SITE INVESTIGATION

J. P. WILLIAMS
Hydrosearch Associates Limited, Chandler House, Anchor Hill, Knaphill, Woking, Surrey GU21 2NL

1. Introduction

Over the last five years a significant amount of effort has been directed towards the use of workstations in conjunction with engineering seismic data for the detection of shallow gas accumulations. In the main, this effort has tended to focus on shallow gas as a drilling hazard and development work has concentrated on improving techniques for detecting shallow gas accumulations in the depth range to approximately 1000m below seabed. Overpressurised gas in this interval can be a major hazard to safe drilling operations and there are a number of cases, both in the North Sea and worldwide, where such accumulations have been encountered unexpectedly with severe consequences.

Recent work in this area has utilised digital multi-channel seismic data acquired during site survey operations. It has concentrated on the use of seismic attribute and amplitude vs offset (AVO) analysis to provide a better prediction of whether or not anomalies seen on the seismic data are due to the presence of gas. The enabling mechanism for this move towards more quantitative data analysis has been the availability of workstations, making it possible to undertake such analysis work efficiently and in a timescale which is compatible with site survey requirements. More recent developments have shown the value of forward modelling and to a lesser extent inversion when used in conjunction with the above techniques as an integral part of the overall drilling hazard detection scheme.

The above work, whilst of significant importance has, to date, been of little direct benefit to the engineering geophysicist and geotechnical engineer concerned with offshore foundation design. The methods developed for shallow gas detection for drilling purposes are available, and in many cases applicable, to detecting gassified sediments or lithological features of relevance for foundation design. However, until recently, the geophysical data appropriate for use in foundation studies was recorded in analogue form primarily on paper records. The fixed display scales,

Volume 28: Offshore Site Investigation and Foundation Behaviour, 229–235, 1993.

limited dynamic range and general lack of flexibility in this form of data display placed fundamental limitations on the amount of true 'analysis' which can be undertaken on data of this type. Although good use of such data has been made by integrating engineering geophysical data with geotechnical borehole information, such correlation and integration has been essentially qualitative with only minimal use having been made of the computer workstation facilities currently available.

However, the advent of digital recording of single channel engineering geophysical data coupled with recent advances in data scanning technology has now made it possible to undertake a much greater level of analysis of such 'analogue' geophysical data. This will inevitably be highly beneficial for offshore foundation studies in general and provides a range of possibilities in the analysis and integration of offshore site investigation data.

2. Integration Techniques – Origins and Future Trends

The majority of the seismic methods outlined in the previous section were initially developed to assist explorationists in the exploration for and development of hydrocarbon accumulations. Good quality seismic data and suites of downhole logs run in exploration wells are standard exploration tools which allow detailed analysis and integration of geophysical and geological information.

In this context the 'integration' is undertaken in a quantitative manner and at a considerably finer level of detail than that currently associated with offshore site investigation programmes. In the latter, correlating seismic reflectors with the boundaries of significant geotechnical units and soil province mapping are two of the key activities in achieving a clear understanding of the foundation zone.

For explorationists, however, integration of well logs, surface seismic data and downhole seismic (VSP) can be undertaken at a level of detail which extracts the maximum amount of information from the seismic data. This is made possible by the availability of downhole electric log data, and possibly VSP data, and the use of such data in generating synthetic seismograms. In this context, the terms 'synthetic seismogram' and forward modelling are inter-changeable. In addition to these techniques of detailed integration, petrophysicists can undertake quantitative analysis of the log suite to determine the values of, or at least place limits on, such parameters as formation water saturation, porosity, permeability, etc., which form key inputs to subsequent calculation of hydrocarbon reserves. Consequently, it must be recognised that the incentive to acquire comprehensive suites of log data in exploration or development wells is significantly greater at the present time than it is in the case of geotechnical boreholes.

Thus, in the exploration sphere 'integration' involves a range of different data sets and, equally importantly, a cross-disciplinary or team effort involving geophysicists, geologists, petrophysicists and reservoir engineers. Whilst this approach has been adopted since the earliest days of hydrocarbon exploration, it is only relatively recently, in the last ten years or so, that the availability of workstation technology has enabled the whole integration process to be streamlined. Although manual correlation of synthetic seismograms with exploration seismic data is still a key element in this process an increasing amount of effort is now being put into the development of analysis software and common data exchange formats. This will ensure that the 'Geoscience Workstation' utilised by a cross-section of explorationists accessing common but multi-disciplinary databanks will be the future platform for this form of data integration.

3. Shallow Engineering Equivalents

So where are the parallels and differences in the way in which offshore foundation design projects are undertaken? There are clearly a number of parallels – in particular in the manner in which the geophysical data are acquired and if appropriate, processed.

It is clear that the engineering geophysicists has benefited considerably from the technological advances which have taken place on the data acquisition side of exploration. Improved seismic source performance and reliability, shorter streamer group lengths coupled with an increase in number of channels available, plus a move towards software-controlled and hence more flexible recording systems have all contributed to improvements in data quality. Although there are exceptions, these advances have generally been developed and proven in the exploration arena before being taken up or adapted for site survey purposes. Similarly, the use of portable or 'in-field' processing systems were originally proven in the exploration sphere. More recently such systems, both multi-channel and single channel, have seen increased use in engineering geophysics.

The same is true with geophysical interpretation workstations. Here the requirement was created by the need for the exploration interpreter to be able to store, access, manipulate and interpret large volumes of seismic data in an efficient manner. The requirement was there when working with large 2D data sets. However, with the upsurge in 3D seismic acquisition in the early 1980's it became mandatory to have access to a suitable workstation if the full benefits of the 3D techniques were to be realised at the data interpretation stage. These requirements resulted in what can only be described as spectacular advances in the facilities and capabilities of interpretation workstations.

This was coupled with equally spectacular reductions in the costs of such systems. The cost of one of the leading interpretation workstations is now approximately one third of the cost of its slower, bulkier and more basic predecessor which was being marketed some eight years ago. Much of this can of course be attributed to advances in computer technology – in particular processor capabilities and large screen graphics. However, geophysicists can take much of the credit for identifying the solution to their needs and harnessing and tailoring the available technology to meet these requirements. Today, interpretation workstations are an integral part of the exploration business with well over 1000 systems or 'seats' operating worldwide.

In contrast, the number of workstations currently being utilised on a regular basis for evaluation of engineering geophysical data and integration with geotechnical information is considerably less than the above figure – probably by some two orders of magnitude. There are a number of reasons for this, the principal ones being:

- Availability of Geophysical Data – until relatively recently much of the geophysical data utilised for foundation zone evaluation was available in analogue form only.

- Availability of Geotechnical Data – limited in content and not available in a form immediately suitable for workstation integration at a detailed level.

- Cost Justification – not an easy task in an environment as volatile as the site survey sector of the industry.

- Data Volume Considerations – even if foundation zone geophysical data were available the data volumes were in some cases so large as to be prohibitive.

- Personnel Skills – although training can overcome this constraint, only a limited number of personnel are available at the present time who are conversant with the vagaries of data loading, interpretation, file handling and workstation operating systems.

Thus, although recent advances have reduced or in some cases removed the impact of some of the above there are still some fundamental differences between the use of workstations for exploration and foundation zone purposes. Perhaps the major difference is that the range and semi-continuous nature of downhole data which forms such a key component in the explorationists' data integration and data analysis scheme is rarely available to geophysicists and foundation engineers engaged on foundation design projects. Where borehole data are available this information is rarely in a form which has well-defined or readily generated relationships with the corresponding geophysical data. This contrasts with the situation with density and compressional wave sonic logs which form the basis of conventional synthetic seismogram generation. In foundation zone boreholes however, even if logging tools were to be used routinely there may be difficulties in obtaining reliable log data due to the unconsolidated nature of the soils.

Therefore, whilst the workstation may be a valuable tool for engineering geophysicists in that it enables them to interpret and map reflectors efficiently making use of attribute displays, optimum colour palettes, display scales, etc., the usefulness of the system for integration purposes is at present curtailed by the limitations on the availability of suitable geotechnical data. This poses two problems – firstly, techniques and systems would need to be developed to provide such data in a cost effective manner and on a routine basis. Secondly, a set of theoretical or empirical relationships would have to be devised to provide the key cross-disciplinary linkages between meaningful geophysical and geotechnical parameters. Whilst both these are feasible to a certain degree and are being progressed in some areas, the key question which needs to be addressed is – would such advances provide sufficient benefits to offshore foundation engineering projects to justify the costs and effort required to develop the technology?

To address this issue in the correct manner will require detailed cross-disciplinary discussions between engineering geophysicists and geotechnical engineers in order to define the capabilities and limitations of individual techniques and to agree the ultimate requirement and output from the integration process. The commercial aspects of both the geophysical and geotechnical phases of an offshore foundation study and the commercial benefits which could accrue from further development of advanced techniques would also need to be considered at this stage.

In summary therefore, a significant proportion of the technology required to develop a sound integration approach particularly with regard to the geophysical aspects, is available but the methodology of integration needs considerable effort and further development if it is to realise its full potential.

4. Use of the Workstation for Data Analysis and Integration

Returning now to the issue of the requirement for such advances. We have only to look at the key role which such detailed data integration techniques play in exploration projects to realise that, on the surface, there are significant further benefits to be gained in developing a similar form of integration methodology for foundation study projects.

Whilst a workstation is not essential in order to undertake a successful geophysical/geotechnical integration project, the flexibility in such basic areas as data access and display which such a system can provide offers significant benefits.

The requirement of a foundation zone 'integration' study may be to obtain a gross correlation between geotechnically significant boundaries marking the top and base of units which are perhaps several metres or even tens of metres thick. Alternately, it may be to identify major features such as infilled channels. Under such circumstances, a workstation, whilst useful, is not essential. However, if the requirement is to 'integrate' the data sets at a level of detail and resolution close to that contained in the geotechnical data set then a workstation may prove to be essential.

The ability of the seismic method to resolve a feature is primarily a function of the frequency content of the wavelet irradiating the feature, the dimensions of the feature and its impedance contrast with the surrounding material. In order to maximise resolution and extract the maximum amount of information from a seismic image then, as minimum, flexibility in display scales and image form (dynamic range attribute options and colour palettes) are required. If 'thin bed' effects are present and resolution below that provided by the dominant frequency in the wavelet is desirable then some form of modelling or synthetic seismogram generation will be required – both to determine unit thickness and to ensure correct interpretation of any attribute analyses.

As indicated above, explorationists use the theoretical relationships between sonic and density wireline or MWD log data and acoustic impedance to construct a reflectivity series which can form the basis for generating synthetic seismograms. In some cases, for example through the top hole section of a well where the sonic data may be unreliable, resistivity data in conjunction with parameter cross-plots may be used to synthesise a pseudo-sonic response which can then be used in subsequent modelling work. The key requirement is to have the ability to generate a theoretical or empirical relationship which can be used to integrate geological and geophysical information. This is the stage to which geophysical/geotechnical relationships need to be developed if integration of these data sets is to be achieved at a fine level of detail.

If only a limited amount of geotechnical data is available, an alternative approach might be to apply all available analysis techniques to the geophysical data. The objective of this would be to 'type' sedimentary units by their seismic response using techniques such as pattern recognition, cluster analysis and 'soft' inversion to identify key parameters which characterise individual units and thus allow their vertical and lateral extents to be determined. Integration with geotechnical data could then take the form of ground truthing using the limited amount of geotechnical information available.

5. Conclusions

Within the UK site survey industry, the role of the geophysical workstation as a tool to both improve efficiency of interpretation and increase the amount of the information which can be extracted from geophysical data is now firmly established. Although such systems have for a number of years been utilised for drilling hazard analysis purposes, to date they have seen only limited use in foundation investigation projects. Now that an increasing amount of single channel high resolution geophysical data is being acquired in digital form, in a format suitable for loading to workstations, more use can be made of the data analysis and data integration facilities available on these workstations.

When considering integration purely from a geophysical viewpoint, workstations enable different sets of geophysical data acquired over the same site to be displayed, manipulated and interpreted on a common system, thus overcoming some of the limitations encountered in correlating events between hard copy displays having differing scales and dynamic range characteristics.

The use of workstations also provides the option for closer integration between data processing and interpretation phases of a project, as a comprehensive suite of processing algorithms can be installed on a number of the workstations which are currently available. Thus, the distinction between interpreter and processor is removed and the combined process becomes better integrated and more interactive.

In order to achieve satisfactory levels of integration between geophysical and geotechnical data, then both the amount and format of the geotechnical data which is routinely available needs further careful consideration. It is clear that this aspect can best be progressed by cross-disciplinary initiatives which address both the current limitations of the detailed integration process and also quantify the technical and commercial benefits which might be expected if development effort was to be focused on this area.

In terms of future potential for workstation-based projects with both geophysical and geotechnical data available on a common system, there is considerable scope for a more automated approach to the interpretation, analysis and integration of both sets of data. This methodology could be developed specifically for foundation zone studies or, alternatively, could draw on established techniques such as pattern recognition, feature analysis and cluster analysis which could be tailored to meet the particular requirements of the foundation engineer.

Finally, as an increasing amount of workstation data becomes available for detailed analysis and subsequent integration studies, the topic of data recording media, and in particular data recording and archive formats, will need to be addressed. As has been shown in the exploration sphere, adoption of agreed data exchange media and formats will help to ensure compatibility at the recording, analysis, storage and retrieval stages of projects, thereby increasing efficiency and enabling optimum use to be made of the acquired data over an extended period of time.

SESSION 4

INTEGRATED INTERPRETATIONS

PROBLEMS ASSOCIATED WITH SEISMIC FACIES ANALYSIS OF QUATERNARY SEDIMENTS ON THE NORTHERN UK CONTINENTAL MARGIN

M. S. STOKER
British Geological Survey, Murchison House, West Mains Road, Edinburgh EH9 3LA, Scotland

F. S. STEWART
GEOTEAM UK Ltd., Regent House, Regent Quay, Aberdeen, AB1 2BE, Scotland

M. A. PAUL
Department of Civil and Offshore Engineering, Heriot-Watt University, Edinburgh EH14 4AS, Scotland

and

D. LONG
British Geological Survey, Murchison House, West Mains Road, Edinburgh EH9 3LA, Scotland

Abstract. Seismic facies analysis is increasingly being used in the interpretation of high-resolution seismic reflection data. Existing depositional systems models, based primarily on seismic data, have been constructed on the basis that acoustic character can be directly correlated with relatively consistent lithologies and geotechnical properties. However, borehole and shallow core data from the northern UK continental margin indicate significant inconsistencies in the use of this technique as a predictive tool. Changes in acoustic texture can occur laterally and vertically, in both lithologically homogeneous and heterogeneous sequences. The successful application of this technique can only be achieved by the integration of seismic data with other subsurface information, with interpretation based on sound geological concepts and models.

1. Introduction

The interpretation of seismic facies in terms of specific lithological and geotechnical characteristics is a problem of very general interest. There is a common expectation that laterally persistent, acoustically homogeneous seismic units are associated with consistent lithologies and geotechnical properties. Conversely, it is generally supposed that lateral or vertical changes in seismic texture necessarily indicate a change of sediment type or character. Although these expectations may be fulfilled in some locations, we believe that they are not infallible, and that reliance on seismic reflection data alone without adequate groundtruth can be shown in certain areas to lead to an unsatisfactory interpretation of the sedimentary succession. For example, much of the Quaternary sediment on the east Canadian and mid-

Volume 28: Offshore Site Investigation and Foundation Behaviour, 239–262, 1993.

Norwegian continental shelves has been interpreted as till deposits, by assuming that all acoustically unstratified units represent till (King and Fader, 1986; King *et al.*, 1987, 1991). Yet borehole data have proven that acoustically similar deposits on the northern UK continental shelf contain glaciomarine sediments, including debris flows and turbidites (Stewart and Stoker, 1990).

The aim of this paper is to demonstrate the variable relationship between acoustic texture and lithology, using examples from the Quaternary succession of the northern UK continental margin (Figure 1). In order to demonstrate this variability, we have focused primarily on acoustically unstratified and acoustically layered deposits from different depositional settings across the margin. These range from locations in the North Sea and on the Hebrides and West Shetland shelves, where sediment deposition has been directly affected by glacial activity, to the adjacent slopes and deep-water areas of the Rockall Trough and Faeroe-Shetland Channel, where glacially-derived, reworked material is interbedded with marine sediments. The study combines high-resolution airgun, sparker and deep-tow boomer data with borehole and shallow core information, collected by the British Geological Survey (BGS) as part of their offshore reconnaissance mapping programme. Our data indicate that successful seismic facies analysis of the Quaternary sediments is dependent on the integration of seismic with other subsurface information to recognise depositional systems.

2. Basic Concept of Seismic Facies Analysis

Seismic facies analysis is the interpretation of sedimentary processes and depositional environments from reflection seismic data (Mitchum *et al.*, 1977; Sangree and Widmier, 1977; Davis, 1984). The technique was primarily developed by the hydrocarbon industry as a predictive indicator of lithology. It involves the description and geologic interpretation of a number of seismic reflection parameters of which configuration, character and external form are the most obvious and directly analysed parameters.

The pattern of the reflections may delineate rock or sediment bodies with distinctive geometries, mappable as discrete seismic units. Their internal stratal configuration is interpreted from seismic reflection configuration, and refers to the geometric patterns and relations of strata within the unit. These are commonly taken to be indicative of depositional setting and processes. The 3-D association of individual seismic units may enable conclusions to be drawn concerning sedimentary processes and depositional environments on both local and regional scales. Limitations placed on the interpretation of seismic data arise primarily from the coarse resolving power of the seismic method, which may lead to oversimplistic interpretation of the depositional environment. This is a particular problem with relatively thin stratigraphic sequences which may be beyond the limit of vertical seismic resolution. Strata, 20 to 50m thick, can be obscured on profiles collected by low-frequency seismic systems such as airgun. In the horizontal dimension, the

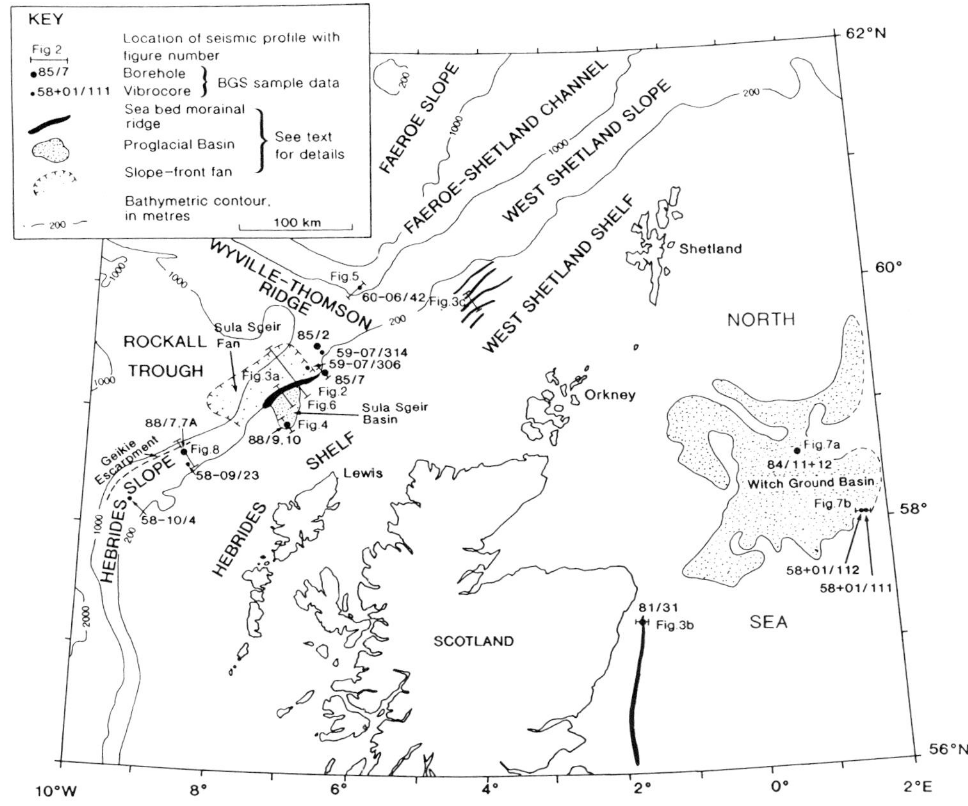

Fig. 1. Northern UK continental margin with location of BGS seismic profiles and sample data, and geographic locations referred to in the text.

interpretation of rapid lateral facies change or stratal discontinuity may be severely constrained by factors such as survey speed and recording speed. Although lateral resolution can be manipulated to some degree using digitally recorded data, analogue profiles represent a permanent record at the fixed survey parameters.

3. Application to High-Resolution Quaternary Studies: The Nature of the Problem

The concept of seismic facies analysis is being increasingly applied to high-resolution studies of Quaternary sequences, particularly on mid- to high-latitude glaciated continental margins. Interpretation techniques are being employed as a means of distinguishing various glacial and glaciomarine facies. The main tools used in such studies involve sparker and boomer systems. The former is capable of up to 500m penetration and 5 to 10m in resolution. The boomer has a penetration of up to 100m and a resolution of 0.25 to 1m. Despite the much greater seismic definition from these systems, the thin-bed problem remains. The complexity of the glacial system, in general, is such that even the resolving power of the high-resolution systems may lead to oversimplistic interpretation of glaciated shelves.

The major problem in high-resolution studies, however, is the lack of adequate geological data to constrain the seismic facies. Yet, this has not discouraged some workers from developing regional sedimentary environment models primarily from seismic interpretation. On the east Canadian and mid-Norwegian continental shelves, changes in sediment lithology, texture and geotechnical properties are being inferred solely on the basis of distinctive acoustic attributes, such as the presence and relative strength both of coherent reflections (stratification) and incoherent backscatter. Acoustically unstratified units with positive relief, and occasional point-source reflections are interpreted largely as subglacial till, whereas silty, glaciomarine deposits have been characterised as displaying low- to high-amplitude, continuous, coherent reflections (King and Fader, 1986; King *et al.*, 1987, 1991). Lithological control in both of these areas is restricted to short piston cores and grab samples which only sample the top few metres of the units. Consequently, this seismic characterisation remains largely untested.

4. The Need for Geological Control

The need for geological control in the interpretation of seismic facies can be well illustrated using examples from the Quaternary of the northern UK continental margin. Acoustically unstratified and layered deposits, from a variety of geographic and depositional settings across the margin, are compared using lithological and geotechnical data derived from numerous boreholes and shallow cores which have penetrated these deposits.

4.1. ACOUSTICALLY UNSTRATIFIED DEPOSITS

Acoustically unstratified deposits are typically preserved as regionally extensive, sheet-like seismic units, up to tens of metres thick, found in shelf, slope-front and deep-water basin settings. Their structureless to chaotic seismic texture reflects the high internal backscatter characteristic of diamicton-dominated facies (Stewart and Stoker, 1990). This consists predominantly of beds comprising a poorly sorted and unlithified admixture of clasts and matrix regardless of depositional environment (Frakes, 1978; Eyles *et al.*, 1983). Although these deposits superficially resemble glacial till, particularly in a restricted core sample, they could equally represent glaciomarine deposits modified by ice loading, debris flow reworking or iceberg turbation. According to Syvitski (1991), there may be little difference in the acoustic properties of glacial till and those of glaciomarine deposits which have undergone some degree of ice loading. The formation of a significant acoustic reflecting or scattering surface, by overconsolidation or iceberg ploughing, may result in no internal acoustic stratification even if the unit is lithologically stratified. This is well illustrated on the northern Hebrides Shelf, where a stacked, prograding series of acoustically unstratified units were proved to contain both lithologically stratified and massive, subaqueous diamictons (Figure 2). BGS borehole 85/7 recovered a hard, massive, distal glaciomarine dropstone diamicton from the upper unit, overlying gravel beds and crudely stratified to massive, proximal glaciomarine debris flow diamictons in the lower units (Stoker, 1988). The former originated mainly from iceberg rafting and dumping of material onto the sea bed, whereas the proximal deposits formed marginal to the ice sheet. The stratification in the proximal glaciomarine deposits is defined by shell fragments consistent with a waterlain, marine origin.

The top of the upper unit displays an iceberg-scoured relief which is highly reflective on the sparker profile (Figure 2). This unit continues on the shelf to the south-east as a thin sheet drape; the iceberg-ploughed furrows locally extending from the sea bed to the bedrock surface. The distal glaciomarine diamicton on this part of the shelf may, therefore, have been partially reworked by iceberg turbation. Often, the resulting diamicton may be coarser than 'normal' glaciomarine sediments as the stirring action of the iceberg must create turbulence and resuspension leading to depletion in fines (Vorren *et al.*, 1983). Acoustically and compositionally, the iceberg-turbated deposit is indistinguishable from glacial till.

Acoustically unstratified units with positive relief have been identified in the North Sea, and on the Hebrides and West Shetland shelves (Figures 1 and 3). They form distinct sea bed ridges ranging from 15 to 60m high and from 4 to 8km wide, and can be traced laterally for up to 130km. They have generally been interpreted as submarine end-moraines (Thomson and Eden, 1977, Stoker *et al.*, 1985; Stewart, 1991a; Stoker and Holmes, 1991). In the North Sea, the recovery of stiff, massive, largely unfossiliferous diamicton in boreholes such as 81/31 (Figures 1 and 3b) has been taken to indicate a subglacial origin (Thomson and Eden, 1977). However, an influx of microfauna near the top of the unit may be indicative of a late-stage glaciomarine component (Stoker *et al.*, 1985).

The main ridges on the Hebrides and West Shetland shelves (Figure 3a, c) have only been partially sampled by shallow cores (3m), which recovered massive diamictons interbedded with sands and gravels. Micropalaeontological data indicate a proximal glaciomarine depositional setting for these deposits (Stoker, 1990a; Stoker and Holmes, 1991), although portions of these ridges could also contain subglacial till (Stewart and Stoker, 1990). On the Wyville-Thomson Ridge, BGS borehole 85/2 (Figure 1) penetrated a smaller, more isolated ridge or mound, and proved hard, glaciomarine deposits with thin turbidite sands (Stoker, 1988). This mound lies beyond the main ice-marginal depocentre on the Hebrides Shelf, and

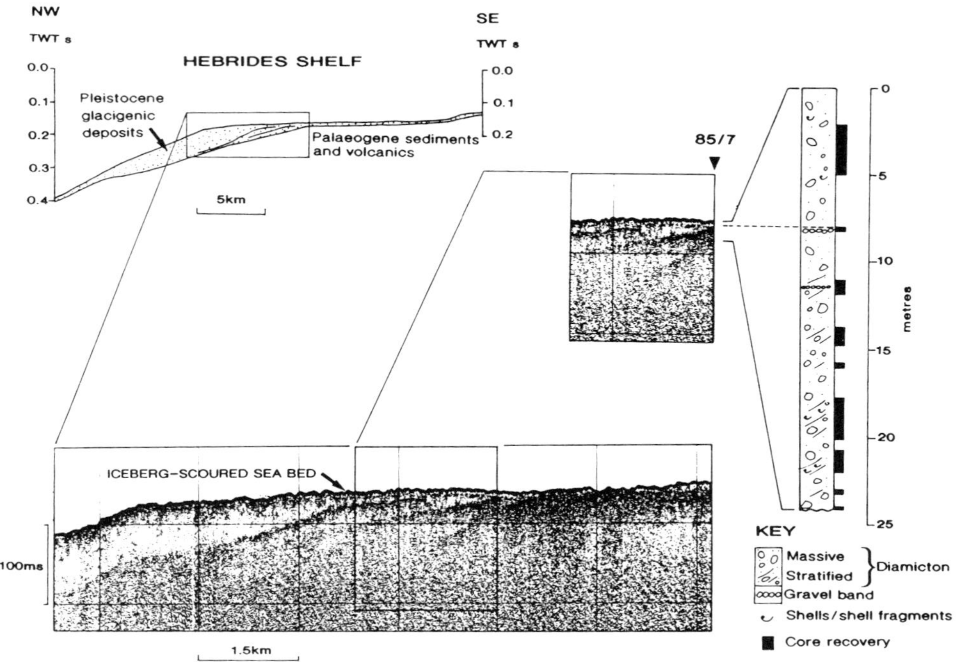

Fig. 2. Seismostratigraphic and lithological characteristics of acoustically unstratified, glacigenic deposits (stippled) on the Hebrides Shelf. Interpreted line drawing based on sparker profile 83/04-12; boxed area shows detail in area of borehole 85/7. Profile located in Figure 1.

an origin such as an iceberg melt dump is possible (Stoker and Holmes, 1991). All of these deposits have been reworked to varying degrees by iceberg scouring.

The lateral relations of the unstratified deposits on the Hebrides Shelf also display evidence for a polygenetic origin. Landward of the main shelf-edge ridge, the acoustically unstratified deposits form a sheet drape covering the floor of the glacially overdeepened Sula Sgeir Basin (Figures 1, 3 and 4). Without geological control, the unit could be interpreted as a basal till deposited by ice occupying the basin, as an undermelt till laid down as the ice began to float, or as a combination of the two. However, BGS borehole 88/9, 10 recovered firm, massive glaciomarine diamicton interbedded in the lower part of the unit with massive muddy sands, partially reworked from the underlying deposits. The sands may represent mass flow deposits and possibly correlate with a series of irregular reflectors observed on the seismic profile (Figure 4) (Stewart and Stoker, 1990). Therefore portions of

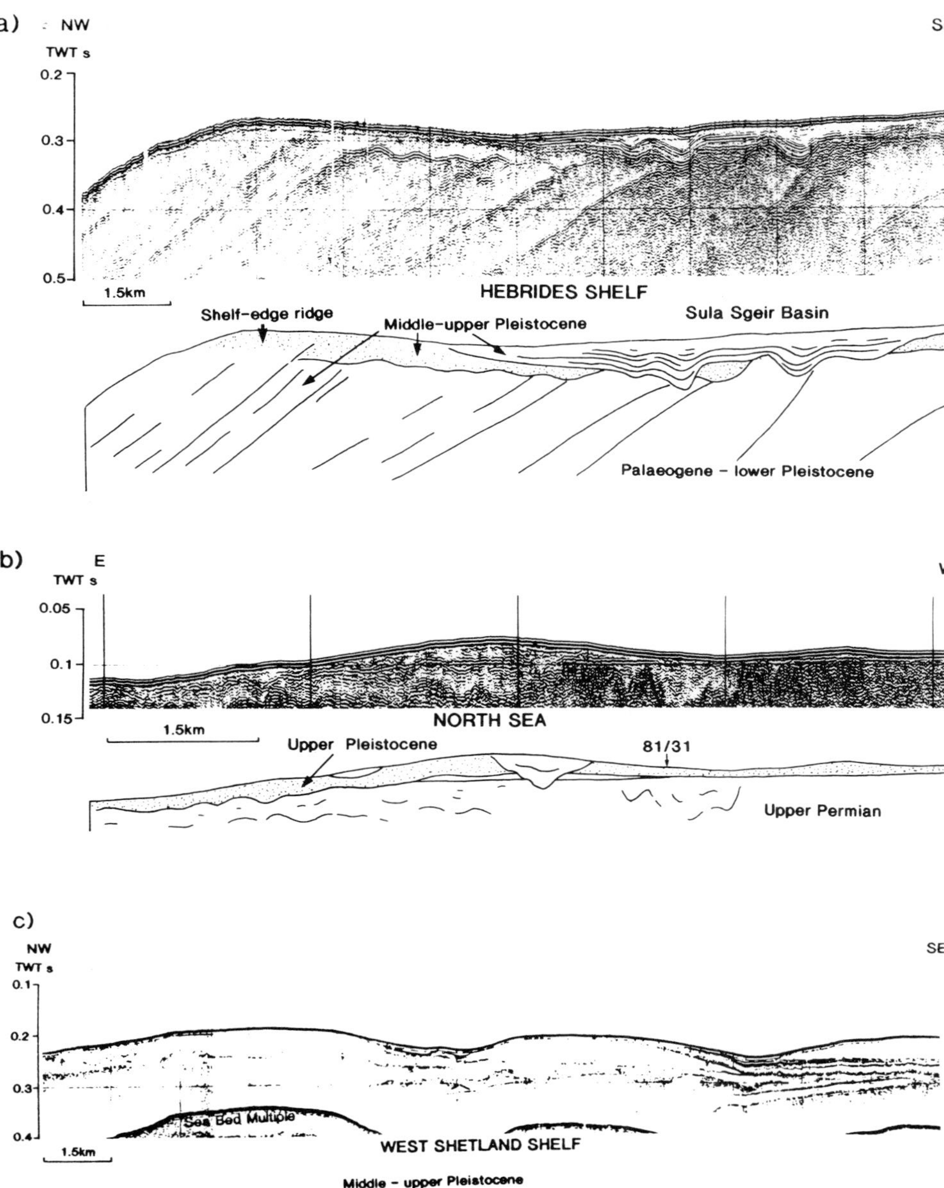

Fig. 3. Characteristics of acoustically unstratified, glacigenic units with positive sea bed relief (stippled). Interpreted line drawings based on (a) sparker profile 83/04-9; (b) sparker profile 80-03/36; (c) sparker profile 83-04/31. All profiles located in Figure 1. For details of borehole 81/31 in (b), see text.

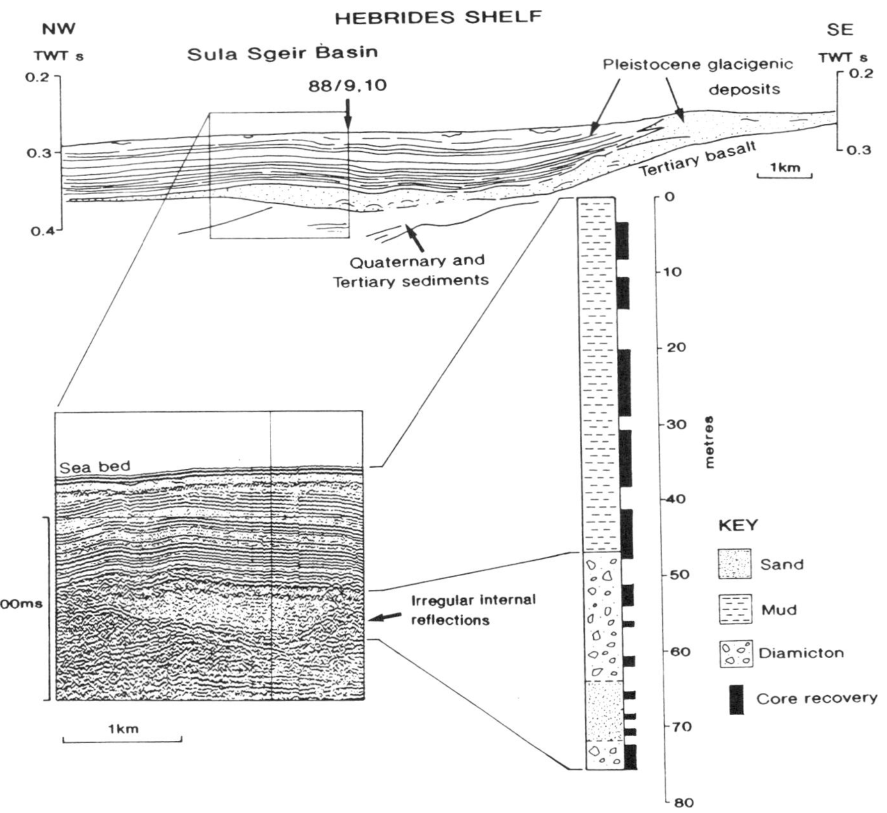

Fig. 4. Seismostratigraphic and lithological characteristics of acoustically unstratified, glacigenic, deposits (stippled) in the Sula Sgeir Basin. Interpreted line drawing based on sparker profile 83/04-15; boxed area highlights detail in area of borehole 88/9, 10. Profile located in Figure 1.

this unit may also include debris flow diamictons.

Significantly, on the margin of the Sula Sgeir Basin, the unstratified unit displays interdigitating relations with the acoustically layered glaciomarine deposits which form the bulk of the basin fill. This type of relationship has previously been interpreted as representing the migration path of the grounding line of ice sheets (King and Fader, 1986; King *et al.*, 1987, 1991; Josenhans and Fader, 1989). The wedge-like deposits were termed 'till tongues' with till deposition envisaged to occur from the meltout of debris immediately proximal to the lift-off point of an ice sheet (King *et al.*, 1987). However, Gipp (1990, 1991) and Stravers and Powell (1991) interpret similar features in terms of reworking by slumping. Syvitski (1991) proposes that such features relate to quasi-stable ice terminus positions wherein a morainal bank is developed, built up largely of debris flows. This is consistent with the evidence from BGS borehole 88/9, 10.

On the north-west margin of the shelf-edge ridge, the ridge sediments clearly interdigitate with prograding, contemporaneous, slope-front deposits; reflectors within the slope-front sequence appear to root into the ridge (Figures 3a and 5). Shallow cores from the upper slope recovered firm to stiff, debris flow diamictons with turbidite sands and muds, representing resedimented glaciomarine deposits (Stoker, 1990a). Their intimate association with the shelf-edge ridge suggest that similar deposits are likely to form a component of the ridge.

Similarly on the West Shetland Shelf, the acoustically unstratified ridges interdigitate on their north-west flank with acoustically well-layered glaciomarine deposits (Figure 3c). Each ridge, in this series of ridges, probably marks a major stillstand during the landward retreat of the ice sheet responsible for their formation. The most eastward of the ridges illustrated in Figure 3c overlies the layered deposits, and this may reflect occasional surging of the ice front (Stoker *et al.*, in press). This may have resulted in ice loading and/or reworking of previously deposited glaciomarine sediments.

The West Shetland Shelf ridges are strikingly similar in geometry and acoustic character to a mounded lensoid body identified on the floor of the Faeroe-Shetland Channel (Figure 6). Yet, the latter occurs at a water depth of 1063m, which rules out a direct glacial origin. The deep-water lensoid package is about 50m thick and traceable for up to 19km. However, the deep-tow boomer profile clearly shows the presence of sporadic, low amplitude and discontinuous internal reflections which suggest that the lensoid package consists of an amalgamation of smaller lensoid bodies. At the south-west end of the lensoid package, small lenses 5m thick and about 1km in profile length are discernible (Figure 6). This is the characteristic acoustic response of debris flow deposits (Nardin *et al.*, 1979; Damuth, 1980), and flow dimensions are consistent with examples from other continental margins such as north-west Africa (Embley, 1976; Simm and Kidd, 1984; Kidd *et al.*, 1987), eastern North America (Embley, 1980), Israel (Almagor and Wiseman, 1982), South Korea (Chough, 1984) and California (Thornton, 1984).

The lensoid package was sampled by vibrocore 60-06/42, which recovered 5.45m of mainly very soft to soft, massive diamicton containing gravel and fossils derived from the West Shetland Shelf (Stoker *et al.*, 1991). These deposits are interpreted as resedimental glacigenic sediments. Material was probably delivered to the outer shelf and slope during glacial cycles when ice sheets reached the shelf edge, and subsequently redistributed down the slope and into the deep-water basin. They form major slope-front accumulations up to several hundred metres in thickness, such as the Sula Sgeir Fan (Figure 5), on both the Hebrides and West Shetland slopes (Stoker, 1990a; Stoker *et al.*, in press). It should be noted that non-glacigenic slope-front debris flow deposits are also acoustically unstratified (Syvitski, 1991, Stoker *et al.*, in press).

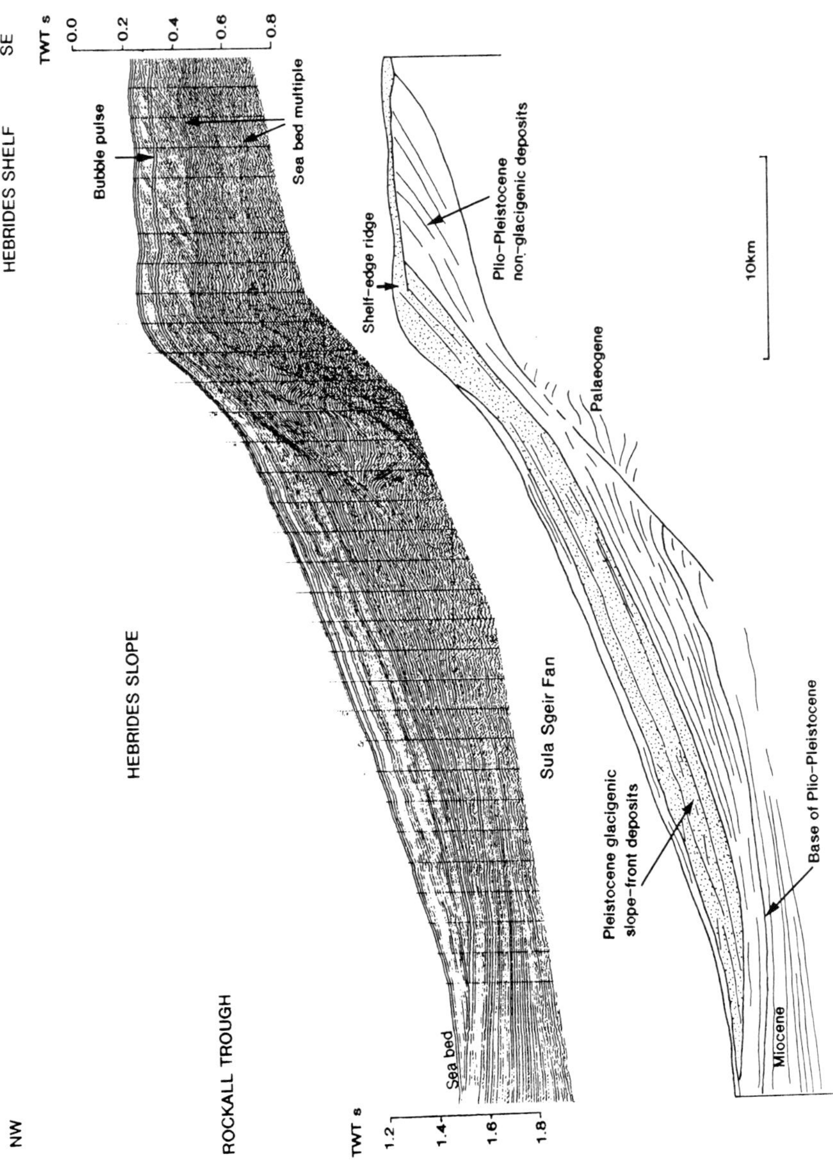

Fig. 5. Airgun profile 83/04-6 and interpreted line drawing across the Hebridean margin, showing the seismostratigraphic relations between acoustically unstratified, glacigenic deposits (stippled) of the shelf-edge ridge and the adjacent Sula Sgeir Fan. Profile located in Figure 1.

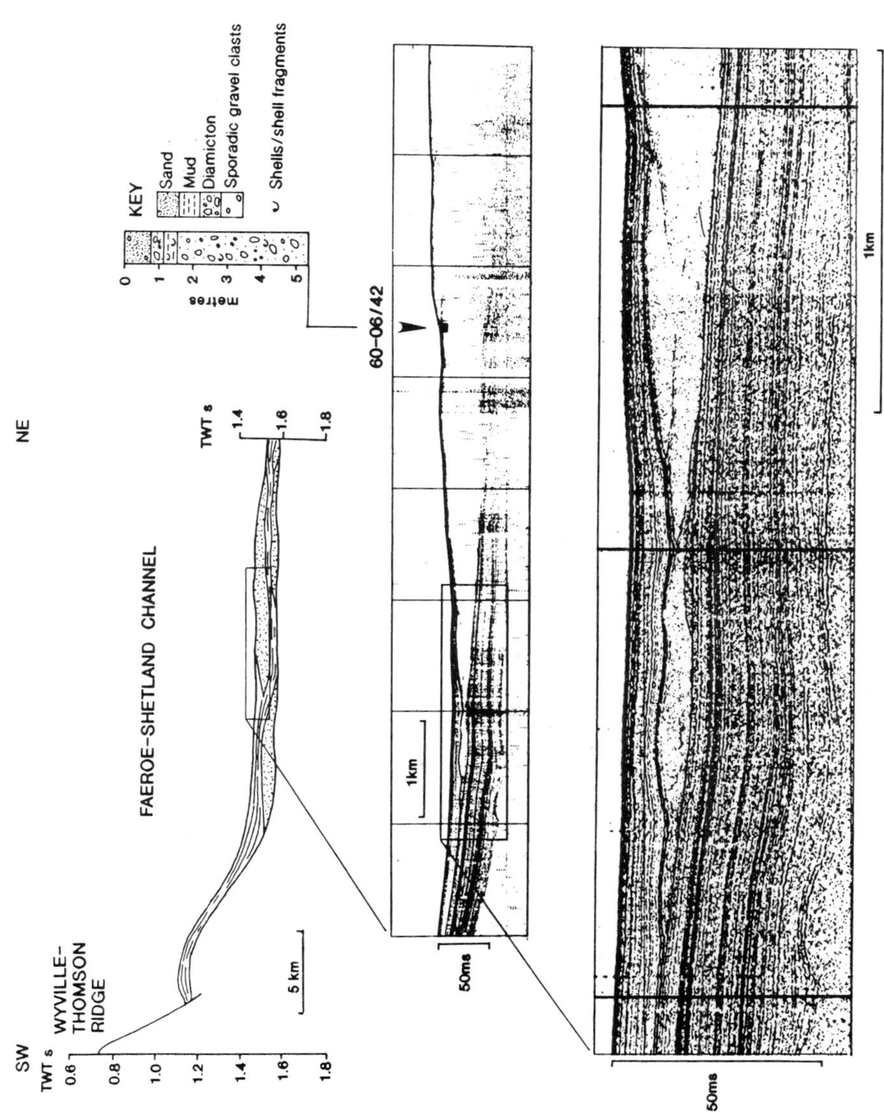

Fig. 6. Seismostratigraphic and lithological characteristics of acoustically unstratified, deep-water lensoid bodies in the Faeroe-Shetland Channel. Interpreted line drawing based on sparker profile 83/04-61 shows several lensoid packages (stippled), with boxed areas showing detail of uppermost package imaged on deep-tow boomer profile 85/07-11. Profile located in Figure 1.

4.2. Acoustically Layered Deposits

On the shelf, acoustically layered deposits display a relatively restricted occurrence largely infilling discrete glacially overdeepened basins and bathymetric lows. Beyond the shelf-edge, these deposits are more widely preserved and form a major component of the slope-front and deep-water basin successions. The change in frequency and amplitude of the layering, both vertically and laterally, is often used as a prime descriptor in interpreting potential lithofacies variation within such deposits. However, the reliability of this interpretive technique has been well tested on the northern UK continental margin, and the results do not support this hypothesis.

In the central North Sea, a sequence of acoustically layered deposits, up to 30m thick, occupies the bathymetric low of the Witch Ground Basin (Stoker *et al.*, 1985; Long *et al.*, 1986). These deposits are interpreted predominantly as the product of glaciomarine sedimentation in a relatively restricted, shallow embayment (Bent, 1986). On seismic profiles (Figure 7), a variable reflection configuration suggests that these deposits may have resulted from a variety of depositional processes. Onlapping reflector terminations may be indicative of a pulsed, turbidite style of deposition, whereas the more draped reflection pattern implies lower energy, basin-wide sedimentation such as suspension fall-out. Reflector cut-outs indicate episodic erosion across the basin. Variation in frequency, amplitude and continuity of reflections is well illustrated.

BGS borehole 84/11, 12 sampled these deposits in the central part of the basin (Figure 1). Although the sediments at the drillsite are predominantly characterised by continuous, high amplitude reflections, the upper 5m of sediment is much more faintly layered and represents a distinct acoustic contrast (Figure 7a). The potential lithological significance of this change was tested by borehole 84/11, 12 which recovered homogeneous muds with occasional thin sand beds across the entire interval. The lack of any discernible visual variation was confirmed by particle size analysis which revealed a uniform sandy, silty clay (Paul and Jobson, 1991). Additionally, the dramatic change in acoustic character at about 26m depth, with unstratified deposits underlying the well-layered section, coincides with minimal lithological and geotechnical change (Paul and Jobson, 1987, 1991: see geotechnical section below).

In contrast, shallow cores to the south (Figure 1) have revealed marked lateral variation in lithology across the uppermost part of the basin-fill, despite the observed continuity of reflections between the core sites. BGS vibrocore 58+01/111 recovered 1.01m of silty, very fine sand near the eastern margin of the basin, whereas vibrocore 58+01/112, located about 3.5km along strike into the basin, proved 5.96m of very soft, silty clay with silt laminae (Figure 7b).

Both reflection seismic and sample data imply some variation in depositional process and hence lithology within the basin-fill deposits. Yet, the two sets of data do not appear to directly correlate. Thus, on present information, seismic facies are not a reliable indicator of specific lithofacies in the Witch Ground Basin.

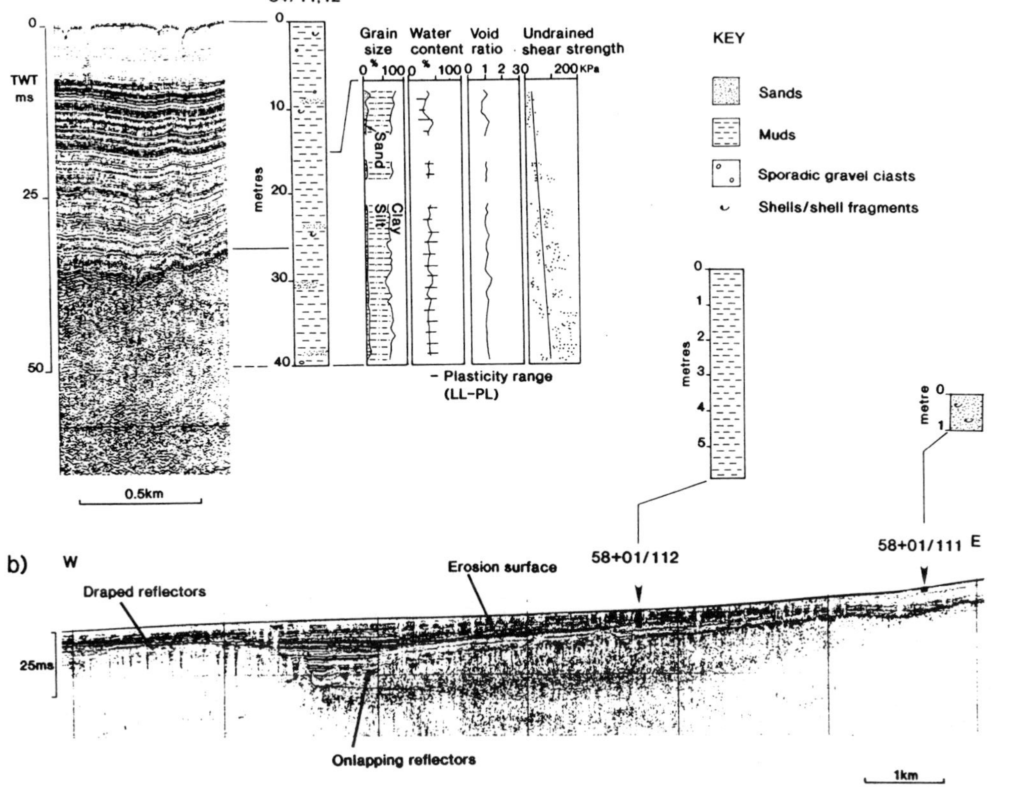

Fig. 7. Seismostratigraphic, lithological and geotechnical characteristics of both acoustically layered and acoustically unstratified, upper Pleistocene, glacigenic deposits in the Witch Ground Basin; (a) boomer profile 81/04-19 in area of borehole 84/11, 12: geophysical data from Paul and Jobson (1991); (b) boomer profile 81/04-11 with vibrocores 58+01/111 and 112. Profiles located in Figure 1.

However, the most striking demonstration of this non-correlation is provided by the sediments preserved on the Hebrides Slope and in the deep-water basins of the Rockall Trough and Faeroe-Shetland Channel. On the upper Hebrides Slope, the entire preserved section of the Plio-Pleistocene succession has been penetrated by BGS borehole 88/7, 7A, landward of the Geikie Escarpment (Figure 1). On seismic profiles, this 89m-thick sequence displays a marked vertical change in acoustic character; a strongly layered lower unit is overlain by a much weaker layered middle unit, in turn overlain by a stronger reflecting upper unit although there is loss of reflection towards the sea bed (Figure 8). The lower part of the lower unit locally downlaps and onlaps onto an irregular, eroded Miocene surface, whereas the upper part of the unit shows a laterally continuous draped reflection configuration. In borehole 88/7, 7A, this unit correlates with sand-dominated, preglacial, marine deposits of Pliocene to early middle Pleistocene age (Stoker *et al.*, 1992). The middle and upper units show a predominantly draped reflection character, and consist of homogeneous, mud-dominated, glaciomarine sediments of middle to late Pleistocene age. Despite the change in acoustic signature there is no discernible change in the lithology of the glaciomarine section. However, the change does appear to have some chronological significance, in that the weakly reflective middle unit broadly correlates with the Anglian glacial stage (about 440 to 350Ka), whereas the upper unit equates with the Devensian glacial stage (about 120 to 10Ka).

Although there is apparent conformity of the units at the borehole site, a disconformity must separate the middle and upper units. Fossil evidence suggests a hiatus of up to 150,000 years (Stoker *et al.*, 1992). Significantly, when these units are traced back up slope evidence of erosion is observed on the seismic profile (Figure 8). Similarly, there is also erosion at the contact of the lower and middle units, which is clearly not observed at the borehole site.

The lack of lithological and acoustic correlation is very evident at this location. The lower and upper units are acoustically similar, but the former is sandy whilst the upper is muddy. However, the homogeneous, muddy, glacial section displays very obvious changes in acoustic signature. Moreover, the identification of at least one intra-Pleistocene erosion surface of substantial duration, within a seismically continuous section, may have some effect on the geotechnical parameters above and below this level.

The erosion on the uppermost part of the Hebrides Slope coincides with the loss of the layered character of the Plio-Pleistocene section, which becomes increasingly structureless and chaotic (Figure 8). Evans *et al.* (1989) suggested that this lateral change in acoustic character reflected the increased influence of glacially-related processes, in shallower water depths, on the deposition of the sediment. Certainly, the irregular nature of the intra-Pleistocene erosion surfaces provides a good wave-scattering surface, which presumably increases the level of backscatter with a loss of acoustic stratification in the units, on this part of the slope. If these surfaces represent buried levels of iceberg ploughmarking (Stoker *et al.*, 1992) then some

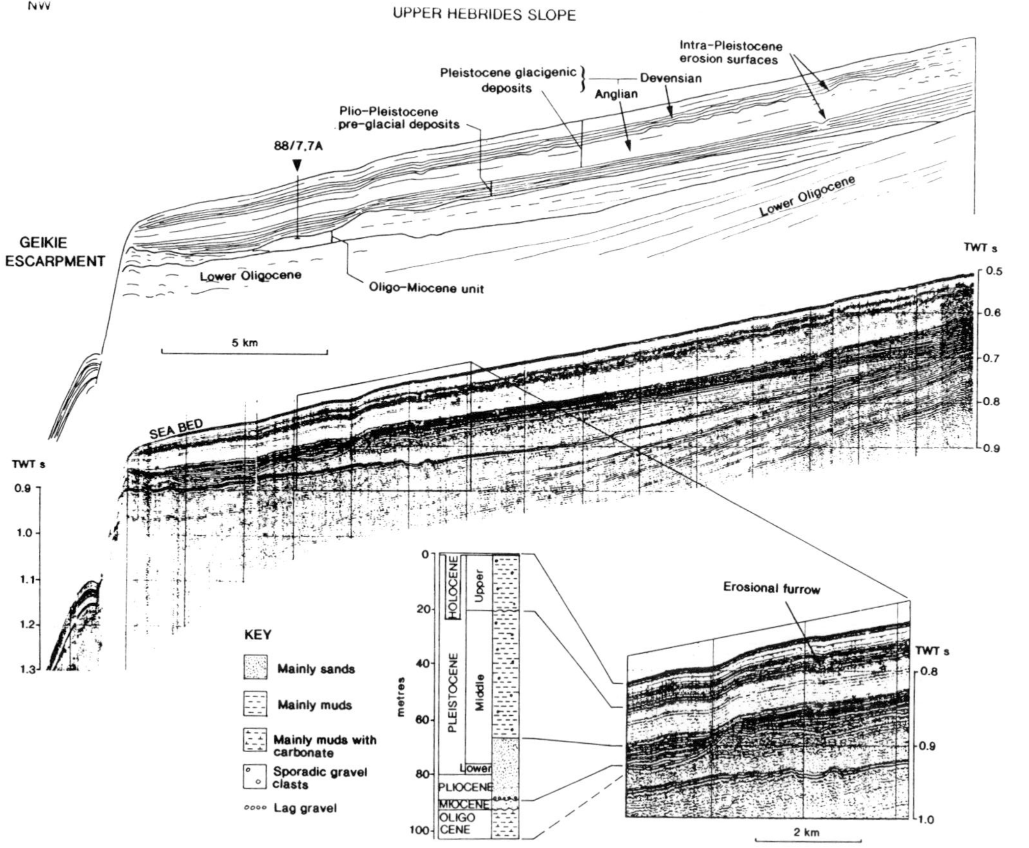

Fig. 8. Seismostratigraphic and lithological characteristics of acoustically layered, glacigenic and non-glacigenic deposits on the upper Hebrides Slope. Interpreted line drawing based on sparker profile 84/06-17; boxed area shows detail in area of borehole 88/7, 7A. Profile located in Figure 1.

physical disruption and turbation of the sediment is also likely.

Lateral change in acoustic character on the lower slope and in the deep-water basins is less readily explained. Numerous shallow cores collected from the area of the Sula Sgeir Fan (Figure 1), have recovered relatively homogeneous, thin-bedded muds and sporadic sands, between the middle slope area and the basin floor. The sediments display no significant visual variation. There is a change, however, from an acoustically well-layered to a weakly layered and transparent/structureless reflection character of the uppermost sediment layers with increasing water depth (Stoker *et al.*, 1989). Similar deposits from the floor of the Faeroe-Shetland Channel are acoustically layered.

The acoustically layered sediments of the Rockall Trough, Faeroe-Shetland Channel and adjacent slopes, have been deposited and modified through the interaction of a variety of downslope and parallel-to-slope depositional processes (Akhurst, 1991; Stoker *et al.*, 1989, 1991, 1992; Damuth and Olsen, in press). Glaciomarine hemipelagites and ice-rafted deposits, and distal turbidites have been reworked to varying degrees by alongslope and basin floor bottom currents. The effects of bottom current activity are well illustrated in Figure 6, where the uppermost layered sediments show modification by erosional pinch-out leading to the localised exposure of an underlying, unstratified debris flow package (Stoker *et al.*, 1991). However, the relative proportion and composition of downslope- and alongslope-derived sediments cannot be determined from the seismic data; the thin-bedded nature of the sequence is beyond the limit of the seismic method.

5. Geotechnical Characteristics and Acoustic Signature

Geotechnical data from the Witch Ground Basin, Hebrides Shelf and upper Slope, and the Faeroe-Shetland Channel allow a further comparison of acoustically unstratified and acoustically layered deposits. We summarise these data here; more detail is provided in Paul and Jobson (1987, 1991), Paul and Talbot (1991) and Paul *et al.* (this volume).

In the central Witch Ground Basin, the acoustically layered deposits described above are underlain by lithologically-similar sediments (silty clays with sporadic sand laminae) which display an unstratified signature (Figure 7a). Measurements of various physical parameters from sediments recovered in BGS borehole 84/11, 12 revealed no significant differences between the different acoustic facies (Figure 7a) (Paul and Jobson, 1991). Variations in the undrained shear strength are gradational with depth as a result of selfweight compression, with some scatter from small compositional changes. There may be secondary effects due to the changing environment of deposition, and to the effects of possible syndepositional disturbance by floating ice (Stoker and Long, 1984). The latter may also explain the unstratified acoustic signature.

In contrast, major variation in geotechnical properties exist between sediments on the northern Hebrides Shelf and upper Hebrides Slope (Paul and Jobson, 1987;

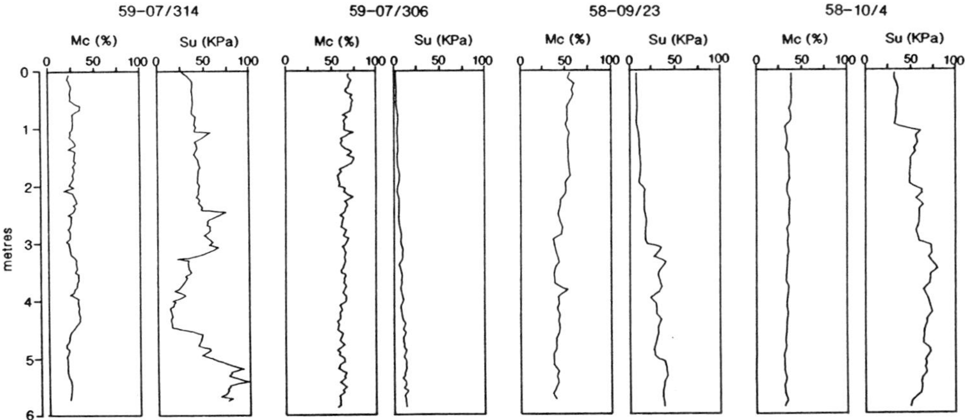

Fig. 9. Profiles of water content (Mc) and undrained shear strength (Su) from selected vibrocores on the Hebrides Shelf and Slope (after Paul and Jobson, 1987). Vibrocores 59-07/314, 58-09/23 and 58-10/4 are taken from acoustically unstratified deposits; vibrocore 59-07/306 sampled acoustically layered sediments. All cores located in Figure 1.

Stoker, 1990b), seemingly reflecting differences in lithology and acoustic signature. For example, samples from vibrocore 59-07/314, which penetrated an acoustically unstratified diamicton on the outer shelf, display low water contents close to the plastic limit and correspondingly high undrained shear strength (Figure 9). These suggest an overconsolidated material subjected to 100 to 200kPa past pressure. Vibrocore 59-07/306 which sampled acoustically layered silty clays on the upper slope proved normally consolidated material, with high water contents and low undrained shear strength (Figure 9). Significantly, the upper part of the acoustically unstratified deposits has been extensively scoured by icebergs, whereas the layered sediments occur below the effective depth of scouring.

However, vibrocore samples from the upper slope above the Geikie Escarpment (Figure 1) invalidate any simple correlation between acoustic facies and geotechnical character. Two lithologically similar cores, 58-09/23 and 58-10/04, were collected about 60km apart at comparable positions (300-400m water depth) on the slope in acoustically unstratified sediments variably disturbed by iceberg ploughing. The muddy sediment recovered in core 58-09/23 was normally consolidated with high water content and low undrained shear strength, whilst that tested in core 58-10/04 was overconsolidated (past pressure about 100 to 150kPa) and had lower water content and higher undrained shear strength (Figure 9). It is tempting to suggest that the first deposit is a glaciomarine sediment that has not undergone ice loading, whereas the latter may have suffered post-depositional ice loading or is a till. Whichever origin applies, the geotechnical distinction between the sediments is not revealed by the seismic facies.

A final example is provided by the sediments in the Faeroe-Shetland Channel. In this area, vibrocore samples highlight consistent differences between the unstratified and layered facies (Paul and Talbot, 1991), but also demonstrates some lithological variation between the unstratified sediments of the deep-water basin. These are described in detail in Paul *et al.* (this volume). In the acoustically layered deposits, vibrocore samples reveal rapid vertical lithofacies changes (Akhurst, 1991; Paul and Talbot, 1991) in which the sedimentological variations are accurately reflected by changes in geotechnical properties. By contrast, the acoustically unstratified deep-water deposits contain uniform sandy and muddy sediment of low plasticity, low compressibility, and moderately high shear strength. However, diamictons with up to 30% gravel have also been recovered from this acoustic facies (Stoker *et al.*, 1991).

In summary, these examples provide a complex picture. Acoustically layered sediments usually appear to be normally consolidated. In some cases they contain important lithological variations, in others they are lithologically uniform with cryptic density layering. Shallow-water, acoustically unstratified sediments are usually diamictons, but are not necessarily overconsolidated, in which case an origin as a subglacial till seems less likely than a glaciomarine origin. However, overconsolidated diamictons are not necessarily tills. Deep-water, acoustically unstratified deposits also include diamictons which clearly cannot be tills; an origin by mass flow seems probable.

6. Discussion

Our geological and geotechnical data illustrate the problem that may arise if too much reliance is placed on the interpretation of high-resolution seismic profiles without adequate groundtruthing. The inconsistency between acoustic signature, lithology and geotechnical properties applies equally to individual seismic units, as well as stacked vertical sequences of different acoustic response. In a hydrocarbon context, the seismic characterisation of units used in a predictive manner on deep commercial seismic data has been founded on an extensive well-log database, and developed over several decades. This underlines the fundamental requirement to include as much data as possible in the interpretations, and to base these interpretations on sound geological concepts and models (Davis, 1984). In high-resolution studies, the potential level of interpretive detail is much greater. Thus, wherever possible, these data should be used in conjunction with observations from modern analogue environments, to infer the possible range of palaeo-environmental conditions operating during the deposition of a particular unit (Syvitski, 1991).

On the northern UK continental margin, which has been subjected to extensive Pleistocene glaciation (Cameron *et al.*, 1987; Stoker *et al.*, in press), an awareness of the nature and style of glaciation is essential to the understanding of the lithological and geotechnic characteristics of the Quaternary sediments. Although the acoustically unstratified facies may be diagnostic of diamicton, it does not necessarily imply a subglacial origin. Regardless of origin, diamictons contain many point sources, such as are obtained from gravel, which reflect the acoustic energy of the sparker or boomer in a disorganised manner. This produces a seismic unit with high internal backscatter, possibly amplified by any surface roughness caused, for example, by surface boulders, iceberg ploughmarks or glacial fluting.

The incorrect identification of subglacial till facies may have significant lithological, geotechnical and palaeo-environmental implications for the Quaternary succession. According to Piper (1988), high-resolution seismic methods cannot adequately distinguish between different types of till and coarse ice-margin deposits. Our data support this view. How would one distinguish a lodgement till interbedded with proximal glaciomarine deposits? The overall lithological similarity would restrict the likely presence of any boundary capable of generating the necessary acoustic impedance contrast to image a reflector. In proximal glaciomarine settings, diamictons often occur in stacked associations with only thin sandy or silty interbeds (Ingolfsson, 1987, Stoker, 1988) which are often of insufficient thickness (< 0.5m) and lateral continuity to produce coherent reflections. If bed thickness increases, reflections may be generated possibly in a manner equivalent to the discontinuous, irregular reflections associated with the mass flow sands in the Sula Sgeir Basin (Figure 4) (Stewart and Stoker, 1990). The significance of these reflections, however, would not have been determined without the borehole control.

The external geometry of a seismic unit is an important aid to interpretation, particularly with the unstratified deposits which often show a mounded relief at sea bed. However, in stacked sequences any diagnostic surface morphology originally associated with a presently buried unit may have been eroded away by the overlying unit, leaving a uniform, sheet-like deposit. In such an instance, seismic facies analysis, alone, is unlikely to resolve the problems of environmental interpretation and lithofacies of the buried sequences. Sheet-like unstratified deposits of seemingly primary origin are preserved at the sea bed on the northern Hebrides Shelf (Figure 2), emphasising the difficulty in the interpretation of these shelf units.

The problems encountered by the variable origin that a diamicton may have within a single unstratified unit are echoed by the layered deposits. Although these deposits superficially give the appearance of a layered lithological sequence, our data indicate that real caution must be exercised in their interpretation. Some vertical changes in acoustic layering do appear to coincide with lithological change, but reflections appear to be equally capable of being generated in a lithologically homogeneous deposit by cryptic change in bulk density. Similarly, lateral homogeneity or variation in acoustic signature is not a reliable indicator to lithological character. No consistent acoustic signature for sandy or muddy deposits is forthcoming from our study; both lithologies seem capable of producing a layered signature. In thin-bedded sequences of polygenetic origin, the nature of the deposit cannot be discriminated by the seismic method.

Discordant reflections suggestive of erosion surfaces do appear to reflect 'real' geological events; the erosional hiatus of up to 150,000 years identified on the upper Hebrides Slope is a significant interval of time. Presumably, units of variable age will possess different geotechnical characteristics. Is it possible that a 'chronological' signature in the sediments is contributing to the reflection pattern?

There is no doubt that certain reflectors have real geological significance. The most unambiguous interpretation revealed by our study concerns the unstratified deposits of the slope and deep-water basins, which seismically, lithologically and geotechnically possess all of the characteristics of debris flow deposits. Perhaps it is no coincidence that similar deposits have been well documented from other continental margins, reiterating the need of an interpretive database. A significant point to be drawn from these deposits comes from their amalgamated nature; a mappable debris flow package is in fact a composite of numerous smaller flows. These packages define major slope front accumulations such as the Sula Sgeir Fan (Figures 1 and 5). In a more dynamic shelf setting, the likelihood must exist that equally uniform and extensive units are in reality composites of various different lithological facies.

Although we emphasize the need for geological control, it should be remembered that the analysis of borehole and shallow core data can also be subjective. Recovery of material from diamicton sequences is typically very poor (Figures 2 and 4). This increases the risk assessment of the interpretation of depositional environment, as diagnostic criteria such as bed contacts and fine scale sedimentary structures may not be preserved (Stewart, 1991b). Moreover, boreholes and shallow cores only provide point-source information. In a glacigenic sequence, which is characterised by rapid lateral facies change, such information remains site specific and is not necessarily regionally applicable. As has been emphasised throughout the paper, hard, sandy gravelly muds do not necessarily represent subglacial till. There is an obvious need for a set of core data from any given area, stratigraphically located to test both vertical and lateral changes in lithology irrespective of seismic character.

7. Conclusions

The main conclusions to be drawn from our study are:

1. Vertical and lateral changes in acoustic signature do not necessarily correspond to changes in lithology or geotechnical character.

2. Acoustically unstratified seismic units are often associated with diamicton lithologies, but this does not necessarily imply a subglacial origin.

3. An acoustically layered seismic reflection configuration does not necessarily represent a real lithological layering.

4. Improved calibration between acoustic data and borehole and shallow core information is essential to the successful development of high-resolution seismic

facies analysis as a predictive tool in the interpretation of Quaternary sediments.

5. In an applied context, our data suggest that extreme caution be exercised, at the present time, in the use of seismic facies analysis as an indicator of specific lithological or geotechnical character.

Acknowledgement

This paper is published with the permission of the Director, British Geological Survey, NERC.

References

1. Akhurst, M. A. (1991), 'Aspects of Late Quaternary Sedimentation in the Faeroe-Shetland Channel, Northwest UK Continental Margin', British Geological Survey Technical Report WB/91/2.
2. Almagor, G. and Wiseman, G. (1982), 'Submarine slumping and mass movements on the continental slope of Israel', in *Marine Slides and Other Mass Movements*, S. Saxov and J. K. Nieuwenhuis (eds.), pp. 95–128.
3. Bent, A. J. A. (1986), 'Aspects of Pleistocene Glaciomarine Sequences in the North Sea', Unpublished Ph.D. Thesis, University of Edinburgh.
4. Cameron, T. D. J., Stoker, M. S., and Long, D. (1987), 'The history of Quaternary sedimentation in the UK sector of the North Sea Basin', *Journal of the Geological Society, London* **144**, 43–58.
5. Chough, S. K. (1984), 'Fine-grained turbidites and mass-flow deposits in the Ulleung (Tsushima) Back-arc Basin, East Sea (Sea of Japan), in *Fine Grained Sediments: Deep-water Processes and Facies*, D. A. V. Stow and D. J. W. Piper (eds.), Geological Society Special Publication 15, Blackwell Scientific Publications, Oxford, pp. 185–196.
6. Damuth, J. E. (1980), 'Use of high frequency (3.5kHz–12kHz) echograms in the study of near-bottom sedimentation processes in the deep sea: A review', *Marine Geology* **38**, 51–75.
7. Damuth, J. E. and Olson, H. C. (in press), 'Neogene-Quaternary Depositional Processes in the Faeroe-Shetland Channel Revealed by High-Resolution Seismic Facies Analysis: Preliminary Observations', Petroleum Geology of NW Europe.
8. Davis, T. L. (1984), 'Seismic-stratigraphic facies models', in *Facies Models, Geoscience Canada Reprint Series*, R. G. Walker (ed.), Ainsworth Press, Ontario, pp. 311–317.
9. Embley, R. W. (1976), 'New evidence for the occurrence of debris flow deposits in the deep sea', *Geology* **4**, 371–374.
10. Embley, R. W. (1980), 'The role of mass transport in the distribution and character of deep ocean sediments with special reference to the North Atlantic', *Marine Geology* **38**, 23–50.
11. Evans, D., Abraham, D. A., and Hitchen, K. (1989), 'The Geikie igneous centre, West of Lewis: Its structure and influence on Tertiary geology', *Scottish Journal of Geology* **25**, 339–352.
12. Eyles, N., Eyles, C. H., and Miall, A. D. (1983), 'Lithofacies types and vertical profile models: An alternative approach to the description and environmental interpretation of glacial diamict and diamictite sequences', *Sedimentology* **30**, 393–410.
13. Frakes, L. A. (1978), 'Diamictite', in *The Encyclopedia of Sedimentology*, R. W. Fairbridge and J. Bourgeois (eds.), Dowden and Ross, Stroudsberg Pa., pp. 262–263.
14. Gipp, M. R. (1990), 'Multiple debris flows on glaciated active and passive continental margins: Evidence from the Yakataga Formation, Gulf of Alaska and Emerald Basin, Scotian Shelf', 13th International Sedimentological Congress, Nottingham, England, Abstracts-Papers, pp. 192–193.
15. Gipp, M. R. (1991), 'Styles of resedimentation of proglacial sediments: Evidence from the Scotian Shelf and the Gulf of Alaska', Geological Association of Canada/Mineralogical Association of Canada/Society of Economic Geologists Joint Annual Meeting, Toronto, Canada, Abstracts-Papers, pp. A45.

16. Ingolfsson, O. (1987), 'The Late Weichselian glacial geology of the Melabakkar-Asbakkar coastal cliffs, Borgafjordur, West Iceland', *Jokull* **37**, 59–81.
17. Josenhans, H. W. and Fader, G. B. J. (1989), 'A comparison of models of glacial sedimentation along the eastern Canadian margin', *Marine Geology* **85**, 273–300.
18. Kidd, R. B., Hunter, P. S., and Simm, R. W. (1987), 'Turbidity-current and debris flow pathways to the Cape Verde Basin: Status of long-range side-scan sonar (Gloria) surveys', in *Geology and Geochemistry of Abyssal Plains*, P. P. E. Weaver and J. Thomson (eds.), Geological Society Special Publication 31, Blackwell Scientific Publications, Oxford, pp. 33–48.
19. King, L. H. and Fader, G. B. J. (1986), 'Wisconsinan glaciation of the Atlantic continental shelf of southeast Canada', *Bulletin of the Geological Survey, Canada*, 363.
20. King, L. H., Rokoengen, K., and Gunleiksrud, T. (1987), 'Quaternary Seismostratigraphy of the Mid Norwegian Shelf, 65°–67°30'N. – A Till Tongue Stratigraphy', Institutt for Kontinental-sokkelundersokelser og Petroleumteknologi A/S, 114.
21. King, L. H., Rokoengen, K., Fader, G. B. J., and Gunleiksrud, T. (1991), 'Till-tongue stratigraphy', *Geological Society of America Bulletin* **103**, 637–659.
22. Long, D., Bent, A., Harland, R., Gregory, D. M., Graham, D. K., and Morton, A. C. (1986), 'Late Quaternary palaeontology, sedimentology and geochemistry of a vibrocore from the Witch Ground Basin, Central North Sea', *Marine Geology* **73**, 109–123.
23. Mitchum, R. M., Vail, P. R., and Sangree, J. B. (1977), 'Seismic stratigraphy and global changes of sea level, Part 6: Stratigraphic interpretation of seismic reflection patterns in depositional sequences', in *Seismic Stratigraphy- Applications to Hydrocarbon Exploration*, C. Payton (ed.), American Association of Petroleum Geologists Memoir 26, pp. 117–133.
24. Nardin, T. R., Hein, F. J., Gorsline, D. S., and Edwards, B. D. (1979), 'A Review of Mass Movement Processes, Sediment and Acoustic Characteristics, and Contrasts in Slope and Base-of-Slope Systems versus Canyon-Fan-Basin Floor Systems', Society of Economic Palaeontologists and Mineralogists Special Publication 27, pp. 61–73.
25. Paul, M. A. and Jobson, L. A. (1991), 'Geotechnical properties of soft clays from the Witch Ground Basin, central North Sea', in *Quaternary Engineering Geology*, A. Forster, M. G. Culshaw, J. C. Cripps, J. A. Little, and C. F. Moon (eds.), Geological Society Engineering Geology Special Publication 7, The Geological Society, London, pp. 151–156.
26. Paul, M. A. and Jobson, L. M. (1987), 'On the Geotechnical and Acoustic Properties of Sediments from the British Continental Margin, West of the Hebrides', Report of MTD Research Grant No. GR/C/71385.
27. Paul, M. A. and Talbot, L. A. (1991), 'Thematic Geotechnical Analysis of Sediments from the Continental Slope Northwest of the British Isles', Report of MTD Research Grant No. GR/E/81289.
28. Piper, D. J. W. (1988), 'Glaciomarine sedimentation on the continental slope off eastern Canada', *Geoscience Canada* **15**, 23–28.
29. Sangree, J. B. and Widmier, J. M. (1977), 'Seismic stratigraphy and global changes of sea level, Part 9: Seismic interpretation to clastic depositional facies', in *Seismic Stratigraphy-Applications to Hydrocarbon Exploration*, C. Payton (ed.), American Association of Petroleum Geologists Memoir 26, pp. 165–184.
30. Simm, R. W. and Kidd, R. B. (1984), 'Submarine debris flow deposits detected by long-range side-scan sonar 1,000 kilometres from source', *Geo-marine Letters* **3**, 13–16.
31. Stewart, F. S. (1991a), 'A Reconstruction of the Eastern Margin of the Late Weichselian Ice Sheet in Northern Britain', Unpublished Ph.D. Thesis, University of Edinburgh.
32. Stewart, F. S. (1991b), 'An evaluation of seismic and borehole data available from onshore and offshore site investigations of relict glaciated areas', in *Quaternary Engineering Geology*, A. Forster, M. G. Culshaw, J. C. Cripps, J. A. Little, and C. F. Moon (eds.), Geological Society Engineering Geology Special Publication 7, The Geological Society, London, pp. 573–578.
33. Stewart, F. S. and Stoker, M. S. (1990), 'Problems associated with seismic facies analysis of diamicton-dominated, shelf glacigenic sequences', *Geo-marine Letters* **10**, 151–156.
34. Stoker, M. S. (1988), 'Pleistocene ice-proximal glaciomarine sediments in boreholes from the Hebrides Shelf and Wyville-Thomson Ridge, NW UK Continental Shelf', *Scottish Journal of Geology* **24**, 249–262.

35. Stoker, M. S. (1990a), 'Glacially-influenced sedimentation on the Hebridean slope, northwestern United Kingdom continental margin', in *Glaciomarine Environments: Processes and Sediments*, J. A. Dowdeswell and J. D. Scourse (eds.), Geological Society Special Publication 53, Blackwell Scientific Publications, Oxford, pp. 349–362.
36. Stoker, M. S. (1990b), 'Sula Sgeir (59°N, 8°W), Quaternary Geology', British Geological Survey, 1:250,000 Offshore Map Series.
37. Stoker, M. S. and Long, D. (1984), 'A relict ice-scoured erosion surface in the central North Sea', *Marine Geology* **61**, 85–93.
38. Stoker, M. S. and Holmes, R. (1991), 'Submarine end-moraines as indicators of Pleistocene ice-limits off northwest Britain', *Journal of the Geological Society, London* **148**, 431–434.
39. Stoker, M. S., Long, D., and Fyfe, J. A. (1985), 'A Revised Quaternary Stratigraphy for the Central North Sea', British Geological Survey Report, 17/2.
40. Stoker, M. S., Harland, R., Morton, A. C., and Graham, D. K. (1989), 'Late Quaternary stratigraphy of the northern Rockall Trough and Faeroe-Shetland Channel, northeast Atlantic Ocean', *Journal of Quaternary Science* **4**, 211–222.
41. Stoker, M. S., Harland, R., and Graham, D. K. (1991), 'Glacially influenced basin plain sedimentation in the southern Faeroe-Shetland Channel, northwest United Kingdom continental margin', *Marine Geology* **100**, 185–199.
42. Stoker, M. S., Leslie, A. B., Scott, W. D., Briden, J. C., Hine, N. M., Harland, R., Wilkinson, I. P., Mitchell, C. J., Kroon, D., Maher, B., Evans, D., and Ardus, D. A. (1992), 'A Multidisciplinary Study of Mid-Tertiary to Quaternary Sediments Recovered in BGS Borehole 88/7, 7A, Hebrides Slope, Northern Rockall Trough Region', British Geological Survey Technical Report WB/92/5.
43. Stoker, M. S., Hitchen, K., and Graham, C. C. (in press), 'The Geology of the Hebrides and West Shetland Shelves, and Adjacent Deep-Water Areas', HMSO for the British Geological Survey, London.
44. Stravers, J. A. and Powell, R. D. (1991), 'Three-Dimensional Seismic Facies of Glacial Deposits on the Southeastern Baffin Island Shelf and the Debris Flow Origin of Till Tongues', Geological Association of Canada/Mineralogical Association of Canada/Society of Economic Geologists Joint Annual Meeting, Toronto Canada, Abstracts-Papers, pp. A120.
45. Syvitski, J. P. M. (1991), 'Towards an understanding of sediment deposition on glaciated continental shelves', *Continental Shelf Research* **11**, 897–937.
46. Thomson, M. E. and Eden, R. A. (1977), 'Quaternary Deposits of the Central North Sea, 3. The Quaternary Sequence in the West Central North Sea', Report of the Institute of Geological Sciences, 77/12.
47. Thornton, S. E. (1984), 'Basin model for hemipelagic sedimentation in a tectonically active continental margin: Santa Barbara Basin, California Continental Borderland', in D. A. V. Stow and D. J. W. Piper (eds.), Geological Society Special Publication 15, Blackwell Scientific Publications, Oxford, pp. 377-394.
48. Vorren, T. O., Hald, M., Edvardsen, M., and Lind-Hansen, O.-W. (1983), 'Glacigenic sediments and sedimentary environments on continental shelves: General principles with a case study from the Norwegian shelf', in *Glacial Deposits in North-West Europe*, J. Ehlers (ed.), A. A. Balkema, Rotterdam, pp. 61–73.

Discussion

Question from J. Noad, BT (Marine) Ltd., Southampton, UK: It seems unlikely that a whole ridge of sediment could have been formed purely by iceberg dumping. Firstly why would this material be selectively dropped in a particular area? Secondly, and more importantly, it is hard to imagine a marine sedimentary deposit formed entirely from dropstones.

Authors' Response: The speaker agreed that iceberg dumping, alone, may only be one of a number of contributory factors influencing the development of the ridges. Although the ridges currently lie beyond the established ice-limits on the Wyville Thomson Ridge, the possibility of grounded ice affecting their formation cannot be discounted, particularly as the ice-limits remain tentative at present.

THE ENGINEERING GEOLOGICAL APPROACH TO THE SITING OF OFFSHORE STRUCTURES IN THE RAVENSPURN NORTH FIELD

S. THOMAS
Fugro-McClelland Limited, 18 Frogmore Road, Hemel Hempstead, Hertfordshire HP3 9RT

Abstract. This paper uses the development of the Ravenspurn North Field by Hamilton Oil Company Limited as a case study to demonstrate the advantages of an engineering geological approach to the siting of offshore structures. In 1986, 1987 and 1988 engineering geological investigations were performed in the Ravenspurn North Field in Blocks 42/30 and 43/26 of the southern North Sea. These investigations comprised engineering geophysical surveys and geotechnical borings followed by geological and geotechnical laboratory testing programmes. These data along with relevant publicly available information were integrated to produce an engineering geological appraisal of ground conditions for the entire field, enabling the selection of optimum platform locations and the planning of detailed engineering, geophysical surveys and site investigations for both concrete gravity base and pile-supported steel jacket structures.

1. Introduction

Ravenspurn North is a natural gas field operated by Hamilton Oil Company Limited and is located approximately 100km east-north-east of Kingston upon Hull in Blocks 42/30 and 43/26 of the UK sector, southern North Sea (Figure 1).

The review of publicly available bathymetric information indicated that the water depth generally ranges from 40m in the south-east to 50m in the north-west. The regional trend in water depths is interrupted by the presence of sandwaves, especially in the north-west of the field, which rise from 1m to 14m above the general seafloor level. The publicly available geological information indicated that the shallow geology in the Ravenspurn North Field consisted of Quaternary deposits of variable thickness overlying bedrock which comprised undifferentiated Middle and Upper Jurassic sediments. The Quaternary deposits were expected to comprise dense sand of Holocene age overlying stiff clay (till) and dense sand of Pleistocene age. The underlying Middle and Upper Jurassic sediments were expected to comprise interlayered very hard clay (mudrock), very dense sand, limestone and sandstone strata.

Volume 28: Offshore Site Investigation and Foundation Behaviour, 263–293, 1993.

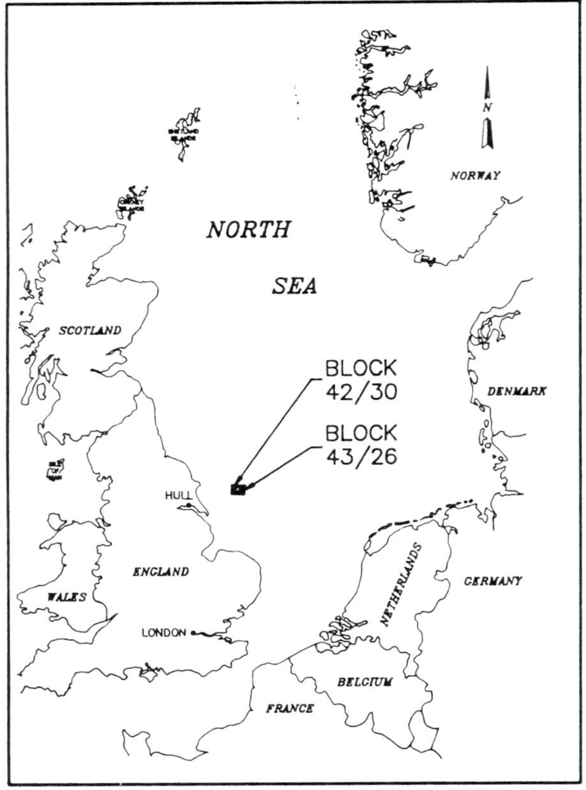

BLOCKS 42/30 AND 43/26

100 km EAST OF FLAMBOROUGH HEAD

Fig. 1. Vicinity map.

2. Field Development Concept

2.1. General

The development concept proposed for the Ravenspurn North Field was based on the requirements set by the evolving reservoir model and on the innovative use of proven technology appropriate to the environmental and ground conditions identified in the field. Based on this information the proposed development concept required the installation of a well-head tower consisting of a conventional pile-supported steel jacket adjacent to ac concrete gravity base structure at the location designated as the Central Complex Site. In addition, the development concept required the installation of two Satellite structures consisting of conventional light-weight pile-supported steel jackets at the locations designated as the ST2 and ST3

Sites (Figure 2) and associated interconnecting and export pipelines. In common with many other field development projects, the assessment of potential platform sites and the design of the proposed structures would be progressed together with the refinement of the reservoir model. It is the reservoir model which ultimately determines the eventual requirements of the field development concept. Fiscal controls and the time constraints imposed by the contract to deliver first gas to the purchaser also exert an influence on the development concept. Clearly, reliable information concerning ground conditions and a degree of flexibility in design or siting of the proposed structures would be requirements for the Ravenspurn North development project.

2.2. Foundation Requirements

The principal foundation requirements of the proposed structures called for by the field development concept are summarised below:

1. **concrete gravity base structure;**
   (a) flat and featureless seafloor to provide maximum base contact for the structure;
   (b) an even thickness of uniformally dense sand to provide a suitable base contact soil that would also permit penetration of the skirts to competent soils at depth;
   (c) an absence of rock layers at shallow depths.
2. **well-head tower and satellite structures;**
   (a) a flat and featureless seafloor to provide an even surface for the placement of the legs of the steel jacket structures;
   (b) the absence of rock layers to below the proposed pile-tip depths required by the conceptual designs of the steel-jacket structures;
   (c) the presence of a sufficient thickness of soils suitable for the installation of driven open-ended steel pipe-piles;
   (d) the presence of a sufficient thickness of cohesive soils in the depth range of pileable strata required in (c) above to provide the necessary axial pile capacity in tension specified by the design of the steel jacket structures.

2.3. Reservoir Zones

In order to identify potential locations that satisfy the principal foundation requirements of the proposed structures both a regional and site specific engineering geological appraisal of ground conditions was required. The regional approach is detailed in Section 3. The site specific approach involved the application of the concept of reservoir zones to the planning of the geophysical and geotechnical

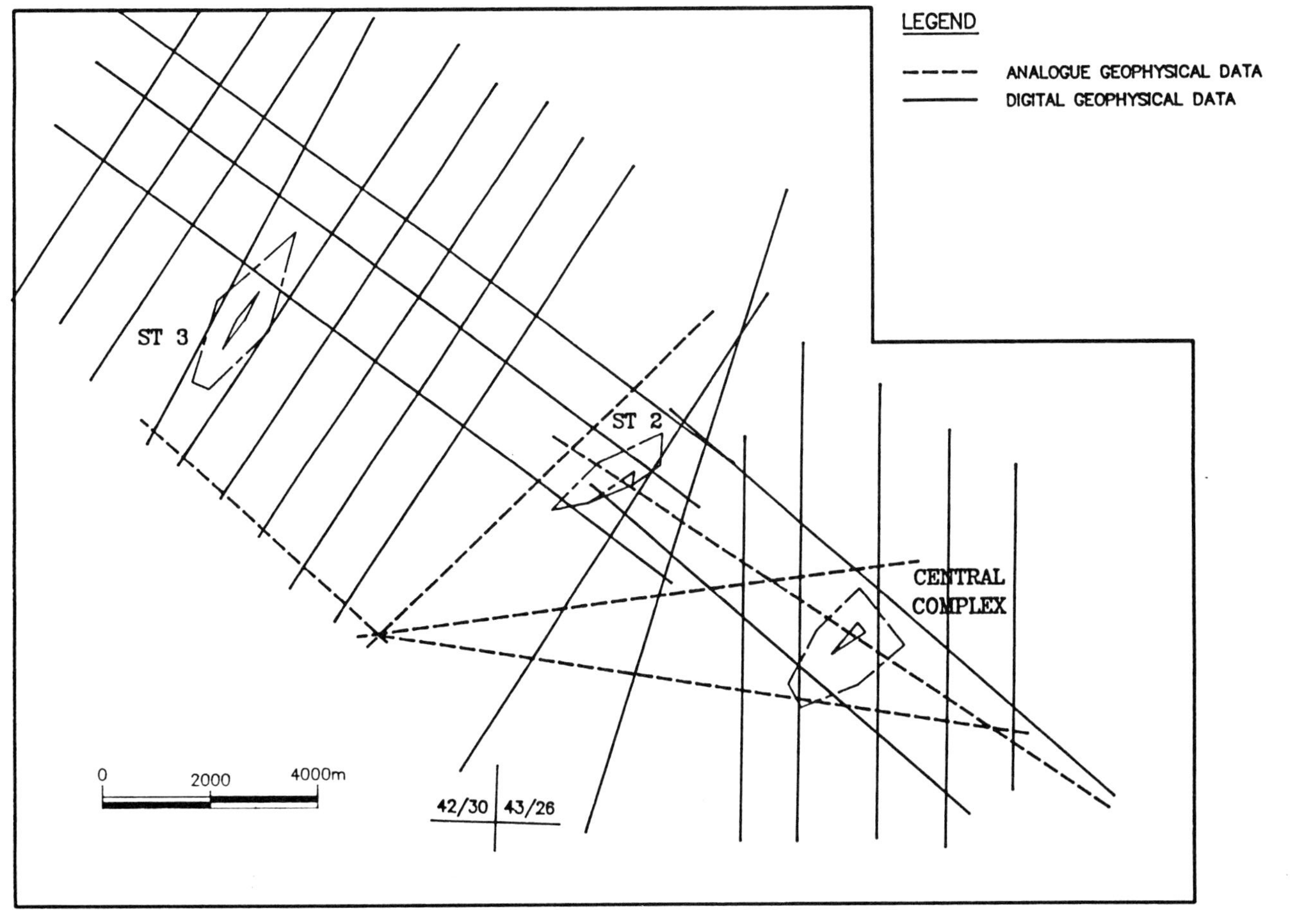

Fig. 2. Summary of engineering geophysical surveys.

investigations. Reservoir zones delineate the area where the production platforms should be located to optimise field development and were provided by the Hamilton Oil Company drilling engineers prior to the commencement of the field investigations. Each reservoir zone comprised a relatively small "inner" target zone defining the optimal area for platform locations and a larger "outer" zone delineating the viable area for platform locations. Significant drilling cost penalties would be incurred for platform locations outside of the reservoir zone. Examples of these reservoir zones are shown on Figure 3.

3. Geological and Predictive Soil Models

3.1. General

This section of the paper describes the methodology used to develop a predictive geological and soil model for the Ravenspurn North Field. The purpose of the predictive geological and soil model is to transform the separate geophysical, geological and geotechnical data sets into a coherent engineering geological framework that makes possible the prediction of the range of the ground conditions at locations throughout the field. In turn, the predictive geological and soil model enables the development of applications maps which consider the specific foundation requirements of the proposed structures called for by the field development concept. In summary, the predictive geological and soil model renders the separate and often unco-ordinated geophysical, geological and geotechnical data sets into readily understandable products relevant to the siting of proposed structures and the planning of engineering geophysical surveys and site investigations.

The predictive geological and soil model for the Ravenspurn North Field is based principally on the original interpretation and synthesis of the available geological, geophysical and geotechnical information. The method and approach adopted for the engineering geological appraisal of ground conditions has been described by Campbell *et al.* (1982) and Campbell (1984) and is outlined briefly in the following sections.

3.2. Geological and Predictive Soil Models

The approach adopted was to develop a geological framework for structural and stratigraphic relationships among the shallow geological formations and to determine their geological history. This framework is described in Section 4 and is summarised on Figure 4. Next, using the geological framework as a basis, a predictive geological and soil model was developed by inferring the types and general properties of sediments likely to comprise each equivalent geological formation. Inferences were refined and calibrated using soil boring information from the Fugro-McClelland investigations and extrapolation of ground conditions using the available seismic data between sites previously investigated.

The predictive soil model consists of a geological features map (Figure 3), stratigraphic correlation diagram (Figure 4), soil province map (Figure 5), predicted soil profiles showing the expected range in soil conditions for each province and the vertical relationships between the geological formations (two examples are presented on Figure 6) and geological cross sections or annotated seismic records.

3.3. Soil Provinces and Subprovinces

A soil province is defined as an area within which soil conditions are generally uniform or are within some specific range. The range is defined on a case-by-case basis depending mostly on the complexity of soil conditions and completeness of the database. The range, in turn, defines a real extent of each province. In general, the more complex the soil conditions and the less complete the database, the fewer soil provinces there are and the wider the range in soil conditions within any given province.

Soil conditions within a subprovince, where delineated, are usually similar to those elsewhere in the province, but differ in some identifiable and mappable respect. Boundaries between provinces or subprovinces are vertical and are generally chosen to be coincident with geological or geotechnically significant boundaries. There are often gradational changes between provinces and the exact location of a province boundary can be arbitrary.

A soil province map differs from a seafloor soils map in that it may be based on a composite of seafloor and underlying soil conditions, typically to some specified depth. The depth limit selected for this study was 80m.

3.4. Predicted Soil Profiles

Two examples of graphical summaries of predicted soil profiles showing the expected range of soil conditions to penetrations of 80m in each soil province is shown on Figure 6. More precisely, each profile shows the stratigraphic sequence and range in thickness of each major geological formation or soil unit likely to be present in the corresponding soil province. Soil conditions shown in the profiles are based on a synthesis of available seismic, soil boring, and seafloor sample data. In provinces where steeply dipping formation boundaries or erosion surfaces are used to define soil stratigraphy the apparent thickness of some formations may be inaccurate.

The seismic data permit an assessment of the extent and thickness of the geological formations and the soil boring information form the Fugro-McClelland investigations allow the geotechnical properties of the sediments comprising each formation to be inferred. Correlating soil boring information with seismic records may be referred to as materials "calibration" of seismic records. However, seismic reflection data do not directly indicate soil type or geotechnical properties, and the nature of soil remote from borings has to be inferred based on the acoustic

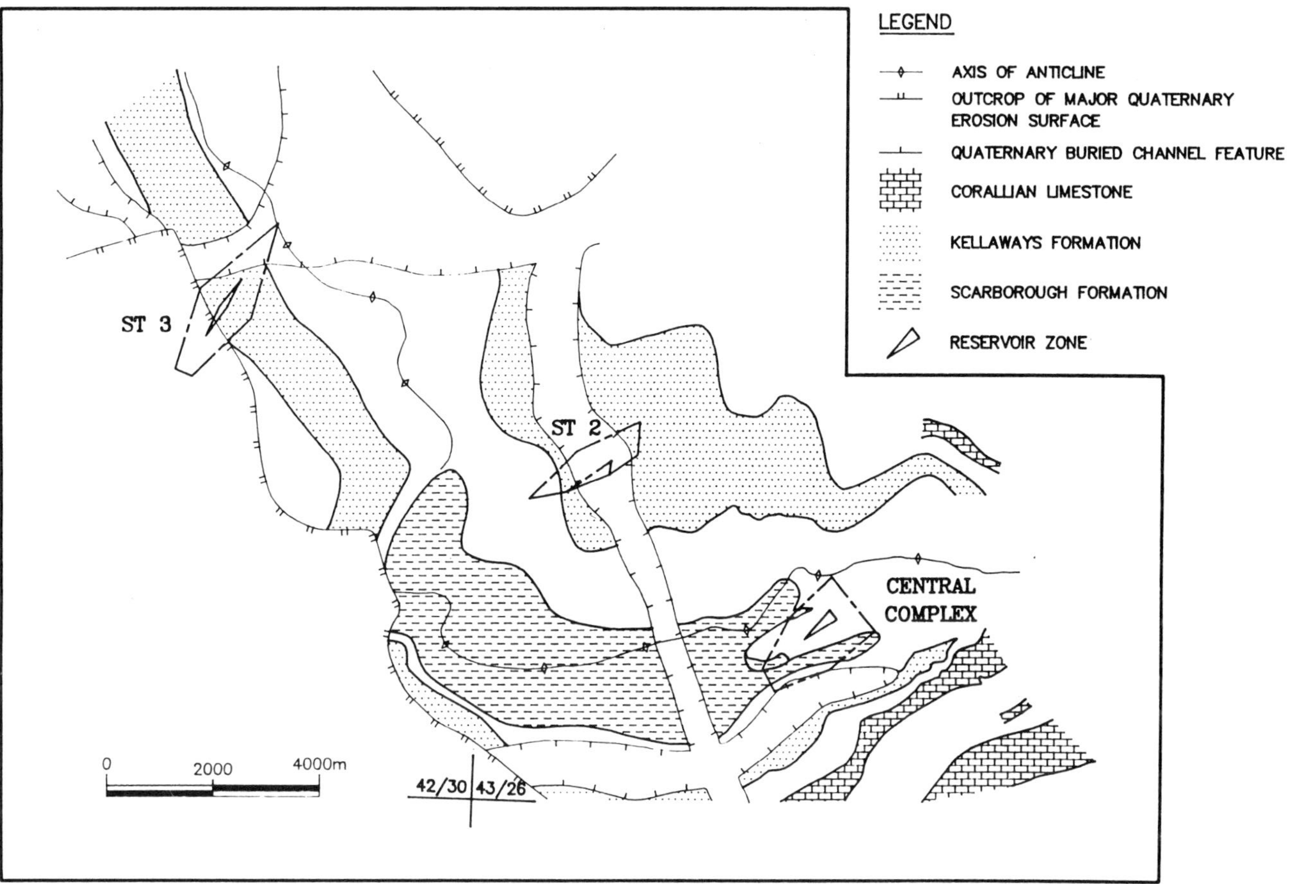

Fig. 3. Geological features map.

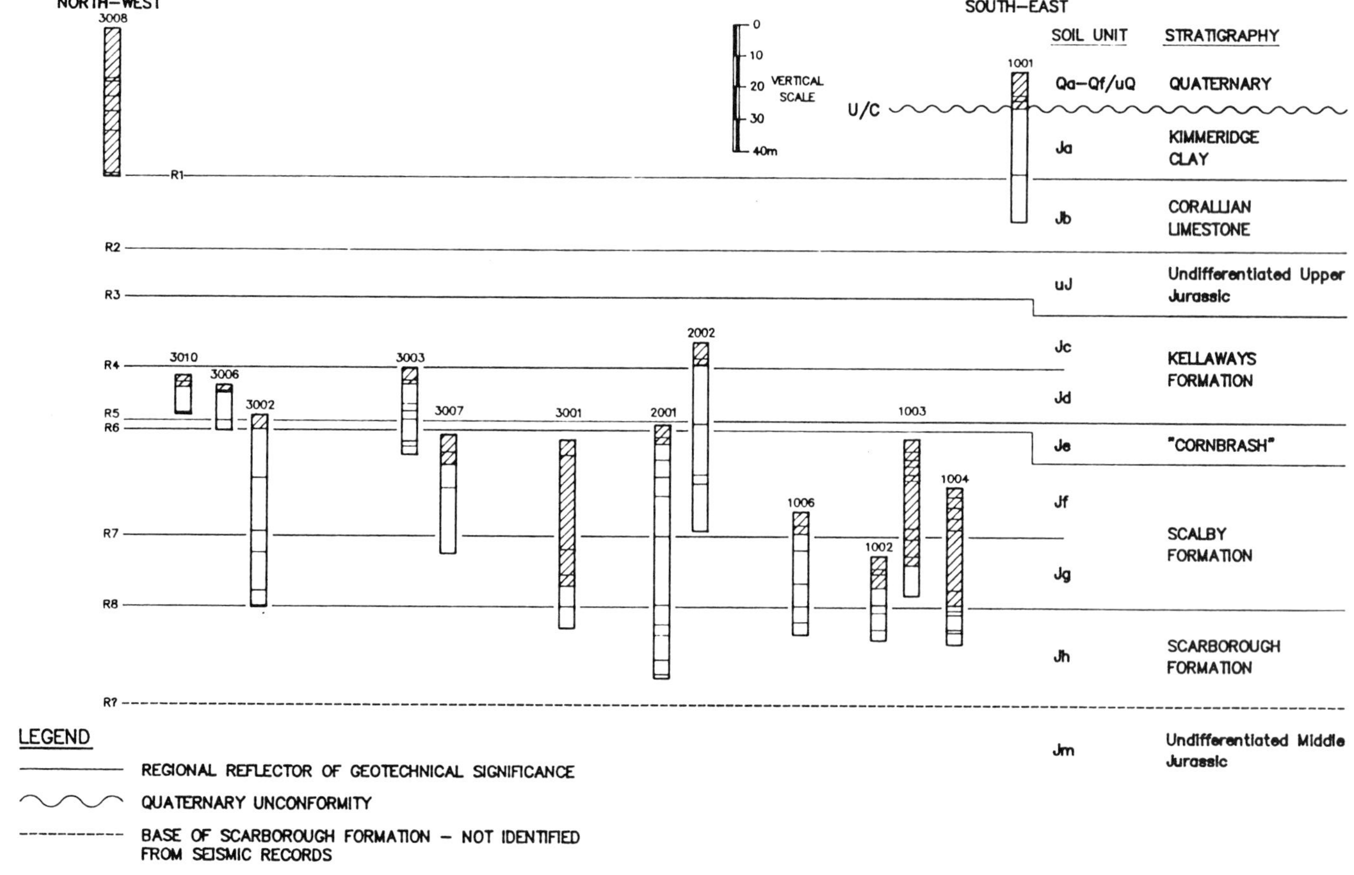

Fig. 4. Stratigraphical correlation diagram.

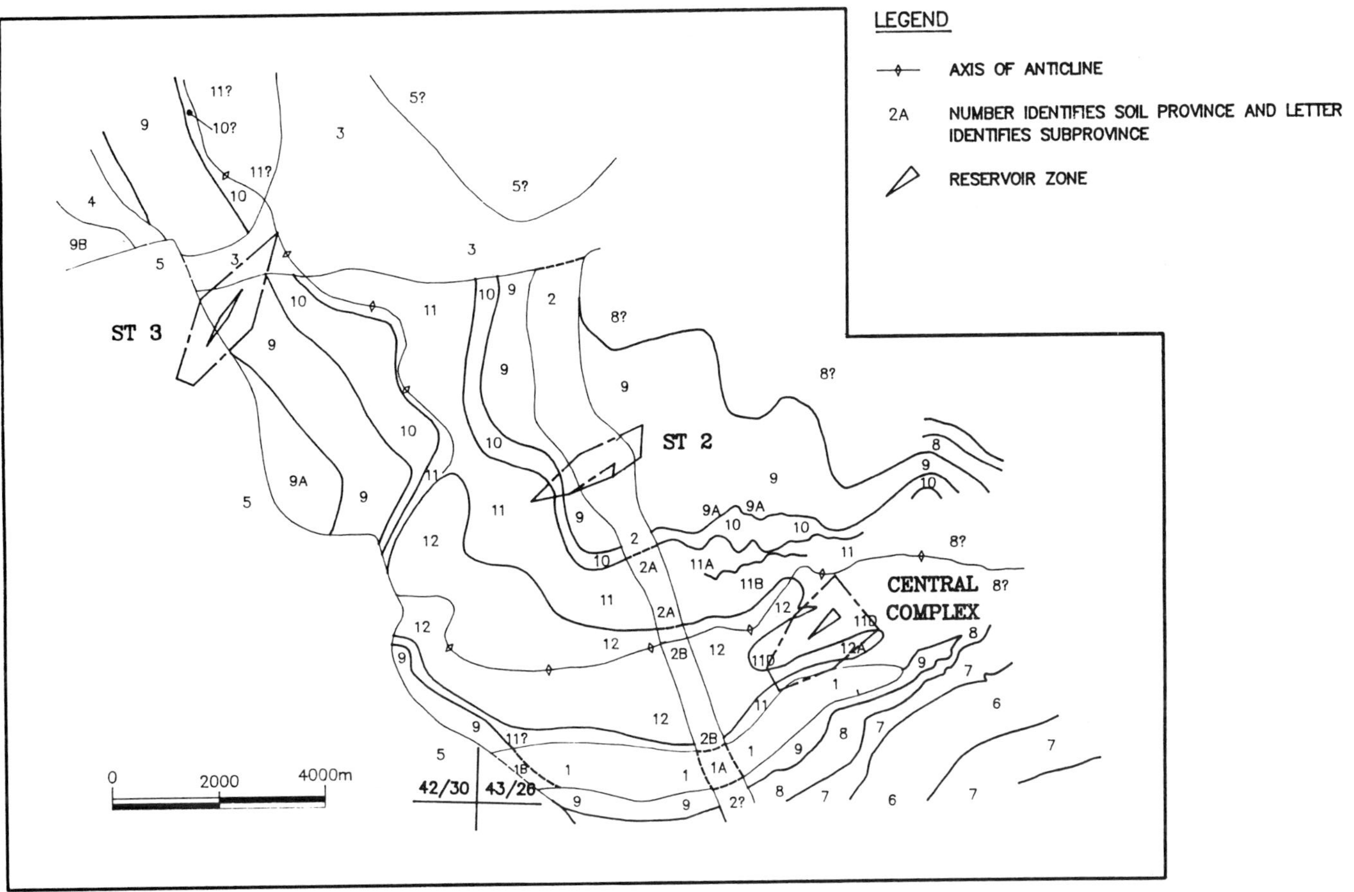

Fig. 5. Soil provinces map.

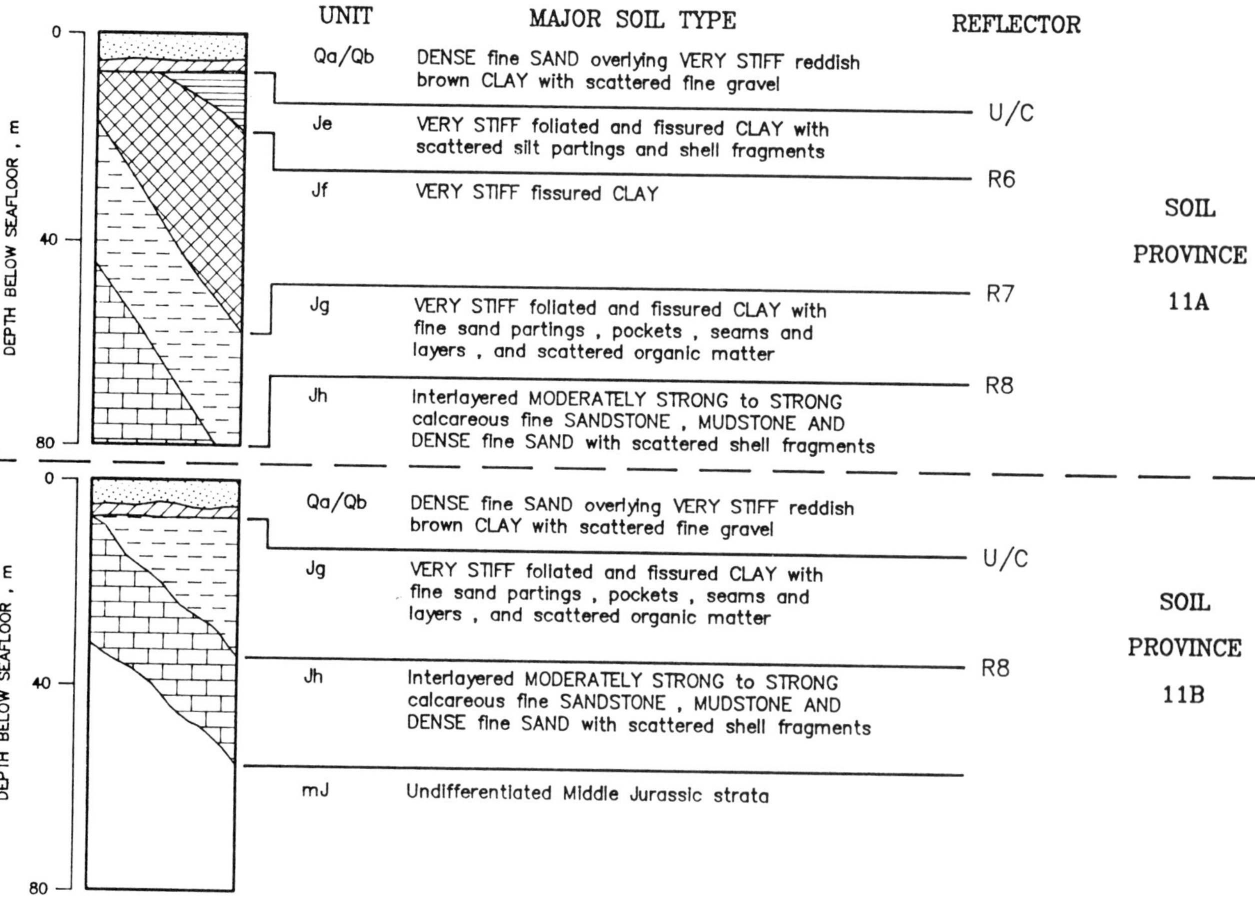

Fig. 6. Predicted soil profiles.

character of the records, information derived from the Fugro-McClelland investigations, the soil model and on geological judgement. Due to this limitation and the complex nature of the soils in the field, extrapolation of geophysical data to obtain the predicted soil profiles should be regarded as a best assessment of general soil conditions in the Ravenspurn North Field.

The predicted soil profiles include only an abbreviated description of the soils expected in each province.

3.5. Use of Predictive Soil Models

The soil model can be used to predict general soil conditions in any part of the area shown on the Soil Provinces map on Figure 5. To facilitate the understanding of soil conditions in any selected area, it is recommended that the Soil Province map (Figure 5), the appropriate predicted soil profiles (e.g. Figure 6) and the geological cross sections be considered together as they are all closely interrelated.

3.6. Applications Maps

Throughout the engineering geological approach to the assessment of ground conditions in the Ravenspurn North Field, maps illustrating geological features that are of particular significance to the siting of structures are produced culminating in the soil provinces map. The information presented on these maps can be further synthesised to produce applications maps which provide information pertinent to one or more of the principal foundation requirements of the proposed structures called for by the field development concept described in Section 2.

In order to select the most appropriate sites and to plan the final geophysical and geotechnical investigations an applications map was produced delineating areas where rock layers were expected to be absent and where in excess of 45m of predominantly cohesive soils suitable for the installation of driven open-ended steel pipe-piles (Figure 7).

3.7. Reliability and Limitations

It should be noted that a predictive soil model can be used to determine the general range in soil conditions for areas covered by the soil provinces map (Figure 5) but cannot be used to define exact soil stratigraphy or geotechnical properties at specific sites. Thus, the information cannot be used for final design and is not a substitute for geotechnical site investigations.

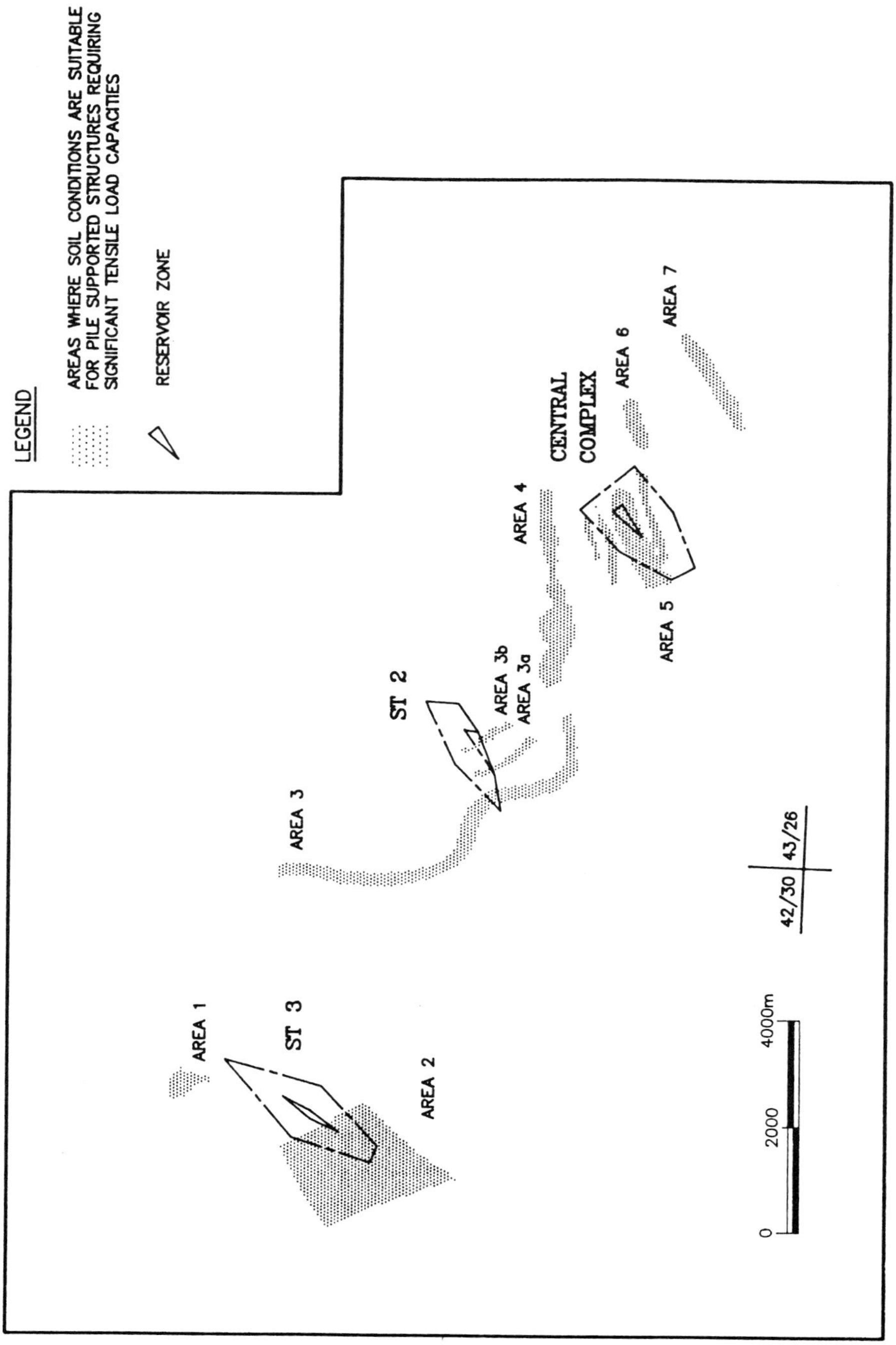

Fig. 7. Foundation considerations map.

4. Regional Investigations

4.1. General

Regional geophysical and geotechnical investigations were performed by Fugro-McClelland Limited (Fugro-McClelland) in 1986 and 1987 to evaluate ground conditions in the field.

The geophysical surveys were performed by Gardline Surveys and Racal Surveys of Great Yarmouth subcontracted and supervises by Fugro-McClelland and comprised the acquisition of bathymetric, sidescan sonar, analogue and digital seismic data, obtained on a grid of survey lines through existing and proposed geotechnical boreholes locations (Figure 2). In addition, vibrocores were taken at ten locations selected to investigated shallow geological features identified from the interpretation of the analogue seismic data.

The geotechnical investigation was performed by Fugro-McClelland from the purpose converted geotechnical drill ships MV Pholas operated by Coe Metcalf Shipping Limited of Liverpool and the MV Mariner operated by Hereema. Several soil borings to between 60m and 80m penetration and a number of surface cone penetration tests to about 10m penetration were performed to investigate ground conditions at strategic locations. The soil samples obtained from the field investigations were subjected to a comprehensive geological and geotechnical laboratory testing programme.

The geotechnical testing programme was performed by Paleo Services Limited of Watford subcontracted and supervised by Fugro-McClelland and comprised lithological classification, and palynological and micropalaeontological analysis (biostratigraphic dating) performed on selected samples to identify the shallow stratigraphic units present in the Ravenspurn North Field.

The soil and rock testing programme was performed by Fugro-McClelland and was devised to provide geotechnical data for both the regional engineering geological assessment of ground conditions and the development of conceptual foundation design information for concrete gravity base and pile-supported steel jacket structures.

The integration of the geophysical, geological and geotechnical findings of the regional investigations provided the basis for the preliminary engineering geological appraisal of ground conditions in the Ravenspurn North Field which is summarised in the following sections.

4.2. Shallow Geology

4.2.1. Structure

The shallow geological structure in the Ravenspurn North Field consists of an antiform with a sinuous fold axis which shows several abrupt changes of direction. In the northern part of the field the fold axis is generally oriented north-west to south-east while in the southern part of the field it is generally oriented east-north-east to west-south-west (Figure 3).

The fold axis of the antiform dips gently to the east in the southern part of the field. The folded strata are Middle and Upper Jurassic in age. The dipping fold axis along with the erosion of the Upper Jurassic strata along the crest of the antiform results in the exposure of Middle Jurassic Strata in the centre of the field surrounded by a 'horseshoe' shaped outcrop of Upper Jurassic and younger strata along its margins. To the north and south of the area surveyed are congruent synclines oriented parallel to the fold axis of the antiform.

4.2.2. Stratigraphy

The identification of the shallow geological strata in the Ravenspurn North Field was made by comparison with Jurassic successions in east Yorkshire and Lincolnshire (Figure 4).

The geological strata encountered in the field can be divided into three broad stratigraphic groups separated by prominent unconformities which represent phases or erosional breaks in deposition. Middle Jurassic strata consisting of lateral equivalents of the Scarborough and Scalby Formations comprise the predominant soils in the centre of the field. Upper Jurassic strata consisting of the lateral equivalent of the Kellaways Formation containing interlayered soil and rock strata (Kellaway Rock) is present at the northern and southern margins of the field. In the extreme south-east of the field, the lateral equivalents of the Corallian Limestone and Kimmeridge Clay comprise the predominant soils (Figure 3).

The Jurassic strata are unconformably overlain by Quaternary sediments which generally form a thin surface layer, typically between 2m and 10m in thickness, except where its thickness is influenced by pronounced seafloor topographic features and the presence of buried channel features.

4.3. Engineering Geological Features

4.3.1. General

This section describes the geological features encountered in the 1986 investigations which are significant for the engineering geological evaluation of ground conditions in the Ravenspurn North Field.

4.3.2. Sandwaves

Sandwaves are undulating bed forms composed of granular material with wavelengths between 30m and 1000m. Wave heights are typically 1m to 10m above the general level of the seafloor and crests may extend for up to 1.5km.

Small sandwaves were identified in the south-east part of the Ravenspurn North Field. The crests of these minor features rise between 0.5m and 1.5m above the general level of the seafloor and are oriented in a north-east to south-west direction.

A group of large sandwaves are identified in the north-west part of the field. The crests of the largest sandwaves rise up to 14m above the general level of the seafloor and are oriented between an east-north-east to west-south-west and a north-east to south-west direction. Separate elongated scour basins up to 3m deep occur along the troughs between the largest sandwaves suggesting that active scour of the seafloor is present in these areas. In addition, the presence of sharp asymmetric forms and symmetrical shapes, perhaps formed by the coalescence of several features, suggests that the sandwaves may be actively migrating across parts of the Ravenspurn North Field.

4.3.3. Buried channels

A number of buried channel features are present in the Ravenspurn North Field (Figure 3). On the basis of their morphology the buried channel features can be divided into three groups. These are: (1) the major Quaternary erosion features which occupy parts of the congruent synclines to the north and south of the antiform previously described; (2) the extensive fan-shaped area of buried channels in the north-west part of the field; and (3) the narrow elongated channel features which are present in the extreme north-west and in the central and south-east parts of the Ravenspurn North Field (Figure 3). The buried channel features described in (2) and (3) above generally extend to between 35m and 60m penetration below seafloor and are infilled by a predominant soil type comprising granular material with subordinate cohesive strata.

4.3.4. Rock outcrops

The Middle and Upper Jurassic strata present at shallow depths in the Ravenspurn North Field consist predominantly of very hard clay with subordinate sand and rock layers. The extent of the formations or parts of formations where soils interlayered with rock strata are most likely to outcrop are presented on Figure 3. The Jurassic strata containing rock layers are described in order of increasing age in the following paragraphs.

The thickest and most uniform rock layer encountered in the investigations is the Corallian Limestone which consists predominantly of a moderately weak to moderately strong medium grained oolitic limestone and is about 20m in thickness. This stratum crops out in the south-east of the field where it forms the margin of the antiform. In the extreme south-east of the field, the limestone is folded into a syncline and overlain by the Kimmeridge Clay which does not appear to contain significant rock layers.

The lateral equivalent of the Kellaways Rock Formation consists predominantly of dense sand interlayered with moderately strong calcareous sandstone and is about 22m in thickness. The rock layers may range from 0.5m to 4.5m in thickness and be separated by 2m to 3m of dense sand which may be locally cemented and contain subordinate rock seams. A rock layer about 4.5m in thickness occur at the top of the formation. The Kellaways Rock crops out extensively along the south-western and north-eastern margins of the Ravenspurn North Field (Figure 3).

The lateral equivalent of the Cornbrash may underlie the Kellaways Rock. This stratum is expected to be extremely variable in thickness and lithology. In some areas it may be cemented or possibly form a rock layer.

The lateral equivalent of the Scalby Formation consists predominantly of very hard clay with subordinate sand layers and does not appear to contain significant rock layers. This is therefore the most suitable stratum from a foundation engineering perspective. The Scalby Formation is about 40m in thickness and crops cut along the centre of the Ravenspurn North Field.

The lateral equivalent of the Scarborough Formation consists predominantly of interlayered moderately weak to strong fine grained sandstones, siltstones, dense sands and very hard clay. The rock layers may range from 0.5m to 6m in thickness and the lithology is laterally extremely variable. The Scarborough Formation is greater than 20m in thickness and is present at shallow depths below the Scalby Formation along most of its outcrop in the centre of the field. The Scarborough Formation crops out extensively in the central and south-eastern parts of the Ravenspurn North Field (Figure 3).

4.3.5. Shallow faults

Shallow faults oriented parallel to the fold axis of the antiform are present in the field (Figure 3). The displacement of sedimentary units along the faults appears to range from between 5m and 10m to between 50m and 55m. The shallow faults do not appear to have displaced the overlying Quaternary strata indicating that significant movement along the faults may not have occurred since about 18,000 a BP (years Before Present). However, the large vertical displacements are significant for the engineering geological appraisal of ground conditions because they can, over a very short horizontal distance, result in a change of soil conditions from predominantly clay to rock bearing strata. In summary, the faults may control the localised occurrence of soils suitable for the installation of driven open-ended steel pipe-piles in the Ravenspurn North Field. This is particularly the case in the central and south-eastern part of the field where the faults with large vertical displacements are more numerous.

5. Site Specific Investigations

5.1. GENERAL

Site specific engineering geophysical and geotechnical investigations were performed by Fugro-McClelland from the MV Dawn Flight chartered to Britsurvey and MV Pholas in 1988 to evaluate ground conditions in the reservoir zones for the Central Complex, Satellite 2 (ST2) and Satellite 3 (ST3) platforms (Figure 2).

The geophysical survey was performed by Britsurvey of Great Yarmouth subcontracted to and supervised by Fugro-McClelland. Bathymetry, sidescan sonar and analogue seismic data were obtained at each of the sites and along the proposed pipeline routes. Also, analogue seismic data were obtained along several tie lines between the Central Complex and the ST 2 Reservoir Zone for regional engineering geological mapping purposes. The analogue seismic data were acquired in a dual pass survey comprising a first pass using a surface two boomer for high resolution shallow geological information and a second pass using water gun or multi-electrode sparker sources for the deeper structural and regional geological information.

The soil samples obtained from the field investigations were subject to a detailed geotechnical testing programme devised to provide the geotechnical data required for both the concrete gravity base and pile-supported steel jacket structures.

The aim of the geophysical survey was to provide sufficient information at the reservoir zones to select platform sites that comply with the principal foundation requirements of the proposed structures.

5.2. CENTRAL COMPLEX RESERVOIR ZONE

5.2.1. General

The Central Complex Reservoir Zone is located in the south-eastern part of the Ravenspurn North Field and corresponds to an area where soils complying with the principal foundation requirements described in Section 2.2 were predicted to occur (Area 5 on Figure 7).

5.2.2. Geological features

In the Central Complex Reservoir Zone, the thickness and presence of geological features in the superficial Holocene sand and Quaternary sediments were mapped (Figures 8 and 9) to ensure the compliance of the predicted ground conditions with the principal foundation requirements of the concrete gravity base structure.

In this reservoir zone, numerous faults are aligned to the axis of the anticline and appear to be down thrown to the north (Figure 10). The displacements along the faults range from 20m to 55m (Figure 11). Displacements of this magnitude are significant from a foundation view point since the maximum observed movements approximate to the interpreted thickness of the Scalby Formation in the reservoir

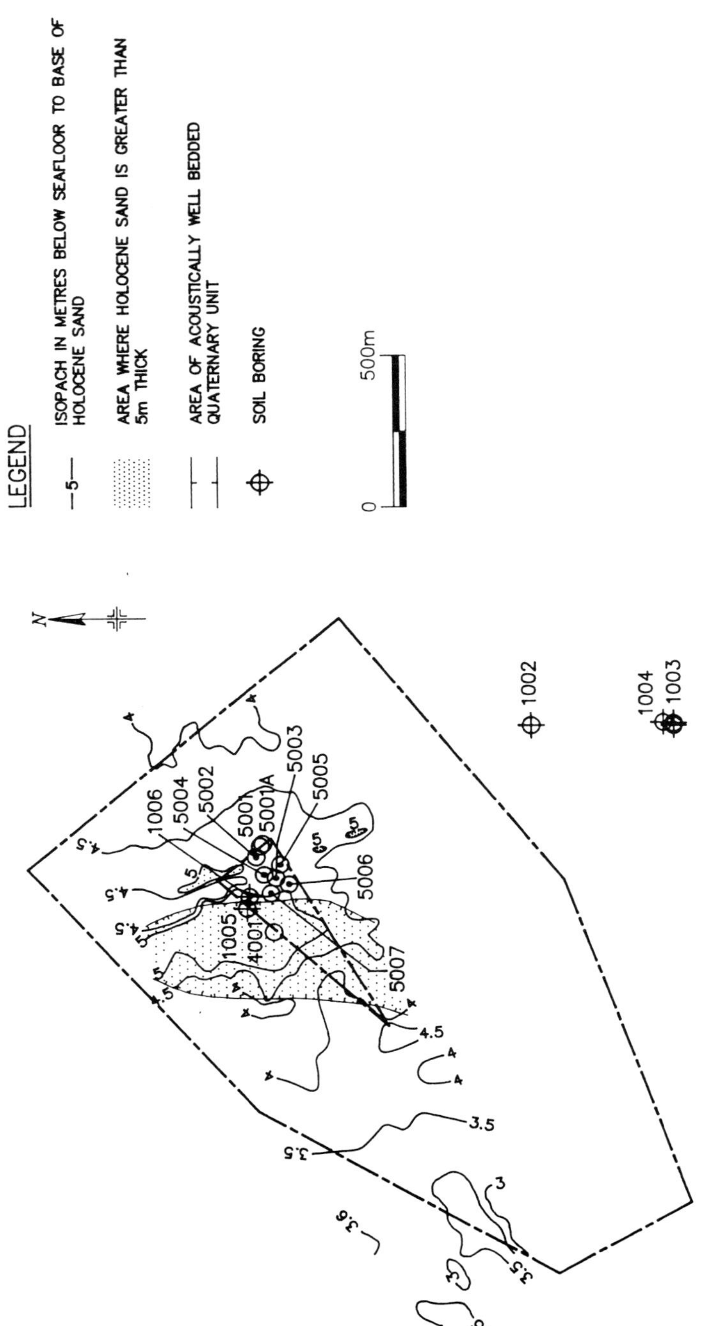

Fig. 8. Isopach map of Holocene sand Central Complex Reservoir Zone.

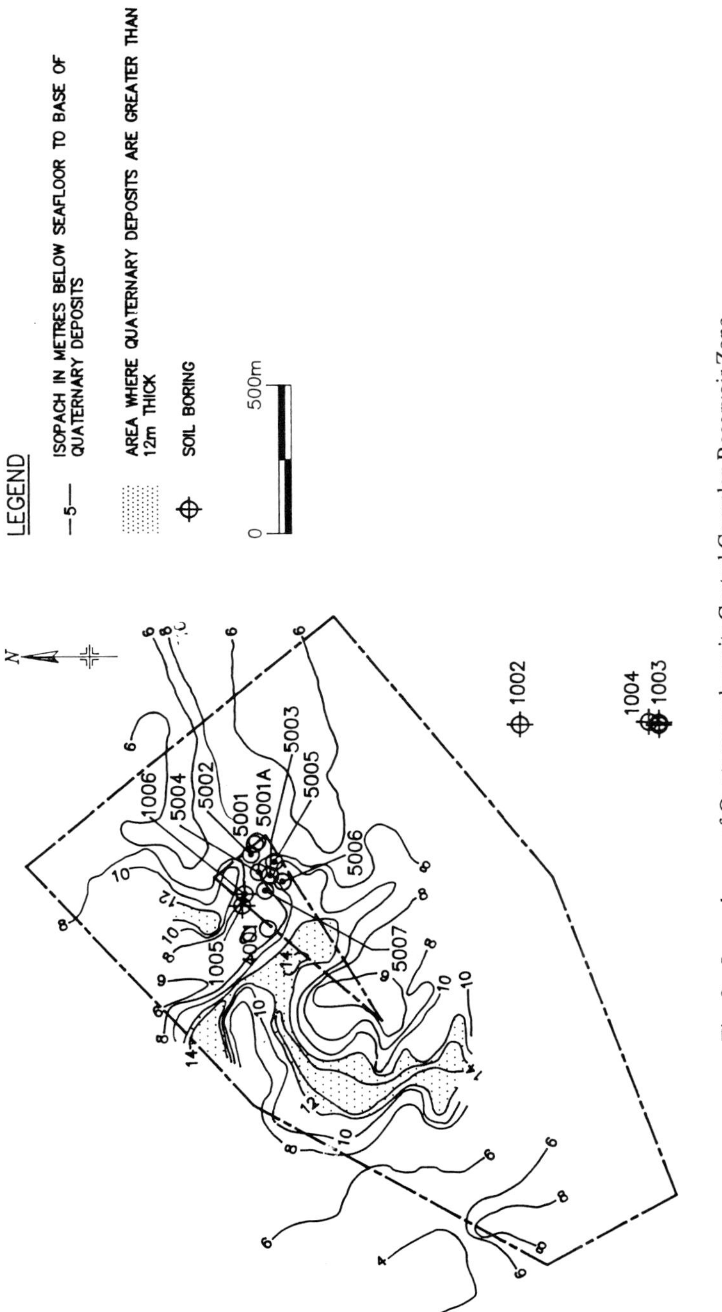

Fig. 9. Isopach map of Quaternary deposits Central Complex Reservoir Zone.

zone. Thus the soil conditions in the reservoir zone may change over a very short horizontal distance from a profile with about 55m of strong cohesive soils (Scalby Formation) to one with interlayered rock strata (Scarborough Formation) outcropping at the seafloor. As shown on Figure 12 this "worst case" scenario occurs in the north-west part of the reservoir zone.

In addition to the faults, a buried channel feature occurs in the extreme south of the reservoir zone. This buried channel is locally significant from a foundation view point because the infill crops out at the seafloor and consists of gravel with subordinate peat and clay layers.

5.2.3. Predicted soil conditions

Based on the interpretation of the seismic records and familiarity with the geology in the Ravenspurn North Field, it was possible to recommend a suitable location for the Central Complex well-head tower and concrete gravity base structures within the "inner" target zone. The soil conditions at the proposed location were subsequently confirmed in the geotechnical site investigation. The following paragraphs discuss predicted soil conditions at depth at the proposed platform location.

The depth contours to the base of the Scalby Formation shown on Figure 10 provide an approximation for the thickness of strong cohesive soils in the Central Complex Reservoir Zone. The reflector configurations suggest that the basal Scalby overlies an erosion surface cut into the underlying Scarborough Formation. Also, the upper part of the Scarborough Formation is acoustically structured, comprising interbedded units which vary between well bedded and structureless, again suggesting vertical lithological variability. The soil model suggests that the upper part of Scarborough Formation comprises interbedded very hard clays and dense silty fine sands with occasional layers of moderately strong to strong calcareous sandstones. Assuming a degree of uniformity across the field these data may suggest the presence of an additional 5-10m thick stratum of very hard clay with occasional layers of moderately strong to strong sandstone up to 0.25m in thickness overlying the more rocky strata below.

5.3. ST2 RESERVOIR ZONE

5.3.1. General

The ST2 Reservoir Zone is located about 4.5km to the north-west of the revised Central Complex Reservoir Zone in the central part of the Ravenspurn North Field (Figure 2).

5.3.2. Geological features

The most significant geological features occurring in this reservoir zone are summarised on Figure 12 and comprise a series of north-west to south-east oriented

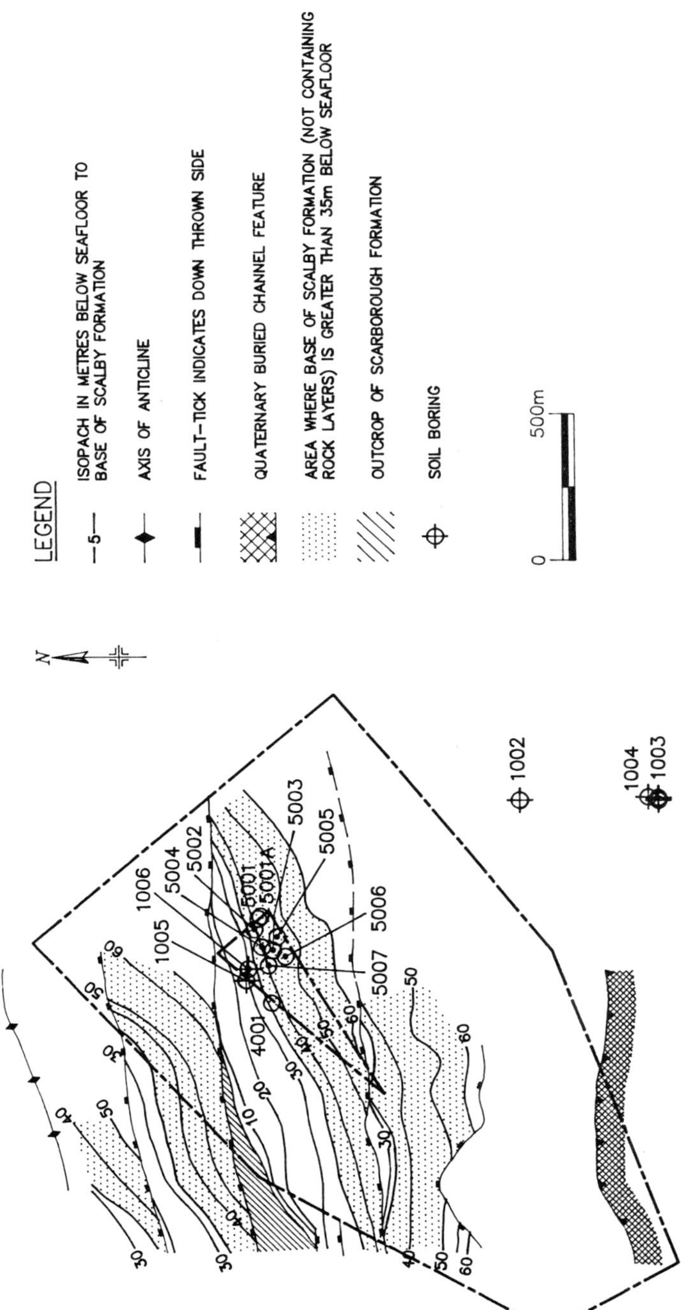

Fig. 10. Isopach map to base of Scalby Formation Central Complex Reservoir Zone.

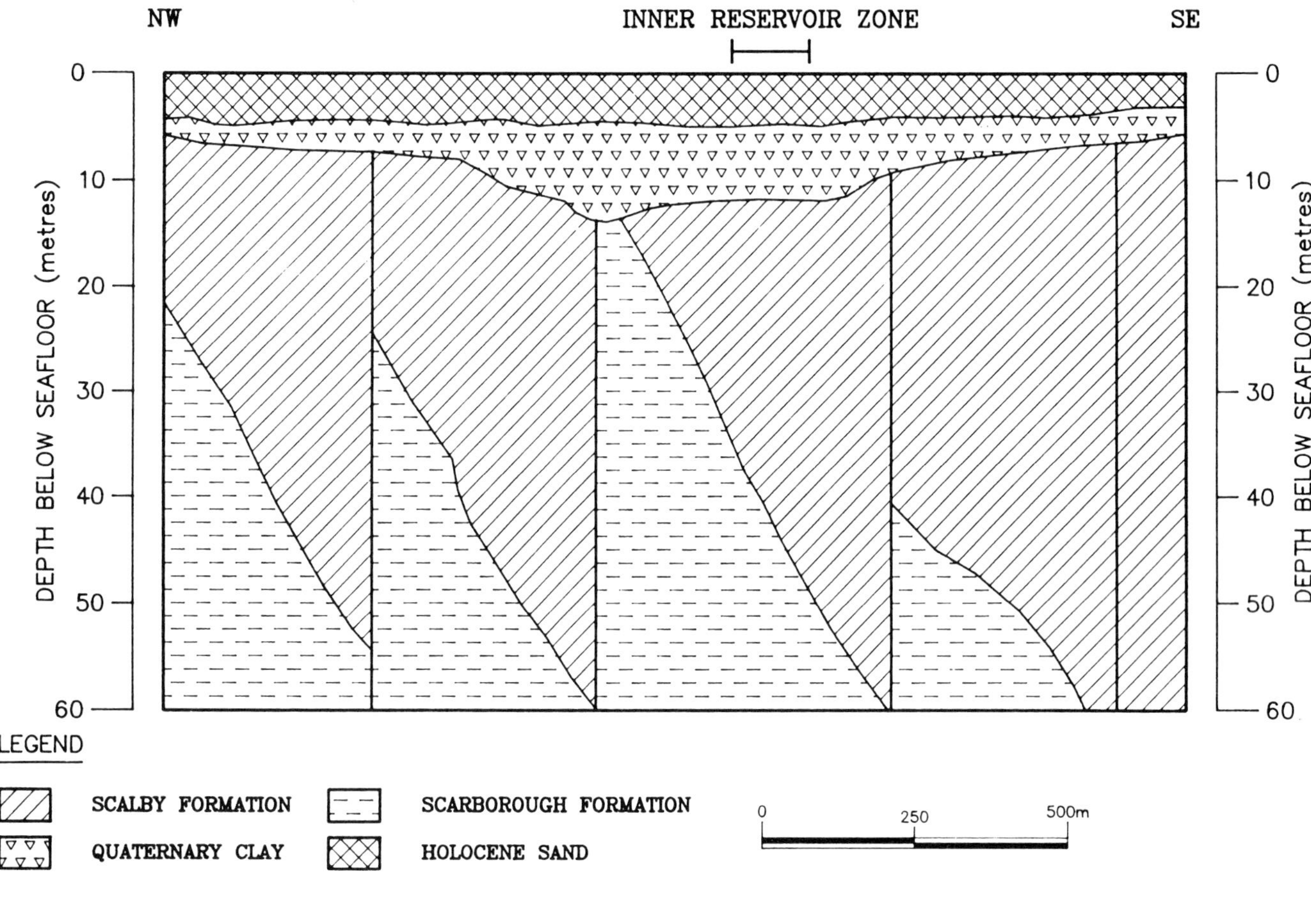

Fig. 11. Geological cross section Central Reservoir Zone.

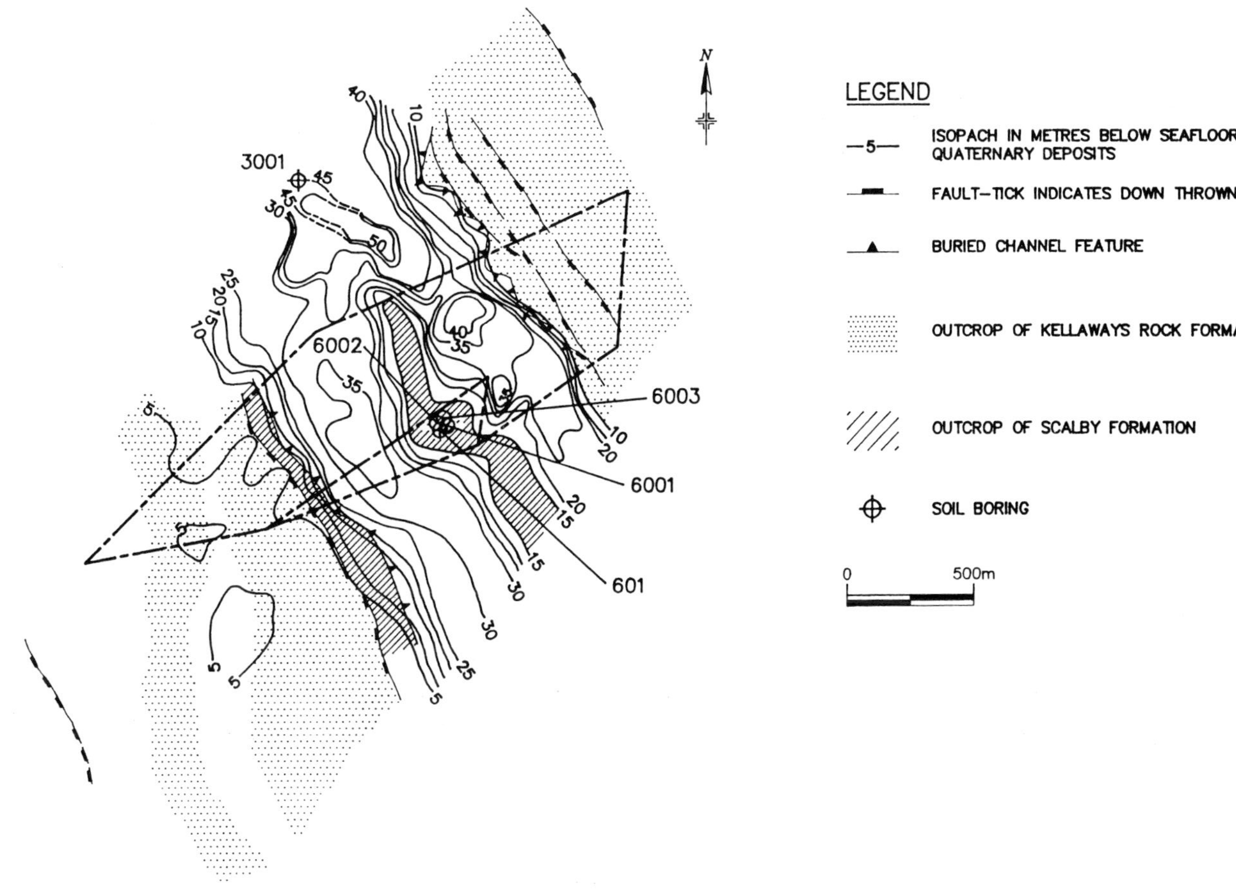

Fig. 12. Geological features map ST2 Reservoir Zone.

faults, a similarly oriented buried channel and the presence of strata containing rock layers. The "inner" or optimum part of the reservoir zone is located in the buried channel.

The main fault is located in the south-west of the reservoir zone and is oriented north-west to south-east parallel to the margin of the buried channel. The fault appears to be down thrown to the south-west with a possible displacement of 50m. This displacement is similar to those observed in the Central Complex Reservoir Zone. The acoustic character of the seismic records suggest that the Kellaways Rock Formation may outcrop to the south-west of the fault and the Scalby Formation to the north-east where it has been preferentially eroded to form the buried channel feature.

The interpretation of the seismic records and the geological and soil model for the field indicate that the Scalby Formation may outcrop on the south-west margin of the buried channel. As shown on Figure 12, this outcrop of strong cohesive soils forms a thin outcrop between the fault and channel margin which range from 50m in the north-west to 150m wide in the south-east. In the "inner" or optimum part of the reservoir zone of the Scalby Formation outcrop occupies about 50 square metres in the south-west apex of the target area.

The Kellaways Rock Formation may outcrop in the north-east of the reservoir zone. It was not clear from the seismic records whether the Scalby Formation outcrops along the north-east margin on the buried channel feature.

The close spacing of the survey lines and the high quality of the seismic records acquired in the site specific survey enabled the identification and mapping of a peninsula or island-like feature within the buried channel feature. The interpretation of the acoustic character of the seismic records and predictions made from the geological and soil model indicates that the peninsula-like feature may be composed or Scalby Formation (Figure 13).

The acoustic character of the channel infill sediments range from well bedded to structureless suggesting possible variability in soil conditions within the channel infill. The channel infill sediments have not been differentiated since a potentially suitable platform location was identified on the peninsula-like feature.

5.3.3. Predicted soil conditions

Based on the available data it was possible to recommend a suitable location for the ST2 structure within the "inner" target zone. The soil conditions at the proposed platform location were subsequently confirmed during the site specific geotechnical investigation.

The interpretation of the seismic records and predictions made from the geological and soil model indicate that the unfavourable interlayered very hard clay, dense sand, limestone and sandstone of the Scarborough Formation is expected between 45m and 55m penetration below seafloor. This suggests that between 35m and 40m of Scalby Formation consisting predominantly of strong cohesive soils

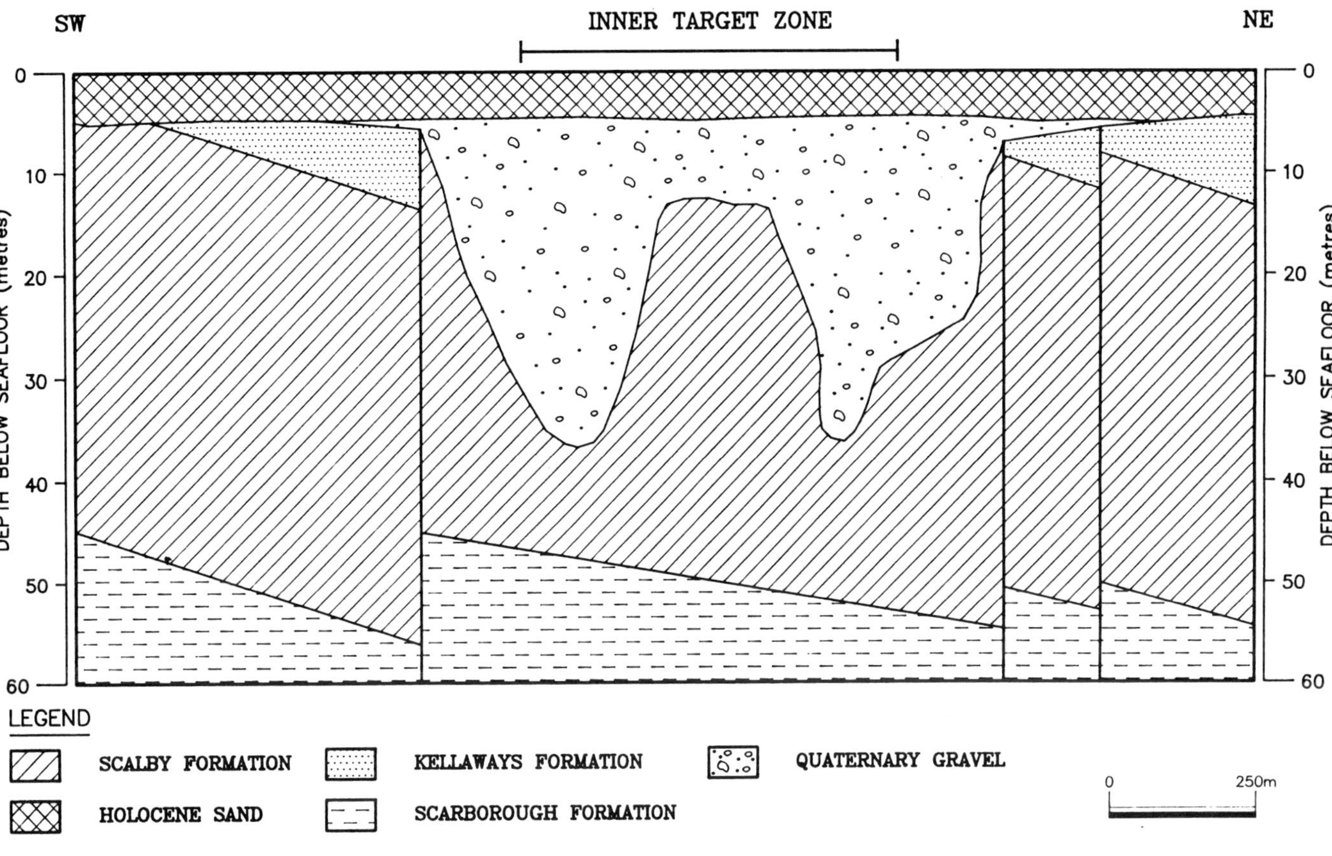

Fig. 13. Geological cross section ST2 Reservoir Zone.

are expected at the proposed platform location.

5.4. ST3 RESERVOIR ZONE

5.4.1. General

The ST3 Reservoir Zone is located about 6.5km north-west of the ST2 Reservoir Zone in the north-west part of the Ravenspurn North Field.

5.4.2. Geological features

The most significant geological features occurring in the ST3 Reservoir Zone comprise sandwaves, the presence of strata containing rock layers (Kellaways Rock Formation) and a major buried channel feature.

As shown on Figure 14, three sandwaves occur in the reservoir zone with axes oriented approximately east-north-east to west-south-west. The sandwaves range from 5m to 12m in height and have wavelengths of between 50m and 80m. The asymmetric shape of the sandwaves suggests that they may be migrating towards the north-north-west. Previous surveys in the study areas indicate that a sandwave field occurs to the south-west and south-east of the reservoir zone.

The ST3 Reservoir Zone is located on the south-west limb of the antiform and on the north-west margin of the major Quaternary erosion feature that occupies the congruent syncline in this part of the Ravenspurn North Field.

As shown on Figure 15, the south-west limb of the anticline is formed by the lateral equivalent of the Kellaways Rock Formation which subcrops at a depth of about 10m beneath Quaternary sediments in the north-east half of the reservoir zone. Where sandwaves are present the depth to the Kellaways Rock Formation increases in proportion to the height of the feature.

In the inner or optimum part of the reservoir zone the Kellaways Rock Formation subcrops at a depth of between 5m and 30m below seafloor. A soil boring was drilled in the inner part of the reservoir zone and proved rock strata at 8.1m penetration below seafloor verifying the ground conditions predicted from the geological and soil model.

Based on the interpretation of the seismic records and lithological information acquired from soil boring information, it was possible to differentiate the Quaternary sediments infilling the major erosion into an upper and lower unit. The depth in metres to the top of the lower unit is presented on Figures 14 and 15. The available data suggests that the upper unit may be predominantly granular while the lower unit may possibly comprise strong cohesive soils. A degree of lateral variability in soil conditions is expected to these Quaternary infill sediments.

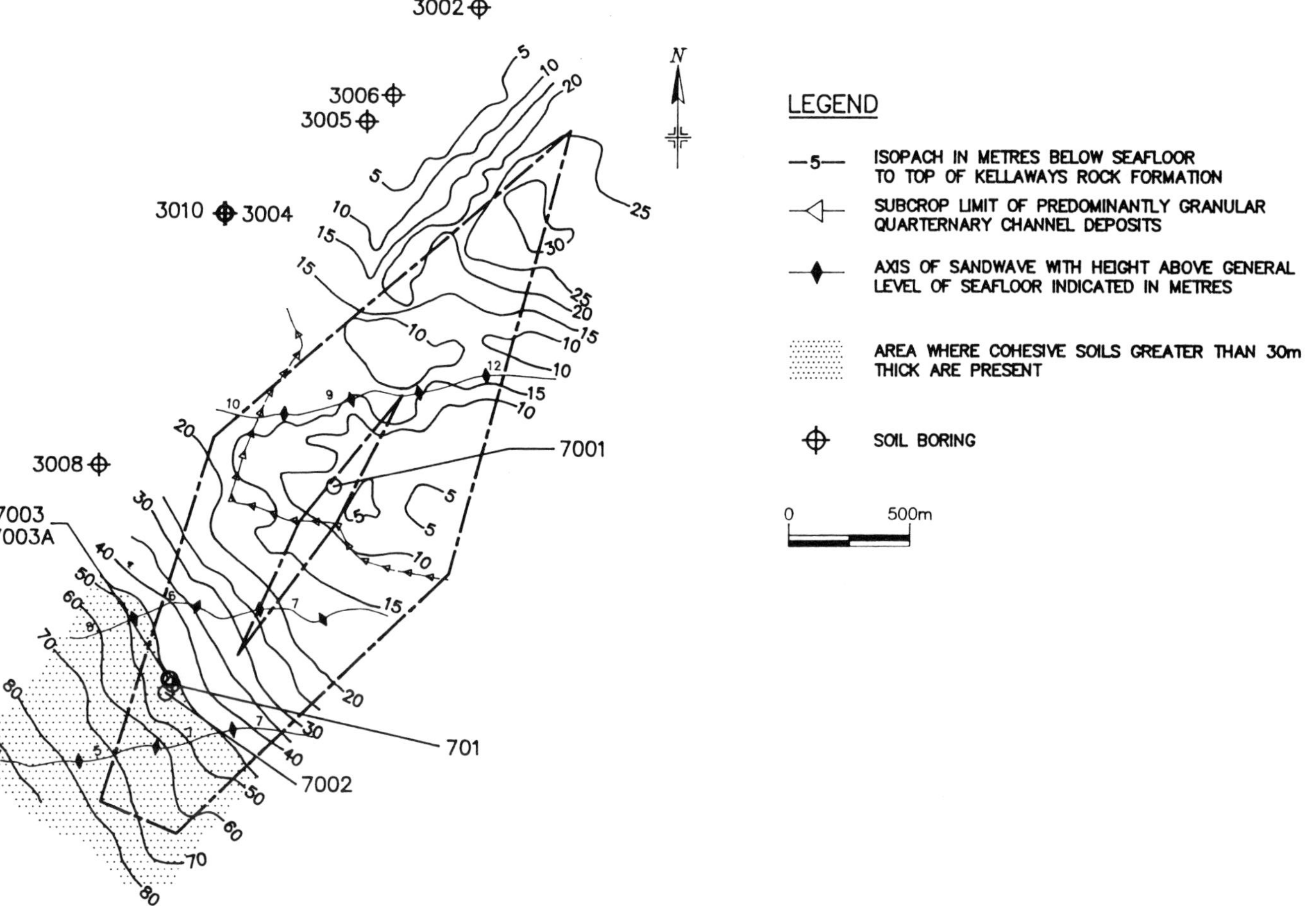

Fig. 14. Geological features map ST3 Reservoir Zone.

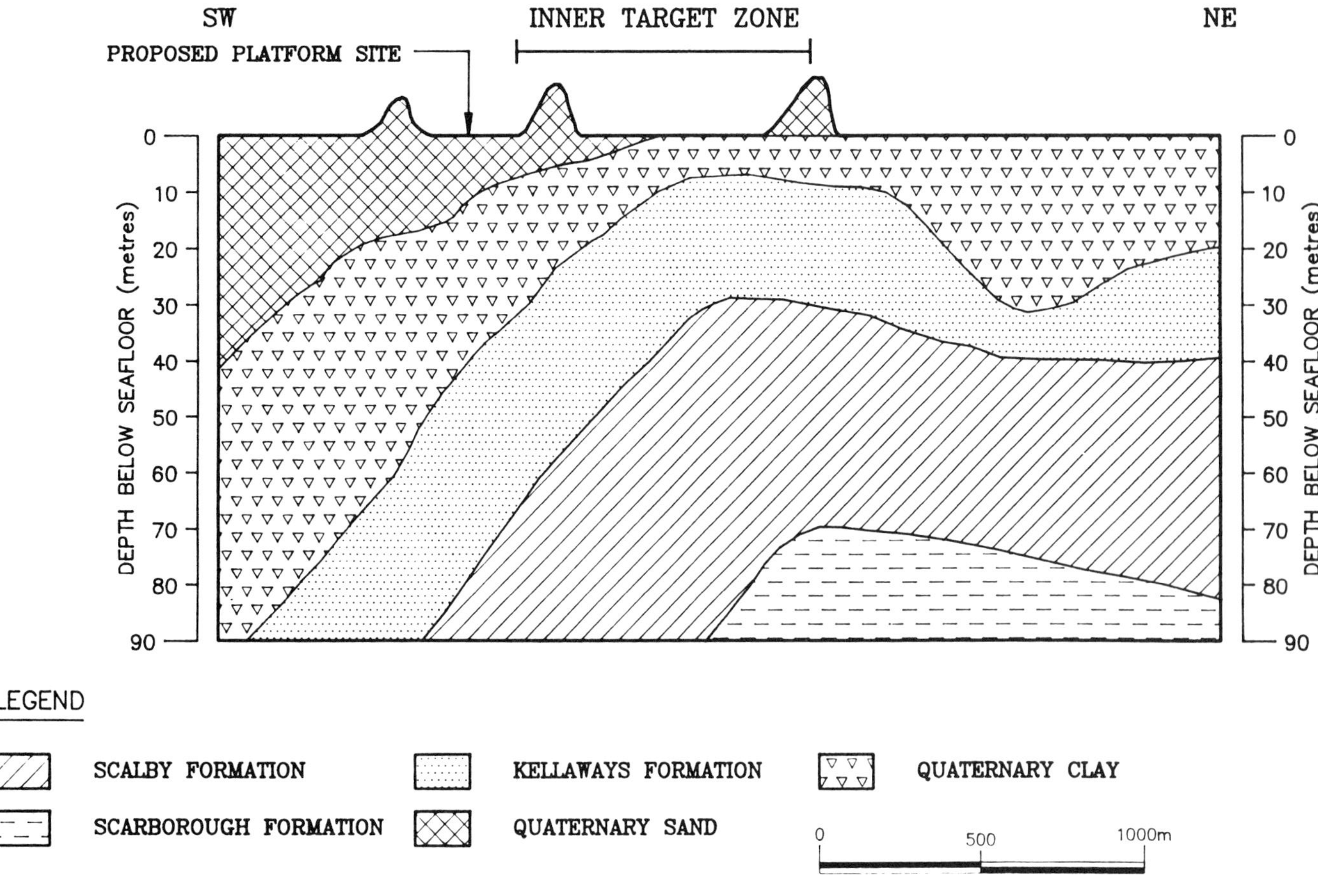

Fig. 15. Geological cross section ST3 Reservoir Zone.

5.4.3. *Predicted soil conditions*

Based on the interpretation of the seismic records and predictions made from the geological and soil model it was possible to recommend of a potentially suitable location for the ST3 platform in the "outer" target zone. The soil conditions at the proposed location were subsequently confirmed by soil borings.

The soil borings verified the depth contours to top of the Kellaways Rock Formation and top of the lower unit of channel infill. However, due to lateral variability in the lower unit of the channel infill between 15m and 20m of interbedded dense silts and fine sands were encountered overlying the very hard clay that we had expected to predominate in the lower unit of sediment infill.

6. Conclusions

The engineering geological approach to the siting of offshore structures exemplified by the development of the Ravenspurn North Field preferably consists of a number of components which are summarised below:

1. **Phase I**
   (a) review of the publicly available bathymetric, geological, geophysical and geotechnical data;
   (b) integration and interpretation of the available data to develop a predictive geological and soil model for the study area;
   (c) develop predicted soil profiles summarising the expected range of soil conditions in soil provinces, designated areas or at specified locations;
   (d) develop predicted design profiles for input to conceptual engineering design studies;
   (e) evaluate the requirements for future work.
2. **Phase II**
   (a) perform items (1a) and (1b) or review the findings of Phase I;
   (b) review the available rig site geophysical survey data;
   (c) perform items (1c) to (1e) or review the findings of Phase I.
3. **Phase III**
   (a) perform items (2a) and (2b) or review the findings of Phase II;
   (b) identify the proposed platform location or preferably define the reservoir zone and design an engineering geophysical survey which provides the relevant information to satisfy the engineering requirements of the development concept;
   (c) perform the engineering geophysical survey and carry out an interpretation of the acquired data which has engineering relevance;

(d) perform item (2c) or review the findings of the Phase II.

4. **Phase IV**

(a) perform items (3a) to (3d) or review the findings of Phase III;

(b) select the preferred platform locations and design a cost effective site investigation programme;

(c) perform the site investigation and carry out the onshore laboratory testing programme to provide the necessary soils data for input to the final engineering design studies;

(d) perform item (3d) or review the findings of Phase III ready for input to additional studies should this become necessary.

The Phase I to IV represent a progression of increasing sophistication and improving levels of confidence in the prediction of soil conditions. Each phase builds on the findings of its predecessor and therefore maximises the benefit obtained from the evolving data set.

The concept of reservoir zones, which delineate the areas where the production platforms should be located to optimise field development, proved to be particularly effective in the planning of the Phase III and IV investigations. In the Central Complex and ST2 Reservoir Zones, the sites of the proposed structures were selected at favourable locations within the "inner" target zone which defines the optimal area for platform locations. In the ST3 Reservoir Zone, the site of the proposed structure was selected in the larger "outer" zone delineating the possible area for platform locations. In each case, suitable ground conditions were limited to specific areas which may not have been located by an expensive "wildcat" soil boring programme and the resultant need to engineer foundation solutions in unfavourable ground conditions may have significantly increased the cost of the development. Thus, the application of an engineering geological approach to the siting of the offshore structures in the Ravenspurn North Field resulted in a cost benefit to the development project.

Hamilton Oil Company Limited are continuing to employ the engineering geological approach to the siting of offshore structures in their Hamilton, Douglas and Johnston Field Development Projects. In addition, this technique is being employed to evaluate ground conditions at proposed well sites to assess the need for, and the type of geotechnical investigations required prior to the commencement of drilling operations.

Acknowledgements

The author would like to thank Hamilton Oil Company Limited and the Partners in the Ravenspurn North Field Development Project for permission to publish this article. In particular, the author would like to thank Dr. A. L. Woodhead for his review, and Liz Fellowes, David Fiddaman and Janet Miller for their help in the preparation of the text and figures.

References

1. Campbell, K. J. (1984), 'Predicting offshore soil conditions', *Proceedings of the 16th Annual Offshore Technology Conference*, Houston, pp. 391–398.
2. Campbell, K. J., Dobson, B. M., and Ehlers, C. J. (1982), 'Geotechnical and engineering geological investigations of deep water sites', *Proceedings of the 14th Annual Technology Conference*, Houston, pp. 25–37.

Discussion

Question from John Arthur and Associates: How much encouragement did the speaker's company have to apply to the client to convince them of the benefit of reviewing ground conditions over the regional area in preference to solely undertaking site specific studies?

Authors' response: We encouraged the client by demonstrating that the Middle and Upper Jurassic strata present at shallow depths in the Ravenspurn North Field could provide both "favourable" and "unfavourable" ground conditions for the installation of conventional open-ended, steel pipe-piles and that it would be possible to identify "favourable" areas by the application of the engineering geological approach to the siting of the proposed structures. In addition, financial analysis of the various foundations options at the conceptual engineering stage revealed a cost benefit to the siting of structures at locations with ground conditions "favourable" to the installation of conventional open-ended, steel pipe-pile foundations.

As indicated in the paper, suitable ground conditions were limited to specific areas which may not have been located by an expensive "wildcat" soil boring programme. The engineering geological approach to the siting of the offshore structures was therefore the most logical and cost effective way to proceed.

GANNET SITE AND PIPEROUTE SURVEYS – AN INTEGRATED INTERPRETATION

J. H. SOMMERVILLE
Geoteam UK Ltd., Regent House, Regent Quay, Aberdeen AB1 2BE

and

P. M. WALKER
Shell U.K. Exploration and Production, Shell Mex House, The Strand, London WC2R 0DX

Abstract. The Gannet Development in the Central North Sea, consists of a platform, seven subsea developments and associated infield and export pipelines.

This paper provides a case history of the site survey aspects of the field development. It describes the advantages of a carefully planned, comprehensive site survey programme and also the advantages of a detailed integration of all available data to provide a reliable interpretation of geophysical data. This is of particular importance in an area such as Gannet where there is shallow gas and highly variable shallow soils.

Several seabed surveys and site investigations have been undertaken in the Gannet area over the last ten years. These commenced with reconnaissance surveys in the early eighties. As the field development plans were finalised, a programme of detailed seabed surveys and site investigations was established, supplemented by a shallow gas pilot hole.

Interpretation of the seismic data required careful integration of all the data sets. The result of this interpretation was a good understanding of the shallow soils and shallow gas within the area.

Shallow gas prognoses were made for all the proposed locations and to date these have been confirmed by wells at five of the locations.

Piling and anchoring conditions have been confirmed at the platform and drilling centre locations and trenching conditions for the pipelines were as predicted.

1. Introduction

The Gannet Field is situated in the Central North Sea in UKCS Blocks 21/25, 21/30, and 22/21. The location of the field is shown on Figure 1. Development of the field is being undertaken by Shell U.K. Exploration and Production, operator in the U.K. sector of the North Sea for Shell and Esso. Field development is well underway and first oil is expected in October 1992.

The Gannet development is in water depths of approximately 95m. It covers three oil and associated gas fields designated Gannet A, C, and D, and a gas field designated Gannet B.

The development is shown in Figure 2. It consists of the Gannet A Platform and seven subsea developments (drilling centres), designated Gannet B1, B2, C1, C2, C3, C4, and D. Pipelines link the subsea developments with the platform. Export of gas is through the Fulmar to St. Fergus Pipeline via the Gannet Diverter and of

Volume 28: Offshore Site Investigation and Foundation Behaviour, 295–332, 1993.

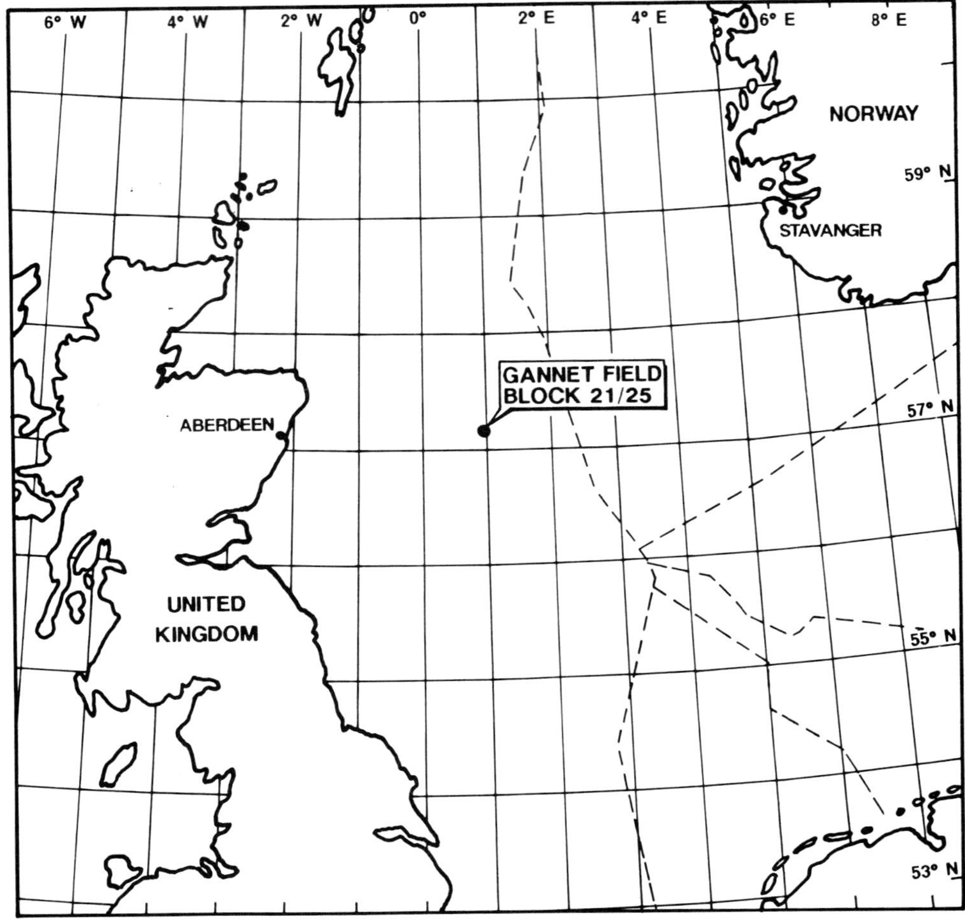

Fig. 1. Gannet Field location map.

oil is via a pipeline to Fulmar Alpha.

Reconnaissance surveys were undertaken prior to completion of the field development plans. These surveys consisted of geophysical surveys in 1982 and 1985 and geotechnical investigations in 1983 and 1985. An integrated interpretation was then undertaken which incorporated the results of these surveys and available well data, to provide a regional overview of the geology of the area. These reconnaissance surveys indicated that the shallow geology was not uniform and considerable variations in foundation conditions were expected. Shallow gas was also observed to be widespread in the Gannet area.

As field development plans were finalised in 1989, detailed geophysical surveys were undertaken, followed later that year by geotechnical investigations. A shallow gas pilot hole was drilled in 1990.

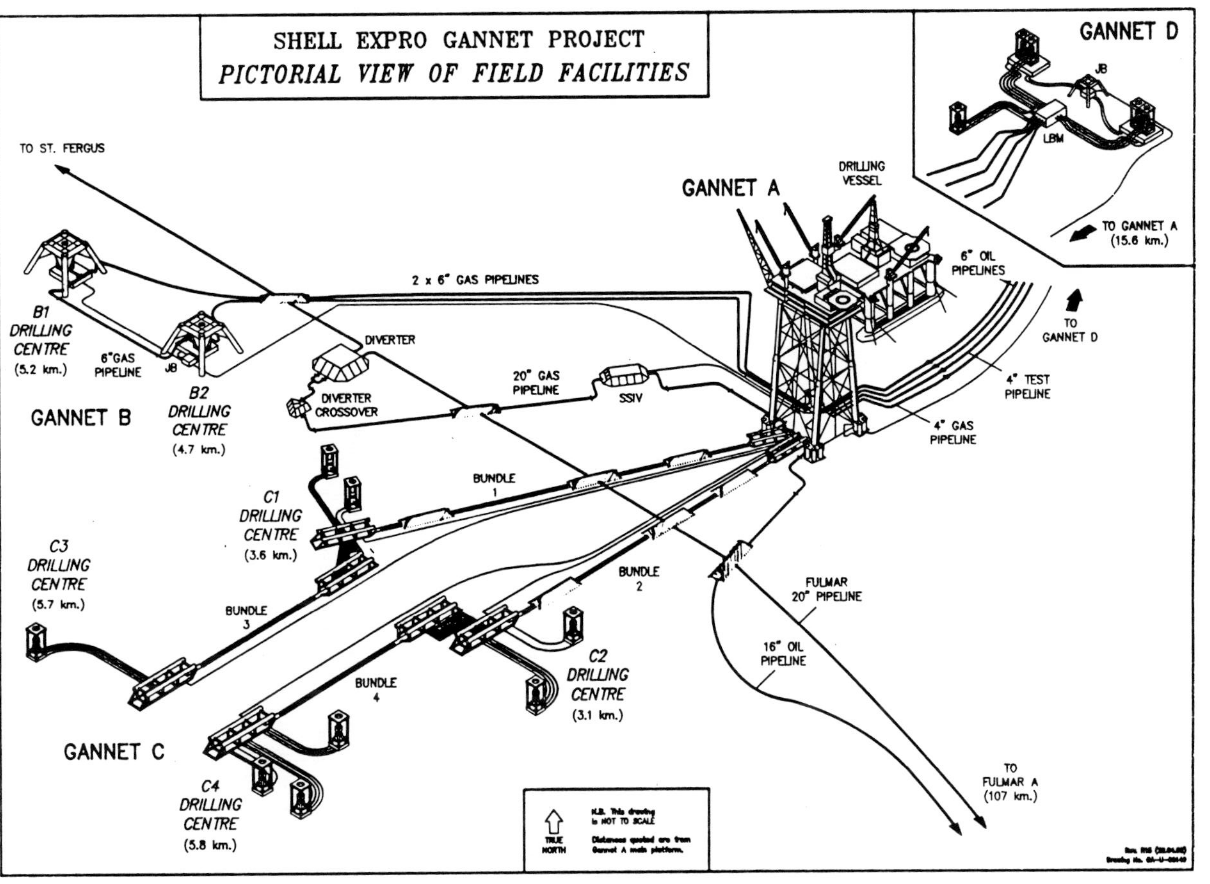

Fig. 2. Pictorial view of field facilities – Shell Expro Gannet Project.

This paper discusses the type of data acquired and the integration of the various data types to produce a reliable interpretation of the shallow geology and potential gas hazards of the area. Although this paper is simply a case history, it attempts to show the advantages of a well planned detailed survey campaign in an area which has highly variable shallow soils and is known to be shallow gas prone.

2. Field Acquisition

2.1. Survey Work Programme

Proposed locations for installations could have been unsuitable due to seabed obstructions, foundation conditions or shallow gas accumulations. To eliminate the need for repeat surveys, due to unsuitable locations, the following schedule of work was set up:

a) Desk study.

b) Reconnaissance seabed surveys.

c) Reconnaissance geotechnical work.

d) Detailed seabed surveys – proposed locations acceptable on the basis of the geophysical data.

e) Detailed geotechnical work.

f) Anchor surveys at drilling centres immediately prior to drilling rigs moving on to the location.

After each phase of the work was completed, adequate time was allowed for interpretation and reporting so that the results could be used to decide suitable locations and scopes of work for the next phase.

A shallow gas pilot hole was undertaken after the detailed geotechnical work as the requirement for pilot holes at platform locations was introduced by Shell U.K. Exploration and Production in late 1989, after the geotechnical work had been completed. Pilot holes at platform locations are now scheduled to be undertaken after detailed seabed surveys and before the detailed geotechnical work. This ensures that the locations are acceptable prior to undertaking such work.

2.2. Geophysical Data

In 1982 reconnaissance geophysical surveys were undertaken in the Gannet area, covering an area of 15km by 8km.

Equipment used was as follows:

- Echo Sounder
- Side Scan Sonar
- Pinger (single channel analogue recording)
- Mini Sleeve Exploder (multi-channel digital recording – 0.5ms sampling)
- Vibrocorer.

In 1985 the pipeline route surveys were undertaken based upon the development plan at that time. These surveys covered 500m wide corridors.

In 1989 detailed site and pipeline route specific geophysical surveys were undertaken. The scope of work for each aspect of the development is listed below:

a) Gannet A Platform location

  (i) 100m by 100m Detailed Bathymetry Survey
  (ii) 1km by 1km Shallow Gas Survey with tie lines to wells
  (iii) 1km by 1km Foundation Survey
  (iv) 4km by 4km Anchoring Survey

e) Gannet B, C and D Drilling Centre Locations

  (i) 1km by 1km Shallow Gas Survey with tie lines to wells
  (ii) Shallow Foundation Survey (covered by c, below)

c) Pipeline routes (A–D, A–B, A–C, A–Diverter)

  (i) 120km of 500m wide corridor with detailed investigations at existing pipeline crossings.

Equipment used was as follows:

- Echo Sounder
- Side Scan Sonar
- Deep Tow Boomer (single channel analogue recording)
- Airgun (multi-channel digital recording – 0.5 and 1ms sampling).

As there was considerable time scheduled prior to semi-submersible drilling at the drilling centres, the seabed survey data would potentially be 'out of date' when the rigs moved to location. Therefore, the anchoring surveys were not undertaken at these locations in 1989. They are undertaken six weeks prior to the rigs moving on to location. The scope of work for these anchoring surveys consists of a survey of an area 3km by 3km using the following equipment:

- Echo Sounder
- Side Scan Sonar
- Deep Tow Profiler – Boomer or Sparker (single channel analogue recording).

2.3. Shallow Gas Pilot Hole

During the development of Gannet, Shell U.K. Exploration and Production introduced a policy that one or more pilot holes will be drilled at any proposed platform location to test for shallow gas. Even if a shallow gas geophysical survey has been undertaken and shallow gas is considered unlikely, the pilot hole will be drilled. A pilot hole is required because interpretation of the geophysical data does not allow the presence of shallow gas to be completely discounted in some areas.

Although pilot holes had been drilled before by Shell U.K. Exploration and Production, the pilot hole at Gannet was the first such hole to be drilled as part of this policy.

These pilot holes are treated as shallow wells and have the same seabed survey and Department of Trade and Industry drilling approval requirements. A standard rig site seabed survey is therefore undertaken prior to drilling the pilot hole to ensure it is drilled at the most suitable location and that it does not encounter any unexpected accumulations of shallow gas. As the pilot hole at Gannet A was close to the platform location, the seabed survey acquired at the platform location was used for the shallow soils and shallow drilling prognosis for this well.

From the interpretation of the geophysical data at the Gannet A Platform location, shallow gas was not expected. However, shallow gas had been encountered in several of the nearby wells. The pilot hole location was chosen to be representative of the shallow geology at the location but at a distance from the location which would ensure it did not influence foundation conditions for the proposed platform. The hole was drilled in April 1990 from a conventional semi-submersible rig normally used for exploration and appraisal drilling. A 12.25 inch diameter hole was drilled without a riser to 600m sub-seabed. The following wireline logs were run:

- Gamma ray
- Sonic
- Density
- Resistivity.

Although log quality was variable due to hole caving, water bearing sand layers were detected. No shallow gas was encountered.

2.4. Geotechnical Borehole and Cone Penetrometer Test (CPT) Data

The boreholes within the Gannet area are shown on Figure 3.

In 1983, 6 boreholes were drilled in the Gannet area to depths of 50m sub-seabed. These were followed in 1985 by a further 14 boreholes to depths of 80–110m sub-seabed and 28 SEACALF CPTs to depths of 30m sub-seabed. These were at locations of significance in the early field development plans.

In 1989, a detailed borehole and CPT programme was undertaken with 2 boreholes at the Gannet A Platform Location to depths of 110m sub-seabed and 50 SEACALF CPTs along the proposed pipeline routes and at the drilling centres to depths of 5m sub-seabed.

3. Additional Data Used

3.1. Exploration Seismic Data

For a regional overview of the geology and shallow gas risk, the available exploration 2D and 3D seismic data was examined.

Figures 4a and 4b are examples of exploration 3D seismic data in the Gannet area showing two of the shallow gas prone horizons. They are 100 millisecond time window displays, where only high amplitude reflections have been highlighted. Also shown on these figures are the high amplitude reflections as mapped from the shallow gas seismic data.

3.2. Seabed Survey Data

Numerous seabed surveys have been undertaken in Blocks 21/25, 21/30 and 22/21 prior to drilling exploration and appraisal wells. These surveys provide additional information on seabed conditions, shallow soils and shallow gas. Where the data was considered useful for correlation purposes, regional geophysical lines were run to tie these sites to the Gannet locations.

3.3. Well Data

There are in excess of 30 exploration and appraisal wells drilled in Blocks 21/25, 21/30 and 22/21. Some of these are shown on Figure 3. Many of these wells have detailed logging within the top hole section. Additional shallow gas seismic lines were run to permit correlation of the shallow geology in the Gannet development area with these wells.

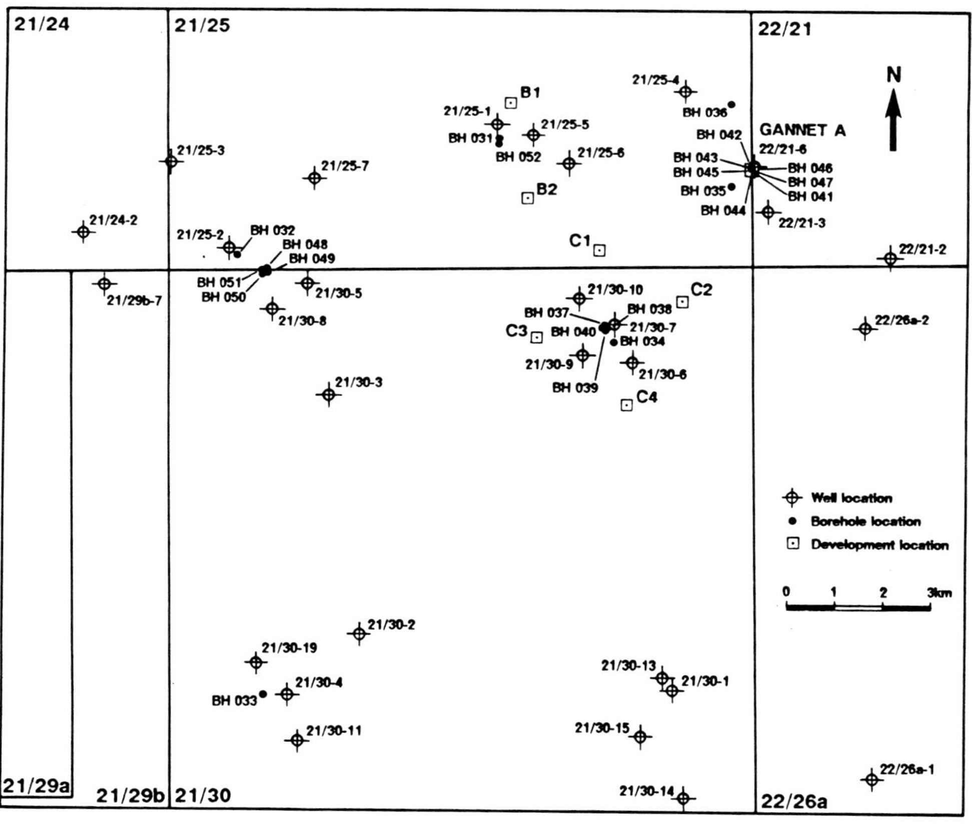

Fig. 3. Location of wells and geotechnical boreholes in the Gannet area.

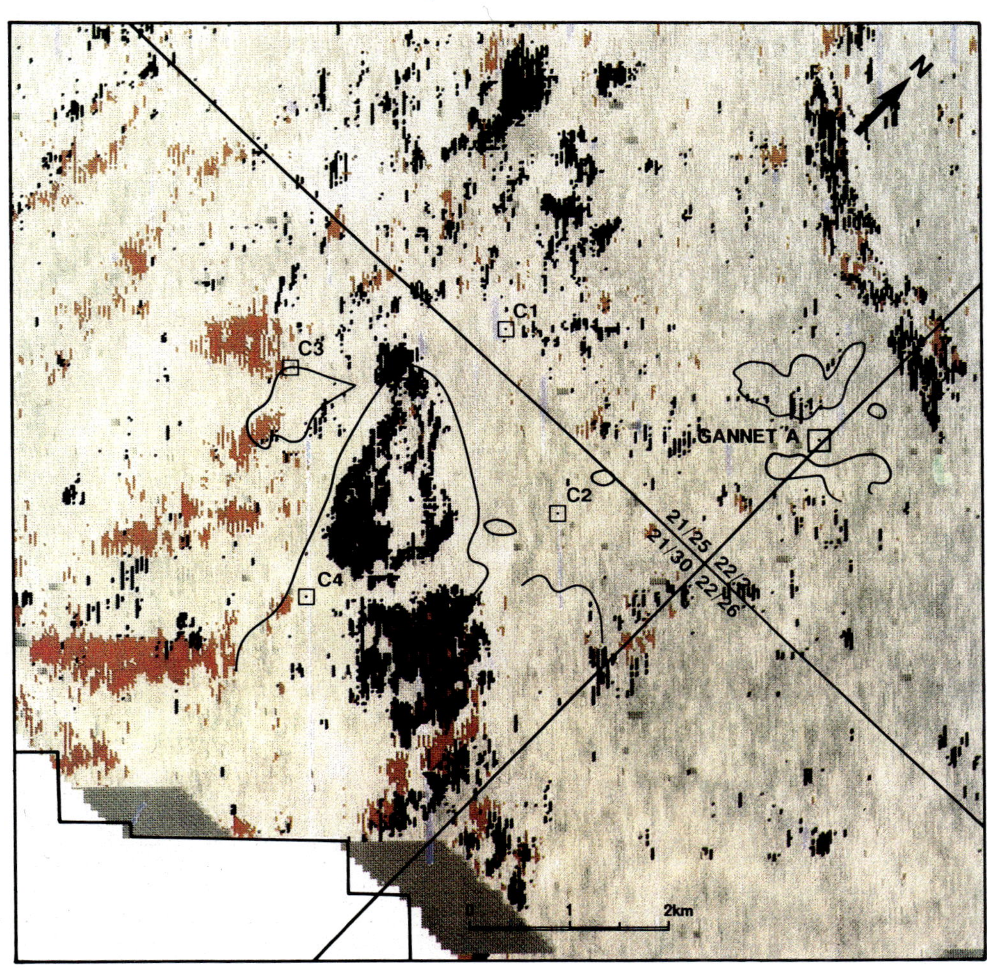

Fig. 4a. 3D seismic data showing potential gas sands in the Gannet area with the shallow gas survey interpretation superimposed. Window 400–500 milliseconds TWTT.

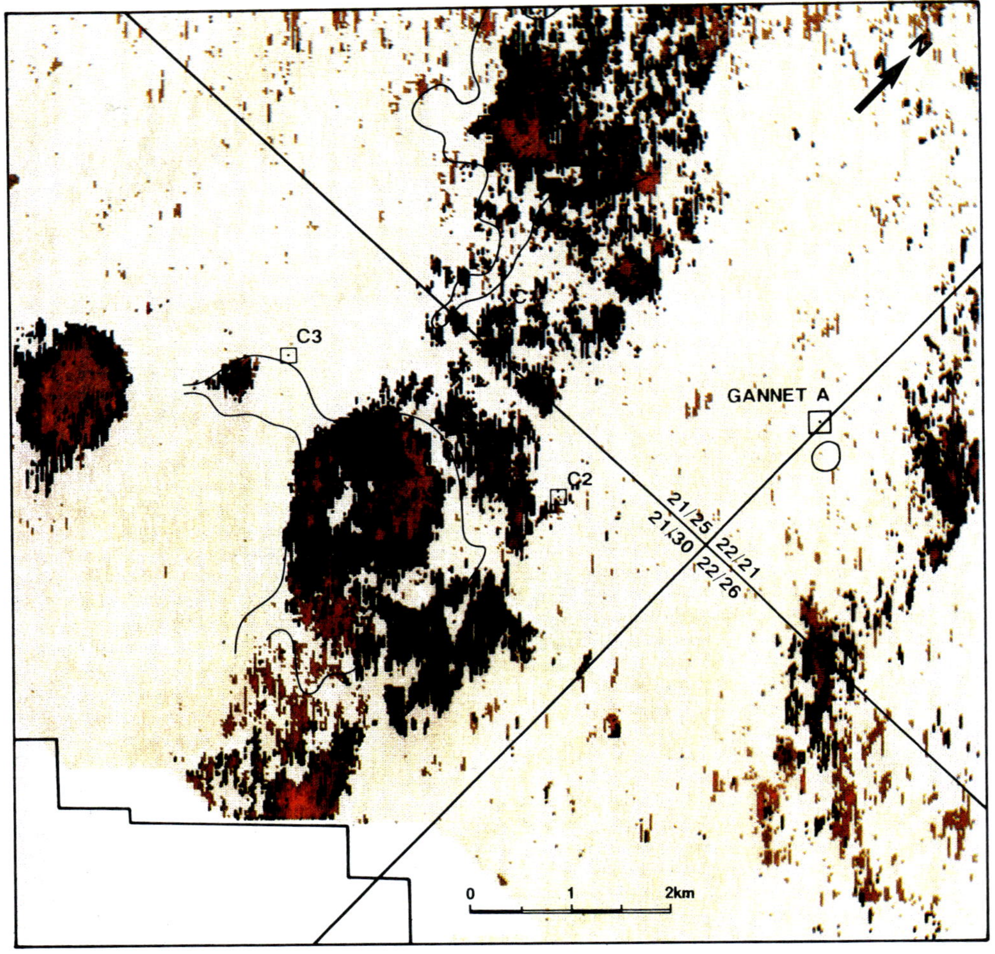

Fig. 4b. See Figure 4a. Window 500–600 milliseconds TWTT.

3.4. BGS DATA

The Gannet Field lies within the area covered by the following BGS Charts:

- Forties Seabed Sediments Sheet (Reference 1)
- Forties Quaternary Geology Sheet (Reference 2).

In addition, the Quaternary geology of the area is discussed in several BGS publications (References 3 and 4).

4. Shallow Geology

An overview of the shallow geology in the Gannet area is shown on Figure 5, an interpreted geological profile through the Gannet A Platform location. In addition, Figures 6 and 10 are examples of seismic data in the Gannet area.

In the upper 1000m sub-seabed, the shallow geology consists of Recent, Quaternary and Tertiary sediments. The total thickness of the Quaternary sediments is in the order of 350m. For descriptive purposes the shallow geology of the area was divided into nine units. Units 1–7 represent the Quaternary geological sequence in the Gannet area.

4.1. RECENT SEDIMENTS

These consists of a veneer of fine sand up to 30cm thick.

4.2. QUATERNARY SEDIMENTS

4.2.1. Forth Formation

The Forth Formation consists of the younger Whitethorn Member and the older Fitzroy Member. The sediments of the Whitethorn Member are very fine grained, often silty sands. The sediments of the Fitzroy Member is interpreted by BGS as post-glacial shallow marine sands, whereas the Fitzroy Member is interpreted as Weichselian late-glacial or early post-glacial, low energy marine sediments laid down on an irregular fluvio-glacial terrain.

As illustrated by Figure 6, the Forth Formation in the Gannet area consists of well laminated sediments infilling irregularities in the underlying erosion surface at the top of the Coal Pit/Fisher Formation sediments.

The Forth Formation in the Gannet area was divided into Units 1–4.

- Unit 1 Medium dense to dense silty fine sand.
- Unit 2 Very soft to soft clay interbedded with clayey sand.
- Unit 3 Soft to firm sandy clay with occasional thin sands.
- Unit 4 Firm to stiff clay.

Unit 1 was interpreted as the Whitethorn Member and Units 2–4 as the Fitzroy Member.

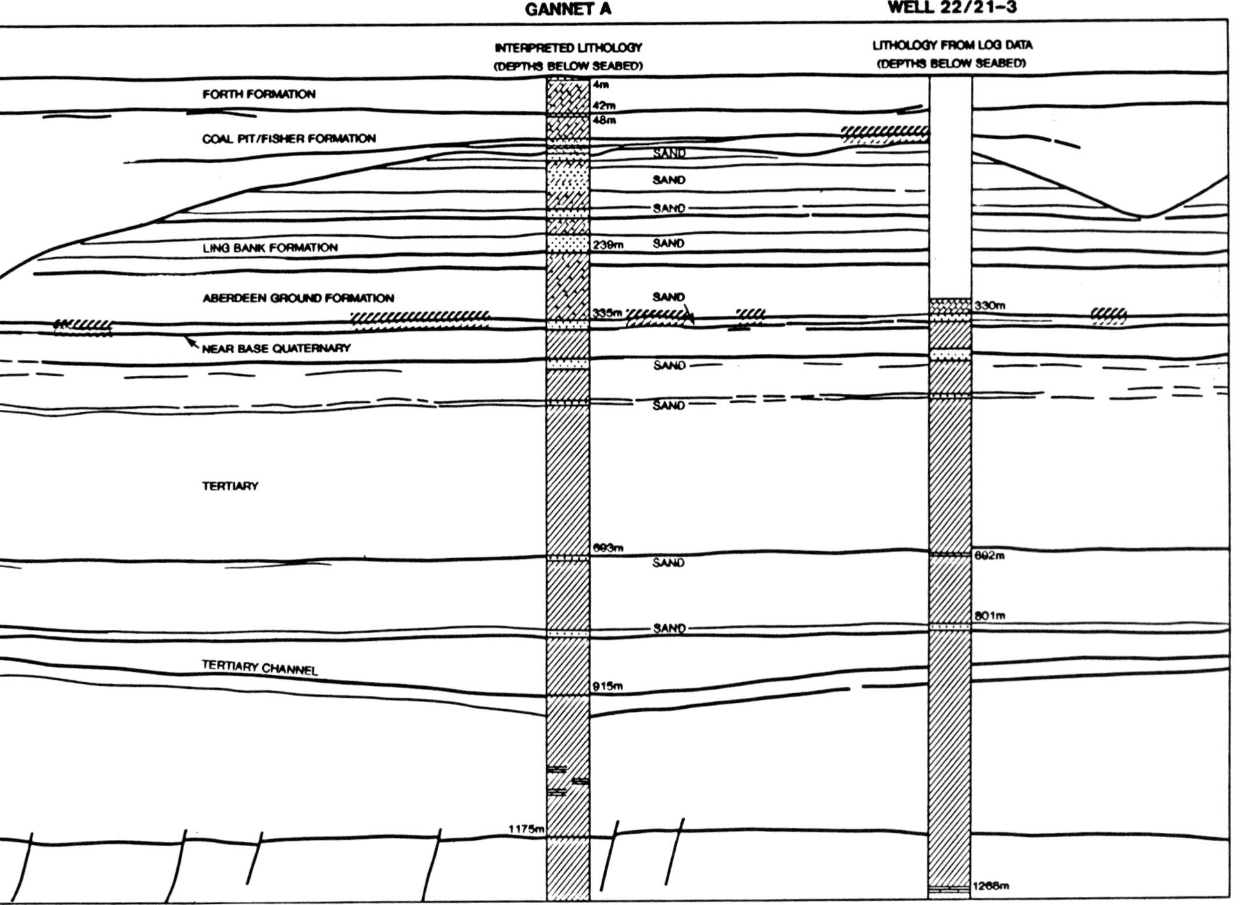

Fig. 5. Overview of the shallow geology of the Gannet area.

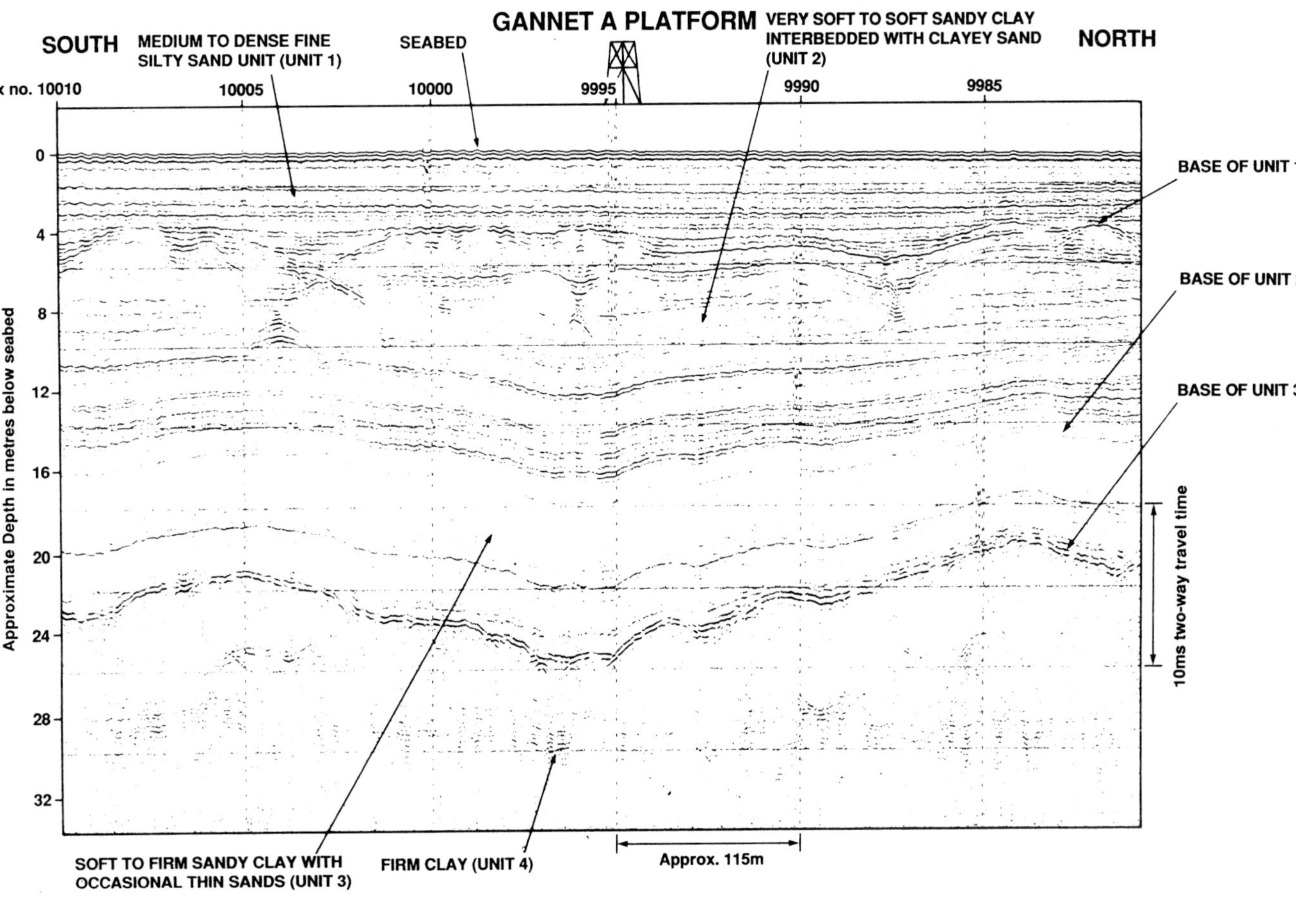

Fig. 6. Deep tow boomer data – Gannet A Platform area.

4.2.2. Coal Pit/Fisher Formation

Coal Pit sediments comprise sand, silty clay and interlaminated clay and silty sand. In general the clay is stiff and overconsolidated. The lower part of the Coal Pit Formation is thought by the BGS to be Eemian interglacial shallow marine deposits laid down on a glacially eroded irregular terrain. The upper part is taken to be Weichselian. The erosion surface at the base of the Coal Pit Formation often forms deep channels.

Fisher Formation sediments comprise very stiff overconsolidated clays with silt and shelly sand. The formation is interpreted by the BGS as glacio-marine, possibly intertidal sediments. These sediments may be Saalian in age.

In the Gannet area, discrimination between Coal Pit and Fisher Formation sediments based on seismic character was difficult. Unit 5 was therefore designated as Coal Pit/Fisher Formation. It was interpreted to consist of firm to very hard normally consolidated sandy silty clays with shell fragments and occasional sand layers. The clays become overconsolidated with increasing depth sub-seabed.

4.2.3. Ling Bank Formation

Ling Bank sediments are dominantly silty with interbeds of sand and clay. These sediments are interpreted by BGS as interglacial and of Holsteinian age. They are shallow marine deposits laid down on a glacially eroded terrain. The base of this formation is generally an irregular erosion surface, often forming deep channels.

Unit 6 was designated as Ling Bank Formation and interpreted to consist of stiff to very stiff silty sandy clays with pebbles. These sediments infill channels eroded into the underlying Aberdeen Ground Formation.

4.2.4. Aberdeen Ground Formation

Aberdeen Ground sediments are varied but according to the BGS are hard, overconsolidated clay which is sometimes fissured. Laminae and lenses of sand are common. These sediments are thought to be glacio-marine, laid down in an inner shelf area, and of Cromerian or older age.

Unit 7 was designated as Aberdeen Ground Formation and interpreted to consist of very hard, normally consolidated to overconsolidated sandy clays with pebbles and numerous sand layers.

4.3. TERTIARY GEOLOGY

Tertiary claystones with thin sandstone and limestone layers underlie the Quaternary sediments. These were divided into Units 8 and 9. The deeper Unit 9 is highly faulted.

Within the Tertiary sedimentary sequence, at approximately 850m sub-seabed, broad channels are observed on the seismic data. These northwest–southeast trending channels are in excess of 1km wide and in the order of 70m deep. The sub-seabed depth to this Tertiary channelled horizon was mapped in the vicinity of Gannet A Platform location and is presented as Figure 7.

As this figure shows, the platform location is situated in the centre of this channel. However due to the sub-seabed depth of this channel it had no impact on the selection of a suitable location for the platform.

5. Shallow Gas

Shallow gas is defined as gas within the open hole drilling section of a well where there is no protection from a blowout preventer. In the Gannet area this would be in the interval up to 600m sub-seabed. Gas at these levels is a potential drilling hazard to drilling rigs or platforms. Shallow gas within the near seabed soils is also a potential hazard to site investigation vessels drilling geotechnical boreholes.

The Gannet area is known to be prone to shallow gas, particularly near to the base of the Quaternary.

5.1. Initial Shallow Gas Evaluation

The reconnaissance shallow gas seismic surveys clearly indicated the presence of high amplitude reflections at two levels within the proposed development area (Reference 5). These potential shallow gas accumulation were mapped and are shown on Figure 8. Comparison of Figure 8 with Figure 4 shows good agreement between the reflections as mapped from this reconnaissance survey and those observed on the exploration 3D seismic surveys. This highlights the potential for the use of conventional 3D data for understanding the regional setting in the Quaternary and Upper Tertiary.

A study was undertaken of the top hole section of the wells drilled in the area. These wells confirmed the presence of shallow gas in sand layers within the Quaternary and Upper Tertiary sequence. Six wells had been drilled through shallow gas accumulations in the Gannet area, as shown on Figures 4 and 8. To illustrate the log response of these sand layers, Figure 9 shows a section of the completion log from well 21/30-6a which was drilled through two of these shallow gas accumulations. No drilling problems were recorded when drilling any of these shallow gas accumulations. The gas was under hydrostatic pressure and the drilling mudweight alone was sufficient to control the gas.

Where possible the proposed development locations were chosen to avoid the high amplitude reflections which were interpreted as gas charged sands. Even though these sands had been drilled by several wells without problem, it was decided that they should be avoided if possible. In particular, the platform location should avoid shallow gas. However, as shown by Figures 4 and 8, it was not practical for all the drilling centres to avoid all the shallow gas accumulations as they were so widespread. Gannet B2 and C4 were situated in areas where shallow

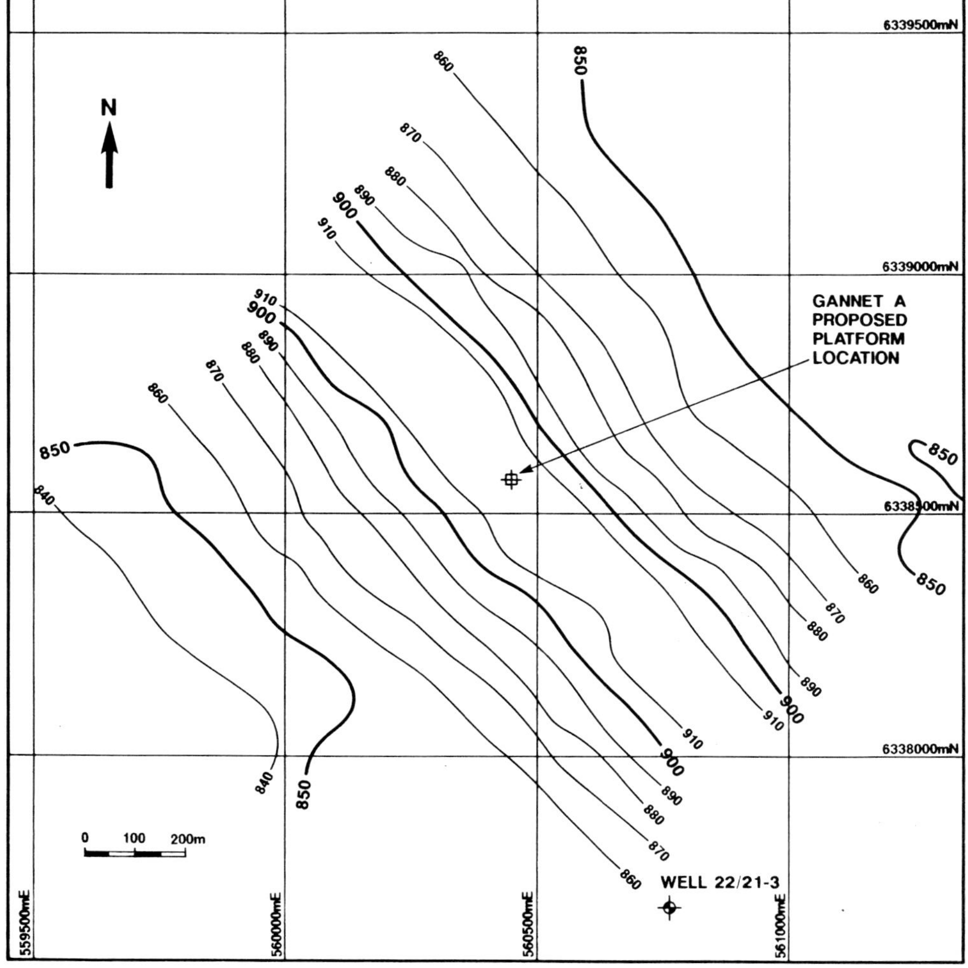

Fig. 7. Depth in metres sub-seabed to the Tertiary channelled horizon.

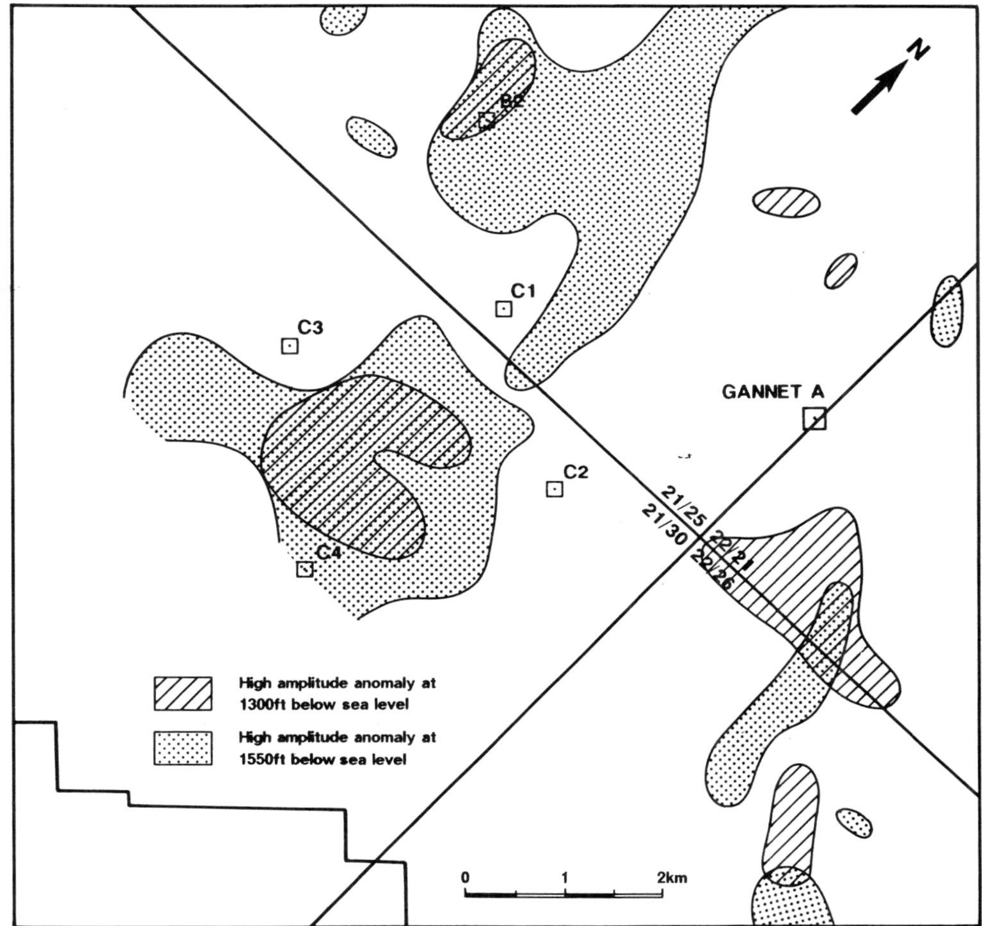

Fig. 8. Potential shallow gas accumulations mapped from the reconnaissance shallow gas seismic surveys.

gas was likely.

5.2. Detailed Shallow Gas Evaluation

Detailed shallow gas geophysical surveys were undertaken at the proposed platform and drilling centre locations. Tie lines were run to nearby wells.

Interpretation was undertaken primarily using seismic data displayed conventionally on paper. In addition, the centre lines and well tie lines were interpreted on a Landmark Workstation. Figure 10 shows two examples of data from the

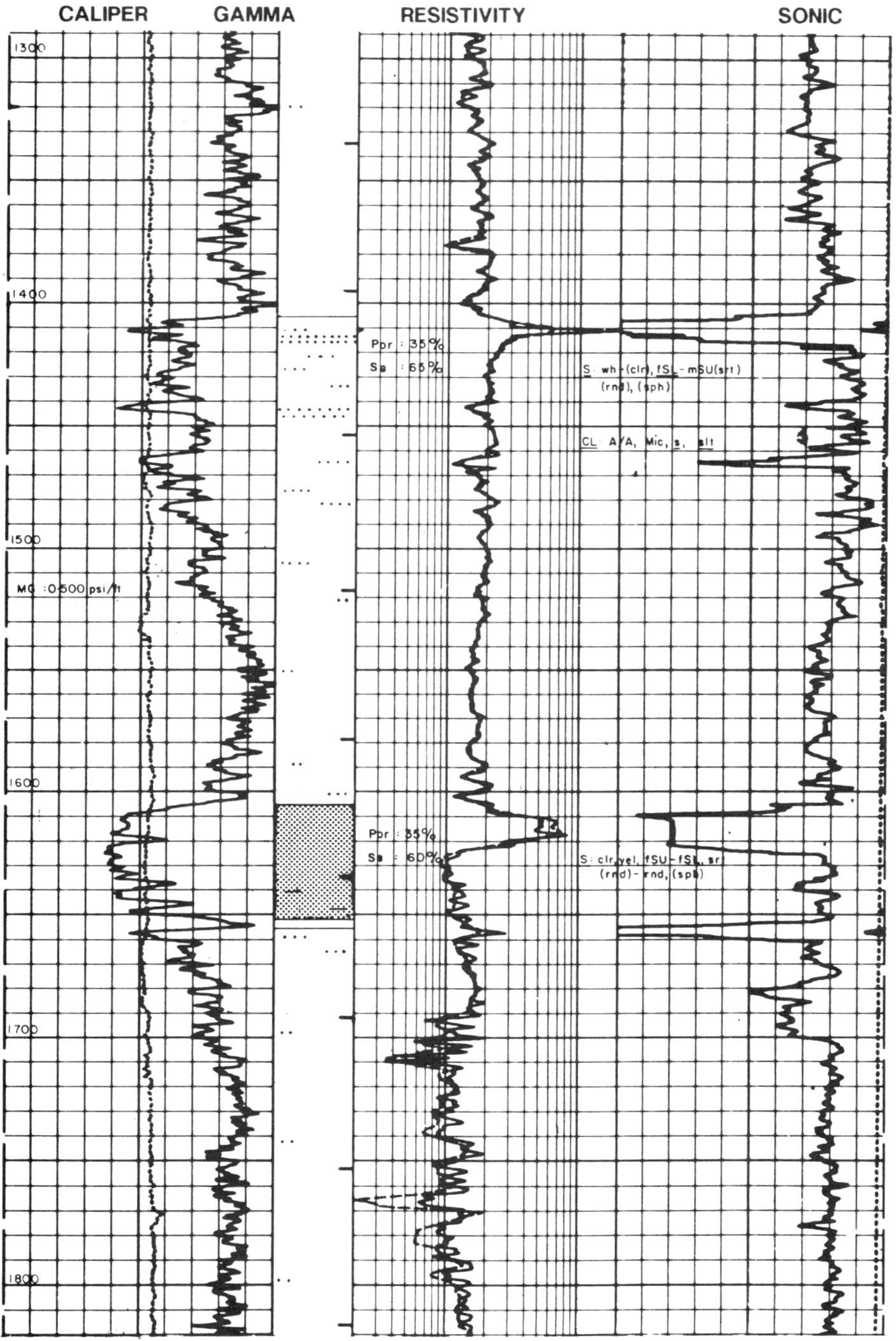

Fig. 9. Completion log from well 21/30-6a showing shallow gas sands.

workstation with high amplitude reflections marked in green. Figure 10a is a data example in the vicinity of the Gannet A Platform location and Figure 10b shows data at two of the well locations where shallow gas was detected. As illustrated by Figure 10, several high amplitude reflections were observed on the seismic data. In general the extent of these reflections were in agreement with those mapped from the reconnaissance survey and those shown by the 3D seismic data.

Correlating the high amplitude reflections observed on the shallow gas geophysical surveys with those observed on the exploration 3D seismic data, allows a regional understanding of the shallow gas prone horizons. This improved confidence in the shallow gas prognoses for the various locations. Many, but not all, of the high amplitude reflections which were interpreted as shallow gas accumulations in the Gannet area on the shallow seismic data were visible on the 3D data.

All the available well data was carefully integrated with the seismic data. Whenever the wells showed shallow gas indications, the seismic data showed anomalous amplitude reflections. This is illustrated by Figure 10b which shows the seismic data in the vicinity of well 21/30-6A, whose log response is shown in Figure 9. There were no anomalous seismic reflections which could not be correlated with shallow gas on the well data. These simple conclusions were confirmed throughout the Gannet area using nearby rig site surveys, producing considerable confidence that the seismic data in this area was a reliable indicator of shallow gas.

Amplitude Versus Offset (AVO) studies, including forward modelling, were also undertaken on the shallow gas seismic data to test the validity of this approach in a known shallow gas area. The shallow gas sands generally showed characteristic AVO affects, but not always. Conventional reflection amplitude analysis proved to be the most reliable gas indicator in this area.

The interpreted shallow geological model was of a predominantly clay sequence with a series of nine regional sheet sands within the Quaternary and Tertiary (designated sands A–I), which were gas prone in certain areas. These sands were up to 15m thick with porosities of up to 35 percent. The sands could be correlated from well to well over considerable distances. Good correlation was possible from well 21/30-12 to well 22/21-5, a distance of 35km. In general, these sands were gas prone near the apex of low relief undulations in these sheet sands.

5.3. Pilot Hole Results

After the interpretation of the data, the Gannet A pilot hole (well 22/21-6) was drilled. No shallow gas was interpreted at the pilot hole location but a series of water bearing sands was expected. Due to the history of shallow gas in the area, care was advised when drilling three of these sand layers. The pilot hole confirmed the interpretation of the seismic data. The regional sand layers were logged and the actual depths of the sand tops agreed to within 2 percent of prognosis. Additional sand layers were logged at two levels but these were thin (less than 3m) and not

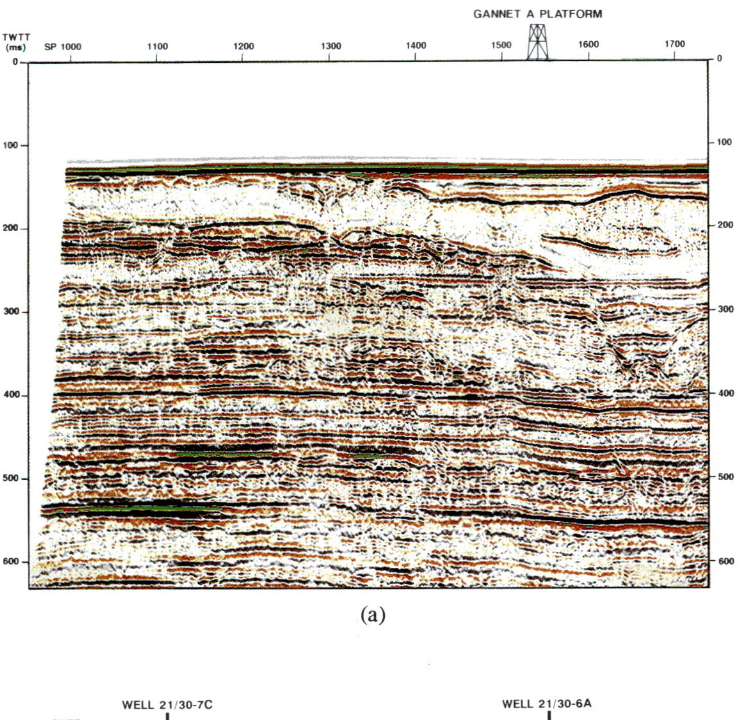

(a)

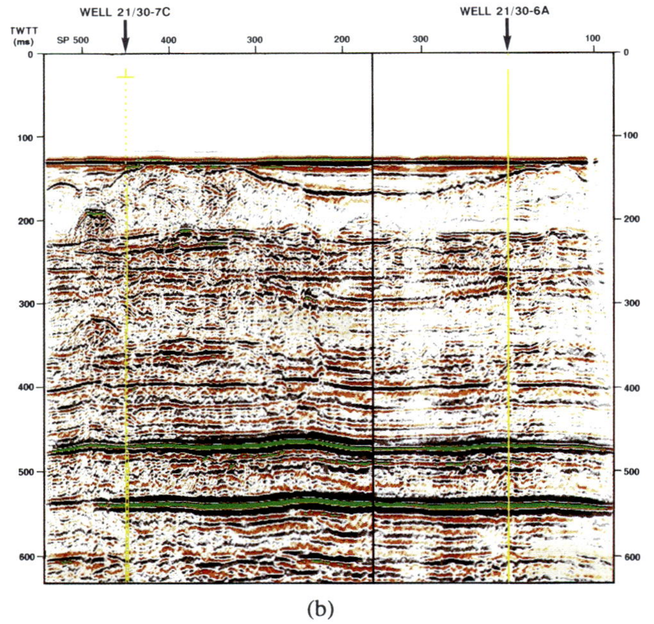

(b)

Fig. 10. Airgun data. (a) Gannet A Platform area. (b) Well correlations.

very clean sands. Within the depths drilled by the pilot hole, all the sands were water bearing and no shallow gas was detected.

5.4. Location Specific Shallow Gas Prognoses

The following shallow gas prognoses were provided to depths of 1000m sub-seabed:

- Gannet A No shallow gas was expected within 500m of the location but care was advised when drilling at three levels.
- Gannet B1 No shallow gas was expected at the location but gas was expected 40m from the location within a glacial channel. This gas accumulation was not expected to represent a hazard to drilling at the location.
- Gannet B2 Shallow gas was expected at two levels and particular care was advised when drilling.
- Gannet C1 No shallow gas was expected at the location but care was advised when drilling at two levels.
- Gannet C2 No shallow gas was expected at the location but care was advised when drilling at four levels.
- Gannet C3 Shallow gas was expected at two levels and particular care was advised when drilling. In addition, care was advised at one other level.
- Gannet C4 Shallow gas was expected at two levels and particular care was advised when drilling. In addition, care was advised when drilling at two other levels.
- Gannet D No shallow gas was expected at the location but care was advised when drilling at three levels.

Based on these prognoses, the well casing schemes and mudweights were designed to combat possible shallow gas, and shallow gas procedures were in place when drilling.

To date, wells have been drilled at Gannet A, B, C and D. With the exception of the pilot hole, well 22/21-6, no mudlogs or petrophysical logs were run over the top hole section of these wells to confirm the presence of gas. Although shallow gas was expected at several of the locations, no shallow gas drilling problems have been encountered in any of the wells. Prognosis of drilling conditions at the proposed locations using the correlation of the seismic character at the proposed locations with the character at the previously drilled wells was proven to be valid. The shallow gas was under hydrostatic pressure and controlled adequately by mudweight alone.

6. Foundation Conditions

Prognosis of shallow soil conditions was required for foundation design at the Gannet A Platform and at the 7 drilling centre locations.

At the Gannet A Platform location, 12 piles of 2.4m outer diameter were to be installed. Soils to depths of 100m might influence the piled foundations. At the drilling centre locations, only the very shallow soils (to 10m sub-seabed) would influence template installations and piling of associated protection structures.

6.1. Platform Location

Within the top 150m sub-seabed, the shallow soils were interpreted to consist of the Forth, Coal Pit, Fisher and Aberdeen Ground Formations. To illustrate the seismic character of these soils, Figure 6 shows deep tow boomer data and Figures 10 and 13 shallow gas airgun data in the vicinity of Gannet A. In addition, Figure 5 shows an interpreted geological profile through the Gannet A Platform location.

As shown in Figure 3, seven boreholes are available within 100m of the platform location to depths of between 71 and 113m sub-seabed. A summary of these boreholes is included as Figure 11. Correlation of borehole and seismic data is very good in the top 80m sub-seabed and reasonable to depths of 110m sub-seabed.

The logs from the pilot hole, 22/21-6, were also useful in the shallow section for interpretation of the shallow soils. Sands and clays were clearly defined.

6.1.1. Forth Formation soils

Within the top 40m sub-seabed at the platform location, Forth Formation sands and soft to stiff clays (Units 1–4) were expected. Of particular interest was the thickness of the medium dense to dense silty fine sands of Unit 1 for the effectiveness of mud mats. The thickness of this unit was carefully mapped in the vicinity of Gannet A and is shown on Figure 12. The base of the Forth Formation at approximately 40m sub-seabed is represented as a seismic reflection on both boomer and airgun data and correlates with a sand or sandy clay layer in all seven boreholes and the pilot hole.

6.1.2. Coal Pit/Fisher Formation soils

Below the Forth Formation soils, overconsolidated firm to stiff clays predominate to depths of 100m sub-seabed. As shown in Figure 11, the four deeper boreholes showed sand layers at 80–100m sub-seabed which were difficult to correlate from borehole to borehole and could not therefore be predicted at the pile locations. These sand layers were also detected on the logs from the pilot hole. They varied from less than 1m to 10m thick and were of concern for pile design.

Figure 11 shows the borehole correlation prior to careful integration with the seismic data. This correlation assumed that the sands were regional sheet sands.

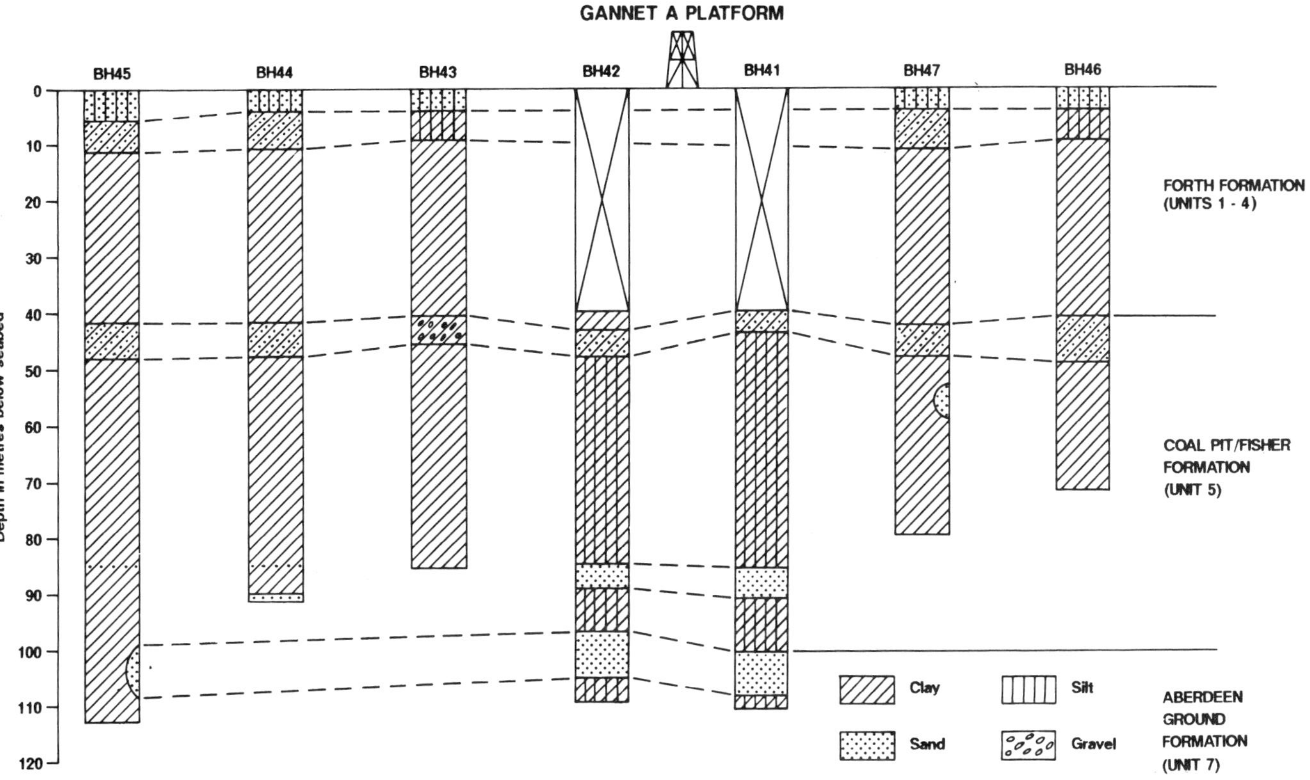

Fig. 11. Geotechnical borehole correlation – Gannet A Platform area.

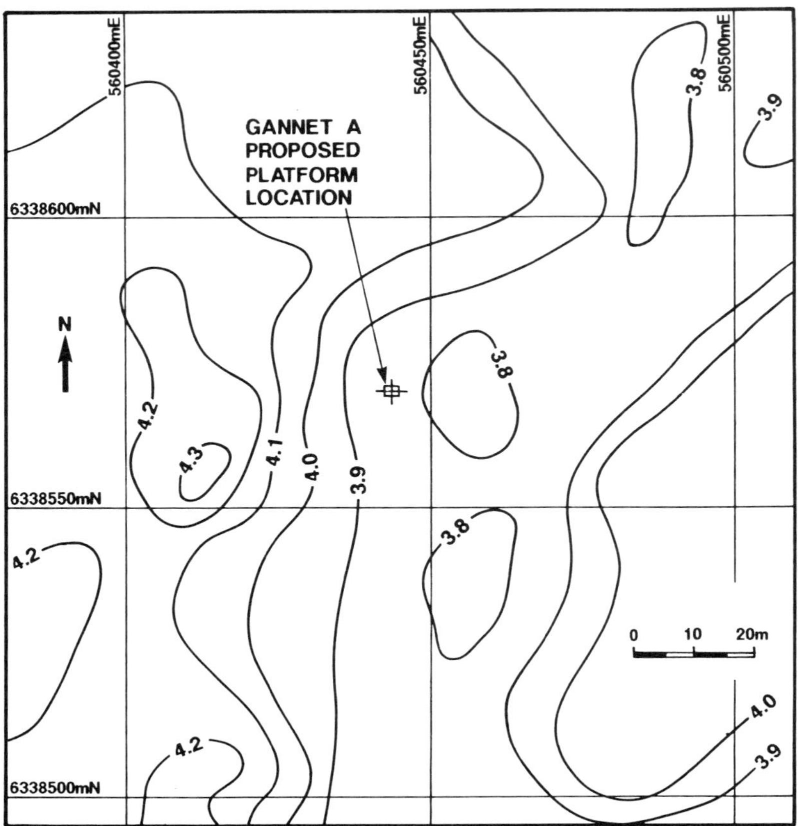

Fig. 12. Depth in metres sub-seabed to the base of Unit 1 – immediate vicinity of Gannet A Platform.

A very careful correlation was undertaken of the borehole and seismic data in the immediate vicinity of the platform location. Seismic line spacing in this area was 25m. Interpretation of the seismic data was undertaken using conventional paper displays and also on a Sierra 2DI and a Landmark Workstation. Integration with the seismic data proved difficult as the sand layers were often of a thickness which corresponded to the limits of the resolution of the available seismic data at these depths.

As shown by Figure 13a, the sand tops could be correlated with seismic reflections on the conventional paper displays. The resulting interpretation of the geometry of the sand layers within the area of the boreholes was considerably different to that produced from using the boreholes alone. Using a workstation provided greater confidence in this interpretation due to the enhanced display options. These displays are illustrated by Figures 13b and 13c. In particular, seismic phase displays show that the deeper sands are not regional sheet sands but are a series of discrete prograding units within a predominantly clay sequence.

The 4–5m thick silty fine sands at approximately 85m sub-seabed in Boreholes 41 and 42 can be correlated with a seismic reflection which dips towards the west of the location. This reflection represents an erosion surface near the base of the Coal Pit/Fisher Formation (Unit 5). In the vicinity of the proposed location this sand layer is generally sub-horizontal in the northeast and southwest but dips towards the northwest and southwest. The seismic data indicates that this sand unit pinches out towards the west and there are only indications of minor sands at this level in CPT records of Boreholes 44 and 45.

Sands were detected in Boreholes 41 and 42 at 101 and 97m sub-seabed respectively. Sandy clay was detected at the same depth in Borehole 45 and a coarsening upwards sand in the pilot hole. As illustrated by Figure 13c, these sands could be correlated with reflections on the seismic data but only locally. They were interpreted as prograding sands at the top of Unit 7.

The top of these deeper sands was interpreted as the base of Unit 5 which was mapped and is presented as Figure 14. This figure in conjunction with Figure 5 shows that the base of Unit 5 forms a significant channel to the north of the proposed platform location. These large Quaternary channels are also clearly seen on the Exploration 3D seismic data which shows that they extend over many kilometres.

6.1.3. Prognosis at the location

The prognosis of soil conditions at the location was:

Depth (m) sub-seabed	**Soil Description**
0 - 4	Medium dense to dense silty fine sand.
4 - 17	Very soft to soft sandy clay. Shear strengths ranging from 5 to 30 kPa.
17 - 25	Soft to firm sandy clay. Shear strengths ranging from 25 to 60 kPa.
25 - 40	Firm to stiff clay. Shear strengths ranging from 50 to 70 kPa. Sand layer at base.
40 - 99	Overconsolidated firm to stiff clay. Shear strengths ranging from 300 to 400 kPa. Dense to very dense silty fine sand layer at 85–90m.
99 - 106	Dense to very dense fine sand.
106 - 246	Overconsolidated hard clay with sand layers.

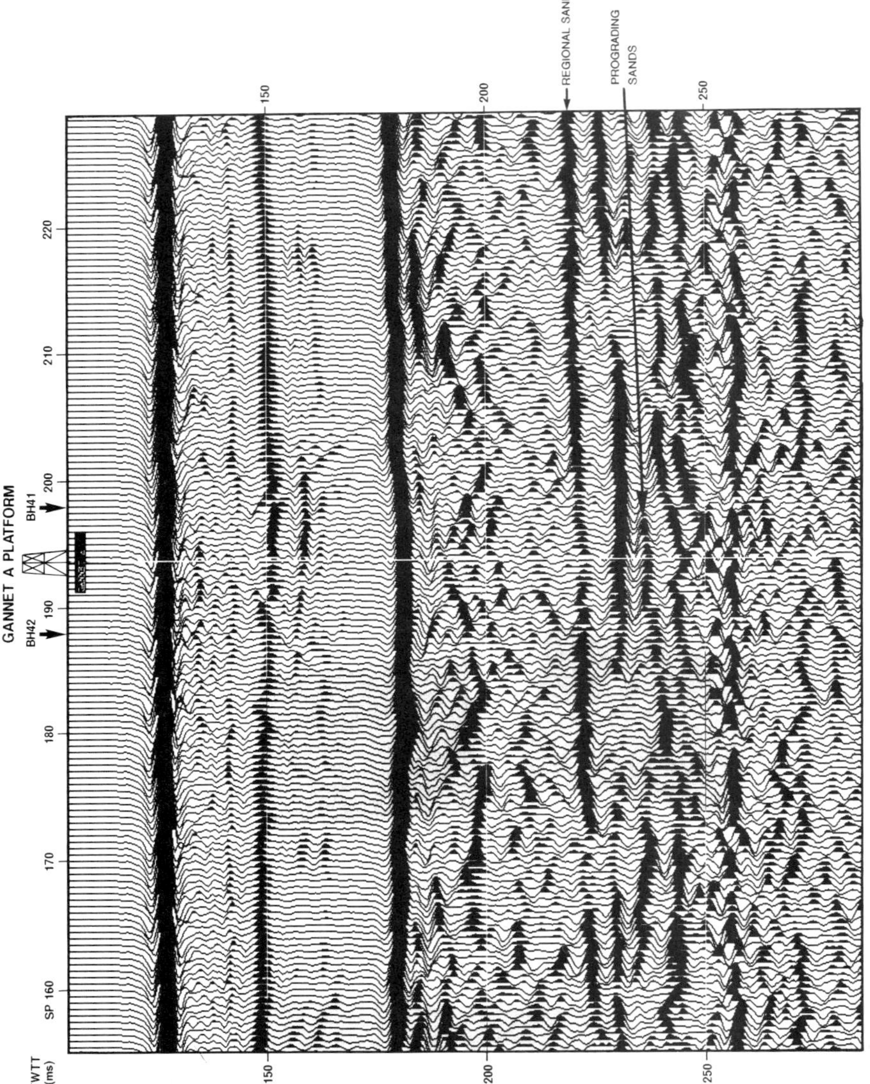

Fig. 13a-c. Airgun data – Gannet A Platform area. (a) Varwiggle display.

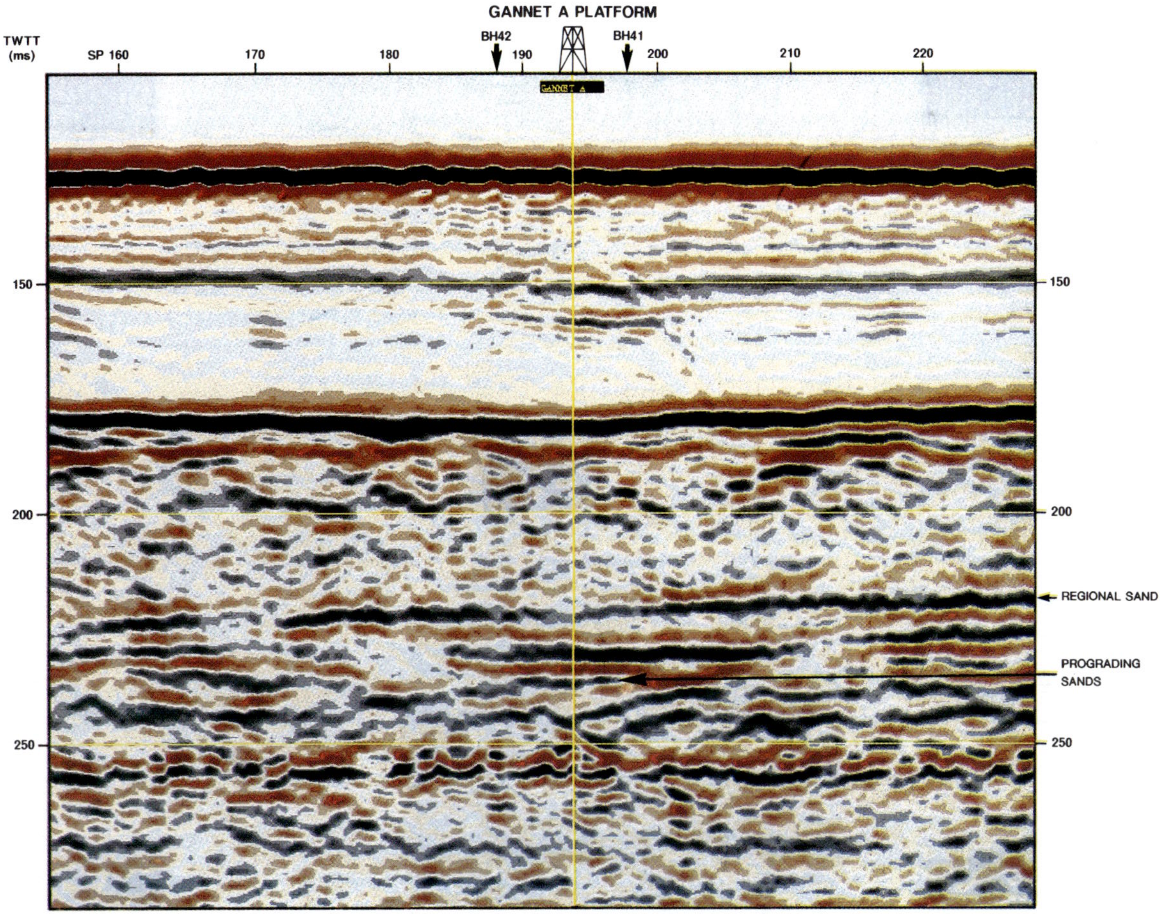

Fig. 13b. Variable area display.

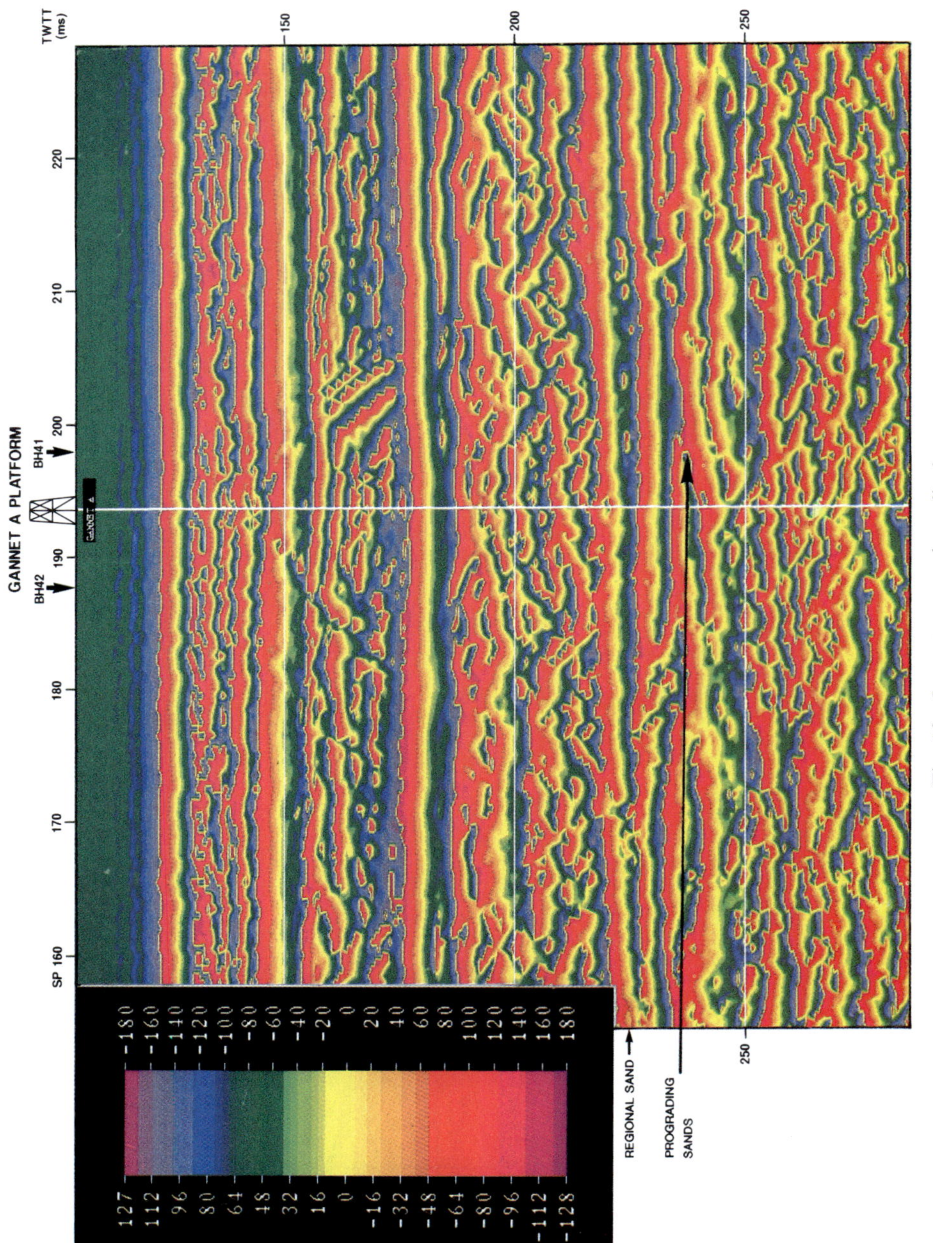

Fig. 13c. Instantaneous phase display.

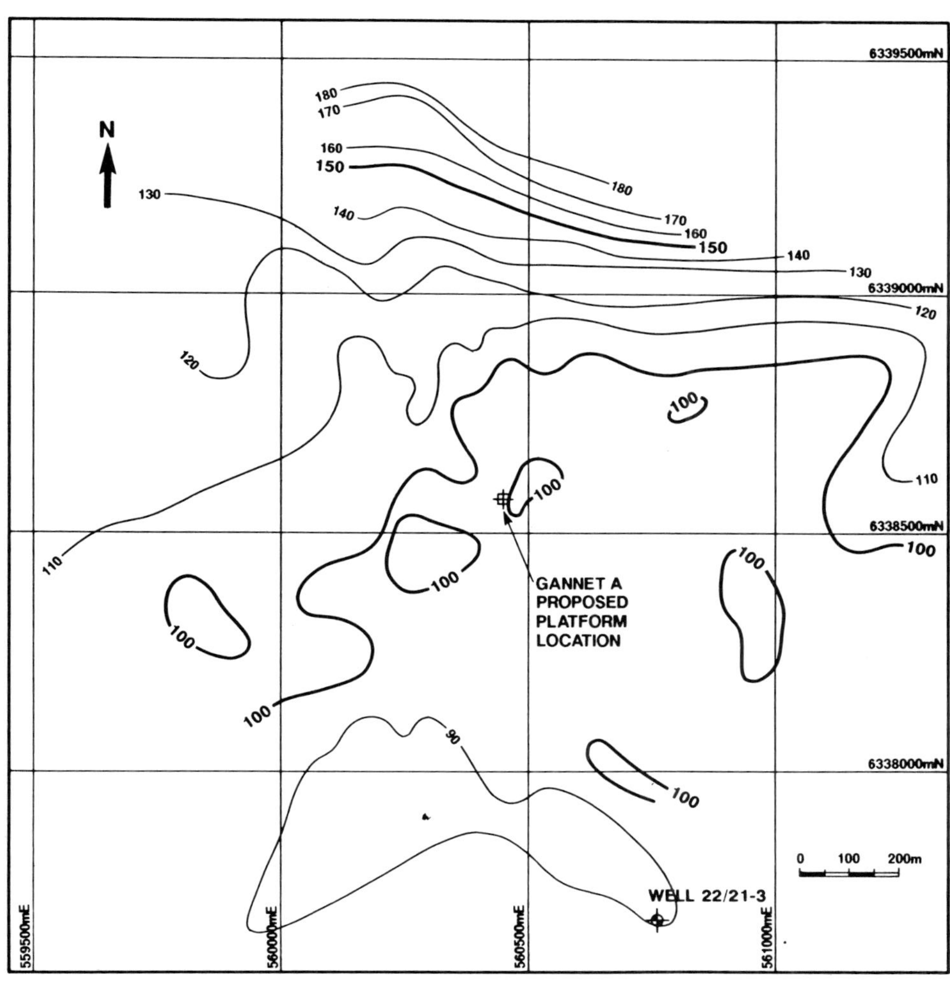

Fig. 14. Depth in metres sub-seabed to the base of Unit 5.

6.1.4. *Field results*

The 12 piles were successfully installed to their target penetrations of either 82m or 85.5m sub-seabed, over a period of 6 days. The measured blowcounts compared very well with the predicted blowcounts.

6.2. Drilling Centre Locations

Within the top 10m sub-seabed at the drilling centre locations, shallow soils were interpreted as either loose sands and soft clays of the Forth Formation or hard to very hard overconsolidated clays of the Coal Pit/Fisher Formations.

As shown by Figure 15, seismic character on the boomer seismic data clearly defined these two soil types. Soil properties were measured with CPTs at the drilling centre locations and in the vicinity of the drilling centres by CPTs acquired along the proposed pipeline routes. There was very good correlation between soil type as interpreted from the boomer data and CPT data.

7. Anchoring Conditions

The requirements for anchoring were different at the platform location compared to those of the drilling centre locations. At the platform location, a semi-submersible rig would be anchored for a period in excess of one year for tender assisted drilling. Eight anchor piles in the order of 30 metres long were required for this. In addition, a semi-submersible rig would be anchored at the location, using conventional anchors, during the hook up and commissioning phase of the platform.

At the drilling centre locations, semi-submersible rigs would install the drilling guidebases and then return at intervals to undertake development drilling through each guidebase.

7.1. Platform Location

The anchoring surveys at the platform location were undertaken in conjunction with the detailed foundation and shallow gas surveys as the results were required to make early decisions on the suitability of soils for long term anchoring in the area.

Seismic character of the boomer data and integration of the boomer data with the borehole and CPT data indicated that the near seabed soils throughout the 4km by 4km area surveyed at the Gannet A Platform location were of the Forth Formation (Units 1–4). The silty fine sand of Unit 1 varies in thickness from approximately 8m to less than 1m. The areas where the silty fine sand was less than 1m thick are shown on a simplified plan view, Figure 16. However, as the thickness of this sand unit was significant for anchor piles, it was carefully mapped throughout the area surveyed and is presented as Figure 17. The underlaying Unit 2 soils comprise soft to very soft clays with some clayey sand and thin sand layers to depths of 18m

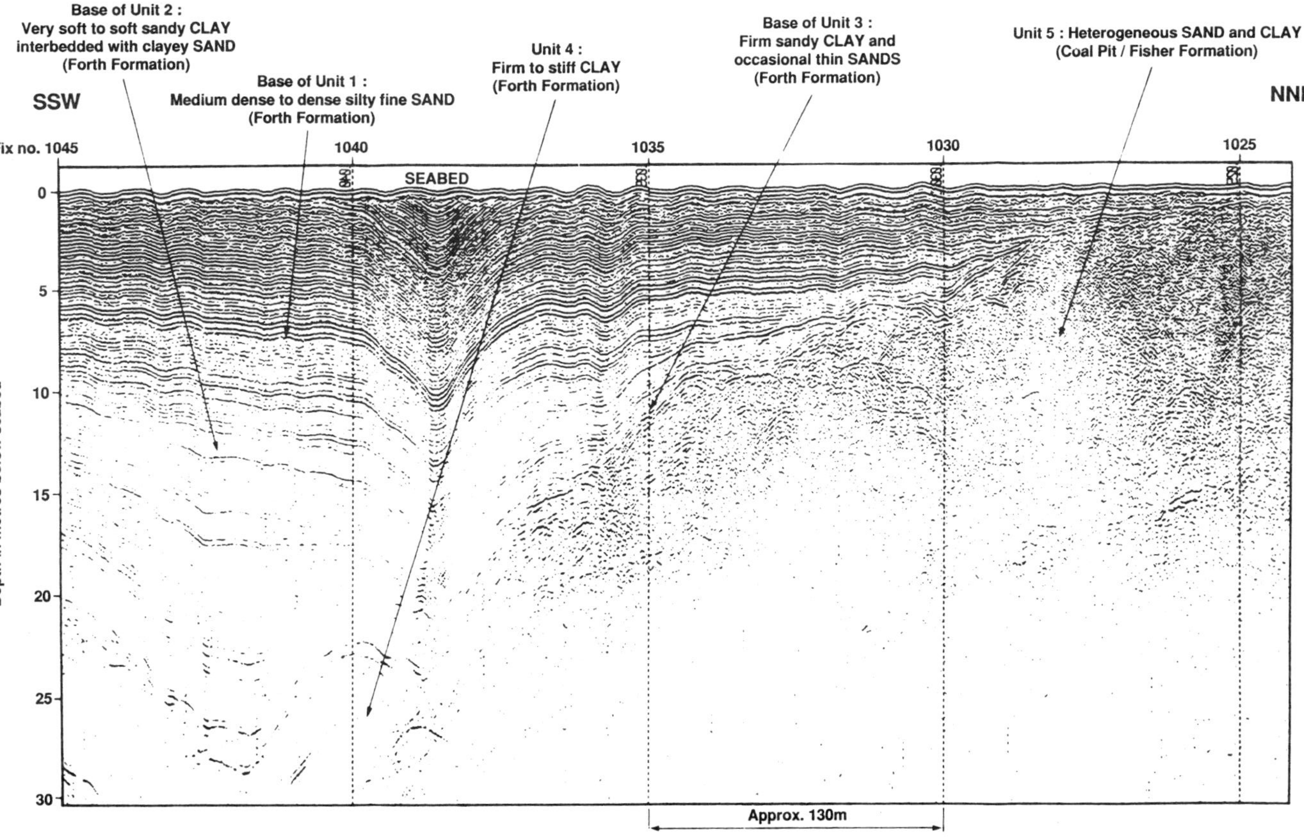

Fig. 15. Deep tow boomer data – in the vicinity of drilling centre locations.

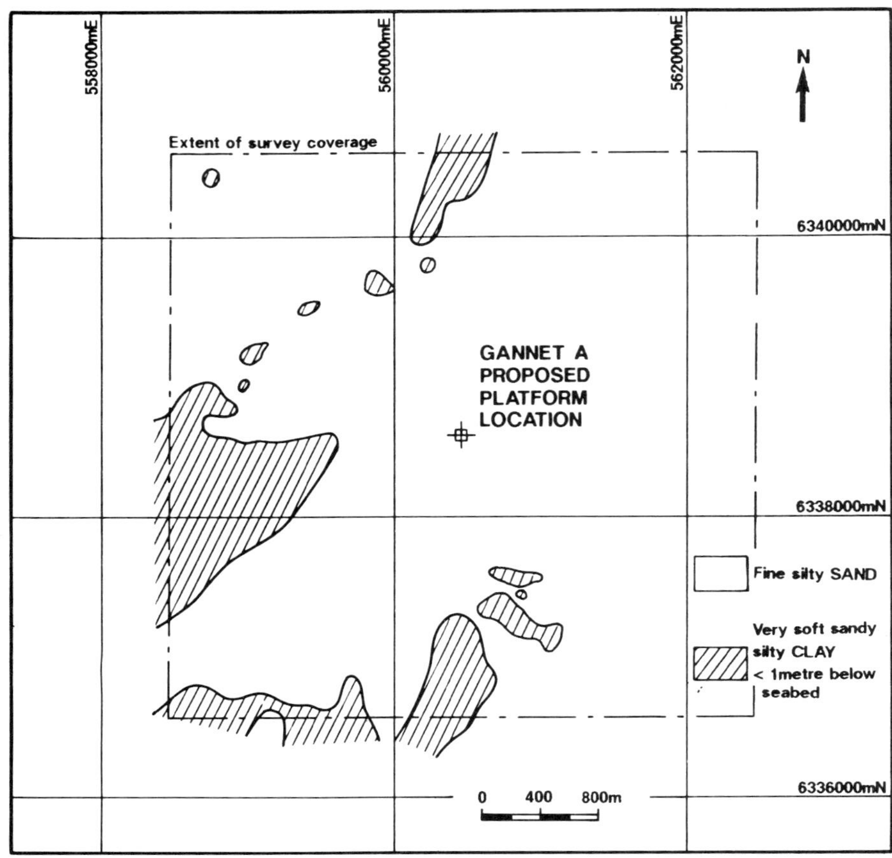

Fig. 16. Simplified plan view of anchoring conditions – Gannet A Platform area.

sub-seabed. These soils in turn are underlain by Unit 3 soft to firm sandy clay with some silt and fine sand partings. Shear strengths of the clays as measured in the Gannet A boreholes vary from 12 kPa at 4m to 75 kPa at 35m sub-seabed.

Beneath Units 1–3, Unit 4 firm to stiff clays are expected throughout the area surveyed. As the thickness of these Forth Formation soils was of significance for anchor pile installation, the thickness was mapped throughout the area surveyed. The thickness of the Forth Formation is shown in Figure 18.

7.2. Drilling Centre Locations

The anchoring surveys at the drilling centres were undertaken immediately (6–8 weeks) prior to the rigs moving onto these locations. Shallow soils were interpreted in the top 10m sub-seabed only. As illustrated by Figure 16, simplified plan views were produced showing the aerial extent of the significant near seabed soil varia-

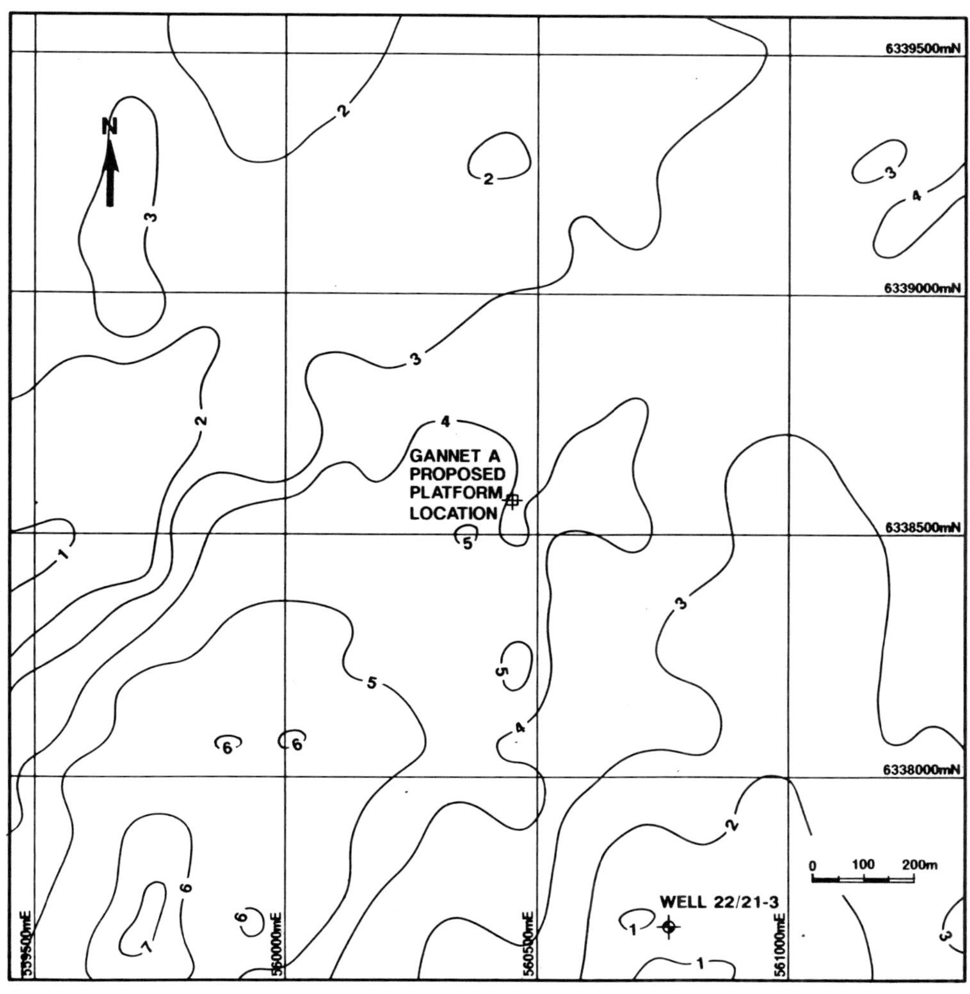

Fig. 17. Depth in metres sub-seabed to the base of Unit 1.

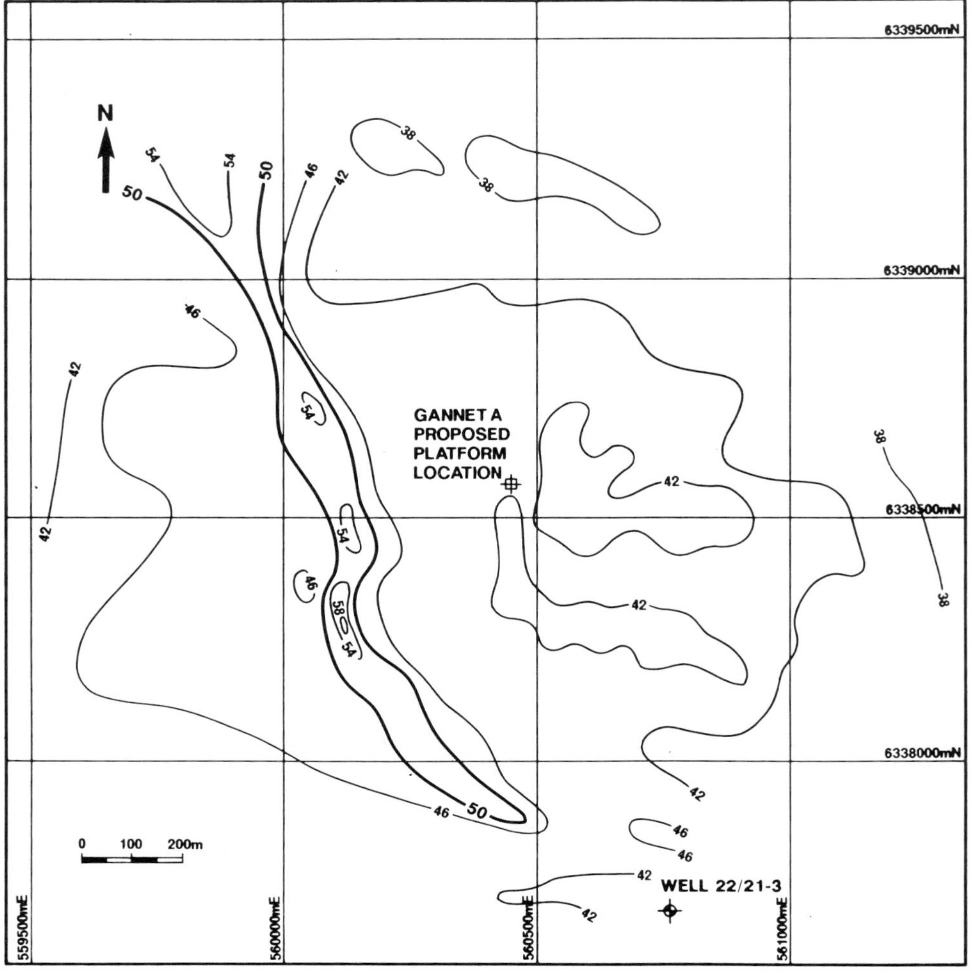

Fig. 18. Depth in metres sub-seabed to the base of Unit 4.

variations i.e. the extent of the Forth Formation or the Coal Pit/Fisher Formation. The variability of anchoring conditions in the near seabed soils is illustrated by Figure 19 which shows the soils in the vicinity of two of the drilling centre locations on an interpreted profile.

8. Pipeline Route Surveys

All the proposed infield pipelines were to be trenched. Mapping of shallow soils variations significant to trenching operations was therefore required along the proposed routes.

CPT data rather than core data was required since the CPT would be more likely to provide the required sub-seabed penetration of 5m sub-seabed. In addition, reliable *in-situ* measurements would be provided from which the lithology, the shear strength of clays, and the density of sands could be assessed.

Shallow soil along the proposed pipeline routes were originally interpreted from the boomer data, correlated to the reconnaissance borehole data. From this interpretation, 50 CPT locations were chosen to ensure near seabed soils along the proposed routes were adequately investigated.

Once the CPT data were available, the results were integrated with the geophysical interpretation. Typical results of this integration are included as Figure 20. Variations in seismic character proved a reliable indicator of soil type. The soil types found within the 0.6m trenching depth along the infield routes were the loose fine sands and soft silty clays of the Forth Formation, and the stiff to hard clays of the Coal Pit/Fisher Formation.

The interpretation of the data was subsequently confirmed by the flowline and umbilical trenching operations. The CPT results were very useful for predicting soil conditions and thus trenchability. However, the pipeline engineering project team concluded that CPT data supplemented with vibrocores in a ratio of approximately 5 CPTs to 1 vibrocore would provide valuable samples to complement the *in-situ* soils measurements. This was considered to be a worthwhile precaution to avoid potential installation contractual difficulties in areas where the soil conditions are variable.

The very soft to soft clays of Unit 2 which occur near the seabed at Gannet D caused traction problems for a trenching spread during the flowline installation.

9. Conclusions

Careful planning of seabed surveys for the Gannet Development was very important over a period of eight years as the development plan evolved.

Seabed survey results, integrated with all relevant data sets provided a reliable interpretation of shallow soils and shallow gas throughout the area.

Careful scheduling of the various seabed surveys, site investigations and the shallow gas pilot hole permitted selection of optimum locations for structures with

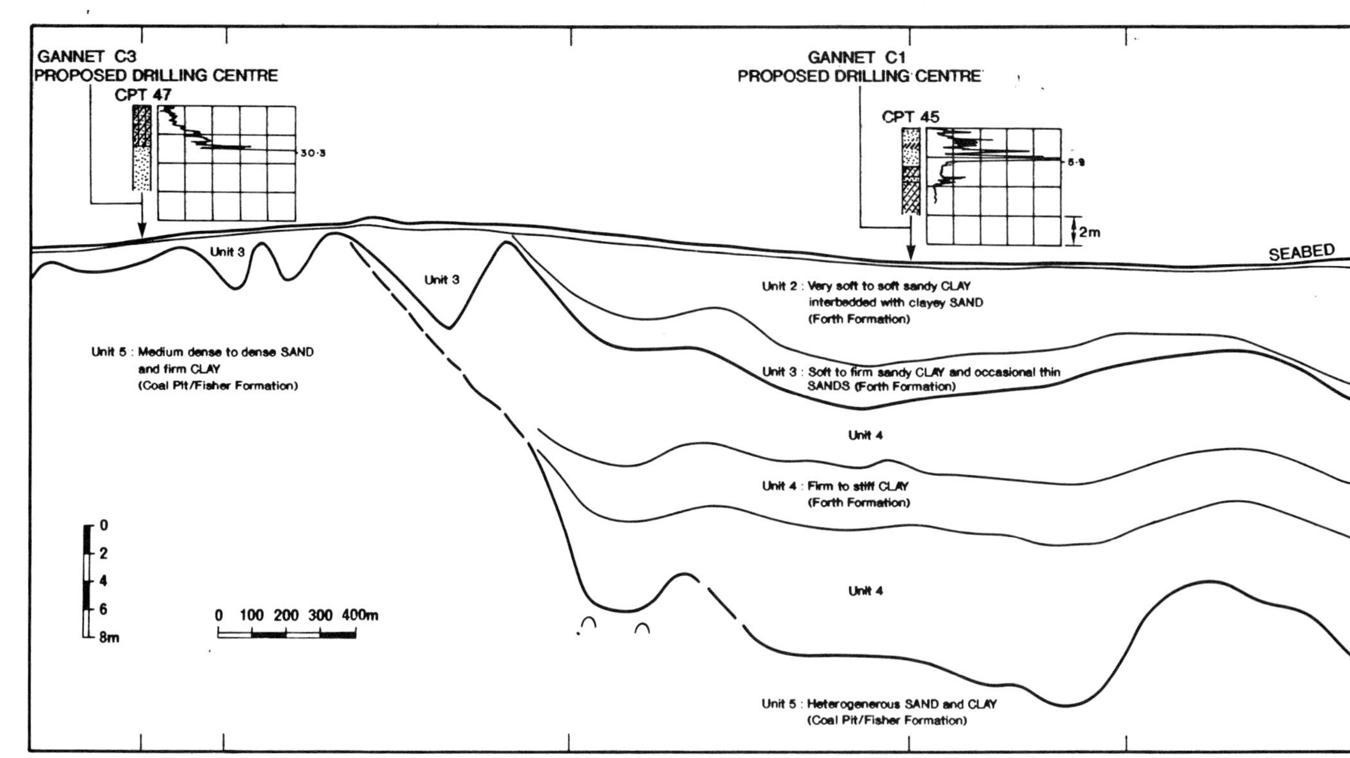

Fig. 19. Interpreted profile – in the vicinity of drilling centre location.

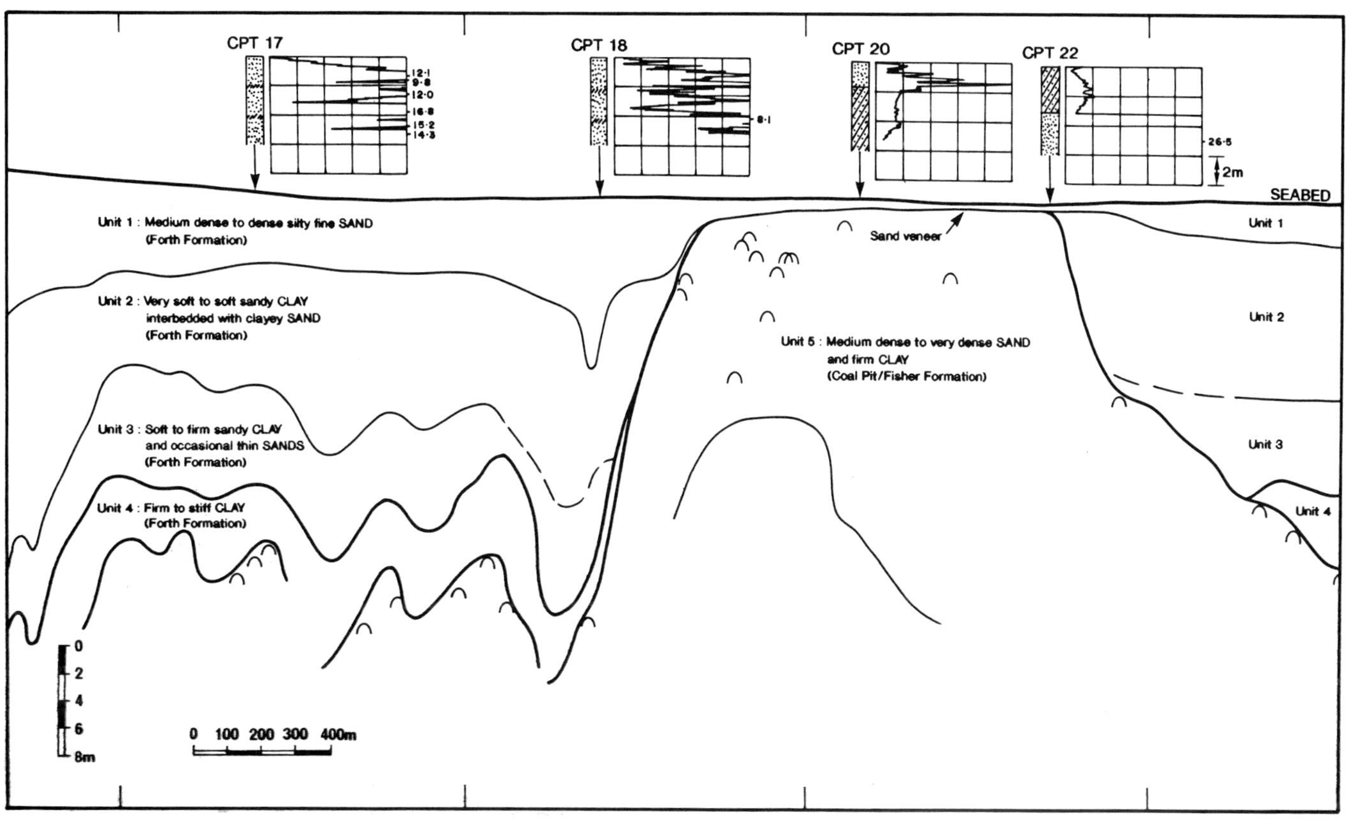

Fig. 20. Interpreted profile – pipeline route.

a minimum of repeat surveys.

The use of seismic workstations enhanced the interpretation of the seismic data.

During installation, conditions were generally as predicted from this integrated interpretation.

Acknowledgements

This paper has benefited from numerous internal reports and reviews by colleagues. The authors would like to thank Shell U.K. Exploration and Production and Esso Exploration and Production UK Limited for permission to publish the paper.

References

1. British Geological Survey (1987), 'FORTIES Sheet 57 N-00, 1:250,000 Series, Sea Bed Sediments'.
2. British Geological Survey (1987), 'FORTIES Sheet 57 N-00, 1:250,000 Series, Quaternary Geology'.
3. Eden, R. A., Holmes, R., and Fannin, N. G. T. (1977), 'Quaternary Deposits of the Central North Sea, 6', Institute of Geological Sciences Report No. 77/15.
4. Holmes, R. (1977), 'Quaternary Deposits of the Central North Sea, 5', Institute of Geological Sciences Report No. 77/14.
5. Kunst, F. and Deze, J. F. (1985), 'The case history of a high resolution seismic survey in the Central North Sea', *Proc. Offshore Technology Conference*, OTC 4968.

Discussion

Question from D. Long, British Geological Survey, Edinburgh: Am I right in understanding that no samples were collected by the pilot hole and that you were relying totally on well log interpretation to aid your seismic interpretation? If so, how near to seabed did you have confidence in the well log signals and that they were not corrupted by borehole instability/disturbance close to seabed?

Authors' response: Yes, no samples were collected in the pilot hole.

Well log data was useful for discrimination of lithology to within 40m of the seabed, even with some collapse of borehole.

INTEGRATED GEOHAZARD STUDY ALONG THE KRISHNA-GODAVARI DELTA SLOPE, EAST COAST INDIA

J. J. A. HARTEVELT and G. L. VAN DER ZWAAG
Fugro-McClelland Engineers B.V., P.O. Box 250, 2260 AG Leidschendam, The Netherlands

Abstract. The continental slope of the Krishna-Godavari delta on the East coast of India exhibits a number of important geological phenomena which should be taken in consideration for offshore foundation design. A detailed geophysical and geotechnical investigation provided the data for an integrated study for the site selection for the foundation of an SBM oil terminal. The geophysical survey consisted of sparker, side scan sonar and bathymetry. Soil borings, including cone penetration testing, were performed at five alternative locations to a depth of 150 m.

The soil profiles generally consist of normally consolidated very soft clays at mudline, becoming stiffer with depth. Geophysical data revealed two phases of faulting and mudsliding; an older phase (over 3000 years old) showing relict structures and a recent phase of mudsliding affecting the seabed topography.

Stable and unstable foundation zones were identified.

Stability calculations based on geotechnical properties of the soil profiles confirmed the delineation of these zones. Wave induced seabed loading analyses showed a mudslide risk for the original proposed SBM location and stable conditions for alternative sites.

1. Introduction

Offshore site investigations for potential structure sites along the East coast of India have for a long time been restricted to the continental shelf. In recent years exploratory activities moved to the deeper water gradually approaching the continental shelf slope. The ONGC oil field developments in the Krishna-Godavari delta area have reached the stage of installation of fixed platforms and pipelines. A number of production platforms will be installed in the shallow water zone near the coast line and will be connected with pipelines to installations on land. A SBM has been planned near the continental shelf edge in sufficiently deep water for safe handling of large oil tankers.

The purpose of this case history is to show how an integrated approach, using a combination of geophysical and geotechnical data, can result in a high degree of confidence for the foundation design of structures in areas which would in general be regarded as potentially unsafe.

Volume 28: Offshore Site Investigation and Foundation Behaviour, 333–346, 1993.

2. Geological Setting

The study area is located on the eastern continental shelf of India (Figure 1), classified as an Atlantic-type margin, close to the shelf break. The continental shelf along the East coast of India has a typical width of about 25 km which only widens in front of the Ganges Delta, some 800 km to the North-east, to a maximum of some 85 km.

The Indian peninsula is a relatively stable platform composed of a basement of Archean and Precambrian igneous rocks, gneisses and schists, and a number of younger sedimentary basins and an important volcanic phase producing the Deccan Basalt plateau.

Seabed sediments of the shelf are reported to be mostly Quaternary in age and on the outer shelf, from water depth of about 60 m onto the shelf edge at about 100 m, relict sediments are a characteristic feature (Closs *et al.*, 1974). These deposits are composed of relict carbonate sands and carbonate rocks in form of algal and oolite limestones of late Pleistocene or early Holocene age, similar to those found on the Western continental margin of India. In areas of active sedimentation as the Ganges and Krishna-Godavari Deltas, the relict sediments are buried below massive Holocene deposits building out beyond the original shelf edge.

To the East of the study area, beyond the continental slope, lies the Bay of Bengal, over 3000 m deep, which is presently floored by the Bengal Deep Sea Fan, the largest deep sea fan in the world (Curray and Moore, 1974). This fan deposit post-dates the first collisions of India and Asia and uplift of the ancestral Himalayas at the end of the Palaeocene, about 54 million years ago. Sediments of the fan have been derived predominantly from the drainage of the Ganges and Brahmaputra rivers, which drain much of the North and South slopes of the Himalayas.

Sediments from the rivers are distributed throughout the fan by a system of turbidity current channels and fan valleys, which extend well beyond the southern tip of India.

The study area is located only 15 km from the shoreline of the Krishna-Godavari Delta complex (Figure 2). The delta has been built from the sediment load of the Krishna and Godavari rivers which drain the central part of the Indian peninsula. The continental shelf, which is about 20 km wide in this area, forms the submarine continuation of this delta complex, which is largely a build-up of very soft to firm Holocene to late Pleistocene clay deposits.

Initial geophysical investigations showed that in the potentially unstable environment of this prograding shelf margin delta, active mudslides of massive dimensions exist (Figures 2 and 6), similar to the features described by Coleman *et al.* (1983) off the Mississippi delta.

This discovery triggered a systematic integrated geophysical and geotechnical investigation over a large area to select a stable SBM location.

Fig. 1. Key map of study area.

3. Elements of Site Investigation and Laboratory Testing

The site investigation covered an elongated area parallel to the SW-NE aligned continental shelf edge. The field work consisted of geophysical and geotechnical elements. The marine geophysical survey included conventional techniques (analogue sparker, side scan sonar and echosounding) to establish the seabed morphology and geological structure of the area. A dense pattern of geophysical lines was surveyed with a line spacing of 100 m perpendicular and 200 m parallel to the shelf edge.

The geotechnical investigation consisted essentially of a number of boreholes

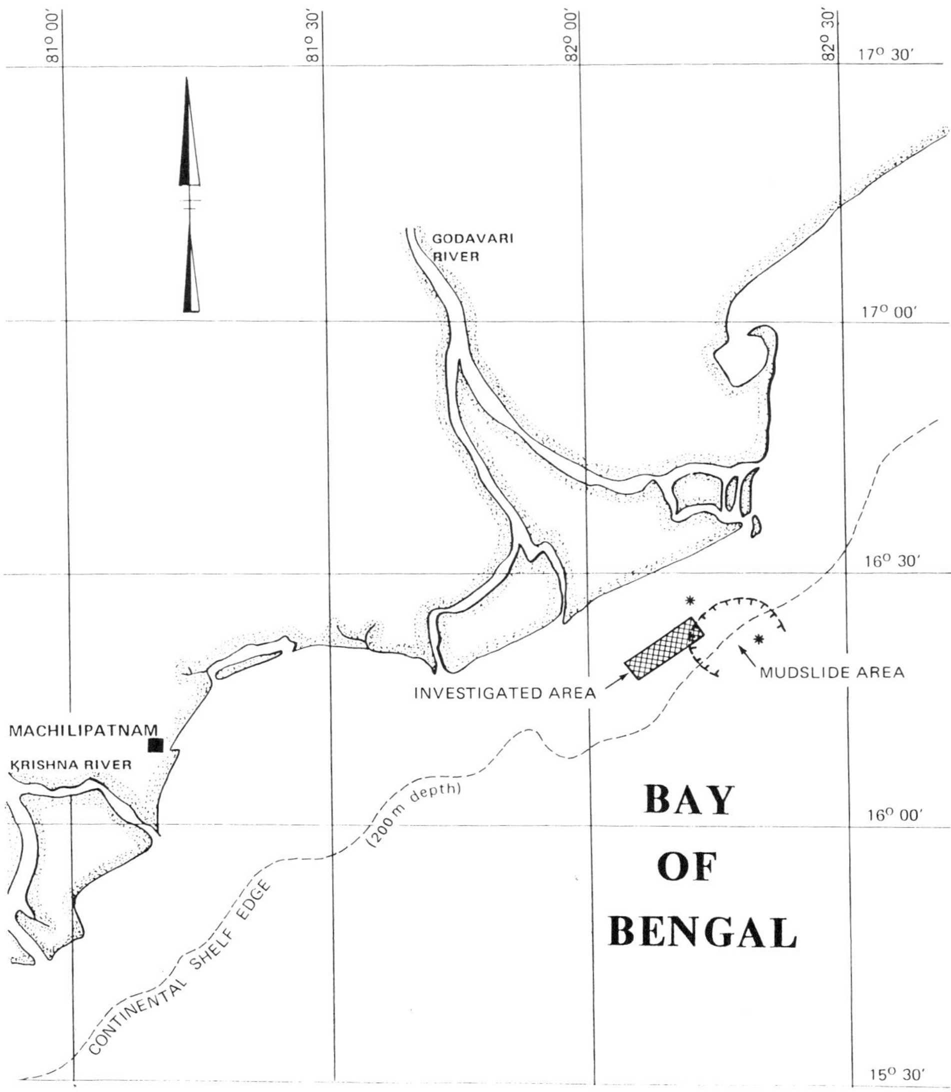

Fig. 2. Map of Godavari Delta area.

to a maximum depth of 150 m, at five alternative locations (Figure 3). Boring operations included alternating push sampling (WIP- Sampling) and cone penetration testing (WISON-testing) (De Ruiter, 1983). An intensive geotechnical laboratory testing program included visual sample inspection, classification testing and determination of undrained shear strength, consolidation data, shear moduli and damping ratios. A paleontological study of the upper 50 m of the soil profile was performed to establish the sedimentation rate of the area.

4. Integrated Study

The integrated study in this area mainly involved correlation between the seabed morphology expressed in the bathymetric map, the geological structure as shown on the sparker records and the soil profiles observed in the boreholes. Stability analysis involved factors for wave induced seabed loading.

4.1. Seabed Morphology

The seabed morphology forms a unique source of information on the distribution of recent seabed disturbances and special attention has been given to the morphologic expression of these features in the bathymetric map (Figure 3). Analyses of this map shows that most of the area investigated (Zone II) displays a smooth seabed sloping down with a very gentle gradient of 1:240 from a shallow depth of 20 m in the NW to 40 m in the SE. Beyond this depth the seabed becomes gradually steeper as it approaches the actual shelf edge at a water depth of about 100 m. In the North-eastern extremity of the area (Zone I), however, the seabed morphology shows a dramatic change. Here a depression is observed with a depth of about 50 m relative to the surrounding topography displaying slope gradients of 1:50. The slopes show a considerable number of scarps which may reach heights of up to 3 m. Further downslope, in water depth beyond 70 m, the seabed shows a hummocky morphology.

4.2. Shallow Stratigraphy and Structure

Two distinct seismo-stratigraphic units can be observed from the geophysical records and can be correlated with the soil profiles established at the five alternative foundation locations (Figures 4 and 5). The top soil unit (Unit A) consists of a very soft to soft clay of about 11 m thickness, which shows up on the geophysical records as a distinct layer with a bedding subparallel to the seabed slope (Figures 8 and 9). Soil Unit A is underlain by a series of similar clays (Unit B) in which the undrained shear strengths increases with depth. On the geophysical records the strata of Soil Unit B show a general steeper slope towards the shelf edge and folding is commonly observed. The interface between the two soil units constitutes an angular unconformity.

Two important observations were made with regard to the clays of Soil Unit B:

1. It was noted that the undrained shear strengths in the top of Soil Unit B decreased towards the NE of the area.
2. The clays of this soil unit typically show zones with fissuring and slickensiding.

The geological correlation profile, shown on Figure 4, shows that the thickness of the upper Soil Unit A is rather uniform. The underlying Soil Unit B shows gentle folding and also a general slope to the NE, which implies that the top of Unit B

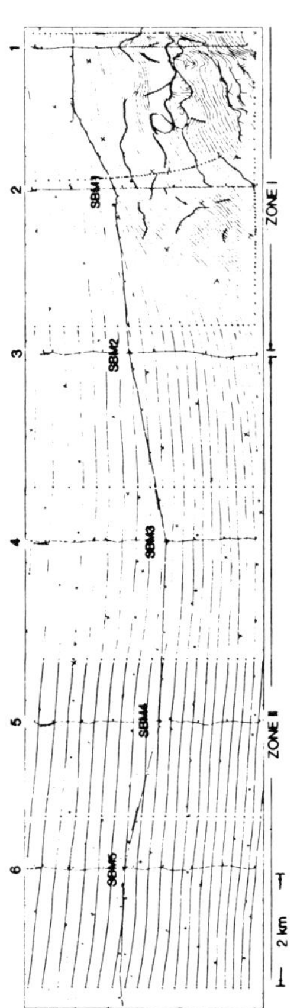

Fig. 3. Seabed morphology and alternative SBM locations.

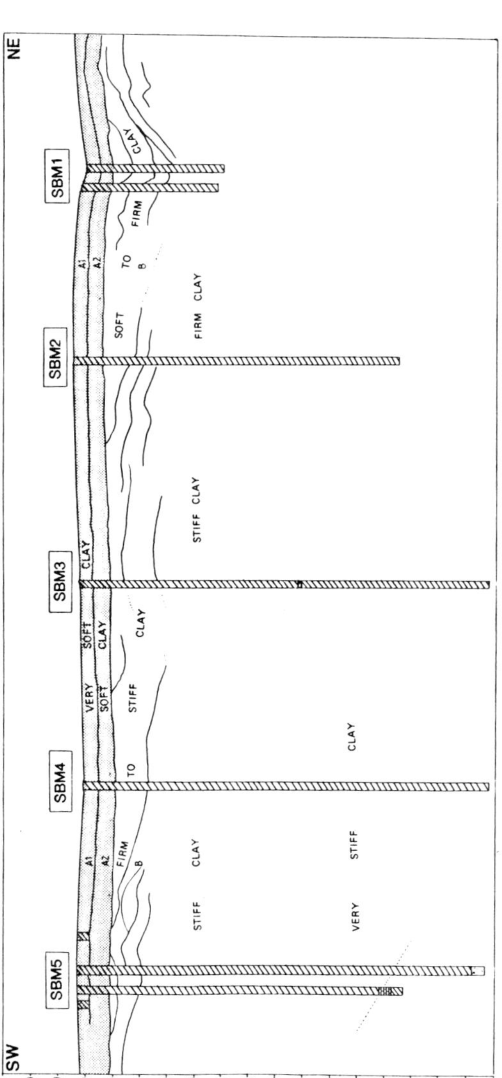

Fig. 4. Geological correlation profile.

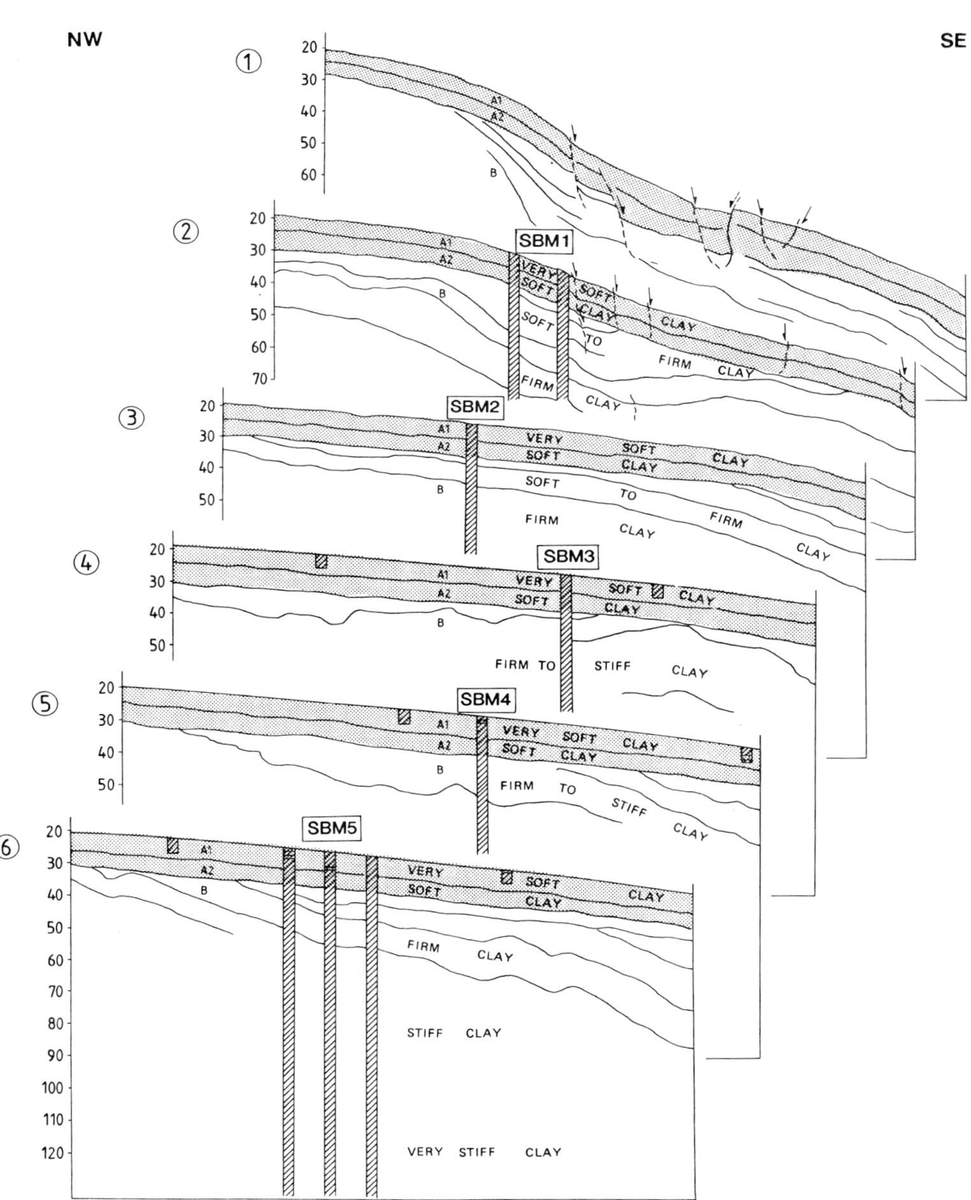

Fig. 5. Geological profiles 1 through 6.

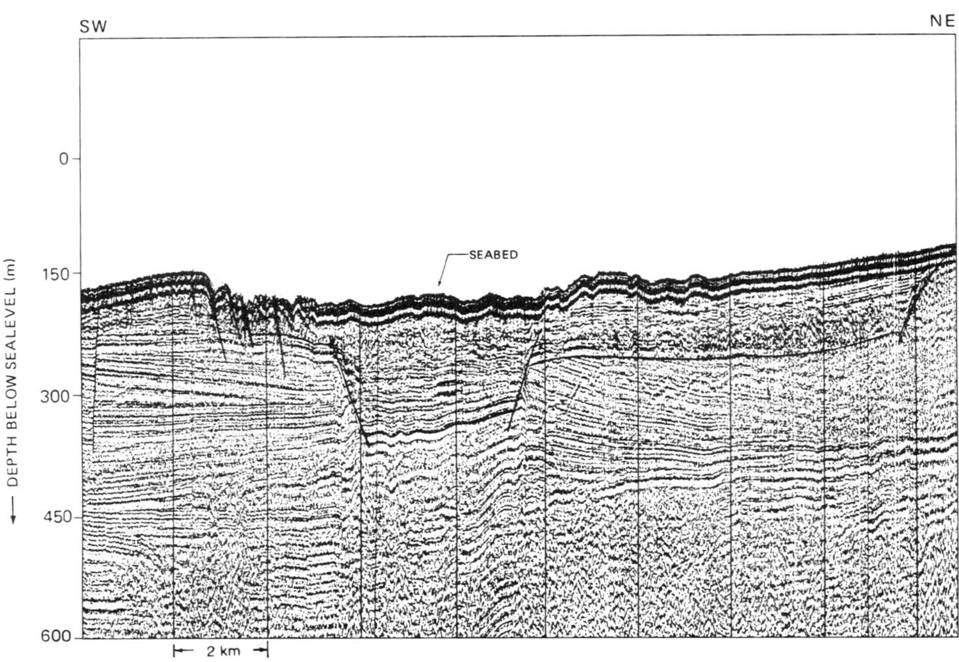

Fig. 6. Sparker record through mudslide channel NE of study area.

becomes gradually younger in that direction. This age difference correlates well with the decrease in shear strengths in the top of Soil Unit B towards the NE.

The structural features observed on the geophysical records correlate well with the observations already made from the seabed morphology. The records clearly show that in Zone II the upper soil unit (A) is undisturbed (Figure 9). Below the unconformity in Soil Unit B, however, the geophysical records show folded strata and clay samples show fissuring which clearly suggests that the clay deposits of Soil Unit B were affected by mudslides. The structural relationship between the two units in Zone II shows, however, that folds in Soil Unit B represent relict mudslides, that no apparent mudslides or other tectonic movements took place after the start of the deposition of the upper soil unit and that the soils of Unit B have since stabilized.

In a paleontological study of the upper 50 m of the soil profile of the SBM 4 location the average sedimentation rate in this area has been assessed as 3.5 m per 1000 years. This would imply that, in Zone II no mudslides took place during the last 3000 years.

Structural zone I shows a dramatic change in character. Here the geophysical records show that the upper soil stratum (Unit A) is disturbed by mudslides which also affect the seabed morphology (Figures 7 and 8). The seabed, which is much

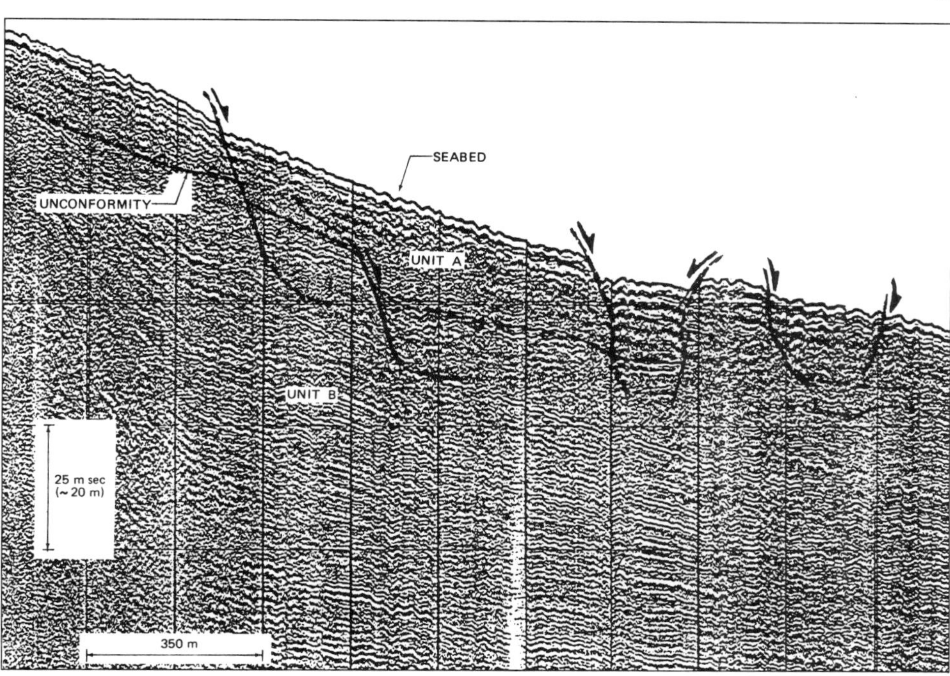

Fig. 7. Sparker record profile 1 in mudslide area.

steeper in this zone, is dissected by slide scarps which may show offsets of up to 3 m.

The fault planes bounding these slides generally penetrate to depths between 10 m and 30 m below seabed and subsequently curve into the bedding plane. Deeper fault planes have been observed locally, but in general the limited acoustic penetration of the sparker system prevented further evaluation of deeper structures.

Further down slope the upper stratum and seabed may become folded due to local compression. Locally, up-slope facing sediment blocks have developed to compensate the down slope movements. The mudslides of Zone I are located adjacent to the large truncated mudflow channel observed just NE of the area of investigation (Figure 6). The approximate location of this channel is shown on Figure 2. The mudslides of Zone I should probably be considered as peripheral slumps of this large mudslide channel.

From the above observations it can be concluded that in Zone I both soil units (A and B) are essentially unstable. The occurrence of steep slump scars in the seabed shows that mudsliding is of recent date and is probably still active. The low undrained shear strength values found in the soil profile of Location A explains this behaviour.

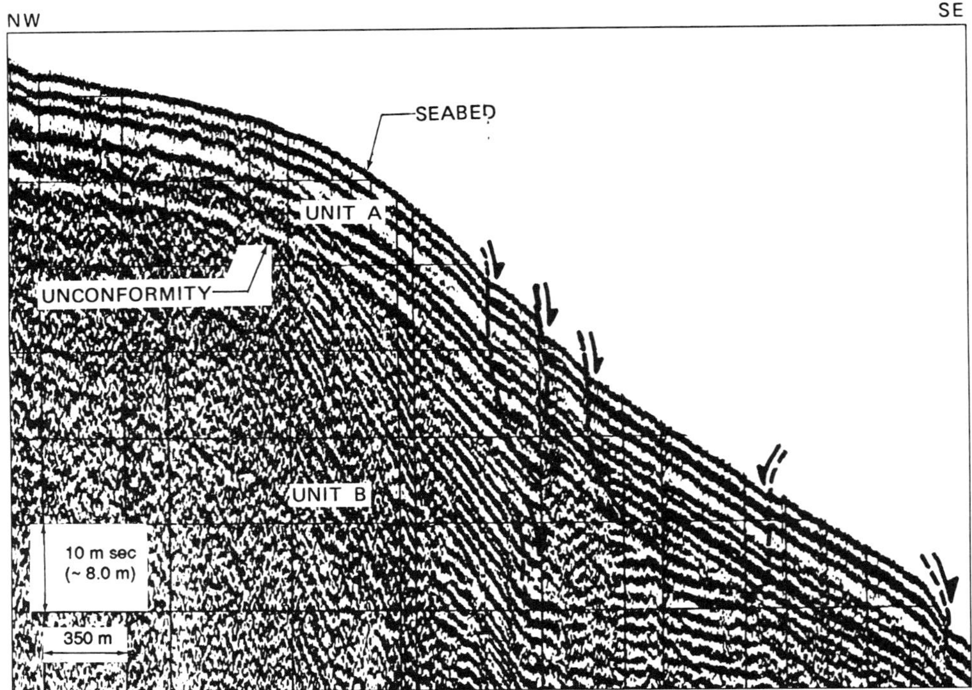

Fig. 8. Sparker record profile 2 through SBM 1 location.

4.3. Seabed Stability

Movements of soft submarine soils may be due to a combination of causes such as gravitational forces, rapid sedimentation, wave induced seabed loading, currents and seismic activity (Kraft *et al.*, 1986).

The movement can be a gradual intermittent mudflow or a mudslide due to shear failure. Mudflows may occur in extremely soft under-consolidated clays in delta areas. This feature has extensively been studied for the Mississippi Delta Area.

Soft clays in a slope may be marginally stable and slides can be triggered by wave induced bottom pressures. The investigation areas close to edge of the continental shelf in Godavari Delta area can be subjected to relatively high wave induced seabottom pressures. The seabed load amplitude for the design 100 year wave was calculated at 66 kPa.

The geophysical profiles indicate that slides have occurred just east of the original proposed SBM-1 location. Studies on wave induced seabed movements show a failure risk at SBM-1 and stable conditions west of the original location.

The soil stability analyses were performed using both simplified limit equilibrium models for circular slip surfaces (Henkel, 1970) and non-linear Finite Element models (Figure 10). The FE runs were performed with programs using static mod-

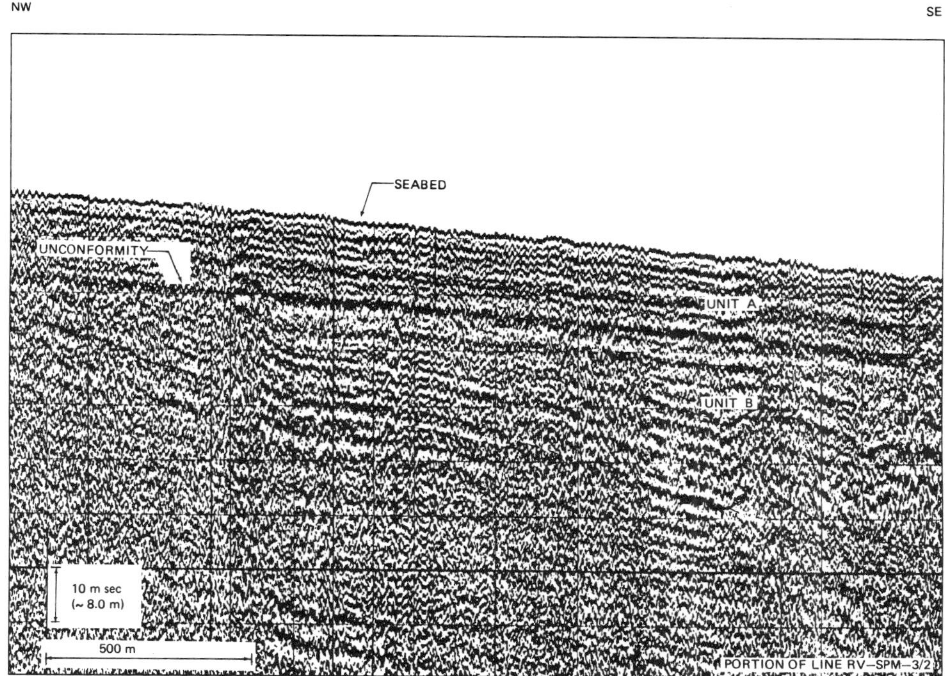

Fig. 9. Sparker record profile 6 through SBM 5 location.

els (FEGRO) and dynamic models (DWAVE/DFOOT, Van der Zwaag, 1988). The main soil parameters are the undrained shear strength c_u, the shear stiffness G, the Poisson ratio μ and the mass density γ. These parameters were derived from the laboratory test program. This included triaxial, oedometer, cyclic triaxial and resonant column tests.

Two design strength profiles were considered for the original SBM-1 location, i.e. an average strength and a minimum strength profile. The stability analyses with the ultimate limit state models show failure for both profiles.

The collapse wave pressure amplitude was 55 kPa for the average strength profile and 41 kPa for the minimum strength for the strength profile. The FEGRO static finite element runs give similar results. The analyses indicate a potential slip surface depth of 20 m. This depth can be substantially deeper as well. The finite element runs with the dynamic model (DFOOT) provided higher collapse pressures. These runs gave failure at a seabottom pressure amplitude of 50 kPa for the minimum strength profile and marginally stable conditions for the average strength profile. A typical example of a deformed mesh is shown in Figure 11.

The soil stability analyses for the alternative locations for design wave loading show stable conditions. The ultimate limit state analyses give safety factors of about 1.5 (Figure 12). The DFOOT finite element analyses give slightly higher safety

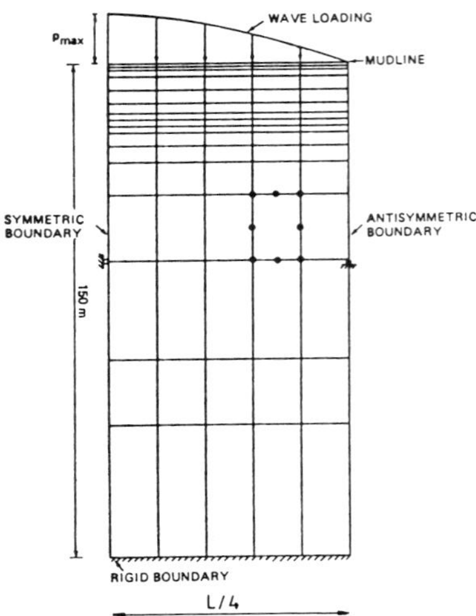

Fig. 10. Finite element model.

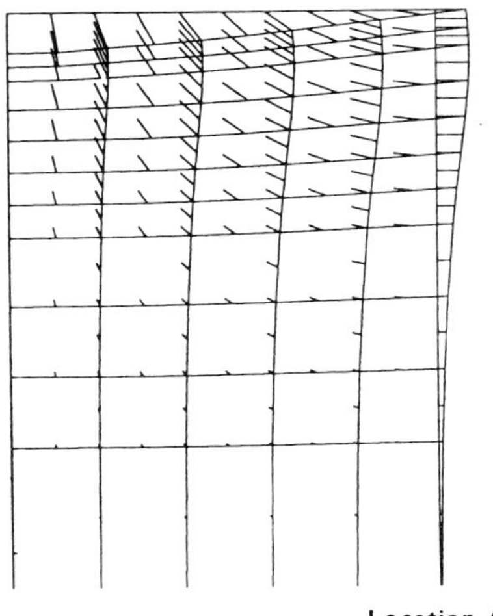

Fig. 11. Deformed mesh DFOOT SBM 1 location.

factors. These runs give elastic movement amplitudes up to 70 mm horizontal and 50 mm vertical.

The conclusion of the soil stability analyses was that slides can be expected at the original SBM-1 location during design wave loading. This conclusion is supported by observed slides just east of the location. Wave induced slides are not expected at the revised locations SBM-3, 4 and 5.

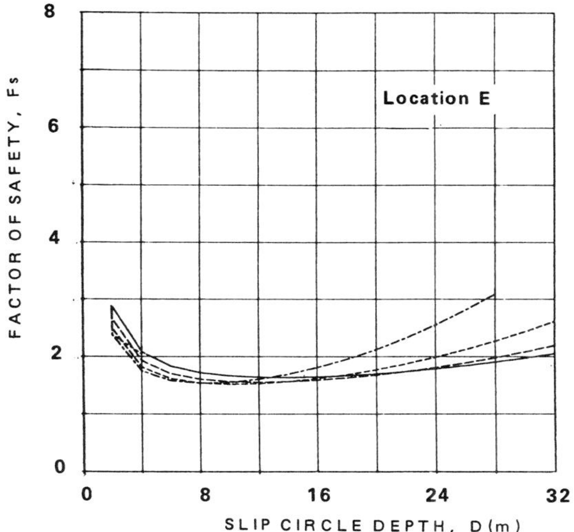

Fig. 12. Safety factors ultimate limit state analysis.

4.4. Foundation Design

The proposed structure involves a buoy which is fixed by chains to six anchor piles. The piles are 42 inch or 48 inch diameter open pipe piles to be located at about 300 m radius. The pipeline runs from the shallow nearshore area to a pipeline end manifold (PLEM). The PLEM has to be fixed with 24 inch pipe piles.

The original SBM-1 location is not suitable because of the slide risk during design wave loading. The depth of the sliding surface may be beyond 20 m. No distinct stronger layers were found at deeper penetrations. Alternative concepts for the anchor piles are not feasible. Moreover the area is not suitable for the pipeline and the PLEM.

The alternative SBM locations 3, 4 and 5 appeared to be suitable. The foundation engineering analyses included studies on ultimate axial capacity and lateral load deformation (p-y curves). The wave induced elastic horizontal seabed movements can be incorporated in the lateral soil response of the piles (Van der Zwaag, 1988). This aspect is of minor importance for the anchor piles but can be relevant for fixed platform foundation piles.

5. Conclusions

The Krishna-Godavari area is essentially an unstable progradational shelf margin delta. This case history shows hat an integrated geophysical and geotechnical study in such areas can result in a high degree of confidence for the selection of safe foundation sites.

Acknowledgements

The authors wish to thank the Oil and Natural Gas Commission (ONGC) for their permission to use data contained in this study.

References

1. Closs, H., Hari Narain, and Garde, S. C. (1974), 'Continental margins of India', in *The Geology of Continental Margins*, C. A. Burk and C. L. Drake (eds.), pp. 629–639.
2. Coleman, J. M., Prior, D. B., and Lindsay, J. F. (1983), 'Deltaic Influences on Shelfedge Instability Processes', Coastal Studies institute, Louisiana State University, Baton Rouge, SEPM Special Publication No. 33, pp. 121–137.
3. Curray, J. R. and Moore, D. G. (1974), 'Continental margins of India', in *The Geology of Continental Margins*, C. A. Burk and C. L. Drake (eds.), pp. 617–627.
4. Kraft, L. M. and Ploessel, M. R. (1986), 'Stability of submarine slopes', in *Planning and Design of Fixed Offshore Platforms*, Part IV Chapter 15, Van Nostrand Reinhold Company, New York.
5. Henkel, D. J. (1970), 'The role of waves in causing submarine landslides', *Géotechnique* **20**(1), 75–80.
6. De Ruiter, J. and Richards, A. F. (1983), 'Marine geotechnical investigations, a nature technology', *Proceedings ASCE Conference on Geotechnical Practice in Offshore Engineering*, Austin, Texas, April 1983.
7. Van der Zwaag, G. L. and Van Seters, A. J. (1988), 'Seabed movements and their effect on offshore structures', *Proceedings of the International Symposium on Modelling Soil-Water-Structure Interaction*, Delft, The Netherlands.

RIGS AND REEF GEOLOGY: A SITE SURVEY IN THE MAFIA CHANNEL OFFSHORE TANZANIA

R. MCELROY, B. P. MEIER, P. M. V. M. GABRIELS and C. D. GREEN
Shell International Petroleum Maatschappij B.V., EPX/233- Seabottom Surveys, Oostduinlaan 75, 2501 AN Den Haag, The Netherlands

Abstract. A suite of high quality, high resolution seabed survey data was used to investigate the flank of the Dira Reef, offshore Tanzania. The reef occurs on the upthrown side of an offshore normal fault related to the East African rift system. The fault is likely to have moved at a rate of approximately 2.5 $mm.yr^{-1}$ for 16,000 years. This extension has opened up migration pathways which allowed biogenic gas to vent into the near-seabed soils sequences. The activity of the fault network also determined the nature and distribution of the shallow overburden sequences. They comprise a single fining-upwards depositional cycle compatible with formation during a relative sealevel rise. The Rufiji Delta drains the rift-flank area of the continental hinterland to the SW, and will eventually enroach the area and bury this sequence with coarser sediment. The relationships are therefore a small-scale geological analogy for features commonly occurring around 'break-up unconformities' buried beneath continental shelves in mixed carbonate/clastic settings. The complexities of the near seabed sediments, the presence of gas accumulations related to fault structures, and the restrictions imposed by the design of the available rig made for a challenging site survey operation.

1. Introduction

A wellsite hazard survey was conducted for Shell Tanzania at the proposed CENTRAL MAFIA-A drilling location in the Mafia Channel, Offshore Tanzania (Figure 1). The data comprised high resolution digital and analogue seismic, side-scan sonar and echosounder records calibrated with soils samples. The aims of the survey were to assess the potential for shallow gas drilling hazards and to provide data to assist in founding the preselected jack-up rig *'Maersk Venturer'*. This contribution describes the unusually detailed and informative geological findings of the survey, outlines the nature of the shallow gas accumulations in the area and assesses typical foundations conditions adjacent to the reef. The application of this information during the drilling operation is discussed and the paper concludes by placing the results in a wider context.

2. Data Acquisition and Processing

The analogue seismic system comprised a surface-tow boomer with a bandwidth of 600–4000 Hz. The data quality is excellent across most of the survey area,

Volume 28: Offshore Site Investigation and Foundation Behaviour, 347–373, 1993.

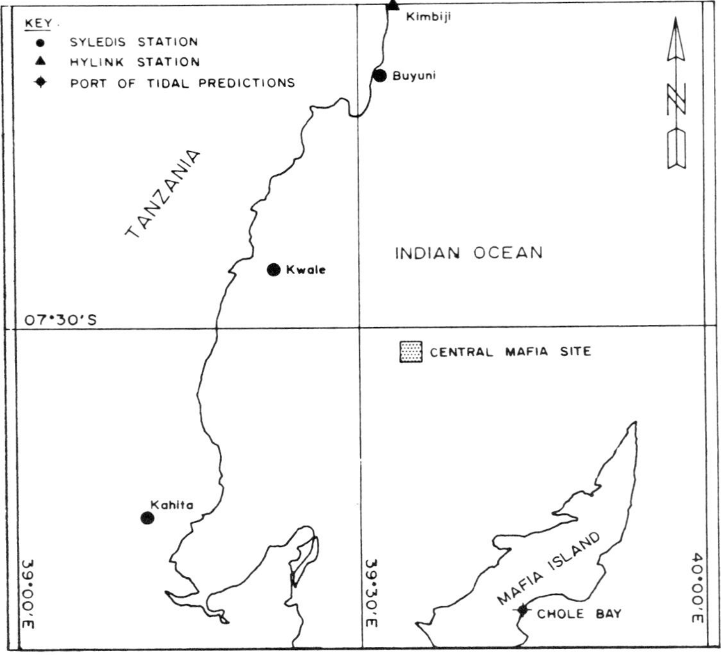

Fig. 1. Location of the study area.

with deep penetration (70 m) characteristic of the northwestern 80% of the survey grid. In this region, the seabed is composed of mudprone overburden deposits or (locally) more granular material. When reef buildups are evident (e.g. the Dira Reef along the SE edge of the survey grid), the energy returns from deeper levels are extremely attenuated or absent. Where present, such acoustic masking limits penetration to less than 10 m. Resolution is of the order of 0.4 m at the top of the section, decreasing to 0.8–1.0 m where maximum penetration is achieved.

For the multichannel digital seismic acquisition a standard high-resolution spread was used. A 4 × 40 cu.inch T.I. sleeve gun cluster provided a bandwidth of 10 hZ/18 dB low cut and 750 Hz/78 dB high cut. The record length was 2.0 seconds. The data quality allowed a standard processing sequence to be followed in-house. On the final displays, the vertical seismic resolution is circa 4 m at the top of the sections gradually decreasing to more than 10 m towards the lower part.

3. Data Interpretation

3.1. Bathymetry and Seabed Conditions

The survey area flanks the steep NW face to the SW–NE trending Dira Reef (see Figures 2, 3 and 4). Water depth is minimal over the top of the reef (the NE end

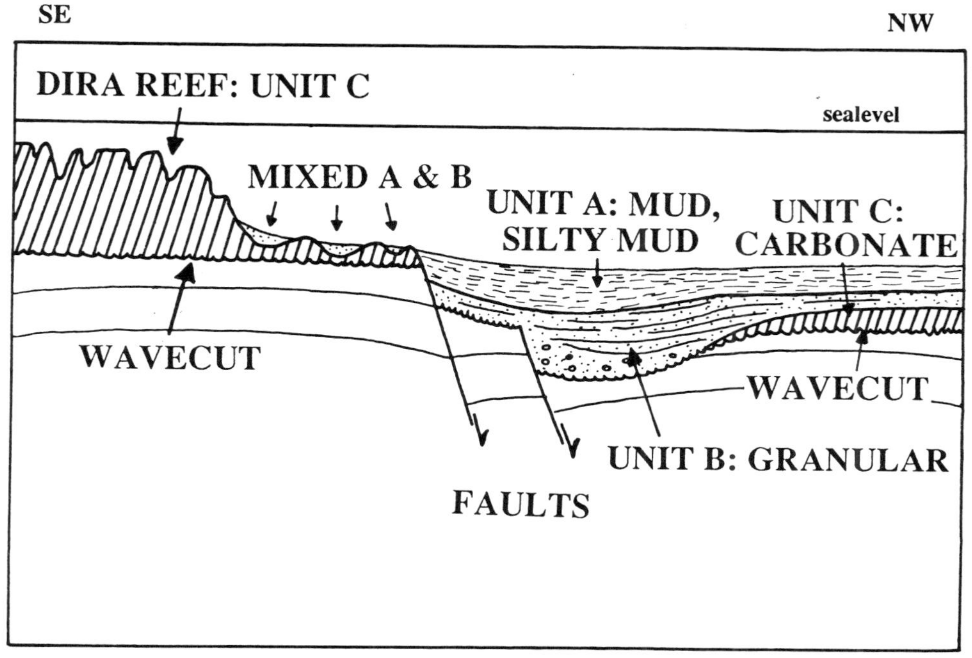

Fig. 2. Schematic illustration of the overburden sequences.

is almost exposed at low tide). Deep notches cut into the reef with extremely steep slopes (up to 1 in 1) can attain up to 30 m in vertical relief. These channels pass carbonate detritus down off the reef into a mud- and silt-filled fault-bounded trough to the NW. In the NE of the survey, a fan-like feature on the bathymetry data (Figure 4) is clearly associated with the mouth of one of the re-entrants in the reef margin.

The northwesternmost seafloor expression of the reef complex is a 15 m high NE–SW trending topographic scarp which marks a fault trace. A series of linear mounds are developed in 35–45 m of water on the upthrown narrow ledge located between the fault scarp and the toe of the main reef to the SE. These reef-edge carbonate buildups are characterised by high sonar reflectivity and acoustic shadowing and have a relief of 1–7 m above the surrounding seabed.

The seafloor away from the reef to the NW is essentially flat over most of the survey area, with water depths of around 55–60 m. In this deeper water realm, the seabed has a low to moderately reflective sonar response and is generally featureless. The surficial sediments were identified as very soft silty clays with some sandy silts from soils samples. However, a rugged bedrock topography is hidden below the flat seabed.

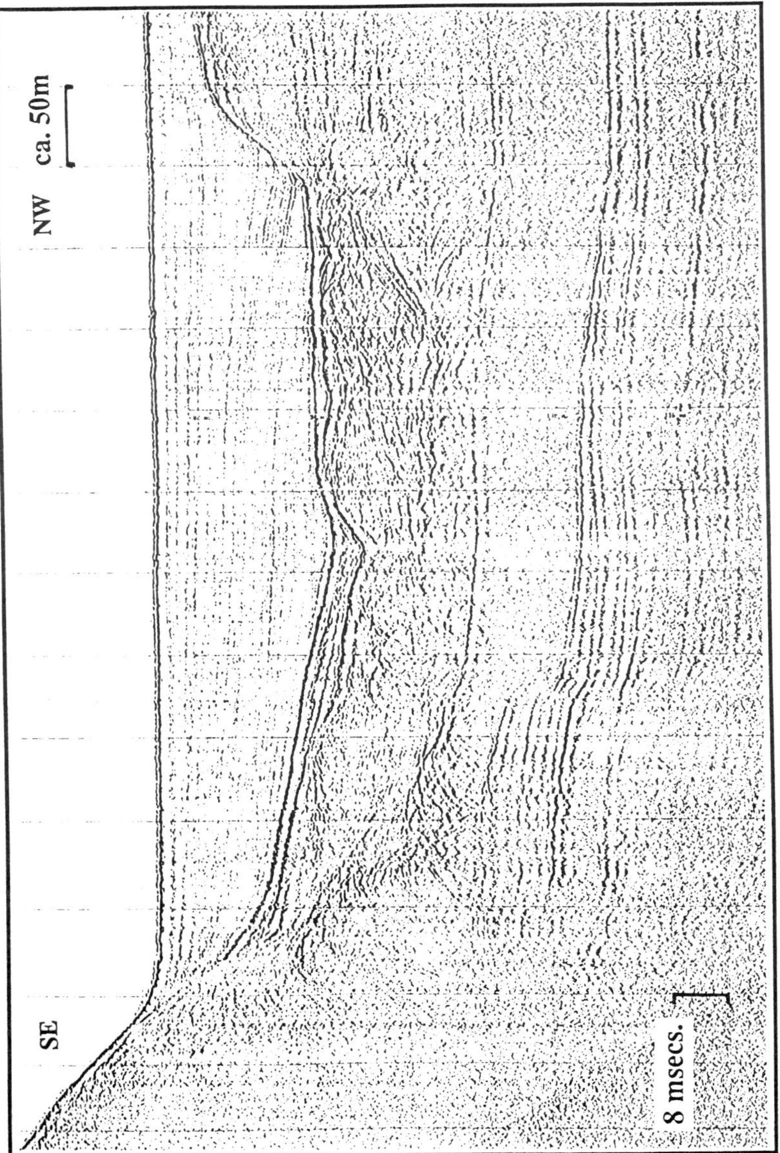

Fig. 3. Line 91-MC-A11. Analogue seismic profile showing the upper mud fill and lower granular facies of the overburden. They fill a channel cut into bedrock which is faulted (lower left centre) and contains low-angle erosion surfaces (lower centre). The reef is located off to the left (compare Figure 2).

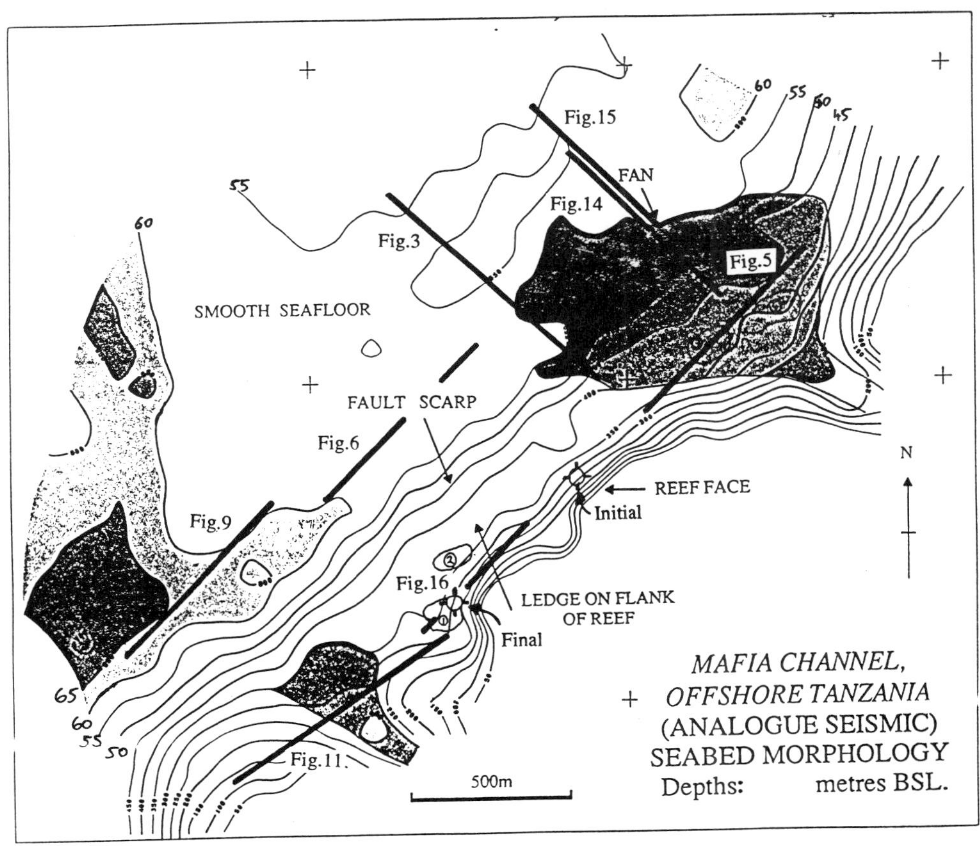

Fig. 4. Seabed morphology based on analogue seismic data.

3.2. Near-Seabed Sequences (Analogue Seismic)

On the analogue seismic data, a distinction is made between unconsolidated overburden and possibly consolidated deposits. The overburden basically consists of three units (A–C), that are characterised by their seismic facies (Figures 2 and 3) and forms a single transgressive sequence above an erosion surface cut into partly lithified deposits, termed 'bedrock',

The shallowest **Unit A** is a mudprone sequence which is largely confined to the deeper water. **Unit B** mainly consists of a series of sandy silt intervals with varying amounts of admixed coral fragments. Unit B is reduced to a veneer on the ledge flanking the reef. The interval is thickest and coarsest in a steep-sided channel buried beneath the flat seafloor. Deposits of **Unit C** facies often form the base to the overburden and persist to the seafloor in the area of the main reef complex in the SE. The facies is inferred to comprise coral buildups, particularly where mounding is expressed: these are separated by and interfinger with reworked debris consisting mainly of carbonate sands with coral fragments. The **Bedrock** is interpreted to consist of a tidally reworked, mixed lithology of coralline limestone and silicic material. The bedrock is interpreted to be at least partly lithified, particularly where its top causes acoustic shielding of the underlying sequence as is the case beneath the reef.

Unit A (Figures 3, 6 and 7). The muddy, silty sequence (from dropcore samples) has a relatively transparent seismic character, with weak continuous reflections defining a lamination of the order 0.5–1 m. At least five overlapping sequences are seen over the interval. They vary in thickness, internal geometry and spatial relationships with other sub-units. Such features represent successive depositional systems which onlapped and buried their precursor bedforms.

The lower boundary of the unit as a whole is also an onlap surface, whereby the deposits drape older erosional and depositional topographies with slopes of up to 15–20 degrees. In addition, a very steep fault scarp bounding the carbonate platform to the Se is onlapped by the unit. The unit is essentially limited to deeper water, but is present with a drastically reduced thickness in a small area in the NE where it occurs on the upthrown side of the fault. A maximum thickness of around 32 m is attained in the adjacent trough which contrasts with the area further to the NW where Unit A thins to zero over a submerged bedrock high.

Unit B (Figures 8 and 9). This interval has a laminated seismic response, with internal foresets, downlap and onlap surfaces. The reflections are high frequency, high amplitude features, and gain continuity upward towards a very high amplitude top boundary to the sequence. The lower boundary of the unit either drapes over coral mounds of Unit C where they occur, or is essentially conformable with inter-mound deposits of Unit C. Where Unit C is completely absent, Unit B banks up over the pre-existing fault- and erosion topographies which define the upper surface of the bedrock.

The upper 3–5 m of the interval has a fairly uniform distribution throughout

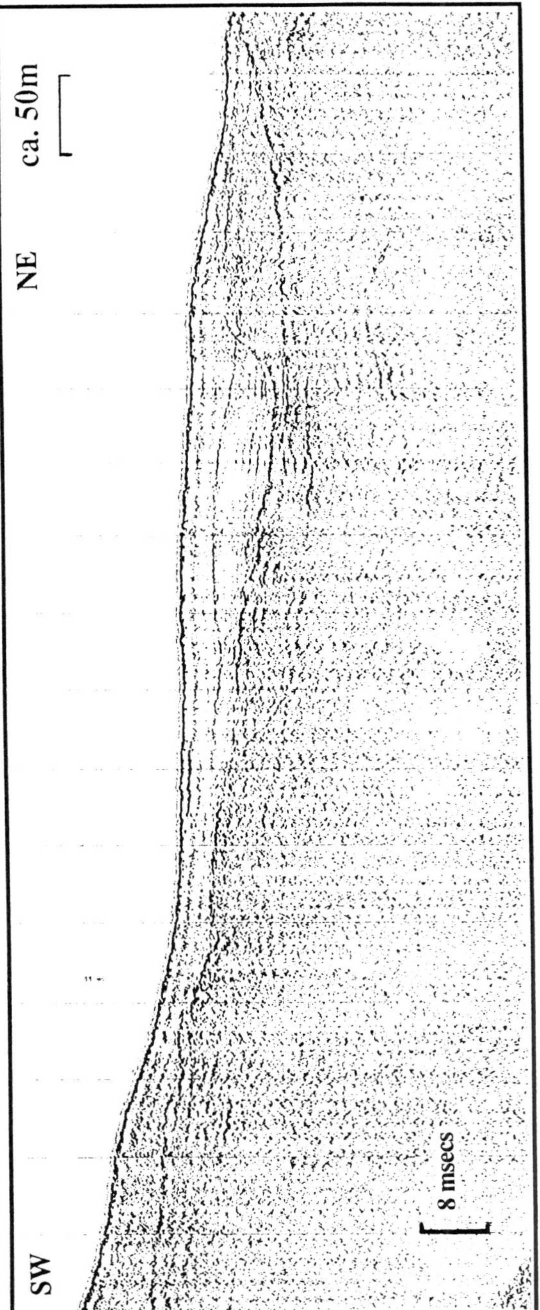

Fig. 5. Line 91-MC-A22. Analogue seismic profile over the upper reaches of the distributary fan in the NE of the area.

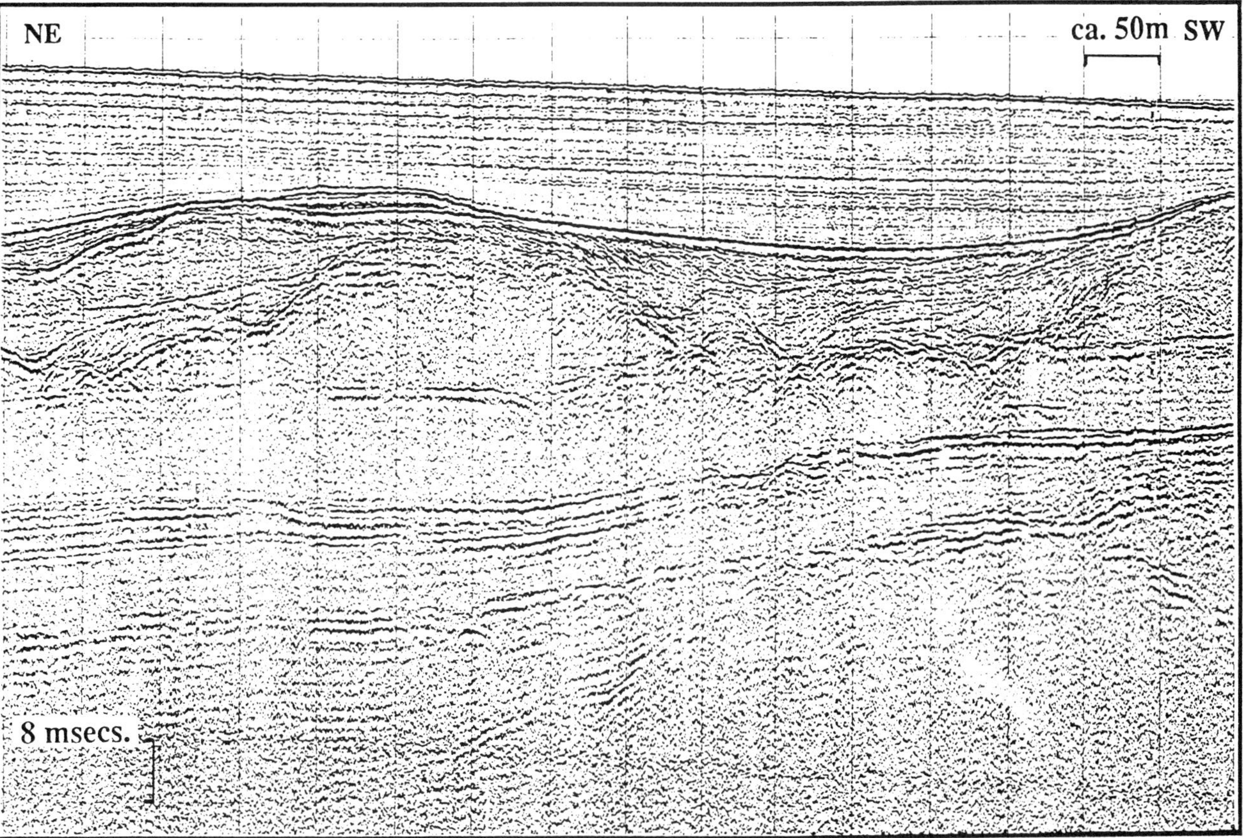

Fig. 6. Line 91-MC-A12. Analogue seismic profile showing small erosional remnants of bedrock separated by filled channels and buried by overburden muds.

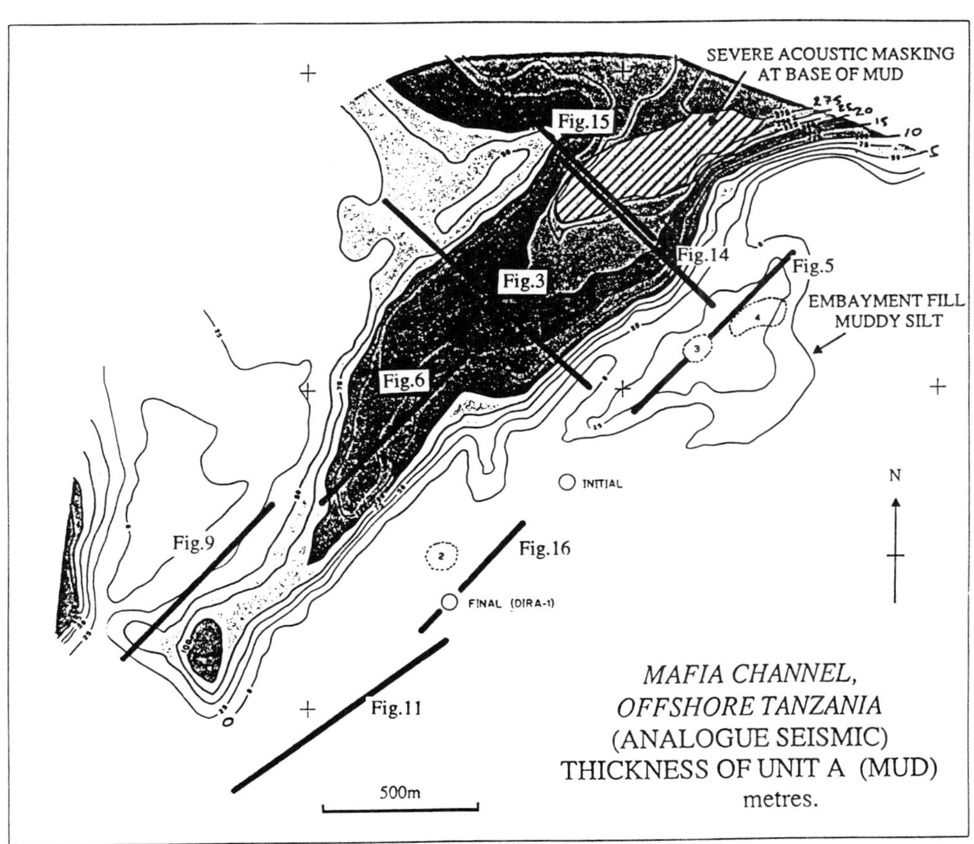

Fig. 7. Thickness map of the overburden mud (Unit A).

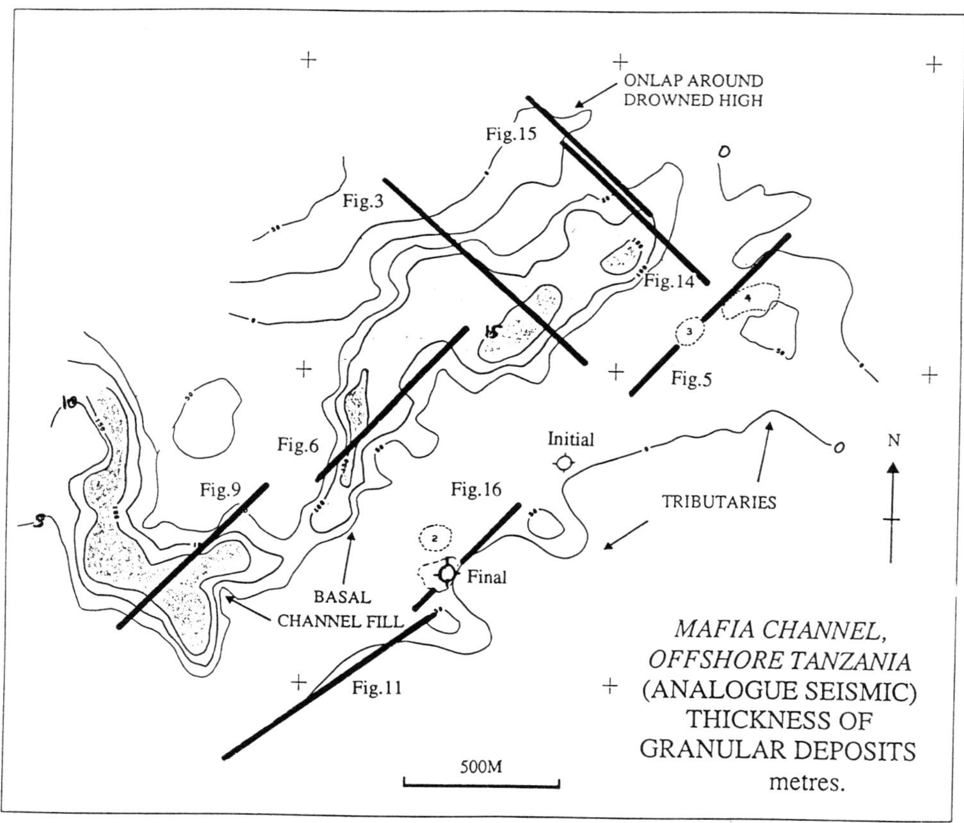

Fig. 8. Thickness of the basal granular deposits of the overburden (Unit B).

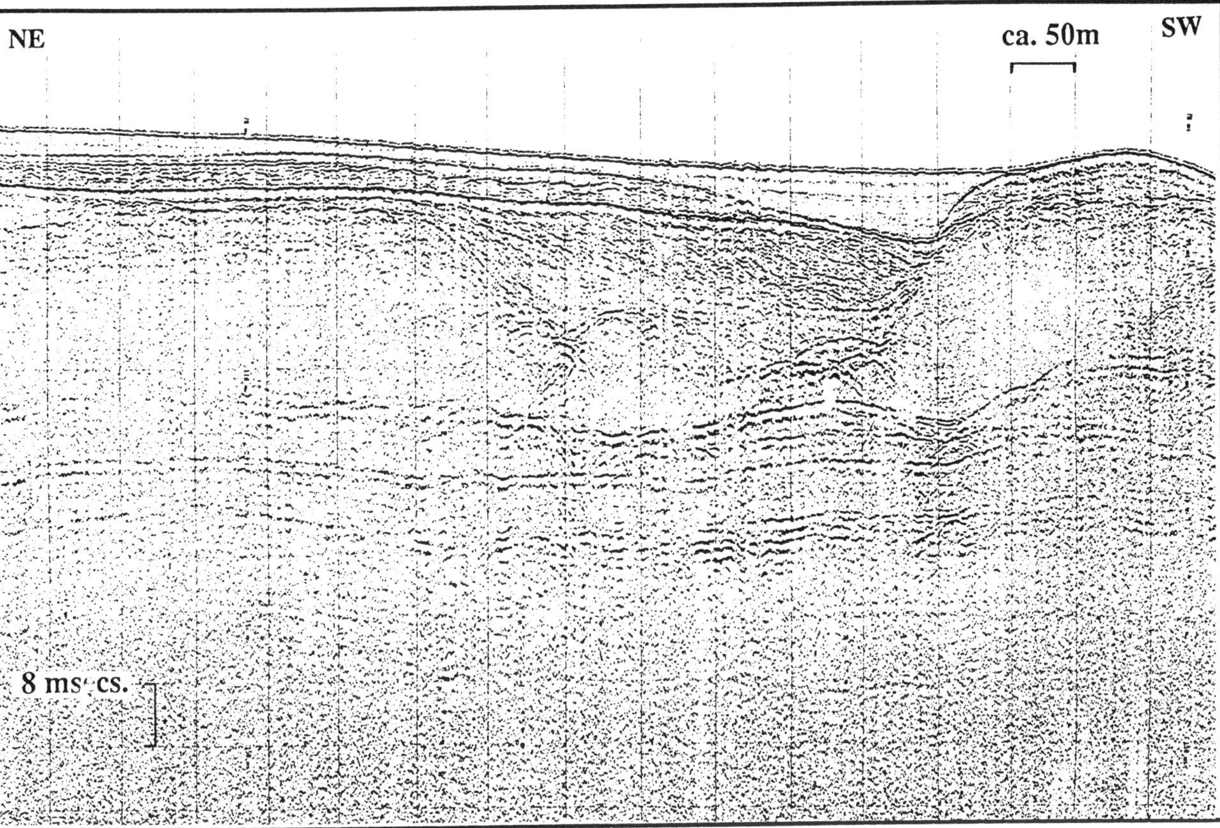

Fig. 9. Line 91-MC-A10. Analogue seismic profile illustrating a channel filled by laterally accreted point-bar sands. They prograde towards the steep bank of the channel where a small bioherm is developed.

the survey area, and is thickest where channels feed off the reef in the SE. The eastern boundary of the interval is a pinchout against the topography of the reef. This upper part of Unit B was sampled by cores and shown to consist of sandy silt with admixed coral fragments.

In contrast, the lower part of Unit B is a 0–10 m thick channel-fill sequence. The seismic response is characterised by high amplitude, high frequency discontinuous reflections, generally displaying clastic depositional bedforms which include prograding channel-margin bars. The basal 5 m of the Unit is rather chaotic, with very steep internal dips and contains high-energy bedforms (locally derived talus slopes, debris flows, collapsed channel banks) on a variety of scales. This lower part of the unit was not directly sampled. However, from the similarity in its seismic response to the overlying deposits, but because of its stronger amplitudes and the characteristic bedforms, the interval is interpreted to consist in part of coarse granular deposits: coral debris with sands and silts.

The base of Unit B is either an erosion surface cut into bedrock (Figure 12) or is a drape over small coral buildups developed on that erosion surface. The isopach pattern filling the topography identifies a series of small tributaries feeding off the reef into a main channel to the NW, where thicknesses of more than 5 m of Unit B are confined. In the NE of the area, the largest of these tributaries is approximately 400 m wide and has cut down into both the reef and the ledge. It passes material from SE to NW and is filled with 5 m of detritus (Figure 4). The base of the channel cuts down to its maximum of 8 m into bedrock where it crosses the ledge. Mounded reef-type features are developed in this area on top of the bedrock erosion surface, and are themselves truncated by younger erosion surfaces evident within the channel-fill. Several subsequences of granular deposits within the fill (both Units B and A) can also be distinguished. Each represents a period of renewed deposition after a minor event of submarine erosion linked to readjustment of the active channel course within its submarine banks.

The channel-fill changes thickness along the length of the channel. This is a result of scarp degradation at both the reef face and the fault. The channel-fill is thinnest where the feature cuts into the top of the fault and reef scarps. Depositional bulges are evident adjacent to the two breaks in slope, one at the base of the reef on the ledge and the other at the base of the fault scarp in the basin to the NW. These bulges are sub-radial and display on a small-scale geometries typical of a submarine fan.

The major channel which collected debris from such tributaries is located in the centre of the survey area, its SE margin defined by the fault scarp. To the NW, however, the channel wall is an erosion surface cut into an isolated bedrock high. The channel floor corresponds geographically with the traces of faults in the underlying bedrock, indicating that the faulted area was the most susceptible to erosion. The channel is interpreted to have been progressively eroded by tidal surges acting on a wavecut platform.

Unit C (Figures 10 and 11). The facies comprising this interval is both distinctive and unusual. Depositional mounds and pinnacles are seen, with high amplitude convex upward reflections in their upper parts. The mounds have a rather opaque seismic character. The few internal reflections that occur tend to be high-frequent and chaotic. The upper and lower bounding reflections in contrast are strong, indicating a marked impedance contrast between the Unit C and the overlying mud or clastic sediment and also the underlying bedrock lithologies. Strong acoustic masking of the underlying sequences occurs where Unit C is thickest and scatters most of the seismic signal. This attenuation relates to the wide range in the size and shape of pore volume and the wide variety of skeletal/shell and coral material expected to form this framework rather than stratified deposit.

The depositional mounds are interpreted as patch-reef bioherm structures. They have a vertical relief of up to 10 metres and a lateral extent of between 30 and 200 m. They are most commonly developed as topographical highs against which the clastic sequences onlap. The facies occurs throughout the area, except for within the channel system in the centre of the area which was probably too deep and subject to periodic burial be debris. Coral growth was most successful on the upthrown side of the fault where the present-day reef occurs. In the downfaulted block to the NW of the Dira Reef, an outlying 'island' of coral facies grew for a time. It developed on a high part of the old wavecut surface, but was eventually smothered by Units B and A.

Thick reef buildups form the seafloor in 9–43 m of water along the extreme SE margin of the area. They are based on a moderate to high amplitude sub-horizontal discontinuous reflection at 37–45 m BSL which is interpreted as a wavecut bedrock surface. The event is strongly attenuated below each of the mounds. The bottom 5–7 m of the distributary channels between such massifs are likely to be filled with granular deposits.

Bedrock (Figures 11 and 12). The bedrock returns are strongly masked beneath the reef in the SE. To the NW, however, the bedrock has a clearer seismic response, and is characterised by a framework of cross-cutting, large-scale, low-angle erosion surfaces. These separate a variety of channel-fill deposits. 'Grainy', opaque, massive seismic units are the most common, and thicken into channels. Occasionally foresets are preserved. The seismic character is interpreted as an indication of well-sorted silty, sandy channel fill sediments, perhaps deposited in a mature alluvial/tidal deltaic environment. The geometries of the low angle, concave upward bounding erosion surfaces suggest the distributary channels were 10–15 m deep and circa 700–1500 m wide. Wedges of moderate to high amplitude reflections with a lamination on a scale of circa 1 m also occur. These sequences are interpreted as low energy interdistributary bay deposits of sandy, silty clay. Erosional truncation and pinch-out against channel bases and margins is common.

The basal fill of some of the larger scale channels within the bedrock is occasionally characterised by a patch reef/inter-reef channel facies (similar to Units B and C of the overburden). These would suggest periodic abandonment of some channels as the focus for sediment dispersal, for long enough to permit local establishment

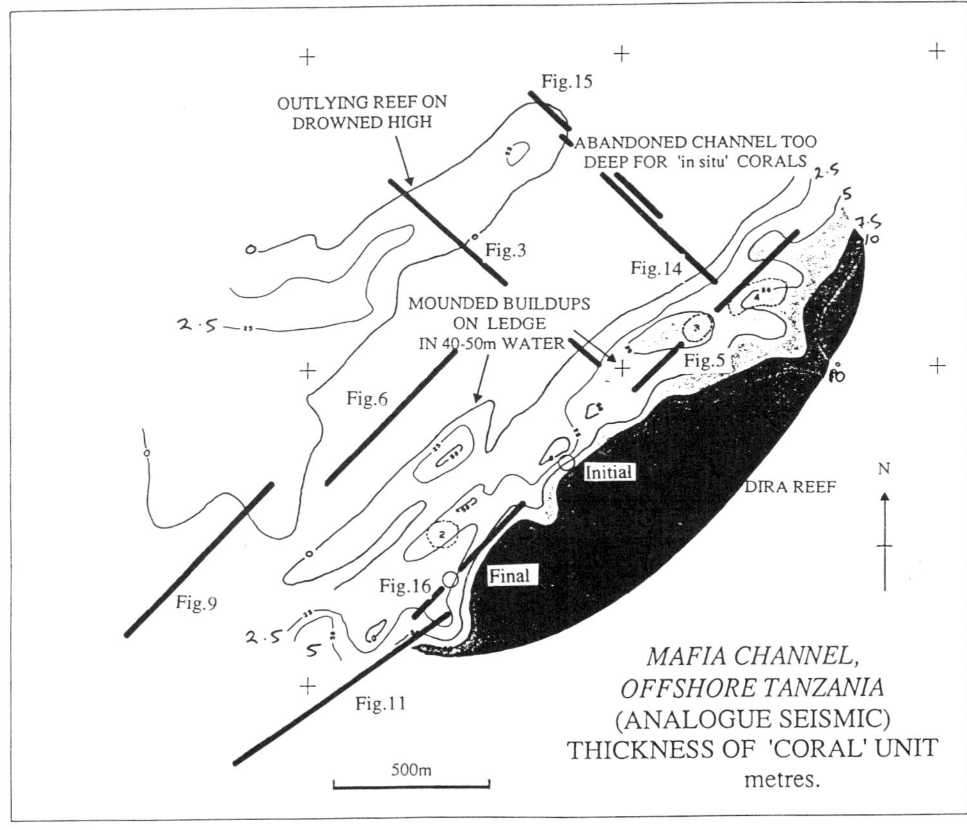

Fig. 10. Thickness and distribution of the coral facies of the overburden (Unit C).

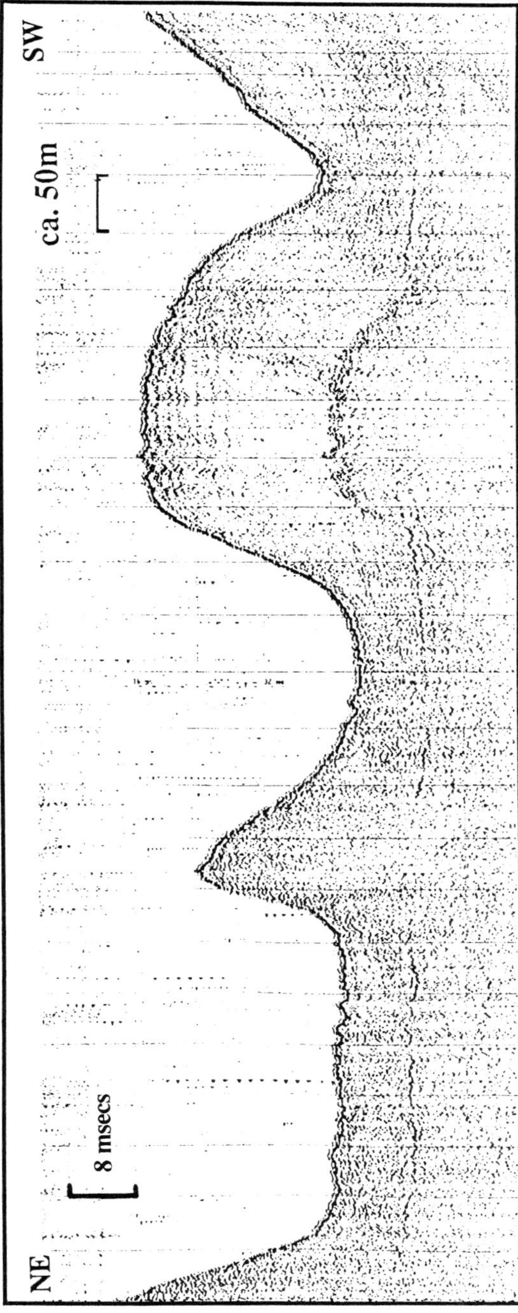

Fig. 11. Line 91-MC-RL3. Analogue seismic profile showing mounds and pinnacles of coral. The sub-horizontal reflector visible beneath them is interpreted as a wavecut surface.

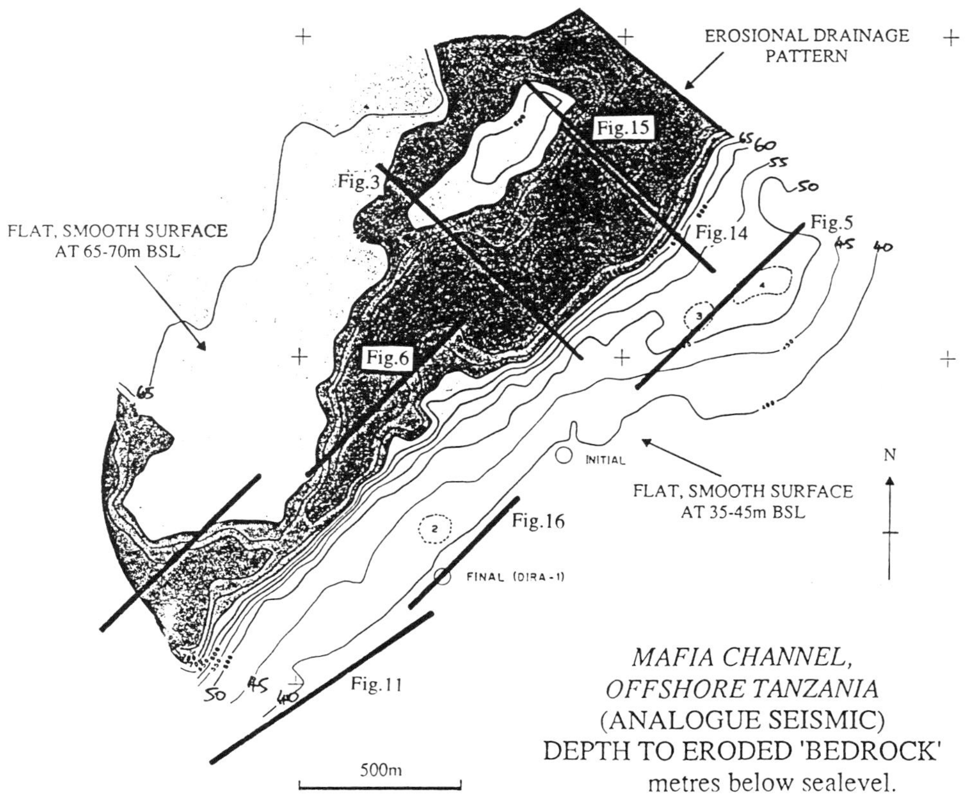

Fig. 12. Depth to the wavecut surface.

of bioherms. They were probably fragile ecosystems, since they can be seen to have been smothered by clastic channel-fill at several stages of development.

3.3. Deeper Levels (Digital Seismic)

Most of the stratigraphic section covered by the high resolution digital seismic is interpreted to comprise mainly Miocene-Recent tidally-influenced delta deposits. These pass up into the mixed clastic/carbonate sediments imaged on the analogue data. In the upper 400 metres, small-scale internal unconformities are developed which suggest syn-depositional fault control of sedimentation. Three faults are identified (e.g. Figure 15). A major deep-seated and linear, NE–SW oriented normal fault appears to control the western flank of the Dira Reef. The throw of this NW-hading fault varies between 20–40 m.

4. Structure

Three sub-parallel NE–SW striking normal faults were identified on both the digital and the analogue seismic data (Figures 2, 3, 14 and 15). They downthrow to the NW to define the margin of the active Dira Reef. The fault nearest the reef has the largest throw: a minimum of approximately 20–25 m is observed from the offset of the buried wavecut platform on the analogue data.

The analogue data contain sufficient information to constrain an investigation of the subsidence history of the areas either side of the fault zone. The geometry of the erosion surface cut into bedrock (Figure 12) is most easily interpreted as an 20 m-deep tidal surge channel incised into a co-existing flat wavecut platform when base-level was lower than present. The flat surface is offset by the fault and is downthrown by 25 m to the NW (although total throw on the fault at deeper levels is circa 40 m from interpretation of the digital data). If the channel/wavecut surface is assumed to have stopped developing at an early stage during the last postglacial sealevel rise (Figure 13), circa 10,000 years ago, its offset suggests a time-averaged rate of motion on the fault of 2.5 mm.yr.$^{-1}$, in good agreement with neotectonic rates. The fault must therefore have begun to move approximately 16,000 years ago in order to develop the total throw of 40 m throw.

The sub-planar wavecut surface is located 40 m and 65 m below present sealevel on the upthrown and downthrown sides of the fault respectively. These give rates of *relative* sealevel rise of 4 and 6.5 mm.yr.$^{-1}$ for the two areas. These rates can be compared with those, for example, over the interval 8,000–6,000 years B.P. in The Netherlands (and therefore the North Sea area in general), where a *relative* sealevel rise at a rate of 6 mm.yr.$^{-1}$ was determined (Reference 1). In conclusion, the assumed ages and derived rates are both realistic and consistent with the geological interpretation of the area and are an unusual bi-product of this type of data.

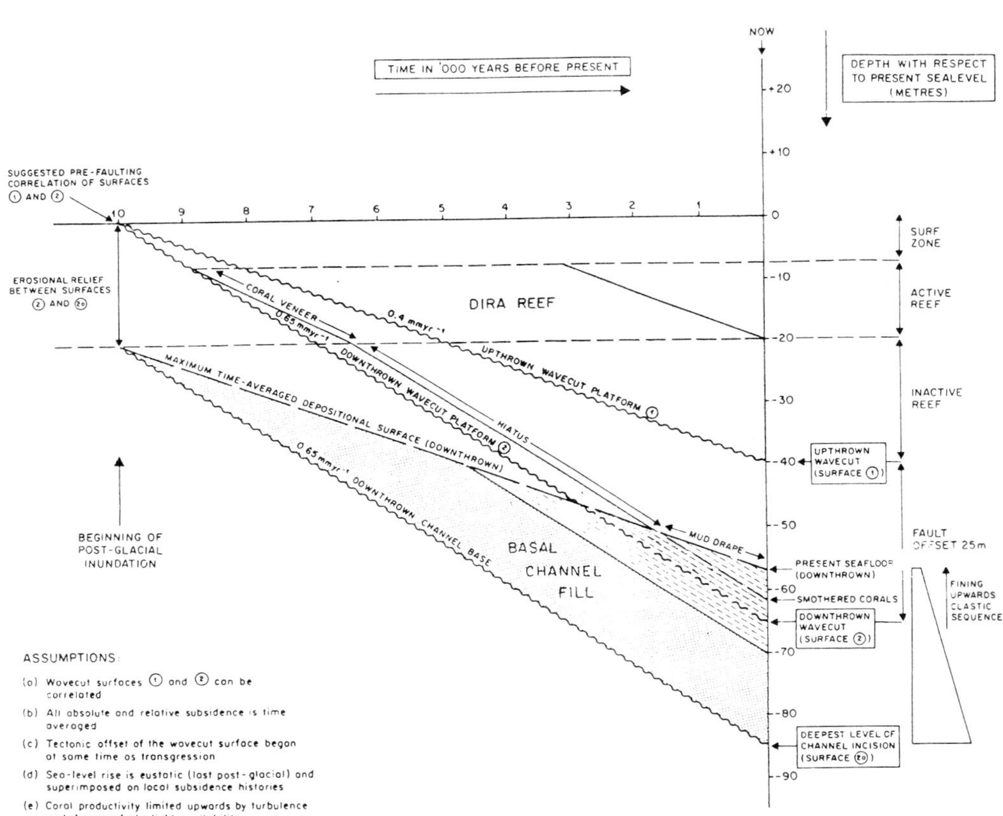

Fig. 13. Sediment accumulation diagram comparing the subsidence history of the wavecut surface either side of the reef-margin fault. Note the listed assumptions.

5. Shallow Gas

Zones of high amplitude chaotic diffractions and acoustic masking were identified at various locations within the survey area. These anomalous zones are associated with faults in the deepest parts of the basin, and are interpreted as shallow gas accumulations. In one case, a patch of diffractions was mapped at the north end of the main survey grid which occupies an area of 200 m × 500 m and trends sub-parallel to the fault pattern. The top of the feature is at approximately 15 m below the seabed, contained within the muds of Unit A. The occurrence reflects the unconsolidated nature of the muds: they still have a high porosity at this shallow level.

Two further types of probable gas occurrence were noted on reconnaissance lines to the north of the main survey grid. In one example, diffractions and masking are initiated at the sequence boundary at the base of Unit A. This boundary is therefore locally sealing. In another case, gas appears to be partly hosted by a small patch-reef in addition to being partly hosted by the muds.

These features were initially identified on the exploration seismic data, and a sampling plan was devised to recover gas for analysis. Reconnaissance analogue lines were shot to calibrate the location and seismic character of the sediments recovered in the series of soils cores. Geochemical results from these samples subsequently indicated that the trace amounts of gas present was composed almost exclusively of methane. Figure 14 illustrates the high resolution digital seismic expression of the feature interpreted as gas on the analogue seismic profile of Figure 15 (note section orientations are reversed). The sections demonstrate a clear fault-association for the occurrence, suggesting that the gas is being vented into the overburden sequence above fracture zones. Considering the seismic data and the results of the geochemical programme, it is suggested that the gas has a biogenic provenance within the upper part of the deltaic succession underlying the shallow carbonate sequence, and underwent focussed migration into the overburden along fractures associated with the fault network.

Results of the digital interpretation indicated that there are no genuine amplitude anomalies present beneath any the four alternative areas discussed below which could indicate a shallow gas-related hazard to tophole drilling. Anomalously high amplitude continuous reflections within the data were interpreted as being caused by strong lithological contrasts. Low-angle convergence of reflections, pinch-outs and wedging is apparent in the bedrock from the boomer data. These appear as amplitude variations when seen on the digital data, and are occasionally very marked. Such features are therefore considered to be a result of loop interference because of the lower vertical resolution of the digital seismic. A similar origin was therefore inferred for uncalibrated amplitude anomalies of this type elsewhere within the digital dataset.

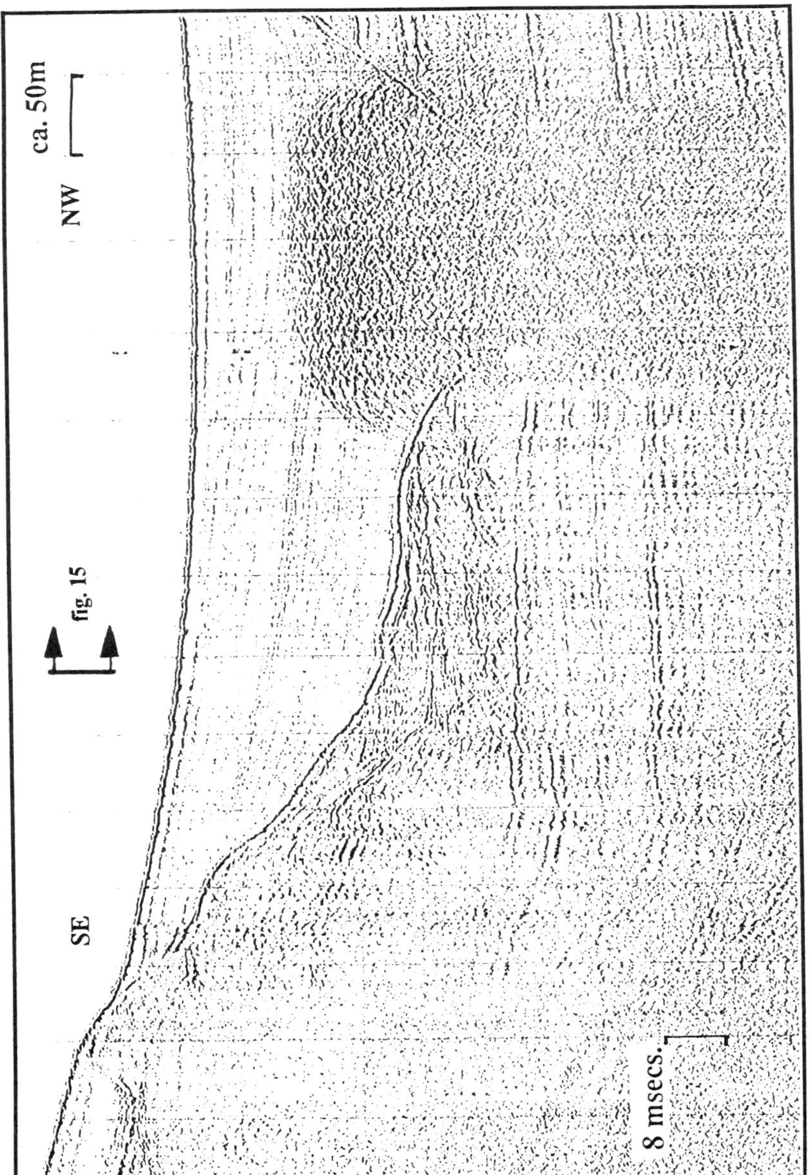

Fig. 14. Line 91-MC-A15. Analogue seismic profile showing diffractions and acoustic masking. The feature is attributed to the presence of biogenic gas venting from a fault at lower right.

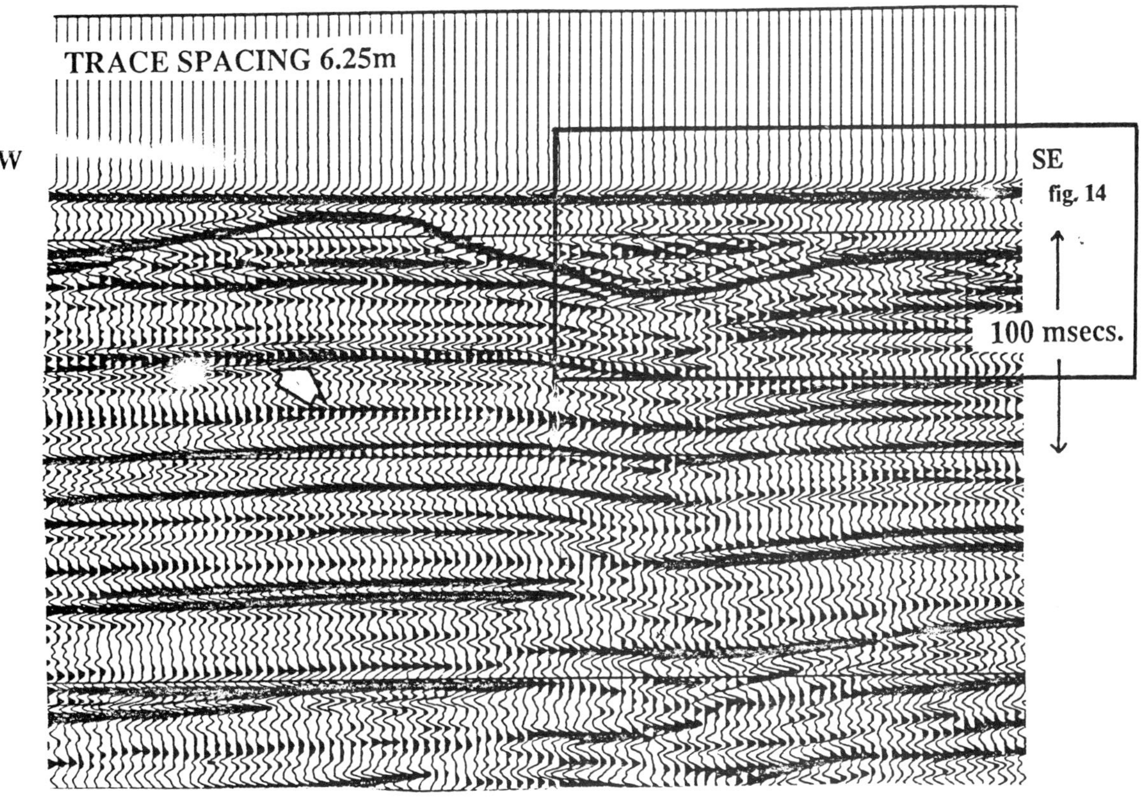

Fig. 15. Line 91-MC-D15. High-resolution digital seismic profile illustrating the gas zone of Figure 14 (box, note reversed section orientation). The feature is directly above a fault expressed to the right of centre. The arrow indicates an amplitude variation that can be demonstrated to relate to terminations and pinchouts of sequences from superior resolution of the analogue seismic data.

6. Site Assessments

General problems in ensuring adequate founding of a rig include determining that there will be no risk of excessive and/or uneven leg penetration nor risk of any sliding of the legs. The bathymetric data indicated that the seabed slopes at the initially proposed site adjacent to the reef are uneven and in excess of 5° whereas the design of the jack-up allowed for dips up to approximately 3°. Alternative areas had to be investigated, taking into account seabed gradients, sedimentary dips, overburden thickness, composition of soils above possibly consolidated 'bedrock', and a water depth constraint of 45 m because of the size of the rig. The only available area with an appropriate depth was the narrow ledge next to the reef. The overburden thickness and composition is extremely complicated because of small coral mounds, submarine channels, a fault scarp and the reef scarp. Further, these features are all developed above a locally rugged and buried bedrock topography. Only areas which had a flat seabed and an essentially uniform soils thickness and composition could be considered.

Considering the above criteria, only four areas could be proposed and tested by computer modelling. The restrictions imposed by the design of the *preselected* rig had drastically reduced flexibility in selecting a final drilling location. Of the four possibilities, Areas 1 and 2 are located on the ledge of the western reef flank SW of the proposed location in water depths less than 40 m. Areas 3 and 4 are located NE of the proposed location in water depths more than 40 m. Each of these areas has advantages and disadvantages which are briefly discussed.

Area 1. This area (Figure 16) is characterised by relatively low seabed gradients and sedimentary dips. The overburden thickness is minimal, and the soils units are considered as well defined. Within this small area, in circa 38 m of water, the location would however remain relatively close (circa 100 m) to the boundary of (steep) coral reefs as established from the side-scan sonar data. The soils consist of a thin upper sequence of Unit B facies circa 2.4 m thick, comprising coral fragments and sandy silts. The lower sequence (Unit C, circa 1.6 m thick), is interpreted to comprise carbonate sands with coral/shell fragments. The bedrock surface occurs at circa 42 m below sealevel and is sub-planar.

Area 2. This area has a water depth of 39.5 m, and the location would be moved further away from the boundary of coral reefs while the soils units remain essentially well defined. The overburden thickness however is slightly increased (to circa 8 m), and the seabed gradients (2.5–3 degrees) are close to the allowable limit. The soils sequences comprise Unit B (circa 3.2 m; coral fragments and sandy silts) and Unit C (circa 4.8 m; carbonate sands with coral/shell fragments and reefal deposits) above the sub-planar bedrock surface at 47.5 m.

Area 3. Within this area (water depth circa 45 m) the location is moved towards the edge of the channel feeding off the reef in the NE. The seabed gradients are relatively low (1–2.3 degrees), and the soils units are considered as relatively well defined. However, the overburden thickness is increased (10.5 m), and although the

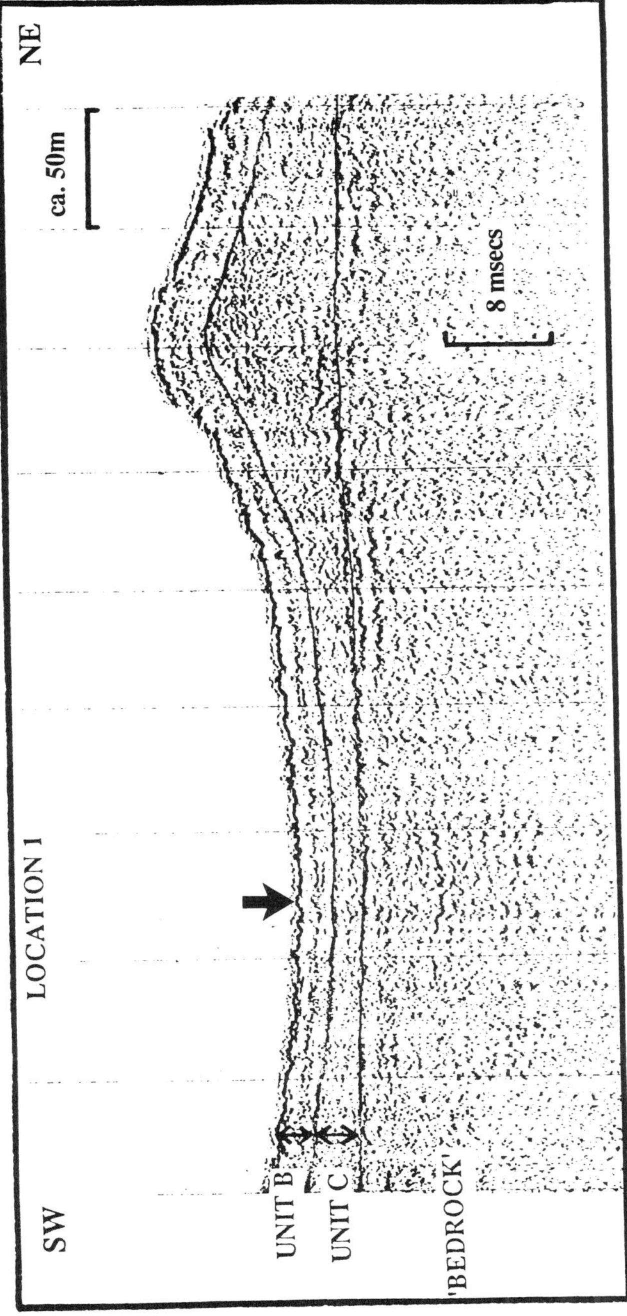

Fig. 16. Line 91-MC-A22. Analogue seismic profile showing the final location in a low area at the toe of the reef.

average sedimentary dips within and at the base of the overburden in this area seem to be low, there is little control on internally varying dips within individual units. The soils sequences is composed of Unit B (circa 5.3 m) sandy silts with coral/shell fragments and is possible coarser at the base. These overlie circa 5.2 m of Unit C carbonate sands with coral/shell fragments and coral reef deposits. The bedrock surface is concave upward (with a channel geometry) and internal sedimentary dips also suggest the bedrock to be laterally variable.

Area 4. This area is located approximately in the axial part of the present day reef-margin channel feature (water depth circa 43.5 m). The seabed slopes in this area are relatively low (0.5–2.2 degrees). However, the overburden is almost 13 m thick, and the seismic units are less well defined and show internal variation in seismic facies. In addition, the top of the bedrock is not clearly expressed. In this area, it is possible that the legs of the jack-up rig may not penetrate down to the interpreted bedrock level. The soils sequence is composed of Unit B (circa 6.4 m) coral/shell fragments with sandy silts and silty sands, overlying Unit C (circa 6.4 m) composed of predominantly reefal deposits. The bedrock has strong internal sedimentary dips.

7. Rig Positioning

The locations were assessed from exploration, engineering and environmental viewpoints. These factors included the possibility of the legs of the rig sliding along the bedrock under large lateral wave loads. The final location chosen was therefore Area 1, where the water is shallowest and the overburden thinnest (Figure 16). At this site, minimal wave-loading and sufficient penetration of the bedrock by the spudcan tip could be ensured (Figure 17). That condition was assisted because the top 1–2 m of the bedrock was likely to have been structurally degraded by exposure during development of the wavecut erosion surface. Although the final site could therefore meet the engineering safety criteria, the area was unfortunately both small and immediately adjacent to the toe of the reef-scarp. Stringent additional operational procedures were required to ease the installation of the rig and to minimise the possible environmental impact of the drilling operation. These requirements determined that:

(a) The rig was ‘walked’ in because of the difficulty in operating tugs so close to the reef face.

(b) Full de-manning would have been required in the event of a cyclone.

(c) The stern faced to the prevailing upwind direction to maximise the turning moment required for instability.

(d) Drill-cuttings were collected to prevent seagrass/coral/benthic fauna being smothered.

(e) Discharge of drilling mud was controlled to minimise any impact on the photic zone ecosystem.

8. Summary and Conclusions

The Dira Reef occurs on the upthrown side of an offshore normal fault related to the East African rift system. The fault is likely to have moved at a time-averaged rate of circa 2.5 $mm.yr^{-1}$ for about 16,000 years. This extension has opened up migration pathways which allowed biogenic gas to vent into the near-seabed soils. The activity of the fault network also determined the nature and distribution of the shallow sequences which comprise a fining-upwards depositional cycle compatible with formation during a relative rise in sea-level.

The cycle began when tidal erosion carved an essentially flat wavecut surface into partly lithified mixed deltaic and carbonate deposits. A deep tidal surge channel was developed where erosion could act most effectively on the weakened substrate above the fault network. Coral communities were established on the adjacent flat areas, while a network of channels began to collect debris and redeposit the material in the fault-related trough. The corals were most successful on the upthrown side of the fault and now form the reef while those on the downthrown block quickly succummed to the effects of subsidence. The downthrown area then became the focus for accumulations of transgressive sands, silts then distal muds.

The Rufiji Delta drains the rift flank area of the continental hinterland to the SW, and will eventually encroach the area and bury these basinal muds with coarser sediment. The entire sequence is therefore a small-scale geological analogy for depositional features which typically occur around the level of 'break-up unconformities' (Reference 2) beneath continental shelves in mixed or transitional carbonate/clastic settings.

The results from the Dira Reef site survey constrain an unusually complex, detailed and comprehensive geological model of a reef-margin environment. The combined datasets, which include soils cores, sonar images and detailed echosoundings in addition to single and multi-channel high resolution seismic data, offer a remarkable degree of internal consistency. Additionally, the survey methods have a resolution at essentially the same scale as the geological relationships and processes discussed. Since the analogue survey investigated the top 80 m of the deposits below the seabed, important seismic details are not obscured by the geophysical effects of a thick overburden as is usually the case when similar but deeply buried sequences are imaged. The results of the survey may therefore have some application in interpreting similar but deeply buried and poorly imaged examples typically encountered in the fields of exploration targeting and reservoir engineering.

The seismic site-survey was completed on 16th July 1991. Dira-1 spudded on 24th October 1991 (leg penetrations of 5.5, 5.1 and 6.0 m, Figures 16 and 17), and T.D. of 3529 m was reached safely on 18th November 1991 with neither shallow gas nor overpressures encountered. However, the complexities involved in deciding

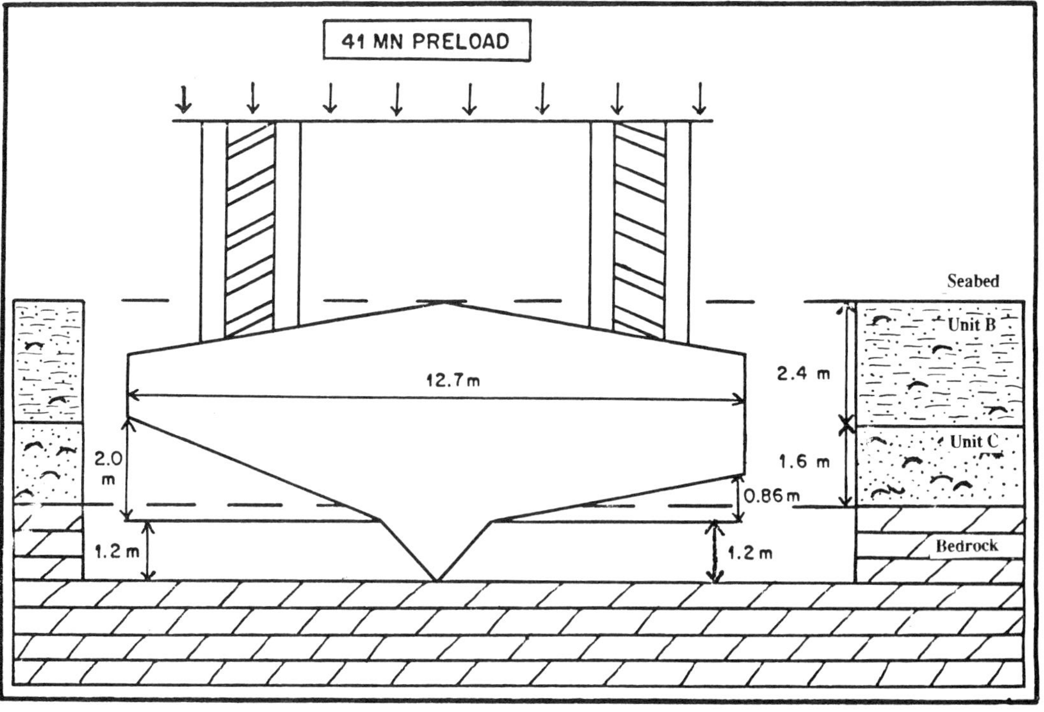

Fig. 17. Diagram showing the final configuration of the spud cans at the base of the rig legs. The sandy silt (upper) and bioclastic sand (lower) overburden sequences were sufficiently penetrated to allow the tips to embed into bedrock and prevent the legs sliding.

on a final location in this case offer a further illustration of why a comprehensive site survey should be conducted *well* in advance of a drilling operation.

References

1. De Jong, J. D. (1967), 'The Quaternary of The Netherlands', in *The Geologic Systems: The Quaternary* 2, K. Rankama (ed.), Interscience Publishers, Wiley & Sons, 477 pp.
2. Boillot, G. (1978), *The Geology of the Continental Margins*, first published by Mason, Paris, 108 pp.

DEEPWATER ENGINEERING GEOLOGY AND PRODUCTION STRUCTURE SITING, NORTHERN GULF OF MEXICO

K. J. CAMPBELL and J. R. HOOPER
Fugro-McClelland Marine Geosciences, Inc., 6100 Hillcroft, Houston, TX 77081, U.S.A.

Abstract. The engineering geology of the continental slope in the northern Gulf of Mexico is probably among the most complex of any offshore area in the world. This complexity is due largely to past and ongoing salt movement and to landsliding and other depositional processes during Quaternary sea-level lowstands. Conditions which can adversely affect siting and design of production structures include: steep slopes and rugged, sometimes rocky, topography; active faults; shallow overpressured zones; both modern and ancient landslides; gas hydrates; seafloor erosion; and variable materials ranging from weak soils to rock. Engineering assessment of deepwater sites may require collection and comprehensive integration of a wide variety of geophysical, geological, and geotechnical data.

1. Introduction

The purposes of this paper are to summarize the wide range and complexity of geologic conditions on the continental slope in the northern Gulf of Mexico, to briefly discuss some engineering concerns related to these conditions, and to summarize some geophysical techniques used to investigate these conditions. Sampling and *in situ* testing of marine soils, including deepwater soils, is discussed in Young (1991).

1.1. Database

Results presented here are based principally on high-resolution geophysical data collected over more than 350 lease blocks (each about 5 km by 5 km) scattered over the Texas and Louisiana slope. Water depths ranged from about 200 m at the shelf edge to nearly 2500 m. Survey line layout was typically 300 m by 900 m. The area of data coverage represents about 900 sq km, or about 4% of the total deepwater lease area off Texas and Louisiana. Most data was collected during the past 6 years in connection with exploratory drilling, structural siting and design, or pipeline routing being planned by major oil companies. Results have also been available from numerous 3- to 6-m-long piston cores along with results from more than 150 rotary soil borings done on the continental slope in water depths up to about 1000 m.

Volume 28: Offshore Site Investigation and Foundation Behaviour, 375–390, 1993.

1.2. Deepwater Production Structures

There are some 4100 production platforms in the Gulf of Mexico, but less than 20 are in water depths greater than 200 m. Among the first truly deepwater structures in the Gulf of Mexico were Shell's Cognac platform and Union's Cerveza platform, which were installed in about 300 m of water in the late seventies. Both are conventional pile-supported jacket structures. Conoco's Jolliet tension-leg platform in about 536 m of water holds the deepwater operating record in the Gulf, but Shell's Auger tension-leg platform is scheduled to be installed in about 872 m of water in 1993. For all of these structures, a thorough understanding of past and ongoing geologic processes and soil conditions was necessary for safe and economical siting and design. Largely as a result of this increasing need, deepwater survey techniques to investigate the seafloor and shallow strata were refined in the mid to late 1980's.

2. Summary of Deepwater Conditions

2.1. Geologic Setting

The shallow geology on the continental slope off Texas and Louisiana is probably among the most complex and dynamic of any offshore area in the world, and is characterized by recent and ongoing geologic processes. In sharp contrast, the shallow geology of the continental shelf, with few exceptions, is simple, uniform, and predictable over wide areas. One of the major factors influencing deepwater geologic processes is the massive Jurassic salt deposits that underlie the northern Gulf of Mexico. The topography of the slope is relatively rugged and irregular (Figures 1 and 2) as a result of past and ongoing flowage and uplift of the buoyant, plastic salt. Average rates of vertical uplift of 2 to 4 m per 100 years have been suggested (Martin and Bouma, 1982) at one location on the slope. Uplift has resulted in faulting, slope oversteepening and landsliding, leakage of gas along faults contributing to formation of authigenic carbonates and gas hydrate accumulations, and other processes.

Another major influence has been sea-level lowering and deposition of massive amounts of sediment on the upper slope by the ancestral Mississippi and other rivers. Local depocenters influenced, and in turn were influenced by, the uplift of the mobile salt. The large volumes of sediment were often deposited rapidly at the shelf edge, causing high pore pressures and instability. Large landslides resulted. In addition, turbidity currents formed the large Mississippi Fan that includes a complex of successive channel systems often filled with permeable sands. Bouma and Roberts (1990) discuss the characteristics of the slope in more detail.

These active or recently active processes, and the resulting conditions, are of clear concern to siting production structures and routing pipelines. Although faulting and other difficult conditions are common in deepwater, many deepwater sites are not affected by them. Difficult areas tend to be most common in the immediate vicinity of salt uplifts. Some of the deepwater geologic conditions and

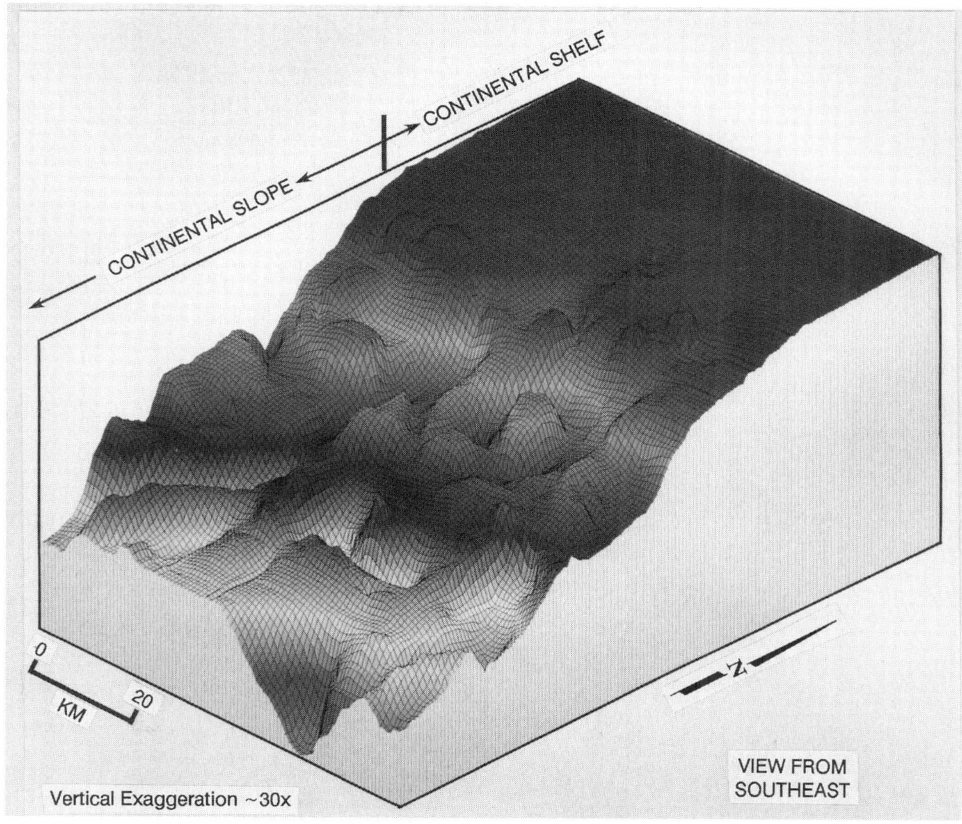

Fig. 1. 3-D perspective view of continental slope Green Canyon area, Gulf of Mexico (from Hooper and Dutt, 1991).

their implications for production structures are summarized here.

2.2. Steep Slopes and Rugged Topography

The most prominent topographic features of the continental slope are the salt uplifts. These features are typically several kilometers across (Figures 1 and 2) and rise as much as 200 meters or more above the surrounding slope. Slope gradients on the flanks of these uplifts can be 15° or more. Locally, on large fault scarps or other features, slopes can be as steep as 45° (Figure 3). Steep slopes are of concern because of possible seafloor instability and for other engineering considerations.

Irregular rocky topography is found on the tops and upper flanks of many salt

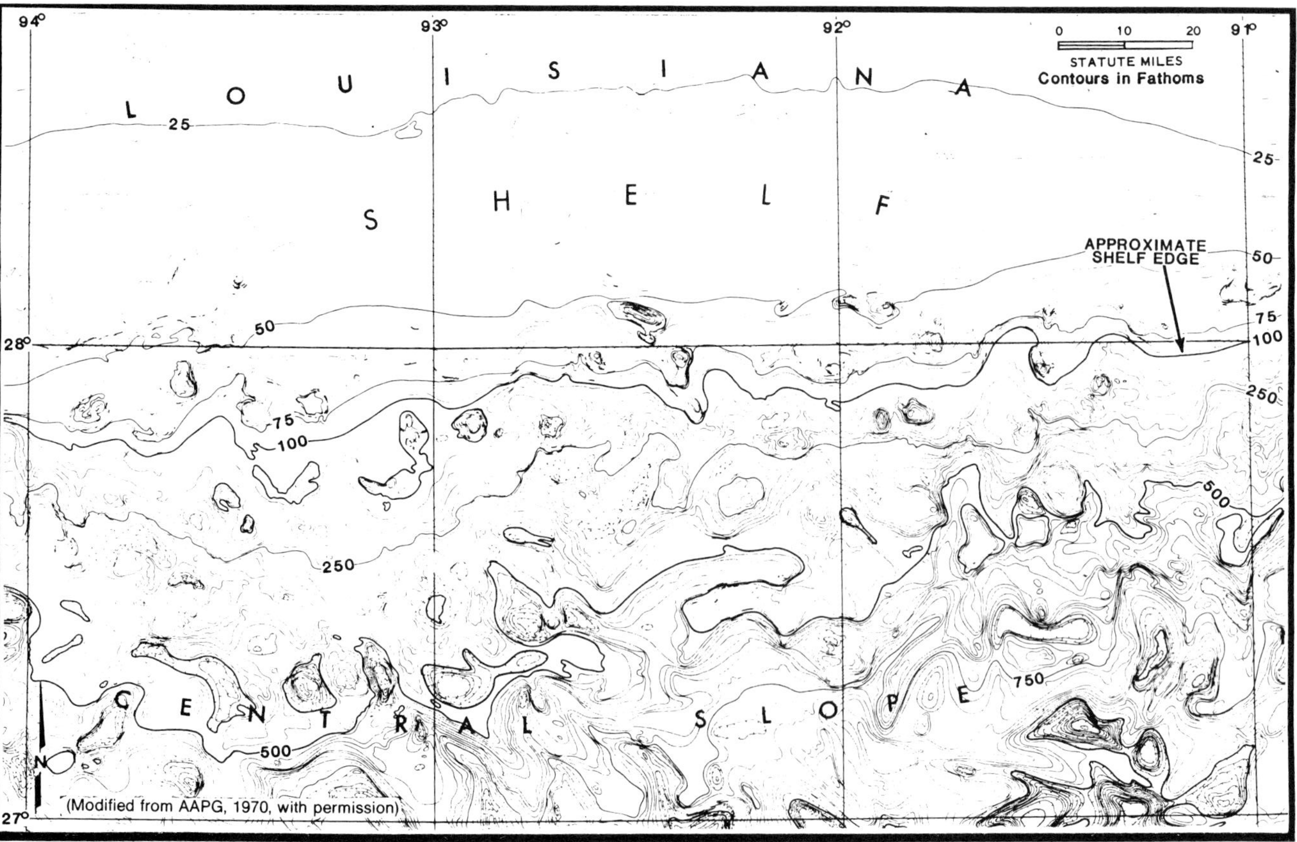

Fig. 2. Bathymetric map contrasting featureless seafloor on continental shelf with complex topography on continental slope.

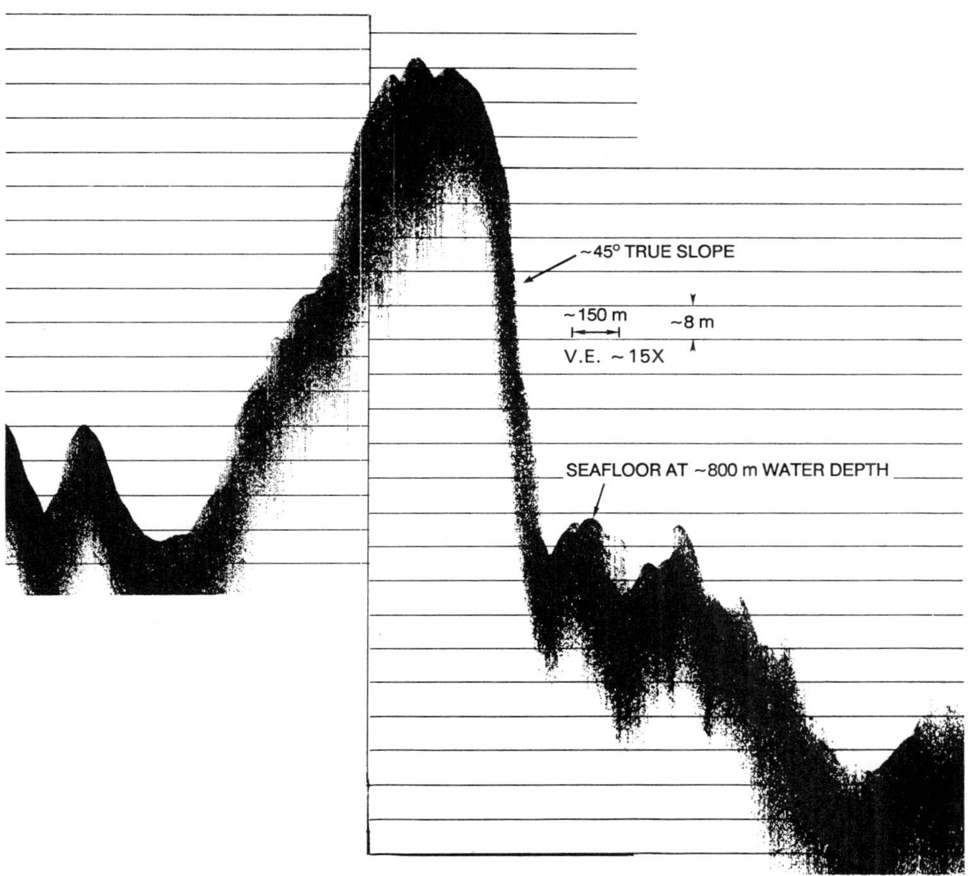

Fig. 3. Narrow-beam echo sounder record showing rugged topography on continental slope.

uplifts. The rocky material is mostly geologically recent authigenic carbonate that results from complex biochemical processes. Bacteria living on seeping hydrocarbons give off CO_2, which then combines with calcium dissolved in seawater, and is deposited as calcium carbonate. Roberts *et al* (1989) and Sassen *et al* (1991) discuss this phenomenon in more detail and believe that this is how caprock forms. Tunnicliffe (1992) gives an interesting description of related venting and biological processes in the deep ocean. Rocky buildups can cover large areas (Figure 4), and individual blocks can be tens of meters or more on a side, and near-vertical relief is known to be as much as 20 m. Overall, blocks can be the size of small office buildings and can present obstacles to pipeline routing and numerous other engineering difficulties.

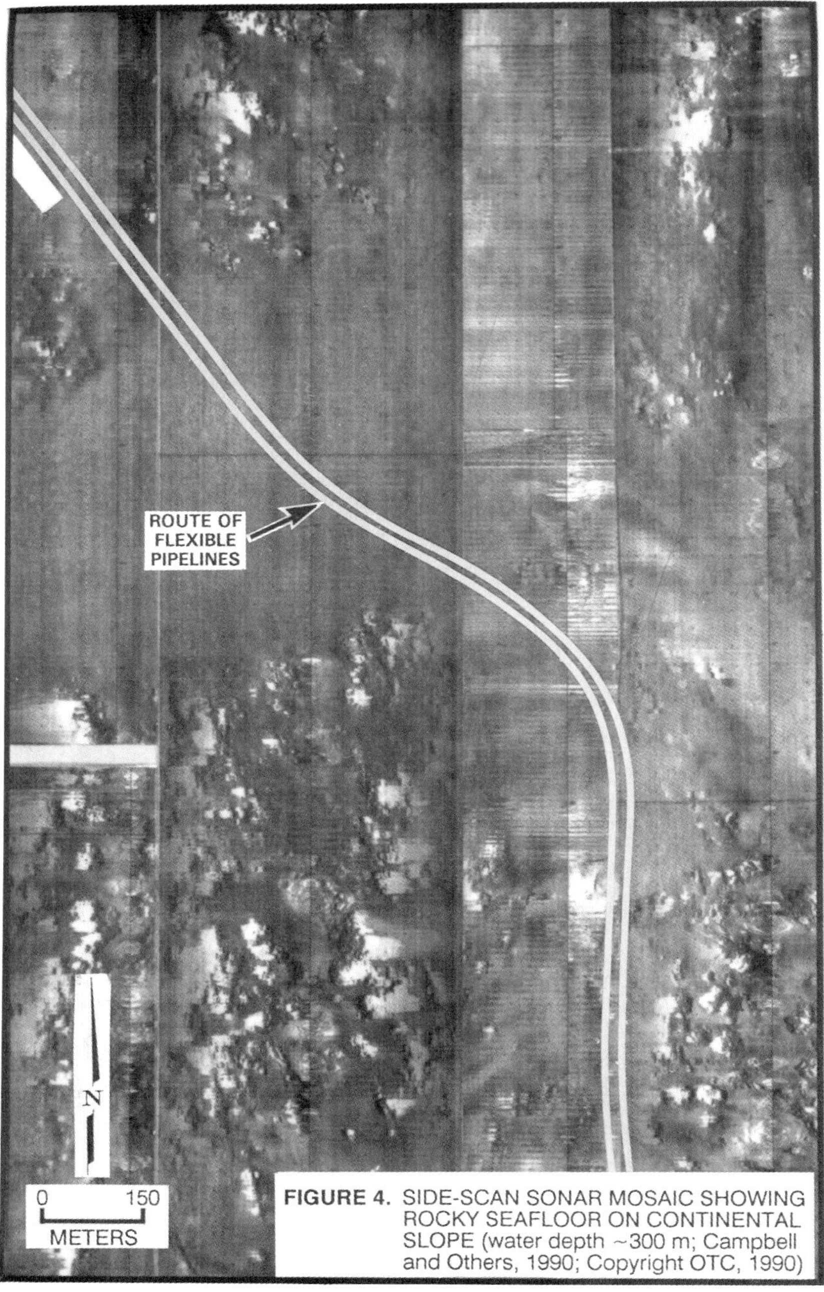

Fig. 4. Side-scan sonar mosaic showing rocky seafloor on continental slope (water depth ~ 300 m; Campbell *et al*, 1990; copyright OTC, 1990).

2.3. Faults

Faults with seafloor expression are common in deepwater. Many are partly buried by a drape of undisturbed sediment, implying that they are now inactive. Others appear 'fresh' and unburied on side-scan and seismic profiler data. These faults are inferred to be active today, although rates of offset are poorly known. Offsets on many may be at imperceptibly slow rates. Fault patterns can be intricate and complex (Figure 5) and result from numerous, closely-spaced faults found in some areas (Figure 6). Seafloor relief across fault scarps is known to be as much as 75 meters, but is more typically less than 20 meters. Fault scarps can present barriers to pipelines by causing spanning and may be locally unstable. Campbell *et al* (1990) give a case history concerning fault scarps and deepwater pipeline routing problems. Production wells that penetrate active faults can be damaged by fault movement.

2.4. Overpressured Zones

Anomalous, overpressured zones have been encountered during exploratory drilling operations at penetrations of 300 to 600 meters. These unexpected conditions have caused drilling problems and delays for several operators. The cause of the overpressured sediments is uncertain but they may be restricted to buried channels which contain permeable turbidite sands. To our knowledge, reliable identification of these anomalous pressure zones using seismic data has not been demonstrated.

2.5. Gas Hydrates

Gas hydrates (solid, ice-like mixtures of gas and water) have been found at numerous sites where water depths are greater than about 500 m. Gas hydrates require a source of free gas and water, along with high pressure and low temperatures. Hydrates accumulate both in seafloor mounds (Figure 7) where gas has seeped upward along faults, and in permeable beds below the seafloor. Hydrate mounds are not unusual and have been sampled by a number of investigators (Campbell *et al*, 1986; Kennicutt *et al*, 1989). Individual mounds typically rise 30 to 40 m above the surrounding seafloor and are 300 to 500 m across. When hydrates melt (as when warm oil flows through a pipeline or upward through a production well), large volumes of gas and water can be produced, and can result in reduced shear strength and vastly changed foundation conditions. One large crater found in more than 2000 m of water may have resulted from the rapid degradation of a hydrate hill (Prior *et al*, 1989).

2.6. Landslides

Landsliding has been common throughout the Quaternary (Hardin, 1986). Many landslides underlie areas kilometers to tens of kilometers long; many others are

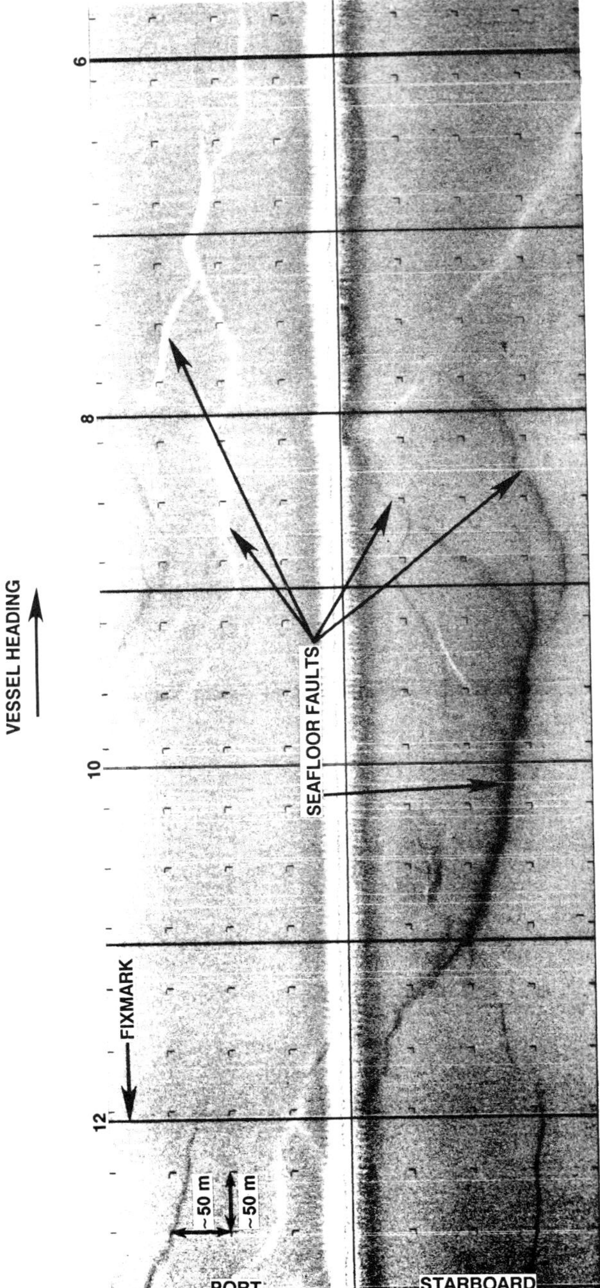

Fig. 5. Deeptow side-scan sonar record showing seafloor faults on continental slope (water depth ~ 1460 m; Shell Offshore Inc.).

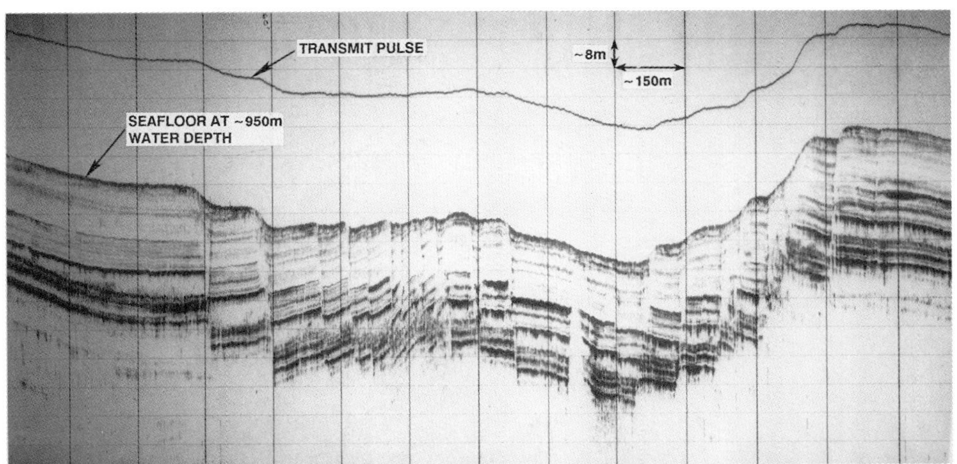

Fig. 6. Deeptow subbottom profiler record showing faults on continental slope (Shell Offshore Inc.).

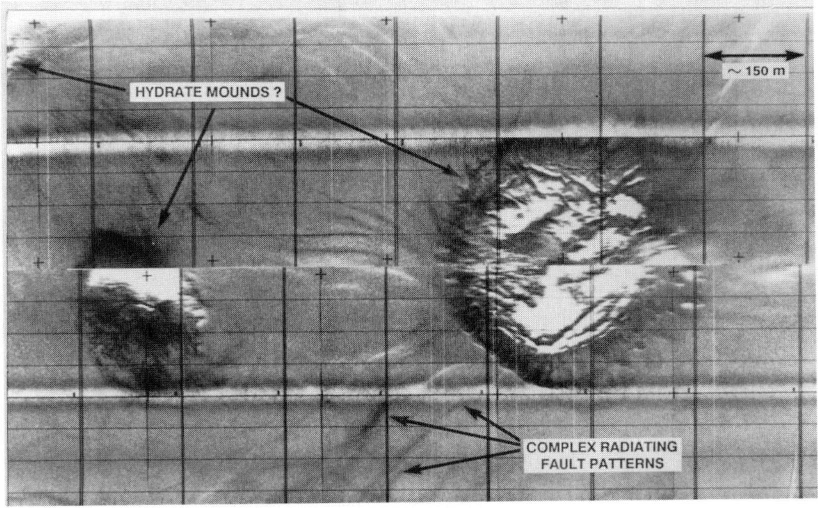

Fig. 7. Deeptow side-scan sonar mosaic showing inferred hydrate mounds on continental slope (water depth ~ 1220 m; Shell Offshore Inc.).

smaller, local features. All of the large ones apparently occurred during Pleistocene lowstands and have been stable for thousands of years. One is about 30 km long and slide deposits are about 150 m thick. Figure 8 shows the headscarp of a large

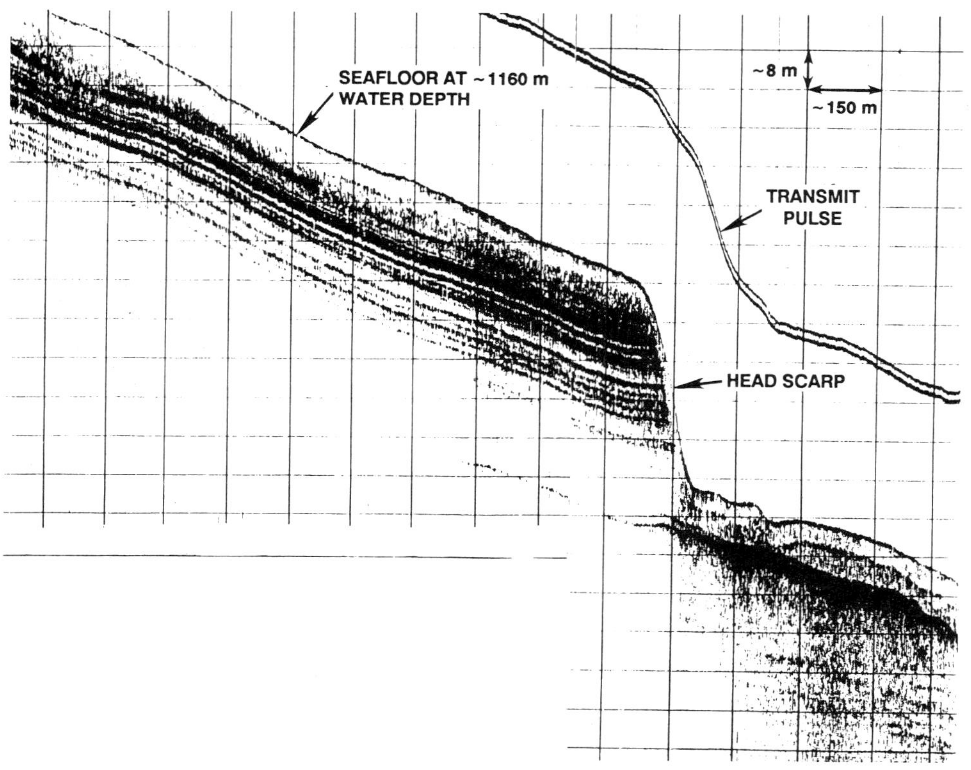

Fig. 8. Deeptow subbottom profiler record showing the headscarp of a major landslide on continental slope (from Campbell *et al* (1988a).

relict landslide that is buried by several meters of undisturbed, post-slide sediment. Relict slide deposits are of concern partly because of possible irregular shear strength profiles and variable foundation conditions.

Some of the smaller slides are believed to be much more recent, and instability of steep slope areas is a particular concern. Many of the smaller slides are less than 10 meters thick and some have formed on slopes of less than 2°. Sensitive soils, apparently resulting from biological activity, are found on the continental slope and modern slides may be most common in areas where these soils predominate.

2.7. Seafloor Erosion

Seafloor erosion of up to tens of meters of sediments is common on the tops and upper flanks of many salt uplifts (Hooper and Dunlap, 1989). Seismic data typically show that significant episodes of erosion also repeatedly occurred throughout geologic time. It is not clear whether seafloor erosion is now ongoing or if erosional processes are relict. Erosion may occur episodically as a result of eddys from the loop current. Where erosion has occurred, seafloor soils are typically overconsolidated and stronger than would otherwise be expected.

2.8. Variable Materials

Materials encountered in deepwater areas can range over relatively short distances from weak, underconsolidated to normally consolidated clays in topographic lows between salt uplifts, to strong, overconsolidated soils and to rock on the upper flanks and tops of salt uplifts. Deepwater soils often originate from much different geologic conditions than shelf soils and the properties and behaviour of deepwater soils may be very different from what we are accustomed to. Because nearly all offshore foundation engineering experience is derived from sites on the continental shelf, empirical soil relationships commonly used in foundation design may not be directly appropriate for deepwater soils (Hooper and Dutt, 1991). For example, the properties of some deepwater clays have been altered by intensive biologic activity such as bioturbation (soil mixing by worms and other organisms) and have become much more sensitive to construction disturbance than shelf clays. High soil sensitivities can create difficult design problems for pile foundations, seafloor templates, and pipelines. Hooper and Dunlap (1989) discuss modeling soil properties on the continental slope.

3. Deepwater Geophysical Survey Tools

3.1. Overview of Survey Tools

The following types of geophysical survey tools are being used routinely for deepwater site investigations:

- Narrow-beam water depth recorder with velocimeter calibration.

- Deeptow side-scan sonar with bathymetric swath mapping combined with 3.5 kHz subbottom profiler (with acoustic navigation and recording of data in digital format) to show the seafloor and geologic conditions to soil penetrations of up to 50 m.

- Intermediate-penetration profiler (using minisparker or small sleeve-gun-array sources, and digital recording) to show conditions within the foundation zone (to soil penetrations of about 200 m).

- Deep-penetration profiler (sleeve guns with digital recording, for example) to show deep-seated faults, buried landslides, and gassy sediments.

In addition to geophysical tools, manned submersibles have been used for visual observation of the seafloor and for collecting cores at visually selected key locations. Manned submersibles are being used to help determine, for example, if faults have recently been active or if landsliding is recent. Visual observations can be critical to a reliable understanding of modern processes. Doyle *et al* (1992) describe using a manned submersible to investigate surface slumps in the vicinity of a proposed structure site.

3.2. DEEPTOW SURVEY TOOL

Routine use of a deeptow side-scan sonar and subbottom profiler system, beginning in 1986 for Shell Offshore, greatly improved the quality of data available for deepwater site assessment. The deeptow tool used for most deepwater surveys to date has been the EDO Western 4075 deeptow system, owned by Shell Offshore and operated by John E. Chance and Associates. Figures 5, 6, 7, and 8 show data collected with this tool. Figure 9 is a picture of an upgraded version of the towfish and traction unit.

The buoyant towfish flies at a nearly constant height of about 30 m above the seafloor as controlled by a chain that drags along the seafloor (for example, the transmit pulse in Figures 6 and 8 indicates the towfish position above the seafloor). Because the towfish stays at a uniform height even in areas of extremely rough topography, the data is of uniformly high-quality. In contrast, data collected with free-flying towfish in areas of rough topography is typically of lower quality because the height of the towfish is difficult to control and varies radically.

In addition, because the subbottom profiler transducer is also near the seafloor, and is not affected by the great water depths, crisp, detailed records are collected (Figures 6 and 8). Data collected with shallow-tow transducers does not always resolve important details because of beam spreading. In other cases, shallow-tow records may include artifacts that unnecessarily cause concern. Prior *et al* (1988) describe the EDO Western 4075 deeptow survey tool and survey techniques in more detail.

4. Geologic and Foundation Engineering Models

4.1. THE INTEGRATED APPROACH

The complex conditions found in many deepwater areas require an integration of studies from several disciplines to provide a reliable site investigation. Typically, geophysical, geological, and geotechnical data are considered collectively to define

Fig. 9. Deeptow towfish incorporating side-scan sonar and 3.5 kHz subbottom profiler transducers; traction unit to right.

site conditions and geologic processes. This requires that geologists, geophysicists, engineers, and other team professionals work interactively to ensure that neither geotechnical nor geological factors are overlooked or misunderstood, and that reasonable assumptions are made to optimize siting and design (Campbell *et al*, 1988a; 1988b).

The integrated approach is in contrast to the traditional approach where survey geophysicists write reports using terms understood only by other geophysicists. The end user, the engineer, often does not get full value out of reports not readily understood. Milliseconds? Unconformities? Anomalies? Quaternary? Reflectors? Mass movement? Terms all potentially confusing to the engineer trying to site or design a structure. In other words, traditional geophysical reports are often not user friendly. As a result, the engineer tends to focus on boring logs and lab test results, and perhaps overlooks or does not recognize in a timely manner important geological factors (such as active faults, landslides, seafloor scour, and so forth) not obvious from borings alone. At most sites on the Gulf of Mexico continental shelf, the use of borings alone for design has been adequate because of the simple, quiescent geologic conditions. Deepwater conditions can be less forgiving.

4.2. Geologic Models

A key component of the integrated approach is to develop a geologic model of the area of interest. Geologic models explain the ongoing processes and past geologic history that help define site conditions in deep water. This understanding then provides the basis for confidently assessing the engineering significance of conditions, developing a foundation engineering model, and allowing safe and economical design. Mapping geologic features without understanding the processes that formed them often will not allow an adequate engineering assessment of conditions. Doyle (1989) discusses the development and use of geologic models for engineering purposes.

5. Concluding Comments

The complex geologic and soil conditions common in the Gulf of Mexico may be unusual, but we think it likely that many other deepwater areas around the world will also have difficult conditions of concern to engineers. Thus, we expect that deepwater site investigation survey techniques developed and successfully used in the Gulf of Mexico will be useful elsewhere as well.

Regardless of the details of the survey tools used, one key to success will be to develop reliable geologic models, before structural siting and design begins, to define past and possible ongoing geologic processes. This will require appropriate, high-quality data, seasoned interpretive skills with an engineering perspective, and the ability and opportunity for the project team of engineering geologists/geophysicists and geotechnical engineers to carefully and methodically synthesize geophysical interpretations with geological and geotechnical lab test results, with results of engineering analyses, and with visual observations of the seafloor. In addition, close teamwork between the foundation engineers, designers, and engineering geologists/geophysicists throughout the site investigation phase will provide the best chance for avoiding costly surprises and project delays.

Acknowledgements

The authors wish to thank our numerous clients on whose behalf we gained our deepwater experience. Shell Offshore, in particular, is acknowledged for commissioning numerous engineering surveys of deepwater sites and for allowing us to use several deepwater data examples. Mr. Earl Doyle, Staff Civil Engineer with Shell Oil Company, has championed the use of geophysical surveys to help develop engineering geologic models as a basis for solving or avoiding deepwater engineering problems. His long-standing support of engineering geophysical surveys, vision, and perspective as an engineer have been instrumental in firmly establishing the critical role of geophysical surveys in deepwater site investigations.

Dr. David B. Prior, Director of the Geological Survey of Canada's Atlantic Geoscience Center, and formerly of the Coastal Studies Institute at Louisiana State University, participated in many deepwater studies and provided the critical geological insight and guidance necessary to interpret seismic data and develop models of deepwater geological processes.

We also would like to acknowledge the significant contributions of our many other colleagues both within the Fugro-McClelland organization and in the wider marine community. Special acknowledgement is due to our sister company, John E. Chance and Associates, Inc., for pioneering deepwater survey work and for collecting, under the dedicated leadership of Mr. Jay Northcutt, most of the geophysical data our deepwater experience is based on. Fugro-McClelland's marine engineering geologists and geophysicists, supervised by Mr. Mike Kaluza, provided the detailed engineering interpretations of the large amount of deepwater data collected. The authors also thank the American Association of Petroleum Geologists for permission to use part of their 1970 bathymetric map (Figure 2), and finally we thank Fugro-McClelland Marine Geosciences for support needed to prepare this paper.

References

1. A.A.P.G. (1970), 'Bathymetric Maps, Eastern Continental Margin, U.S.A.', 1:1,000,000, sheet 3 of 3, Northern Gulf of Mexico, compiled by W. C. Holland, American Association of Petroleum Geologists, Tulsa, Oklahoma.
2. Bouma, A. H. and Roberts, H. H. (1990), 'Northern Gulf of Mexico slope', *Geo-Marine Letters* **10**, 177–181.
3. Campbell, K. J., Hooper, J. R., and Prior, D. B. (1986), 'Engineering implications of deepwater geologic and soil conditions, Texas-Louisiana slope', *Proc. 18th Annual Offshore Technol. Conf.*, OTC Paper 5105, pp. 225–232.
4. Campbell, K. J., Quiros, G. W., Machado, C., and de Castro, J. R. (1988a), 'Integrated geophysical and geotechnical deepwater site investigations', in *Offshore Engineering*, F. L. L. B. Carneiro, A. J. Ferrante, and R. C. Batista (eds.), Proc. 6th Intl. Symp. Offshore Engineering, Rio de Janeiro, Aug. 1987, Pentech Press, London, pp. 125–190.
5. Campbell, K. J., Quiros, G. W., and Young, A. G. (1988b), 'The importance of integrated studies to deepwater site investigation', *Proc. 20th Annual Offshore Technology Conference*, Houston, OTC Paper 5757, pp. 99–107.
6. Campbell, K. J., Tillinghast, W. J., Roulstone, J. A., and Hoffman, J. S. (1990), 'Geohazards surveying and complex seafloor conditions along deepwater Jolliet pipeline routes', *Proc. 22nd Annual Offshore Technol. Conf.*, Houston, OTC Paper 6370, pp. 223–234.
7. Doyle, E. H. (1989), 'The usefulness of high-resolution geophysical surveying to the offshore oil and gas industry – Gulf of Mexico', *Conf. Proc. Marine Engineering Geology for Petroleum Development in Developing Countries*, sponsored by the United Nations and the People's Republic of China, Guangzhou, 30 Nov.–6 Dec.
8. Doyle, E. H., Kaluza, M. J., and Roberts, H. H. (1992), 'Use of manned submersibles to investigate slumps in deepwater Gulf of Mexico', *Proc. Civil Engineering in the Oceans*, Texas A&M Univ., 2–5 Nov., Am. Soc. Civil Engineers, New York, N.Y.
9. Hardin, N. S. (1986), 'Mass transport deposits of the upper continental slope, northwestern Gulf of Mexico', in *Late Quaternary Facies and Structure, Northern Gulf of Mexico*, H. L. Berryhill (ed.), AAPG Studies in Geology, No. 23, Tulsa, Oklahoma, pp. 241–284.
10. Hooper, J. R. and Dunlap, W. A. (1989), 'Modeling soil properties on the continental slope, Gulf of Mexico, *Proc. 21st Annual Offshore Technol. Conf.*, Houston, OTC Paper 5956, pp. 677–688.

11. Hooper, J. R. and Dutt, R. N. (1991), 'Geotechnical and geological research challenges of the deepwater frontier', Fugro-McClelland Marine Geosciences, Inc., proprietary Research Report, 194 pages, 136 plates, 11 tables.
12. Kennicutt, M. C., II, Brooks, J. M., and Burke, R. A., Jr. (1989), 'Hydrocarbon seepage, gas hydrates, and authigenic carbonate in the northwestern Gulf of Mexico', *Proc. 21st Annual Offshore Technol. Conf.*, Houston, OTC Paper 5952, pp. 649–654.
13. Martin, R. G. and Bouma, A. H. (1982), 'Active diapirism and slope steepening, northern Gulf of Mexico continental slope', *Marine Geotechnology*, **5**(1), 63–91.
14. Prior, D. B., Doyle, E. H., Kaluza, M. J., Woods, M. M., and Roth, J. W. (1988), 'Technical advances in high-resolution hazard surveying, deepwater Gulf of Mexico', *Proc. 20th Annual Offshore Technol. Conf.*, Houston, OTC Paper 5758, pp. 109–117.
15. Prior, D. B., Doyle, E. H., and Kaluza, M. J. (1989), 'Evidence for sediment eruption on deep seafloor, Gulf of Mexico', *Science* **243**, Jan., pp. 517–519.
16. Roberts, H. H., Sassen, R., Carney, R., and Aharon, P. (1989), 'Carbonate buildups on the continental slope off central Louisiana', *Proc. 21st Annual Offshore Technol. Conf.*, Houston, OTC Paper 5953, pp. 655–662.
17. Sassen, R., Grayson, P., Cole, G., Roberts, H. H., and Aharon, P. (1991), 'Hydrocarbon seepage and salt-dome related carbonate reservoir rocks of the U.S. Gulf coast', Trans. Gulf Coast Assoc. Geol. Societies, vol. XLI, pp. 570–578.
18. Tunnicliffe, V. (1992), 'Hydrothermal-vent communities of the deep sea', *American Scientist* **80**(4), 336–349.
19. Young, A. G. (1991), 'Marine foundation studies', in *Handbook of Coastal and Ocean Engineering, Vol. 2. Offshore Structures, Marine Foundations, Sediment Processes, and Modeling*, J. B. Herbich (ed.), Gulf Publishing Co., Houston, pp. 445–596.

SESSION 5

GRAVITY FOUNDATIONS

REVIEW OF THE DESIGN DEVELOPMENT OF A HIGH PERFORMANCE ANCHOR SYSTEM

J. J. OSBORNE, R. D. COLWILL and D. ROWAN
Noble Denton Consultancy Services Ltd., Noble House, 131 Aldersgate Street, London EC1A 4EB

and

D. PHILLIPS
Formerly Tarmac Construction Ltd., now MTRC Hong Kong

Abstract. Particularly onerous positioning restraints, performance criteria and ground conditions called for a design review of available anchor systems and a consequential detailed ground investigation and design. High quality geophysical and geotechnical surveys were performed to establish the nature and characteristics of the ground conditions. The data acquired was used for the selection and design of an anchor system. The soil-structure interaction of the anchor was modelled and analysed for installation, operation and survival conditions.

1. Introduction

Tarmac Construction, with Dowty and Vosper Thorneycroft, were contracted by the M.O.D. to design and construct a floating magnetic treatment facility in the Gareloch, Scotland. Noble Denton Consultancy Services were sub-contract designers to Tarmac for the berth moorings and anchor system.

The site location is indicated on Figure 1. The initial ground investigation data and design criteria allowed the use of anchor piles. Further to subsequent site investigation and amended design criteria an additional anchor system review identified the most economic solution to be a modified deadweight anchor.

This paper describes the ground conditions, anchor system design requirements and design decision procedure. The additional site investigation results are discussed and the final analyses and design of the modified deadweight anchor are described.

Throughout the design process unusual restraints were imposed on the anchor design. Such restraints included a requirement to provide a non-magnetic mooring leg within a zone adjacent to the facility, and a maximum yaw of $\pm 1°$ about a heading which had to be accurately achieved.

Additionally the floating structure, manufactured in glass reinforced resin, imposed restrictions upon both the location and magnitude of the anchor connections.

Volume 28: Offshore Site Investigation and Foundation Behaviour, 393–416, 1993.

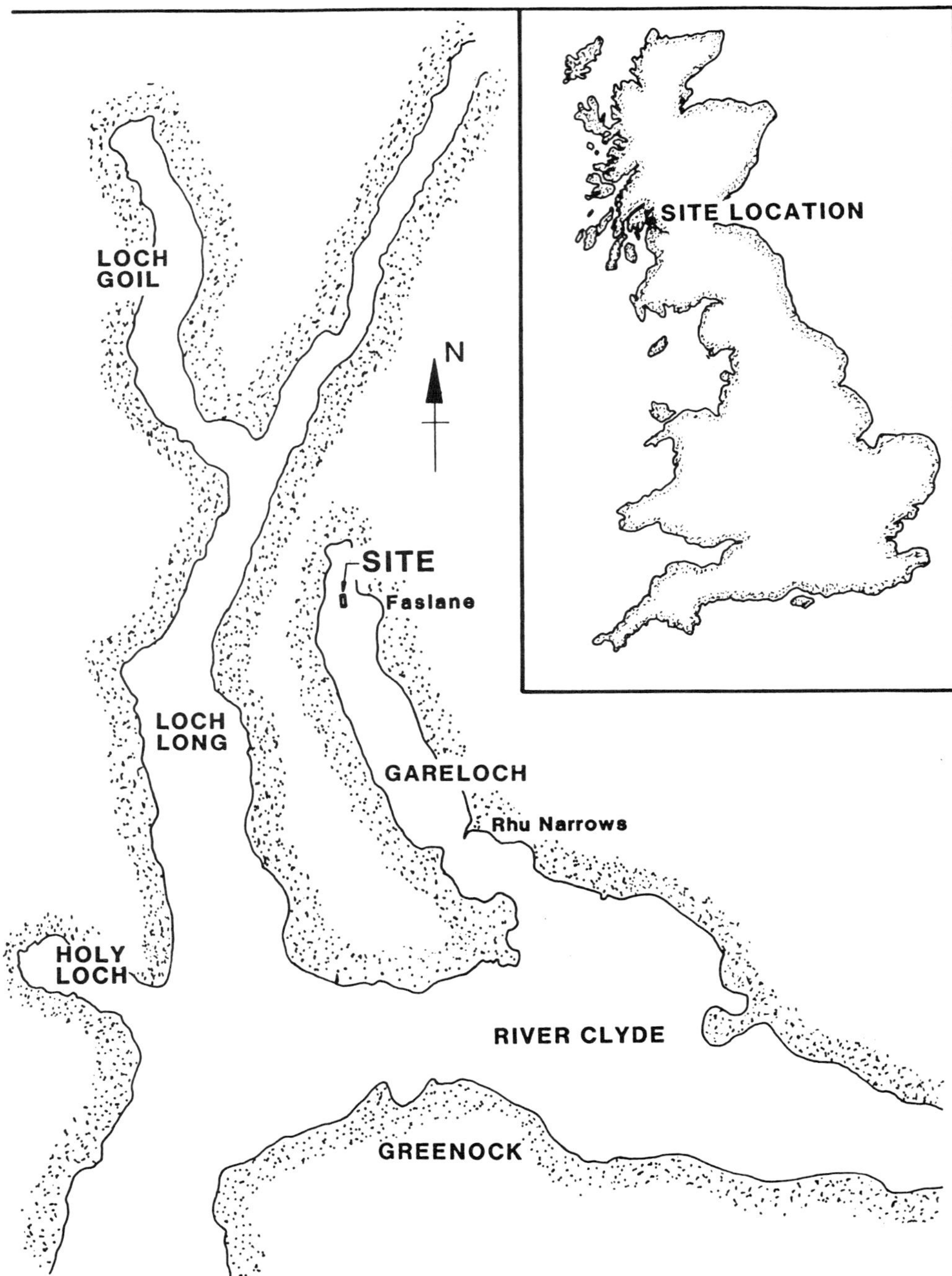

Fig. 1. MTF site location – Gareloch.

2. Original Anchor Scheme

2.1. Anchor Design Premise

The anchor system was required to conform to the following requirements:

- Design life of 25 years.
- Maintain position and stability under design conditions.
- Provide a platform for a chain tensioning device.
- Exhibit minor displacements throughout design life.
- Facilitate uncomplicated installation procedures and have a weight of less than 220 tonnes in air.
- Have no magnetic components within a specified magnetic exclusion zone.

An integrated design procedure was defined, a flow diagram illustrating this procedure is shown in Figure 2. Of necessity this procedure does not define the magnetic and structural restraints imposed on the anchor system, though they are implicit within the procedure.

2.2. Initial Review of Site Conditions

The proposed site is located in a loch on the west coast of Scotland, formed with a glacial "U" shaped valley. A review of the British Regional Geology [1] indicated that the bedrock comprises schists and grits laid down during the early Ordovician which were subsequently deformed during the late Devonian period. Post-glacial activity has overlain the bedrock by glacial till, comprising sandy boulder clays, and alluvial clays.

At tender stage, and during the initial stages of the design, the results of two existing site investigations were available ((1984), References [2] and [3]). These data indicated that the soil over the nominal berth area comprised very soft to soft silty alluvial clays between 17 to 27 metres depth which overlay glacial till, of upto 5.5 metres thick. These sediments overlay bedrock of Ordovician schists.

The marine borehole locations are illustrated on Figure 3 and a SW to NE cross section through four of the boreholes is shown on Figure 4. Due to the sampling techniques and possible sensitivity of these soft materials the recorded shear strengths were considered to be weaker than would be anticipated for the *in-situ* soil conditions. During preliminary design stage the following undrained shear strength design profile was assumed:

$$s_u = 8 + 0.179\gamma'(kPa).$$

2.3. Mooring, Anchoring and Installation Constraints

The MTF mooring and anchor system was required to withstand a number of onerous design conditions during the 25 year service life:

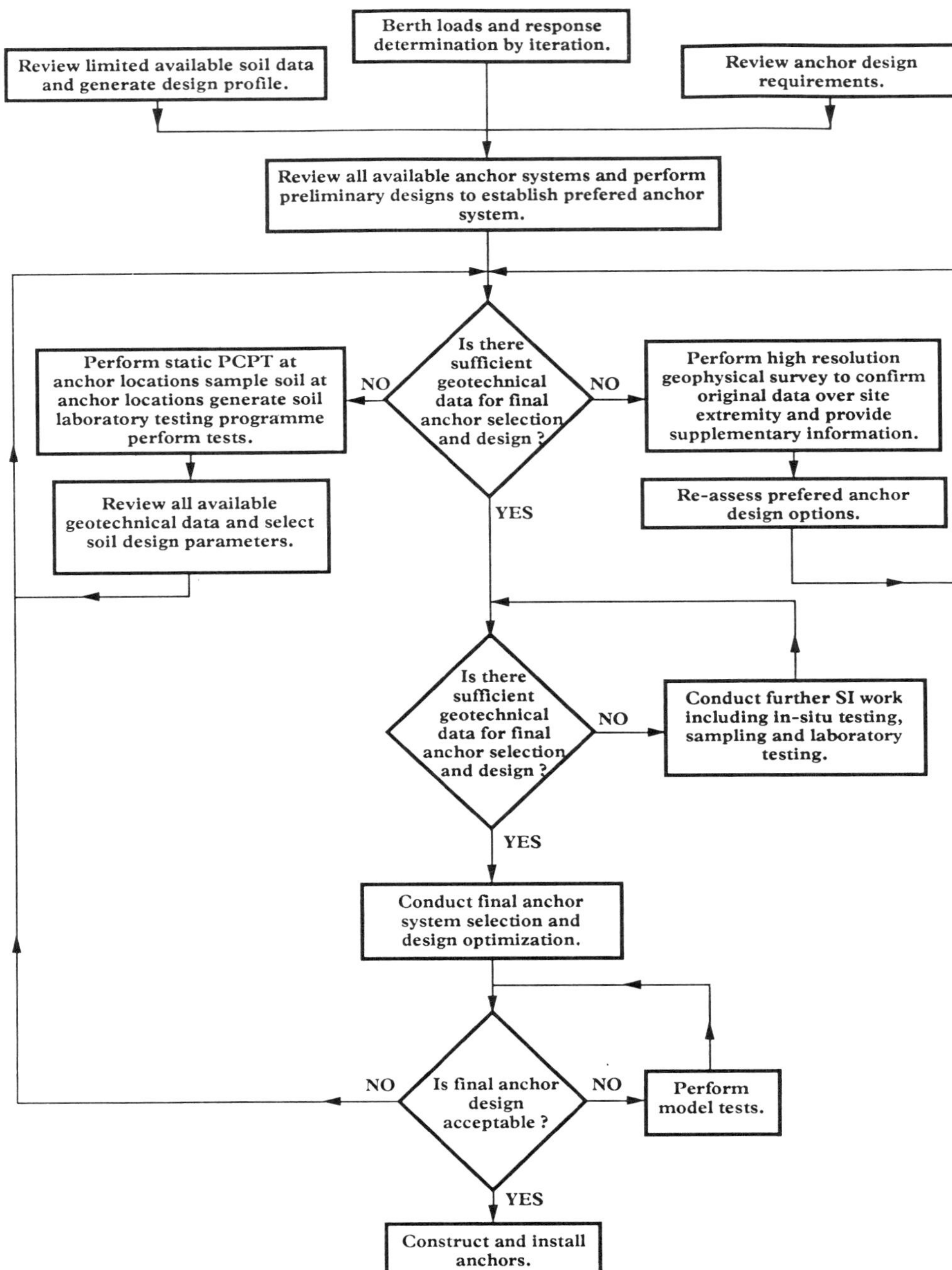

Fig. 2. Anchor system design procedures flow diagram.

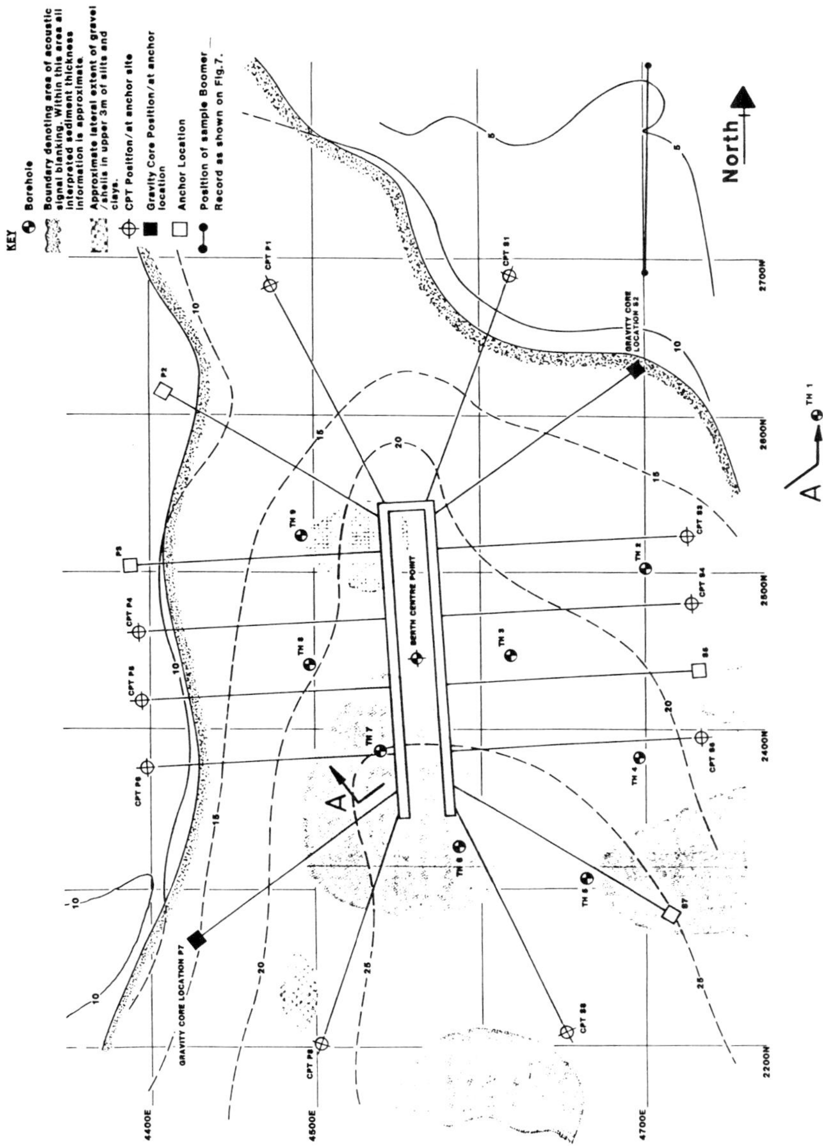

Fig. 3. Silt isopach with borehole and CPT locations.

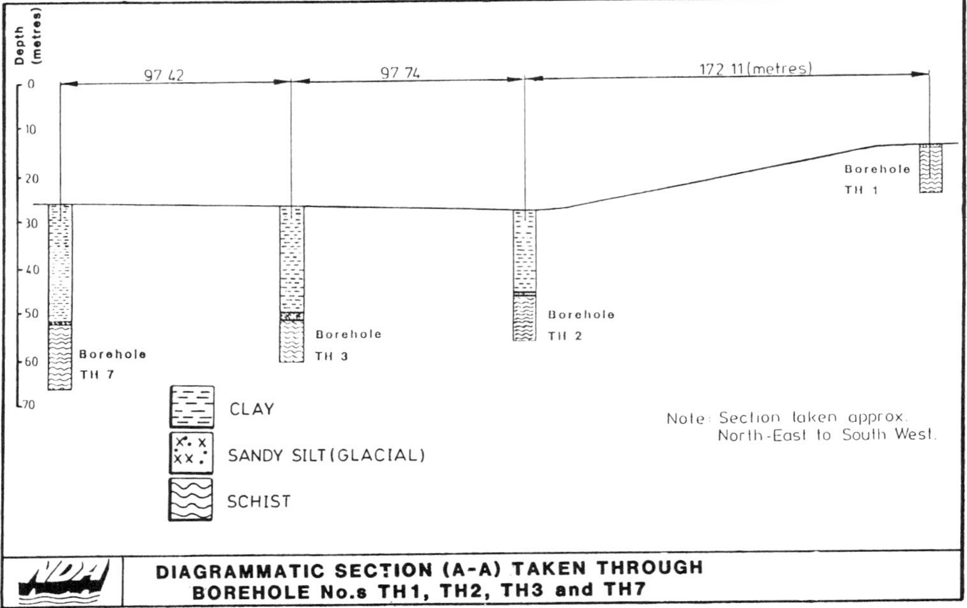

Fig. 4. Diagrammatic section (A–A) taken through borehole Nos. TH1, TH2, TH3 and TH7.

- Nuclear safety – 10,000 year return to environmental event.
- Nuclear safety – earthquake (M.O.D. specified ground accelerations).
- Safe mooring and survival conditions – intact mooring system.
- Safe mooring and survival conditions – one line failed.
- Submarine berthing/unberthing.
- Yaw movement of $\pm 1^\circ$ during operation.

The design specification restricted the movements of the berth during operations to less than $\pm 1^\circ$ yaw, while the mooring system was required to accommodate a 7.15m tidal range in only 25 metres water depth and to have no magnetic components in the immediate berth vicinity.

The above criteria, coupled with the structural limitations of the glass reinforced resin berth, resulted in a sixteen line mooring system. The berth structural design was determined, by the berth structural designer, to be highly sensitive to environmental current induced loads. Design current values were subsequently produced on completion of a current buoy survey which, while critical for berth structural considerations, did not significantly influence the anchor line design philosophy, though the number of anchors and their locations were defined by these factors.

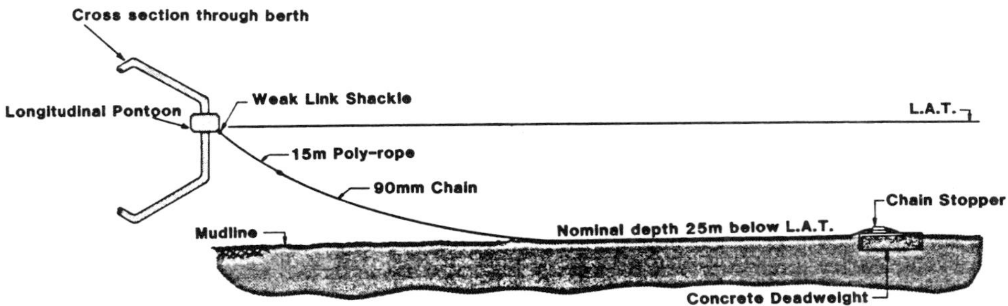

Fig. 5. Typical mooring line profile.

A typical mooring line comprising 90mm diameter chain and 104mm polyester rope is illustrated on Figure 5. Polyester rope was used as the berth specification called for a 13m magnetic exclusion zone while the chain component was necessary to reduce the mooring leg stiffness in shallow water. A chain tensioner was located outside the magnetic exclusion zone positioned on, or close to, the anchor.

Evaluation of the environmental loads was conducted according to commonly accepted design practice (CP3 (1972) Reference [4], BS6349 Reference [5]), using drag values derived from model tests (Belcher (1988) Reference [6], H.R. (1986) Reference [7], MIRA (1989) Reference [8]).

Anchor design factors of safety were selected to be compatible with all other mooring system components. For intact operating conditions an F.O.S. of safety of 3 had been determined, and 2 for the one line damaged conditions. These factors satisfy BS6349 (1986) Reference [5].

The maximum line loads, at the anchor, are as follows:

$$\text{Design Survival Load} = 68 \times 3 = 204 \text{ tonnes.}$$
$$\text{Design 1 Line Damaged Case} = 93 \times 2 = 186 \text{ tonnes.}$$

A weak link is incorporated in each mooring leg which is designed to fail at 204 tonnes. To accommodate the fabrication tolerance of -0, +2% for this weak link an additional 3% is added to this load which increases it to 210 tonnes. The maximum factored vertical load at the mooring line/mudline touchdown point was found to be some 4 tonnes.

2.4. Anchor System Options

During the pre-tender design stage the presence of a thick surface clay layer, which had been identified in the original ground investigation data, focused attention on the following anchor system options:

- Drag embedment anchor.
- Deadweight anchor.
- Single pile.
- Pile groups (2 or 4).
- Composite pile and clumpweight system.

A summary of these system advantages and disadvantages is given in Table 1. At this stage the preferred anchor system comprised a composite "Anchortech" type anchor pile and clumpweight as illustrated on Figure 6. This system requires a depth of soft material in excess of 15 metres.

The purpose of the small clumpweight anchor was to provide a platform for the chain tensioner. A composite system required pre-tensioning of the chain lying between the anchor pile and the clumpweight to ensure that the mooring line loads would be transmitted directly onto the anchor pile without causing clumpweight displacement as the chain/soil inverse catenary flattens under load.

3. Further Site Investigation and Subsequent Anchor Review

3.1. Geophysical Investigation

Further to the preliminary design stage (post-tender) a geophysical site survey was conducted (Reference [9] 1990) in order to:

- Establish bathymetry (echo sounder).
- Confirm site geology (high resolution boomer analogue sub-bottom profiling).
- Identify surface/buried magnetic anomalies (magnetometer).
- Identify seabed characteristics and surface obstructions (side scan sonar).

TABLE 1

MFT Faslane comparison of anchoring systems.

SYSTEM OPTION	ADVANTAGES	DRAWBACKS	APPROX. INSTALLATION TIME	COMPARATIVE COSTS
Drag embedment/mud anchors	Simple design. Proved load capacity during installation.	Clump weight required for chain tensioner. Anchor designed to drag under ultimate load. Anchors and connections not easily inspected if embedded. System would require high proof-tension load.	10 days	High
Deadweight anchor	Requires limited soil depth. Low lateral displacements under working load. Provides platform for chain tensioner.	Installation surcharge to ensure sufficient skirt penetration. Large construction – heavy lift capability required. High quality soil data essential.	14 days	Low to Medium
Single pile	Simple installation concept and design. Pile cap provides platform for chain tensioner.	Pile diameter > 2.85m large installation equipment required.	12 days	Medium
Pile group – four vertical piles	Small pile diameter – more easily handled. Large capacity against lateral loads. Small lateral displacements under working loads. Pile cap provides platform for chain tensioner.	Complicated installation requires pile cap construction and underwater installation. Cap pile connection must be grouted.	29 days	Medium to High
Pile group – two vertical piles	Comparatively small piles. Good response with fixed heads. Small lateral displacements under working loads. Simpler installation than four pile group. Pile cap provides platform for chain tensioner.	Complicated installation requires pile cap construction and underwater installation. Alignment of piles with mooring line is important. Hydralok type tool needed for pile-cap connection.	18 days	Medium
Composite system – Anchortech pile and clump weight	Reduced pile length as padeye located at centre of earth pressure. Simple installation commonly used offshore. Clump weight would be designed to take a proportion of the load.	Complicated installation of pile and clump weight. Pile pre-loading required prior to hook up to clump weight. Greater mooring spread due to chain section in soil. Requires significant soil depth.	20 days	Low

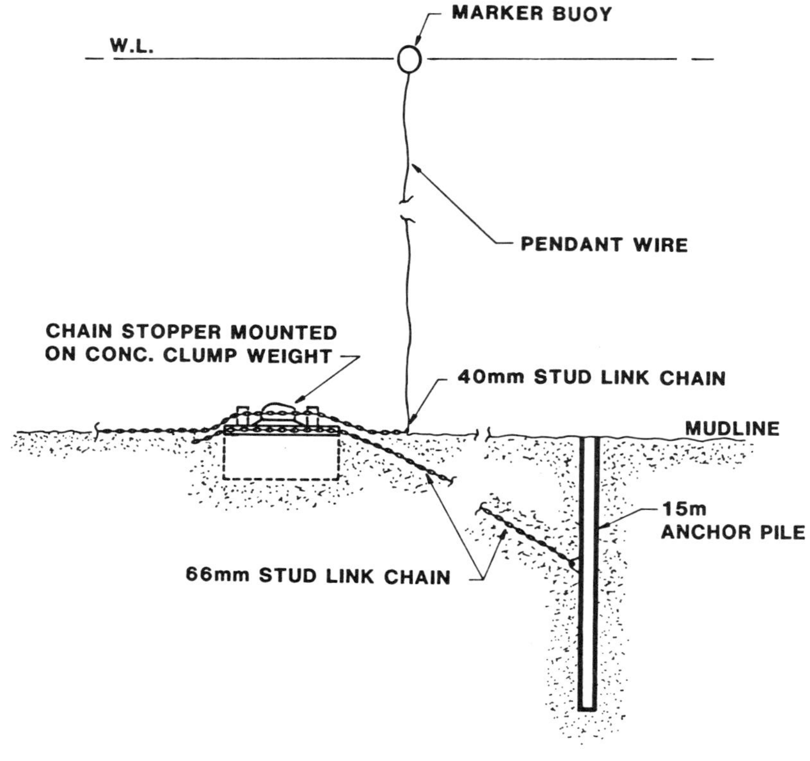

Fig. 6. Original mooring leg elevation for combined anchor and Anchor tech pile.

Interpretation of the sub-bottom analogue boomer data was severely hindered as acoustic signal blanking occurred over most of the site, presumably due to the presence of biogenic gas present within the clay. This blanking rendered a detailed interpretation of both clay thickness and depth to bedrock difficult, however an interpretation was attempted with the aid of the available borehole data. It was also considered that layers with increased granular content, within the upper 3 metres of the clay, may have contributed to the signal attenuation. A sample boomer record is shown on Figure 7 (with its corresponding position shown on Figure 3).

3.2. Geotechnical Site Investigation and Soil Parameter Selection

The clay depth limitation identified during the geophysical survey precluded the use of the proposed anchor pile systems. Further assessment of the available anchor scheme options was conducted and a deadweight anchor system was considered the most appropriate. However, the design and optimisation of this system relied on the

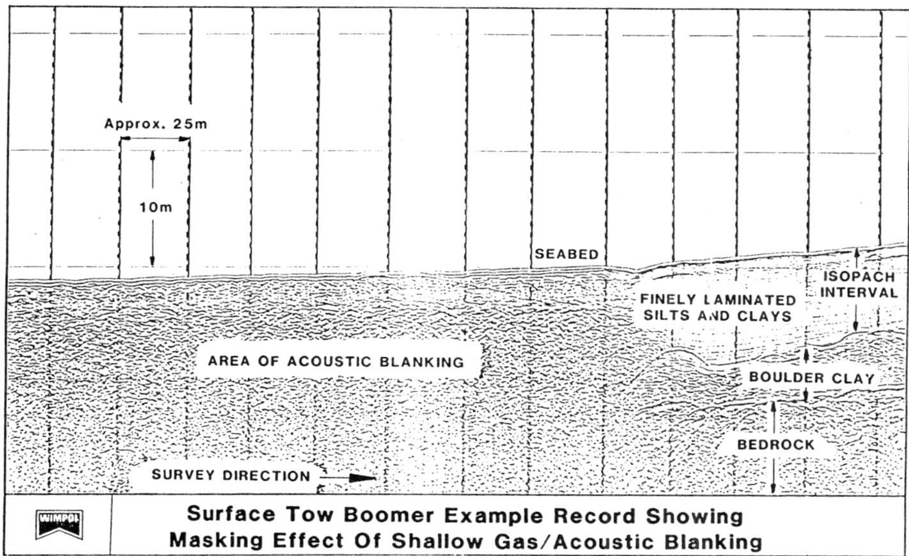

Fig. 7. Surface tow boomer example record showing masking effect of shallow gas/acoustic blanking.

engineering parameters of the extremely soft surface and near-surface sediments. This necessitated the acquisition of high quality *in-situ* and laboratory soil test data for the near surface soft alluvial clays.

A programme of quasi-static piezocone penetrometer tests (PCPTs), sampling and laboratory testing was therefore conducted (Reference [10], 1990). The investigation indicated extremely consistent soil conditions across the site with a very soft clay overlying a very stiff sandy silt/silty sandy clay (glacial till). Soil samples were taken using a 6m gravity drop core sampler and a laboratory testing programme was conducted. Figures 8(a), (b) an (c) illustrate typical particle size distribution curves, bulk density values and plasticity data.

The undrained shear strength of this clay was determined using torvane, fall cone and unconsolidated undrained triaxial tests. These data were used to correlate the *in-situ* cone penetrometer test data. A typical set of PCPT field data is shown on Figure 9.

The clay is extremely plastic with a low undrained shear strength at mudline of about 1 kPa which increases with depth at a rate of between 1 to 2 kPa/m. The design profile of $s_u = 1.0 + 1.92z$ was adopted which is significantly weaker than that previously assumed. The thickness of this material was seen to vary from 7.7 metres to greater than 22.0 metres.

Of significance was the high degree of correlation between the PCPT recorded clay layer depths and the interpreted sub-bottom profiling data illustrated on Figure 3.

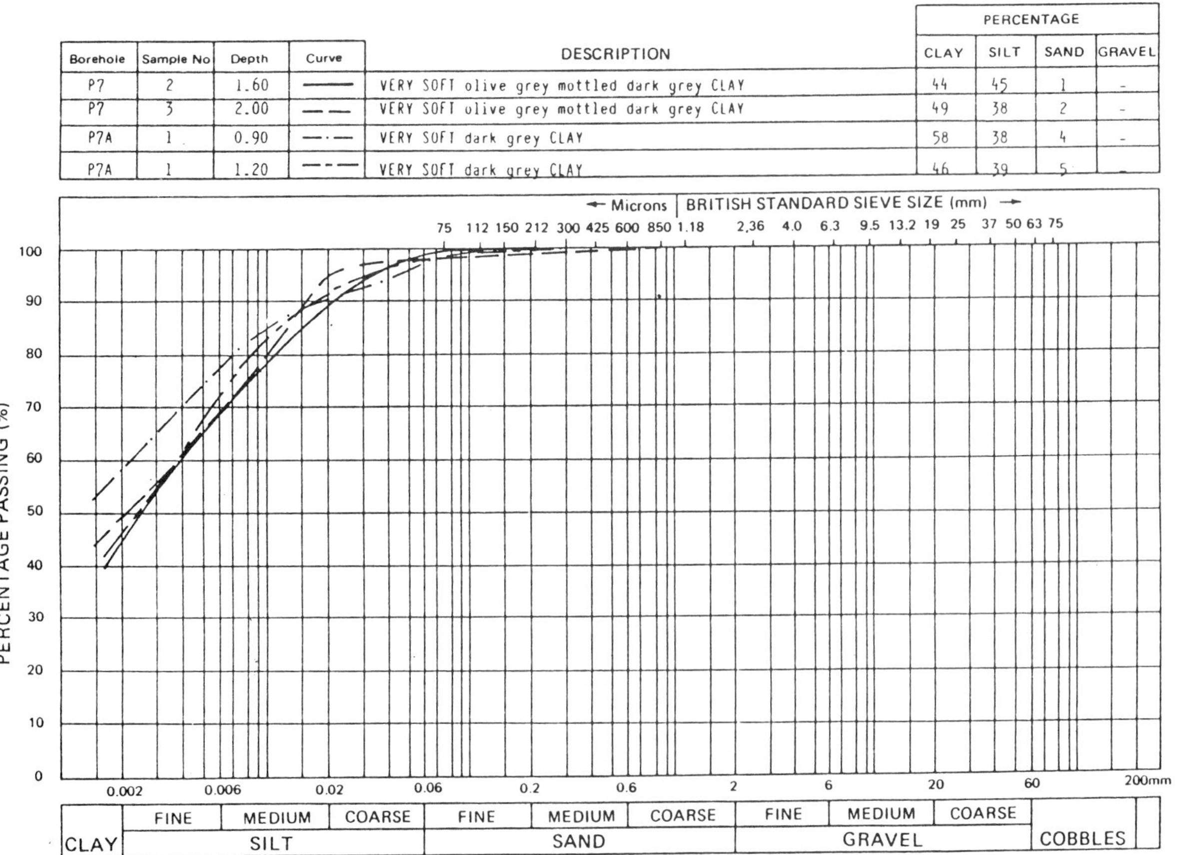

Fig. 8a. Particle size distribution.

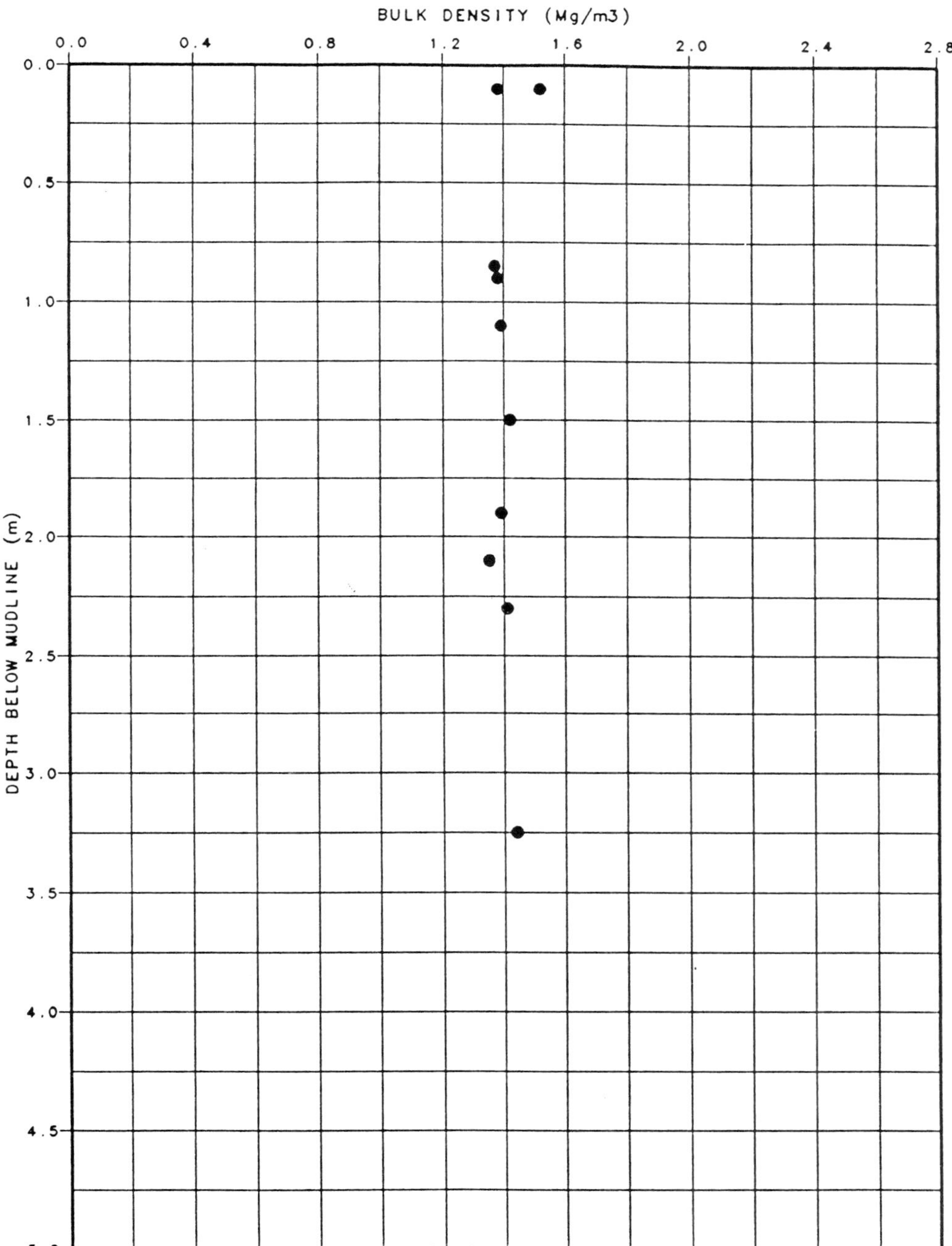

Fig. 8b. Bulk density profile (0 – 5 metres).

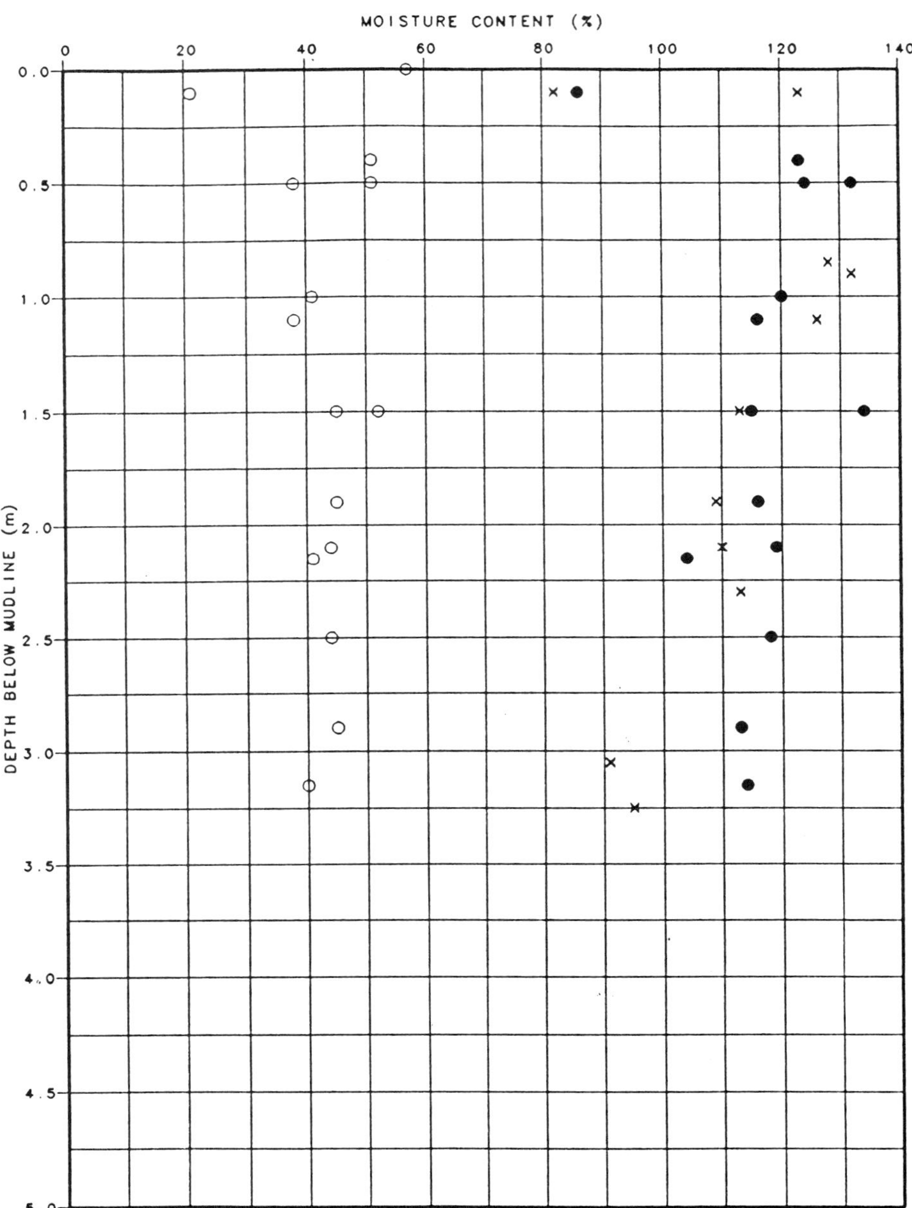

Fig. 8c. Plasticity data (0 – 5 metres).

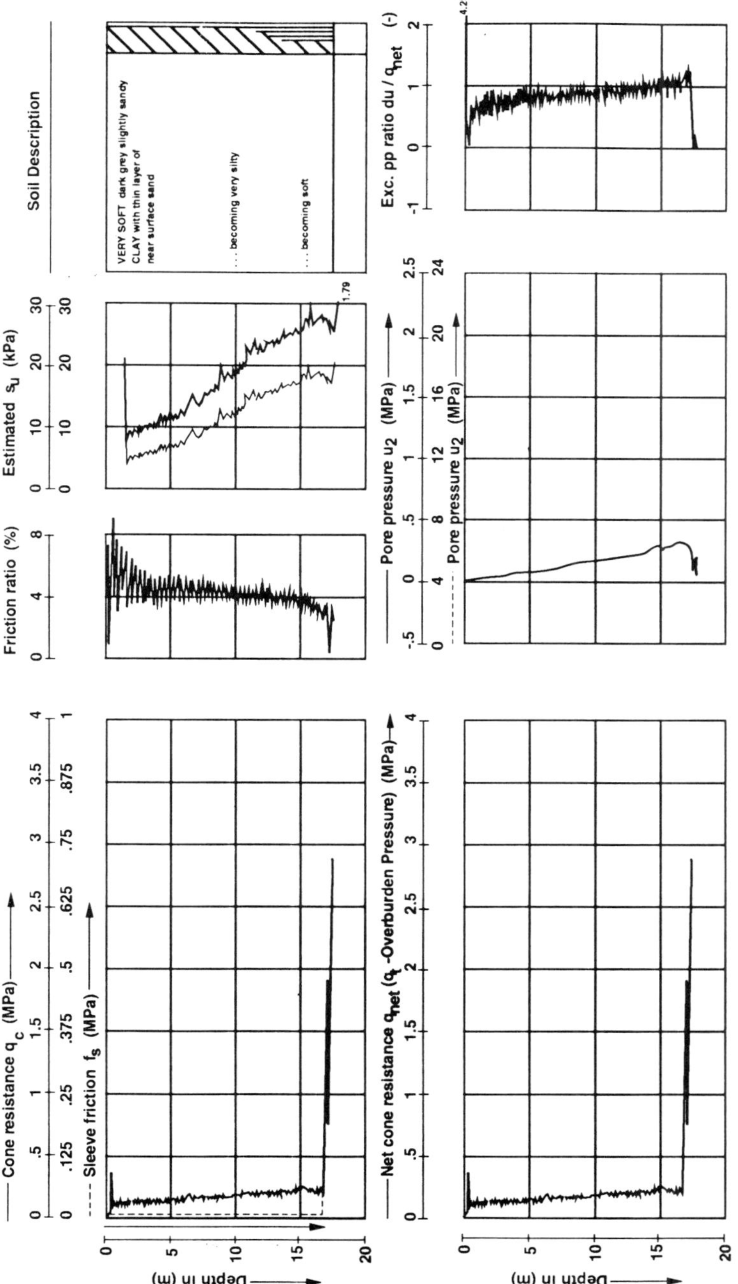

Fig. 9. Seasprite cone penetration tests S4 (Tarmac-Faslane MTF, preliminary).

4. Anchor Redesign

4.1. Deadweight Anchor Design

It was therefore necessary to review the anchor system options. This further review resulted in the selection of a modified deadweight anchor design.

Traditionally simple deadweight block anchors are designed to efficiencies (load/weight) of only 0.3–0.5, BS6349 (1989) Reference [5]. In this instance the size of block required would not provide an economic or practical solution. The optimisation of a skirted box type structure was therefore necessary. The engineering characteristics of the extremely soft plastic clay into which the anchors would be installed also complicated the anchor design.

The chosen deadweight anchor design was formed in reinforced concrete from two "skirt" walls (front and rear) and a "spine" wall in an "H" formation with a top covering slab, Figure 10. The major dimensions of the sixteen identical anchor blocks are 13.2m $\times$ 13.2m by 3.2m deep, weighing approximately 210 tonnes in air (with an additional 5 tonnes for the chain tensioner).

Under ultimate load the theoretical restraining forces developed by the soil on the anchor are mobilised by the ultimate soil shear stresses in the side and base failure planes, and the combination of active and passive forces developed on the skirts.

The theoretical (conservative) distribution of load, under ultimate design conditions, was evaluated as:

- Ultimate undrained soil shear stress at anchor base – 1220 kN.
- Ultimate undrained soil shear stress along side faces – 765 kN.
- Resolved soil active and passive forces on front and rear skirts – 380 kN.

In order to derive the distribution of these loads for structural design purposes finite element (F.E.) soil and structural models were created using the PAFEC 7.2 F.E. system (Woodford (1992), Reference [11]).

A 3D F.E. soil model examined the distribution of forces through the soil onto the anchor structure. The rectangular block of soil between the skirts and spine walls, beneath the top slab, was modelled with 240 twenty noded elasto-plastic brick elements. The vertical variation of soil properties was modelled; soil response properties being specified according to initial (small strain) stiffness data derived from the results of the consolidated undrained triaxial test programme. An undrained Mohr-Coulomb relationship was adopted, Houlsby (1983), Reference [12]).

Variations in specification of the boundary restraints of the soil block were found to have a significant influence on the critical back skirt stress distribution. Elastic horizontal and vertical restraints were examined on the "free" side and base, respectively. However, rigid restraints orthogonal to the loading direction were adopted as they, together with the stiff soil response, combined to provide a

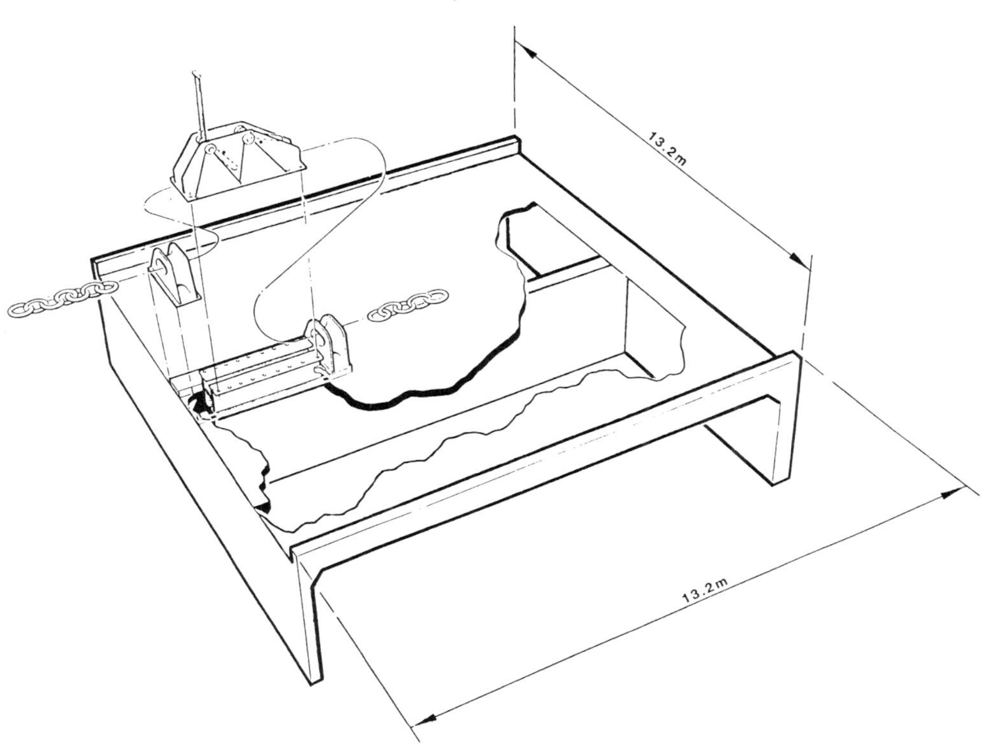

Fig. 10. Schematic representation of deadweight.

conservative loading criteria for the back skirt stress distribution.

For structural design purposes, it was assumed that no shear load was mobilised by the slab and spine on the soil block. The stress distribution evaluated for ultimate loading is illustrated in Figure 11. The front skirt distribution is a conventional 2D passive distribution.

Soil parameters leading to conservative loading distribution combinations on the structure were adopted throughout the design process. This philosophy was adopted as the concrete anchor block was exceptionally slender for the spans and loading considered.

As the soil bearing capacity, acting on the anchor base, and side wall friction would resist full penetration under selfweight, slab surcharges were designed to ensure that the slab soffit rested at mudline. The installation of a surcharge block is illustrated in Figure 12.

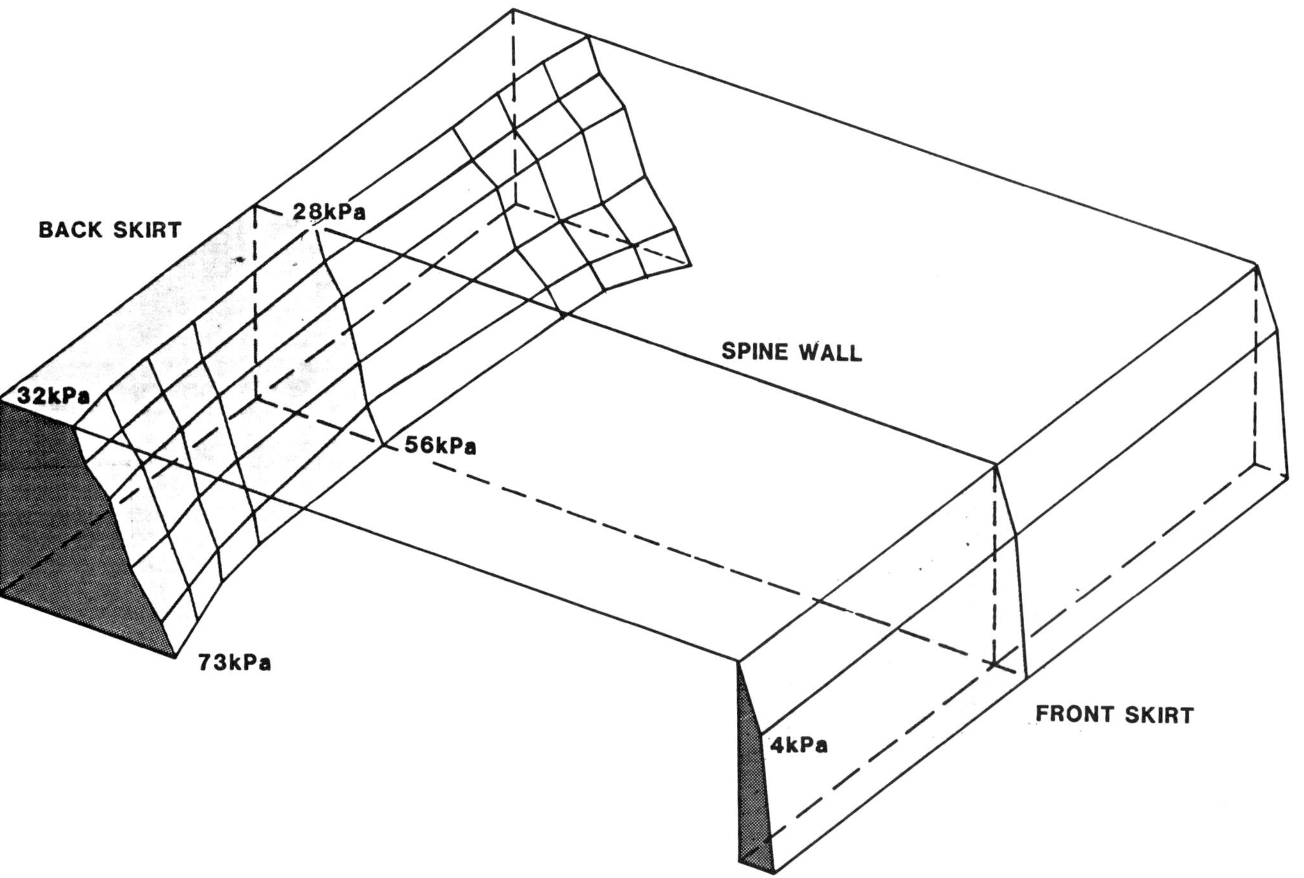

Fig. 11. Pressure distribution used for structural model.

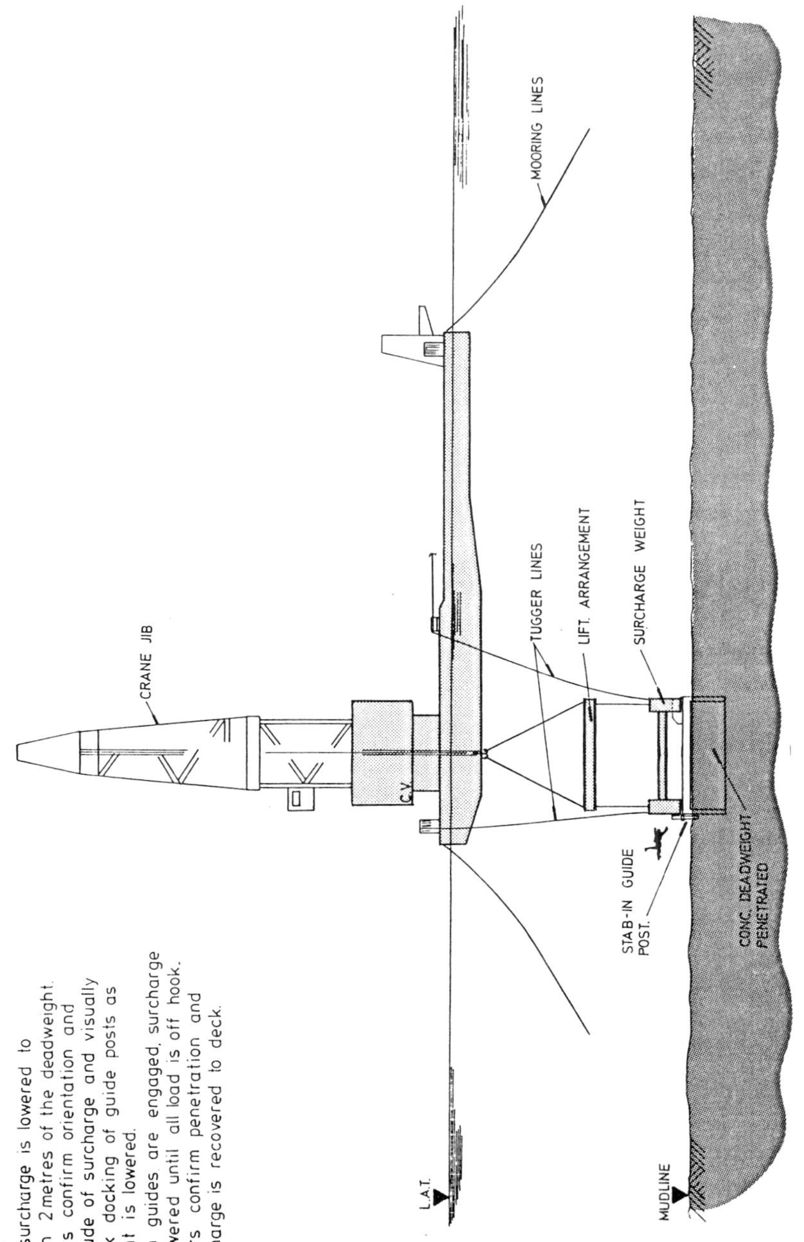

Fig. 12. Surcharge installation.

4.2. ANCHOR STABILITY

The chosen deadweight design was evaluated for overturning under ultimate load. Although remoulding of the soil will occur in the immediate vicinity of the skirts during installation the shear strength will largely recover with time. The laboratory soil tests indicated that a completely remoulded clay sample had a substantially reduced strength, approximately 30% of its undisturbed value. For design purposes a shear strength reduction of 50% was adopted. Conservatively, suction acting on the base of the structure and shearing of the material along the spine wall was ignored. The lowest factor of safety for the ultimate design load was for line P4, for which the mudline gradient was 5.2°, with a F.O.S. against overturning of 1.15.

The slope stability of the concrete anchor under ultimate load conditions for the anchor resting on the highest slope was addressed. Undrained analysis, based on Bishop's method of slices, were performed giving a minimum F.O.S. against slope failure of 3.0. Drained failure conditions were assessed using triaxial compression strength (s_u^c) in the active zone, triaxial extension strength (S_u^E) in the passive zone and direct simple shear strength $s_u^{\rm DSS}$) for the horizontal shear zone for potential failure surfaces. The direct simple shear tests were stress controlled under cyclic conditions which simulated the operating case. The results indicated that after a simulated four year exposure to these cyclic loads degradation of the near surface materials would not be expected.

Using the results from the laboratory oedometer tests, performed on samples both vertically and horizontally taken, the short and long term vertical and lateral anchor displacements were considered for settlement and creep. The total maximum vertical displacement of the deadweights expected after 25 years was approximately 440mm. Under normal cyclic load mooring conditions after 25 years the lateral displacement would be expected to lie between say 35 to 675mm.

The effects of horizontal and vertical ground accelerations due to earthquakes were also assessed and the induced lateral loads were found to be less than the ultimate capacity of the anchor.

4.3. DEADWEIGHT ANCHOR STRUCTURAL ANALYSIS

The deadweight anchor was assessed for eleven separate load cases, covering the construction, installation, operation and survival conditions. Several iterations were performed to optimise the member dimensions of the basic deadweight anchor and ensure anchor stability was maintained, Figure 13. A load case, combining four of the individual load cases, was investigated to determine the most onerous realistic load combination on the anchor block during the design storm condition.

The combined load cases adopted for *in-situ* structural design consisted of:

- Ultimate in place lateral load (219T including the maximum down- slope selfweight component)

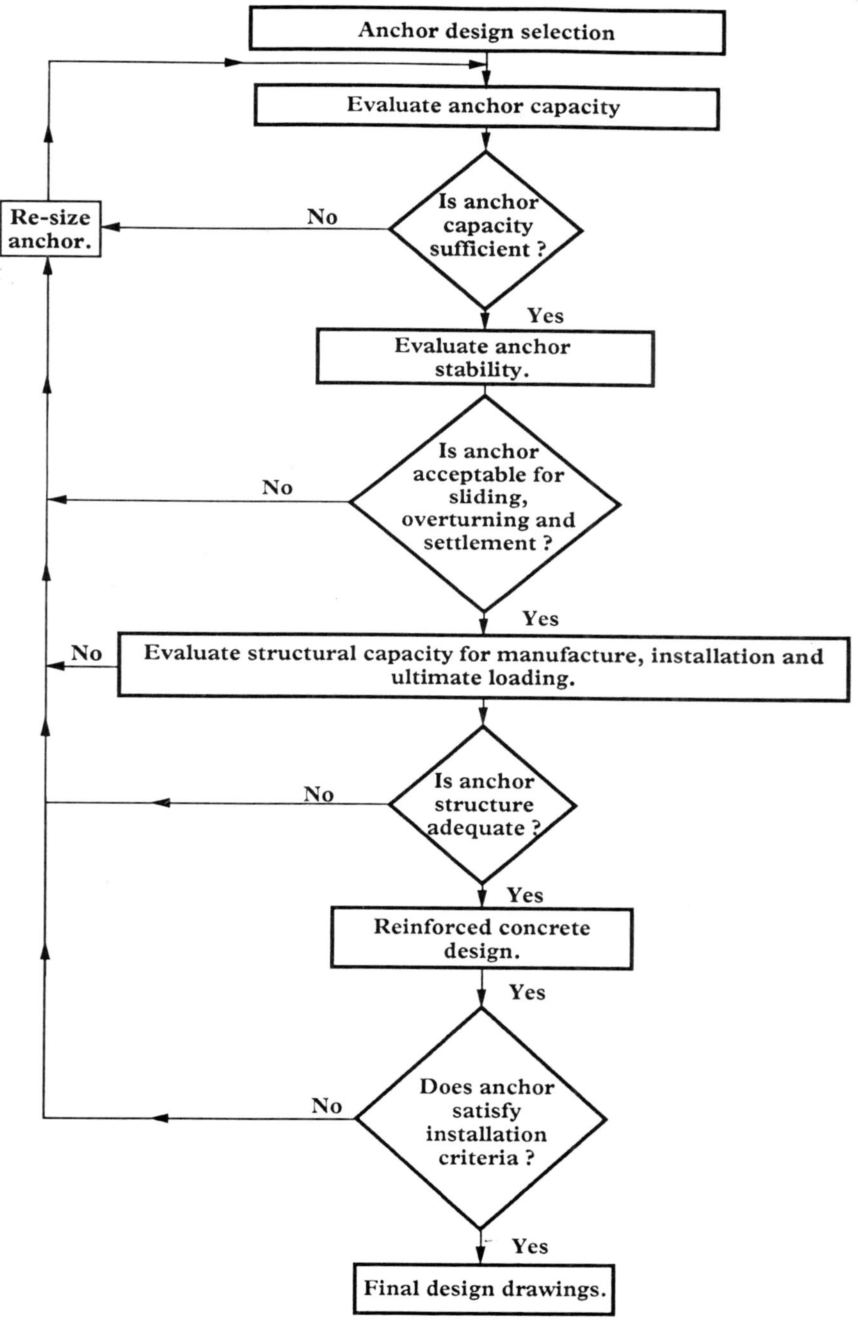

Fig. 13. Anchor design optimization.

– + Skew chain load
– + Non-symmetrical soil resistance
– + Silt accumulation above slab.

The PAFEC(7.2) finite element system was used to build a 3-D model of the deadweight anchor from elastic 20 noded isoparametric brick elements. The deflected shape of the anchor during ultimate loading is illustrated in Figure 14. The back skirt soil stress distribution produced by the "brick" soil model, Figure 11, was applied as point loads to the structural model, which was restrained at the spine and chain tensioner.

Variations of this model were employed to examine the behaviour of the block during lifting, installation, and embedment. Elastic surface stresses produced by the F.E. model were developed into bending moments and the concrete reinforcement was detailed in accordance with BS8110 (1985), Reference [13].

5. Conclusions

High quality geophysical and geotechnical ground investigation information is critical for successful anchor system design.

The use of the modified deadweight anchor necessitated an adequately designed ground investigation programme in order to optimise the anchor system design.

The application of soil/structure F.E. techniques allowed for conservative structural design of the anchor under installation and ultimate load conditions.

It is concluded that the design of this high performance anchor system required the close integration of geophysical and geotechnical information with mooring and structural design to produce the most appropriate anchoring solution.

Acknowledgements

The teamwork and support of the members of Tarmac Construction, Vosper Thorneycroft and Dowty involved in this project is gratefully acknowledged. The M.O.D. has kindly given the authors their permission to publish this paper for which they are indebted. The authors would also like to specifically acknowledge their appreciation to Mr. T. Swift (of Tarmac Construction Ltd.), Mr. A. Jackson (of Wimpol Ltd.) and Dr. T. Henderson (formerly of Fugro-McClelland Ltd., now G.C.G.) for their technical input during this project.

Footnote

The M.O.D. programme to build the MTF facility was originally budgeted at £50 million. Escalating costs, primarily due to the structural implications of the recognition of a more severe current regime than originally specified, caused the abandonment of the project in October 1991, Reference [14].

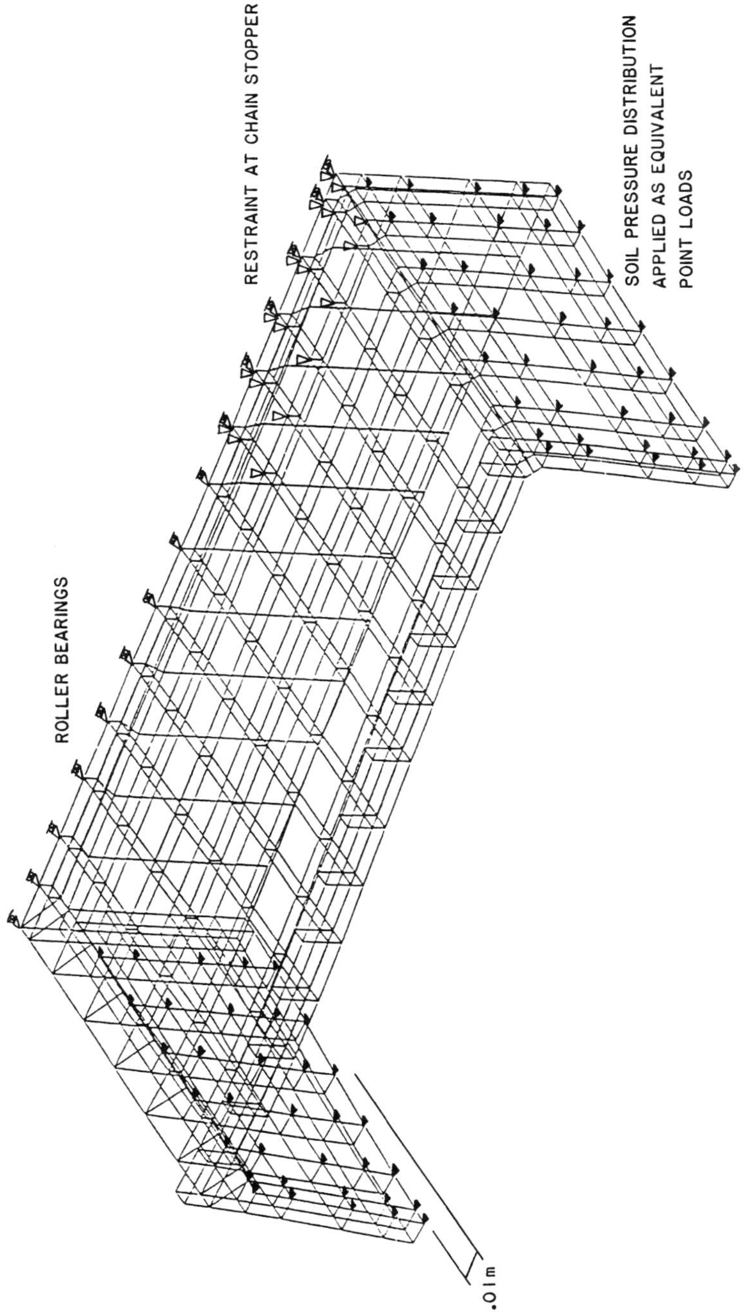

Fig. 14. Structural finite element model deflections under ultimate load (Figure 11).

References

1. British Regional Geology (1971), Grampian Highlands, HMSO.
2. Wimpey Laboratories Limited (1984), 'Report of Marine Geotechnical Investigation – Faslane CSB – Proposed MTF', Lab Ref. 5/20543.
3. Wimpey Laboratories Limited (1984), 'Proposed MTF at Faslane Submarine Base, Clyde Schotland. Task No. FGE/1899, Report on Site Investigation', Lab Ref. 5/20543 (for PSA).
4. British Standards Institution (1972), 'Code of Basic Data for the Design of Buildings CP3, Chapter V: Loading, Part 2: Wind Loads'.
5. British Standards Institution (1989), 'British Standard Code of Practice for Maritime Structure, Part 6: Design of Inshore Moorings and Floating Structures, BS6349', Ref. 86/12351 DC.
6. Belcher, R. E. and Wood, C. J. (1988), 'Wind Tunnel Calibration of the Faslane Anemometer Site', Oxford University Engineering Laboratory, Report No. 1728/88.
7. Hydraulics Research Limited (1986), 'Wind Analysis and Wave Hindcasting for Faslane, Gareloch, Schotland', H.R. Report No. EX. 1482.
8. MIRA (1989), 'Magnetic Treatment Facility Half Scale Wind Tunnel Test of a Section of Hoop Member at MIRA, Nuneaton', VT Report No. D/89-490.
9. Wimpol Limited (1990), 'Tarmac Construction Geophysical Survey M.T.F.', Wimpol Limited Reference S1489. (Main author: A. Jackson.)
10. Fugro-McClelland Limited (1990), 'Faslane MTF Geotechnical Site Investigation Laboratory and Parameters Report', Report Reference No. 90/1162. (Main author: T. Henderson.)
11. Woodford, C. H., Passaris, E. K., and Bull, J. W. (1992), '*Engineering Analysis Using Pafec Infinite Element Software*, 1st edition, Blackie and Son, London.
12. Houlsby, G. T. and Wroth, C. P. (1983), 'Calculation of Stresses on Shallow Penetrometers and Footings', *Seabed Mechanics, Proc. Symp. IUTAM*, pp. 107–112.
13. British Standards Institution (1985), 'Structural Use of Concrete: Part 1. Code of Practice for Design and Construction', BS8110.
14. The Engineer (1991), 'MOD Scraps Subs Facility', 12th Dec. 1991 edition, p. 7.

OPTIMIZATION OF UNDERBASE DRAINAGE SYSTEMS FOR GRAVITY STRUCTURES ON SAND

N. J. O'RIORDAN and J. W. SEAMAN
Ove Arup & Partners, Consulting Engineers, 13 Fitzroy Street, London W1P 6BQ

1. Introduction

For many off-shore sites, gravity structures can provide cost-effective solutions for the provision of support to topside facilities. Uncertainties in drivability and long-term performance of piled foundations, particularly in new and unusual ground conditions, can lead to the adoption of relatively shallow foundation systems to carry the forces collected by the superstructure. Installation of a gravity structure can be undertaken without the requirement for an expensive heavy lift vessel in attendance while a steel jacket structure is set onto the site seabed and piled. The need for such a vessel at the platform site can be completely avoided if the topsides are mated to the structure in the construction yard or inshore. Furthermore, the large volumes associated with gravity structures can be utilised for oil storage off-shore.

Experience has shown that to avoid bearing capacity failure due to storm loads on predominantly clayey sites, a combination of a light structure and a large base area is to be preferred. However, for sand sites, heavier structures are normally needed to prevent sliding.

Sliding resistance of sandy soils can be reduced due to increased pore pressures caused by cyclic loading. This can be avoided by maintaining a sufficiently large positive pressure under the base of the structure. The heavier the structure, the costlier it is to build, float out and install; therefore additional ballast may be needed after installation. The placing of ballast to ensure short term stability is costly and increases the length of the weather window required for the sub-structure installation. An alternative solution is to provide an underbase drainage system to encourage the dissipation of excess pore pressures during and after a storm. The design of the foundation system is usually driven by the optimisation of structure weight and ballasting.

Codes and other guidance documents for the design of gravity structures on sand require the designer to take due account of accumulated excess pore pressures during cyclic storm loading. However, the designer is given very little guidance on

Volume 28: Offshore Site Investigation and Foundation Behaviour, 417–432, 1993.

the manner in which this is to be done. Det norske Veritas (1990) provides some helpful comments on the philosophy of carrying out stability analysis using the cyclic strength of sand, but concludes by observing that no documented detailed procedure has yet been published. Conceptually, Det norske Veritas admits the possibility of performing analyses including models for pore pressure build-up and dissipation.

Various authors, notably Bjerrum (1973), Eide (1974) and Dobry *et al* (1982), have set down procedures for the testing of soils and the calculation of excess pore pressures during undrained loading of sands. Drainage of sand underneath structures can occur during storm loading and excess pore pressures normally dissipate between storm, Andersen *et al* (1982).

Observations of the behaviour of gravity foundations on sand from Ekofisk, installed in 1973, through to Ravenspurn, installed in 1989, have shown that large movements, accompanied by high excess pore pressures, have not been encountered. Table 1 lists weights and dimensions of North Sea gravity structures built predominantly on sand. The horizontal load effect from storm loading on the platforms listed in Table 1 falls generally in the range of 0.15 to 0.3 times the submerged weights, with the exception of the Ekofisk Tank and Frigg CDP-1, for which the horizontal storm load effect is closer to 0.4 times the submerged weight.

Drainage ports are normally required to pass through the base to remove trapped water during installation. If left ungrouted, these ports can be used as part of a permanent stability control system for the structure in service. In the years between Ekofisk and Ravenspurn, there has been greater employment of underbase drainage systems in order to reduce excess pore pressures in the soil and thus the requirement to provide dead weight purely to avoid liquefaction.

2. Design Process

When a gravity structure sustains storm loading, the forces are transferred into the soil. The soil deforms and there is a tendency for the void spaces to close up, inducing excess pore pressures, higher than hydrostatic pressures, in the ground. The strength of the sand is governed by the mean normal effective stress. During a storm, excess pore pressures accumulate during successive cycles of loading, so the soil progressively loses strength. The designer has to ensure that, at the height of the design storm, the remaining soil strength is sufficient to withstand the environmental loads with an adequate reserve of safety.

The stages in the design of a gravity foundation system on sand are well established and are encapsulated in Figure 1. Andersen (1991) describes a procedure for analyzing foundation stability for undrained conditions. The method has the merit that asymmetric shearing is allowed for in design, indicated schematically in Figure 2. For sands, the highest excess pore pressures are generated during full shear reversals, with significantly lower excess pore pressures generated when τ_{cy} is offset by τ_a. The process leads to definition of zones of high pore pressure, as

TABLE 1.

Summary of gravity structures founded on sandy soils in the North Sea, based on data from Eide and Andersen (1984).

Name	Operator	Sector	Year installed	Water Depth m	Submerged Weight 10^6 kN	Foundation area m^2	Soil Properties
Ekofisk Tank	Phillips	Norway	1973	70	1.9	7,400	Fine dense silty sand
Beryl A	Mobil	U.K.	1975	120	1.7	6,200	Fine dense silty sand (0-10m) overlying very stiff silty clay
Frigg CDP-1	Elf	U.K.	1975	98	1.8	5,600	Fine dense silty sand (8m) overlying stiff silty clay
Frigg TP-1	Elf	U.K.	1976	104	1.8	5,600	Fine dense silty sand (3-7m) overlying stiff silty sand
Frigg Manifold	Total	U.K.	1976	94	1.8	5,600	Fine dense silty sand
Frigg TCP-2	Elf	Norway	1977	102	1.6	9,300	Fine dense silty sand (3-6m) overlying stiff silty clay
Gullfaks B	Statoil	Norway	1988	143	3.0	8,700	Dense sand
Ravenspurn North	Hamilton	UK	1989	43	0.41	3,400	Dense sand (3.6-4.5m) over very stiff silty clay
F3	NAM	Netherlands	1992	42	-	-	Dense sand (7m) over interbedded sands and clays

shown diagrammatically in Figure 3.

In order to simplify the calculation of drainage during the storm, it is suggested that the soil and excess pore pressure systems are divided into representative zones beneath the platform. The underbase drainage system is then designed so that the desired rate of dissipation in each zone is obtained.

3. Pore Pressure Generation

The development of excess pore pressures in sands during cyclic loading has been discussed by many authors, including Lee and Focht (1975) and Muira and Toki (1982). Castro (1975) and others have observed, particularly for dense sands, that

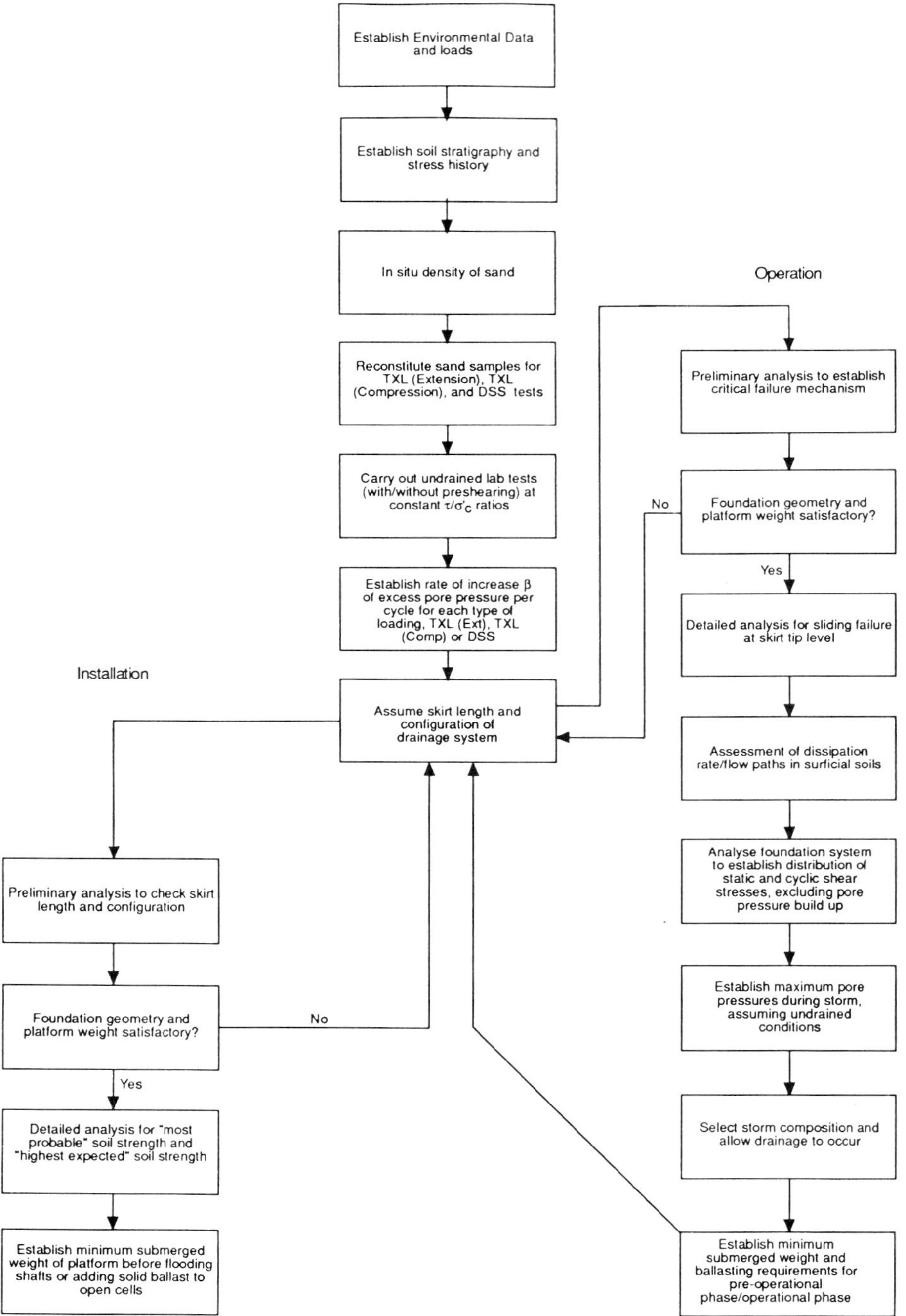

Fig. 1. Flow diagram showing foundation design methodology.

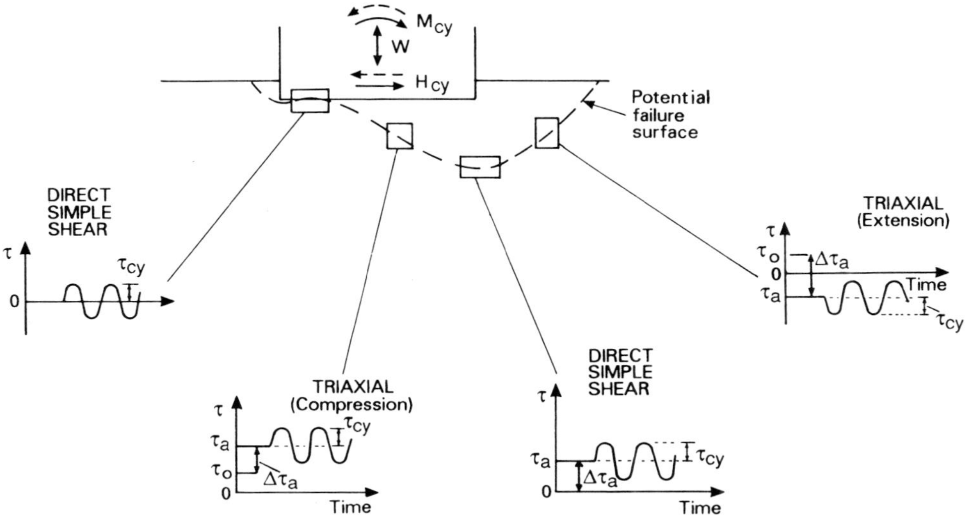

Fig. 2. Simplified storm condition for some elements along a potential failure surface, after Andersen (1991).

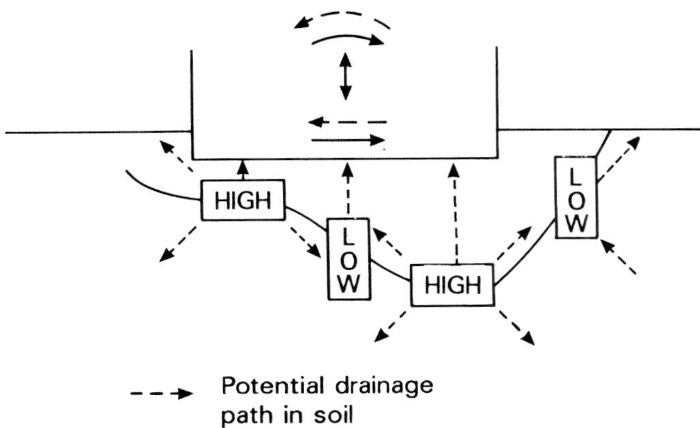

Fig. 3. Zones of high and low excess pore pressure during storm loading.

the development of either large permanent strains or large cyclic strains as liquefaction is approached in laboratory tests may be a function of sample preparation. Care should therefore be exercised when attempting to match excess pore pressures and cumulative strengths obtained from different tests with different methods of sample preparation.

Generally, and this is recognised explicitly in Det norske Veritas (1990), the relationship between excess pore pressure, Δu, number of cycles, N, stress level, τ_{cy}, and mean effective stress, σ'_c, is semi-logarithmic, as shown in Figure 4. At high cyclic shear ratios, τ_{cy}/σ'_c, greater than about 0.35, the excess pore pressure generated per cycle β, normalized with respect to the mean effective stress, becomes depressed, possibly due to dilatancy effects.

β is often estimated using a linear logarithmic function of the form:

$$\beta = 10^{(A(\tau_{cy}/\sigma'_c)-B)}$$

where A and B are constants.

The adoption of a linear function for all values of τ_{cy}/σ'_c, weighted towards values obtained below $\tau_{cy}/\sigma'_c = 0.35$, will avoid over-reliance on the beneficial effects of dilatancy on foundation stability.

Having established values of A and B from laboratory testing in compression, extension and simple shear, the increase in pore pressure for one cycle at a given stress level can be calculated.

4. Design Storm

A storm contains a wide range of waves that individually apply a load cyclically to the structure and thence to the foundation. Corresponding cyclic shear stresses in the underlying soil lead to the generation of excess pore pressures which can be estimated for each cycle using the method discussed above.

The modelling of the design storm is important when considering excess pore pressure build-up and associated drainage. Conventionally the peak 6 hours of a design storm are composed of 1800 waves of varying intensity, Hansteen (1981). The wave forces are assumed to follow a Raleigh distribution as shown in Figure 5. The largest wave of the storm typically represents the 100 year return wave for the ultimate limit state condition. 50% of the waves are assumed to induce 20% of the force effect of the maximum wave.

It is often assumed in geotechnical analyses that the wave force increases in magnitude throughout the storm, following the order shown on Figure 5. The probability of waves arriving in this order is, however, very remote (about 10^{-34}) and is incompatible either with material failure or load partial factors with exceedance probabilities of around 10^{-2} to 10^{-4}, implied elsewhere in design codes.

The ordering of the waves during the storm does not greatly influence the build-up of excess pore pressures in a clay soil, where little drainage is to be expected before the arrival of the largest wave, assumed to occur at the end of the storm.

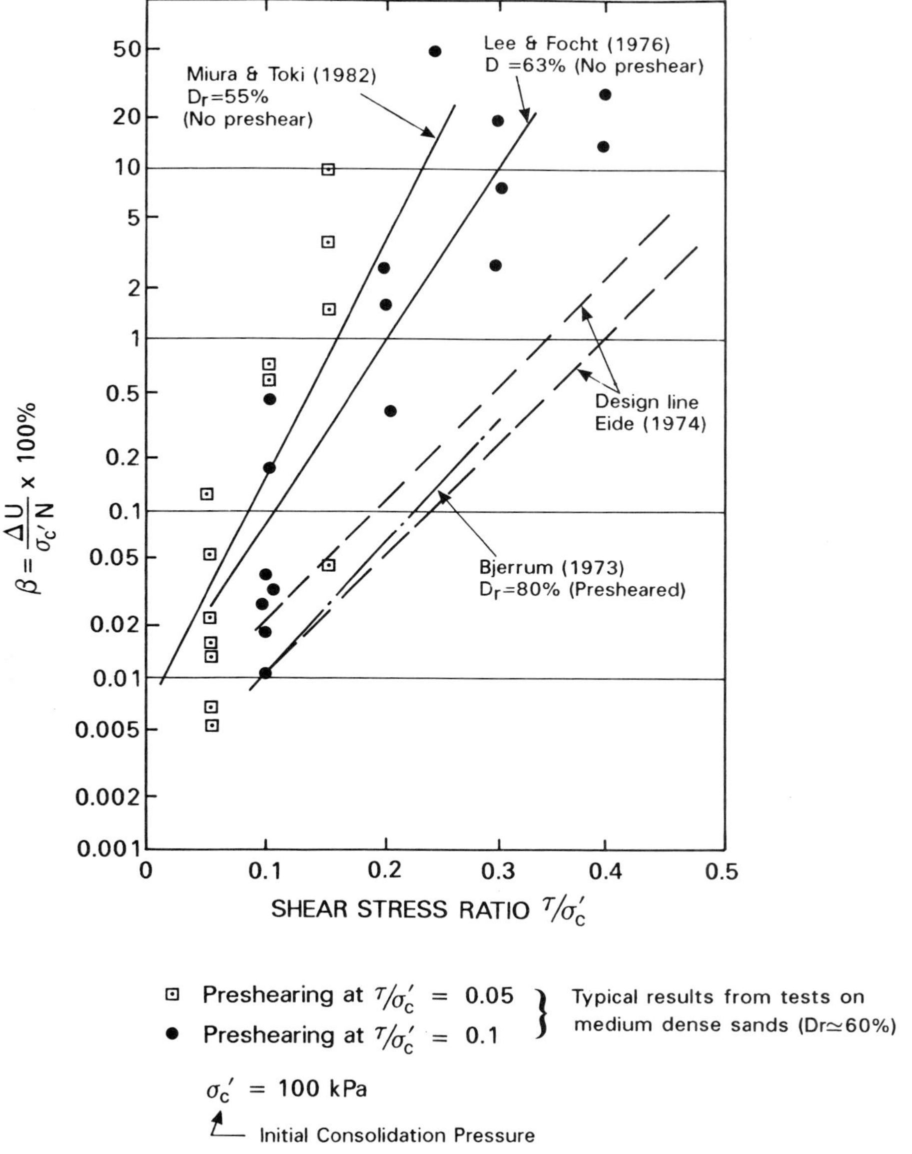

Fig. 4. Generation of excess pore pressure, Δu, in sands.

However, for sand soils the assumption of a progressive build-up in wave force can cause excess pore pressures to be overestimated, resulting in an uneconomic foundation design. Therefore, the distribution of wave forces throughout the design storm should better reflect the randomness observed from field observations.

Rahman *et al* (1977) suggested that the distribution should reflect a build-up

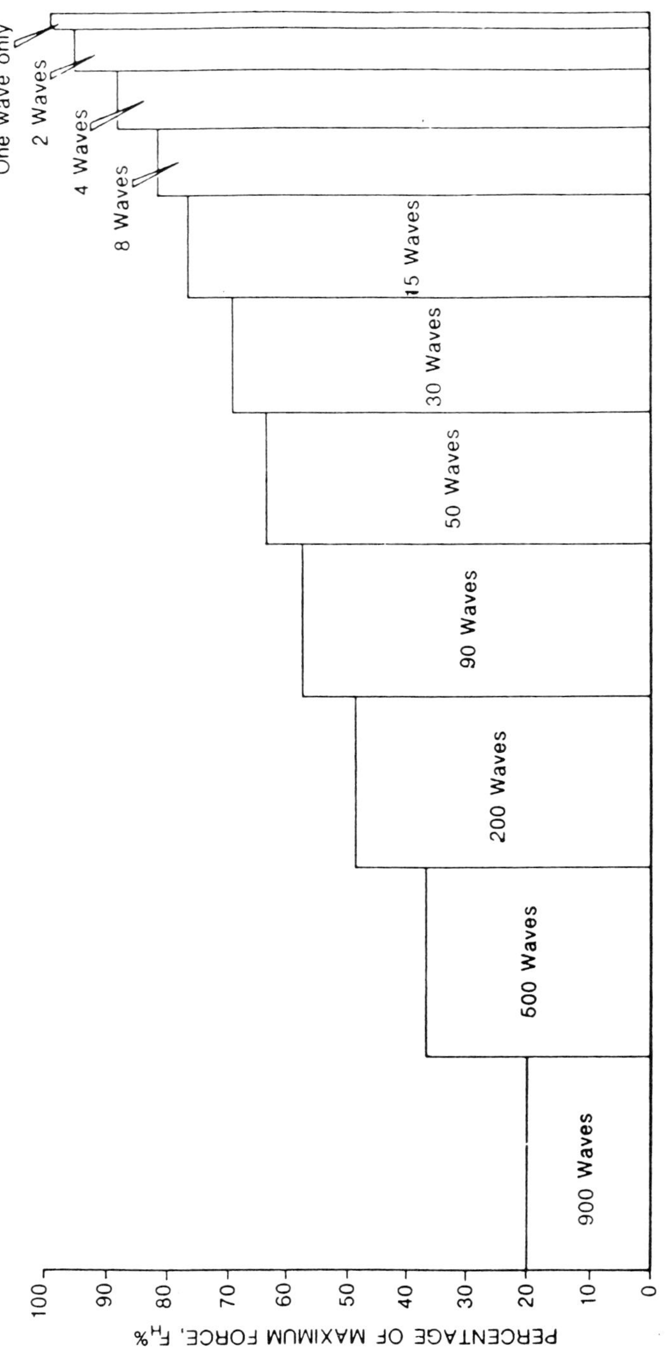

Fig. 5. Composition of design storm, after Hansteen (1980).

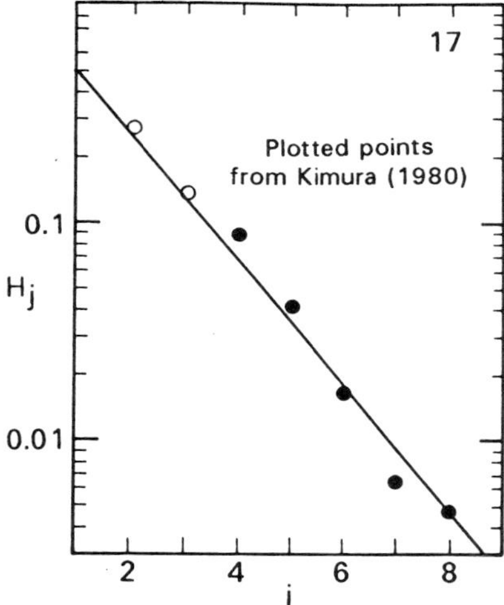

Fig. 6a. Probability H_j of large waves occurring in a run of j waves, after Longuet-Higgins (1984).

to the maximum wave force half-way through the storm, dying down towards the end. Wu *et al* (1984) recognised the randomness of storm behaviour in their back-analysis of the Ekofisk tank, but permitted the maximum wave to occur randomly at any time in the storm. Then the number of waves that may appear prior to the highest wave was assumed to be normally distributed. The results presented by Wu *et al* (1984) showed calculated excess pore pressures either significantly higher or close to values measured in the field and their analysis suffered from a very high coefficient of variation. In all cases, however, excess pore pressures calculated by Wu *et al* (1984) were always less than those established using a progressive build-up, deterministic approach, as might be expected.

Longuet-Higgins (1984), using real wave spectra from Kimura (1980), has provided a theoretical framework from which it can be concluded that a Hansteen type storm, applicable in the North Sea, can be characterised by a series of random Gaussian events in which, to a probability of about 10^{-2}, an average wave group length of twelve waves is associated with no more than six waves greater than the mean wave height. This is illustrated in Figure 6 taken from Longuet-Higgins (1984). Goda (1985) reached a similar conclusion using observed data.

In order to encapsulate Longuet-Higgins' (1984) work into a design method, the computer program RANWAVE has been developed, generally following a procedure described conceptually by Marex (1988). In this program the wave force spectrum of Hansteen (1981) is re-ordered pseudo-randomly, ensuring that the largest wave forces occur within groups of 12 lesser waves. Three simplifying but pessimistic assumptions using the basic framework suggested by Longuet-Higgins are made:

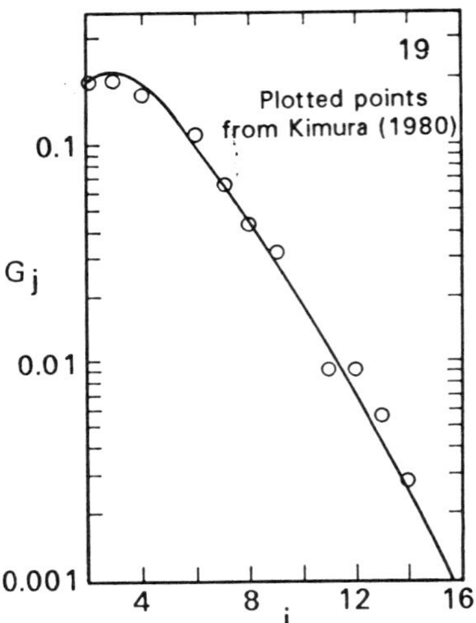

Fig. 6b. Probability of G_j of a group of j waves occurring together, after Longuet-Higgins (1984).

1. The maximum wave force occurs in the final wave group at the end of the six hour period of the storm.

2. The group containing the maximum wave force is preceded by groups of twelve waves containing the other highest waves in descending order.

3. Each group of twelve waves can contain more than six waves in the upper half of the wave population.

These assumptions combine to produce design storms that fall between Longuet-Higgins and an ordered Hansteen storm in terms of overall probability.

An example of the composition of the last hour of a design storm produced by RANWAVE is given in Figure 7, which can be compared with an ordered Hansteen storm given in Figure 5.

5. Drainage

Having obtained a storm history, the next step is to model drainage during the storm by superposition of pore pressure generation and dissipation curves for each individual wave. The rate of dissipation is governed by the geometry of each representative drainage zone beneath the base caisson and by the coefficient of consolidation of the soil. The selection of this coefficient is discussed in the following section.

Figure 8 shows the dissipation curves for two arrangements of drainage: to a single port and to a continuous drainage blanket on the underside of the base

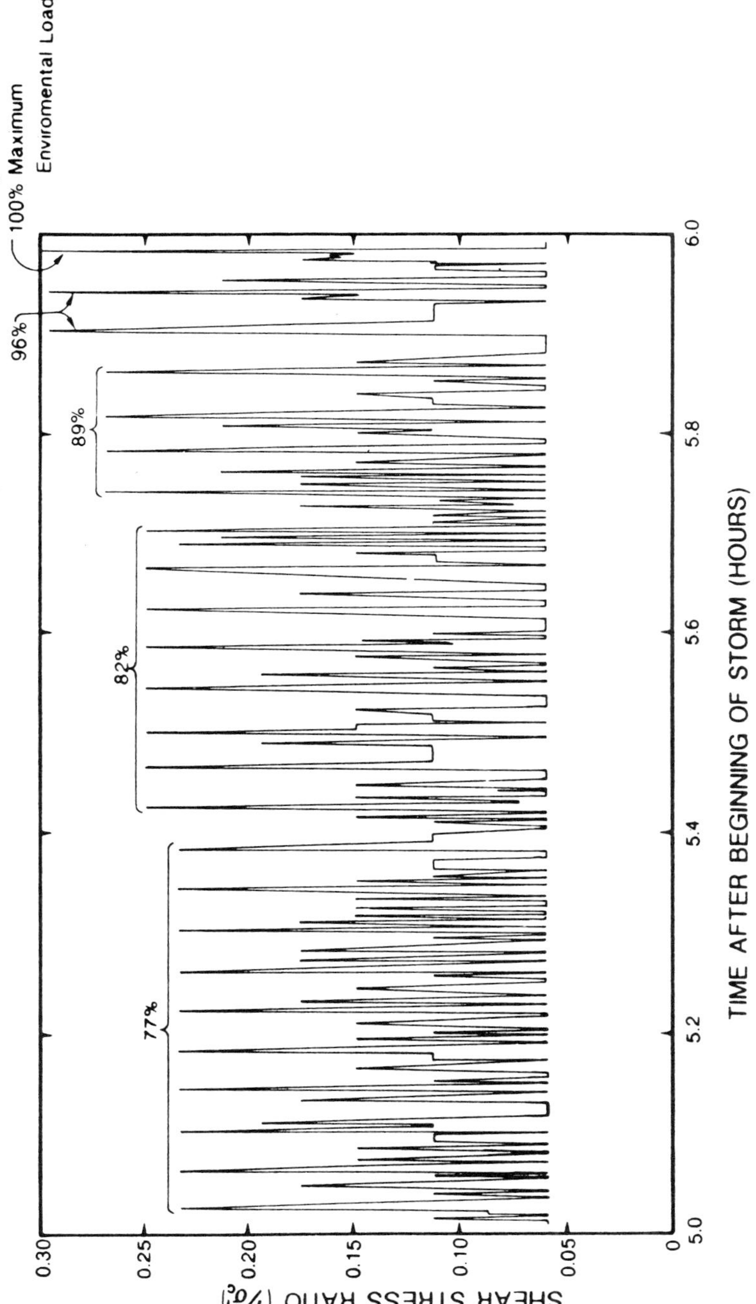

Fig. 7. Shear stress ratio applied during final part of a six hour pseudo-random design storm.

caisson. In this example, finite element analyses demonstrated that the soil would deform in simple shear, so that within each drainage zone, the cyclic shear stress τ_{cy} is constant with depth. Furthermore, the coefficient of consolidation, c_v, is maintained at a constant value.

The provision of a continuous drainage blanket that demonstratively produces one-dimensional dissipation of excess water pressure can be extremely onerous and costly. An alternative is to provide a sufficient number of isolated drainage ports. The number and location of these ports will depend on the magnitude of excess pore pressure that can be tolerated at the end of the storm and the consolidation characteristics of the soil.

It is found that the excess pore pressure at the end of a number of design storms, calculated by the above approach, approximately fall into a normal distribution. Figure 9 shows the results of over 200 RANWAVE runs, in which drainage has been permitted. High excess pore pressures are found outside the range of a theoretical normal distribution curve; however, inspection of these individual storms confirms that the important final hour of the storm contains a sequence of waves closely conforming to Hansteen (1981). This is the result of assumptions (2) and (3) above. As such occurrence would be rejected as improbable by Longuet-Higgins' theory, these high values of excess pore pressures are considered to be unrealistic.

In order to select a "design" excess pore pressure for use in stability analyses when the final, largest wave arrives, a value is taken at 3 standard deviations above the mean values obtained from successive RANWAVE runs. Figure 10 shows the result of RANWAVE calculations for a structure on 4m of medium dense sand at values of maximum τ/σ'_c in the range of 0.2 to 0.35. It may be seen that theoretical liquefaction, $\Delta u/\sigma'_c \geq 1$, is calculated to occur at about $\tau/\sigma'_c = 0.31$ with the ordered Hansteen storm and 0.34 with the RANWAVE storm. Thus, for a given imposed τ_{cy}, the required submerged weight of the structure would be at least 10% lower using pore pressures calculated using RANWAVE than if Hansteen's storm history was assumed to give waves of increasing magnitude.

Clausen *et al* (1975) report excess pore water pressures beneath the Ekofisk Tank of up to 35 kPa during the passage of a major storm shortly after installation of the tank. The Ekofisk Tank is founded on 21m of dense sand over very hard clay. It has a flat base slab with 0.4m deep concrete ribbed skirts. There is no underbase drainage system. Taking the average consolidation stress beneath the tank as 260 kPa, this gives an excess pore water pressure to consolidation stress ratio of about 13%. Clausen *et al* estimate that the horizontal wave force during this storm was a maximum of 690,000kN, about 85% of the 100 year wave force. The corresponding shear stress ratio is about 0.36. Using the RANWAVE procedure with underbase drainage the calculated pore pressure ratio at the end of a 6 hour storm with a maximum shear stress ratio of 0.36 would be greater than 100%. As there has been no evidence reported of a foundation failure of the Ekofisk Tank it is concluded that the procedures discussed so far result in a conservative assessment of foundation stability on sand soils.

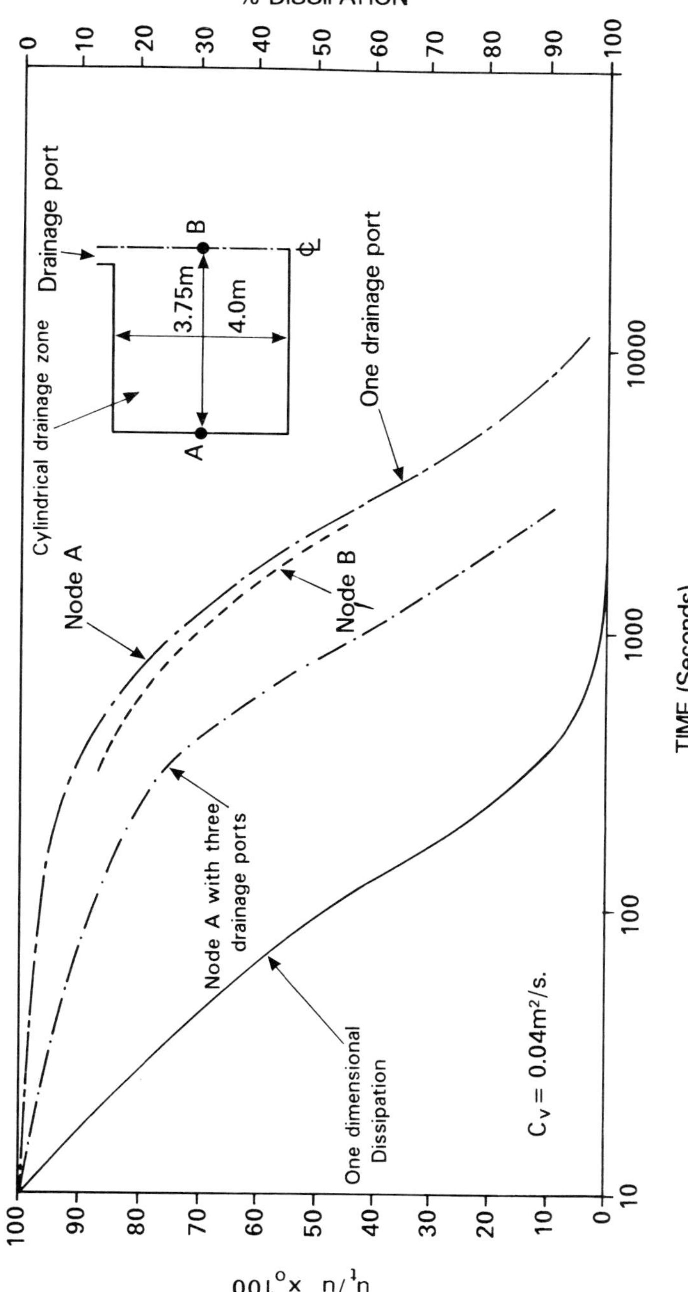

Fig. 8. Dissipation rates for excess pore pressures following a single wave for different drainage conditions.

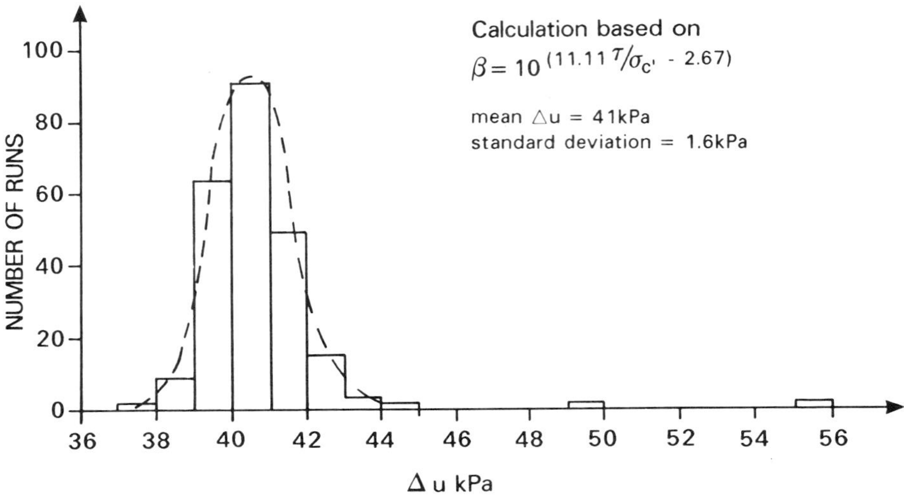

Fig. 9. Excess pore water pressure at the end of a series of six hour pseudo-random design storms.

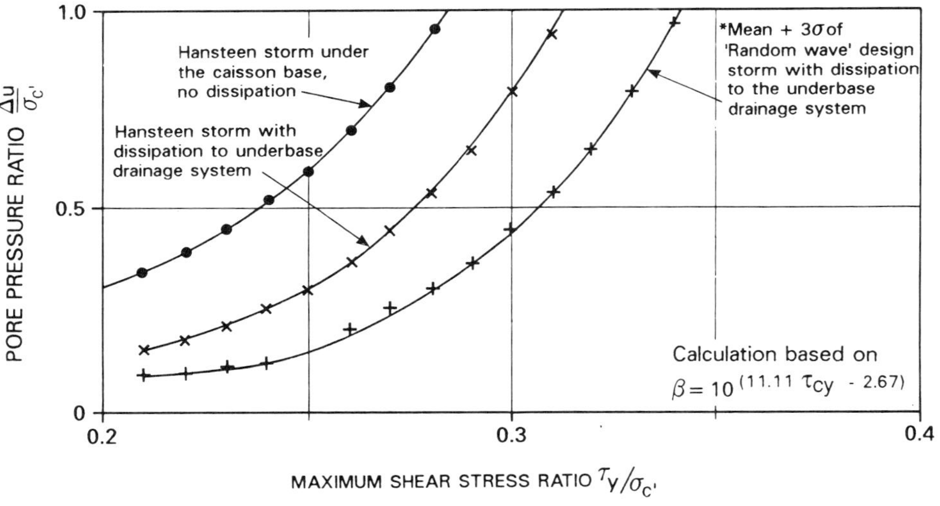

Fig. 10. Excess pore pressure ratios at the end of a six hour design storm.

6. Choice of Coefficient of Consolidation

In the above examples a constant c_v of 0.04m^2/s has been adopted for the sand under consideration. Such small values of c_v are difficult to demonstrate in the laboratory. Therefore, c_v is normally calculated from linear elastic consolidation theory using

$$c_v = \frac{kD}{\gamma_w}$$

where γ_w is the density of sea water, k is permeability and D is uniaxial modulus and

$$D = \frac{2G(1-\nu)}{(1-2\nu)}$$

where ν is Poisson's ratio and G is shear modulus.

G is normally evaluated from empirical relationships such as those suggested by Robertson and Campanella (1983), Stroud (1971) and Det norske Veritas (1990).

It is well known that secant shear modulus G is a function of shear strain, falling by an order of magnitude between shear strains of 10^{-6} and 10^{-2}. Provision for this behaviour is made in many design codes. A corresponding dramatic drop in permeability is unlikely over this range of strain. It may be conjectured that c_v is then also a function of shear strain. Dissipation of excess pore pressures during the early part of the storm, when cyclic shear strains are likely to be relatively small, will therefore be more rapid than dissipation under large waves, where cyclic shear strains are correspondingly large. Therefore as a further refinement RANWAVE can include for this non-linearity in dissipation rate.

7. Conclusions

By using a theoretical framework developed by Longuet-Higgins (1984), based on real wave data, a pseudo-random "design" storm has been established. This is a more reasonable representation of real storms than that obtained by other methods adopted in foundation design which assume a progressive increase in wave force during the period of a storm.

Combining the pseudo-random storm with drainage characteristics of the structure and soil, the program RANWAVE calculates the cumulative excess pore pressure at the worst point in the storm. This excess pore pressure is used in stability calculations. By iteratively adjusting drainage zones, drainage outlet design, and skirt geometry, it is possible to optimise the structure weight so that costs are minimised while ensuring an adequate factor of safety under operational conditions. A worked example shows how at least 10% of the submerged structure weight can be saved using this approach.

It is considered that this approach provides a closer representation of the performance of gravity foundation systems by using the characteristics of real storms rather than approaches based on fatigue theory, in which the order in which the cyclic loads are applied is unimportant.

The framework for calculation discussed above has been used routinely by the authors on many structures, from feasibility through to full implementation on projects.

Acknowledgements

The authors record their appreciation of the help and encouragement of their colleagues, in particular Michael Long, Dr. Barry Lehane and Dr. David Henkel in developing this approach to gravity foundation design on sands.

References

1. Andersen, K. H. (1991), 'Foundation design of offshore gravity platforms', in *Cyclic Loading of Soils*, M. P. O'Reilly and S. F. Brown (eds.), Blackie, pp. 122–173.
2. Andersen, K. H., Lacasse, S., Aas, P. M., and Andenæs, E. (1982), 'Review of Foundation Design Principles for Offshore Gravity Platforms', Norwegian Geotechnical Institute Publication No. 143.
3. Bjerrum, L. (1973), 'Geotechnical problems involved in foundations of structures in the North Sea', *Géotechnique* **23**(3), 319–358.
4. Castro, G. (1975), 'Liquefaction and cyclic deformation of sands', *Journal of the Geotechnical Division, ASCE* **101**, 551–569.
5. Clausen, C. J. F., De Biagio, J. M., Duncan, J. M., and Andersen, K. H. (1975), 'Observed behaviour of the Ekofisk oil storage tank foundation', *Offshore Technology Conference*, Houston, vol. 3, pp. 399-413.
6. Det norske Veritas (1990), 'Rules for the Design, Construction and Inspection of Offshore Structures'.
7. Dobry, R., Ladd, R. S., Yokel, F. Y., Chung, R. M., and Powell, D. (1982), 'Prediction of Pore Water Build-Up and Liquefaction of Sands during Earthquakes by the Cyclic Strain Method', Nat. Bureau of Science Series 138, Washington.
8. Eide, O. and Andersen, K. H. (1984), 'Foundation Engineering for Gravity Structures in the Northern North Sea', Norwegian Geotechnical Institute Publication No. 154.
9. Goda, Y. (1985), *Random Seas and Design of Maritime Structures*, University of Tokyo Press.
10. Hansteen, O. E. (1981), 'Equivalent Geotechnical Design Storm', Norwegian Geotechnical Institute Report 4007-16.
11. Kimura, A. (1980), 'Statistical properties of random wave groups', *Proc. 17th Conf. on Coastal Engineering*, Sydney, pp. 2955–2973.
12. Lee, K. L. and Focht, J. A. (1975), 'Liquefaction potential at Ekofisk Tank in North Sea', *Journal of the Geotechnical Division, ASCE* **101**(GT3), 1–18.
13. Longuet-Higgins, M. S. (1984), 'Statistical properties of wave groups in a random sea state', *Phil. Trans. R. Soc. London*, Vol. 312, 219–230.
14. Marex Ltd (1988), 'Review of the Report: Equivalent Geotechnical Design Storm', Report to Ove Arup and Partners.
15. Muira, S. and Toki, S. (1982), 'A sample preparation method and its effect on static and cyclic deformations – Strength properties of sand', *Soil and Foundations* **22**(1).
16. Rahman, M. S., Seed, M. B., and Booker, J. R. (1977), 'Pore pressure development under offshore gravity structures' *Journal of the Geotechnical Division, ASCE* **103**(GT12), 1419–1436.
17. Robertson, P. K. and Campanella, R. G. (1983), 'Interpretation of Cone Penetration Tests: Parts 1 and 2', *Can. Geot. J.* **20**, 718–745.
18. Stroud, M. A. (1971), 'The Behaviour of Sand at Low Stress Levels in the Simple Shear Apparatus', PhD Thesis, Cambridge University.
19. Wu, T. H., Martinez, R. E., and Kjekstad, O. (1984), 'Stability of Ekofisk Tank: Reliability analysis', *ASCE* **110**(GT7), 938–956.

THE NON-PILED FOUNDATION SYSTEMS OF THE SNORRE FIELD

HANS PETER CHRISTOPHERSEN
Saga Petroleum a.s., P.O. Box 490, 1301 Sandvika, Norway

Abstract. A new milestone has been passed in offshore development with the completion of the Snorre (Phase I) field installations this summer. The installation are located in more than 300m water depth supported on very soft soil. The foundation systems represent a major breakthrough for their simplicity, cost-effectiveness, and technical reliability.

The Tension Leg Platform (TLP) and the worlds largest Subsea Production System (SPS) are secured to the seabed without the use of piles. Skirted gravity systems are used which have minimized construction costs and installation time. Stricter requirements to removal can more easily be met with these systems. Future subsea wells in the area are planned to be fully protected within subseabed silo structures, minimizing impact on and from the fishing activities.

1. Introduction

The Snorre field is the first true deepwater North Sea development and is located within Block 34/7 north of the Statfjord fields. Block 34/7 was awarded to Saga Petroleum a.s. as operator in 1984. The first oil was produced in July 1992 from a Tension Leg Platform (TLP) located in 310m water depth. The seabed comprises very soft clay and is influenced by local depressions (pockmarks). The first installation being the TLP Well Template was placed on location in the summer 1990. Production wells have since been drilled from a semi-submersible rig until the TLP was installed in May 1992. Drilling was interrupted only for a period in May–June 1991 when the TLP concrete foundations were installed. The Subsea Satellite Template and Manifold (SPS), the worlds largest subsea installation, was installed during May 1992 in 335m water depth. The oil and gas is transported through 28km long pipelines to the Statfjord A Platform.

The foundation systems of both the TLP and SPS comprise skirted gravity systems without the use of piles for load-bearing. The selected foundation systems have considerably reduced foundation costs compared to conventional piled solutions.

The Tordis and Vigdis fields are located south of the Snorre field within Block 34/7. The Tordis field will be developed with single subsea wells connected to a central manifold structure. The oil and gas will be transported through pipelines to the Gullfaks C platform. The manifold structure will comprise a skirted foundation system similar to that of the Snorre SPS. The production wells will have a conventional protection structure which will be supported on skirts. One of the water

Volume 28: Offshore Site Investigation and Foundation Behaviour, 433–447, 1993.

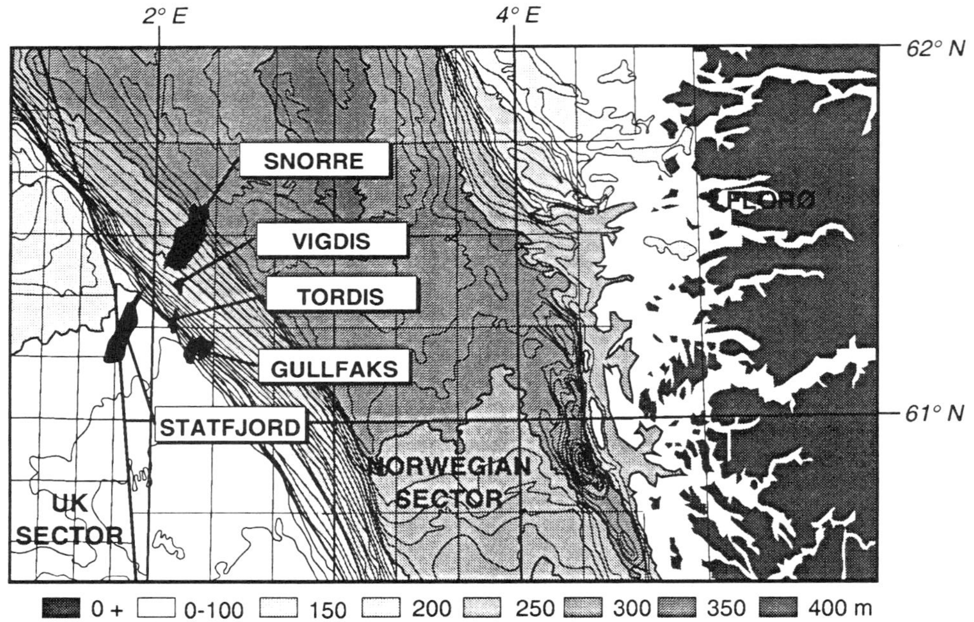

Fig. 1. Field lay-out area.

injection wells will be placed in a subseabed silo structure completely protected from trawl impact loads. This well will be the first in the North Sea to be placed in a subseabed silo. The subseabed silo concept is also considered for the next development in Block 34/7 – the Vigdis field.

2. Seabed Conditions

The Snorre (Phase I) field installations are located in 310 to 335m water depths within the "Norwegian Trench". The soil conditions at these depths consist of an upper layer of very soft clay gradually increasing in strength with depth. At around 60m depth the clay turns into a hard glaciated till containing stones.

Figure 2 gives an overview of the water depth, and seabed conditions from the Oseberg and Statfjord area ("North Sea Plateau") at around 110–150m water depth, through the "Slope" area (where the Gullfaks C platform and the Tordis field are located), and into the "Norwegian Trench" with the Snorre field installations in more than 300m water depth.

Figure 3 summarizes the soil conditions at the Snorre TLP locations. Coming into the "Norwegian Trench", beyond some 250m water depth, the seabed becomes irregular containing "pockmarks" or depressions. These pockmarks may be a potential hazard to installations if not properly identified.

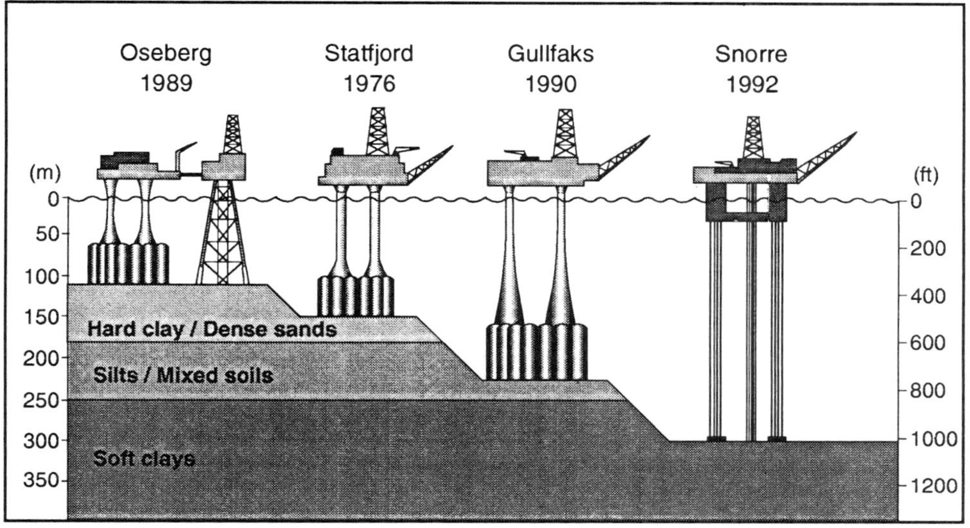

Fig. 2. Water depth/Soil conditions – Year of installation.

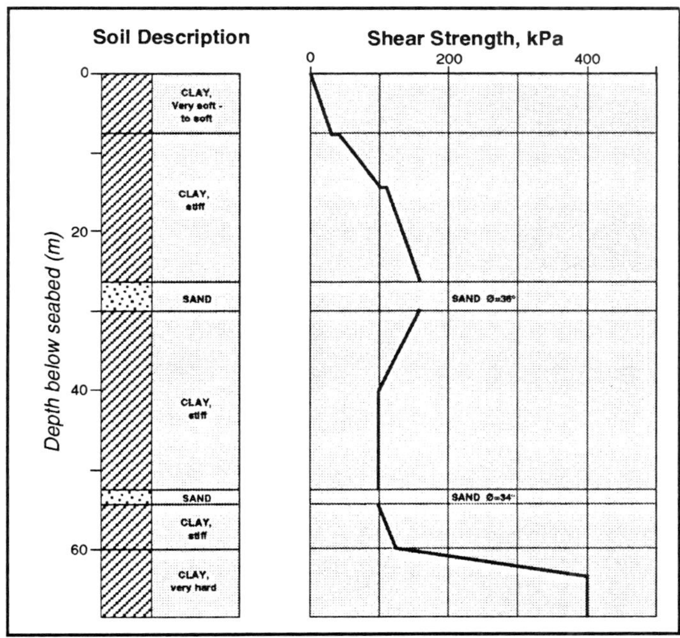

Fig. 3. Soil conditions Snorre TLP location.

3. Foundation Systems

3.1. Background for the Selected Systems

The award of Block 34/7 to Saga Petroleum a.s. as operator in 1984 represented a breakthrough for Saga as an oil company. The technical challenges in developing the field were tremendous. This was certainly true also for the foundation designers. There was no operational experience in the North Sea with the soft clays and deep waters as at Snorre.

A close liaison between the engineers involved in conceptual design development and foundation research programmes made it possible to propose cost-effective solutions, to verify and "mature" these solutions, and finally to achieve acceptance for these by the Snorre licence.

Two platform concepts were considered in the later phase of conceptual development. These were a "Condeep" type concrete gravity structure and a Tension Leg Platform (TLP). Both represented a considerable challenge with respect to foundation design, and many consultants and institutions were involved in solving various aspects of the foundation design of these concepts.

3.2. Piles or Gravity Foundations?

Relevant operational experience with tension loaded foundations in soil conditions as at Snorre is not available. Extensive research was carried out at the time of conceptual engineering for the Snorre development. Saga participated in Research and Development (R&D) projects related to both gravity foundation and pile foundations in soft clay sites. Two independent projects being conducted at about the same time could be mentioned: model tests on a gravity skirted foundation on soft clay conducted by J&W Offshore (Reference 1) and model tests on tension piles conducted by the Norwegian Geotechnical Institute (NGI) (Reference 2). These R&D programmes led us to the conclusions that existing analyses methods confidently predicted the behaviour of a skirted gravity foundation undergoing dynamic loading. The pile testing, however, indicated far lower skin friction capacities in silty clays (as found at Snorre) than "state-of-the-art" methods would predict.

Figure 4 indicates range of static skin friction and bearing capacities using various "state-of-the-art" methods for a 2.1m diameter pile at the TLP location. The range in predicted static (compression) capacity is high. For the TLP, the piles would undergo cyclic loading with shear stress reversals (alternate tension and compression). Further, the databases for the "state-of-the-art" methods did not contain soft, silty clays as found at Snorre. Model tests on tension piles being conducted by NGI (Reference 2) at the time included relevant loading conditions and soil conditions for Snorre. This programme concluded that a dramatic reduction in pile capacities is seen for piles undergoing alternate tension and compression loading. The capacities in this situation is about 40 per cent of that of the static capacities. Further, the static capacity was in general seen to be less than "state-

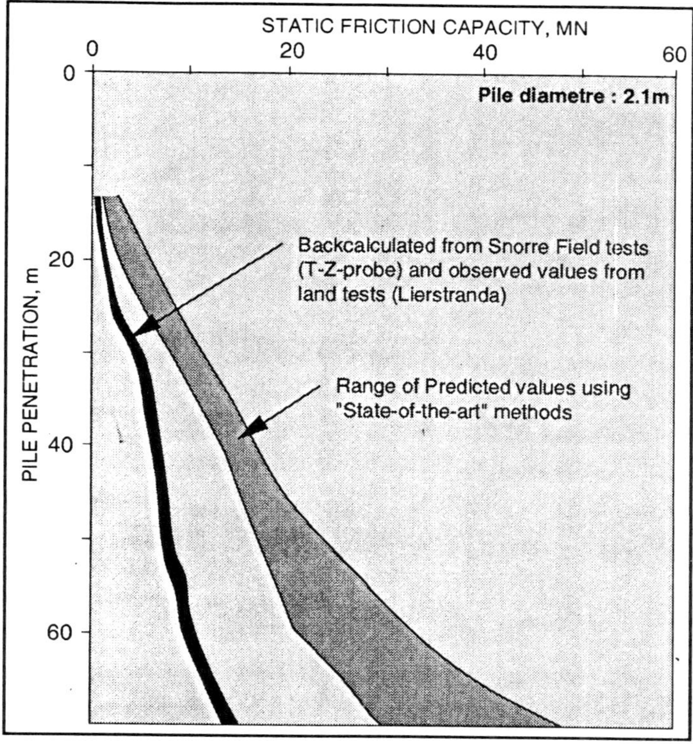

Fig. 4. Friction pile capacities at Snorre.

of-the-art" (Figure 5). In particular for the "Lierstranda" site, containing the most silty clay which resembles that at Snorre, very low frictional capacities were seen.

The concerns about low pile capacities at Snorre were strengthened by field observations. Field model tests at the Snorre field ("T–Z"-probe) gave frictional resistances of only 0.1 to 0.2 of the shear strength after 60 per cent consolidation. Backcalculated frictional resistances from the installation of the TLP Well Template piles indicated the same low ("unconsolidated") frictional resistances. It could take 1 to 2 years for the piles at Snorre to reach full consolidation which meant that design loads needed to be resisted for only partially consolidated soil.

Static skin friction capacities backcalculated from the Snorre field model tests and onshore tests ("Lierstranda") are included in Figure 4 together with the "state-of-the-art" predictions. The large discrepancy of these results gave serious concern to our ability to confidently predict pile capacities. It is emphasized that the curves in Figure 4 have not been reduced for cyclic degradation.

Being faced with the above conclusions and uncertainties related to the design of tension piles in soft, silty clays, a considerable and costly upgrading of a piled foundation system would be necessary.

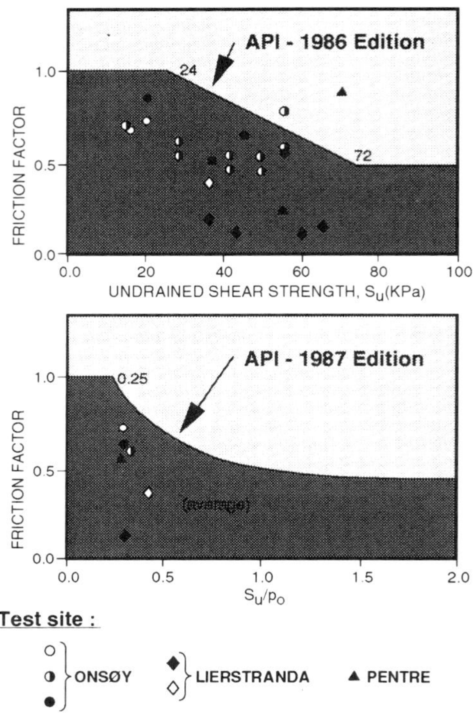

Fig. 5. Observed skin friction on model piles.

Having gained confidence in our ability to predict capacities for gravity foundations, conceptual development of a TLP gravity foundation was undertaken by Peconor (with J&W Offshore as geotechnical consultant) and Norwegian Contractors (with NGI as geotechnical consultant). By 1987 we were confident of having technically attractive alternatives to a piled foundation system. Cost comparisons were made which turned out favourable for the gravity foundations which were selected late 1987.

4. Concrete Foundation Template (CFT)

Each Concrete Foundation Template (CFT) weighs about 5700 tonnes in air, and 3500 tonnes submerged (unballasted). The structures penetrated 12m into the seabed, partially by selfweight and partially by suction force created by pumping water out of the skirt chambers. The top compartments of each CFT were filled with 3500 tonnes of heavy ballast (iron ore and olivine), giving a total submerged weight of 7000 tonnes. This equals the average static pull from the TLP. In the 100 year design storm, the tension on each CFT varies between 0 and 14200 tonnes.

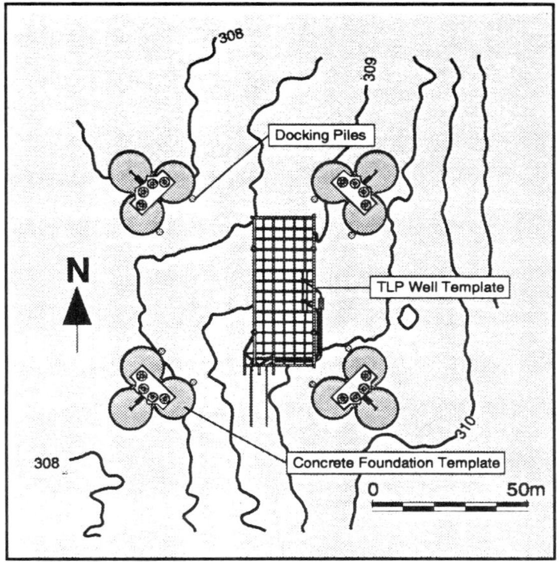

Fig. 6. Field layout Snorre TLP area.

Thus, accounting for the CFT weight, the average long term loads experienced by the soil is zero. During a storm, the TLP will offset some 8.5 degrees from the foundations which results in a moment on the CFTs which gives the most critical load situation. The criteria for soil material coefficient was set to 2.0 for the 100 year design wave condition. A high level of safety was applied because of the novelty of the structure, and consequence of failure. It may be noted that the concept design made it possible to increase safety level with only moderate cost impact by slightly increasing skirt depth.

5. SPS Foundation

Having selected the foundation concept for the TLP, the conceptual design of the Snorre Subsea Satellite Production System (SPS) was advancing. The SPS, being 48m long and 32m wide and weighs 2500 tonnes in air, would be worlds largest subsea installation (Reference 3).

The SPS foundation system would need to support the structure on the very soft soils and to withstand trawl impact loads. Further, a levelling system was required to bring the structure within the level tolerance of ± 0.5 degrees. The foundation would conventionally comprise mudmats to support the structure in a temporary phase, piles to support the structure in a permanent phase, and mechanical jackets connected to the piles to level the structure.

The experience with the foundation design of the TLP was now considered for the SPS. It was concluded that all the design requirements would be fulfilled by

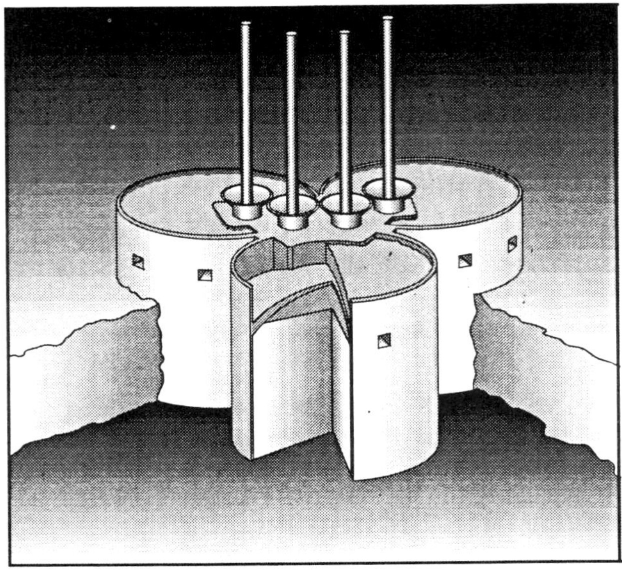

Fig. 7. Artist view of one of the four Snorre TLP foundations (CFT).

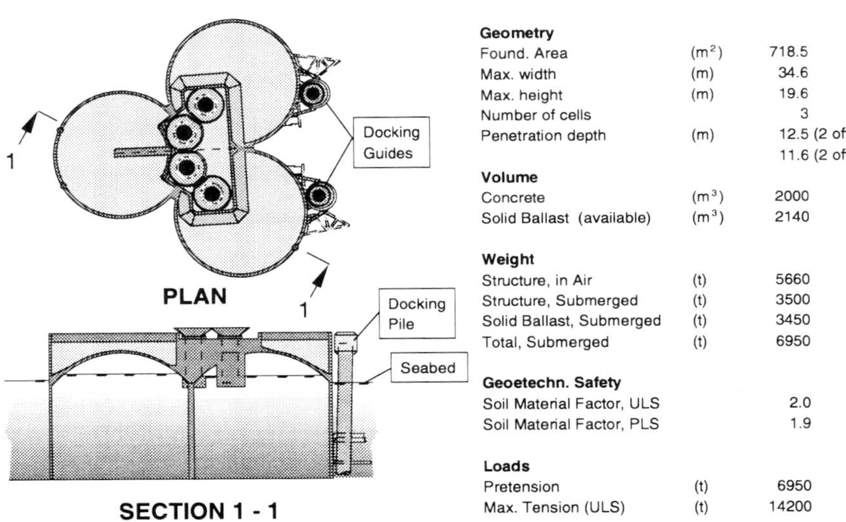

Geometry		
Found. Area	(m^2)	718.5
Max. width	(m)	34.6
Max. height	(m)	19.6
Number of cells		3
Penetration depth	(m)	12.5 (2 off)
		11.6 (2 off)
Volume		
Concrete	(m^3)	2000
Solid Ballast (available)	(m^3)	2140
Weight		
Structure, in Air	(t)	5660
Structure, Submerged	(t)	3500
Solid Ballast, Submerged	(t)	3450
Total, Submerged	(t)	6950
Geoetechn. Safety		
Soil Material Factor, ULS		2.0
Soil Material Factor, PLS		1.9
Loads		
Pretension	(t)	6950
Max. Tension (ULS)	(t)	14200

Fig. 8. CFT Geometry.

adding 3m skirts to the mudmat areas such that each end of the SPS foundation comprised 3 water tight skirt chambers with a total foundation area of $720m^2$. Butterfly valves (72 inch) ensured quick escape of air and water from the skirt chambers when the SPS was lowered through the splash zone, and when penetrated

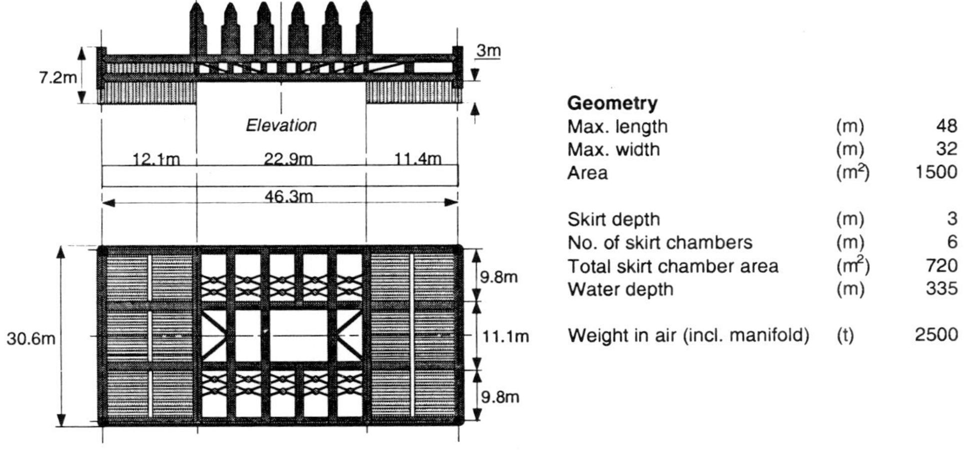

Geometry		
Max. length	(m)	48
Max. width	(m)	32
Area	(m^2)	1500
Skirt depth	(m)	3
No. of skirt chambers	(m)	6
Total skirt chamber area	(m^2)	720
Water depth	(m)	335
Weight in air (incl. manifold)	(t)	2500

Fig. 9. The Snorre Subsea Satellite Production System (SPS).

into the seabed. The SPS penetrates the seabed fully by its own weight with large margins on skirt resistance. If this had not been the case, additional penetration force could have been achieved by applying suction to the skirt chambers.

A water pressure control system (operated by an ROV) made it possible to adjust the level of the Template when installed. A low end of the SPS would simply be raised by pumping water into the relevant skirt chambers. The "water tight" clay at the site would ensure the SPS to be kept in level for several months. The permanent foundation principle is to secure the structure to the 10 drilling well conductors. The expected skirt chamber pressures required to raise the structure were relatively small, only around 2.5m water head, with a design value of 7m.

6. Model Tests

Because of the novelty of the foundation concepts, it was decided to verify the analyses methods with a series of model tests.

A 1:13 scale model of a TLP gravity foundation was selected for field tests in an onshore site with similar soil conditions as that at the Snorre field. The tests programme, contracted to the Norwegian Geotechnical Institute (NGI) was undertaken in 1989. The description and results of the tests are thoroughly described in References 4 and 5.

Four tests were performed relating to the CFT behaviour. First, one static "pull-

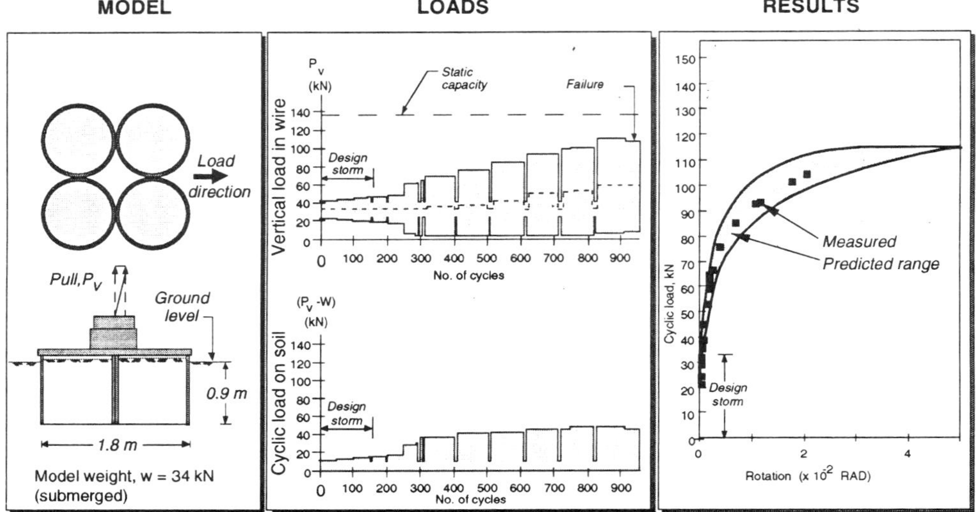

Fig. 10. Snorre CFT model test.

out" test was done to get a "reference" capacity. Three dynamic tests simulating storm conditions of a prototype structure were then performed. Figure 10 illustrates the loads, observed capacities and predicted capacities of one of these tests.

As can be seen from Figure 10, the observed behaviour was remarkable consistent with the predictions giving good confidence in the analyses methods used. Figure 11 illustrates one of the tests after having been brought to failure.

The same model was used to investigate various aspects of the SPS levelling system design such as:

- Controlling level of the model by pumping water into/out of various skirt chambers.
- Checking capacity of soil to withstand excessive water pressures.
- Investigate settlements with time when the model was resting on a water column.

These tests confirmed the validity of "state-of-the-art" methods for estimating fracturing pressures and seepage rates.

Fig. 11. Model after failure.

7. Installations

7.1. TLP Concrete Foundation Templates (CFTs)

The four CFTs were successfully installed in the field during June 1991.

Each CFT, weighing some 5700 tonnes in air, was lowered to seabed using the heavy lift crane barge DB102. Preinstalled docking piles ensured the correct positioning of each CFT. The CFT docking guides were lowered onto the docking piles prior to seabed touch-down. The docking guides were released from the CFT structures, being left on top of the docking piles, thus avoiding any possible open channels which could be the result if the docking guides had followed the penetrating CFT structure. After seabed touch-down water was evacuated in a controlled manner at a rate which gave a penetration of about 1m/hr. At 2m penetration all the CFT weight could be supported by the soil. The exact level of the structure was achieved by adjusting water flow from each of the three skirt compartments.

Judging from the various friction tests performed on the site during previous soil investigations, installation of the CFT docking piles, and from experience with similar clays (References 1, 2, 7) expected selfweight penetration of the CFTs was 10m using an average soil/skirt friction factor of 0.2. This was exactly what was

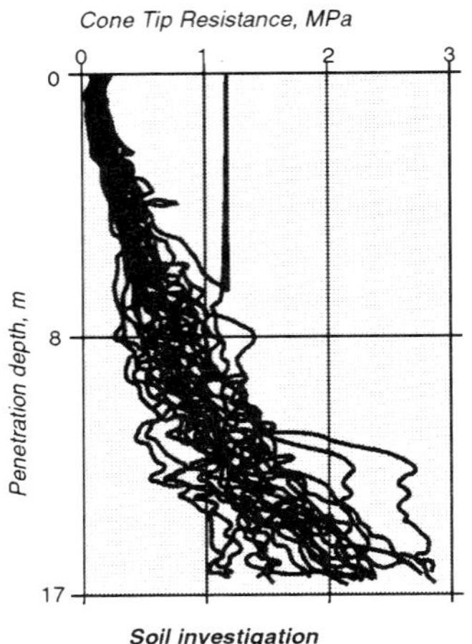

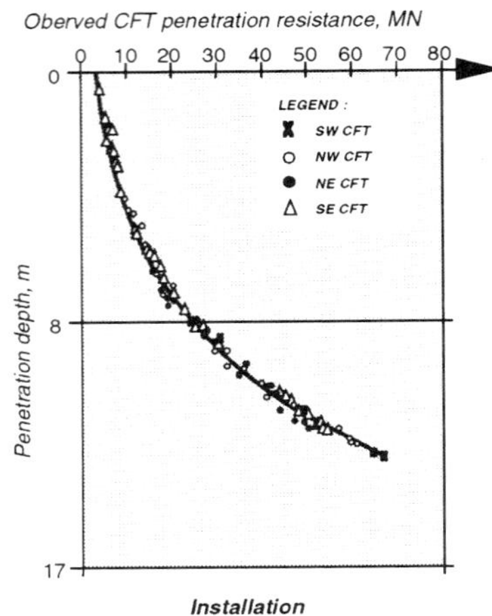

Fig. 12. CPT results and observed CFT penetration resistance.

observed.

Cone Penetrometer Tests (CPTs) performed over the whole TLP foundation area indicated a significant scatter in the data. From these tests soil/skirt friction factors were judged to lie between 0.1 and 0.3 with 0.2 as an average.

It is interesting to see that all four CFTs experienced the same penetration resistance, not reflecting any of the scatter as indicated by the CPTs. It is likely that very local variations in soil properties are reflected by the CPTs, however not by the far larger CFTs. Some of the variations in the CPT results can also be explained as instrument inaccuracies considering the very soft clay.

7.2. Subsea Satellite Production System (SPS)

The 2500 tonnes SPS was lifted off the heavy lift barge M7000 17th May 1992. The structure was completed installed on the seabed within a level position after some 7 hours! The skirt system design, avoiding a lengthy piling operation, made this short installation time possible. The SPS was sitting perfectly on the seabed, with each skirt chamber in full soil contact, within a level of 0.34 degrees. As this was well within the level tolerance, there was no need to perform levelling operation by pumping in water. However, if levelling had been required, this operation could have been done within very few hours.

8. Foundations on Next Developments in Block 34/7

The next developments within Block 34/7 include the Tordis and Vigdis fields. The Tordis field is located in the "Slope" area at 210m water depth with the upper

Fig. 13. SPS being lowered through the splash zone.

soil layer comprising a very soft clay. The development includes 5 single subsea production wells and 2 water injection wells connected to a central subsea manifold structure (Figure 14). The production wellheads and the manifold structure will be equipped with a protection structure with a skirted foundation using the same principles as used for the Snorre SPS.

A new concept for wellhead protection is introduced to the North Sea with one of the water injection wellheads being placed in a subseabed silo structure (Figure 15). The silo concept (Reference 8) involves an efficient protection of the wellhead which almost eliminates impact on or from the fishing activities. Further, the costs for construction and installation of a subseabed silo is less than for a conventional protection structure. Subseabed silos have been used offshore Canada to protect exploration wellheads from drifting icebergs (Reference 8).

The Vigdis field, located between the Snorre and Tordis fields, is planned to be developed with the wellheads protected within subseabed silo structures, highly reducing impact on or from the fishing activities in the area.

9. Conclusions

The development of the Snorre field, being the first true North Sea deepwater development, has resulted in many new technological achievements. Amongst these are the new cost-effective and simple foundation systems for anchoring of floating structures, foundations and levelling systems of subsea units, and the introduction of the subseabed silo wellhead protection structures to the North Sea.

Adaption of this technology to the future offshore development worldwide will

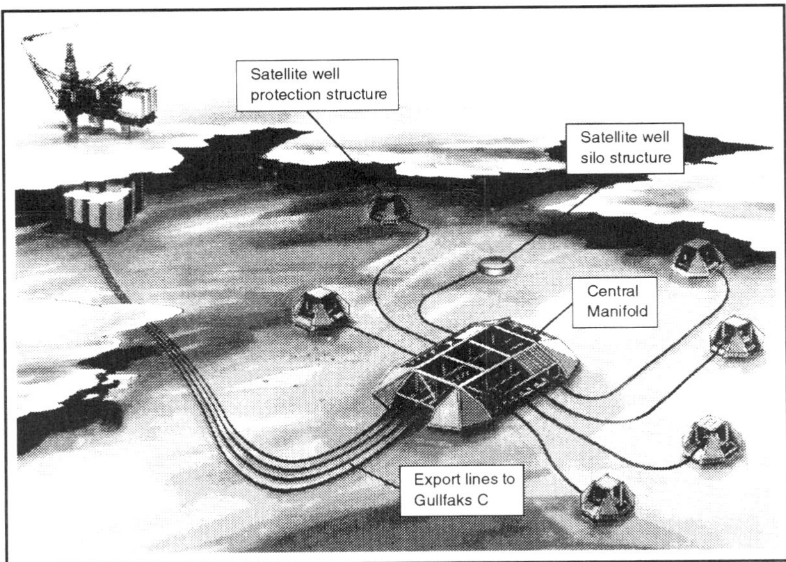

Fig. 14. Tordis development layout.

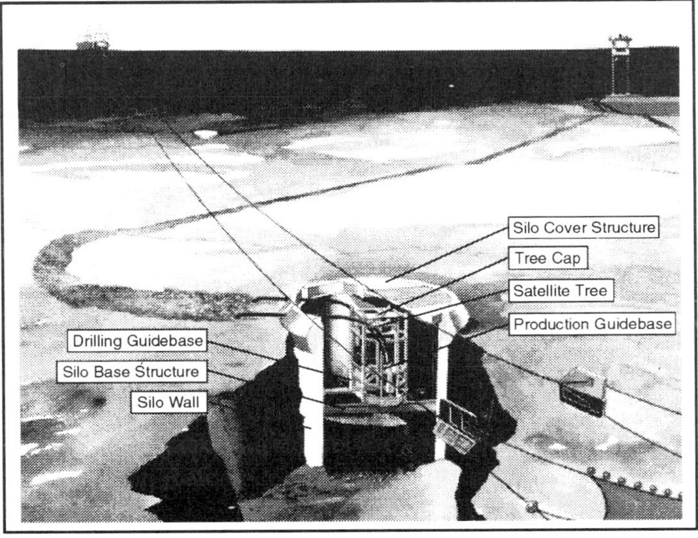

Fig. 15. Tordis subseabed silo protection structure.

reduce development costs and improve foundation safety. Requirements to removal can more easily be met at lower cost with skirted structures.

Conflict of interests with fishing industry will be reduced with the protection of wellheads in subseabed silos.

References

1. Andreasson, B., Christophersen, H. P., and Kvalstad, T. J. (1988), 'Field model tests and analyses of suction installed long-skirted foundations', *Proc. Behaviour of Offshore Structures Conference*, Trondheim.
2. Karlsrud, K., Nowacki, F., and Kalsnes, B. (1993), 'Response of piles in soft clay and silt deposits to static and cyclic loading based on recent instrumented pile load tests', *Proc. Int. Conf. on Offshore Site Investigation and Foundation Behaviour*, Society of Underwater Technology, London.
3. Childers, T. W., Øvergaard, I., and Helle, E. (1991), 'Snorre Subseabed Production System', *Proc. Offshore Technology Conference*, OTC 6625.
4. Dyvik, R., Andersen, K. H., Hansen, S. B., and Christophersen, H. P. (1992), 'Field tests of anchors in clay – Part I: Description', *Journal of Geotechnical Engineering, ASCE*, submitted for publication.
5. Andersen, K. H., Dyvik, R., Schrøder, K., Hansteen, O. E., and Bysveen, S. (1992), 'Field tests of anchors in clay – Part II', *Journal of Geotechnical Engineering, ASCE*, submitted for publication.
6. Andersen, K. H. 1992), 'Bearing capacity analyses of suction anchors for tension leg platforms', *Proc. Behaviour of Offshore Structures Conference*, London.
7. Tjelta, T. J., Guttormsen, T. R., and Hermstad, J. (1986), 'Large-scale penetration test at a deepwater site', *Proc. Offshore Technology Conference*, Houston, Vol. 1, pp. 201–202.
8. Meadows, B. W. and Gilbert, D. C. (1989), 'Drilling and installing large diameter caissons for wellhead protection', *Proc. Offshore Technology Conference*, OTC 6628,.
9. Jonsrud, R. and Finnesand, G. (1992), 'Instrumentation for monitoring the installation and performance of the concrete foundation templates for the Snorre Tension Leg Platform', *Proc. 6th Int. Conf. on Behaviour of Offshore Structures (BOSS)*, London.

Discussion

Question from R. Hobbs, Lloyd's Register:

1. In calculating the pile friction in Figure 4 using API RP2A procedures, what shear strength was used to define S_u? The API procedure is used, in U.K. North Sea practice, with unconsolidated undrained triaxial test data.
2. How did the sleeve friction in cone penetrometer tests compare with the frictions in Figure 4?

Author's response:

1. In calculating pile friction in Figure 4 using API RP2A procedures, the shear strength used was as that given in Figure 3 which is based on Unconsolidated Undrained (UU) triaxial shear strength tests.

2. There was some scatter in Cone Penetrometer (PCPT) sleeve friction. The ratio of the PCPT sleeve friction to undrained shear strength were:

   0–8m depth 0.2–0.4

   8–17m depth 0.3–0.6

   When comparing these values with the frictions in Figure 4, they would fall in between the results from the field test data and the 'state of the art' data sets.

SESSION 6

FOUNDATION PERFORMANCE MONITORING

FOUNDATION BEHAVIOUR OF GULLFAKS C

TOR INGE TJELTA
STATOIL, P.O. Box 300, 4001 Stavanger, Norway

Abstract. The Gullfaks C platform is a heavy concrete gravity structure resting on soft soils. To resist the design storm loads both a solution with circular concrete walls (skirts) to great depth and active improvement of soils were necessary. This paper describes the solution and gives some results from a continuous monitoring programme of the structure and foundation.

1. The Gullfaks C Concrete Gravity Structure

The Gullfaks C concrete gravity base structure was installed in 218 m of water on a soft soil site in 1989. It is a combined production, drilling and quarter platform with a total topside weight of approximately 55000 tons. With a displacement of 1,5 million tons it is far the largest and heaviest offshore structure ever installed.

With the combination of a heavy structure and soft soils, Figure 3, this did not only call for a new foundation solution with skirt piles, Figure 1. It also necessitated an *active* foundation improvement scheme, i.e. accelerated consolidation of the foundation soils through a suction driven drainage system, see Figure 2 and References [1] and [2]. Since these features occurred for the first time, they each and all called for a monitoring programme through instrumentation.

The Norwegian Petroleum Directorate (NDP) regulations, and more specifically the guidelines (NS 3481) states that:

Structures shall be fitted with permanent instrumentation, i.a. in cases when

- *the safety or behaviour of the foundation is dependent on active operation, e.g. the use of drainage systems; data shall then be immediately accessible to the user;*

- *the foundation solution, the soil conditions or the load conditions differ substantially from those with which experience has been gained;*

- *there is a need for monitoring of the whole foundation with regard to penetration, settlement, tilt or other items.*

It is results from this instrumentation programme that are presented in this paper, together with some interesting observations on effective stress control, dynamic

Volume 28: Offshore Site Investigation and Foundation Behaviour, 451–467, 1993.

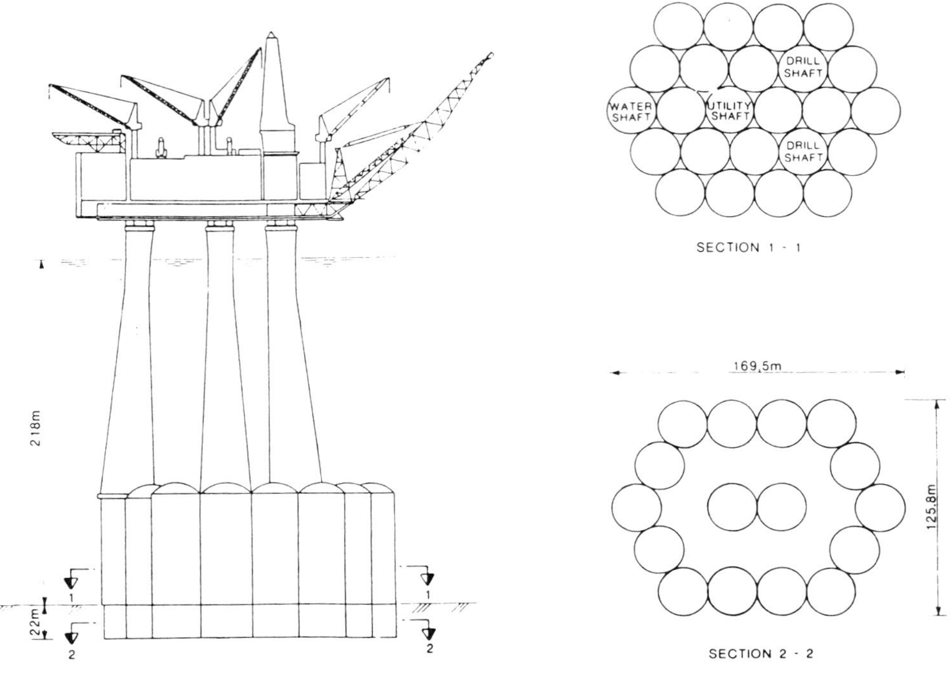

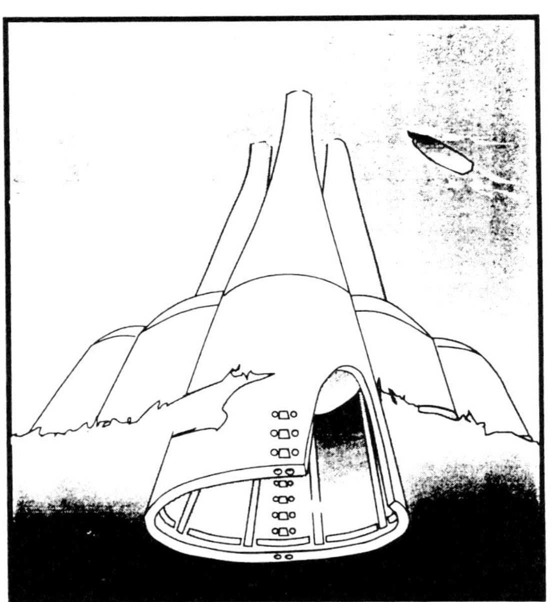

Fig. 1. The Gullfaks 'C' structure.

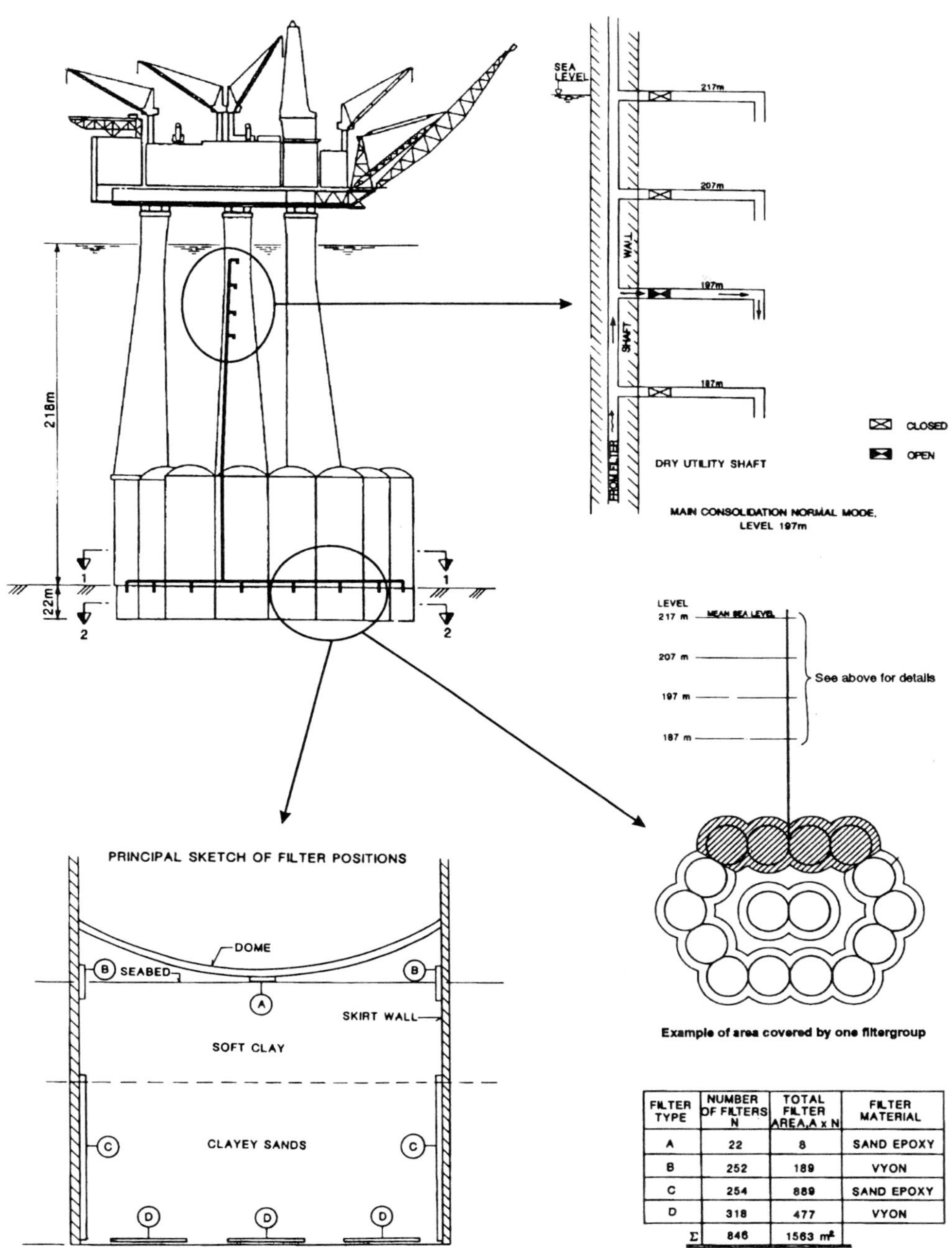

FILTER TYPE	NUMBER OF FILTERS N	TOTAL FILTER AREA, A x N	FILTER MATERIAL
A	22	8	SAND EPOXY
B	252	189	VYON
C	254	889	SAND EPOXY
D	318	477	VYON
Σ	846	1563 m²	

Fig. 2. Principles for the Gullfaks 'C' drainage system.

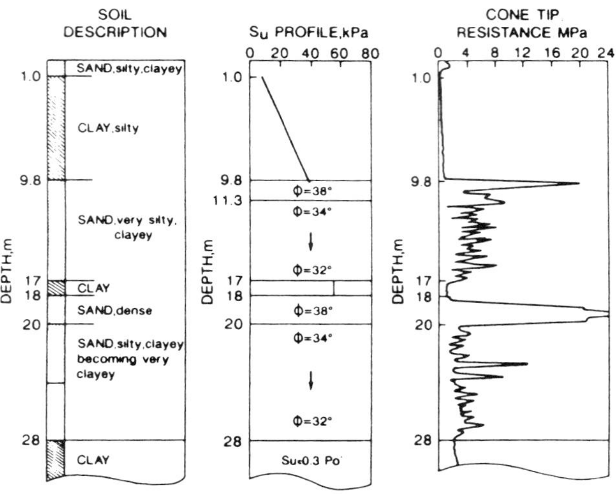

Fig. 3. Summary of soil conditions.

pore water pressures and displacements, long term settlements and load distribution between base and 22 meters long skirt piles.

2. Results from the SEM Instrumentation

The structural and environmental (SEM) instrumentation systems on the Gullfaks C platform provide three principal types of observations: oceanographic/meteorological measurements, data needed for operation, and load/response data. The first group includes measurements of wave data, tide, ocean current and standard meteorological data. The latter two groups included data needed both for installation of the structure, and performance monitoring data during the subsequent period of the platform life. This paper will concentrate on performance monitoring.

The type and relative location of different types of instruments are shown in Figure 4, and is further detailed in Reference [3], where also the data acquisition system is described. The SEM system comprises a total of 191 instruments.

Typical results from the instrumentation system are shown on Figure 5. Excess pore pressure as average values of 5 sensors in the depth interval 0–20 meters, 3 sensors at 23–25 meters and 8 sensors in the interval 28–40 meters. Figure 6 shows long term settlement, Figure 7 inclination and Figure 8 load distribution between base and skirts.

The significance of these results will be commented upon in the following sections. As can be seen, most of the plots are influenced by something that looks like variations or fluctuations in signals or in the measured parameter and consequently the lines in the figures are not as smooth as would be expected, e.g. for pore pressures at -40 meters level. This phenomenon is due to incomplete

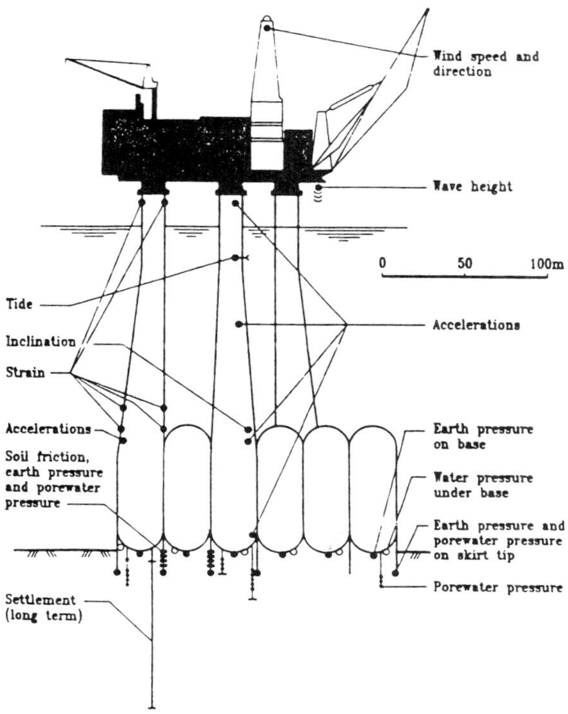

Fig. 4. The Gullfaks 'C' instrumentation for long term performance monitoring.

corrections for tidal variations, and is explained below.

3. Influence from Tidal Variations on Measurements

All earth pressure sensors and most of the pore pressure sensors are measuring pressures relative to absolute zero. This is an inconvenient reference level to work with for our profession, which normally tends to relate everything to a water table or ground surface. In this case the sea level is not a fixed level since waves, tide and atmospheric pressure are three amongst many other factors affecting this level. The sea bottom is then the only proper level to work with, also since the total pressure at 218 m water depth is significant and masks the pressure variations. The recorded pressures consequently have to be corrected for atmospheric pressure and water pressure at sea bottom in order to give reasonable and familiar values. The water pressure at sea bottom is recorded for this purpose.

However, if all pressure readings are corrected for water pressure at sea bottom, the basic assumption is that pressure variations due to tide and atmospheric conditions are the same at any depth and equal to what is measured at sea bottom. This assumption is based on the theory that tidal variations have a large areal extent and

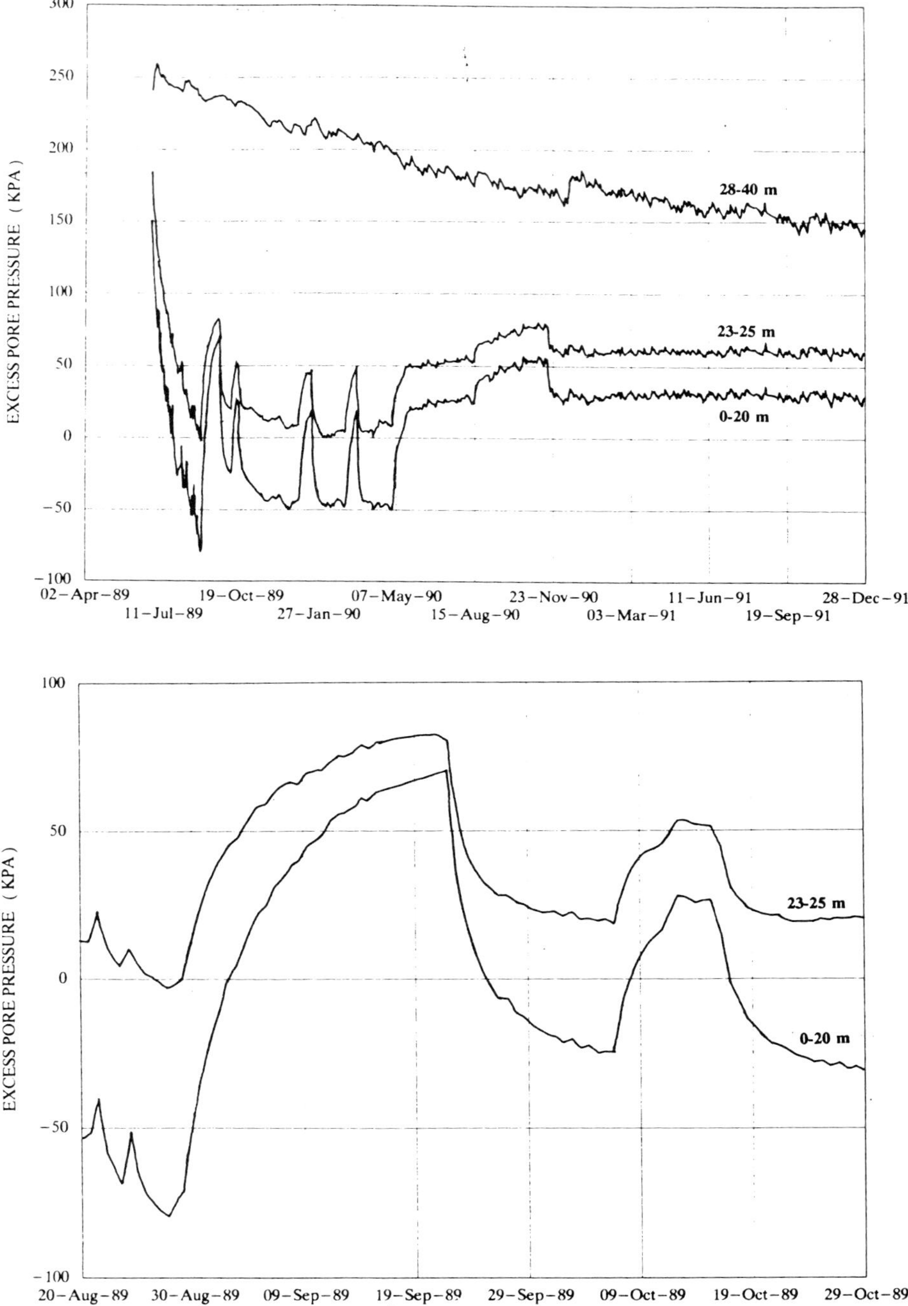

Fig. 5. Excess porewater pressures at various levels.

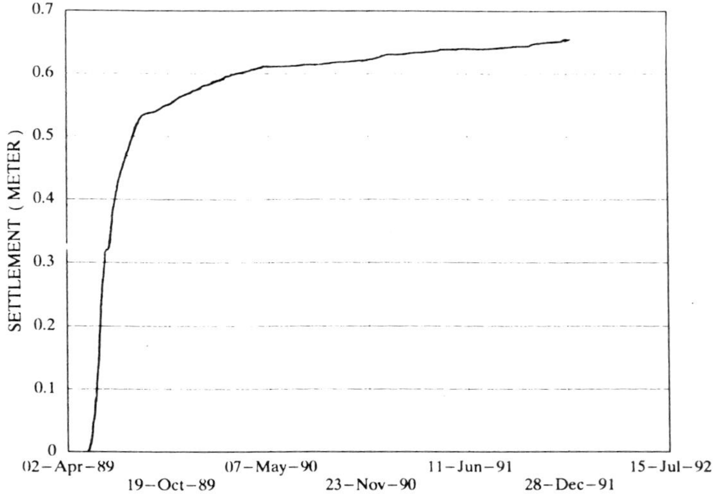

Fig. 6. Long term settlement.

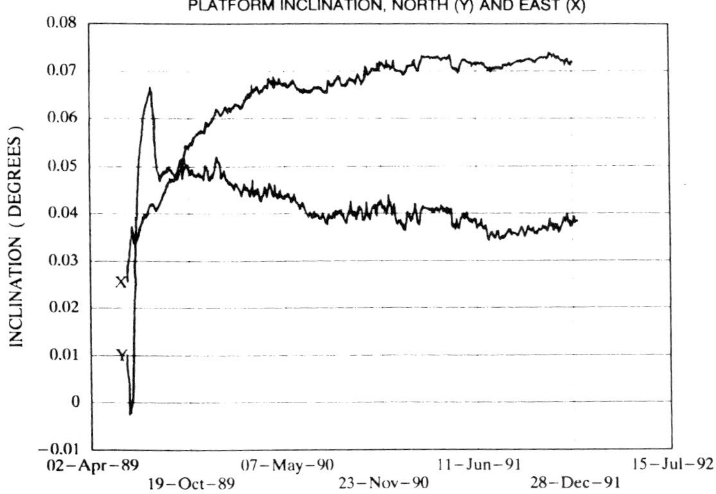

Fig. 7. Platform indication.

in a saturated sediment will only affect the total pressure and pore water pressure while the effective stresses will remain constant.

Figure 9 shows the total pressure variation at sea bottom and the pore pressure variations at 40 meters below sea bottom for the same recording period in July 1990. It is obvious from comparison between the two that pressure variations in the soil are less than at sea bottom. Figure 10 shows that the difference between

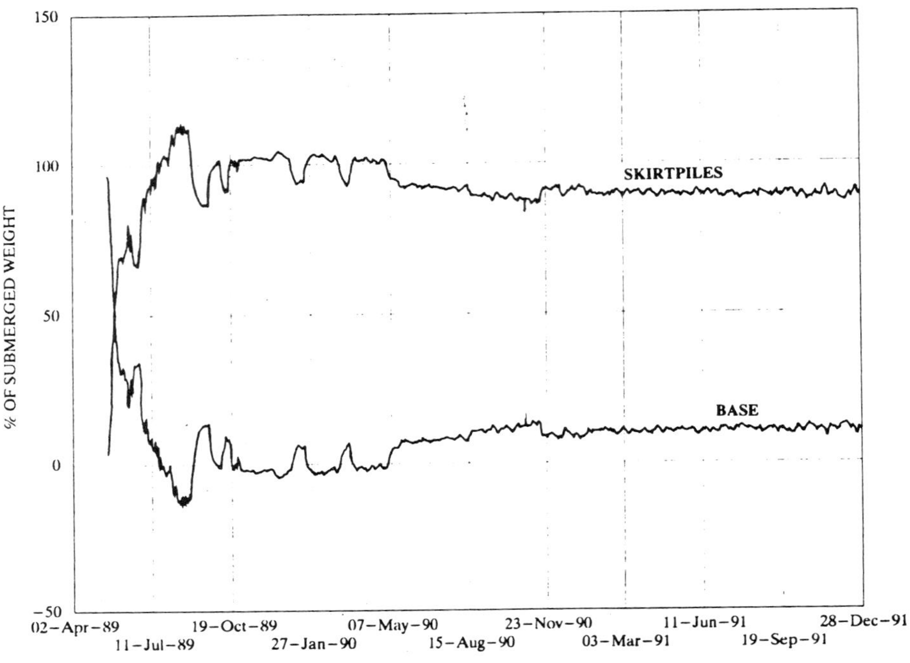

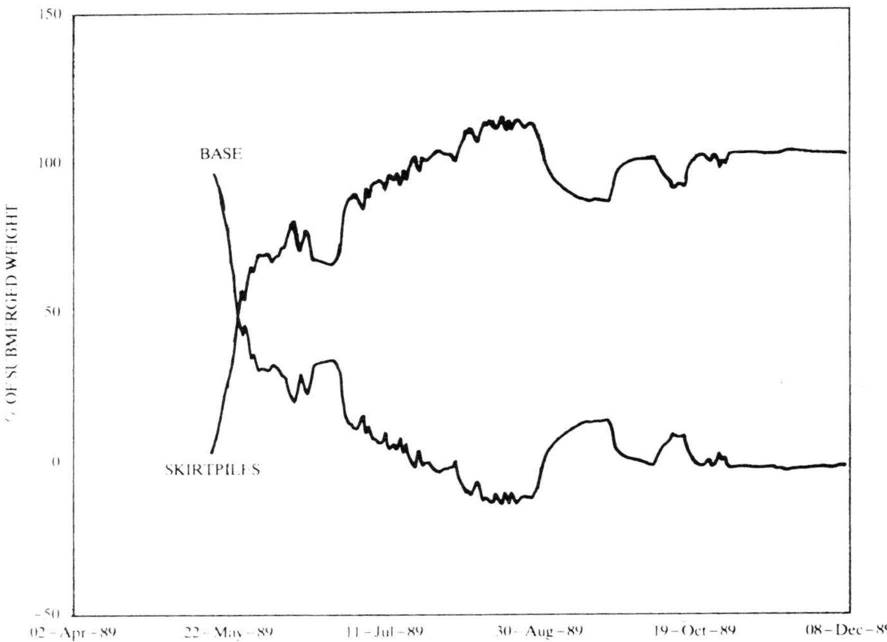

Fig. 8. Load distribution.

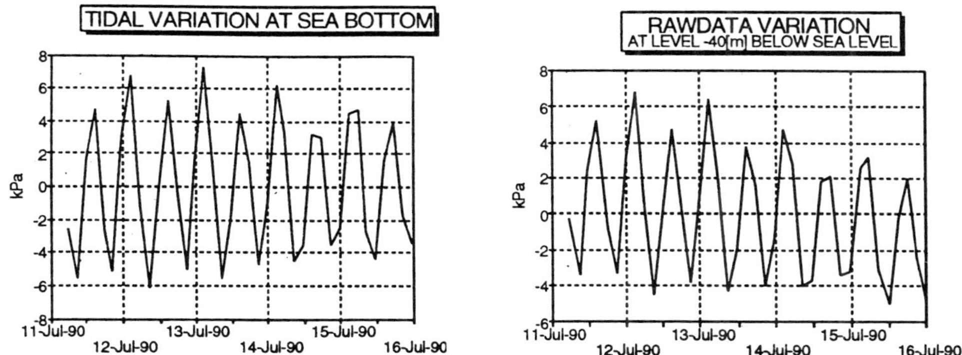

Fig. 9. Total pressure variations at sea bottom and 40 meters below sea bottom.

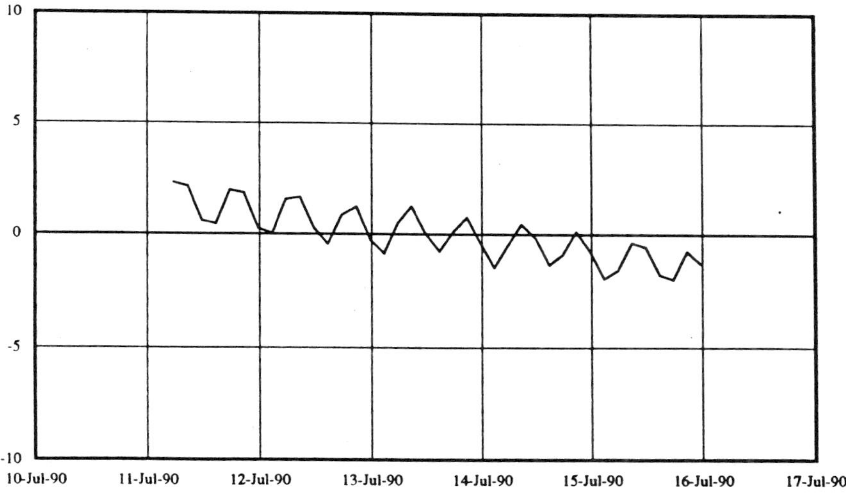

Fig. 10. Porepressure at -40 meters corrected for tide variation.

the two is approximately 80% or that the ratio between the load and the pore water pressure response is 0.8, i.e. $B_q = \Delta u/\Delta\sigma = 0.8$. This indicates that only 80% of the undrained load caused by tidal variation is carried by the pore pressure, and that the remaining 20% is carried as effective stresses in the soil skeleton.

The effect of this could easily be accounted for in the presented graphs, but the situation is even more complicated. Since the drainage system has in some periods been actively used or as for the last almost two years been open to hydrostatic level, this is a fixed level inside the utility shaft of the structure. Consequently the boundary condition at the skirt tip and skirt wall is a constant level not influenced by any environmental action like tide, atmospheric pressure, etc.

The immediate effect of this is that all pore water pressure and earth pressure readings are fluctuating, even when corrected for what was believed to be the cause for the fluctuations. Since most of the data presented here are based on one reading per day (average of a 20 minutes recording period), and the tide is out of phase with the 24 hours day, the graphs will appear as they do.

4. Pore Pressure Response below Platform, Static

The impression one gets by a quick glance at the pore water pressure history since platform installation 11 May 1989 and to the end of 1991 is that the platform and the soils have had a turbulent period the first year and a half. Since then almost nothing has happened.

This is a correct observation as far as the operation of the soil drainage system is concerned, Reference [3] and [4]. The first three months after platform installation, the soil drainage took place with an underpressure of 300 kPa. From August 1989 and through to May 1990 the skirt drainage was used at an underpressure of 100 kPa. In this period, the system was operated at hydrostatic level in several short intervals which is clearly observed in the immediate rise in excess pore water pressure, e.g. in January 1990.

The drainage system was also closed down completely at the end of August 1989, but since pore water pressure increase was higher than expected, drainage resumed three weeks later with a moderate underpressure of 100 kPa through all skirt drains. The base drains have all been closed since late August 1989, Reference [3].

Pore water pressures below 30 meters depth are little affected by drainage operations. The four sensor levels at 30, 33, 35 and 40 meters are all in clay and it is not expected either that any response to what happens with boundary conditions at skirt tip and wall would take place at these sensors. What did influence the pore pressures in clay below 30 meters was a storm in early December 1990. With a significant wave height of 13.6 m it accounted for approximately 80% of the 100 year design storm. During this storm an accumulated pore water pressure build up took place in the clay. The accumulated pore water pressures are 15 kPa as an average, but in some positions 20–25 kPa build up took place. It has taken this storm induced permanent pore water pressure up to one year to dissipate at the deepest locations. Several storms of almost similar magnitude have hit the platform since this early December storm in 1990, but with no such effect on pore water pressures.

It is of some interest to notice the very stiff response once a change in soil drainage level takes place. It took three months to bring down the very high pore water pressures due to platform installation, but it only takes a few days to change the pore water pressures up or down with almost similar magnitudes. Actually it only takes a few hours from operation of the drainage system to see a significant response in pore water pressure. The water flow that is involved is insignificant and demonstrates that pore water pressure at some remote point in the soil is dependent on pressures at the skirt wall where the drainage filters are mounted. This is an indication that the soil skeleton is very stiff in the "preconsolidated" sand layers.

The main observation from the almost three years with pore water pressure measurements underneath the platform is that operation of the soil drainage system is more important and have a larger effect on bearing capacity than anything else. The pore water pressure in the important sand layer below the skirt tip can be manipulated with as much as 200 kPa, and this clearly shows what a powerful tool such a soil drainage system is. This fact is further demonstrated in the next section.

5. Settlement

Settlement occurred primarily in the first three months of the platform life, i.e. when the most active soil drainage operation took place. Since then, only limited settlements have taken place. This was also aimed at in the design. Both pipelines tied in to a platform at sea bottom and all production and injection wells can only take a limited amount of settlement. By accelerating consolidation settlements in these first three months in the softest layers within the skirt compartments and immediately beneath the skirt tip several objectives were achieved;

- majority of settlements completed before wells and pipelines are in operation;
- consolidation and associated strength improvement obtained before the winter storms arrive;
- transfer of load from base to skirts and by this action actually reducing total expected settlements.

After the main drainage period, the level of underpressure on the system was kept at 100 kPa for almost one year. Settlement rate in this period was approximately 10 cm/year. For the last year and a half, the drainage system has not been actively used. The skirt drains have been open, but no active underpressure has been used. The settlements in this period have dropped to 1.5–2 cm/year. Considering this huge platform on relatively soft soils, this is quite remarkable and really shows the benefit of the skirt piles.

The measurements of settlements take place in several layers, Reference [3], and clearly indicate that settlements now only take place in the clay layers below 30 meters. This is further confirmed by the pore water pressure measurements.

6. Inclination

The inclination of the platform is for all practical purposes vertical, and seems to remain so. Experience from the very early drainage period shows that if required, the inclination of the platform can be influenced by operating the drainage system over only one half of the platform area. Again this is an indication of the power of the drainage system in combination with skirt piles.

7. Load Distribution Mechanism: Is Gullfaks C a Piled or a Gravity Based Structure?

Immediately after platform installation all weight was carried as contact pressure between platform base and soil. The effect of soil displacement when pushing solid concrete skirts into the soil and the following underbase grouting was that very high pore water pressures were set up inside the skirt compartments and below the skirt tip. Effective stresses at skirt wall were zero and no skirt wall friction could develop until excess pore water pressures had dissipated. During the first three months after platform installation, the soil drainage system brought these pore water pressures under control and the effect on the skirt piles bearing capacity was dramatic. They now carried 100% of the platform submerged weight, and the Gullfaks C is in no way different from a traditional piled platform, other than pile dimensions are larger: 16 huge 28 meters diameter concrete piles with 40 cm wall thickness, 22 meters long.

Again, when looking at Figure 8, it is the operation of the soil drainage system that dominates the behaviour of the platform. Load is shed back and forth between skirt piles and base depending on the level and mode of drainage operation. If an underpressure of 100 kPa is applied to the drainage system, all dead load is taken by the skirt piles. If the drainage system is passive but still open, i.e. no active underpressure, 90% of the dead load is still carried by the skirt piles with the remaining 10% now taken as contact pressure at the base. With the drainage system closed and given some time, the distribution between skirt piles and base is approximately 80/20.

The transfer of load from base to skirts or vice versa is similar to the change in pore water pressure every time the drainage conditions change. This is also very well documented on Figure 11 which shows effective earth pressure and soil friction, measured at four different levels on the skirt wall. (Total earth pressure and pore water pressure are measured to give effective pressure.) These measurements also indicate that the soil plug inside the skirt compartment at the top actually hangs on the skirt wall, possibly because the platform base (dome) push the upper soil downwards, and since the top 9 meters is a soft normally consolidated clay it is less susceptible to consolidation.

In total, the behaviour of the skirt piles has given rise to increased confidence in such foundations and have triggered development of skirt type foundations

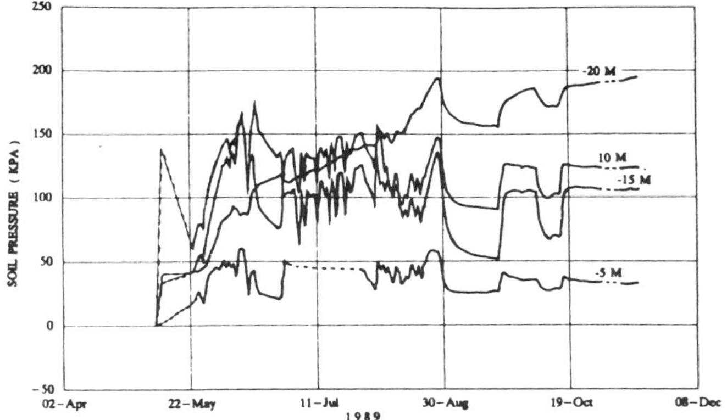

Fig. 11a. Effective earth pressure against the inner side of a skirt wall.

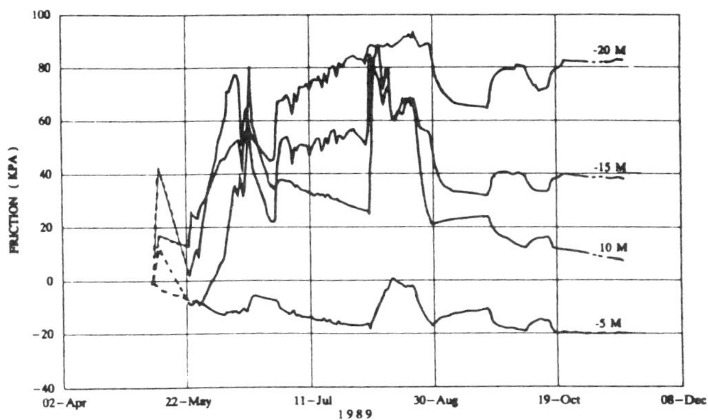

Fig. 11b. Measured soil friction at skirt wall (same position as Figure 11a).

for many applications. Traditional pile foundations for jackets are likely to meet competition from skirt foundations in the time to come, Reference [5].

8. Pore Pressure Response, Dynamic

Although this paper almost exclusively deals with static behaviour in the way that only statistical data have been presented so far, i.e. average values from a 20 minutes recording period, this section will be an exception. This will cover the dynamic pore water pressure response during a moderate storm, and compare this response with waves, accelerations in shaft and derived displacements both for wave dominated response as well as for the resonant response.

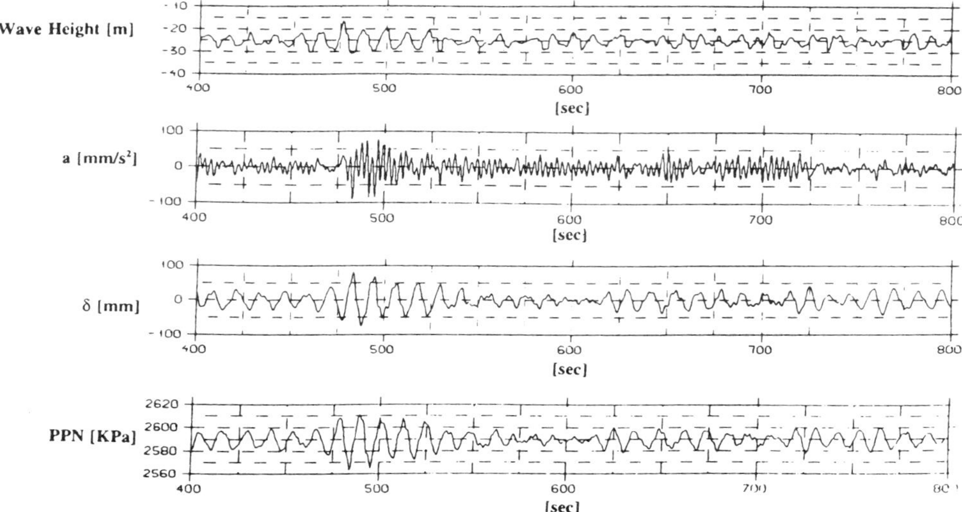

Fig. 12a. Time series, total response (wave dominated and resonant).

The purpose of this is to indicate what is possible to achieve from the instrumentation system and also to give an example of what benefits that can be the outcome of a good instrumentation system but which cannot be planned for since nobody was aware of it at the time of planning and design of the instrumentation. This is true for the so-called "ringing", a transient resonant vibration of a structure caused by an impact type of load.

Figure 12a shows four time series from a 20 minutes recording period in a moderate storm with a significant wave height of 7.4 meters. The data are from the period 400–800 seconds into the recording period. The interesting part to look at is the wave at 475 seconds, which is a steep, non-symmetrical wave. it apparently gives an impact load on the platform since high, resonant accelerations are set up. Also the derived displacements (by double integration of acceleration signals) are clearly visible, as are the dynamic pore water pressures. The pore water pressures are shown as raw data, i.e. show the total pressure at the depth of the sensor which is 28 meters below the sea bottom at the periphery of the platform base.

The accelerometer is located at the top of the utility shaft. The displacements of ±60–70 mm is caused by the wave dominated displacement of the structure.

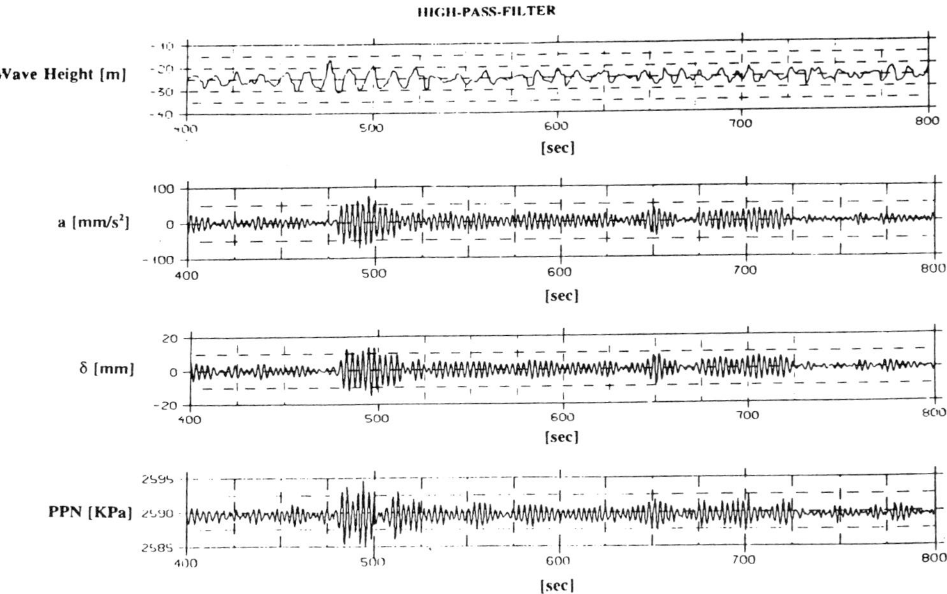

Fig. 12b. Time series, resonant response.

Figure 12b shows the resonant response of the structure and the reaction of the soil. These time series have a period of 2.7–3 seconds and are due to the natural periods of the platform. The acceleration, displacements and pore water pressures are high pass filtered, i.e. only signals with a frequency higher than 0.25 Hz are shown. This means that the quasi-static wave dominated reactions are filtered out, since they are already studied in Figure 12a, and we can look closer at the resonant part of the reactions.

It is now very clear what the wave at 475 seconds has created. A strong resonant signal is set up and is similar in form for accelerations, displacements and pore water pressures. The characteristic shape of this response signal is similar to an electric ringing signal, hence the name *ringing*. The remarkable observation from a geotechnical point of view is that a pressure sensor detects this effect in the pore water, and that the pore water pressure resembles the whole time series in shape and period from the acceleration and displacement.

The magnitude of displacement and associated forces in the shaft from this ringing effect is for the shown example increasing the quasi-static wave response with some 10%. This is also true for the pore water pressure. For a larger sea state this seems to diminish to a few percent for the Gullfaks C, and for the largest storm less than one percent increase.

9. Conclusions

The behaviour of the Gullfaks C structure is possible to study and learn from since an instrumentation system is monitoring the full scale behaviour. Only through measurements is real structural behaviour learned. The instrumentation system has provided:

- Database for design verification;
- Monitoring of soil strength improvement;
- Assistance in operation of the soil drainage system;
- Contributions to a better understanding of geotechnical and structural mechanisms;
- Information on mechanisms not specifically looked at in design, i.e. the effect of wave impact loads and the associated resonant behaviour (ringing).

The most important lesson learned through the period of three years monitoring is that we have an increased *confidence in active control of foundation behaviour*.

It is believed that the instrumentation will continue to increase our understanding of structural and foundation behaviour and will do so for a long time.

Acknowledgements

The author appreciate the opportunity given by Statoil to work with challenging projects like Gullfaks C. Both colleagues at Statoil and friends at NGI, GCG, NTH and Imperial College have given support in this project. Special thanks to Professor Lars Grande, NTH.

References

1. Tjelta, T. I., Aas, P. M., Hermestad, J., and Andenæs, E. (1990), 'The skirt piled Gullfaks C platform installation', *Proceedings of 22nd Annual OTC*, May 1990, Vol. 4, pp. 453–462.
2. Tjelta, T. I., Skotheim, A. Å., and Svanø, G. (1988), 'Foundation design for deepwater gravity base structure with long skirts on soft soils', *Proceedings International Conference on Behaviour of Offshore Structures, BOSS '88*, Trondheim, Vol. 1, pp. 173–192.
3. Myrvoll, F. (1992), 'Instrumentation of the skirt piled Gullfaks C platform for performance monitoring', *Proceedings International Conference on Behaviour of Offshore Structures, BOSS '92*, London.

4. Tjelta, T. I., Grande, L. O., Janbu, N. (1992), 'Observation on drainage control effects on Gullfaks C gravity structure', *Proceedings International Conference on Behaviour of Offshore Structures, BOSS '92*, London.
5. Tjelta, T. I. (1993), 'Novel foundation concept for jackets finding its place', *Proceedings International Conference on Offshore Site Investigation and Foundation Behaviour*, Society for Underwater Technology (SUT), London.

Discussion

Question from C. H. Price, Kent, UK: Can you give some idea of the cost of the instrumentation programme in relation to the project cost?

Author's response: The cost of the total instrumentation programme is approximately 1% of concrete structure or 0.2% of the total project cost.

FOUNDATION MONITORING ON THE HUTTON TENSION LEG PLATFORM

P. J. STOCK
CONOCO (U.K.) Limited, Rubislaw House, Anderson Drive, Aberdeen AB2 4AZ

R. JARDINE
Department of Civil Engineering, Imperial College, Imperial College Road, London SW7 2BU

and

W. MCINTOSH
CONOCO (U.K.) Limited, Rubislaw House, Anderson Drive, Aberdeen AB2 4AZ

Abstract. The Hutton tension leg platform (TLP) was installed in the North Sea in the summer of 1984. This structure was a world first in adopting the tension leg principle for a full scale floating oil production platform. The platform is held in place by 16 vertical tension legs, four at each corner, which are attached to foundation templates piled into the sea bed. The tension legs are held taut by the inherent buoyancy of the platform.

This type of loading places unique conditions on the foundations as they are required to work almost exclusively in cyclic tension. As design methods were unproven for these conditions in the early 1980s, higher factors of safety were adopted in their design than would have been used for the equivalent foundation under compression. A monitoring system was installed to identify any deformations of the foundation system.

This paper describes briefly the salient features of the monitoring system and summarises the data obtained between 1984 and 1991. Reference is also made to recent reanalyses of the piled foundation, which have been performed using a more advanced Effective Stress method.

1. Introduction

As numerous papers have been published on the various details of the Hutton TLP (Tetlow, Ellis and Mitra (1983), Bradshaw, Barton and McKenzie (1984) and others) it is only necessary to present a brief summary here.

The Hutton field lies 90 miles north east of the Shetland Islands in block 211/28 on the United Kingdom continental shelf (see Figure 1). The water depth at the site is approximately 145 m. The platform consists of six circular columns, connected at their bases by pontoons of rectangular cross section. The mass of the TLP is approximately 50000 t and its displacement is 615000 t. A general arrangement of the platform is shown in Figure 2.

The platform was installed in 1984. The hull and deck were fabricated in separate yards in the north of Scotland and mated in the Moray Firth in the summer of 1984. The platform was then towed complete to the field.

The tension leg platform principle is described graphically in Figure 3. The

Volume 28: Offshore Site Investigation and Foundation Behaviour, 469–491, 1993.

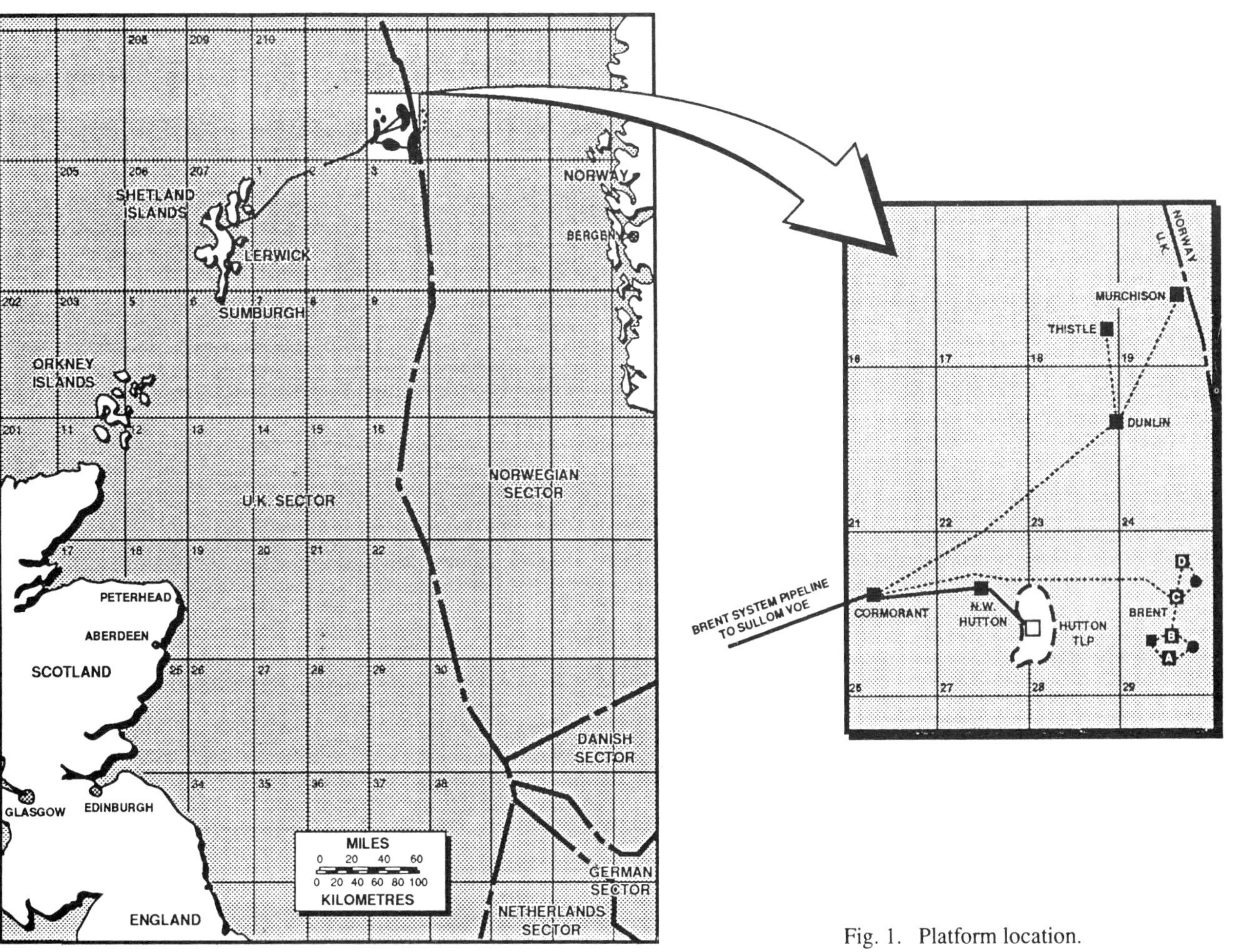

Fig. 1. Platform location.

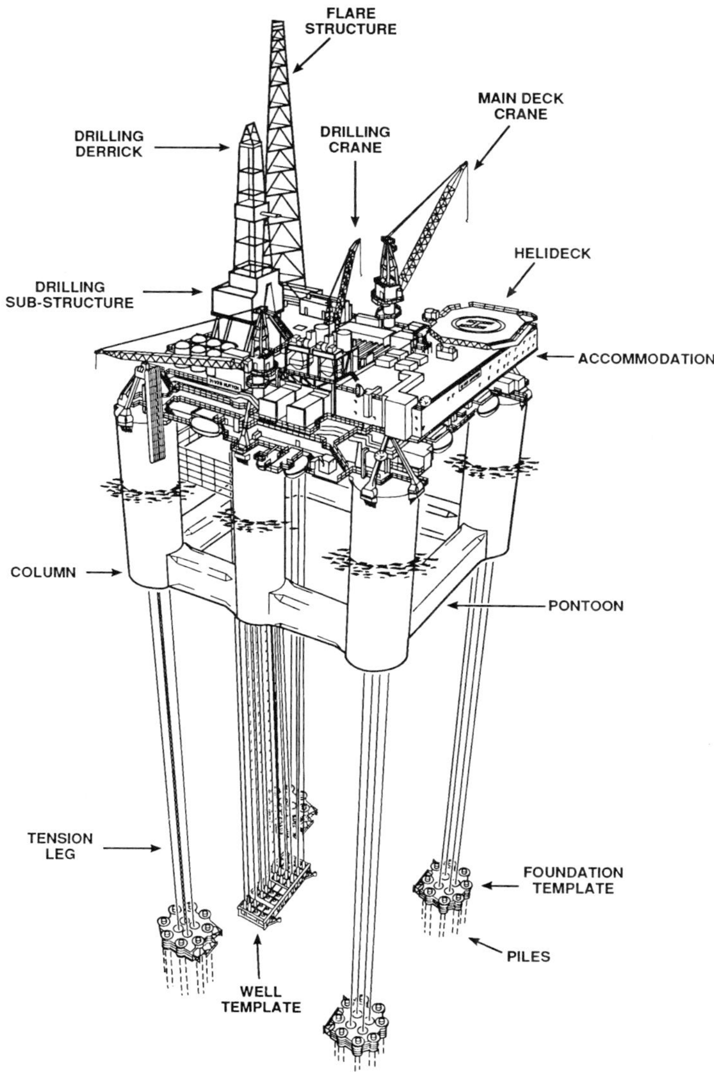

Fig. 2. General arrangement of Hutton TLP.

platform is designed to float, and was transported to the field in this condition. Once on station, a series of four tethers in each corner column were deployed and latched into the pre-positioned foundation templates. Once the first tether in each corner was engaged, the hull was deballasted to generate approximately 1000 tonnes of tension in each tether. The remaining three tethers in each corner were then deployed and latched and further deballasting occurred to complete the operation. These tethers are 260 mm diameter with a 70 mm diameter bore and

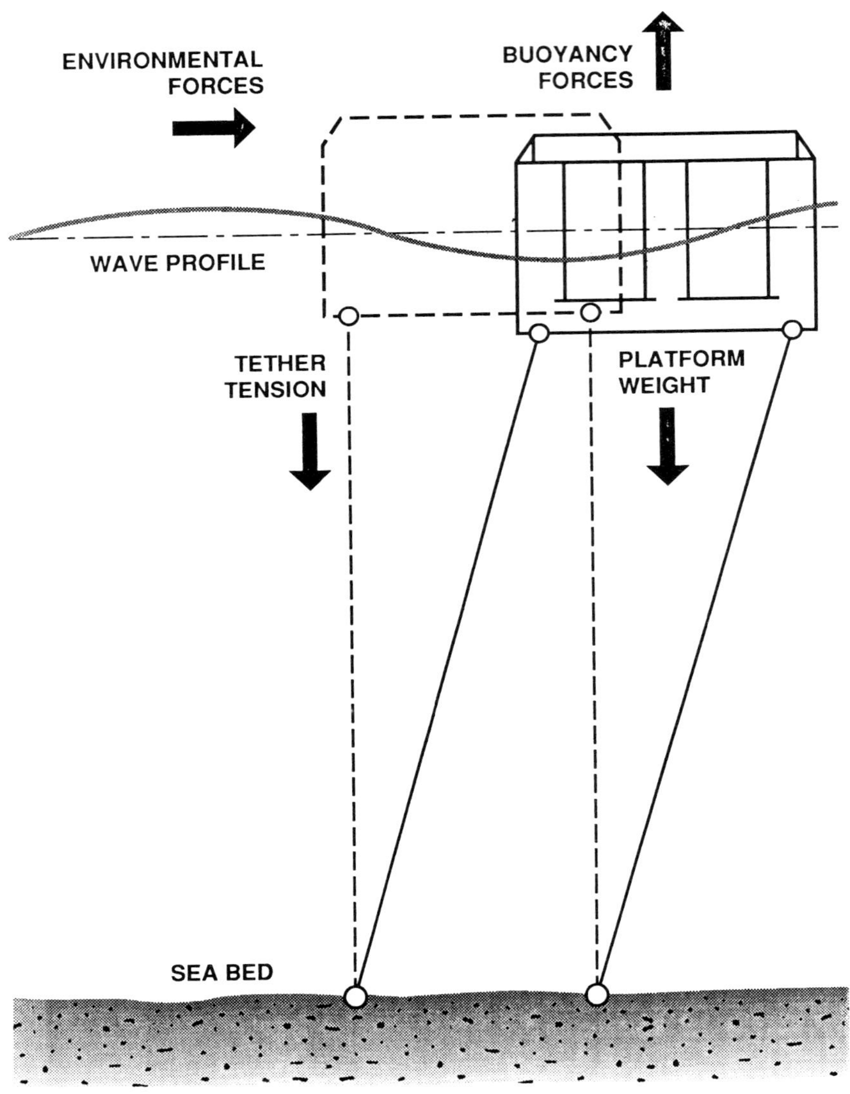

Fig. 3. Tension leg principle.

are made of a very high quality nickel chromium alloy steel. The tethers provide a very stiff anchorage to the platform and heave motion is virtually eliminated.

Lateral motion does occur under the action of waves, winds and currents. Lateral offsets of the order of 20 metres can occur under extreme weather conditions. To accommodate these motions, flexible bearings are fitted into each tether where it leaves the hull (Cross Load Bearing) and in the foundation template (Anchor Connector). The foundations provide reactions to the resulting tension and horizontal loads.

It is highly unusual to have to design pile groups to sustain the combination of a mean tensile load at all times and cyclic axial and horizontal forces variations about this mean, depending on the weather conditions. As there was little experience in designing for these conditions in the early 1980s, two approaches were used to ensure that the foundations would be adequate for the life of the platform. Firstly, more conservative assumptions than would be applied to a compression foundation were adopted. Secondly, and the topic of this paper, a foundation monitoring system was designed and installed to gain reliable knowledge and confidence in the long term performance of the foundations under these tensile loading conditions.

Subsequently, additional theoretical studies were undertaken to interpret the field data and up-date the foundation analysis.

2. Hutton TLP Soils and Foundations

Borehole and CPT site investigations performed by Fugro have shown that the Hutton site is underlain by a series of stiff glaciomarine tills and dense sands. Deposition probably took place during the past 10,000 to 100,000 years and, although variations in stratigraphy were noted between different boreholes, the first 100 metres of quaternary soils can be summarised by the simplified layering system and scheme of soil properties given in Figure 4.

The ground conditions at Hutton are typical of many locations in the Northern North Sea and, in particular, there are many similarities with those at the Magnus site (Rigden and Semple, 1983 and Jardine and Potts, 1992). The correspondence between the sites was fortunate as detailed studies of the Magnus soils had been carried out previously at Imperial College, which allowed parameters to be developed for the nonlinear finite element analyses described by Jardine and Potts (1988).

The layout of the foundations is given in Figure 5. There are four foundation templates, one under each of the four corner columns of the TLP. Each template has 8 piles, 1867 mm in diameter, with a wall thickness of 50 mm. These were driven to a typical penetration of 60 m (see Figure 6). A 23 m long pin pile was also installed for each template.

The foundation templates and piles were installed in the Hutton field in 1983, one year before the platform was towed into position.

3. Monitoring Systems

The Hutton TLP is fitted with two distinct monitoring systems, one for the foundations and one for the platform response.

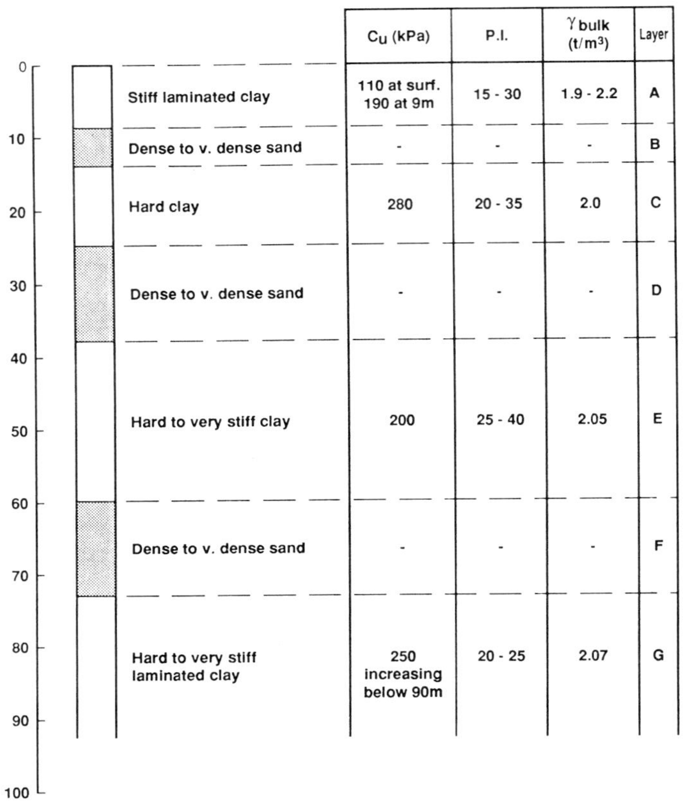

Fig. 4. Simplified 'typical' Hutton profile (after Jardine and Potts, 1988).

3.1. Foundation Monitoring System

The foundation gauge system is described in detail by Jardine *et al* (1985) and (1988) and shown schematically in Figure 7. The layout at the time of platform installation is shown in Figure 5. The gauges were designed specifically for the TLP project by Imperial College.

The gauges work like aneroid barometers; two flexible reservoirs (one containing oil, the other mercury) act as manometers or U tubes. One end of each tube is attached to a foundation template (the active pot) and the other to a pot on the sea bed (the passive pot). As the foundation deflects, the ends of the U tubes move relative to one another. A highly sensitive pressure transducer monitors the changes in pressure in the two U tube fluids, from which the displacement can be derived. The dual fluid systems allows the affects of tides, waves and barometric changes to be cancelled out.

There are two transducers on each gauge, giving fine and coarse ranges of

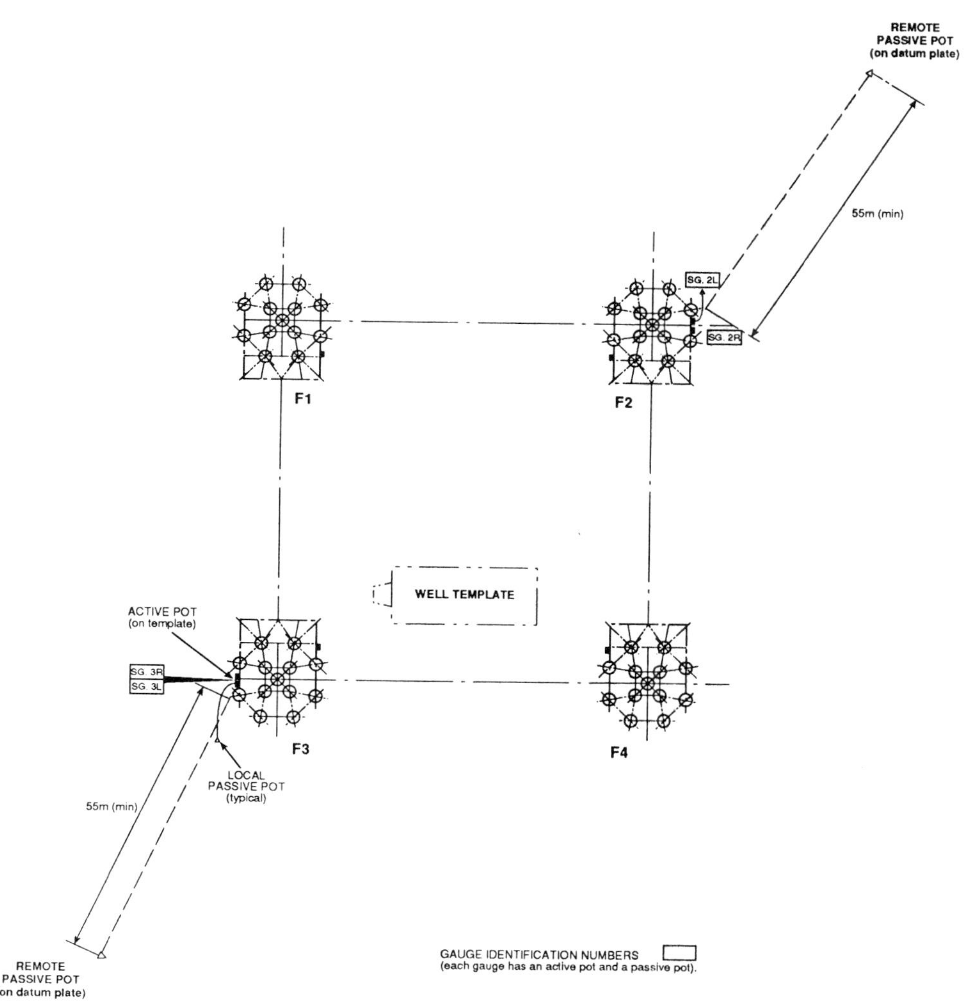

Fig. 5. Hutton seabed plan. Location of original settlement gauges.

±60 mm and ±300 mm respectively. Calibration tests performed in special hyperbaric vessels proved accuracies and resolutions of 0.1 mm and 0.03 mm for the course and fine transducer respectively.

The nature of the TLP precluded a conventional cable data transmission link between the sea bed and the platform. Therefore, an acoustic telemetry system, based on a micro processor slave system on the sea bed fitted with a long life battery, and a master on the under side of the hull, was developed in association

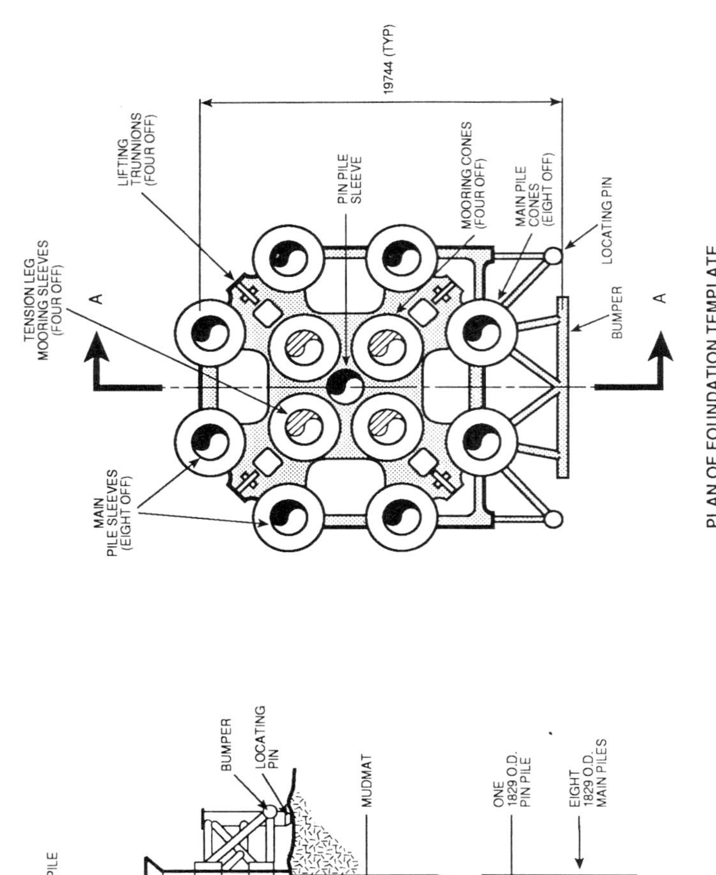

Fig. 6. Foundation template general arrangement.

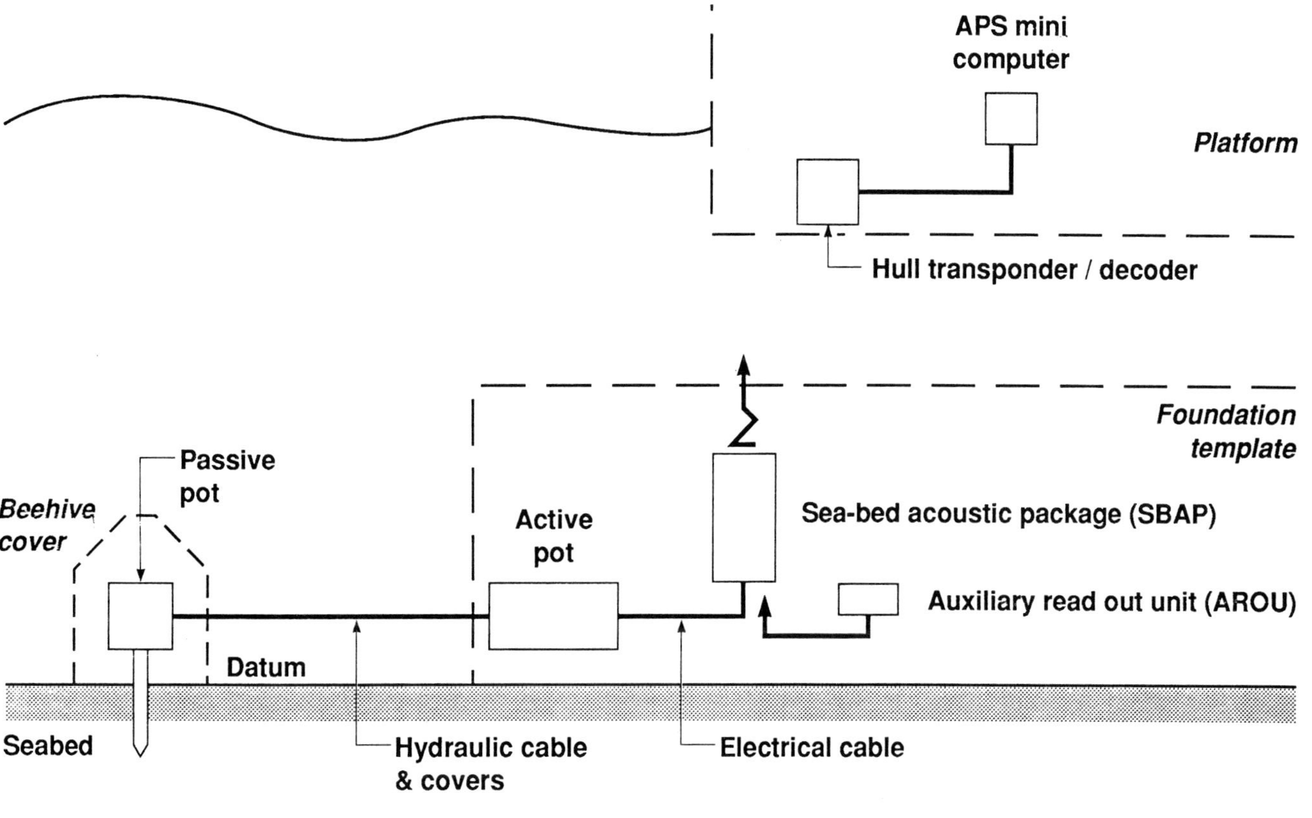

Fig. 7. Foundation settlements monitoring system.

with a specialist sub-contractor (Reference Jardine, Dore and McIntosh, 1985).

The settlement gauges can be read at rates of up to one measurement per second. They can also be read by divers operating an auxiliary underwater voltmeter. This proved to be highly valuable as the first design of telemetry system was not reliable in the long term. Replacement telemetry equipment produced by Sonardyne Ltd in 1986 has performed much more reliably. After several years in service, some of the settlement monitoring transducers have also shown drifts or malfunctions, but the gauges have self checking ability and faults can be identified relatively easily. It is estimated that the typical working life of a gauge is 5 years.

3.2. Platform Response Monitoring

The Hutton TLP is fitted with a very sophisticated platform monitoring system called the Performance Monitoring Verification (PMV) system. This system gathers over 100 channels of data at a collection rate of 1 hertz and can switch to a higher data collection frequency in heavy water.

Of particular interest to us is the monitoring of the tether tensions. Each tether is mounted on an array of four load cells and the variations of tether loading are continually recorded. Wave heights and platform offsets are also recorded.

4. Platform Installation

The platform installation presented a unique opportunity to monitor the performance of the foundations under the applications of known loadings. Each of the four tethers in each corner were installed in turn, and a known tension applied by the deballasting of the platform. The tensions were monitored as part of the on-board tether tension load monitoring system. An average load of 900 t per tether was finally achieved after the four day installation period.

During this period, the foundation displacements were taken at regular intervals (see Figure 8). The effects of the eccentricity of the loads applied to the template as individual tethers were installed and tensioned were accounted for in the interpretation. Accurate load-displacement plots were produced for pure axial loading (see Figure 9) and pile group moment–rotation stiffnesses were also calculated (see Table 1) (Reference Jardine, Hight and McIntosh, 1988).

The data gathered during the installation were very valuable for several reasons. The behaviour of the foundations was much stiffer (approximately 400%) than the conventional design calculations had predicted but was consistent with non-linear finite element predictions based on the Imperial College 'small-strain' approach (see Jardine and Potts, 1988). The load–displacement plot was only slightly non-linear, suggesting that most of the displacement will be recoverable at the sustained working load, and not prone to loss of load carrying capacity. Finally, the resolution aimed for when designing the gauges (± 0.03 mm) proved to be both achievable and necessary to monitor the small displacements experienced.

TABLE 1.
Rotational stiffnesses calculated for pile groups as deduced from installation measurements.

Gauge	Change in Eccentricity m	Change in Displacement at 3000 t total (-ve = uplift) mm	Rotational Stiffness $x10^6$ tm/rad
F1 Local	2.30 - 0.77	-0.28	146
	0.77 - 0.00	-0.13	150
F2 Remote	2.30 - 0.77	-0.35	112
	0.77 - 0.00	-0.21	93
F2 Local	2.30 - 0.77	-0.32	122
	0.77 - 0.00	-0.20	98

5. In-Service Foundation Performance

The foundation displacements have been recorded almost continually since platform installation. As noted earlier, there were some periods when data was not available due to telemetry malfunction.

Every year, tension data, wave height measurements and the settlement gauge readings are collated and sent to the Civil Engineering Department at Imperial College, London, for interpretation. The extraction and presentation of that data has evolved over the life of the platform and is now available in a PC spreadsheet manipulatable form.

The principal conclusion drawn from the long term monitoring is that the foundation movements are very small. Figure 10 shows a typical annual record for a local datum settlement gauge that had been positioned on Template F2. Features to note include:

1. The coarse and fine transducers give essentially parallel traces.

2. The template appeared to move up by 1 to 2 mm during October to January but then returns to its original level.

3. Storm loading had little or no effect on the regular weekly measurements.

4. There was no significant overall change in template elevation over the 12 month period.

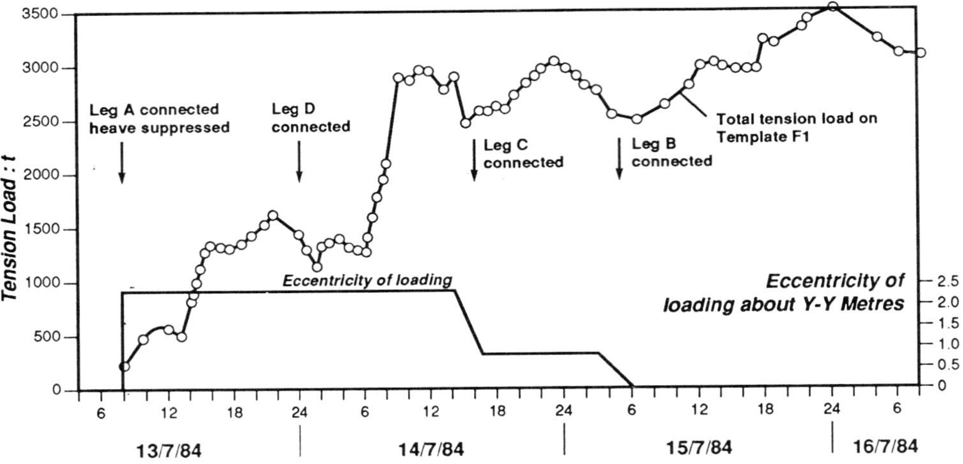

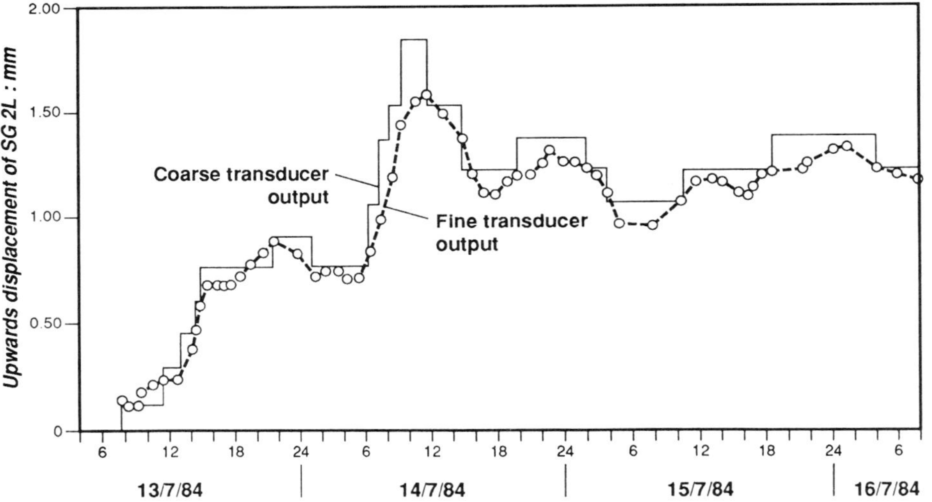

Fig. 8. Measurements taken during platform installation, July 1984.

The remote datum gauges tend to show larger apparent movements during winter than the local datum instruments. However, the true foundation movements are smaller than they appear to be because of the effects of temperature changes. Correlating the known seasonal changes in Northern North Sea sea bed temperature with temperature calibrations of the settlement gauge systems proves that almost all

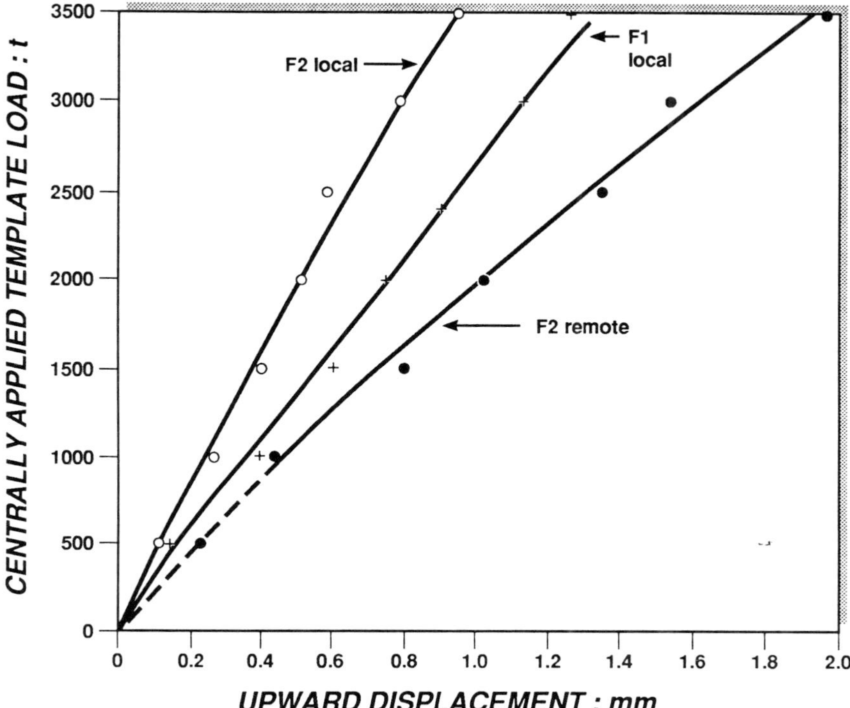

Fig. 9. Pile group axial load displacement curves deduced from measurements made during platform installation.

of the (very small) apparent movements can be ascribed to the effects of temperature on the settlement gauges themselves.

One way of eliminating temperature effects from the long term records is to plot the mid-summer readings alone. Figures 11 and 12 present these data as bounded scatter diagrams for the remote and local datum gauges respectively. Part of the scatter is due to some instruments being changed out after telemetry malfunctions. Instrument drift also contributes to the deviations between different gauges although the annual drift rates of working gauges are typically less than 0.5 mm per year.

Overall, there is no tendency for the locally measured upward displacement to increase with time: the remote gauge data suggest that global pull-outs might have increased by perhaps 2 mm over seven years. Comparing the measurements made before and after individual severe storms shows that the environmental loads experienced to date, including the extreme event of 12 December 1990 cause no measurable increase in permanent displacement.

As mentioned earlier, the settlement gauge system can be used to assess the dynamic foundation response during continuous storms. The data obtained by cor-

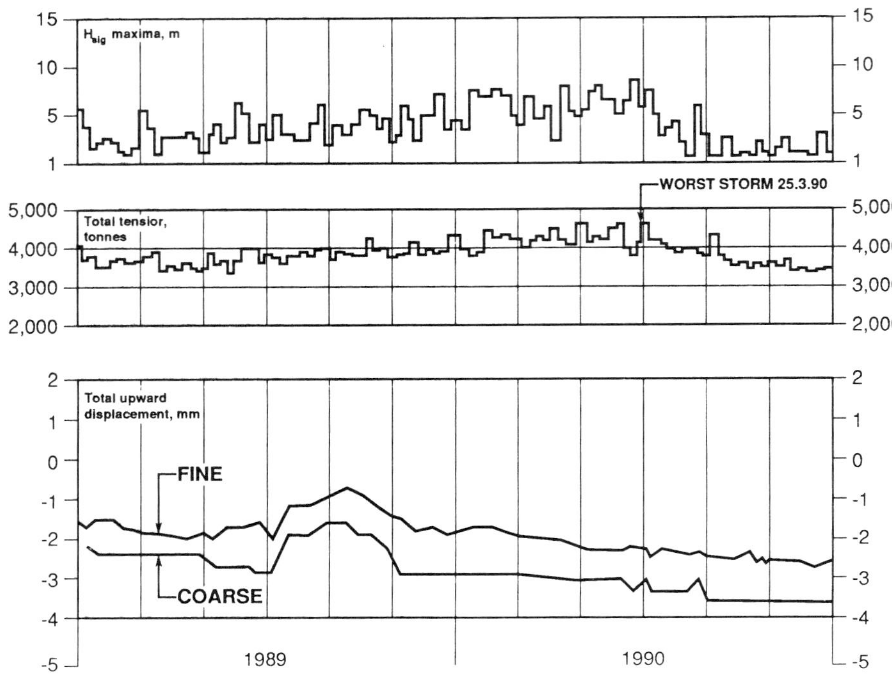

Fig. 10. Typical annual record of measurements made on local gauge 2L.

TABLE 2.
Load and deflection values for storm in November, 1984.

Gauge	Peak to trough Upward Displacement (mm)	Change in Tension (tonnes)	Equivalent Global Stiffness tonnes/mm
Local	0.16	475	1000/0.34

relating peak to trough load-cell measurements with the corresponding settlement gauge minima, maxima and standard deviations, obtained by taking 180 measurements over a three minute interval area are given in Table 2 for a storm which was monitored in November 1984.

Comparing this foundation stiffness with that measured during TLP installation, excellent agreement is noted. The slow dynamic response of the remote gauges makes them unsuitable for measuring the response to storm loading.

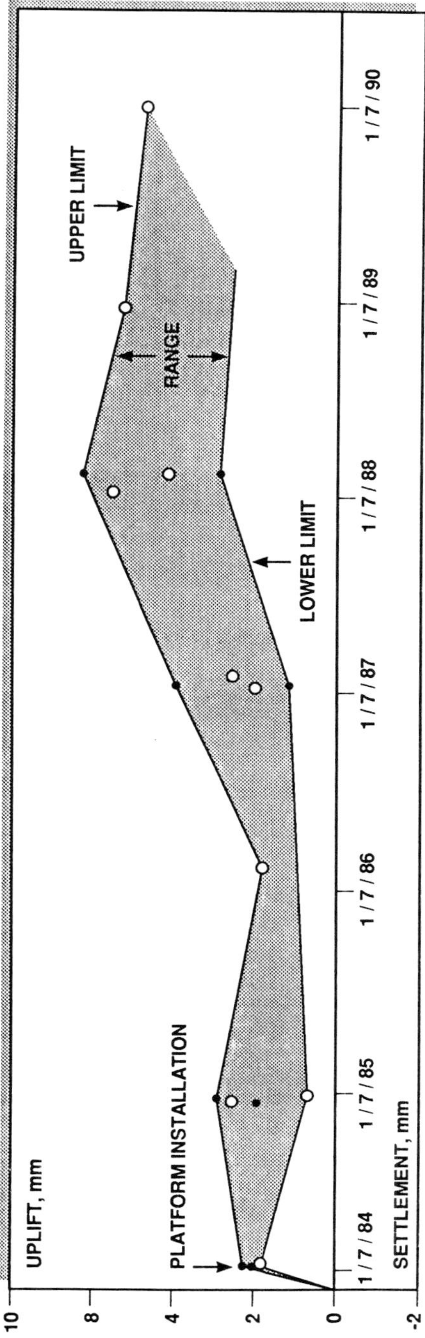

Fig. 11. Summary of data from remote gauges and mid Summer readings.

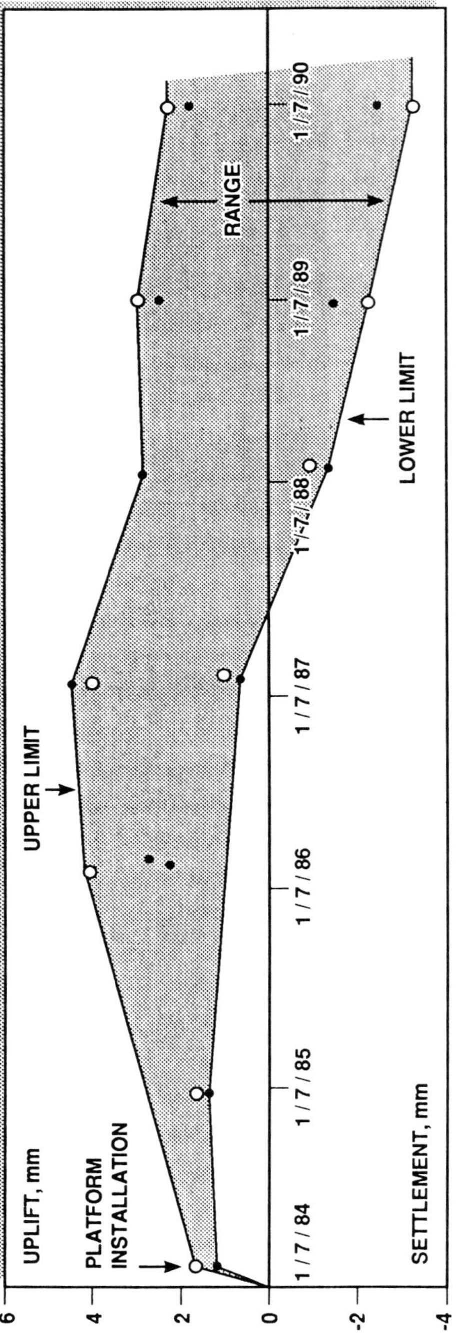

Fig. 12. Summary of data from local gauges: mid Summer readings.

6. Foundation Reassessment

6.1. Load Derivation

The performance verification monitoring carried out over the early years of the TLP's service life revealed that the platform movements and foundation loads developed by environmental loading were considerably smaller than the original design analyses had predicted. The original loads were derived from both model tests and early analytical methods. It was recognised that, at the time of the TLP design, the theoretical understanding and analytical capability to predict the performance of the platform was in its infancy; designing the platform was a pioneering activity and safe judgements had been made at each stage of the process.

Since that time, both theory and analysis methods have been developed. In 1989, it was possible to reanalyse the response of the TLP to environmental loadings, using the PMV results as benchmarks for known storms. These analyses confirmed that the TLP responds in a less extreme way than had been predicted at design. This study went on to demonstrate that the platform could carry larger deck loads, thus opening up opportunities for the tie-in of adjacent marginal fields.

As a spin-off from this analysis, several other components of the TLP were now open for reassessment. The performance of risers, tethers and foundations had also shown much lower forces and responses than predicted. All three of these components were given considerable attention during design and have been monitored as part of the PMV design. In each case, it has been possible to confirm safety factors that are larger than those demonstrated in design.

The maximum design foundation loads have been revised using data generated from two sources. The first source made use of the platform environmental response numerical model described above. For a 100 year return wave height of 27.5 metres, the analysis gave a total load in four tethers of approximately two thirds of the original design value of 9100 tonnes.

The second method of deriving the 100 year return foundation load was by performing a statistical analysis on the PMV records of the total column loads and extrapolating to 100 years. This also gave a total corner load of roughly 6000 tonnes. The 100 year return maxima derived by the two independent methods were within 2%, and gave considerable confidence in the load derivation method.

To determine the maximum pile loading, account was made of the weights of the tethers, foundation templates and piles plus accounting for a certification conditions of only 3 tethers present. This condition reflects the concern and uncertainty felt by both Conoco and the certifying authority in the reliability of the unproven tether system.

The new design maximum single pile load is now approximately 50% lower than the original single pile load used in design. The removal of one tether has the effect of increasing the eccentricity of the loading and hence generating an extra moment effect which is uncoupled into the piles and accounts for 60% of the above pile design load.

6.2. Static Foundation Assessment

The foundation assessment methods have also developed considerably in the intervening years, including the work at Imperial College on applying effective stress techniques to the analysis of load deflection behaviour. The developments have been described by Jardine and Christoulas (1991), Bond *et al* (1992) and others and will not be repeated here. However, the principle steps in the calculation methods developed for assessing pile shaft capacities in clays and sands are shown graphically in Figures 13 and 14 respectively. The Hutton foundations have been reanalysed by Imperial College with the objective of determining the current factor of safety against single pile failure.

The latest reanalysis followed the approach shown schematically in Figures 13 and 14, with parameters being assessed from the Hutton TLP site investigations and research at other nearby sites. A series of special ring-shear interface tests were also undertaken at Imperial College on a range of North Sea clays and sands, to help assess appropriate interface friction angles for the Hutton clays.

The current estimate for the load-deflection curve of a single pile is given in Figure 15 with the earlier prediction by Jardine and Potts (1988) using the non-linear finite element method and the original design assessments of group stiffness and axial capacity (from Barton, 1984). Also shown are the original design loads and those that have since been assessed, taking into account the performance monitoring data. When the initial foundation stiffnesses deduced from the installation and storm settlement gauge records are plotted on the same diagram, they are indistinguishable from the non-linear finite element predictions.

6.3. Cyclic Effects

In addition to static calculations, an assessment was made of the effects on capacity of the revised extreme cyclic environmental loads. A series of cyclic triaxial tests were performed at Imperial College which allowed the approach proposed by Jardine (1991) to be applied to the Hutton site. The analysis indicated that a small reduction in shaft resistance (approximately 5%) should be expected to result from the loading cycles currently anticipated for a 100 year return period storm.

6.4. Factor of Safety

As can be seen from the results presented, both the applied load has decreased and the pile capacity has increased. As a result, significant increases in the factors of safety have been realised which satisfy both the API (API, 1987) and UK guidance (HMSO, 1990) requirements.

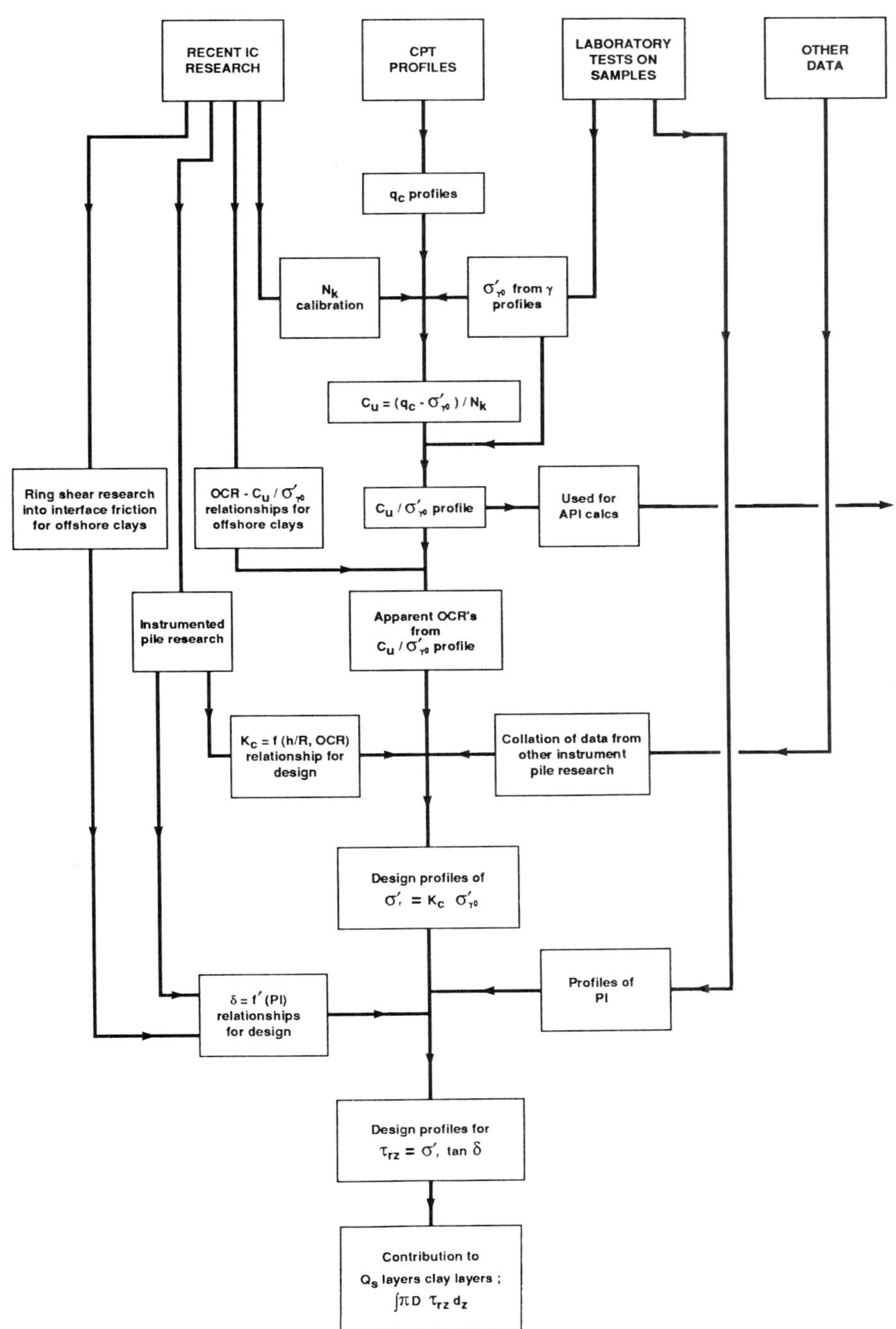

Fig. 13. Revised effective stress approach adopted for clay layers.

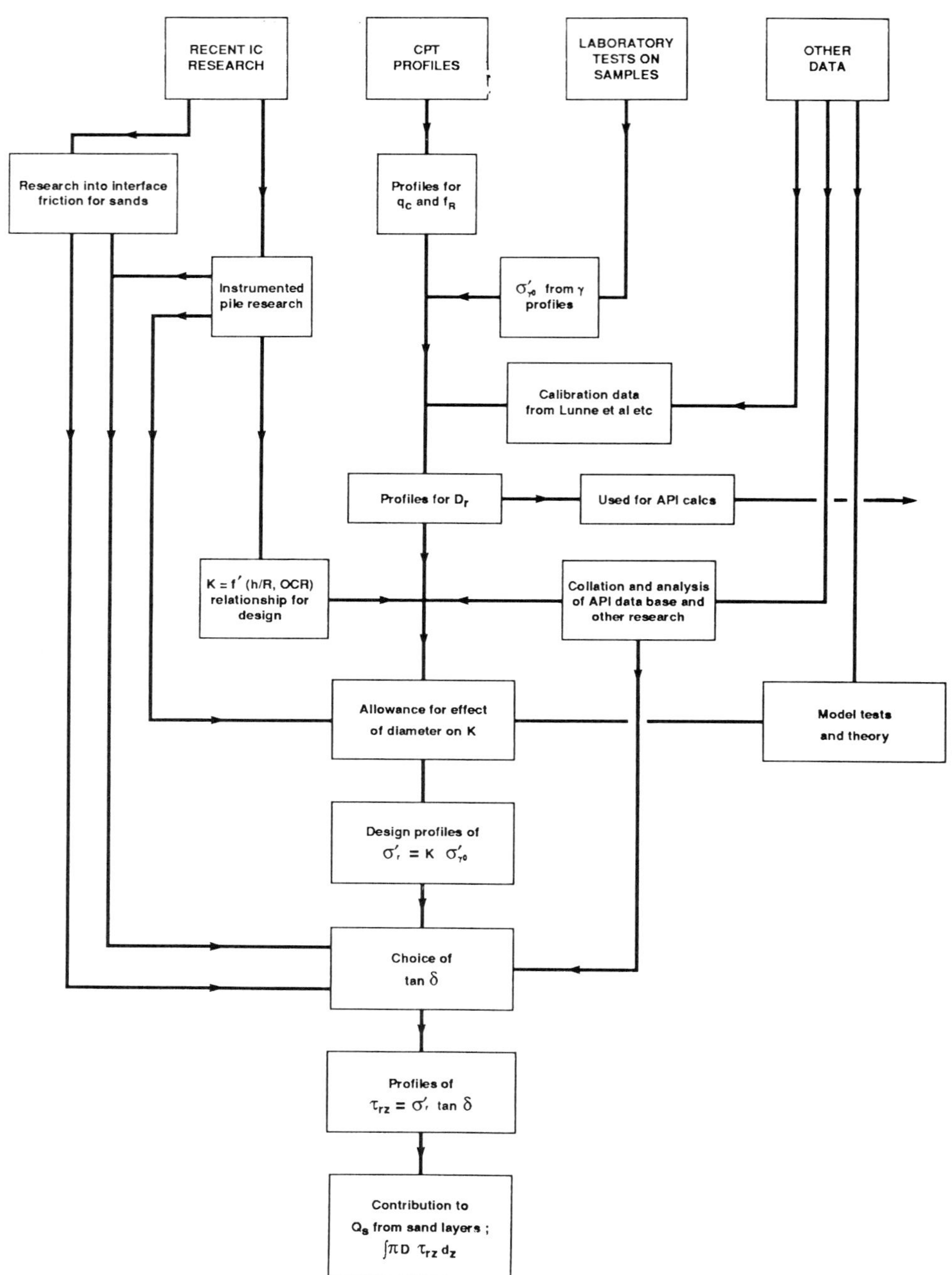

Fig. 14. Revised effective stress approach adopted for sands.

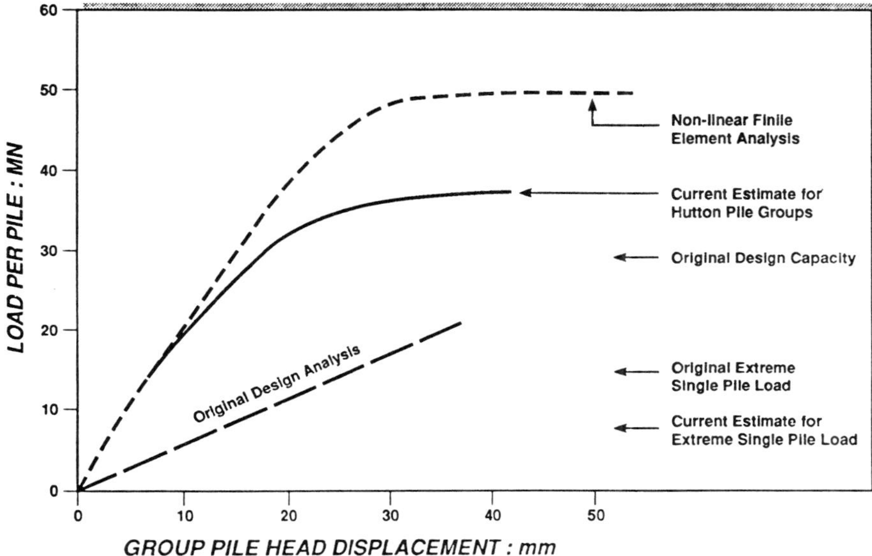

Fig. 15. Group load-displacement curves (displacement is upward).

7. Conclusions

1. The service performance of the Hutton TLP has been verified by a comprehensive programme of long-term monitoring.

2. Field measurements of the tensions applied to the piled foundations show that the extreme environmental loads are far smaller than those allowed for in design. The maximum tether tensions expected from the combined effects of extreme tides, the 100 year storm and the static loads have been significantly downgraded.

3. The foundation settlement gauge records prove that the pile groups have developed very small displacements under the applied loads.

4. Whilst the initial foundation response was much stiffer than predicted by conventional analyses, non-linear 'small-strain' analyses of the pile group behaviour have proved to be accurate.

5. Long term monitoring has shown that storm loading and creep have caused no further significant, permanent, long term movements. The dynamic foundation stiffness interpreted from in-storm measurements are consistent with the initial characteristics proven during platform installation.

6. With the exception of the first design of acoustic telemetry system, the foundation monitoring system has performed very well and has been able to resolve movements of 1 to 2 mm over long periods of time. Given that the displacements are so small, any new system should be designed to have a greater degree of temperature compensation.

7. In parallel with the in-service monitoring, the Hutton TLP piles have been reassessed using an improved effective stress approach. The recent analyses all indicate that the design capacity calculations were conservative.

8. Combining the revised estimates of critical loads and capacities indicates that the Hutton TLP pile groups' factors of safety are far more substantial than was specified at the design stage. This conclusion is supported by the high quality field measurements made since the platform was installed.

9. Recent papers on other North Sea foundations which are acting in compression (Reference Jardine and Potts, 1992) have presented similar increases in strength and stiffness which are also supported up by full scale measurements results.

Acknowledgements

The authors are grateful to the management of partner companies of the Hutton field for permission to publish this paper. Partner companies are Conoco (UK) Ltd (operator), Oryx UK Energy Company, Chevron (UK) Ltd, Amerada Hess Ltd, Amoco (UK) Exploration Co., Enterprise Oil plc and Mobil North Sea Ltd. Thanks are also extended to Steve Bultema and Paul Erb at Conoco Houston for their work on the load derivation and to all colleagues past and present at Conoco, Imperial College, Sonardyne and Custom Design Mouldings who have contributed to the success of the project over the past decade.

References

1. API (1987), 'Document API RP2T – Recommended Practices for Planning, Designing and Constructing Tension Leg Platforms', American Petroleum Institute, Washington DC.
2. Bond, A. J., Jardine, R. J., and Lehane, B. M. (1993), 'Factors affecting pile capacity in clay soils', *Proc. Int. Conf. on Offshore Site Investigations and Foundation Behaviour*, SUT, London, September 1992, in press.
3. HMSO (1990), 'Offshore Installations: Guidance on Design, Construction and Certification (4th Edition)', Department of Energy, HMSO, London.
4. Jardine, R. J., Dore, P. M., and McIntosh, W. (1985), 'A settlement monitoring system for the foundations of the Hutton Tension Leg Platform (TLP)', *Electronics in Oil and Gas Conf.*, January 1985, Cahners, London, pp. 217–230.
5. Jardine, R. J. and Potts, D. M. (1988), 'Hutton Tension Leg Platform foundations: An approach to the prediction of pile behaviour', *Géotechnique* **38**(2), 231–252.
6. Jardine, R. J. and Christoulas, S. (1991), 'Recent developments in defining and measuring static piling parameters', Gen. Report, *Int. Conf. on Deep Foundations*, Paris, Presse de l'École de Ponts et Chausées, pp. 713–746.

7. Jardine, R. J. (1991), 'The cyclic behaviour of large piles with special reference to offshore structures', *Cyclic Loading of Soils*, O'Reilly and Brown (eds.), Blackie, Glasgow, pp. 174–248.
8. Jardine R. J. and Potts, D. M. (1992), 'Magnus foundations: Soil and properties and predictions of field behaviour', *Proc. Conf. on Recent Large Scale Fully Instrumented Pile Tests in Clay*, ICE, London.
9. Rigden, W. J. and Semple, R. M. (1983), 'Design and installation of the Magnus foundations: Prediction of pile behaviour', *Proc. Conf. on Design and Construction of Offshore Structures*, ICE, London pp. 29–52.
10. Tetlow, J. H. and Ellis, N. (1983), 'The Hutton Tension Leg Platform. Developments in the design and construction of offshore structures', ICE, London, pp.
11. Bradshaw, H., Barton, R. R., and McKenzie, R. H. (1984), 'The Hutton TLP Foundation Design', *Proc. Offshore Technology Conference*, OTC 4807.
12. Barton, R. R. (1984), Personal Communication.

MAGNUS FOUNDATION MONITORING PROJECT – SUMMARY OF STATIC AND DYNAMIC BEHAVIOUR

D. E. SHARP
BP Engineering, Uxbridge One, 1 Harefield Road, Uxbridge, Middlesex UB8 1PD

and

R. M. KENLEY
Structural Monitoring Division, Fugro-McClelland Limited, Glasgow G20 0XA

Abstract. The Magnus Foundation Monitoring Project (FMP) was a major instrumentation project involving the measurement of the actual performance of the foundation system of one leg of the Magnus platform under service conditions. This paper describes some of the methods used in analysing the data collected during the project and summarises key aspects of the behaviour of the platform foundations.

1. Introduction

BP's Magnus oil-field was discovered in Block 211/12 of the UK sector of the North Sea, as shown on Figure 1, and was developed by a single production platform installed in 186m of water in 1982.

In 1979, when BP's Magnus structure was being designed, methods for the design of offshore structural foundations were extrapolated conservatively from onshore experience although pile sizes employed offshore were very much larger than land piles. Much of the required capacity of offshore piles is attributable to wave loading but there were few measurements to confirm how much of this loading is actually transmitted to the piles. Furthermore, little work had been done onshore and none offshore on the distribution of loads within a pile group. There was a need to confirm pile group capacity, the actual loads imposed on offshore pile groups, the distribution of those loads within pile groups and the conservatism inherent in design. As a result BP proposed to monitor the performance of the foundation system of one leg of the platform (leg A4) in order to obtain some basic information to verify current design methods and to compare actual foundation behaviour with design.

The lower section of one of the four legs of the Magnus platform and the piles supporting that leg were instrumented to determine the actual loads imposed on the piles and seabed by the structural and environmental forces.

Volume 28: Offshore Site Investigation and Foundation Behaviour, 493–510, 1993.

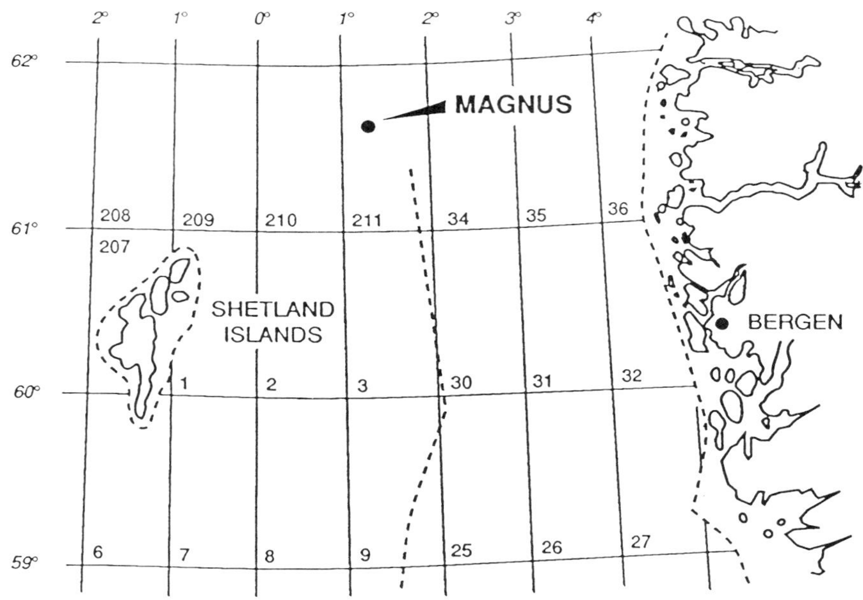

Fig. 1. Vicinity map.

Objectives of the project included measuring:

- the variation of static load distribution with time;
- the environmental load acting on a pile group during storms;
- the proportion of this load taken on the mudmat and on the pile group;
- the distribution of this load within the pile group and along the length of one pile;
- the effective stiffness of a pile group during storms.

Details of the Magnus platform, its foundations, the soil conditions at the platform location and the instrumentation system and its performance are described by Sharp (1992). Figure 2 shows the principal instrumentation installed on the platform and piles of leg A4 which comprised:

- wave sensors and accelerometers at deck level;
- strain gauges and accelerometers on the structure and near the base of leg A4;
- total pressure cells on the underside of the mudmat;
- strain gauges at one level on seven piles and at three levels on another pile.

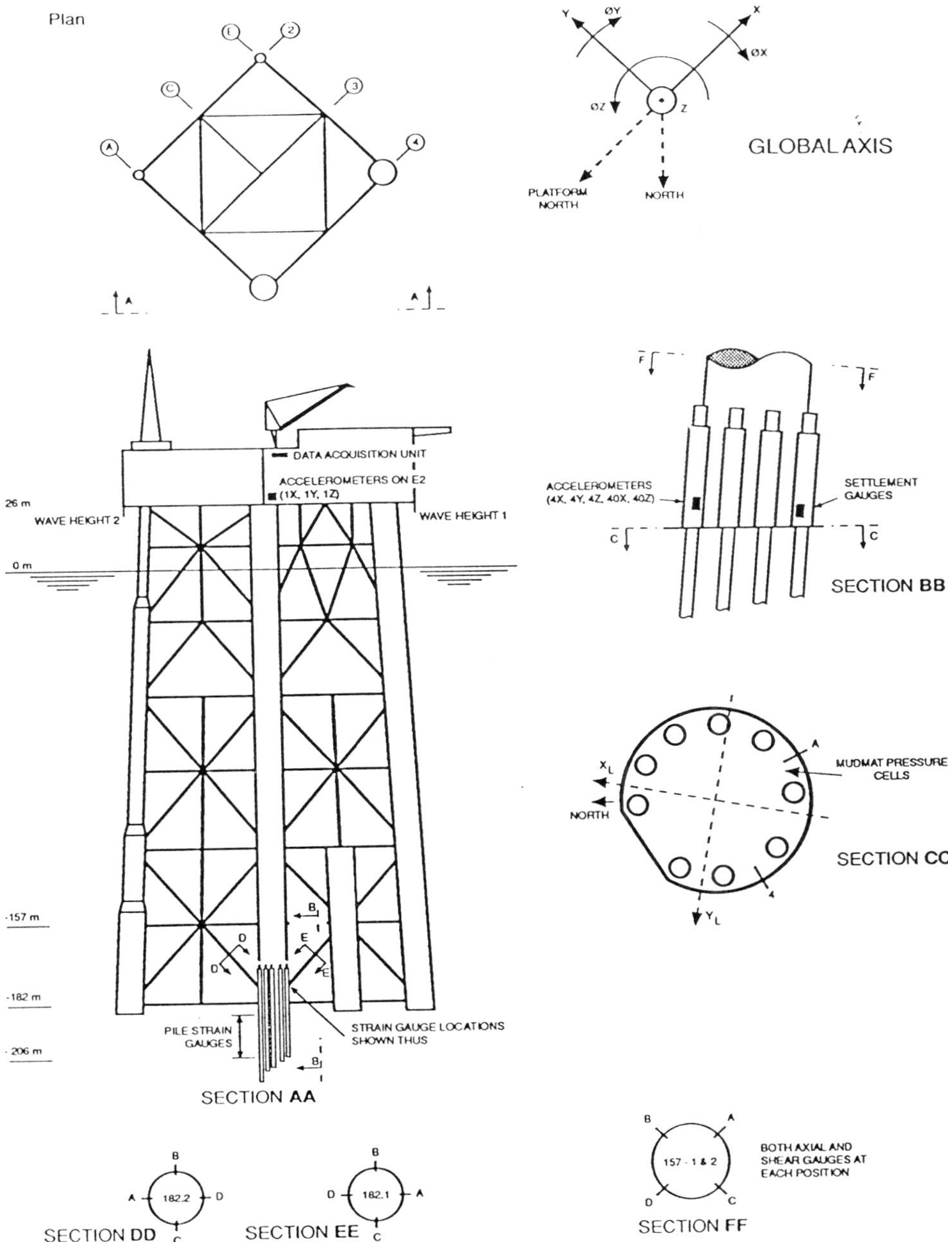

Fig. 2. Location of sensors.

This paper concentrates on data collection, the analysis procedures and actual behaviour of the platform as disclosed by the measurements.

2. Data Collection

Sensors were connected to recording units located on the Magnus platform. Initially, a Temporary Data Logger (TDL) was installed inside the top of leg A4. Following upending of the platform this collected and processed data from pressure cells on the underside of the mudmat and later, following pile installation, also from strain gauges on the piles. It sampled data for 5 minutes every six hours from all connected sensors, computed mean, maximum, minimum and standard deviation values and stored the results on tape.

Some 18 months after platform installation, the TDL was replaced by a more sophisticated permanent Data Acquisition Unit (DAU). The DAU consisted of signal conditioning for all sensors, a micro-processor to sample data from all the sensors on the platform and piles and a tape recorder to store time series data. Normally, data were recorded for 40 minutes every six hours but during stormy weather recording was continuous.

3. Data Analysis Procedures

Tapes from the TDL were replaced onshore and the statistical values stored in a database. The data was used to study long term behaviour mainly by plotting values against time.

Time series data from the DAU were analysed by replaying data into a computer onshore. Initially, a routine analysis was undertaken to obtain mean, maximum, minimum, standard deviation and period values for each recording. In addition, the acceleration signals were integrated to give displacements and wave direction was estimated from the direction of platform motion. Strain signals were combined to give axial and bending loads in the structure and piles. The mudmat gauges were analysed to give axial forces and bending moments carried by the mudmat.

A routine analysis was undertaken on at least one recording each day. Additional recordings were analysed during stormy conditions. Results were stored in a database and examined primarily by plotting one parameter against another or against time. They were used to determine general platform behaviour and to identify recordings for more detailed analysis. Nine recordings were selected for detailed analysis

The long term behaviour of the foundations was examined by plotting the mean values of sensor signals from both the TDL and DAU against time and the results are discussed below.

The dynamic response of the structure and its foundations was examined by undertaking a number of separate but related analysis methods:

- manual analysis of single selected individual waves;
- statistical analysis of all waves in one recording;
- analysis of all waves in one recording;
- frequency analysis of selected recordings;
- long term analysis of natural frequencies and mode shapes.

Initially, the analysis concentrated on a detailed analysis of selected single waves. This is the simplest approach and can be related to a specific applied static load. It yielded a considerable understanding of the behaviour of the foundations. The general applicability of the information obtained was then demonstrated by undertaking statistical analysis of all individual waves in selected recordings and by examining the largest wave in all recordings. Finally, a frequency analysis examined the variation with frequency and the natural frequency analysis gave an independent assessment of the foundation stiffness.

4. Summary of Major Results

The results have demonstrated that it is important to distinguish between the behaviour of the foundation in the long term, i.e. over months or even years, and that in the short term, i.e. during wave loading. This was most obvious with the loading carried by the mudmat.

4.1. Long Term Load Variations

Initially, during platform installation, the mudmat had to carry all the applied load. Figure 3 shows most of this load being transferred to the piles over a period of about a year from when the mudmat carried only about 7% of the static (still water) load. It is shown later that during storms the mudmat carries about 15% of the load attributable to environmental loading. Thus in the short term the mudmat carries more load than in the long term.

A continuous record of loads in five piles was obtained from just before module installation and from a further three piles soon after. Figure 4 shows that pile loads increased on module installation and later showed small increases due to both load transfer from the mudmat and increases in loads imposed on the platform. From measurements taken on the piles in the fabrication yard indications of the absolute loads in the piles were obtained. These loads and the load changes measured in the piles are consistent with expectations yielding values which are comparable with the applied loads. The absolute static load on the pile group is about 130MN, an average of about 15MN per pile.

The long term (static) data suggest no obvious decrease in axial load in the piles with depth down the pile over the upper 25m. This contrasts sharply with the distribution of axial load measured down the piles during environmental loading.

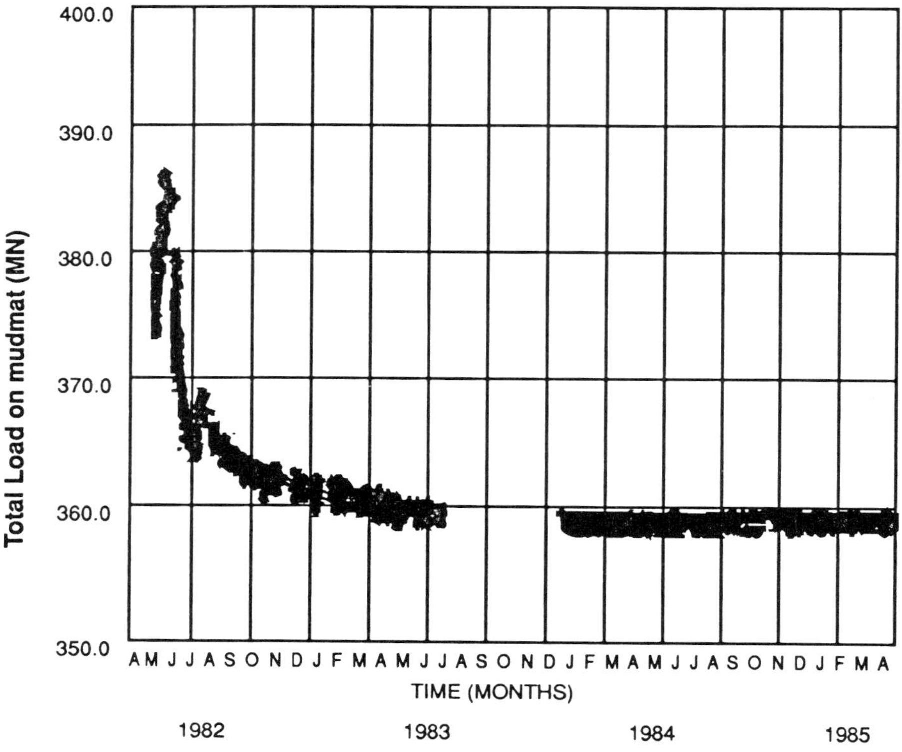

Fig. 3. Variation in total mudmat load.

4.2. General Dynamic Response Characteristics

Typical examples of time series signals, from a storm recorded on 10th January 1986, are shown on Figure 5. This storm, the most severe experienced over the measurement period, had a significant wave height of 12.5m. The highest individual wave measured was about 21m which compares with the 100 year design wave of 27.5m. The plots show the dynamic portion of a simultaneous 200 second extract from the signals which includes the largest wave. The upper plot is the wave signal: it is random in form with a zero-crossing period of about 12 seconds. The next plot is the deck displacement of the platform. This is similar in form to the wave signal, indicating that most of the motion of the platform is in direct response to wave action. Peak to peak displacement was about 230mm. The third plot is the axial load in leg A4 just above the pile sleeves. It is virtually identical to the displacement signal and has a maximum peak to peak value of about 72MN. When the loads in the diagonal brace members are included the total axial load on the leg was 89MN and the maximum peak to peak shear load was about 17MN. These are considerably less than the equivalent design axial and shear loads of about 390MN

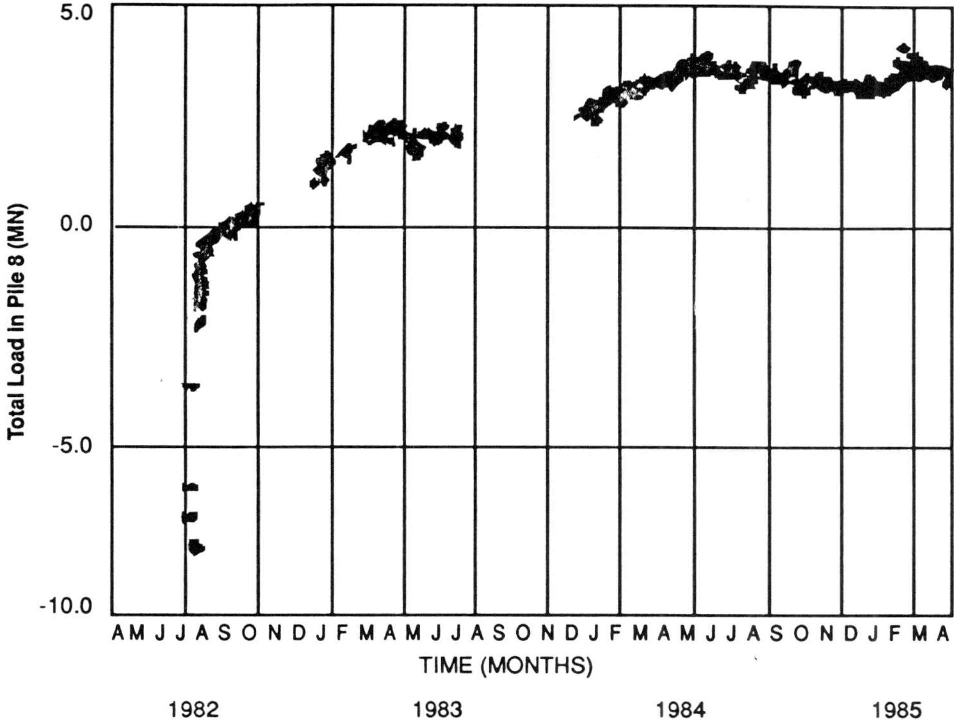

Fig. 4. Variation in axial load in Pile no. 8.

and 60MN, respectively. Whilst this is in part because the 100 year wave did not occur during the measurement period there also appears to have been considerable conservancy in the estimation of structural loading. The fourth plot, the mudmat load, is also similar with a maximum peak to peak value of about 17MN. The final plot is axial force in one of the piles which has a maximum peak to peak value of 13MN.

The dynamic response characteristics of the Magnus platform and its foundations were dominated by direct response to wave action. Response amplitudes at platform natural frequencies can be ignored for most practical purposes. The signals were all random in form but had simple and well-defined statistical properties. Amplitude distributions were Gaussian in form for all locations at all times. The distributions of individual wave heights and load cycles were approximately Raleigh in form. Consequently, the response characteristics could be accurately described by a single amplitude parameter (either standard deviation or maximum peak to peak value) and a single period or frequency parameter (e.g. zero-crossing period).

Clear relationships were found between wave height, platform displacements,

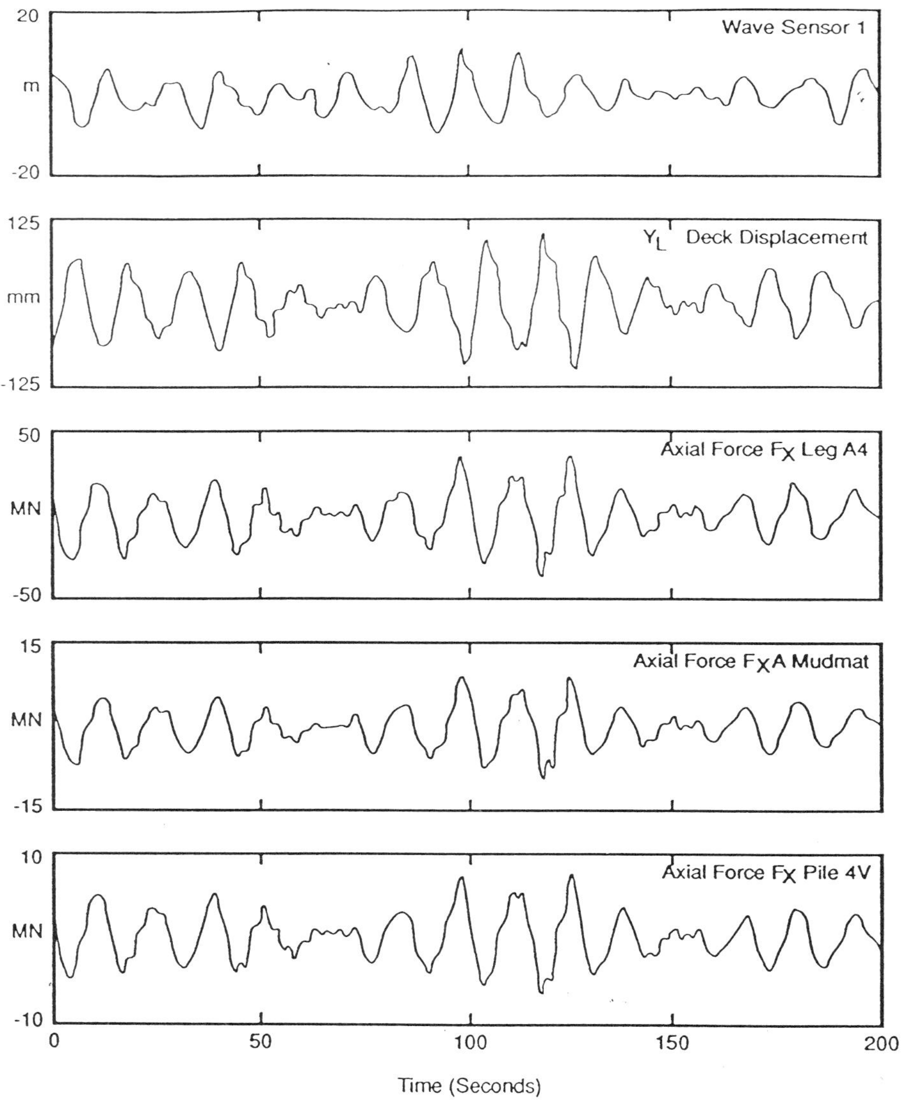

Fig. 5. Examples of time history signals.

platform loads, mudmat loads and pile loads. Relationships were generally linear with no resonant type effects within the foundations. The complete behaviour could be described in a quasi-static manner.

4.3. General Individual Wave Analysis

The analysis of selected individual waves from the Magnus recordings has been described in some detail by Kenley and Sharp. However, the inevitable variability of the wave configuration in a real sea within an individual wave may restrict the accuracy of the single wave analysis. Extending the analysis to an examination of all waves removes the variability and allows a more precise understanding of foundation behaviour to be obtained. For example, it will be shown that it was possible to identify slight differences in the behaviour of the structure in two horizontal directions and a slight difference in the lateral response of the pile group between high and low energy environments. This section focuses on that additional analysis. Overall, the underlying feature was that all the methods gave similar results and were entirely consistent with each other. This gives confidence that the findings are valid. Two separate additional analyses were undertaken.

1. The relationship between pile, mudmat and leg forces over the long term were examined by plotting peak to peak values obtained from the routine analysis results, i.e. by reviewing the largest wave from each recording.

2. An individual wave analysis was carried out on every wave in the selected recordings and the relationships between piles, mudmat and leg forces obtained, i.e. by reviewing all the waves in the selected recordings.

In addition to the foundation loads, relationships between the structural loads, overall platform displacements and wave height were also examined enabling foundation behaviour to be related back to wave conditions.

4.3.1. Long term relationships

Figure 6 shows some examples of results from the long term plots. The upper left graph plots standard deviation of deck displacement measured in the same direction as the wave direction against significant wave height. Each + symbol represents data from one 40 minute recording. The points virtually all lie on a straight line. This suggests that, over the range of the measurements, the applied loading on the structure is linear. There is a very slight indication of possible non-linearity for the largest storm.

The upper right graph plots maximum axial force cycles against deck displacement cycles in the direction of the leg A4 to leg E2 diagonal. Each point is obtained from the maximum measured cycle in each recording. There is a strong linear relationship between leg force and displacement.

The central graphs show mudmat load cycles and peak to peak axial force in one pile against axial force cycles. Once more, all points lie close to a straight line. All results suggested that under axial loading the foundations behave in a linear

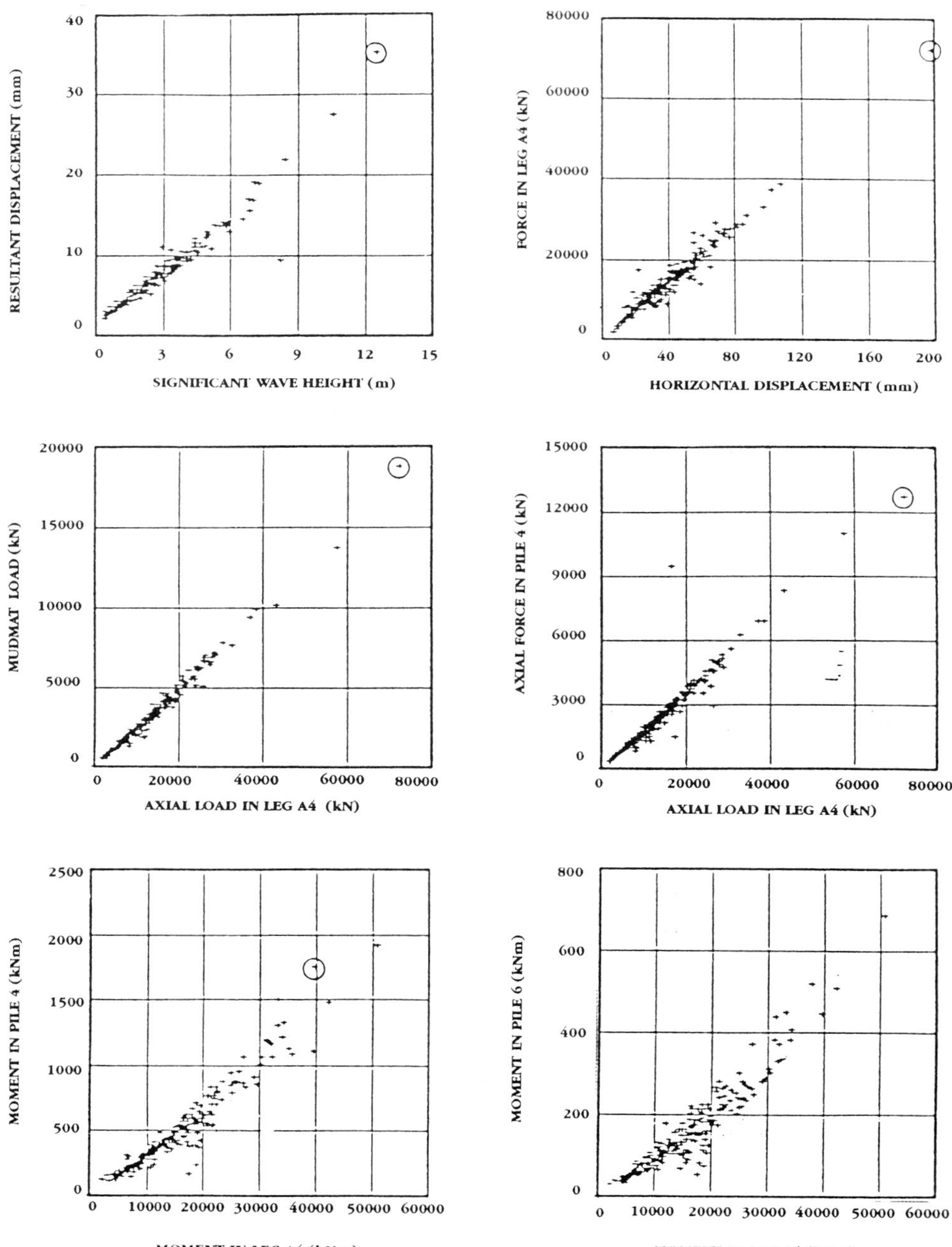

Fig. 6. Long term dynamic relationships.

manner for all conditions measured.

The lower graphs plot the maximum bending moments in two of the piles, piles 4 and 6, against maximum bending moment in leg A4. Whilst all points lie close to a line in this case the line is slightly curved. This suggests some slight non-linear behaviour of the piles under lateral loading.

From these plots, and other similar ones, the long term relationships between wave height and platform displacements, and between platform displacements and forces and bending moments in the structure, mudmat, and piles and foundation forces were obtained.

Bending moment distributions. Values of bending moments, normalised relative to the leg A4 bending moments, were obtained for each pile from the long term plots for low and high wave height conditions. They were used to estimate bending moment distributions down a pile in a similar manner to the selected cycle values described in Kenley and Sharp. Figure 7 summarises the results. Each plot shows the low and high wave height distribution for one direction. Also presented are the estimated shear force, bending moment, rotation and displacement values at mudmat level. The shear force values were obtained by differentiation of the bending moment curves. Rotations and displacements were obtained by integration.

The bending moment distributions are similar in shape although higher bending moments (per unit applied force) occur in the 2 to 15m depth range for the higher wave condition. This is consistent with non- linear behaviour of the soil over that depth range. The rotations and lateral displacements derived from the curves are about 30% greater per unit applied force in the higher wave conditions. It should also be noted that there are slight differences between the two directions.

Axial load distributions. Values of axial loads, normalised relative to applied loads, were obtained for each pile for motion across the platform diagonal. The loads were linear over all wave heights measured. These values were used to derive the pile group axial load distribution and an indication of the distribution down a pile obtained, again in a similar manner to that described by Kenley and Sharp for the selected cycle analysis. Figure 8 plots estimated axial load in each pile at mudmat level against pile position relative to the axis normal to the load direction. Loads were estimated from the loads at each gauge level assuming a linear decrease with depth of 2%/m. A straight line was fitted by eye to give the best fit to the data points and overall axial load and bending moments derived from it. Results proved to be very similar to those obtained from the selected individual wave. This enhances confidence in the results from the selected cycle analysis.

Overall force balance. A check was carried out on the balance of forces and displacements at the base of leg A4 for data extracted from the long term plots, again in a similar manner to that described by Kenley and Sharp for the selected cycle analysis. Table 1 summarises the force balance. The data is normalised with

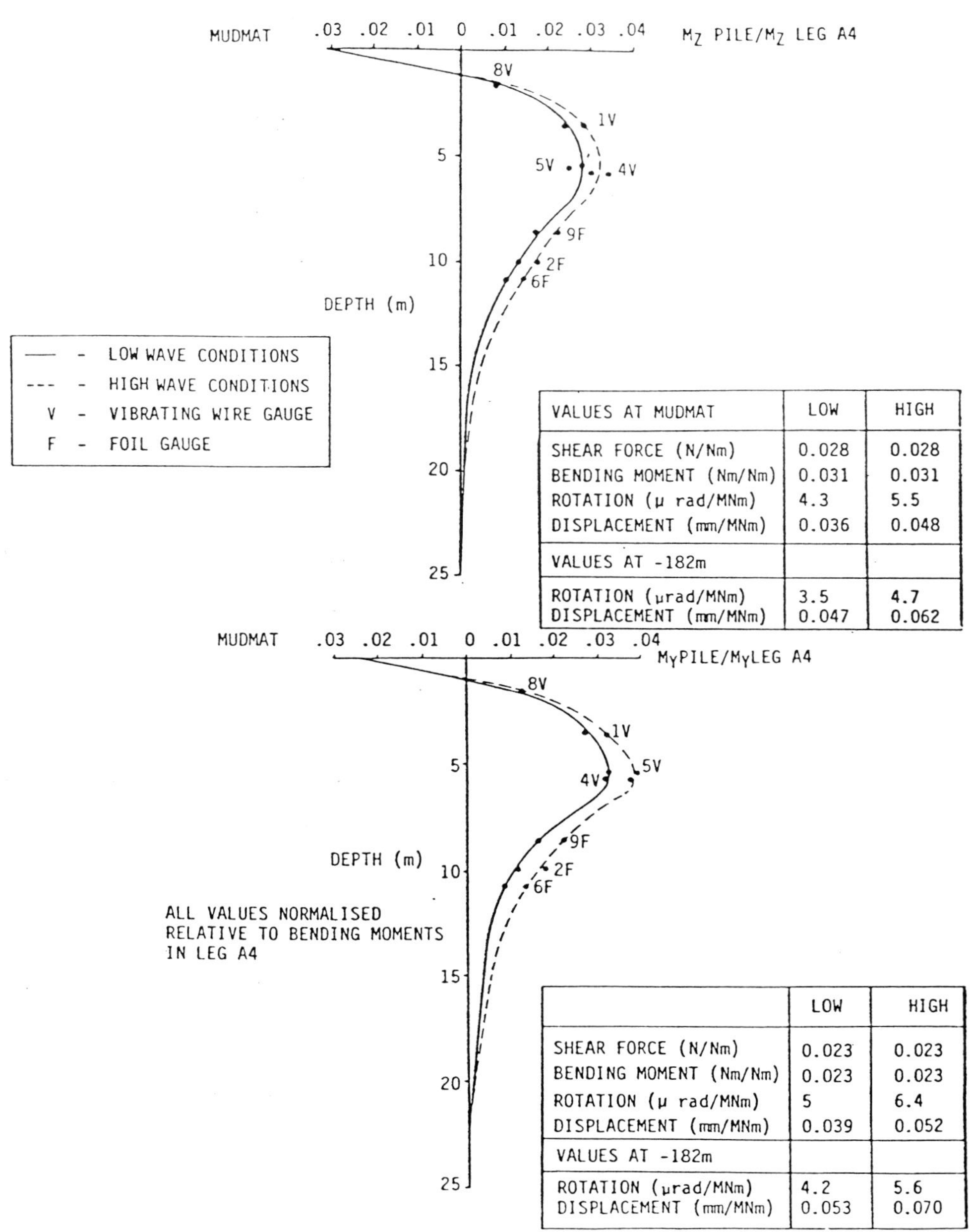

VALUES AT MUDMAT	LOW	HIGH
SHEAR FORCE (N/Nm)	0.028	0.028
BENDING MOMENT (Nm/Nm)	0.031	0.031
ROTATION (μ rad/MNm)	4.3	5.5
DISPLACEMENT (mm/MNm)	0.036	0.048
VALUES AT -182m		
ROTATION (μrad/MNm)	3.5	4.7
DISPLACEMENT (mm/MNm)	0.047	0.062

	LOW	HIGH
SHEAR FORCE (N/Nm)	0.023	0.023
BENDING MOMENT (Nm/Nm)	0.023	0.023
ROTATION (μ rad/MNm)	5	6.4
DISPLACEMENT (mm/MNm)	0.039	0.052
VALUES AT -182m		
ROTATION (μrad/MNm)	4.2	5.6
DISPLACEMENT (mm/MNm)	0.053	0.070

Fig. 7. Bending moment distributions from long term data.

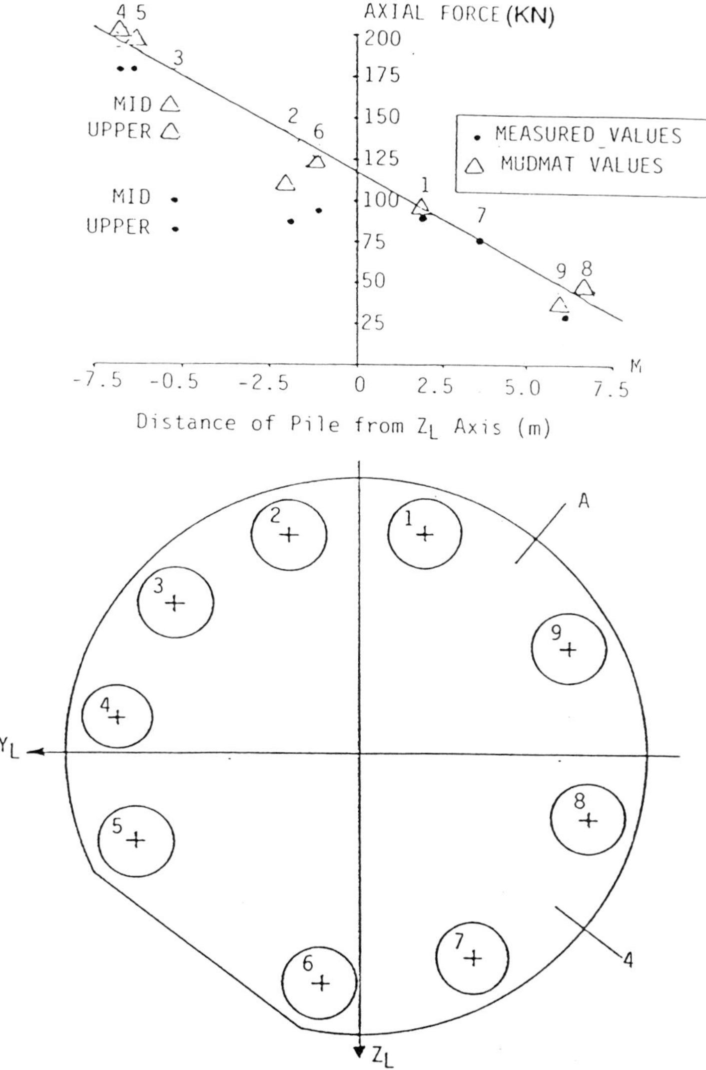

Notes:

1. Values are normalised relative to the axial force in leg A4.
2. The first estimate of axial load at mudmat level was obtained assuming that the load distribution decreased at 2%/m with depth.
3. The second estimate was obtained from straight line which was drawn by eye to give a 'best fit' to the first estimate value.

Fig. 8. Axial force distribution from long term plots.

TABLE 1.
Force balance on leg A4 (from long term plots).

	STRUCTURE		FOUNDATION		% DIFF.
AXIAL FORCE (MN)	MAIN LEG	1.0	MUDMAT	0.22	
	DIAGONALS	0.3	PILE GROUP	1.1	
	TOTAL	1.3	TOTAL	1.32	2
LATERAL FORCE (MN)	MAIN LEG	0.03	PILE GROUP	0.25	
	DIAGONALS	0.21			
	TOTAL	0.23	TOTAL	0.25	9
BENDING MOMENT (MNm)	LEG MOMENT	1.04	MUDMAT	0.29	
	LEG SHEAR FORCE	0.87	PILE GROUP	2.89	
	DIAGONALS	0.8	INDIVIDUAL PILE MOMENTS	-0.28	
	TOTAL	2.71	TOTAL	2.9	7

respect to applied axial leg loads associated with loads applied diagonally across the platform from leg E2 to A4.

The results in Table 1 are very similar to those obtained from the selected cycle analysis. The forces and bending moments from the structure agree with the forces and bending moments from the foundation within 10%. The agreement is excellent, especially considering the number of different measurement parameters that have to be combined. It confirms earlier conclusions about the high quality of the data and the validity of the assumptions made in its analysis.

Data were also extracted from the long term plots to compare displacements and rotations calculated from the pile bending moment and axial load distributions with those obtained by integration of the acceleration signals. Results were again normalised with respect to applied leg loads and the results were similar in form to those obtained from the selected cycle analysis.

4.3.2. Individual wave analysis on all waves in a recording

Analyses were undertaken to obtain the relationships between pile forces or moments and leg force or moments from all the individual waves in each of the recordings selected in the single wave analysis described by Kenley and Sharp.

A statistical approach was used based on individual load cycle distributions. An individual load cycle was considered to be the difference between each trough peak and the next crest peak or each crest peak and the next trough peak (hence strictly a half cycle). For each signal in each recording the number of cycles exceeding particular values of cycle range were found. Values of each pile force or moment and each leg force and moment were then obtained for equal numbers of cycles and plotted against each other. The resulting plots indicate the mean relationship between various signals over each recording and should provide a more accurate definition of these relationships than can be obtained from the analysis of individual cycles.

Figure 9 plots some of the relationships obtained for the axial forces and bending moments for the piles for a storm on 22nd January 1984. In the cases presented and all other cases for all nine recordings analysed in detail the relationships are linear, even up to the largest cycle value. Furthermore, the relationships obtained from this analysis is similar to that obtained from the long term plots. There are similar differences in the two directions and similar differences between moderate and storm conditions. The long term plots presented above suggest a slight non-linear relationship between some pile bending moments and applied bending moments. The short term plots, such as those presented on Figure 9, indicate that over the duration of a recording the relationship is linear but that there is a change in the slope of this relationship between stormy and moderate conditions. This is consistent with the stiffness of the soil being dependent on the general wave conditions at the time rather than the severity of any individual wave.

4.4. Dynamic Behaviour of Mudmat

The dynamic signals from the pressure cells on the mudmat were correlated with the applied platform loads indicating that wave loading was being transmitted through the mudmat. The signal amplitudes varied with the position of sensors and were consistent with the application of an overall axial load and bending moment. Results indicate that 15% of total axial load and 10% of the bending moment was carried by the mudmat. Consequently, the mudmat makes a significant contribution to the dynamic load carrying capacity of the foundation. No contribution was assumed during design.

4.5. Behaviour of Piles under Dynamic Axial Loading

The dynamical axial loads in the piles were consistent with the application of an overall load and bending moment to the pile group. This meant that some piles always experienced higher dynamic loads than others, in particular crotch piles experienced lower dynamic loads than piles on the leading edge of the pile group. The results obtained were consistent with a decrease in pile load with depth of about 2%/m. This is a much greater decrease than deduced from the long term (static) data.

No evidence of any non-linear behaviour of the pile group in axial loading

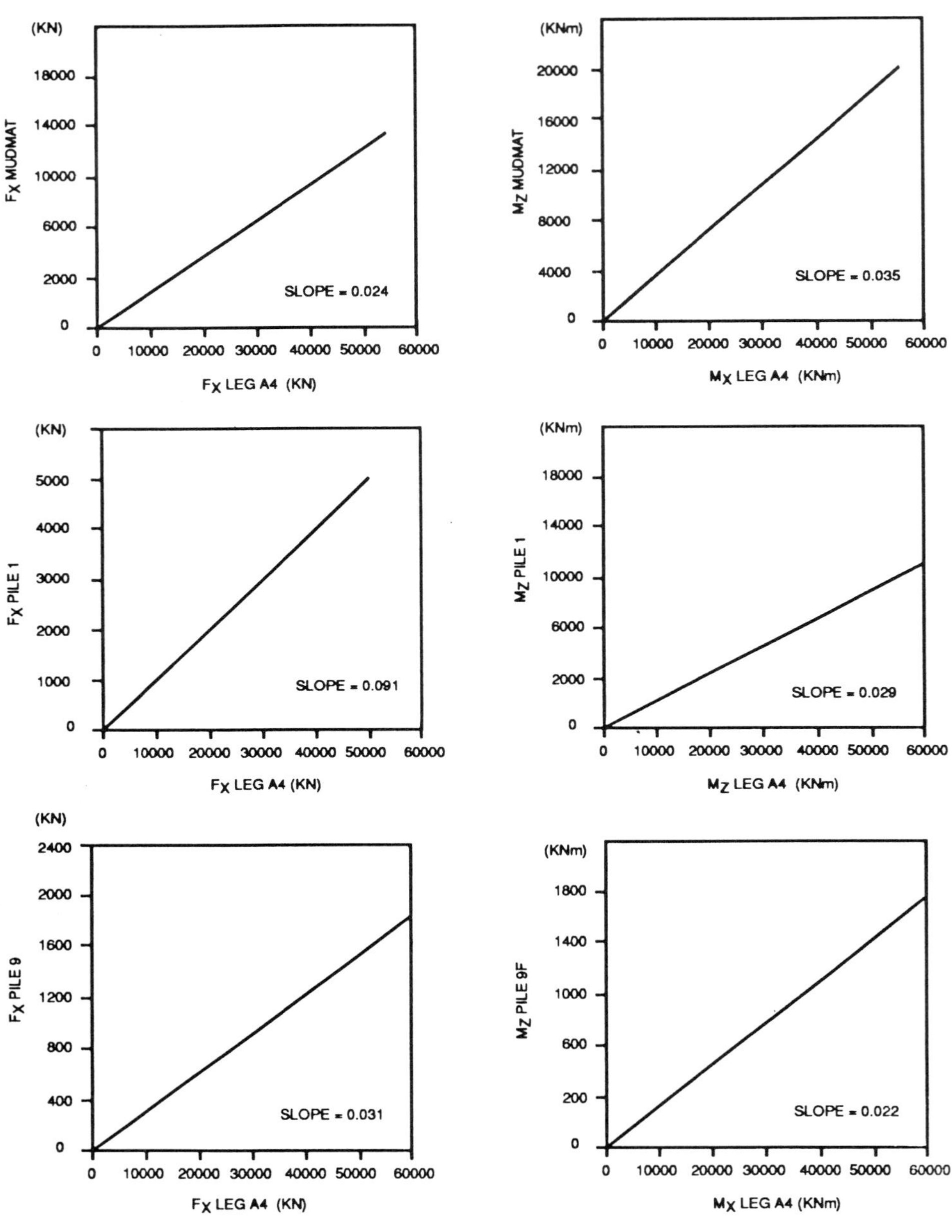

Maximum cycle range in leg A4: Force = 60000 KN
Moment = 55000 KNm

From 08:00 hr recording on 22/1/84

Fig. 9. Individual wave analysis: force and bending moment relationships.

for wave conditions encountered was noted. The maximum peak to peak force measured in an individual pile was about 14MN giving a maximum absolute load level, taking into account the static load, of about 25MN. This is modest when compared to the ultimate design load for each pile of 60MN and the load levels may not be large enough for non- linear behaviour to occur.

The measured dynamic axial stiffness of the pile group was about 32MN/mm which is about four times that assumed during design.

4.6. Lateral Loading on Piles

Assuming that all piles would behave in a similar manner, an estimate of the bending moment distribution down a pile was obtained by combining results from all piles. The bending moment distribution was used to derive shear force, rotation and lateral displacement distribution. The results were consistent with displacement, rotation and shear force measured on the platform. The bending moment distribution reached a maximum about 6m below the mudmat and was approximately zero about 1m below the mudmat.

The foundation appeared to behave in a linear manner over the duration of a recording but there was a small change in bending moment distribution per unit applied load in stormy conditions compared with in calm conditions. However, the load levels measured were only about 20% of the design loads and lower lateral stiffnesses may occur at higher load levels.

The measured lateral stiffness was about two or three times that predicted at the design state. Partially as a result of this the measured natural frequencies were 20% higher than those predicted. Whilst this had little influence on foundation behaviour, it had an influence on the dynamic characteristics of the structure itself. Because of the predicted frequencies, it had been anticipated that there would be more response at the platform natural frequencies and this had an influence on stress levels and fatigue lives. As the natural frequencies were higher, the actual response at the natural frequencies was insignificant.

Acknowledgements

The Authors wish to thank the British Petroleum Company plc for permission to publish this paper.

FMP was a joint industry funded project and the contribution, both financially and technically, of the following organisations is gratefully acknowledged: BP Petroleum Development Limited, Britoil plc (both of the above are now part of BP Exploration), Brown & Root Inc., Engineers India Limited, Exxon Production Research Co., Total CFP, Marathon Oil Co., Den Norske Stats Oljeselskap (Statoil), John Brown Offshore Structures Limited, UK Department of Energy.

We would like to acknowledge the contribution of the myriad of consultants and contractors, and specific individuals, whose talents and commitment were so crucial in ensuring firstly that FMP was initiated and secondly that it was implemented so successfully. In particular, the dedication of staff from Queen Mary College, London, the consistent involvement of key staff from Structural Monitoring from conceptual planning through to data analysis and the contribution of many colleagues within BP was invaluable.

References

1. Kenley, M. and Sharp, D. E. (1992), 'Magnus foundation monitoring project – Instrumentation, data processing and measured results', in *Recent Large Scale Fully Instrumented Pile Tests in Clay*, Thomas Telford Services, London.
2. Sharp, D. E. (1992), 'Magnus foundation monitoring – An Overview', in *Recent Large Scale Fully Instrumented Pile Tests in Clay*, Thomas Telford Services, London.

Discussion

R. Hobbs, Lloyd's Register: One of the findings of the Magnus instrumentation was that calculated pile loads due to environment were less than anticipated. In reviewing pile loadings for some platforms we have also found that unrealistic or unlikely combinations of deck load have been assumed, or that allowance has been made for future conductors or risers which are unlikely to be installed.

SESSION 7

PILING RESEARCH

LARGE DIAMETER PILE TEST PROGRAMME – SUMMARY

J. CLARKE and M. D. LAMBSON
BP International, Uxbridge 4B8 1PD

1. Introduction

This paper summarises the large diameter pile test programme undertaken between 1985 and 1988 in the UK on behalf of a consortium of oil companies. The purposes of the work were to improve the axial capacity design method for offshore sized pile foundations and to gain more insight regarding the fundamental behaviour of long piles in cohesive soils. The test programme was recently reported in detail at a specialist conference (Clarke, 1992).

Initially a comprehensive study of the ultimate capacity of offshore piled foundations in cohesive soils was carried out. This recognised that differences in pile size, the magnitude and type of loading, and the soil conditions that can be encountered offshore required significant extrapolation of empirical design criteria beyond the existing database. Two important soil conditions not included in the API RP2A database are overconsolidated soils with shear strengths greater than 400 kPa, and stiff essentially normally consolidated soils.

Following the initial study an outline scope of work was prepared to test two 0.762m diameter tubular steel piles. One pile, up to 60m long, would be tested at a site termed NC which comprised uniform soft becoming stiff essentially normally consolidated silty clays. The second pile, up to 40m long, would be tested at a site termed OC which consisted of uniform hard, heavily overconsolidated silty clay. With the work basis established funding was obtained from an EEC loan, the UK Department of Energy, Lloyd's Register, McClelland Limited and seven major oil companies. The project was managed on behalf of the participants by BP International with McClelland Limited being the managing consultant. The total cost of the project was £3.99 million plus £0.34 million for additional axial tension and lateral load tests at Tilbrook Grange. The overall programme time was about three years and is shown by Figure 1. Reference to this shows that three axial load tests were performed. Compressive load tests were carried out at the NC and OC sites, and an additional tensile load test undertaken at the OC site.

The site selection and soil conditions are described first, followed by the in-

Volume 28: Offshore Site Investigation and Foundation Behaviour, 513–547, 1993.

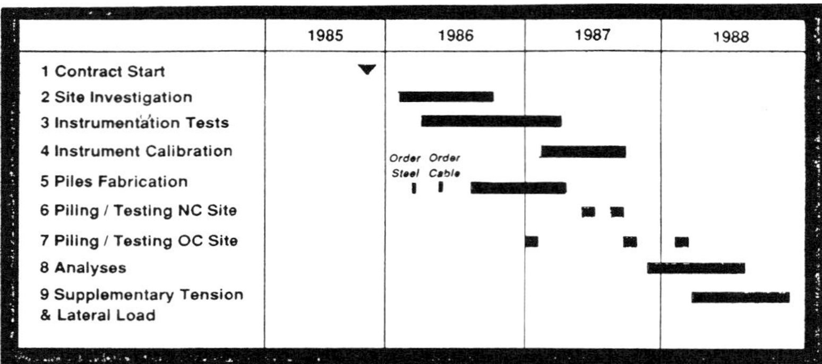

Fig. 1. Test programme.

strumentation and data acquisition system development. The pile installation and testing procedures are then presented before the interpretation of the results and design implications are discussed.

2. Site Selection and Soil Conditions

A full description of the soil conditions and properties at the Pentre and Tilbrook Grange sites is given by Lambson *et al* (1992).

2.1. Site Selection

A thorough literature review of UK Quaternary deposits was undertaken to identify potential onshore sites similar to North Sea sediments. This identified eight regions where thick sequences of cohesive Quaternary sediments are found. Borings were undertaken to complement existing borehole information. Thomas (1990) summarises the literature review and reports on the ground conditions at sites within the regions of interest. The Quaternary deposits reported divide into two categories, NC and OC clays. Table 1 summarises the sites considered. Reference to this shows that of the five regions reviewed for the NC site, one was considered to have too thin a deposit combined with plasticity which was generally too high, two sites had an insufficient mean strength, and at another the clay was thought too plastic to be truly representative of most North Sea soil conditions. Hence, the site at Pentre was selected as being the best compromise. The borings at Edinburgh and Tyneside revealed insufficient depth of strong clays with gravel and cobble inclusions being more predominant than investigations disclosed in the Bedford and Huntingdon region soils. The site at Tilbrook Grange was chosen from this region. Figure 2

TABLE 1.
Sites considered for pile test.

	REGION	SITE	PI %	Su/σ'_v	Depth (m)	Comments
NC	Forth Valley	Alva	30	0.2 – 0.1	70	Insufficient mean strength
		Grangemouth	50–20	0.2 – 0.1	73	
	Team Valley	Lady Park	25–50	0.7 – 0.3	30+	Deposit too thin, plasticity generally too high
		Ouston	25–50	0.7 – 0.3	26+	
		Chester-le-Street	25–40	0.9 – 0.3	15+	
	Aire Valley	Silsden	35–55	0.14	57	Too plastic
	Shrewsbury	Pentre	10–30	0.16 – 0.32	60	Chosen
	North West Norfolk	Watlington	20–40	0.6 – 0.8	48	Insufficient mean strength
OC	Edinburgh	Leith	10–20	5 – 1.5	25	Insufficient depth and gravel/ cobbles more predominant
	Tyneside	North Shields	20	4 – 1.5	28	
	Bedford and Huntingdon	Tilbrook	20–30	10 – 1	41	Chosen
		Chelveston	15–30	10 – 1	34	Insufficient depth

shows the location of the Pentre and Tilbrook Grange sites with UK National Grid references of SJ395 177 and TL 009 710, respectively.

2.2. Fieldwork and Laboratory Testing

Detailed site investigations were carried out at both sites. The soil conditions were investigated by a combination of soil sampling and *in situ* testing. The drilling and sampling methods were chosen to be of a similar high standard to those employed offshore. Hence, the borings were advanced by rotary drilling methods with mud flush and the soils push sampled at close intervals. Sampling and *in situ* test equipment was wireline based. At both sites the *in situ* testing techniques included:

- Cone penetrometer testing
- Downhole vane tests (NC site only)
- Axial load-displacement (T-Z) probe tests

Fig. 2. Site locations.

- Pressuremeter testing
- Dilatometer testing (OC site only)
- Crosshole and surface to downhole seismic testing

At the Pentre and Tilbrook Grange sites, the soils were investigated at 6 and 7 locations respectively. Piezometers were installed in some of the boreholes.

A comprehensive laboratory testing programme was undertaken to evaluate the pertinent physical properties of the foundation soils. A geological testing programme was also undertaken to examine the microfabric and mineralogy of the sediments. Some in-field geotechnical testing, including triaxial compressive strength tests, was performed. All other geotechnical tests, including additional triaxial tests, were conducted in fully equipped soil mechanics laboratories. Most emphasis was given to soil strength measurement, both in terms of total and effective stress. Undrained shear strength was assessed from a wide range of triaxial tests using both undisturbed and remoulded samples. Direct shear tests were performed to measure undrained strength and the shear strength at the interface of soil and the pile steel material. Ring shear tests were carried out on remoulded specimens to determine residual soil to soil and soil to steel interface shear strengths. For the Pentre soils, results from resonant column tests were used to augment soil stiffness measurements obtained from the triaxial tests. This enabled stiffness to be assessed over a range of strains of practical interest. The compressibility characteristics were investigated by one-dimensional consolidation tests on undisturbed specimens. Soil permeability was measured in the laboratory by the constant head permeability test, using undisturbed specimens within a triaxial cell.

2.3. Pentre Ground Conditions

A representative borehole log for the Pentre soils is given by Figure 3. The test stratum comprises a soft becoming very stiff, very silty clay. The silt and clay content varies with depth resulting in occasional layers of clayey silt of low to intermediate plasticity and silty clay of intermediate to high plasticity. The clayey silt layers are between 0.5m and 3.0m thick and generally occur between depths of 36m and 41.5m. The layers of silty clay generally occur between depths of 40.5 and 48.5m. Typically for the Pentre test stratum the plasticity index ranges between 10 and 30 with the majority of results plotting above the 'A' line on the plasticity chart. Particle size distribution analyses reveal the clay sized fraction is generally below 20%. The apparently high PI compared with the relatively low clay fraction is reconciled by the geological studies.

Microfabric and mineralogy studies of the test stratum suggest the sediments were deposited following the input of seasonal pulses into a glacial lake. They may be described as laminated to very finely laminated. The clay minerals occur generally as simple aggregates forming silt sized particles. The microfabrics examined are generally not flocculated and do not show any evidence of glacial over-riding. The overconsolidation ratio, OCR, was estimated from oedometer tests and from plasticity based correlations. At Pentre the OCR was judged to decrease from 1.8 at 15m to 1.2 at 60m depth. Hence the test stratum is essentially normally consolidated. Groundwater conditions at Pentre are summarised by Figure 4 and show little seasonal variation. The site is underlain by a sandstone aquifer which is estimated to lie at a depth of 80m. An artesian head up to 5m above ground level reportedly is present within the aquifer. Figure 5 summarises the undrained shear strength as measured by unconsolidated undrained (UU) triaxial compression tests on 'undisturbed' specimens. However, shear strengths determined from *in situ* remote vane tests are similar to those obtained with the self-boring pressuremeter and both are higher than the UU shear strength. It is thought that partial drainage influenced these field measurements. Laboratory testing revealed consolidation and permeability characteristics which were generally similar to clayey silt. Effective stress testing gave stress paths during shearing at *in situ* stresses which reflected clayey silt type behaviour at large strains although positive pore pressures were generated initially. Figure 6 summarises the bulk unit weight profile at Pentre.

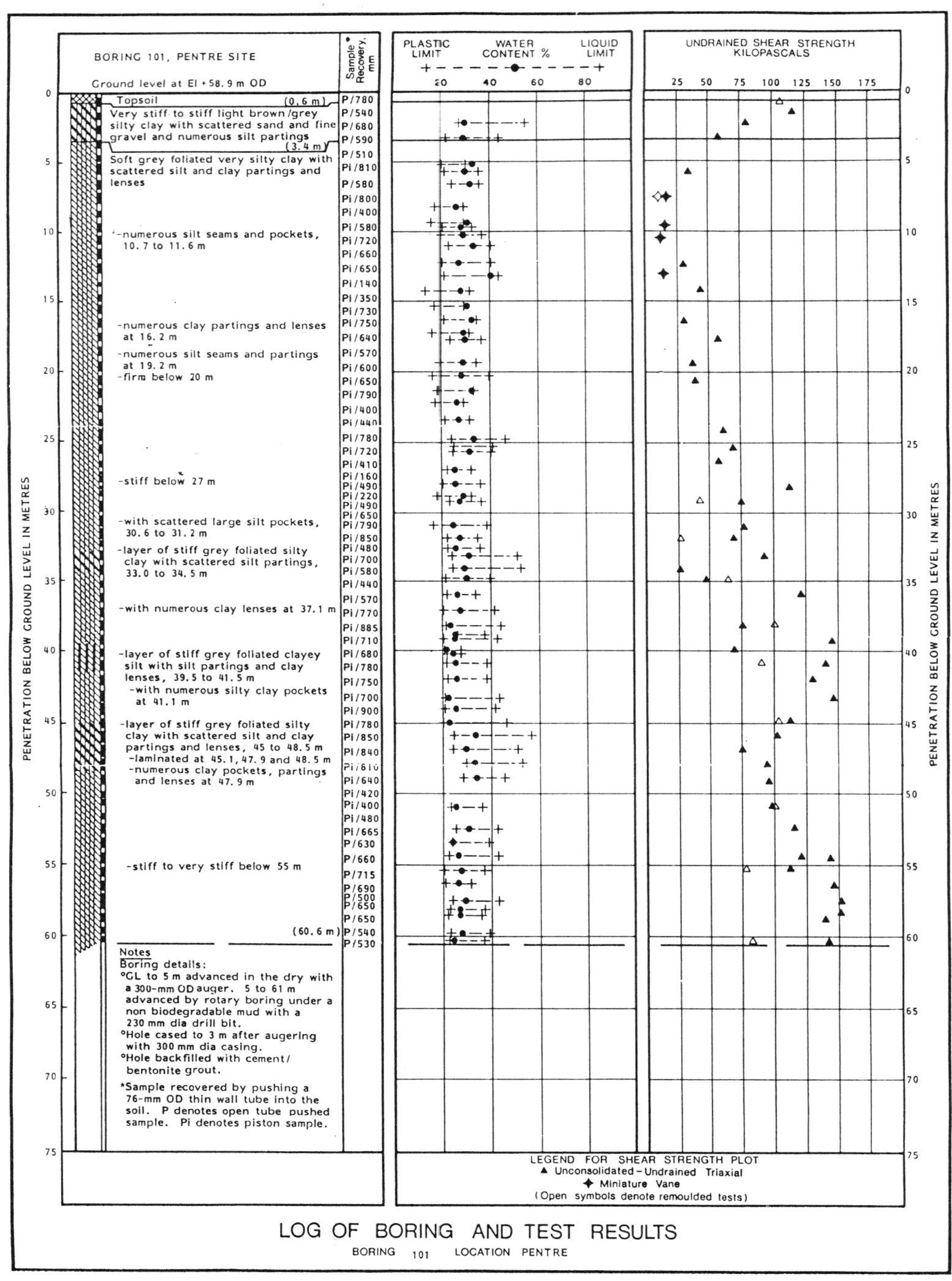

Fig. 3. Log of boring and test results (boring 101, location Pentre).

2.4. Tilbrook Grange Ground Conditions

Figure 7 presents a representative borehole log for the Tilbrook Grange soils. The stratigraphy at the site is uniform with some 18m of Lowesoft Till overlying at least 22m of Oxford Clay. The Lowestoft Till is a very stiff to hard dark grey silty clay with numerous sand to medium gravel sized chalk fragments and with occasional gravel to cobble sized flints. It was considered that these inclusions did not adversely affect the performance of the pile since the material is clay matrix dominated. The Lowestoft Till is of intermediate plasticity with plasticity index in the range of 20% to 30% and natural water content close to the plastic limit. The Oxford Clay stratum comprises a hard fissured clay with occasional silt partings and shell fragments. It is of high to intermediate plasticity with liquid and plastic limits of about 55% and 20%, respectively. Natural water content is about 18%. The OCR at Tilbrook Grange, as judged from a combination of *in situ* and laboratory tests, is about 17 close to the surface and decreases with increasing penetration to have values of approximately 10 at 18m and 5 at 30m penetration. Groundwater pressures at the site were monitored for about 18 months prior to pile testing. The pressure profile is given by Figure 8 which shows a reduction from hydrostatic pressure below 25m-GL and reflects the probable underdraining of the Oxford Clay associated with water extraction from the deeper Cornbrash strata. Figure 9 summarises the UU shear strength data for Tilbrook Grange. There was good agreement between laboratory and *in situ* measurements of strength. Laboratory testing revealed effective stress behaviour was typical of many North Sea soils. The bulk unit weight profile is summarised by Figure 10.

3. Instrumentation and Data Acquisition System

3.1. Instrumentation Development

The procedures undertaken to obtain a system of instruments that would withstand the rigours of being driven with a pile, and yet be stable and sensitive enough to record the long term static behaviour of the pile, are discussed in detail by Solomon *et al* (1992). The key parameters required to be measured using instrumentation installed on the pile were:

- Stresses in the pile wall during driving and load testing
- Axial and lateral acceleration of the pile during driving
- Soil pressure on the pile wall during driving and load testing
- Pore water pressures at the pile-soil interface during driving and load testing

An extensive testing programme was carried out prior to selecting suitable instruments capable of surviving the harsh environment envisaged (range of ±3500 microstrain and accelerations of up to 850g). This was done to ensure a high level of confidence in the ability of each type of instrument to operate throughout the

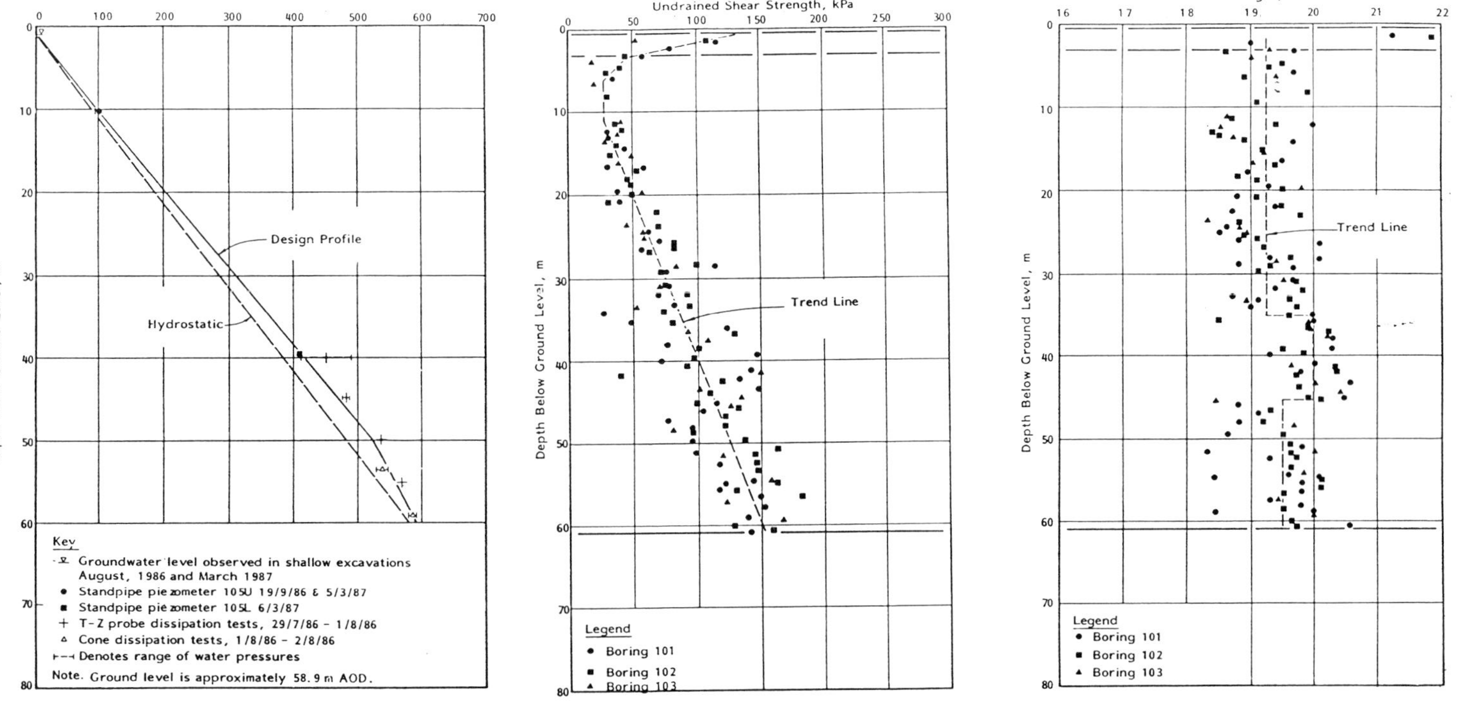

Fig. 4. Pore water pressure profile (Pentre).

Fig. 5. Undrained shear strength (unconsolidated-undrained triaxial tests, Pentre).

Fig. 6. Bulk unit weight (Pentre).

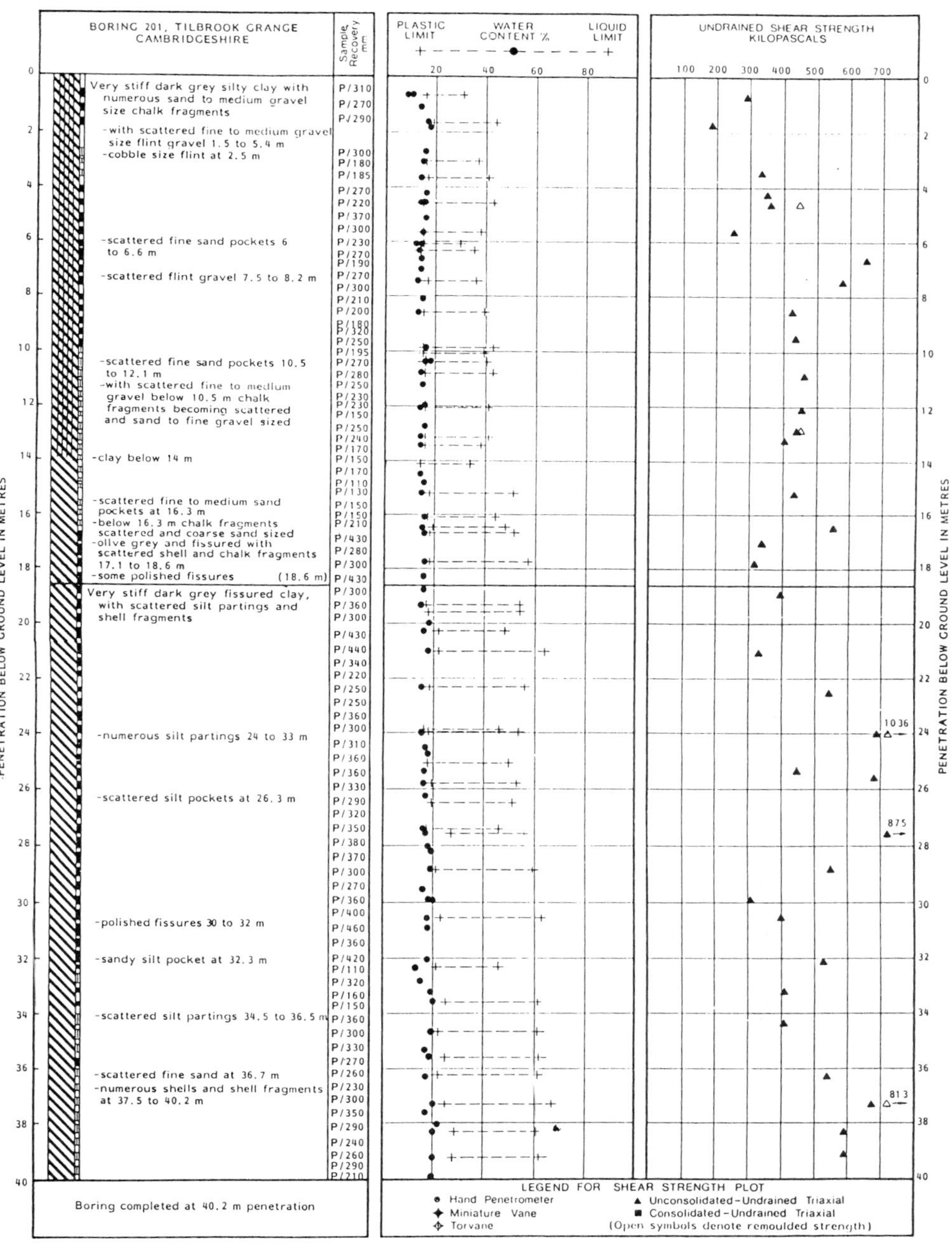

Fig. 7. Log of boring and test results (boring 201, location Tilbrook Grange).

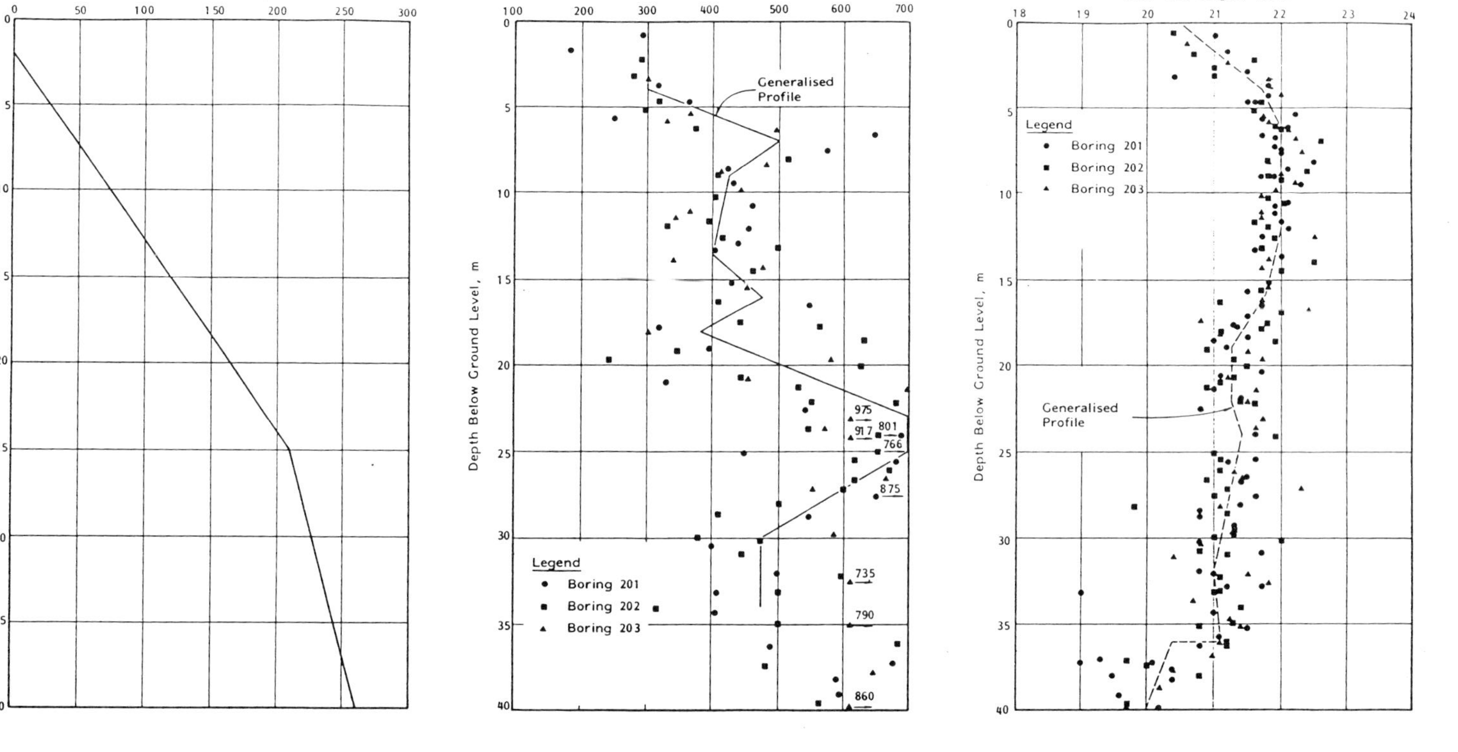

Fig. 8. Ground water pressure profile (Tilbrook Grange)

Fig. 9. Undrained shear strength (unconsolidated-undrained Triaxial Tests, Tilbrook Grange)

Fig. 10. Bulk unit weight (Borings 201, 202 and 203, Tilbrook Grange)

pile driving, and remain operational through the set-up and load testing phases of the project. Selection and testing of a system of instrumentation for the project was carried out in conjunction with the Civil Engineering Department at Queen Mary and Westfield College, part of London University. The testing regime to which the instruments, cable bundling and securing techniques were subjected included static, flexure and dynamic tests.

Wherever possible, system redundancy was enhanced by developing more than one type of instrument to measure each parameter of interest. Piezometers and standpipes were also installed in the surrounding soil to measure pore pressures prior to and after pile driving, and to monitor their dissipation after driving.

3.2. Pile Fabrication and Calibration

The length and wall thickness for each test pile were selected to have similitude with piles used in North Sea offshore platforms. The selected wall thickness required the piles to be fabricated from high-strength steels. The NC pile was selected to have a diameter of 762mm (30 in.) and a wall thickness of 15mm. The OC pile was also selected to have a diameter of 762mm, but with a wall thickness of 30mm.

Fabrication of the two compression test piles began in September 1986 and was completed in August 1987. Steels for the piles were supplied by The British Steel Corporation. The two test piles were fabricated from roller-quenched and tempered (RQT) steels. RQT 501 steel with a minimum yield strength of 470 N/mm^2 was used for the NC pile. RQT 701 steel with a minimum enhanced yield strength of 770 N/mm^2 was used for the OC pile.

The layouts of instruments selected for the NC and OC test piles are shown in Figures 11 and 12 respectively. Instruments were installed around the inside circumference of the pile at 90 degrees spacings. Installed instrumentation included strain gauges, strain modules, total pressure and pore pressure cells together with axial and lateral accelerometers. Channels were installed inside the piles to protect instruments and cables.

Pile section calibration was carried out for four main reasons:

- to stress-relieve the pile prior to driving and load testing in order to minimise stress relief that could cause offset changes in instruments during driving and load testing;
- to verify correct operation of strain sensors;
- to calibrate strain sensors against applied axial load, in order to obtain a sensitivity value in volts per kN that could be used during load testing;
- to measure the cross-sensitivity of the total pressure cells and pore pressure cells to axial load, i.e. to measure the spurious output caused by compression of the cell body, in order that a correction factor could be applied in the field.

Details on the method of performing pile section calibration are contained in Cox *et al* (1992). The conference discussion (Clarke, 1992) considered the effect

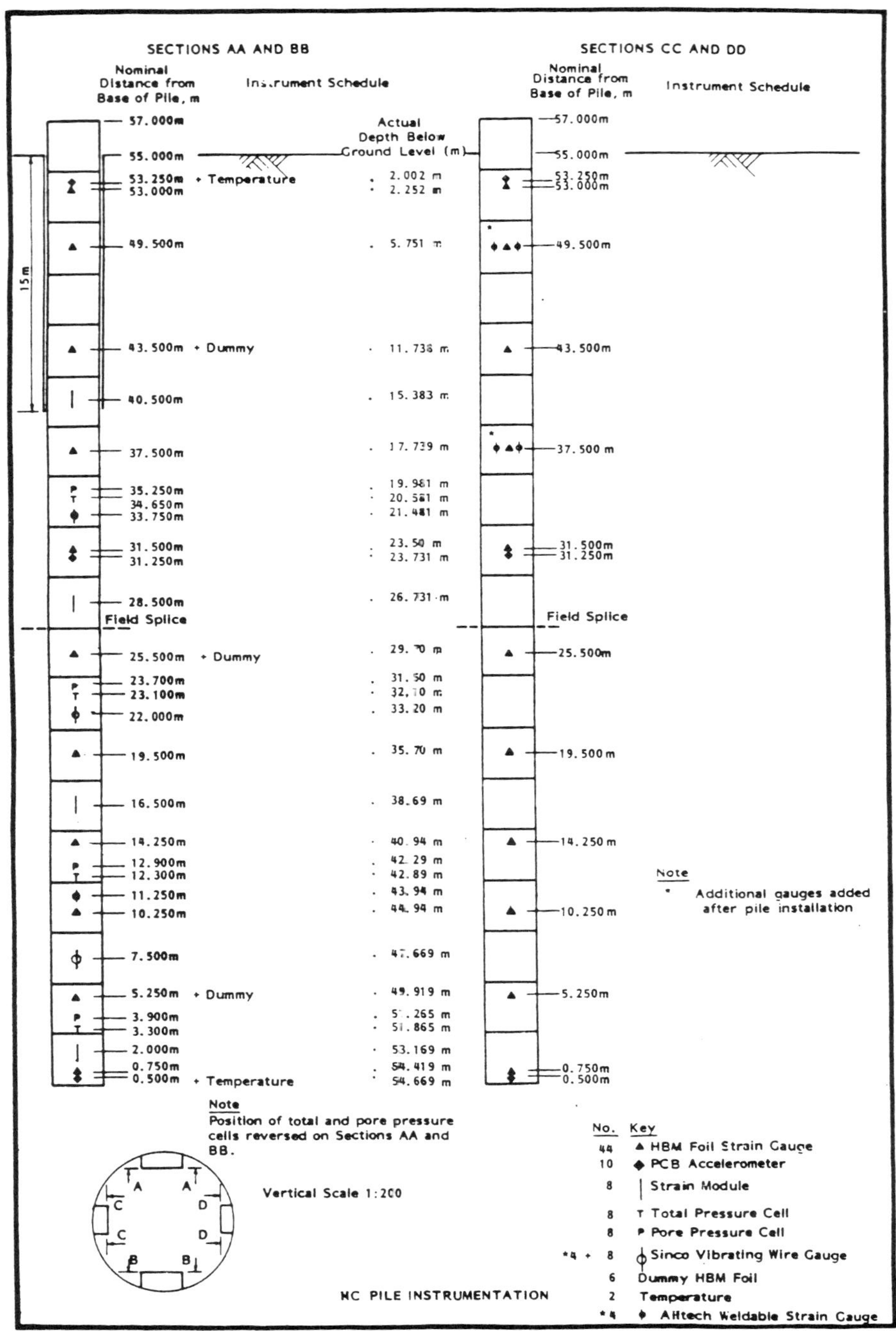

Fig. 11. NC pile instrumentation.

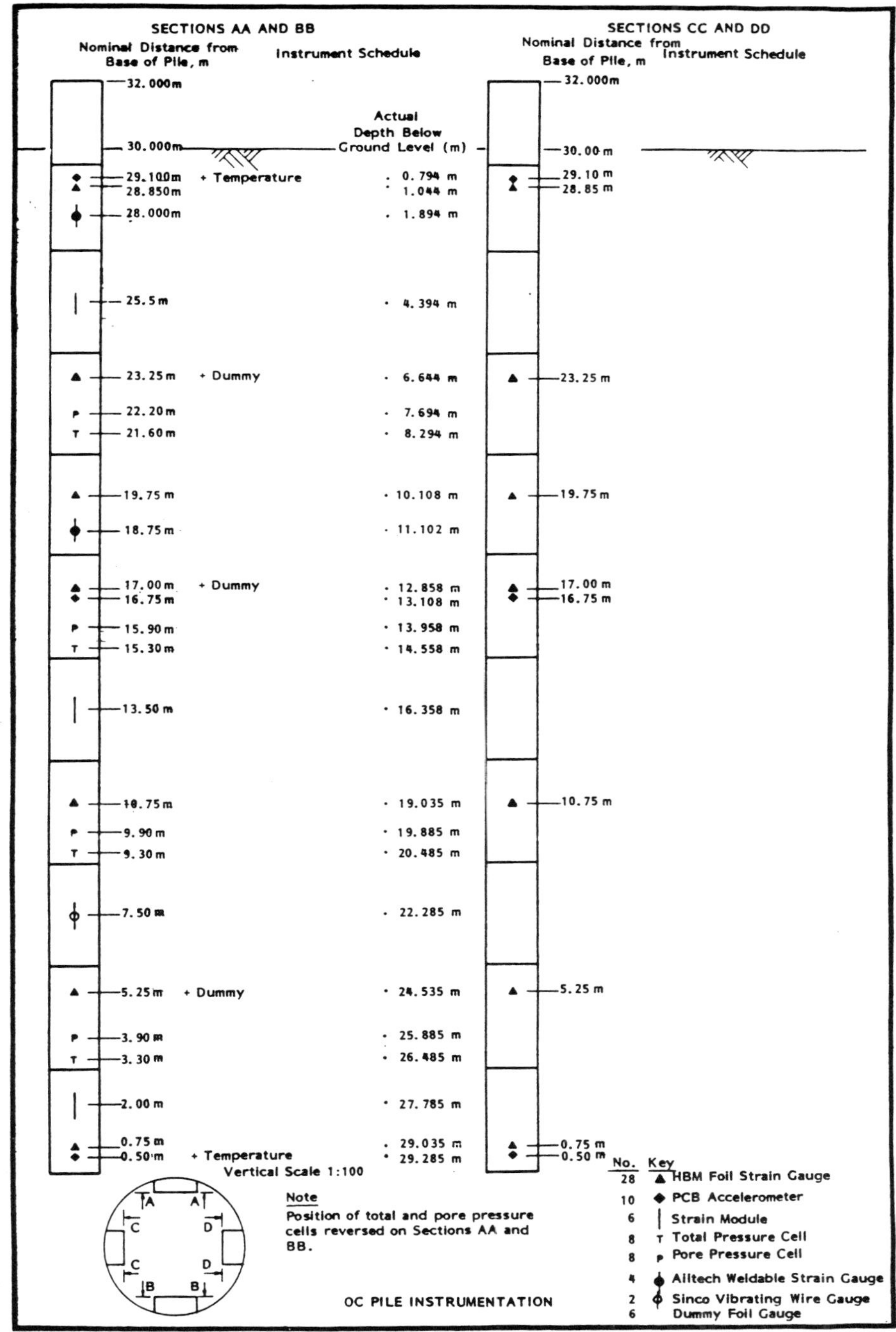

Fig. 12. OC pile instrumentation.

of the total pressure cell cross-sensitivity correction on the results, since this was significant for the NC pile.

3.3. Data Acquisition System

The same data acquisition system was used at both sites for all three piles to collect data from instrumented piles during driving, to record pressures and stresses during the set-up period, and to monitor short term trends during load tests. The system consisted of signal conditioning units, patch units, tape recorders and a computer system. It operated successfully and proved to be reliable and adaptable to the requirement of all three tests, Currie *et al* (1992).

4. Installation and Test Set Up

4.1. Installation

The first pile driven was the tension test pile at the Tilbrook Grange (OC) site. The dimensions of this pile were 762mm OD, 35mm WT and 33m long. It was driven in two sections, the first of which was 15m long. The second section of length 18m was spliced to the first section in the field. This pile was installed as a pile driving trial and therefore had no instrumentation. The compression test pile at the OC site was 32m long. It carried 72 transducers attached internally, with the signal cables being carried in channels. The pile was driven to 30m penetration. The NC pile at Pentre was 57m long. It carried 92 transducers and again had internal channels for the signal cables. The pile was installed through a 15m sleeve to a penetration of 55m. The piling hammer used to install both piles at the OC site was the BSP HA40, with a rated energy of 470 kJ and ram weight 40 tonnes. At the NC site, the BSP HH7, with a rated energy of 84 kJ and a ram weight of 7 tonnes was used. In both cases the driving was only moderately difficult. Details on behaviour of pile and sensors during driving can be found in Poskitt *et al* (1992).

4.2. Test Frame

The test frame used for all three tests was designed for a maximum test load of 30MN to be consistent with the maximum likely load at the OC test site. The anchorage systems (barrette walls NC site, bored piles OC site compression test) were designed such that the horizontal clearance between the anchorages and the test piles were at least five pile diameters. The reaction frame was connected by the use of debonded macalloy bars. Two spread footings 12m long by 2m wide were used as reaction for the tension test at the OC site. The piles were loaded by hydraulic jacks reacting against the load frame.

At both sites, piezometers were installed adjacent to the compression test piles and together with pore pressure gauges on the piles themselves used to monitor pore water pressures. Details on response of these gauges, the reaction frame and

Fig. 13. 30-MN-capacity reaction frame at NC site.

Fig. 14. 30-MN-capacity reaction frame at OC site.

the loading system used are contained in Cox *et al* (1992) and Clarke *et al* (1992). Figures 13, 14 and 15 show the general arrangement of the load reaction system used for all three tests.

5. Axial Static Test Results

5.1. Pentre

The NC pile was driven on 20 July 1987 with the static load test being conducted some 44 days later on 2 September 1987. At the time of initiating the static load test, there was complete dissipation of the excess pore pressures generated by pile driving. Peak load measured at the pile head during the static test was 6.03MN at a pile head movement of 36.1mm. Loading of the pile continued to about 115mm displacement where a post peak load of 5.48MN was recorded. Upon unloading,

Fig. 15. Tension test set up

the pile head rebounded by 21mm. Pile head load versus deflection for the static test is presented on Figure 16. At the peak load, the end bearing was 0.85MN or about 14% of the total capacity and was mobilised at a tip movement of 10mm.

5.2. Tilbrook Grange Compression Test

The OC compression test pile was driven 5 October 1987 and the static load test carried out 130 days later on 16 February 1988. It is estimated that excess pore water pressure dissipation in the Oxford Clay layer was only around 73% at the time of the test, although it had reached 90% in the upper Lowestoft Till layer. Peak load measured at the pile head during the static test was 16.13MN at a pile head movement of 28.1mm. Loading of the pile continued to 128.7mm movement where a post peak load of 14.21MN was measured. Upon unloading, the pile head rebounded by 17.4mm. Pile head load versus deflection for the static test is presented on Figure 17. At the peak load, the end bearing was 1.45MN or about 9% of the total capacity and was mobilised at a tip movement of 8.6mm.

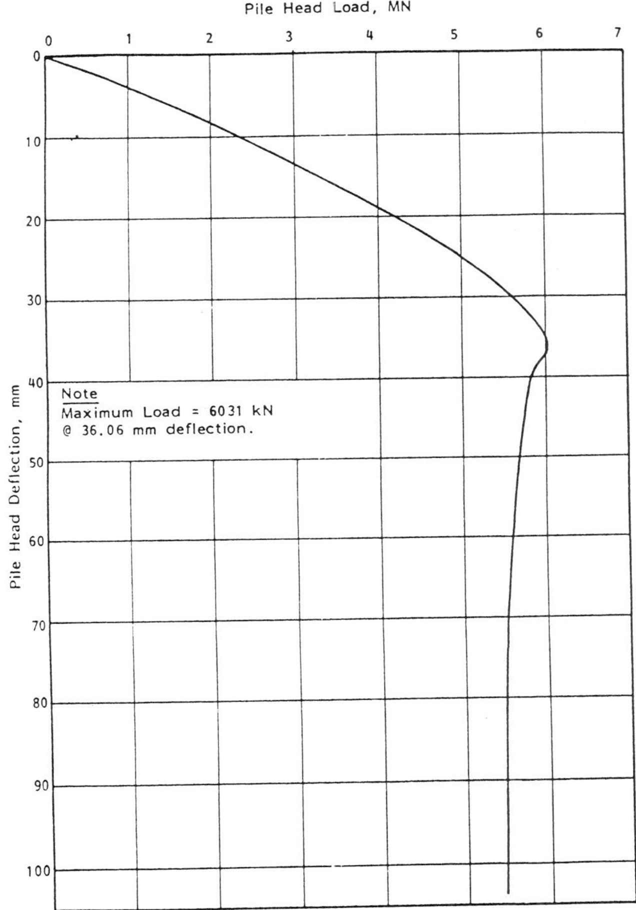

Fig. 16. Load versus deflection at pile head (NC pile, Pentre site, static test).

5.3. Tilbrook Grange Tension Test

The OC tension test pile was installed to a penetration of 30.5m on 19 January 1987. It was driven a further 0.5m on 5 October 1987 and load tested on 22 September 1988, some 20 months after the initial installation. Peak load measured at the pile head during the static tension test was 16.20MN at a corresponding pile head movement of 41.5mm. Loading of the pile continued to 96mm movement by which time the pile head load had decreased to 15.2MN. On removal of the load, a residual displacement of 77.3mm remained, giving a pile elastic recovery of 18.7mm. Pile head load versus deflection for the tension test is also presented on Figure 17.

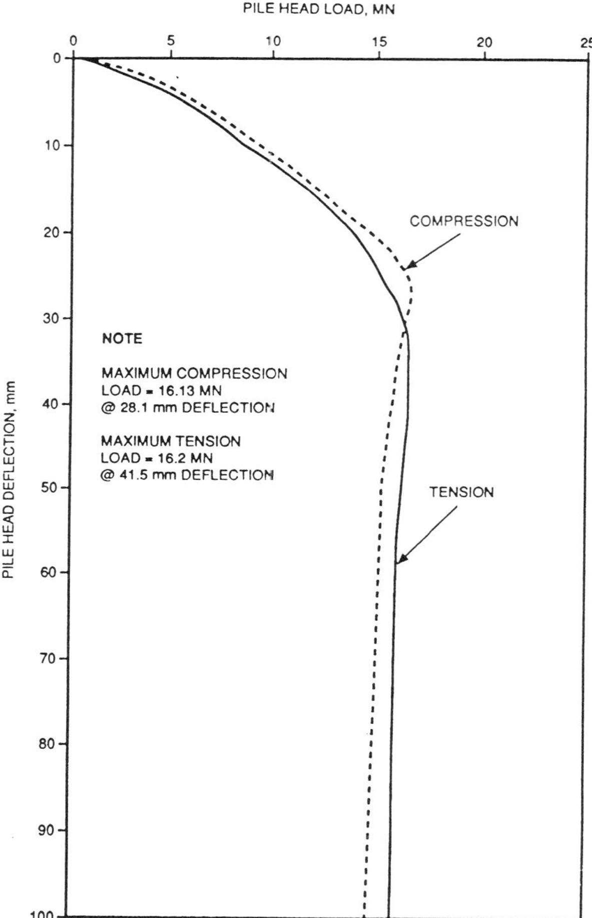

Fig. 17. Load versus displacement (Tilbrook Grange).

6. Interpretation and Analyses

A more detailed appraisal of the compression load test results is given by Gibbs *et al* (1992). It is recognised that they presented the project viewpoint at that time and subsequent analyses by others present alternative interpretations. These opinions will be recorded in the published proceedings of the Recent Large Scale Fully Instrumented Pile Tests in Clay Conference (Clarke, 1992).

6.1. Pentre Compression Test

The pile was driven through a 15m deep 914mm OD casing from which the soil had been removed. At the end of driving the soil plug was 1.73m above the tip of

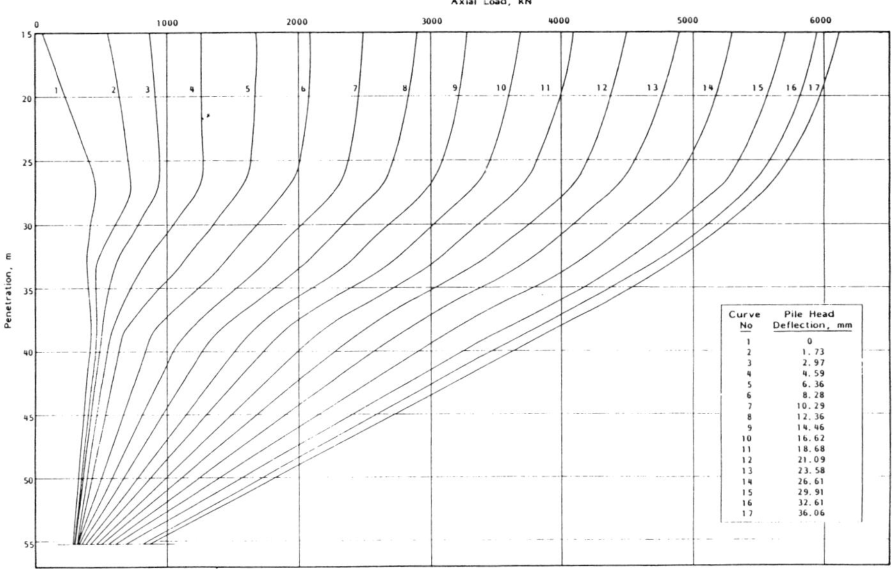

Fig. 18. Axial load distribution NC pile (static load test, pre-peak loads).

the casing.

Total earth pressure measurements indicated that the total radial stress was more or less constant throughout the consolidation stage down to 32m. Below this depth reductions of up to 35 percent from the value immediately after installation were measured. However, at the end of consolidation, the radial effective stresses were much smaller than expected except in the more plastic silty clay zone at 43m, where the ratio of the effective radial to the original effective vertical stresses was about 0.54. This is in reasonable agreement with the K_0 value for normally consolidated soil. Above 32m the effective radial stress was nearly zero, and at 52m, the effective stress was also smaller than expected. It is not known why the final effective radial stresses were so small. However, it is considered that the total stress measurements warrant further investigation since the pore pressures returned sensibly to their expected values. Low radial stresses in the upper section of driven piles has also been recorded by Lehane and Jardine (1992).

The axial load distribution curve is given by Figure 18. The corresponding unit skin friction curves are shown by Figure 19. This figure presents the skin friction distributions for load levels up to the maximum pile head load of 6.03MN. In addition, the skin friction is shown for the residual load which existed at the start of the test. Residual top load was 0.02MN, corresponding to the dead weight of the load cells and bearing plate.

The measured unit skin friction profile at peak load is shown on Figure 20 and

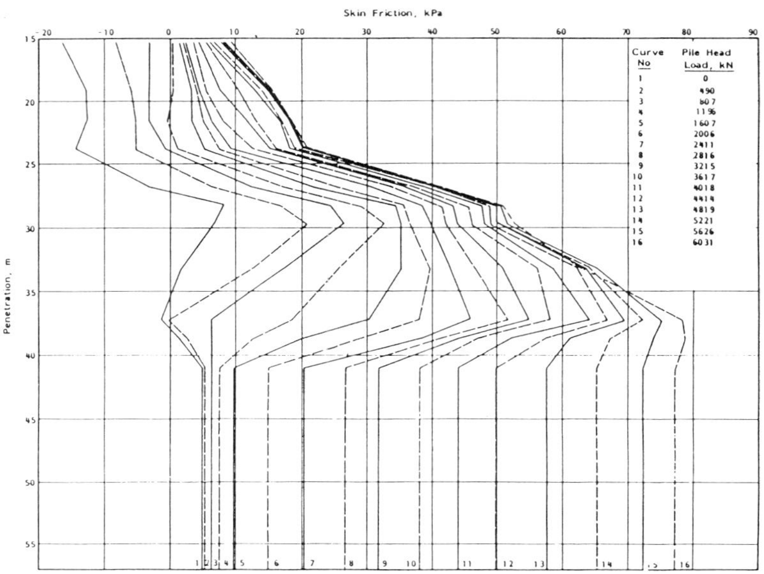

Fig. 19. Skin friction versus penetration NC pile (static load test, pre-peak loads).

compared with UU profiles from the three sample borings. Selection of the UU profiles rather than any other test type is consistent with the recommendation of API RP2A (1987). The undrained shear strength profile corresponding to a rate of increase of 2.5 kPa per metre is superimposed on the UU profiles. It is generally considered that this idealised profile fits the UU data reasonably well although variations above and below the trend are apparent below about 40m.

The average unit skin friction at peak load was about 54 kPa, corresponding to a conventional friction/strength alpha value of 0.62. The skin friction profile at peak load consisted of a linearly increasing portion between 15 and 37m followed by a constant portion between 37 and 55m. While this latter portion appears at variance to the selected design s_u line it is in better agreement with the individual UU profiles.

Over the lower third of the pile length, the measured axial load – displacement (t- z) curves exhibited distinct peak values followed by a strain softening residual stage. They are in accordance with models for non-linear work hardening behaviour prior to yield followed by strain softening after peak, (Randolph 1983, 1985, Kraft *et al* 1981b), see Figure 21.

The peak friction is mobilised at a displacement of 10 to 12mm (1.5% pile diameter) and the ratio of residual to peak skin friction is about 0.75, with the residual values being reached after an additional displacement of 40mm. In contrast to the well defined t-z curves in the deeper soil, the curves in the shallower soil

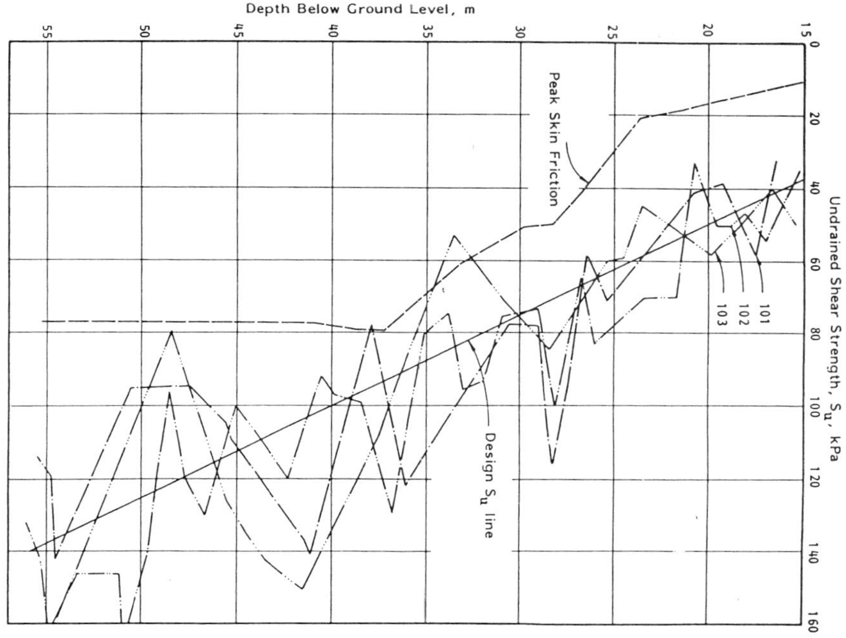

Fig. 20. Comparison of peak skin friction and undrained shear strength, NC pile.

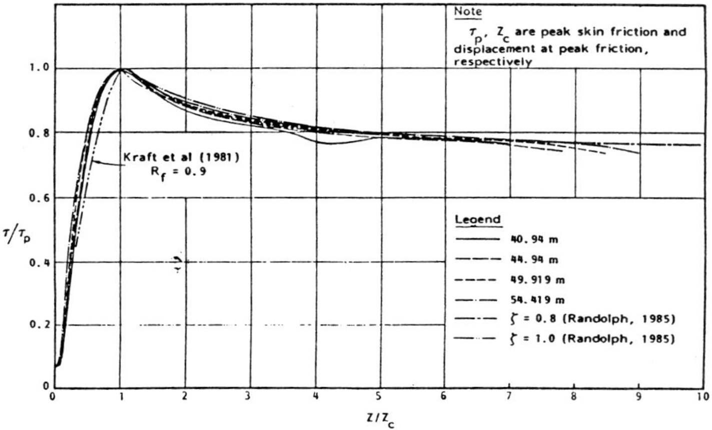

Fig. 21. Normalised t-z curves NC piles (comparison with analytical procedures).

did not exhibit peaks but values comparable in magnitude to residual friction. Here the maximum friction appears mobilised at displacements of 25 to 30mm. Over the middle third of the pile there is a transition zone where the peak is less well defined than in the deeper zone. Possible reasons for the lower frictions include soil fatigue caused by pile driving forces.

Gibbs *et al* (1992) employed several analytical procedures when comparing the measured peak frictions with predictions. These included total as well as effective and quasi-effective stress methods. Seven predictive methods were used. Both rigid and compressible pile based methods were used in comparisons. However, in this instance only the results from the rigid pile methods are outlined since these are indicative of those initially employed by designers. It is recognised that compressive (t-z) based methods are more representative of long offshore piles. Table 2 summarises the results of the analyses. The peak pile capacity corresponds to the load transfer only when the pile is infinitely rigid. The peak capacity is affected by the shaft resistance at each point along the pile and by the soil-pile interaction as defined by the load transfer (t-z) response. In addition, the amount of softening in the t-z curve (residual skin friction) also affects the capacity, Murff (1980), Kraft *et al* (1981a), Randolph (1983), as well as the pile-soil stiffness ratio. Therefore, for a long compressible pile, peak frictional capacity may not correspond to the integration of the peak load transfer curves over the pile length since each level down the pile may be at a different point on the load transfer curve. It is clear that for long compressible piles in normally consolidated soils the frictional capacity is best estimated using load transfer based methods.

The NC pile end bearing response was stiffer than predicted by most methods. Ultimate pile tip load was mobilised at 2.6% pile diameters. The corresponding bearing capacity factor increased throughout the load test, from a value close to 10 up to a maximum of 15.8. There was strong evidence of consolidation resulting from partial drainage.

6.2. Tilbrook Grange

This section first discusses the results from the compression test pile at the OC site. The findings from the tension test pile are then outlined before a brief comparison of the two axial load tests is given.

6.2.1. Compression test

At the end of pile installation internal soil plug level measurements indicated that the pile was continuously coring but approached plugging at 13m, although full plugging was not achieved. It was estimated that the displacement ratio below 13m penetration was 80%. Excess pore pressures at the pile wall generated by pile driving were generally 0.5 to 0.7 s_u in the Lowestoft Till and 2 s_u in the Oxford Clay. At the end of driving, the ratio of maximum excess pore pressure to effective

TABLE 2.
Predicted "rigid" peak friction capacity, NC pile.

Method	Predicted Frictional capacity MN	Average peak skin friction, kPa	Predicted/ Measured
API (1984) Para 2	4.47	46.6	0.86
Kraft et al λ	6.59	68.6	1.27
Meyerhof β (remoulded interface strength)	6.92	72.3	1.34
Randolph and Murphy	7.09	74.0	1.37
Semple and Rigden	8.41	87.8	1.63
Meyerhof β (remoulded soil strength)	8.66	90.5	1.68
Esrig and Kirby	9.46	98.8	1.83

overburden pressure was between 1.6 and 3.9 in the Lowestoft Till and between 3.9 and 4 in the Oxford Clay.

The OC pile load test was performed 130 days after driving. At that time, the percentage dissipation of excess pore pressures, assessed from the pore pressure measurements on the pile wall, was between 80% and 95% in the Lowestoft Till and around 73% in the Oxford Clay. During consolidation, the measured total earth pressure was nearly constant for all cells except those near the pile tip, which showed reductions of about 15% to 20% after the set-up period of 130 days. At the end of consolidation, the effective radial stress was found to be between 0.85 times the UU strength at 14m increasing to 1.8 times at 26m.

The measured profile of residual stresses for the Tilbrook Grange compression pile prior to pile load testing is presented in Figure 22. Both the mobilised skin friction and distribution of load have been plotted versus penetration. As shown, large negative residual stresses occur over the upper 13m of the pile. These negative friction values range between 40 and 65 kPa with an average value of 45 kPa corresponding to about 0.12 times the average undrained shear strength over this depth interval. As shown subsequently, the residual stresses in the upper part of the pile lead to reduced friction values throughout the test. The measured residual load at the tip is about 0.5MN, or approximately 25 percent of the computed ultimate

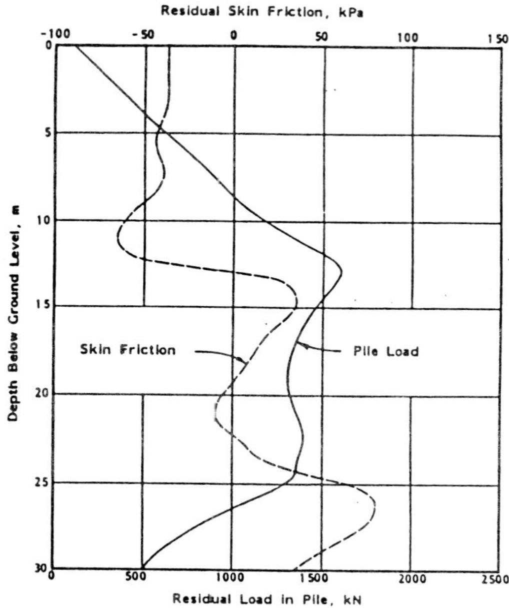

Fig. 22. Residual pile load prior to load test, OC pile.

end bearing value assuming an undrained shear strength of 475 kPa at the pile tip.

The measured peak load was 16.13MN and the pile top displacement was 28mm. At the final residual (post-peak) stage the load reduced to 14.21MN (88 percent of peak) and the top displacement was 129mm. The field load versus pile head deflection curve is shown on Figure 17. At the peak load the load carried by the tip was estimated to be 1.45MN or about 9% of the total load. The average measured unit skin friction is 209 kPa, giving a conventional friction/strength α value of 0.44 based on the UU data. At the final residual (post-peak) load, the average unit skin friction is 173 kPa giving an α value of 0.37 and the load carried by the tip is 2.07MN (14.6% of residual capacity). The axial load distribution curve is given by Figure 23. Figure 24 presents the skin friction distributions for the maximum pile head load.

Figures 25 and 26 show representative unit skin friction displacement t-z curves for the Lowestoft Till and Oxford Clay materials, respectively. Comparison of the two characteristics shows differences in behaviour. Over the lower part of the Lowestoft Till the critical displacements are between 15 and 20mm. Overall maximum frictions appear mobilised at displacements of 10 to 20mm (1.3% to 2.5% of pile diameter) larger than expected in heavily overconsolidated clays for piles of this size. In the Oxford Clay, the curves exhibit distinct peaks with residual to peak values typically in the range of 70 to 80 percent. The maximum skin friction throughout this stratum appears to be mobilised at displacements of 8 to 10mm

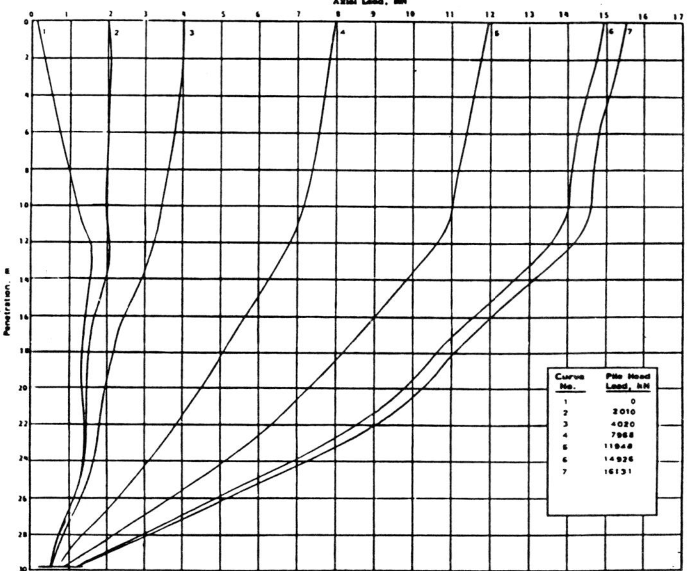

Fig. 23. Axial load distributions, OC pile (static load test, pre-peak loads).

(1.3% of pile diameter).

Predictions of pile axial capacity were made using several recognised methods and compared with measured capacity. These predictions were computed on the basis of a compressible pile. Table 3 from Gibbs *et al* (1992) presents the results of the comparison. It shows that the majority of the methods predict the measured peak capacity to within 15 percent. The scatter in the predicted residual capacity, however, is greater, varying between overpredictions by 13 percent and underprediction by 25 percent.

The OC pile end bearing response was close to that normally predicted by offshore pile design methods. At peak pile top load the measured tip load was 9% of the end total with a corresponding tip movement of 1.1% pile diameter. However, the ultimate end bearing load was achieved at about a 10% diameter movement and corresponded to a bearing capacity factor of about 9.

6.2.2. *Tension Test*

A lower than anticipated static ultimate axial capacity from the compression test, combined with concerns regarding the degree of excess pore water pressure dissipation, resulted in an additional axial load test being performed. Advantage was taken of the existing pile-driving trial pile which was tested under tensile loading conditions. Consequently, the soil plug within this pile was augered out and a range of instrumentation installed. Further details are given by Clarke *et al* (1992). Owing to the post installation of the instruments no information was available for

TABLE 3.
Comparison of measured/predicted load deflection response, OC pile.

Predictive Method	Peak Capacity MN	Pile Head Deflection mm	Residual Capacity MN	Predicted/Measured	
				Peak	Residual
API (1984) Para 2	17.4	24.0	14.6	1.08	1.03
Meyerhof - β (1)	13.6	20.7	11.5	0.84	0.81
Meyerhof - β (2)	19.1	26.7	16.0	1.18	1.13
Parry and Swain	14.8	22.2	12.2	0.92	0.86
Esrig and Kirby 1. ϕ' from Direct shear tests 2. δ from Direct shear tests	 17,3 18.8	 23.7 25.3	 14.2 15.4	 1.07 1.17	 1.00 1.08
Kraft et al λ β	 12.3 17.2	 17.3 20.7	 10.7 14.8	 0.76 1.07	 0.75 1.04
Randolph and Wroth 1. $D_R = 1.0$ 2. $D_R = 0.22$	 15.5 14.1	 22.2 20.6	 12.7 11.6	 0.96 0.88	 0.89 0.82
Randolph and Murphy	14.0	21.3	11.7	0.87	0.82
T-Z Probe	13.8	23.2	15.7	0.86	1.11
Measured	16.1	28.1	14.2	-	-

effective stress considerations. However, high quality load and strain information was obtained.

The measured peak load was 16.2MN at a displacement of 41.5mm. Figure 17 gives the load deflection response. At peak load the load distribution down the pile is given by Figure 23, while Figure 24 presents the unit skin friction distribution. The mobilisation of soil resistance with pile movement (t-z) has been calculated from the strain gauge data at six levels as shown by Figure 27. The measured profiles for the Lowestoft Till and at the top of the Oxford Clay stratum show no degradation with increasing displacement. However, the t-z characteristic measured deeper in the Oxford Clay exhibits a distinct peak at a smaller displacement with a residual to peak value of approximately 80%.

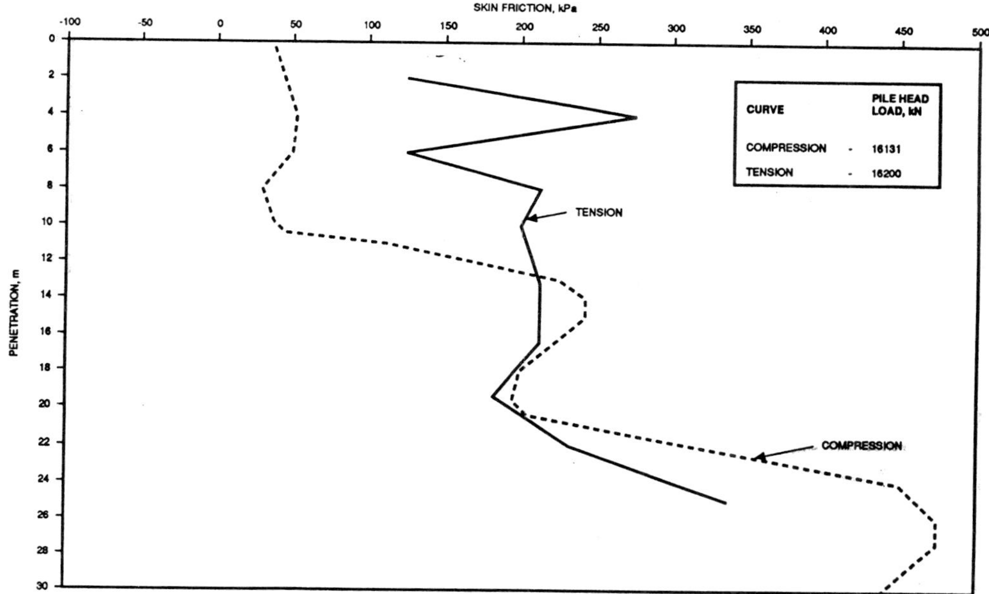

Fig. 24. Skin friction in piles at peak load (Tilbrook Grange).

6.2.3. Comparison

Reference to Figure 17 shows that the peak loads measured in the compression and tension tests were similar (16.13MN and 16.2MN). However, about 50% more displacement was required to mobilise the maximum load in the tension pile. Comparison of the magnitude of shaft resistances indicates that mean unit skin frictions were very similar. The end bearing component of the compression test pile (9% of total load) essentially was balanced by the additional skin friction from the deeper tension test pile.

Figure 23 shows that 'load take out' was substantially greater in the Lowestoft Till layer for the tension test, i.e. an extra 2.5MN over the upper 10 to 11 metres. This difference is shown more clearly in terms of skin friction, Figure 24. In contrast 'load take out' and hence skin friction for the bottom two thirds of the pile are less for the tension test than for the compression test. The reason for this may be due to residual stress effects. Measurements detailed Gibbs *et al* (1992) showed that after set up residual friction is negative over the upper 12.5m of the compression test pile.

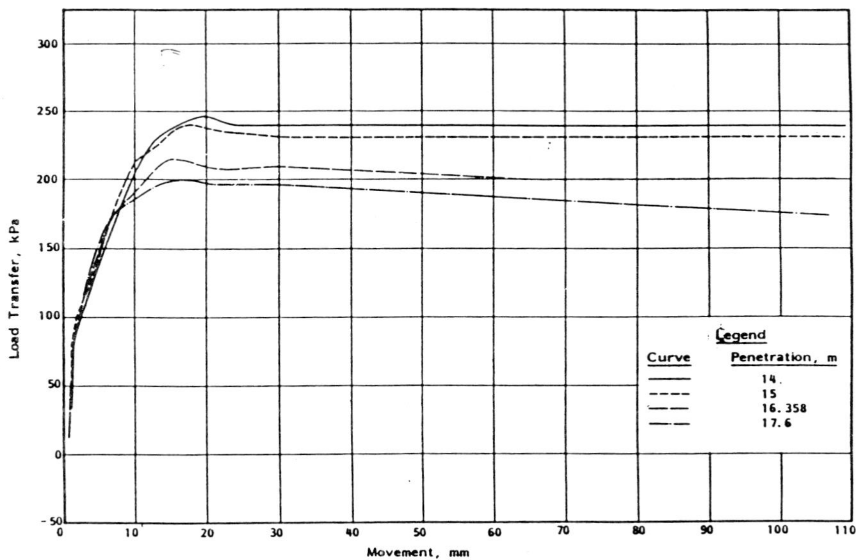

Fig. 25. Load transfer curves OC pile (static load test).

This probably results from the soil resisting rebound of the pile. During the static compression test, the skin friction over this upper portion of pile is reversed. This strain reversal may account for the very low skin friction observed in this section of pile at the conclusion of the compressive pile test. In contrast, the bottom section of the pile starts in compression and remains so during the test. Other possible reasons for the low unit skin friction in the upper third of the compression test pile include pile 'whip' during installation and a friction fatigue effect related to the distance behind the pile tip. The latter is addressed briefly in section 6.1 while the former was discussed in some detail at the conference on Recent Large Scale Fully Instrumented Pile Tests in Clay (1992), with the consensus being that this was unlikely.

Comparison of the t-z characteristics (Figures 25 to 27) reveals that for the tension test pile the peak frictions in the Lowestoft Till are mobilised at movements of some 2.5 to 4.5% of pile diameter, i.e about double that for the compression test pile. As noted earlier the characteristic shape of the t-z curve differs for the Lowestoft Till and Oxford Clay. Much greater similarity of t-z curves occurs in the Oxford Clay soil for both the compression and tension test piles.

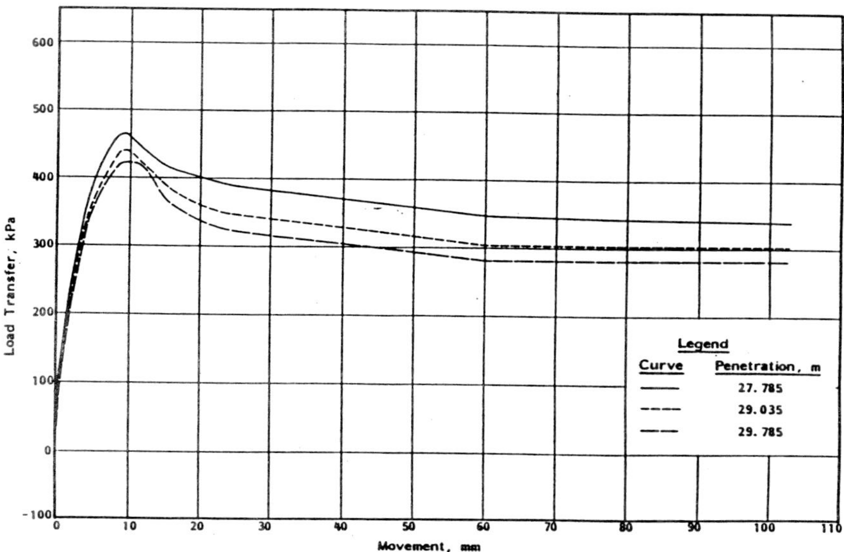

Fig. 26. Load transfer curves (static load test, OC pile).

7. Design Implications

The three load tests were carried out on fully instrumented piles, which were carefully selected to reproduce the compressibility characteristics of most offshore piles. Control was exercised to ensure continuity of the measurements. Installation stresses were measured prior to load testing for both compression test piles. Hence the measured pile load distributions are considered to be reasonably accurate. For the tension test pile the residual stress profile from the similar compression test pile was used when calculating load distributions. The data from the three tests are therefore considered to be of high quality, and the quality is believed to be better than on any other test in the API database.

Figure 28 presents the Pentre and Tilbrook Grange pile test results against load test data from the API database. Also shown is the current API RP2A (1991) recommended design line and the design line proposed by Semple and Rigden (1984) for piles with length to diameter ratios less than 60. For more flexible piles they proposed a length factor with potential to reduce axial capacity up to 30%. Clearly, there are differences between the high-quality test results presented here and the API (1991) main text method. These are briefly discussed in the following paragraphs.

At Tilbrook Grange the API recommended method is shown to be valid, albeit slightly conservative. However, Hobbs (1992) recommends the adoption of a limiting skin friction approach for the upper portion of the pile combined with the

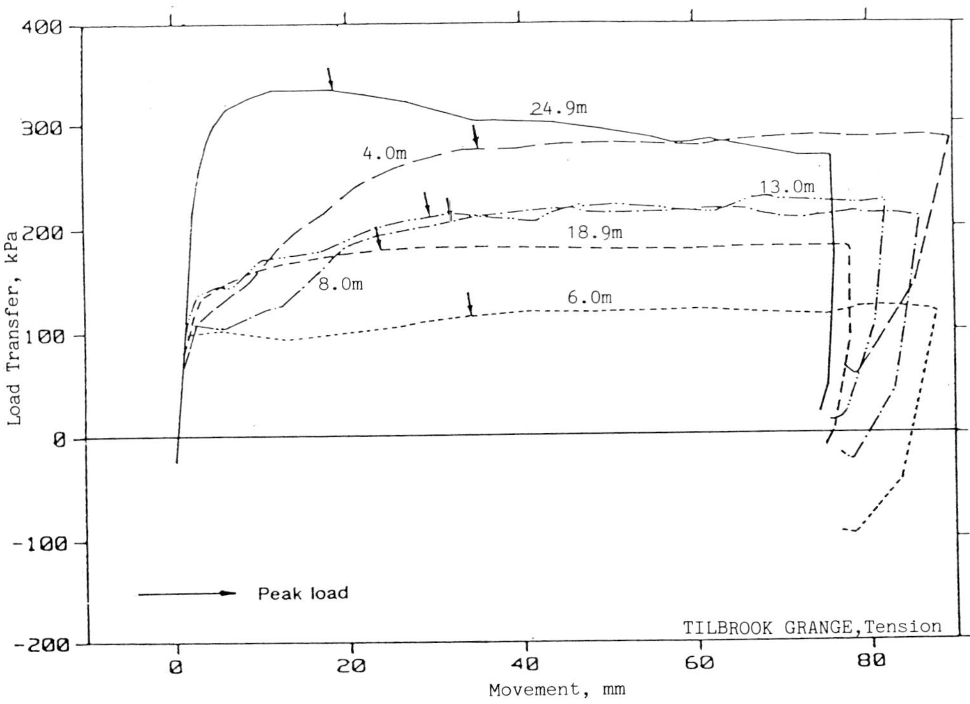

Fig. 27. Load transfer curves (tension load test, OC pile).

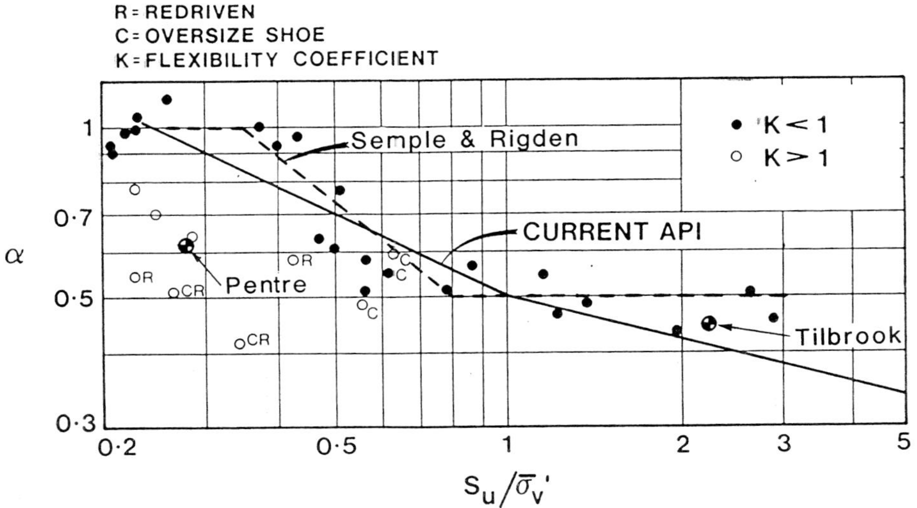

Fig. 28. Load test data from API database (after Randolph and Murphy).

former API-2 Method. This approach correctly predicts the low skin frictions on the upper third of the compression test pile but significantly underestimates skin frictions on the upper part of the tension pile. An alternative approach was presented by Nowacki *et al* (1992), which modified the API criteria for s_u/σ'_v, values greater than 0.7. The suggested change provides a better fit to the data, especially at values of the strength ratio in excess of 2.

At Pentre the API main text method is shown to be unconservative. Reference to Figure 28 reveals that the design line is weighted towards relatively rigid piles with a pile-soil stiffness ratio (Randolph, 1983) close to unity. Offshore sized piles in essentially normally consolidated soils are relatively flexible with typical K values of the order of 2. In recognition of this discrepancy API are considering the adoption of an approach which recognises relative pile-soil flexibility effects. This is based on a degrading t-z curve criteria. Some preliminary work on this is presented by Aldridge and Schnaid (1992). For offshore piles in essentially normally consolidated soils, Hobbs (1992) recommends use of the API-2 method (API RP2A, 1987). Reference to Table 2 indicates that at Pentre this method underpredicts axial capacity by some 14%. It should be recognised that other factors also influence the capacity of offshore piles. The majority of the API database relies upon the observed performance of piles which have been load tested over a relatively short period, and usually only a few days to weeks after installation. Generally offshore piles must withstand their design loads for very long periods and are very subjected to a combination of static and cyclic loads. The effectively viscous nature of the response of cohesive soils to stress changes has long been recognised with implications for rapid loading. Usually time effects can be divided into two main categories, which may both lead to increased capacities.

- consolidation effects generally occur over a relatively long time period due to the dissipation of installation pore water pressures,

- viscous rate effects occur over short time periods due to high rates of loading.

Consolidation effects are not relevant for the Pentre pile since excess pore pressures were shown to have dissipated prior to testing of the pile. Cyclic loading can have two potentially counteractive effects on pile axial capacity.

Load degradation may occur at combinations of static and cyclic load which approach the static ultimate capacity. In normally consolidated clays failure under cyclic load is caused primarily by excess pore pressure generation. Fundamentally this is a strain dependent rather than a stress dependent process. Since soils are stiffer at smaller strains than at larger strains there is a threshold level for cyclic loading, below which the pile is stable and above which there is degradation and it moves to failure. Typically this threshold occurs at 70% to 80% of the pile ultimate capacity. Hence there is strong evidence that provided the single pile minimum factor of safety is above about 1.3 cyclic load degradation effects should be negligible.

Loading rate effects are of particular interest in predicting pile performance for offshore structures where load rise times may be rapid. This is in contrast to the quasi- static, monotonic pile load test or conventional laboratory strength testing with durations at least three orders of magnitude longer. A review of field data by Briaud *et al* (1984) found that the total dynamic monotonic load resistance could be as high as twice the 'static' capacity with comparable increases in stiffness of the pile response. There is a consensus that these rate effects are greatest in soft plastic clays at high liquidity indices. Centrifuge model test work has also shown that the rate effects around the pile shaft are typically an order of magnitude greater than those around the pile base (Lambson, 1987). The Pentre pile carried over 80% of the applied load on the pile shaft. API RP2A recognises that rapidly applied loading can cause increased resistances but gives no specific indication on how to take them into account. Both laboratory element and model test work indicate that viscous rate effects around a pile shaft typically cause a capacity increase of 15 to 20% per log cycle increase in loading rate, referenced to the initial loading rate. Quantification of these rate effects is beyond the scope of this paper. However, it should be recognised that they do occur, may be substantial, and always increase the pile axial capacity.

Hence, the axial capacity of long offshore piles in essentially normally consolidated clays is likely to be underpredicted by the API-2 Method, especially once time effects are taken into account. However, Dunnavant *et al* (1990) have shown that some piled foundations in these clays can have greater calculated axial capacity when cyclic loading effects are considered explicitly than when static procedures are used. The current API main text method overpredicts capacity. It is clear that soil type, stress history, relative pile-soil flexibility and time all have an influence. However, it is recognised that the current API methods, while offering a practical approach, are empirically based. The pile test programme summarised here has added significant high quality data allowing interpretation in terms of effective stresses. It is considered important that further interpretation is undertaken to increase understanding and to derive a pile axial capacity approach which is more fundamentally based. In particular, the discussion on pile axial capacity considered soil plasticity as an important factor. There are new and fundamental insights in the papers presented to this conference which may offer a way forward.

Acknowledgements

The authors wish to thank BP International for permission to publish this paper.

References

1. Aldridge, T. R. and Schnaid, F. (1992), 'Degradation of skin friction for driven piles in clay', *Proc. Conf. on Recent Large Scale Fully Instrumented Pile Tests in Clay*, ICE, London, June.
2. API (1987, 1991), 'RP2A: Recommended Practice for Planning, Designing and Constructing Fixed Offshore Platforms', 17th, 19th editions.
3. Bond, A. J. and Jardine, R. J. (1990), 'Research on the Behaviour of Displacement Piles in an Overconsolidated Clay', (prepared by the Imperial College, London, for the Department of Energy), Department of Energy Offshore Technology Report, OTH 89296.
4. Briaud, J. L., Garland, E., and Felio, G. Y. (1984), 'Rate of loading parameters for vertically loaded piles in clay', *Proc. 16th OTC*, Houston, Vol. 1, pp. 407–412.
5. Clarke, J. (ed.) (1992), *Recent Large Scale Fully Instrumented Pile Tests in Clay*, Thomas Telford, London.
6. Clarke, J., Long, M. M., and Hamilton, J. (1992), 'Tilbrook Grange tension test', *Proc. Conf. on Recent Large Scale Fully Instrumented Pile Tests in Clay*, ICE, London, June.
7. Cox, W. R., Cameron, K., and Clarke, J. (1992), 'Static and cyclic axial load tests on two 762mm diameter pipe piles in clay', *Proc. Conf. on Recent Large Scale Fully Instrumented Pile Tests in Clay*, ICE, London, June.
8. Cox, W. R., Solomon, I. J., and Cameron, K. (1992), 'Instrumentation and calibration of two 762mm diameter pipe piles for axial load tests in clay', *Proc. Conf. on Recent Large Scale Fully Instrumented Pile Tests in Clay*, ICE, London, June.
9. Currie, C. and Soloman, I. (1992), 'Large diameter pile tests data acquisition system', *Proc. Conf. on Recent Large Scale Fully Instrumented Pile Tests in Clay*, ICE, London, June.
10. Dunnavant, T., Clukey, E. C., and Murff, J. D. (1990), 'Effects of cyclic loading and pile flexibility on axial pile capacities in clay', *Proc. 22nd OTC*, Pater 6378.
11. Gibbs, C., McAuley, J., Mirza, U., and Cox, W. R. (1992), 'Reduction of field data and interpretation of results for axial load tests of two 762mm diameter pipe piles in clay', *Proc. Conf. on Recent Large Scale Fully Instrumented Pile Tests in Clay*, ICE, London, June.
12. Hobbs, R. (1992), 'The impact of axial pile load tests at Pentre and Tilbrook on the design and certification of offshore piles in clay', *Proc. Conf. on Recent Large Scale Fully Instrumented Pile Tests in Clay*, ICE, London, June.
13. Kraft, L. M., Focht, J. A., and Amerasinghe, S. F. (1981a), 'Friction capacity of piles driven into clays', *Journal of Geotechnical Engineering Division, ASCE* **107**(GT11), 1521–1541.
14. Kraft, L. M., Ray, R. P., and Kagawa, T. (1981b), 'Theoretical T-Z curves', *Journal of Geotechnical Engineering* **107**(GT11), 1543–1562.
15. Lambson, M. D. (1987), 'The Behaviour of Axially Loaded Piles in Clay', PhD Thesis, University of Manchester.
16. Lambson, M. D., Clare, D. G., Senner, D. W. F., and Semple, R. M., (1992), 'Investigation and Interpretation of Pentre and Tilbrook Grange Soil Conditions', *Proc. Conf. on Recent Large Scale Fully Instrumented Pile Tests in Clay*, ICE, London, June.
17. Lehane, B. M. and Jardine, R. J. (1992), 'The behaviour of displacement pile in glacial till', *Proc. BOSS '92 Conference*, Vol. 1, pp. 555–556.
18. Murff, J. J. (1980), 'Pile capacity in a softening soil', *International Journal of Numerical and Analytical Methods in Geomechanics* **4**, 185–189.
19. Nowacki, F., Karlsrud, K., and Sparrevik, P. (1992), 'Comparison of recent tests on OC clay and implications for design', *Proc. Conf. on Recent Large Scale Fully Instrumented Pile Tests in Clay*, ICE, London, June.
20. Poskitt, T. J., Yip-Wong, K. L., and Cox, W. R. (1992), 'Measurement of axial strain and pile wall displacement and the determination of skin friction during the driving of instrumented piles', *Proc. Conf. on Recent Large Scale Fully Instrumented Pile Tests in Clay*, ICE, London, June.
21. Randolph, M. F. (1983), 'Design considerations for offshore piles', *Proc. Conf. on Geotechnical Practice in Offshore Engineering, ASCE*, Austin, Texas, pp. 422–439.
22. Semple, R. M. and Rigden, W. J. (1984), 'Shaft capacity of driven pipe piles in clay', *Proc. Symp. on Analysis and Design of Pile Foundation, ASCE*, San Francisco.

23. Solomon, I. J., Cox, W. R., Clarke, J., and Poskitt, T. J. (1992). 'The development of instrumentation for drivability testing and load testing of large, diameter piles', *Proc. Conf. on Recent Large Scale Fully Instrumented Pile Tests in Clay*, ICE, London, June.
24. Thomas, S. (1990), 'Geotechnical Investigation of UK Test Sites for the Foundations of Offshore Structure', Department of Energy Offshore Technology Report, OTH 89294, HMSO, London.

Discussion

Question from C. H. Price, Kennington, Ashford, Kent:

(a) How were temperature-change effects taken into account in respect of its effect on the steel reference beams?

(b) How do the authors relate the rate of loading in the test (from no load to failure in one hour) back to practice, where rates of loading are different?

Authors' response:

(a) Temperature change effects on the aluminium reference beam were given proper consideration. Two main measures were undertaken to minimise them. Firstly the beam was shielded from any direct sunlight. Secondly the beam was pinned at one end and free to slide at the other so that linear expansion did not translate into vertical movement. Hence only the expansion due to the depth of the beam was relevent. It should also be recognised that the pile head vertical displacements to peak load were between about 28 and 36mm and were measured to a system accuracy of 0.1mm. Thus any small variations due to temperature fluctuations were negligible and did not detract from the overall validity of the vertical movement measurement.

(b) The compression tests were undertaken at a constant rate of penetration, typically 1mm/min, to be compatible with other pile load test results. Additionally, for the NC pile, two fast static tests were conducted at a rate of 14.1mm/min. These recorded a load increase of some 3% over the initial compression test. The interpretation of the pile load test results (Gibbs *et al*, 1992) did not explicitly account for rate of loading effects since comparison was made with various pile design methods which also did not account for them explicitly.

The authors accounted for rate effects using a variety of methods since it is understood that there is not yet one accpeted method for estimating them. Craig (1983) and Lambson (1987) report results from centrifuge model pile tests which examine load changes over 3 log cycles of loading rate. More recently Bea (1992) describes a procedure to estimate strain rate effects related to cyclically loaded offshore piling, and Tika- Vassilikos *et al* (1992) report the work at Imperial College which correlates fast jacked model pile test results with laboratory interface shear tests at different rates.

Question from R. Hobbs, Lloyd's Register: With regard to the comparison made in the paper between predictions using the API-2 method and the results of the Tilbrook Grange tension test, it is worth noting that the differences in friction distribution are less significant if only the changes in friction from the start of the test are considered (Hobbs, 1992).

Reference is made to work by Aldridge and Schnaid (1992) on degrading *t-z* criteria. My understanding is that the peak unit frictions used with their proposed criteria were *not* those calculated from the API "main text" procedure, which explains any apparent discrepancy with my own findings for Pentre, as presented at this conference.

Authors' response: The authors agee with Dr. Hobb's comments. However, it is considered that the pile tests at Pentre and Tilbrook Grange model offshore piling very well. It should be recognised that part of the reason for undertaking the test programme was to examine two important soil conditions not included in the API RP2A database. These were OC clays with shear strengths greater than 400 kPa and stiff, essentially NC soils. The selected sites were representative of soil conditions found beneath the North Sea and expanded the range of soil conditions in the database.

Open ended steel tubular piles were used. The relative pile soil stiffness was selected to be representative of full sized offshore piling. The pile diameter was 0.76m with typical North Sea platform piles being 1.37 to 2.34m diameter. The piles were installed by driving, using hammers comparable to those used offshore.

RESPONSE OF PILES IN SOFT CLAY AND SILT DEPOSITS TO STATIC AND CYCLIC AXIAL LOADING BASED ON RECENT INSTRUMENTED PILE LOAD TESTS

K. KARLSRUD, B. KALSNES and F. NOWACKI
Norwegian Geotechnical Institute, P.O. Box 40 Taasen, 0801 Oslo, Norway

Abstract. The Norwegian Geotechnical Institute (NGI) has carried out an extensive series of static and cyclic axial load tests on well instrumented piles, respectively in a soft plastic clay deposit at Onsøy (Norway), a silty clay deposit at Lierstranda (Norway) and a mainly clayey silt deposit at Pentre (UK). Three types of test piles were used. Test piles type A were a closed-ended 219 mm o.d. steel pipe pile with length of 10 m. They were driven through cased boreholes to tip penetrations of 15 to 37.5 m. Test Pile B was a tubular open-ended steel pile with o.d. of 812 mm driven to tip penetration of 15 m. Test Pile C, only used at Onsøy, was of the same type as Pile A, but was 30 m long and driven to 35 m tip penetration.

Each pile was subjected to a series of static (monotonic) and cyclic axial load tests after the excess pore pressures generated during pile installation had essentially fully dissipated. The paper summarizes the main aspects of the test results and how these tie in with various theoretical models and design methods and some previous pile tests in similar clay deposits.

1. Introduction

This paper presents results of axial pile load tests recently carried out by the Norwegian Geotechnical Institute in three soft clay deposits, respectively at Onsøy and Lierstranda in Norway and at Pentre in the UK. The piles were all steel pipe piles driven into the ground and were heavily instrumented to measure distribution of axial load and shaft friction, and earth and pore pressures against the pile shaft.

Three types of piles were tested. Pile type A was closed-ended with a diameter of 21.9 cm and embedded length of 10 m. They were driven through cased boreholes to reach tip penetrations ranging from 15 to 37.5 m. Pile type C was a 37.5 m long flexible pipe with same dimension as Pile A (only tested at Onsøy). Pile type B was open-ended pile with diameter of 81.2 cm and 10 m embedded length, driven through a 5 m casing to 15 m tip penetration (only tested at Onsøy and Lierstranda).

Each pile was subjected to a series of static and cyclic load tests. The static (monotonic) tests were all tensile (pulling) type tests, whereas the cyclic tests included both one-way tensile loading and reversed tension/compression loading.

A vast amount of data has been collected from these pile tests, and the main results have been summarized in References 1, 2, 3, 4. This paper will only highlight

Volume 28: Offshore Site Investigation and Foundation Behaviour, 549–583, 1993.

the test results and try to show how the results tie in with analytical models and some design rules. Most emphasis is put on observed response during the pile installation and reconsolidation phase, the state of effective stress after reconsolidation, and the ultimate shaft friction and how that relates to effective stresses. The load-displacement response and effects of cyclic loading will only be briefly referred to. The results are also compared to similar and very extensive pile tests carried on 5 m long and 15.4 cm diameter instrumented piles at the Haga site (References 5, 6, 7).

2. Test Piles and Soil Conditions

The Onsøy site is located approximately 100 km to the south-southeast of Oslo, Norway. The soil deposit at Onsøy consists of about 45 m of marine clay overlying moraine and bedrock. The upper 8–9 metres consist of a thin desiccated crust followed by soft plastic clay with some black spots of iron sulphide, organic matter and fragments of shell. The deeper clay deposit is grey with occasional shell fragments.

The Lierstranda site is located outside the city of Drammen, 35 km southwest of Oslo. The Lierstranda test site is partly a reclaimed area. About 2.5 m of fill material was placed in the test area in the early 1970's. Below the fill materials there is more than 50 m of soft normally consolidated marine silty clay.

The Pentre site is located approximately 10 English miles northwest of Shrewsbury in England. The upper 3.5 m at the site consists of a stiff weathered clay crust with scattered sand and fine gravel. Below the surface crust is mainly clayey silt with variable content of clay.

Index soil properties for the sites including soil description, water content and content of clay material are presented in Figure 1. At Onsøy the clay is plastic, plasticity index varying from 30 to 50%. At the two other sites the soil is more silty and less plastic, with plasticity index from 10 to 20%.

The *in situ* stress conditions are shown on Figure 2. At all three sites measured *in situ* pore pressures are somewhat larger than hydrostatic pore pressures.

The undrained shear strength at the sites have been determined by a number of methods, both *in situ* tests and laboratory tests. At Onsøy these include field vane tests, unconfined compression tests, consolidated undrained triaxial compression and extension tests, and direct simple shear tests (DSS) on samples consolidated to *in situ* effective stresses. The s_u profile used for interpretation of the pile tests are mainly based on the average laboratory strength, $(s_u)_{\text{lab}}$ in which

$$s_{u,\text{lab}} = \frac{1}{3}(s_u^C + s_u^{\text{DSS}} + s_u^E),$$

which also corresponds closely to s_u^{DSS} and $(s_u)_{\text{vane}}$.

At Lierstranda, the recommended s_u as shown in Figure 14 is also identical to $s_{u,\text{lab}}$. The triaxial and DSS tests on Lierstranda material showed a fairly strong

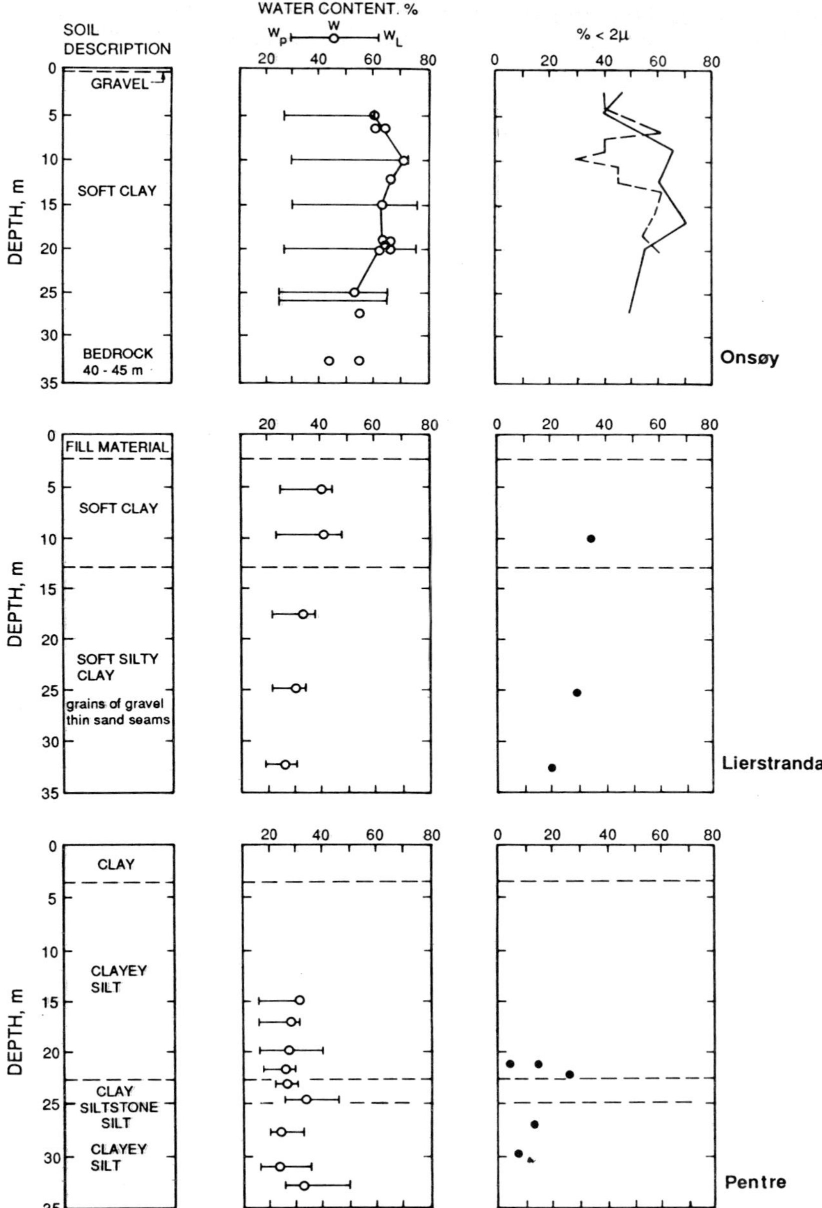

Fig. 1. Typical soil profiles, Onsøy, Lierstranda and Pentre test sites.

tendency for dilation, which started when the effective stress paths approached the Mohr-Coulomb failure envelope. The dilating nature of some of the specimens may be due to sample disturbance and volume changes during consolidation to *in situ*

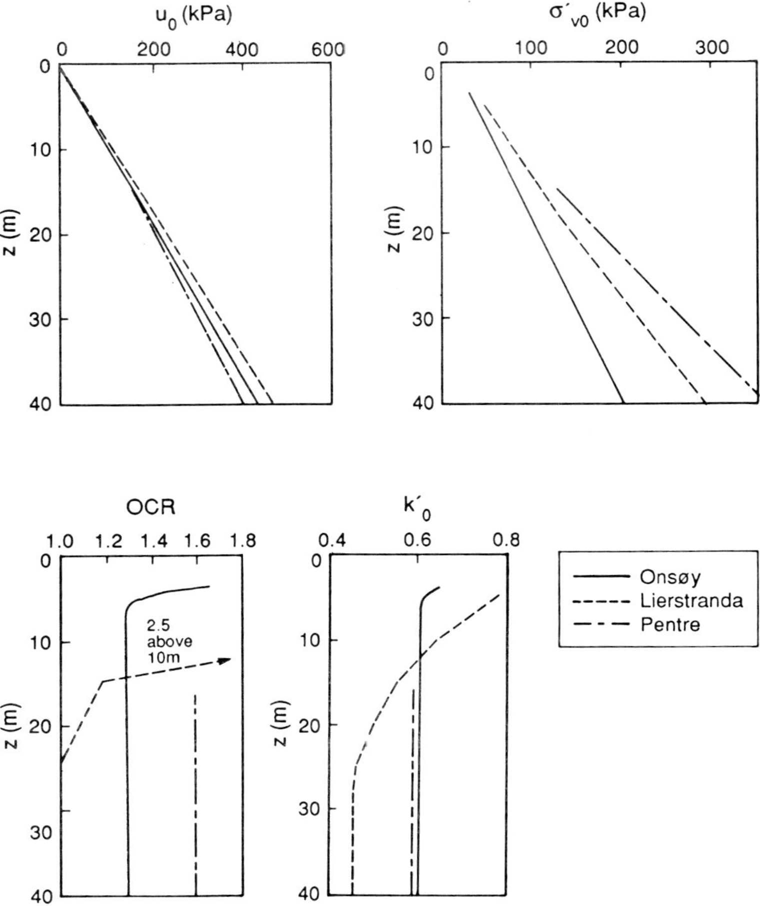

Fig. 2. In situ stress conditions, Onsøy, Lierstranda and Pentre.

effective stresses. Truly undisturbed clay may not show such tendency for dilation. The representative profile as presented in Figure 14 is therefore determined at a strain of about 6%, neglecting the dilation effects at larger strains. This strength profile varies somewhat from *in situ* vane measurements. Above 15 m depth the *in situ* vane strength is somewhat larger than $s_{u,\mathrm{lab}}$, while below it is opposite. The proposed profile corresponds well with s_u interpreted from CPT tests using $N_K = 15$.

Most of the site and laboratory tests at Pentre were carried out for the Large Diameter Pile testing project (LDP-project) carried out at the same site, and have been summarized in Reference 8. The various types of undrained tests show considerable variation, corresponding to normalized undrained shear strength ratios ranging from $s_u/\sigma'_{v0} \approx 0.28$ for UU tests to $s_u/\sigma'_{v0} \approx 0.70$ for *in situ* vane tests. The vane

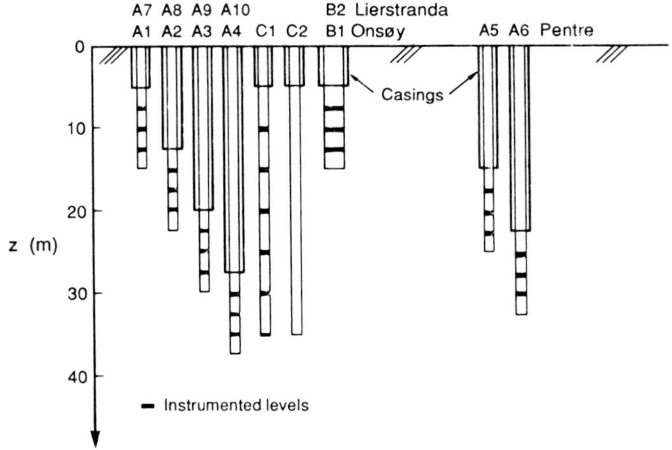

Fig. 3. Illustrations of test piles and embedded depths.

tests are probably not representative due to pore pressure dissipation effects after penetration and during shearing (Reference 8). Interpretation of undrained shear strengths from CIU and CAU triaxial tests are here even more difficult due to strong dilation at large strains. The CIU/CAU triaxial compression tests typically gave s_u/σ'_{v0} of 0.31 at small strains when the Mohr-Coulomb line is reached, climbing to $s_u/\sigma'_{v0} \approx 0.50$ at large strains (20%) as a result of dilation. Two direct simple shear tests gave an average $s_u/\sigma'_{v0} = 0.35$ at large strain and 0.29 at moderate strain. A representative average *in situ* undrained shear strength of $s_{u,\text{lab}} = 0.33\sigma'_{v0}$ has been used in most of these interpretations, except for calculation of α-values from measured shaft friction, where UU-strengths were applied.

In the area close to both the Onsøy site and the Lierstranda sites extensive field and laboratory tests have been carried out. The different tests made at Onsøy suggest that the soil conditions are very uniform in the area. At Lierstranda the soil properties seem to vary more locally. In addition to the shear strength tests on undisturbed clay, NGI has on all deposits carried out DSS-tests on remoulded reconsolidated clay as will be referred to later. Cyclic DSS tests have also been carried out as well as oedometer tests. These results will only be briefly referred to when relevant.

3. Test Piles and Test Program

Figure 3 illustrates the test levels and type of piles tested at each site. Piles A and C had 8.2 mm wall thickness and Pile B 9.5 mm.

Four A-segment piles and one open-ended B-pile were tested both at Onsøy and Lierstranda. The two C-piles were tested only at Onsøy. At Pentre only two A-segment piles were tested. The piles were all driven through pre-emptied casings

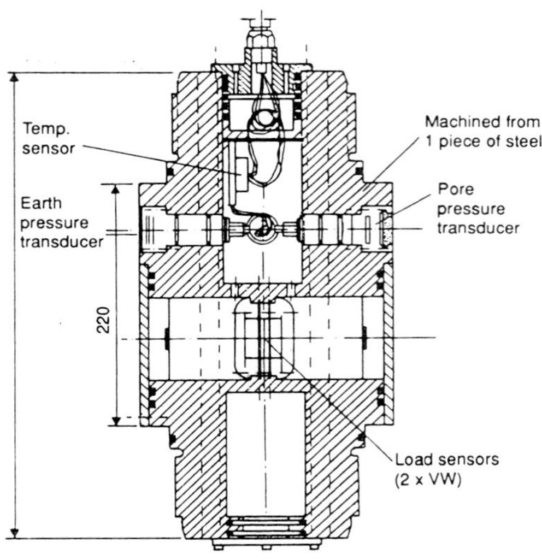

Fig. 4. Instrumentation unit, A- and C-type piles.

extended to the top of the test piles. However, one did not always succeed in getting the casings completely emptied, as some remoulded clay slurry often remained.

The instrument units used for the A-piles and C-piles were identical. Each instrument unit contained 2 earth pressure sensors, 2 pore pressure sensors, 2 axial load sensors and a temperature sensor. All sensors were of the vibrating wire type. Figure 4 shows a typical arrangement of instrumentation for Piles A and C. Distance between instrument units was 2.5 m for Pile A and 5.0 m for Pile C.

The instrumentation of the B-pile was different from the A- and C-piles. The transducers were fixed to the inner pile wall. At each of the three instrumentation levels there were 2 earth pressure sensors and 2 pore pressure sensors. The axial forces in the pile were measured with 8 sealed strain gauges fixed directly to the inner wall of the pile. The location of the strain gauge was offset 0.5 m relative to the pressure transducers to avoid local stress concentration during penetration.

The pore pressure gauges were vacuum saturated in the laboratory and protected with a gasket to prevent air entry until the sensor entered into the ground. In addition to the above one measured loads and displacements at the pile top. All sensors were logged three times per second using a Fastscanner system. The data tapes were run through a specially written computer program to produce time charts, load-displacement curves etc.

The instrumentation of the piles functioned very well. All sensors were carefully calibrated in laboratory before pile testing. Zero-readings were taken both before pile installation and after pile removal to check whether the pile driving had made any influence on the instrumentation. The zero-readings after the pile removal were used. Normally, the zero-readings before and after installation were quite similar.

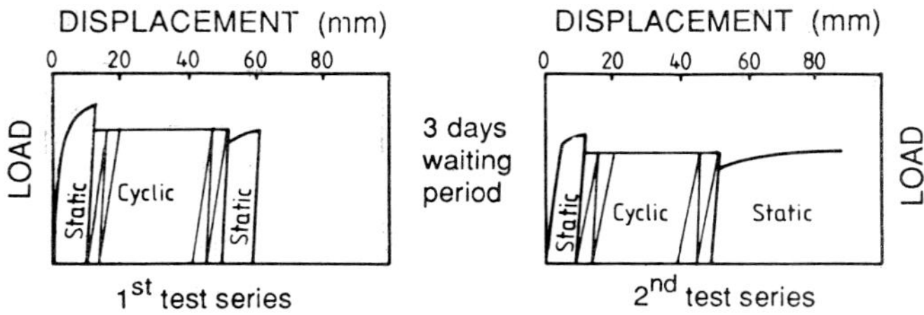

Fig. 5. Typical loading sequence.

There were also very few drop-outs. Generally the two earth- and pore pressure sensors at each instrumentation level gave almost identical readings.

Both the casings and the piles were driven into the soil using a drop hammer of either 1 ton, 2 tons or 3 tons weight. The time required for installation was normally a few hours.

The piles were tested after a reconsolidation period ranging from 30 to 50 days after installation. The load testing program on each pile generally consisted of two test series with a waiting period of some days in between.

In each test series the pile was first subjected to a static (monotonic) load test where the load was increased stepwise to failure in about one hour in a tensile (pulling) mode, Figure 5. Immediately after unloading the pile was subjected to a cyclic load test. These were also load controlled, with the cyclic load levels generally set at a constant predetermined level, with the intention of achieving failure within less than 500 cycles. The cyclic period was 9.2 seconds. The cyclic load levels were in most tests set so that Q_{ave} and Q_{cy} were about the same (e.g. $Q_{\min} \approx 0$), and with $Q_{\max}$ in tensile direction. There were, however also three symmetrical two-way tests on A-piles, two at Onsøy and one at Lierstranda (all as part of a second or third test series).

In addition to the above, a few piles were subjected to a "final static" test immediately after cyclic loading, one pile was loaded statically to failure within a few seconds, and one pile was subjected to a sustained static creep load. Thus, altogether 55 different load tests were carried out on the 14 test piles.

4. The Installation Phase

4.1. Measured Results

Both total earth- and pore pressures against the pile shaft increased significantly compared to the *in situ* stresses during driving of the piles. The excess pressures at Onsøy and Lierstranda were very similar. At Pentre the measured pressures were much lower, due to significant pore pressure dissipation during the driving process itself. The Pentre deposit is more layered and contains more granular soils than at Onsøy and Lierstranda and has about 100 to 200 times larger coefficient of consolidation than the other sites.

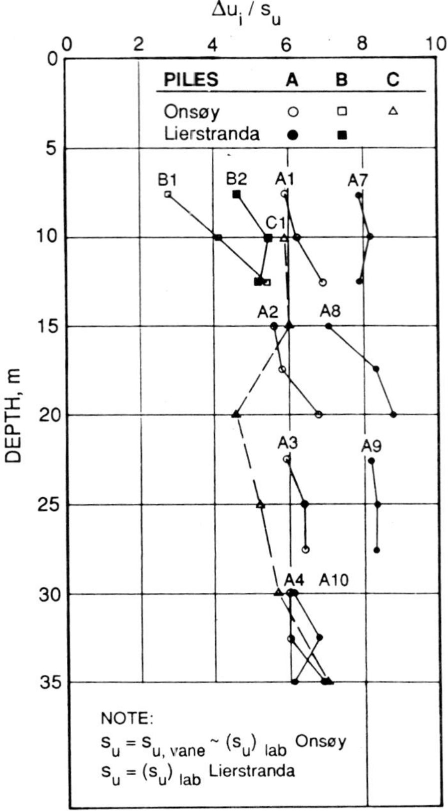

Fig. 6. $\Delta u/s_u$ *vs* depth, Onsøy and Lierstranda.

Figure 6 presents measured excess pore pressures distributions at the end of pile driving, Δu_i, normalized to undrained shear strengths, and Figure 7 shows $\Delta u_i/s_u$ and $\Delta u_i/\sigma'_{v0}$ (σ'_{v0} = *in situ* vertical effective stress), versus overconsolidation ratio, OCR. Included in Figure 7 are also measured values at Haga. The measured data are rather consistent, suggesting that $\Delta u_i/\sigma'_{v0}$ increases with increasing OCR. The Pentre data are not included due to the significant pore pressure dissipation that occurred during pile driving.

The excess pore pressures induced by the closed-ended piles were higher than for the open-ended piles. While typical values of $\Delta u_u/s_u$ and $\Delta u_i/\sigma'_{v0}$ at depth 12.5 m were 7–8 and 1.8–2.2, respectively, for closed-ended piles, corresponding values for the open-ended piles were 5 and 1.45.

The measured total earth pressures after installation were almost identical to the measured pore pressures, which means that the effective horizontal earth pressures measured at the end of installation were very close to zero. Typical effective horizontal earth pressures measured at the end of installation were -10 to 0 kPa.

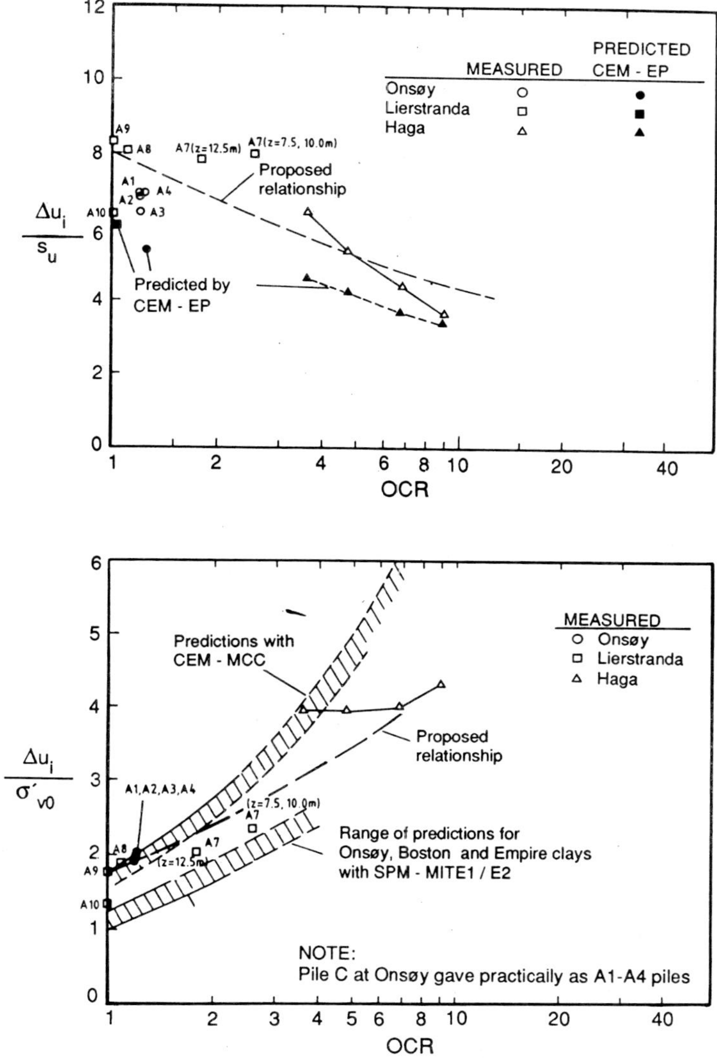

Fig. 7. Normalized excess pore pressures induced by pile installation in relation to OCR.

Note that effective horizontal earth pressures have been taken as the difference between two large values of earth pressure minus pore pressure (which were up to 1800 kPa), which can introduce some inaccuracies.

4.2. Analyses and Discussion

The effect of pile installation on stresses induced in the clay surrounding the piles, can analytically be modelled by coupling an *installation model* for predicting displacement patterns and strain fields, and a *constitutive soil model* for analysing stress changes. Two installation models have been considered:

- The Cavity Expansion Method, CEM (e.g. References 9, 10, 11) which assumes that all displacements during pile penetration occur in a radial direction like expansion of a cylindrical cavity.
- The Strain Path Method, SPM (Reference 12) that also accounts for vertical soil displacements and shear strains during steady penetration of a pile into the ground.

Constitutive soil models that have been applied to analyse these problems are the following:

- A linear elastic- perfectly plastic (EP) total stress model for predicting stresses during pile installation.
- A linear elastic model for predicting stress changes during reconsolidation.
- The effective stress based Modified Cambridge Clay model (MCC) which can be used to predict both stresses during installation and reconsolidation. This is a non-linear work-hardening soil model which assumes that the yield surface is symmetric and undrained shear strength is isotropic.
- Soil models developed at the Massachusetts Institute of Technology (MIT). The MIT-models (MITE1, MITE2 and MITE3) can be considered an extension of the MCC-model. They account for shear strength anisotropy and strain softening in addition to other aspects of the MCC-model.

The simplest CEM-EP model predicts a pore pressure change at the pile shaft given by:

$$\Delta u = s_u ln(G/s_u).$$

For these pile analyses s_u as presented in Figure 14 is used. The modulus G is selected in line with previous recommendations (e.g, Reference 10) as the secant modulus at 50% of the applied shear stress to reach failure (G_{50}). This modulus was determined from direct simple shear tests on undisturbed clay.

Figure 7 shows that this CEM-EP approach consistently underpredicts the installation pore pressures. For practical purposes the data as shown in Figure 7 suggests that the simple CEP-EP model can give a good assessment of the excess pore pressure against the pile shaft if the predicted results are upgraded by a factor of about 1.3. Figure 7 also presents typical predictions of excess pore pressures using CEM-MCC model and SPM-MITE1/E2 (based on References 13, 14, 15, 16).

The CEM-EP, as well as the CEM-MCC approach, predict significant positive lateral effective stresses against the pile shaft during pile penetration both for open- and closed-ended piles. This strongly contradicts the field observations of close

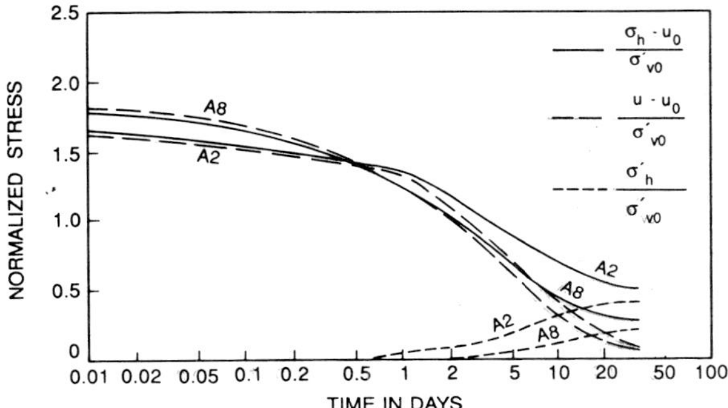

Fig. 8. Typical earth and pore pressure development during the reconsolidation phase, A2-Onsøy and A8-Lierstranda.

to zero effective stresses. The SPM-MITE1/E2 model predicts, however, more correctly the low effective stresses measured. This is partly due to the installation model and partly due to the strain-softening model in MITE1/E2.

5. The Reconsolidation Phase

5.1. Measured Results

The measurements made at Onsøy and Lierstranda were generally quite similar with respect to the reconsolidation process. Figure 8 shows typical measured evolutions of normalized total and effective earth pressures, and pore pressures. Both the total earth pressures and the pore pressures decreased by the same amount during the first period after installation. Thus, the effective horizontal pressures stayed close to the values at the end of installation until 30–50% of consolidation had taken place. At that time the effective horizontal stresses started to increase.

Figure 9 shows typical curves of pore pressure dissipation with time. The shape of the consolidation curves was very similar for all piles; a very fast pore pressure dissipation during the first time after installation and a very slow dissipation towards the end of the consolidation process. The variation in consolidation times were generally in accordance with variations in coefficient of consolidation. Due to the very slow pore pressure dissipation towards the end of the consolidation, load testing was normally initiated before 100% consolidation had taken place (from 90 to 98% after 30 to 50 days). At Pentre the general evolution of earth- and pore pressures was much the same, but the excess pore pressures dissipated completely after maximum 10 hours.

Figure 10 shows ratios between effective horizontal pressures and *in situ* effective vertical stress, $k'_c = \sigma'_{hc}/\sigma'_{v0}$ for all piles. Note that the values shown in Figure 10 are extrapolated values which correspond to 100% consolidation.

The k'_c values measured at Onsøy were somewhat lower than the *in situ* stress

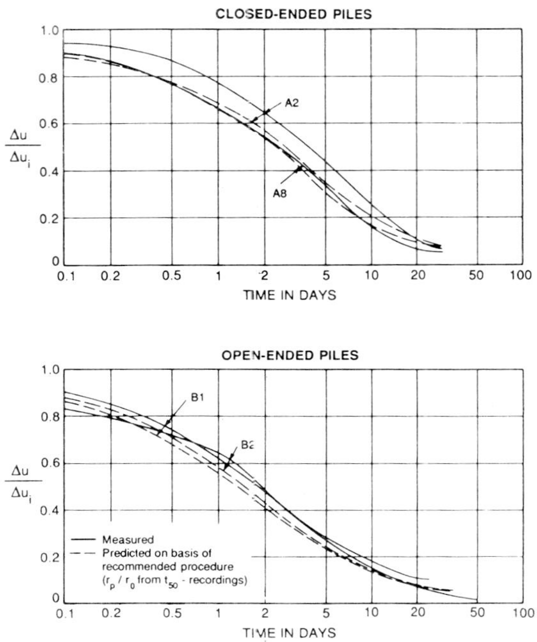

Fig. 9. Typical measured and calculated pore pressure dissipation curves.

ratio, k_0'. At Lierstranda, however, the measured k_c' values were extremely low, with k_c' of only 0.10 for the deepest pile. This falls way on the low side of what has previously been reported from laboratory "pin" model tests (Reference 17), small scale *in situ* tests (References 18, 19), and experiences from Haga and Onsøy. Low k_c'-values were also measured at Pentre, especially for the upper pile (Figure 10). Both at Lierstranda and at Pentre these low horizontal effective stresses were reflected in low axial capacity. This will be further discussed in the next section.

5.2. Analyses and Discussion

Three factors that have major influence on predictions of time-rate of pore pressure dissipation:

- The shape of the initial excess pore pressure field.
- The radial extent of the excess pore pressure field, r_p/r_0.
- The operating coefficient of consolidation.

CEM theories predict an initial excess pore pressure which decrease logarithmically with normalized radial distance from the pile wall, r/r_0, whereas SPM predicts an essentially linear decrease with r/r_0 (Reference 4). The influence these

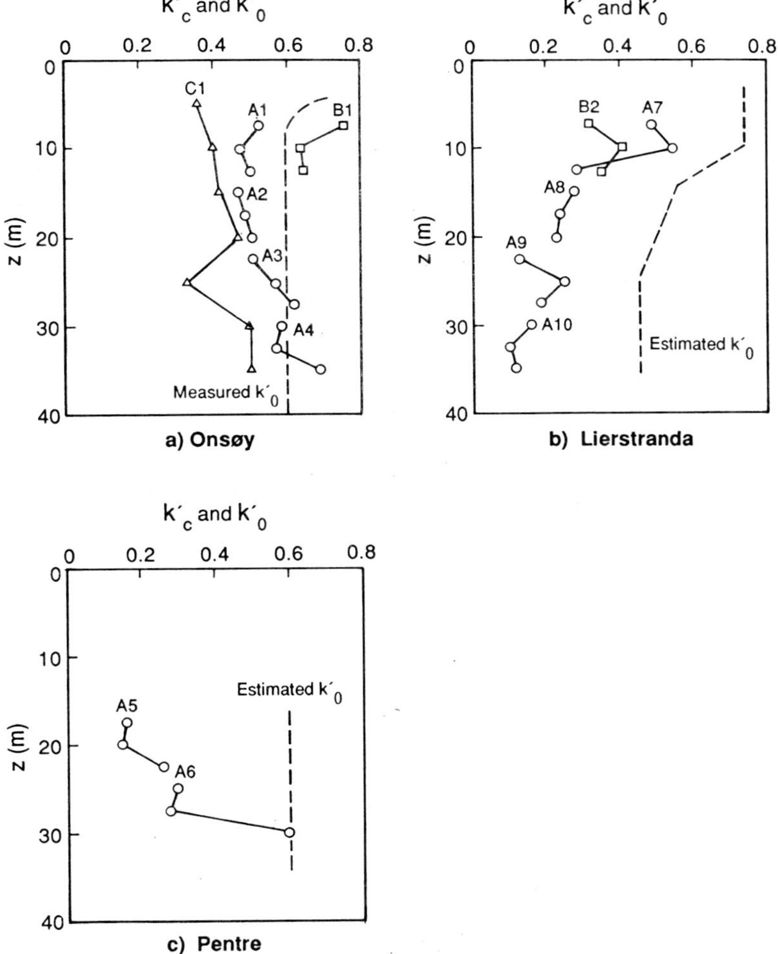

Fig. 10. Horizontal effective stress ratio, k'_c *vs* depth after 100% consolidation.

two types of initial pore pressure fields have on the pore pressure dissipation curves are visualized in Figure 11, for various values of $\lambda = r_p/r_0$ (r_p is the extent of the initial excess pore pressure field and r_0 is pile radius). The curves were taken from Reference 20.

The analyses of the pile tests showed that a very good prediction of the complete time-rate of pore pressure dissipation can be made with a simple linear radial consolidation analysis when the following assumptions are made:

– Use of a coefficient of consolidation corresponding to the true *in situ* recompression value of the coefficient of consolidation, c_r, determined in the stress range of about 0.5 to 1.0 σ'_{v0}. To arrive at the correct c_r-value, may require

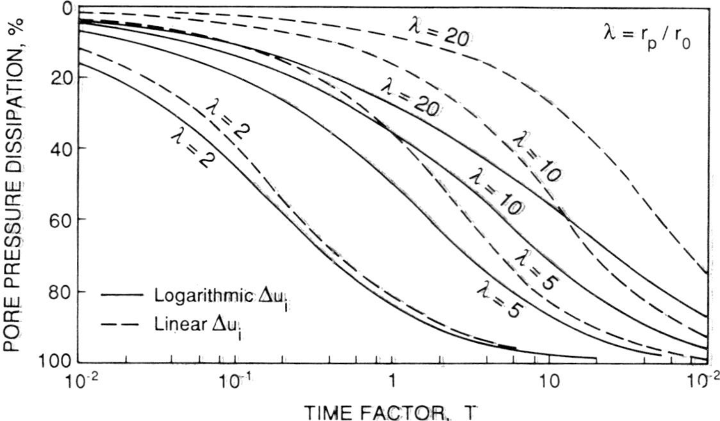

Fig. 11. Time factors from linear consolidation theory for logarithmic and linear initial excess pore pressure distribution (based on Levadoux, 1980, Reference 2).

that the results of oedometer tests are corrected for sample disturbance effects. This generally requires that the oedometer tests include an unloading/reloading branch. If the clay is layered one must also consider the possibility of a higher permeability in the horizontal than the vertical direction. Thus, the oedometer tests may have to be carried out on vertically trimmed specimens.

- The initial excess pore pressure field should be assumed to decrease linearly with r/r_0.

- The radial extent of the excess pore pressure field, r_p/r_0, is computed on basis of CEM-EP theory, i.e. that

$$\frac{r_p}{r_0} = \sqrt{\frac{G_{50}}{s_u}} \quad \text{for closed-ended piles}$$

$$\frac{r_p}{r_0} = \sqrt{\frac{G_{50}}{s_u} \cdot \frac{(r_0^2 - r_i^2)}{r_0^2}} \quad \text{for open-ended piles}$$

$$r_0 = \text{external pile radius}$$

$$r_i = \text{internal pile radius}$$

Figure 12 compares r_p/r_0 predicted with the above assumptions with values that have been backfigured to give a perfect match with the measured dissipation times. These results suggest that to fit t_{90} (time to reach 90% consolidation), r_p/r_0 from CEM-EP should be slightly increased, on average by a factor of 1.1. Because the consolidation time is about proportional to $(r_p/r_0)^2$, this means that the consolidation time should be increased by a factor of about 1.2.

Predicted pore pressure dissipation curves based on the recommended procedure, agree very well with the observed pore pressure dissipation for the different

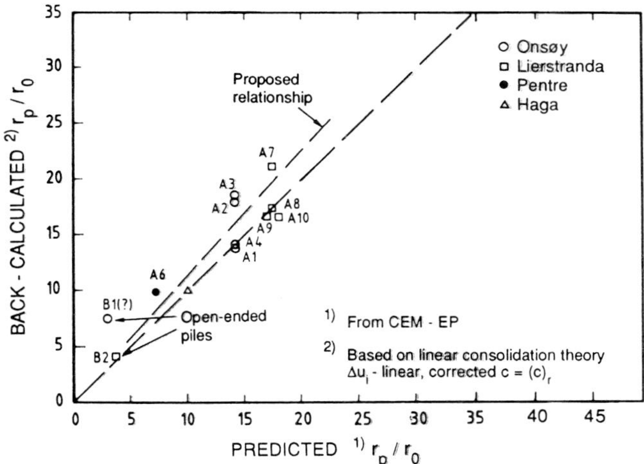

Fig. 12. Theoretical r_p/r_0-values compared to values back-calculated to match t_{90} (90% pore pressure dissipation).

piles, as shown by Figure 9.

Pile plugging will have a dramatic effect on the dissipation times. For the open-ended B-piles, for instance, plugging would theoretically increase the consolidation time by a factor of 21.6. Thus, the aspect of plugging may be of great importance for offshore piling practice.

A linear type of consolidation theory will, no matter what the initial stress condition is, predict no change in total earth pressure against the pile shaft during the reconsolidation phase. This is in direct conflict with the measurements which showed a significant reduction of total horizontal stress during reconsolidation. As a consequence of this, the CEM-MCC approach grossly overpredicts the final effective stresses as illustrated in Figure 13.

The SPM-MITE1.E2 models predict the final effective horizontal stresses reasonably well for piles in plastic clays as at Onsøy. The typical range of predicted values in Figure 13 was based on analyses presented in References 13, 14, 15, 16, 20, 21. For the silty, low plastic clays at Lierstranda and Pentre, however, no analytical models are able to predict so low final effective horizontal stresses as was measured at those sites.

Direct explanations for the very low final effective horizontal stresses measured at Lierstranda and Pentre are not easy to find but are discussed more in the next section. However, the observations suggest that there is a connection between the low stresses and soil characteristics such as:

– Plasticity index, I_p. The final effective horizontal stresses seem to be very low when the plasticity index, I_p, is lower than 15-20%. From literature it is

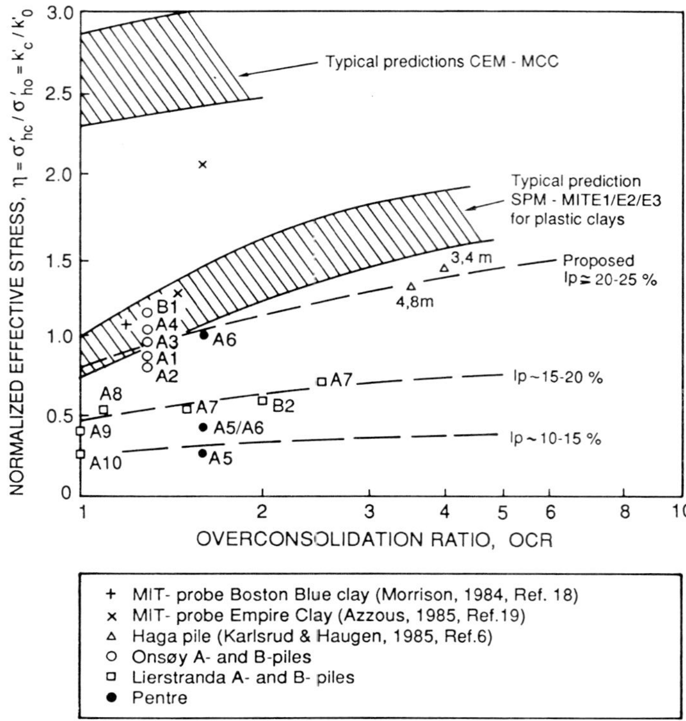

Fig. 13. Measured and typical theoretical relationship between k'_c and k'_0.

also found very low static capacities for wooden piles in low plastic clays (see Section 6).

– Silt content. Both Lierstranda and Pentre are silty clays with clay content 10–20%.

– NGI has performed a number of dilatometer tests at different clay deposits. Dilatometer tests at Lierstranda revealed extremely low material index, I_D (References 22 and 32). The material index I_D is defined as

$$I_D = \frac{p_1 - p_0}{p_0 - U_o}$$

p_1 = pressure at 1.1 mm expansion

p_0 = lift-off pressure

U_o = *in situ* pore pressure

The I_D-factor in the silty clay below 15 m at Lierstranda was in the range $I_D = 0.05$ to 0.08, whereas in the upper more plastic clay and at Onsøy $I_D = 0.15$ to 0.25.

Also effective stresses after consolidation had been allowed were low at Lierstranda. Thus, dilatometer tests may give information of extreme soil conditions with respect to the effective stress state at the end of consolidation and axial pile capacity.

One has also carried out cone penetration tests with pore pressure measurements (CPTU-type) in these deposits, but those results did not show any major differences at Onsøy and Lierstranda.

6. Ultimate Limit Shaft Friction

6.1. Measured Results

Figure 14 and Table 1 present measured values of ultimate lift shaft friction, τ_{us}, for the first static load test on each pile. For A- and B-piles the values given are average values deduced from the axial loads measured at 2.5, 5.0 and 7.5 m below the bottom of the casings. For the Pentre A-piles it is also shown the distribution of shaft friction from 2.5 m to the tip of the pile, because one here had fairly large local variations. The shaft friction along the lowest 2.5 m was calculated by assuming a theoretical tip resistance (suction) corresponding to $q_{\text{tip}} = 9 \cdot s_u \cdot$ Area. As for the k_c'-values in Figure 10, the τ_{us}-values in Table 1 and Figure 14 were extrapolated to correspond to 100% pore pressure dissipation. The extrapolation was made by assuming the τ_{us} would increase in proportion to σ_{hc}'. The extrapolated values are typically 10 to 25% larger than the values actually measured for the Onsøy and upper three Lierstranda piles, whereas 100% dissipation was reached for the others.

For direct comparisons, Figure 14 and Table 1, also present in situ undrained shear strength values, s_u, as discussed in Section 2 above. Subsequent comparisons are based on the s_u^{lab}-values for Onsøy and Lierstranda and s_u^{UU} for Pentre. On that basis apparent measured α-values ($\alpha = \tau_{us}/s_u$) are presented in Table 1 together with β-values ($\beta = \tau_{us}/\sigma_{v0}'$).

The following main observations are made from these results:

- In the plastic Onsøy clay the measured apparent α-values for the A-segment piles were generally close to unity, but decreased with depth to a minimum value of 0.77 for Pile A4. β-values correspondingly decreased with depth from 0.28 to 0.20.

- The corresponding A-segment piles in the silty Lierstranda clay showed extremely low α- and β-values, with α-values also in the case decreasing with depth from 0.50 to 0.22, and β from 0.15 to 0.05.

- The upper A5-pile in the clayey silt at Pentre showed α- and β-values which were of the same order as the upper Lierstranda piles, whereas the lower Pen-

TABLE 1.
Summary of measured pile data.

SITE	PILE	SOIL DATA						MEASURED DATA						
		Depth (m)	po' (kpa)	OCR	ko'	su (KPa)	su/po'	sighc' (kpa)	kc'	kc'/ko'	tus (kpa)	tus/po'	tus/sighc'	tus/su
Ons¢y	A1	10.0	62	1.30	0.60	17	0.27	32	0.52	0.86	17.7	0.29	0.55	1.04
	A2	17.5	100	1.30	0.60	24	0.24	48	0.48	0.80	21.8	0.22	0.45	0.91
	A3	25.0	138	1.30	0.60	35	0.25	79	0.57	0.95	29.2	0.21	0.37	0.83
	A4	32.5	176	1.30	0.60	45	0.26	108	0.61	1.02	34.8	0.20	0.32	0.77
	C1	7.5	51	1.30	0.60	15	0.29	20	0.39	0.65	10.5	0.21	0.53	0.70
	C1	12.7	75	1.30	0.60	1) 19	0.25	32	0.43	0.71	13.8	0.18	0.43	0.73
	C1	17.5	100	1.30	0.60	24	0.24	47	0.47	0.78	21.0	0.22	0.46	0.90
	C1	22.5	125	1.30	0.60	31	0.25	49	0.39	0.65	22.7	0.18	0.46	0.73
	C1	27.5	150	1.30	0.60	38	0.25	64	0.43	0.71	25.3	0.17	0.40	0.67
	C1	32.5	176	1.30	0.60	45	0.26	89	0.51	0.84	36.0	0.20	0.40	0.80
	B1	10.0	62	1.30	0.60	17	0.27	42	0.68	1.13	17.2	0.28	0.41	1.01
Lier– stranda	A7	10.0	78	2.50	0.70	23	0.29	33	0.42	0.60	11.4	0.15	0.35	0.50
	A8	17.5	126	1.35	0.52	29	0.23	34	0.27	0.52	11.3	0.09	0.33	0.39
	A9	25.0	181	1.00	0.46	1) 38	0.21	36	0.20	0.43	12.9	0.07	0.36	0.34
	A10	32.5	237	1.00	0.45	50	0.21	28	0.12	0.26	11.0	0.05	0.39	0.22
	B2	10.0	78	2.50	0.70	23	0.29	32	0.41	0.59	13.5	0.17	0.42	0.59
Pentre	A5	20.0	174	1.60	0.60	2) 50	0.29	30	0.17	0.29	22.3	0.13	0.74	0.45
	A6	27.5	243	1.60	0.60	69	0.28	96	0.40	0.66	60.1	0.25	0.63	0.87
Haga		1.3	27	9.50	1.45	42	1.56	34	1.26	0.87	18.0	0.67	0.53	0.43
		2.7	51	5.00	1.00	40	0.78	69	1.35	1.35	18.0	0.35	0.26	0.45
		3.9	72	3.20	0.80	1) 42	0.58	93	1.29	1.61	40.0	0.56	0.43	0.95
		4.5	82	3.70	0.88	48	0.59	106	1.29	1.47	49.0	0.60	0.46	1.02

1) Su taken as average lab. strength Su^{lab}
2) Su taken as UU strength

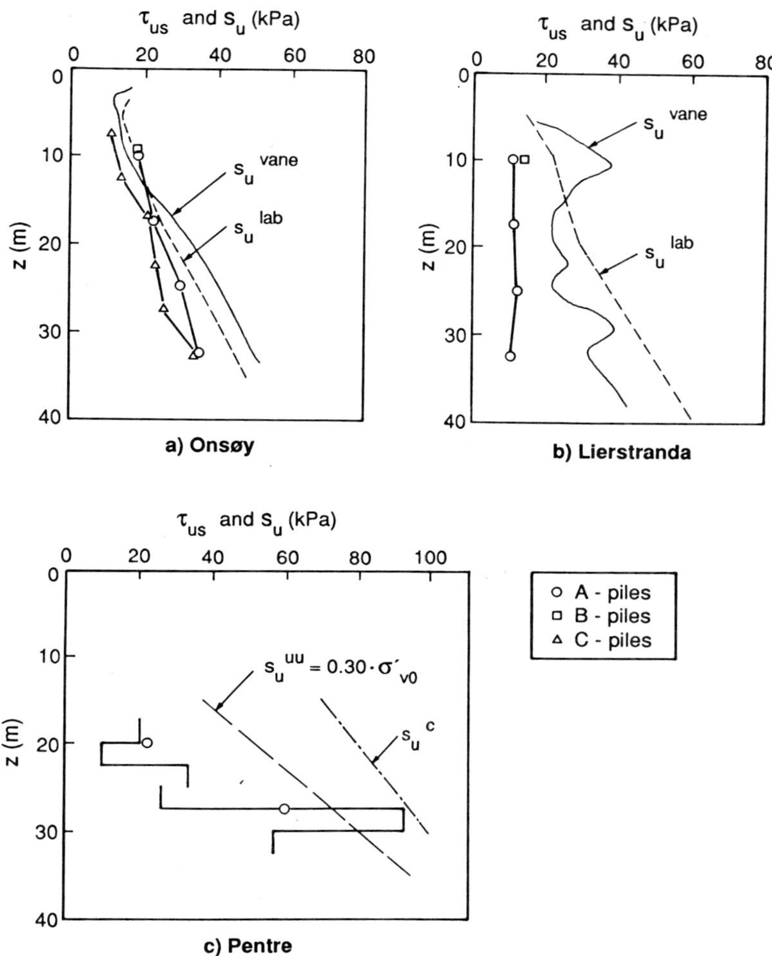

Fig. 14. Measured (partly extrapolated values) of ultimate shaft friction at 100% consolidation.

tre pile was more in the line with the Onsøy piles. Note, however, the large variation along this pile, which may be associated with the very layered nature of this deposit, and a locally very stiff seam described as "siltstone" found in some soil borings around this depth.

- The very low shaft friction values for the Lierstranda and upper Pentre piles are apparently at least partly related to the very low k'_c-values also measured for these piles. Notice also the low plasticity index for these deposits, and in particular at Lierstranda (Figure 1).

- The long Pile C1 at Onsøy showed consistently smaller τ_{us}-values than the

A-segment piles at similar depths. Measured in per cent, the difference to the A-piles tends to be larger along the top half than the lower half, but is on average 20% for the entire pile. There were no indications in the observed relationships between mobilized shaft friction and local pile displacements (τ–z data) that suggested strain softening had anything to do with the lower shaft friction along Pile C1 compared to the A-segment piles. Thus, the only possible explanation is that driving of a long pile causes more direct disturbance of the clay, and/or more "whipping" effects. The result is, as was also measured, lower effective earth pressures against the long flexible Pile C1 compared to the stiffer A-piles (Figure 8).

– At both Onsøy and Lierstranda the large diameter open-ended B-piles showed nearly the same τ_{us}-values as the corresponding closed-ended A-piles.

Most of the piles were subjected to a second test series some days after the first series of static and cyclic loading. In these second static tests the average shaft friction was typically 10-15% and 25 to 45% larger than in the first tests for the Onsøy and Lierstranda piles respectively. At Haga it was observed a similar increase of about 25% from a first to a second test, and after a third test on the same pile the capacity increased to as much as 63% above the first. It is only the Pentre piles that showed no such increase.

6.2. Analyses and Discussion

For essentially normally consolidated clays the current API(89) design guideline (Reference 23) gives α-values close to unity and therefore, grossly overpredicts the shaft friction of all Lierstranda piles and the upper Pentre pile, Figure 15. There are no other conventional design rules (α, β or λ-approaches) that would come near to predicting such low values of shaft friction.

As already stated above, the low τ_{us}-values are to some extent related to very low σ'_{hc} or k'_c-values. This is directly evident from the normalized values of τ_{us}/σ'_{hc} presented in Table 1, which show considerably less variation than either α- or β-values.

Based on NGI's pile tests at the Haga site (References 5, 6, 7) the first author started development of a new method for predicting the ultimate shaft friction along piles in clay. The method, for simplicity named the NGI-method, tries to resolve and take into account the influence of the following factors on the shaft friction:

i) How are the basic undrained shear strength properties of the clay influenced by the severe remoulding and large shear distortions caused by the pile installation?

ii) What is the state of effective stresses in terms of σ'_{hc}, $\sigma'_{\theta c}$ and σ'_{vc} when all excess pore pressures have dissipated (Figure 16)?

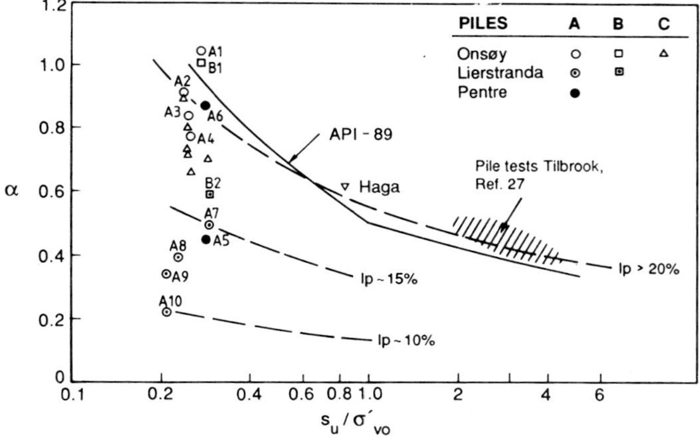

Fig. 15. Results of pile tests in $\alpha - s_u/\sigma'_{vo}$ diagram.

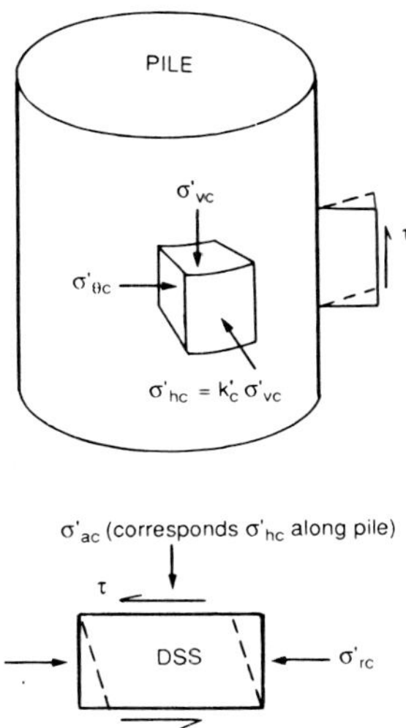

Fig. 16. Stress states along pile and in direct simple shear tests (DSS).

iii) How does the shear strength vary with distance from the pile surface and the level of effective stresses after complete reconsolidation (Figure 17)?

Based on strains predicted by SPM and CEM, and field observations from block samples taken next to the Haga piles (References 5 and 6), one can identify three distinct zones in the soil after pile installation (Figure 17).

– Zone A, or the "Remoulded Reconsolidated" (RR) zone, is a rather thin zone

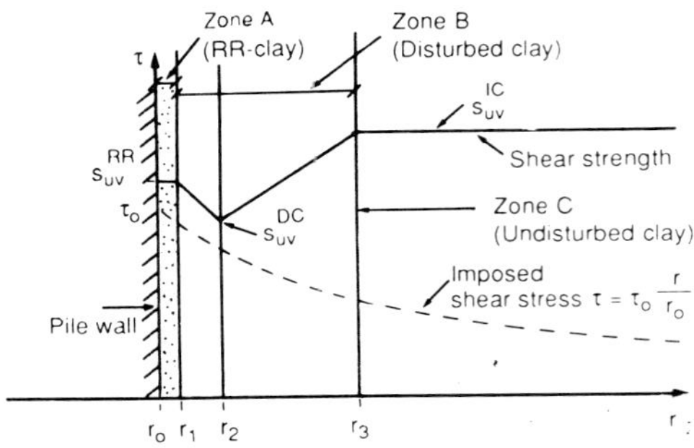

Fig. 17. Schematic variations of undrained shear strength away from the pile wall.

closest to the pile wall (5–10 mm at Haga) which is subjected to extremely large shear distortions and severe remoulding during pile penetration. In this zone the original fabric of the clay is completely wiped out.

– Zone B, or the “Disturbed” zone, is a zone which experiences large but “finite” strains, more or less as predicted by SPM. The shear strains, and thus, the extent of Zone B, are much larger for a closed-ended pile than for an open-ended pile.

– In Zone C, the stresses and strains caused by the pile installation are so small that one may assume that the undisturbed *in situ* shear strength is not influenced.

The types of shear stresses and shear strains induced by pile loading correspond to that of direct simple shear (DSS). Therefore, the NGI-method was developed on basis of DSS tests. This agrees with the suggestion in Reference 24. To model the shear strength in Zone A, undisturbed clay is severely remoulded in the laboratory at its *in situ* water content. It is then stamped into a consolidometer and consolidated to the appropriate effective stress. The specimen is then taken out, trimmed and built into the DSS-apparatus.

NGI has carried out a fairly large number of such DSS tests on remoulded reconsolidated (RR) clay from various onshore and offshore sites. The type of clays range from normally consolidated to moderately overconsolidated (OCR = 1.0 to 5.0) with undisturbed undrained shear strengths ranging from about 20 kPa to 500 kPa. The different clays all show that the shear strength is essentially proportional to the axial effective consolidation stress, σ'_{ac}, applied in the DSS test, e.g.

$$\left(\frac{\tau_f}{\sigma'_{ac}}\right)^{RR}_{DSS} = \text{constant}.$$

Most clay models imply that the undrained shear strength of a normally consolidated clay is proportional to the mean effective consolidation stress at onset of undrained shearing. Therefore, it is proposed that the ultimate limit shaft friction, τ_{us} in the RR Zone A is computed from:

$$\tau_{us}^{\mathrm{RR}} = \xi\left(\frac{\tau_f}{\sigma_{ac}}\right)_{\mathrm{DSS}}^{\mathrm{RR}} \cdot \sigma'_{hc}. \tag{1}$$

The correction factor ξ accounts for the difference between the mean effective stress for a clay element close to the pile wall, and the mean effective stress in the DSS apparatus for a specimen consolidated to an axial effective stress, σ'_{ac}, equal to σ'_{hc} against the pile shaft. The mean effective stress in the DSS apparatus for a severely remoulded specimen is assumed to be given by

$$(\sigma'_{mc})_{\mathrm{DSS}} = \frac{1}{3}(1 + 2 \cdot k'_{onc})\sigma'_{ac} \tag{2}$$

where $k'_{onc} = k'_o$ for truly normally consolidated material. The mean effective stress against the pile shaft is given by:

$$(\sigma'_{mc})_{\mathrm{pile}} = \frac{1}{3}(\sigma'_{hc} + \sigma'_{vc} + \sigma'_{\theta c}). \tag{3}$$

Only σ'_{hc} is really known from the pile tests. If, however, one assumes $\sigma'_{vc} = \sigma'_{v0}$ (in situ vertical effective) and that $\sigma'_{\theta c} = k'_{onc} \cdot \sigma'_{hc}$, one gets the following expression for the correction factor, ξ:

$$\xi = \frac{(\sigma'_{mc})_{\mathrm{pile}}}{\sigma'_{mc})_{\mathrm{DSS}}} = \frac{\left(1 + \frac{1}{k'_c} + k'_{onc}\right)}{(1 + 2k'_{onc})}. \tag{4}$$

For a typical value of $k'_{onc} = 0.5$,

$$\xi = 0.75 + \frac{1}{2} \cdot k'_c.$$

As implied by Figure 17, and discussed in more detail in References 4 and 7, the critical shear plane may not necessarily lie within the most severely remoulded Zone A, but could also lie within the Zone B where the clay has undergone large but finite strains. Attempts have been made to assess the shear strength in this zone (Reference 4). The strength can definitely be smaller than in the remoulded Zone A, which undergoes larger volume changes during reconsolidation, and especially in clays which are normally to lightly overconsolidated. A lower shear strength in Zone B is, however, partly compensated for by the decrease in imposed shear with distance from the pile surface. Thus, assuming failure to occur within the RR-zone has been found to overpredict the shaft friction by maximum 10–20. For overconsolidated clays (OCR $\geq$ 1.5 to 2.5) there are no doubts that failure will occur in the RR-zone A.

Table 2 presents predicted values of shaft friction for the different piles, also including Haga, based on Equations (1) and (4); $(\tau_f/\sigma'_{acc})_{\mathrm{DSS}}^{\mathrm{RR}}$-values as measured

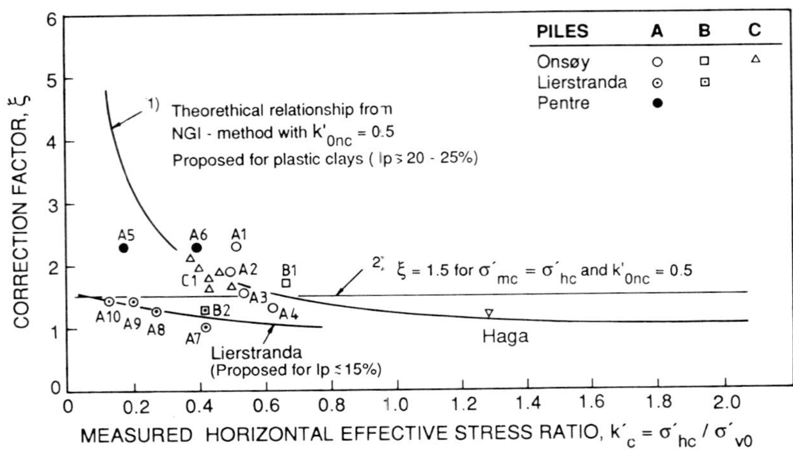

Fig. 18. Measured *vs* predicted ζ *vs* k'_c relationship.

in laboratory tests (Table 2), and the values of σ'_{hc} and k'_c actually measured against the pile shafts. The ratio of measured to predicted values of τ_{us} are also given. The agreement is generally very good for all piles except the Lierstranda piles and the upper Pentre pile. This is more clearly seen from Figure 18, where the backfigured values of the correction factor, ξ, is compared to predicted by Equation (4) and assuming for simplicity $k'_{onc} = 0.5$ for all the clays.

Figure 19 shows actual measured and predicted distributions of shaft friction with depth for the Haga tests. The very good agreement with the NGI-method for this case, that is, a clay deposit where the in situ shear strength is nearly constant with depth, but OCR decreases considerably with depth, gives strong support to the basic concepts of the NGI-method. Also API(89) (Reference 25) gives fair agreement with the Haga pile, whereas the previous API(82) fails completely to predict the variation in τ_{us} with depth.

The low backcalculated values of ξ for the Lierstranda and the upper Pentre piles (Figure 18) can, according to the basic principles of the NGI-method, only be explained by vertical "silo" or "arching" effects, making σ'_{vc} considerably smaller than σ'_{v0}. As an illustration, to get down to $\xi = 1.0$ requires that

$$\sigma'_{vc} = \sigma'_{\theta c} = k'_{onc} \cdot \sigma'_{hc},$$

which would correspond to σ'_c only 20% of σ'_{v0} for Pile A7. The value of $\xi = 1.5$ for Pile A10 can be obtained with σ'_{vc} and $\sigma'_{\theta c} = \sigma'_{hc}$. That would imply a σ'_{vc}-value of only 12% of σ'_{v0}.

The very low effective stress states, and as a consequence shaft friction, along the Lierstranda and upper Pentre piles is an issue of major concern. Similar mechanisms of silo or arching effects and very low shaft friction have been made for driven piles in loose silica sands and cemented calcareous sands.

TABLE 2.
Summary of predicted shaft friction.

SITE	PILE	'PREDICTED NGI–METHOD						'PREDICTED API–89		
		Depth (m)	tf/sigac'	konc'	ksi	tus (kpa)	tusm/tuspr	alfa	tus (kpa)	tusm/tuspr
Ons¢y	A1	10.0	0.24	0.55	1.66	12.8	1.39	0.95	16.2	1.09
	A2	17.5	0.24	0.55	1.73	19.9	1.09	1.02	24.5	0.89
	A3	25.0	0.24	0.55	1.57	29.8	0.98	0.99	34.7	0.84
	A4	32.5	0.24	0.55	1.51	39.2	0.89	0.99	44.5	0.78
	C1	7.5	0.24	0.55	1.95	9.4	1.12	0.92	13.8	0.76
	C1	12.7	0.24	0.55	1.85	14.2	0.97	0.99	18.9	0.73
	C1	17.5	0.24	0.55	1.75	19.8	1.09	1.02	24.5	0.88
	C1	22.5	0.24	0.55	1.95	23.0	0.99	1.00	31.1	0.73
	C1	27.5	0.24	0.55	1.85	28.5	0.89	0.99	37.7	0.67
	C1	32.5	0.24	0.55	1.68	35.9	1.00	0.99	44.5	0.81
	B1	10.0	0.24	0.55	1.44	14.5	1.18	0.95	16.2	1.06
Lier–stranda	A7	10.0	0.34	0.48	1.96	22.0	0.52	0.92	21.2	0.54
	A8	17.5	0.26	0.46	2.69	23.8	0.48	1.04	30.2	0.37
	A9	25.0	0.25	0.46	3.38	30.4	0.42	1.09	41.5	0.31
	A10	37.5	0.27	0.46	5.17	39.1	0.28	1.09	54.4	0.20
	B2	10.0	0.34	0.48	2.00	21.7	0.62	0.92	21.2	0.64
Pentre	A5	20.0	0.32	0.48	3.71	35.7	0.63	0.93	46.6	0.48
	A6	27.5	0.27	0.48	2.05	53.0	1.13	0.94	64.7	0.93
Haga		1.3	0.33	0.48	1.16	13.0	1.38	0.40	16.8	1.07
		2.7	0.33	0.48	1.13	25.8	0.70	0.56	22.6	0.80
		3.9	0.33	0.48	1.15	35.3	1.13	0.65	27.5	1.45
		4.5	0.33	0.53	1.12	39.1	1.25	0.65	31.4	1.56

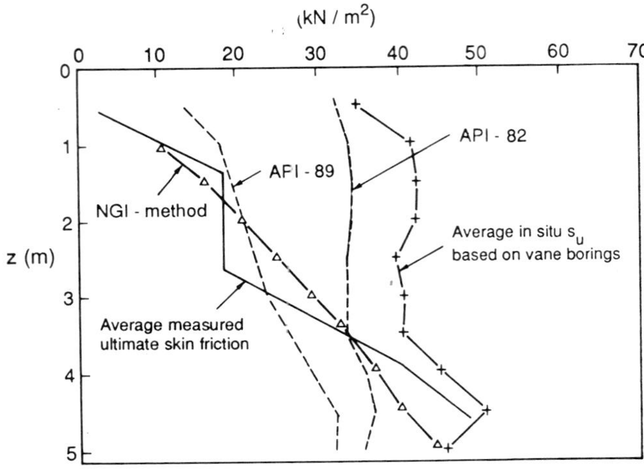

Fig. 19. Measured and predicted shaft friction, Haga pile.

There are previously also reported very low shaft friction values for other pile tests in soft normally, or lightly overconsolidated clays. Within the API database (References 27 and 30) there are pile tests with average α-values down to about 0.4.

Even more striking are results of a large number of load tests summarized in Reference 26, mostly on wooden piles from Norway and some other countries. Most piles were load tested several times after installation, typically 1 month and 3 months. Figure 20 shows reported α-values at these two time intervals as function of plasticity index of the clay. Only tests on wooden piles in Norway have been included for simplicity. The piles were from 10 to 24 m long, with average diameter in the range of about 21 to 27 cm. The clay deposits were normally to moderately overconsolidated, with water contents ranging from 30 to 60%. The average undrained shear strengths ranged from 18 to 70 kPa, and the corresponding s_u/σ'_{v0} ratio from about 0.25 to 0.70.

Figure 20 shows quite clearly that at both 1 and 3 months the α-value drops off rapidly from close to or even above 1.0 when $I_p \geq 20\%$, to respectively α of about 0.15 and 0.35 after one and three months when $I_p \leq 12\%$. For the 1 month tests Figure 21 shows α-values against approximate average s_u/σ'_{v0}-values. The numbers given for each point represent the I_p-values. Note that the σ'_{v0}-values have only been roughly estimated as we have not had access to *in situ* pore pressure data from these old case records. For comparison, the current API (1989) design line is also shown in Figure 21.

Figures 20 and 21 confirm fully the impression from the pile tests reported herein, that low plastic silty clays can have very low shaft friction (α-values), and that the plasticity index is an important design parameter at least for normally or lightly overconsolidated clays.

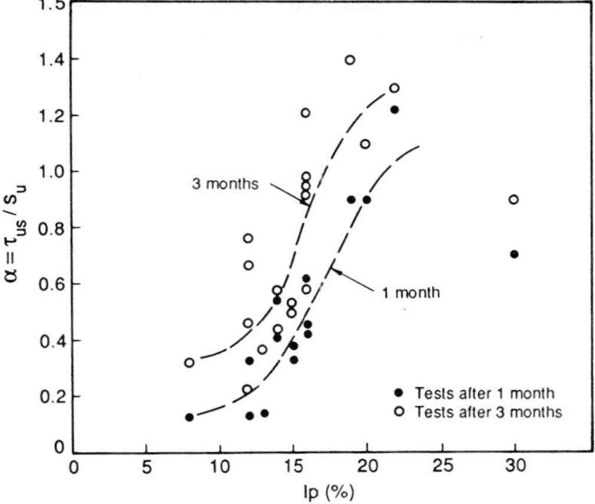

Fig. 20. Relationship between α-factor and plasticity index from load tests on wooden piles in Norway (based on data in Reference 26).

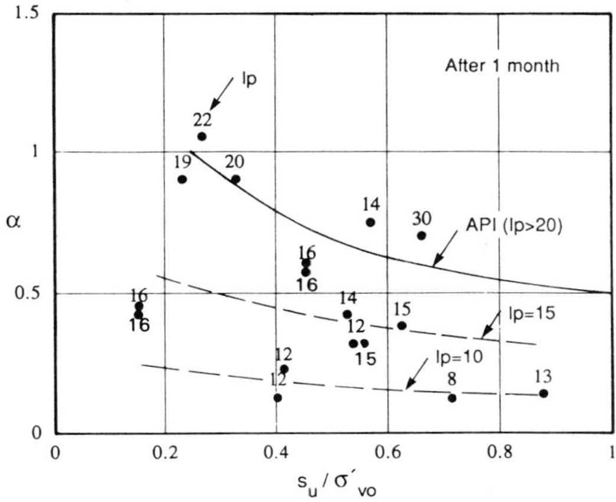

Fig. 21. Relationship between α and s_u/σ'_{vo} for piles from Figure 20 load tested after 1 month.

The increase in capacity from 1 to 3 months shown by Figure 20 can only to a very limited extent be explained by pore pressure dissipation. The reason is that for the pile diameters and clay types in question, a time of 1 month is sufficient to reach about 80–90% pore pressure dissipation. The relative increase from 1 to 3 months is generally largest for the low plastic clays, and up to a factor of about 3.0. It is quite possible that as time goes there is a gradual restoration of total and effective stresses back towards the *in situ* values. The effects of this is quite naturally largest in the low plastic clays where the effective stresses apparently are much lower than the *in situ* stresses when 100% pore pressure dissipation is reached.

As presented above, most piles (except Pentre) showed an increase in pile capacity from the first to the second test series (10 to 45%). This increase was not found to be related to any corresponding increase in horizontal effective stress against the pile shaft. It was, on the other hand, for these clays found a very similar increase in undrained shear strengths in direct simple shear tests on RR-clay, when the same specimen was sheared twice but allowed to reconsolidate in between, as in the pile tests. This "preshearing" effect of typically 10 to 45%, contributes to some of the increase in capacity from 1 to 3 months in Figure 19, but not the up to 3-fold increase in same cases. Thus, a gradual restoration of *in situ* stresses must be the major cause of the increase with time. It is conceivable that such a gradual restoration of effective stresses and increase in shaft friction could go on over a long period of time.

The implications of this study when it comes to design of friction piles in silty low plastic clays are quite serious. It suggests that the simple API-design guidelines should consider the plasticity index as an additional parameter for selecting α-values. The dashed lines in Figure 15 tentatively outline a relationship for low-plastic clays. This is based on the current pile test results and the data presented in Figure 21. The curves may be conservative, as they do not reflect the potential for gradual restoration of total *in situ* stresses and consequently build-up of pile capacity with time, as indicated by Figure 20. Note that for large s_u/σ'_{v0}-values (OCR's) the API curve for plastic clays have been raised somewhat in line with the suggestion in Reference 28 based on recent pile tests in very stiff clays. Again, these values may be conservative due to possible time effects as discussed above.

In line with the API-method, the NGI-method can also be adopted to account for the effects of clay plasticity on the k'_c and ξ-values as tentatively suggested in Figures 13 and 18.

The results of the load test on the long Onsøy pile also suggest that design methods should account for some effects of length of pile driven into the ground. It is not at present clear whether this length effect should only be related to length of pile driven and/or pile flexibility (stiffness). These pile tests results do not suggest that strain softening in terms of τ-z behaviour contributes significantly to length effects. Pile "whipping effects" discussed earlier will on the other hand be influenced by pile flexibility. There is, however, at present not sufficient data to separate the pure length and the pure flexibility or whipping effect. Presently, the most realistic approach is to use a correction (reduction) factor based on the length/diameter ratio as proposed in References 27 and 29. For the Onsøy piles that would give 30% reduction for the C-pile compared to the A-piles, which is somewhat higher than the measured 20%.

Most offshore piles are open-ended pipe piles, and are commonly reported to not plug during driving. The tests reported herein, do not suggest any significant difference in ultimate shaft friction compared to closed-ended piles.

The pile tests reported herein were all carried out under tensile loading conditions. It has been a much debated issue whether or not the pile capacity in

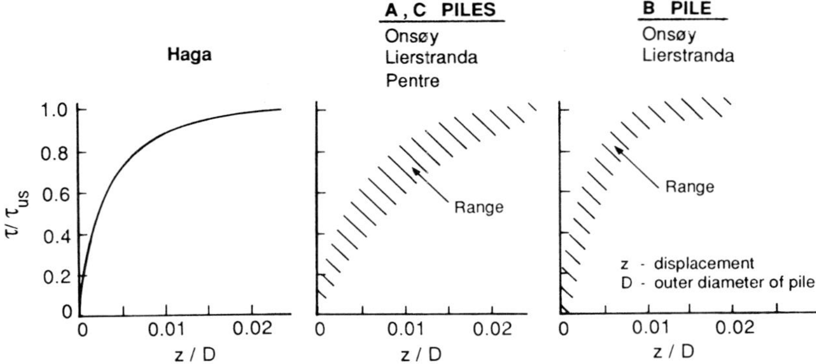

Fig. 22. Typical measured t-z curves.

compression is larger than in tension. The results of tests at Haga clearly showed no difference (Reference 5), and the same conclusion was drawn from the recent large diameter tests in very stiff clay reported in Reference 28.

7. Load-Displacement Response, Static Loading

7.1. Measured Results

From all the pile tests it has been computed local shear stress versus local pile displacement curves (τ-z curves) down along the pile. This was done automatically on the computer. Figure 22 presents typical average curves, where the local shaft friction has been normalized with respect to the ultimate value, τ_{us}, and the local displacement to the pile diameter, D.

These normalized curves show quite similar responses, the ultimate shaft friction being mobilized at a displacement corresponding to 1.5 to 2.5% of the pile diameter. The A and C-piles at Onsøy, Lierstranda and Pentre showed the softest response.

7.2. Analyses and Discussion

There exists a number of empirical suggestions for development of local τ-z behaviour along a pile shaft.

The τ-z behaviour can also be computed theoretically for instance as described in References 7 and 31. The required input is:

- The radial distribution of shear stress, which is simply given by $\tau_r = \tau_{ro} r / r_o$.
- The radial distribution of shear strength, which can simply be assumed estimated as outlined in Reference 4 on basis of the NGI-method.
- The normalized stress-strain behaviour of the soil under a simple shear type mode.

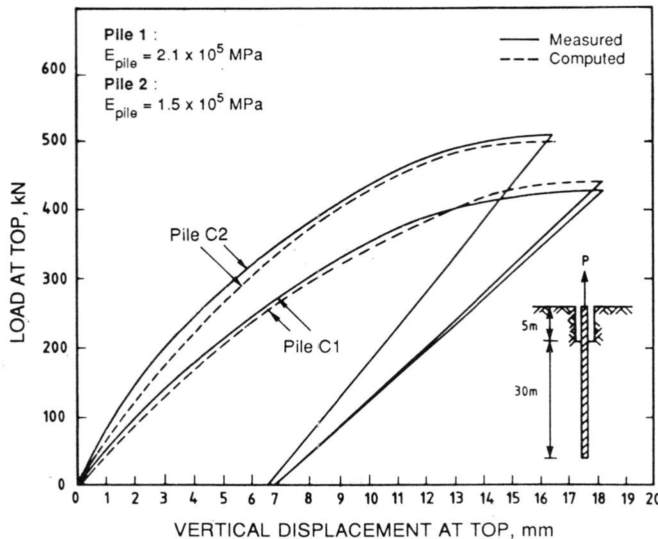

Fig. 23. Example of computed *vs* measured load-displacement response, piles C1 and C2 Onsøy.

For a given imposed shear stress on the pile wall the radial shear strain distribution is first determined from the radial degree of strength mobilization and the normalized stress-strain curve for the soil. The local displacement is then found by simple integration of strains out to a selected cut-off distance.

When t-z springs have been developed it is a fairly straight forward task to analyse a long flexible pile. Figure 23 shows a comparison between measured and computed load-displacement response at the pile top for the long C1 and C2 piles. The computed curves were established with the program PAXCY described in Reference 31 based on the principles outlined above. As input one used stress-strain behaviour as directly measured in DSS tests, but the ultimate shaft friction values were adjusted to give a reasonable fit to the measured values. The lower stiffness of the instrumented Pile C1 compared to the non-instrumented Pile C2 is due to the flexibility of the threaded couplings of the instrument units.

The PAXCY program has also been found to give very good agreement with most of the other pile tests reported herein (References 4 and 31).

8. Pile Response to Cyclic Loading

8.1. Measured Results

At the Haga test site 26 individual piles were tested first under static then under cyclic loading. Most cyclic tests were constant load-amplitude tests, but the cyclic load values varied over a large range of combinations of cyclic load amplitude,

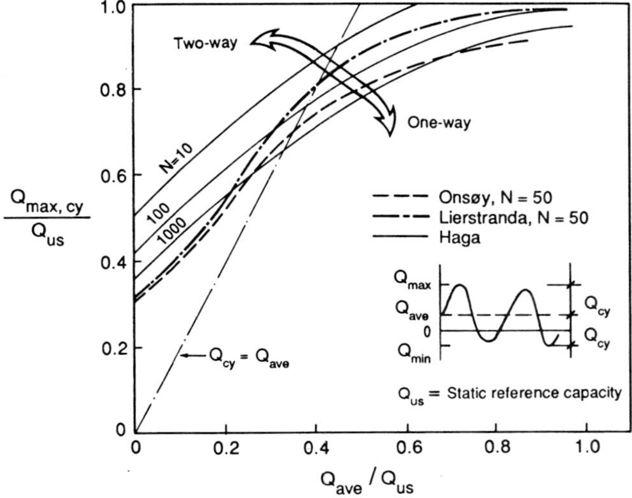

Fig. 24. Cyclic capacity diagram.

Q_{cy}, and average bias load, Q_{ave}. The Haga tests have been thoroughly treated in References 7, and have given a very fundamental picture of the influence of cyclic loading effects on axially loaded piles in clay.

Figure 24 gives a simplified representation of the Haga test results. It relates the peak cyclic failure load amplitude, $Q_{max,cy}$, to the average bias load, Q_{ave}, and the number of load cycles, N_f, the piles could sustain before failure was reached. Both $Q_{max,cy}$ and Q_{ave} are normalized with respect to the static capacity, Q_{us}. Thus, Figure 24 gives a direct picture of the influence of cyclic loading on the pile capacity. As seen, the cyclic capacity drops rapidly off when $Q_{cy} > Q_{ave}$ and thus, one gets a full reversal on imposed load direction (tension to compression).

The cyclic pile load tests at Onsøy, Lierstranda and Pentre cover a smaller range of cyclic loads. They were generally either purely one-way or symmetrically purely two-way, but the results fit into the general picture of the Haga results as illustrated in Figure 24, emphasizing the very detrimental effects of two-way type cyclic loading.

8.2. Analyses and Discussion

It is beyond the scope of this paper to go into any detailed discussion on how to analyse and model cyclic loading effects. As described in References 7 and 31, however, the cyclic pile response can be rather well understood and predicted on the basis of cyclic DSS laboratory tests. A complete representation of cyclic as well as accumulated displacement under any arbitrary cyclic loading history can be made by the computer program PAXCY described in Reference 31.

Parametric studies in Reference 31 show that a typical 6-hour North Sea storm may be represented by an equivalent of about 35 to 100 constant amplitude load cycles under the peak wave amplitude (lowest for two-way loading). Thus, the cyclic loading effects are indeed quite substantial. For symmetrical two-way cyclic loading, Figure 24 suggests a capacity of only 35 to 50% of the static capacity, increasing to 70–80% for pure one-way loading ($Q_{cy} = Q_{ave}$).

It was somewhat surprising that the Onsøy, Lierstranda tests showed lower two-way cyclic capacity than the Haga tests. Cyclic DSS laboratory tests show that the relative effects of cyclic loading should generally increase somewhat with OCR, whereas these pile tests suggest the opposite effect (Haga has OCR = 4 compared to 1.0 to 1.5 for the others).

In any case the results strongly suggest that the capacity under cyclic loading should be accounted for in offshore design practice. This applies in particular to light weight fixed structures which may experience load reversal at the pile top.

9. Summary and Conclusions

The following main conclusions are drawn from the comprehensive instrumented axial pile load tests reported herein. The tests were carried out in three different deposits of respectively soft plastic clay (Onsøy), low-plastic silty clay (Lierstranda) and clayey silt (Pentre, UK).

- The excess pore pressures induced by pile installation are consistent and typically correspond to $\Delta u_u/s_u$ in the range 7–9, but tend to decrease with overconsolidation ratio. It agrees fairly well with simple cavity expansion (CEM-EP) models.

- The installation effective stresses are very close to zero in all cases, which is in strong contrast to CEM-EP theory, but agrees better with SPM.

- The dissipation of excess pore pressures are well predicted for both the closed ended and open-ended piles with the proposed linear radial consolidation theory, provided that one uses the recommended radial distribution of initial excess pore pressures and a coefficient of consolidation corresponding to unloading/reloading (typically 5–10 times larger than in normally consolidated state).

- During the reconsolidation phase the total horizontal earth pressure reduces dramatically, and there is little build up of effective stresses before 30–50% pore pressure dissipation is reached. After 100% dissipation the radial effective stress approached the *in situ* k'_C-value in the plastic Onsøy clay, but very much lower values were measured at the two other sites. This phenomenon seems to be related to the low plasticity index of these clays.

- The ultimate shaft friction corresponds to α- values approaching unity for the plastic Onsøy clay, but was very low in the low plastic deposits, down to $\alpha = 0.22$ at Lierstranda. These extremely low α-values were clearly related to the low effective stresses. Previous pile tests confirm extremely low α-values in soft low-plastic clays. A tentative correction of the $\alpha - s_u/\sigma'_{v0}$ relationship of API(89) for plasticity index is suggested.

- The basic principles of the NGI-method, relating shaft friction to the state of effective stresses and the undrained shear strength of remoulded reconsolidated (RR) clay, seems to give a good representation of and explanation for the measured shaft friction. Semi-empirical relationships are proposed to determine k'_c- values and shaft friction based on OCR, I_p and normalized shear strength of RR clay.

- In the low plastic deposits it is suggested that there must occur silo or arching effect, causing a significant reduction in total and effective vertical stress as well as in horizontal direction close to the pile shaft. There are indications from previous pile tests that there may be a gradual increase and restoration of these low effective stresses back towards *in situ* stress levels. Thus, the shaft friction may increase considerably with time after 100% pore pressure dissipation is reached in such low-plastic deposits.

- Comparisons between the long Pile C1 and the A-segment tests confirm that there exist some length effects. In contrast to previous suggestions, the length effect is primarily related to length of pile driven and/or "whipping" effects. Strain softening effects seem to be negligible.

- Comparisons between the open-ended B-piles and the corresponding closed-ended A-piles suggest no major difference in consolidation effective stresses and shaft friction.

- The load-displacement response (τ-z response) of the piles was fairly consistent and can be well predicted on basis of the stress-strain behaviour from simple shear tests on remoulded and undisturbed clay.

- The capacity of the piles under two-way cyclic loading was only 35 to 50% of the static capacity, increasing to 70–80% under pure one-way loading. The results tie in with the results of the very comprehensive cyclic load testing program previously carried out by NGI at the Haga site. Furthermore, the cyclic pile response can be well predicted on basis of an analytical model where the cyclic τ-z response is developed on the basis of cyclic simple shear tests.

Acknowledgements

NGI and the authors would like to thank the many companies that gave financial and technical support to these pile testing programs. These included: Agip, Amoco, BP, the Department of Energy (UK), Conoco, Elf, Esso, Marathon (except Pentre tests), Mobil, Norsk Hydro, Saga, Shell and Statoil.

The authors would also like to thank the many persons within NGI's staff who contributed to this work. Special thanks goes to Svein Borg Hansen and Knut Solheim who were responsible for much of the instrumentation and field work.

References

1. Norwegian Geotechnical Institute (1988), 'Summary, Interpretation and Analyses of the Pile Load Test at the Onsøy Test Site', NGI Report 52523-23, Rev. 15 March.
2. Norwegian Geotechnical Institute (1988), 'Summary, Interpretation and Analyses of the Pile Load Test at the Lierstranda Test Site', NGI Report 52523-26, Rev. 1 July.
3. Norwegian Geotechnical Institute (1987), 'Summary, Interpretation and Analyses of the Pile Load Test at the Pentre Test Site', NGI Report 52523-27, Rev. 16 August.
4. Norwegian Geotechnical Institute (1989), 'Design of Offshore Piles in Clay-Field Tests and Computational Modelling, Final report, Summary and Recommendations', Report 52523-28, April.
5. Karlsrud, K. and Haugen, T. (1984), 'Cyclic Loading of Piles and Pile Anchors – Field Model Tests – Phase II. Final report. Summary and Evaluation of Test Results and Computational Models', NGI Report 40018-11, 1 June.
6. Karlsrud, K. and Haugen, T. (1985), 'Axial static capacity of steel model piles in overconsolidated clay', *Proc. Int. Conf. on Soil Mech. and Found. Eng.*, 11, San Francisco.
7. Karlsrud, K., Nadim, F., and Haugen, T. (1986), 'Piles in clay under cyclic loading: Field tests and computational modelling', *Proc. 3rd Int. Conf. on Num. Meth. in Offshore Piling*, 165-90, Nantes, France, May 21–22.
8. Lambson, M. D., Clare, D. G., and Semple, R. M. (1992), 'Investigation and interpretation of Pentre and Tilbrook Grange soil conditions', *Conference on Recent Large Scale Full Instrumented Pile Tests in Clay*, Institute of Civil Engineers, London, June.
9. Hill, R. (1950), *The Mathematical Theory of Plasticity*, Clarendon, Oxford.
10. Ladanyi, B. (1963), 'Expansion of a cavity in a saturated clay medium', *American Society of Civil Engineers, Proceedings*, Vol. 89, No. SM 4, pp. 127-161.
11. Carter, J. P., Randolph, M. F., and Wroth, C. P. (1979), 'Driven piles in clay – The effects of installation and subsequent consolidation', *Géotechnique* **29**, 361–393.
12. Baligh, M. M. (1975), 'Theory of Deep Site Static Cone Penetration Resistance', Massachusetts Institute of Technology, Cambridge, Mass., Department of Civil Engineering. Research report, R75-56, No. 517, p. 113.
13. Norwegian Geotechnical Institute (1987), 'MIT's Prediction of Pile Shaft behaviour at Onsøy', NGI Report 52523-25, 26 November.
14. Kavvadas, M. (1982), 'Non-Linear Consolidation around Driven Piles in Clay', D.Sc. Thesis, Massachusetts Institute of Technology, Cambridge, Mass.
15. Baligh, M. M. (1985), 'Predictions of Pile Shaft Capacity. Recent Developments in Measurement and Modelling of Clay Behaviour for Foundation Design', Massachusetts Institute of Technology Special Summer Course 1.60S, August, Lecture 11, p. 73.
16. Chin, C. T. (1986), 'Open-Ended Pile Penetration in Saturated Clays', Thesis Massachusetts Institute of Technology, May.

17. Francesçon, M. (1982), 'Model Pile Tests in Clay, Stresses and Displacements due to Installation and Axial Loading', Ph.D. Thesis, University of Cambridge.
18. Morrison, M. J. (1984), '*In situ* Measurements on Model Piles', Ph.D. Thesis, Massachusetts Institute of Technology, Cambridge, Mass., Department of Civil Engineering.
19. Azzouz, A. S. (1985), 'The Piezo-Lateral Stress (PLS) Cell. Recent Developments in Measurement and Modelling of Clay Behaviour for Foundation Design', Massachusetts Institute of Technology Special Summer Course 1.60S. August, Lecture 9, Part 2, p. 48. bibitem20. Levadoux, J.-N. and Baligh, L. M. (1980), 'Radial Consolidation Solutions', Research report, Massachusetts Institute of Technology, Cambridge, Mass., Department of Civil Engineering.
20. Chin, C. T. (1988), 'Open-Ended Pile Penetration in Saturated Clays', Thesis, Massachusetts Institute of Technology, Cambridge, Mass.
21. Mokkelbost, K. H. (1988), 'Application of Dilatometer for Pile Design', NGI Report 521610-5.
22. American Petroleum Institute (1989), 'Planning, Designing and Constructing Fixed Offshore Platforms', API, Recommended practice, RP2A, 18th edition.
23. Randolph, M. F. and Wroth, C. P. (1981), 'Application of the failure state in undrained simple shear to the shaft capacity of driven piles', *Géotechnique* **31**(1), 143–157.
24. American Petroleum Institute (1982), 'Planning, Designing and Constructing Fixed Offshore Platforms', API, Recommended practice, RP2A, 13th edition.
25. Flaate, K. (1968), *Capacity of friction piles in clay* (in Norwegian), Published by Norwegian Road Research Laboratory.
26. Semple, R. M. and Rigden, W. J. (1984), 'Shaft capacity of driven pipe piles in clay', *Proc. Symp. on Analysis and Design of Pile Foundations, ASCE Nat. Convention*, San Francisco.
27. Nowacki, F., Karlsrud, K., and Sparrevik, P. (1992), 'Comparison of Recent Tests on OC Clay and Implications for Design'.
28. Kraft, L. M., Ray, R. P., and Kagawa, T. (1981), 'Theoretical t-z curves', *American Society of Civil Engineers, Proceedings*, Vol. 107, No. GT 11, pp. 1543–1561.
29. Randolph, M. and Murphy, S. S. (1985), 'Shaft capacity of driven piles in sand', *Offshore Tech. Conf., OTC*, 1985, Paper 4883.
30. Karlsrud, K. and Nadim, F. (1990), 'Axial capacity of offshore piles in clay', *Offshore Tech. Conf., OTC*, 1990, Paper 6245, pp. 405–416.
31. Gabr, M. A., Lunne, T., Mokkelbost, K. H., and Powell, J. J. M. (1991), 'Dilatometer soil parameters for analysis of piles in clay', *Proc. X ECSMFE*, Florence, May 1991, Vol. 1, pp. 403–406.

FACTORS AFFECTING THE SHAFT CAPACITY OF DISPLACEMENT PILES IN CLAYS

A. J. BOND
Geotechnical Consulting Group, London, formerly Imperial College, London

R. J. JARDINE
Imperial College of Science, Technology and Medicine, London

and

B. M. LEHANE
Arup Geotechnics, London, formerly Imperial College, London

1. Introduction

There has been concerted effort in recent years to investigate the fundamental processes that govern the behaviour of displacement piles in clay soils. Attempts have been made by various research organizations (e.g. MIT, NGI, Oxford University, and Imperial College) to measure the effective stresses acting on pile shafts using high quality instrumented model piles. In addition, a joint industry group led by British Petroleum has tested two large diameter (Ø762mm) instrumented driven piles at sites within the UK (Mullis, 1992).

The ultimate aim of these experiments has been to explain displacement pile behaviour in terms of effective stresses. The benefits of such an approach have long been recognized (Chandler, 1966; Burland, 1973; and Meyerhof, 1976) even if a satisfactory theory of driven pile behaviour has so far proved elusive. In the absence of reliable theoretical analyses, field experiments have assumed a major role in identifying the key parameters that govern the ground's response.

This Paper presents a brief overview of the Imperial College research into pile behaviour in clay soils and summarizes the principal findings. The Imperial College research has focused almost exclusively on closed-ended jacked piles, and some of the results may not apply to open-ended or driven piles.

2. Programme of Research by Imperial College

The programme or research by Imperial College involved installing instrumented piles at the four sites listed in Table 1. The experiments at Canons Park identified several aspects of the behaviour of displacement piles in clays that had hitherto

Volume 28: Offshore Site Investigation and Foundation Behaviour, 585–606, 1993.

TABLE 1.
List of sites for Imperial College instrumented pile tests.

Site	Location	Soil type	Reference
Canons Park	North London	London Clay	Bond & Jardine (1991)
Labenne	South-West France	Loose and medium dense sand	Lehane *et al.* (1993)
Cowden	North Humberside	Stiff glacial till	Lehane & Jardine (1992a)
Bothkennar	Grangemouth, Scotland	Soft marine clay	Lehane & Jardine (1992b)

been neglected. The experiments at Cowden and Bothkennar have extended the research to cover different soil types and have confirmed or clarified many of the original findings. The tests at Labenne revealed some marked similarities between the behaviour of displacement piles in clays and sand, thereby adding a degree of assurance to the overall thrust of the research.

Figure 1 shows the general arrangement of the instrumented piles that were used at the four sites. The instruments record – at a minimum of three levels along the pile – the shear stresses (τ_{rz}), radial total stresses (σ_r) and pore pressures (u) that act at the pile wall, plus the axial load that passes through the pile. The position of the instruments has been characterized by their height (h) above the pile tip. *Leading* instruments are located at about eight times the pile's radius (R) above the tip (i.e. at $h/R \approx 8$); *following* instruments at $h/R \approx 25$; and *trailing* instruments at $h/R \approx 50$.

The instruments have performed remarkably well at all the test sites, showing virtually no zero shift in signal or change in sensitivity during each test programme. The design and performance of the instruments is described by Bond *et al.* (1991).

3. Factors Affecting Shaft Capacity

The findings of the Imperial College research are described separately for each site in the papers listed in Table 1. In this Paper, we attempt to summarize the research and to identify the factors that govern the shaft capacity of piles in clay. As will be shown later, these factors include:

- The overconsolidation ratio, sensitivity, grading, and mineralogy of the clay;
- The pile's surface texture, slenderness (i.e. length to diameter) ratio, and compressibility;
- The method and rate of pile installation;

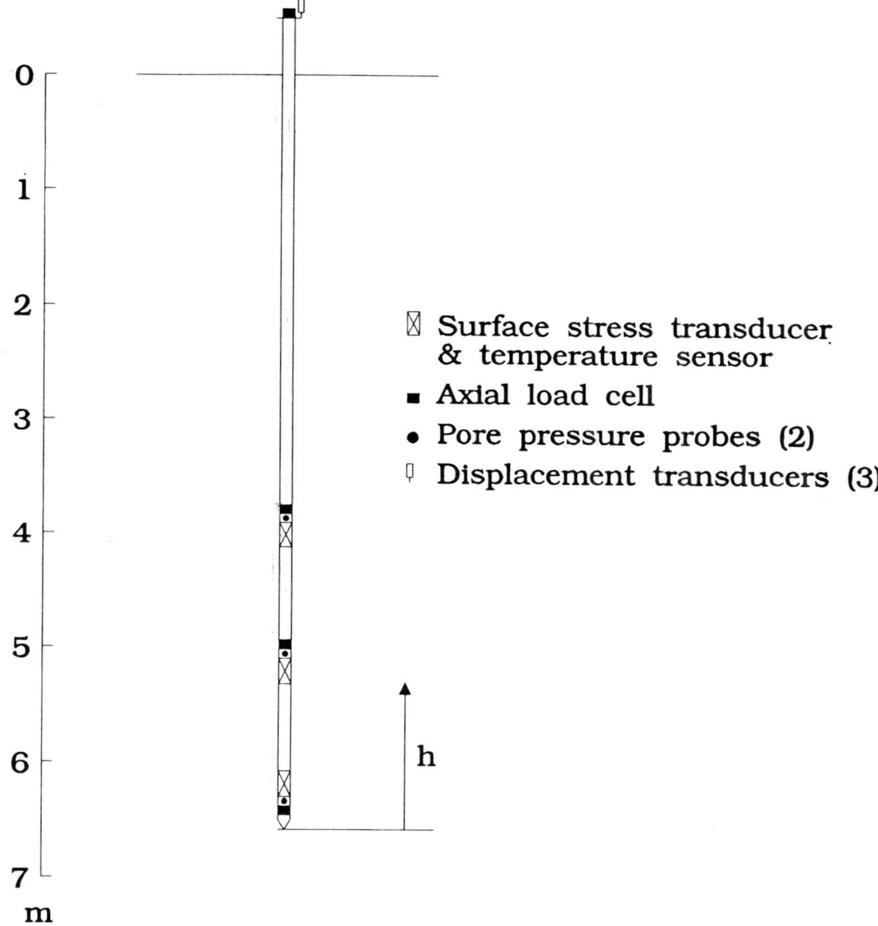

Fig. 1. The Imperial College Instrumented Pile.

- The time left between pile installation and loading (the set-up period);
- The direction of loading.

3.1. Effective Stress Formulation

The parameters we use to present our findings are derived from Coulomb's effective stress failure criterion, in which the shear stress mobilized along the pile shaft (τ_{rz}) is related to the radial effective stress (σ_r') by an equation of the form:

$$\tau_{rz} = \sigma_r' \tan \delta \tag{1}$$

where δ is the angle of shaft friction at the pile/soil surface. Re-writing this equation non-dimensionally and denoting conditions at failure by the subscript f, we obtain:

$$\tau_{rzf} = f_L K_c \sigma'_{vo} \tan \delta_f \tag{2}$$

where f_L $(= \sigma'_{rf}/\sigma'_{rc})$ is termed the "pile loading factor"; K_c $(= \sigma'_{rc}/\sigma'_{vo})$ is an earth pressure coefficient representing the radial effective stresses acting on the pile at the end of equalization (σ'_{rc}); and σ'_{vo} is the original vertical effective stress in the ground.

These non-dimensional parameters are used throughout this Paper to illustrate trends in displacement pile behaviour and to identify factors that are important in determining the shaft capacity of piles in clay.

3.2. Progressive Failure

Equations (1) and (2) represent the *local* failure criterion for soil in contact with the pile. Different elements of soil may reach their local failure criterion at different times owing to the pile's compressibility, and this leads to progressive failure along the pile shaft. When this occurs, it is the *operational* shear stress ($[\tau_{rz}]^{op}$) – i.e. that which acts when the pile reaches its maximum capacity – which contributes to the overall shaft capacity of the pile (Q_{shaft}):

$$Q_{\text{shaft}} = 2\pi R \int_0^L [\tau_{rz}]^{op} dz. \tag{3}$$

Operational values of shear stress may be different from the limiting values given in equation (2) if the local values of δ or σ'_r change with post-peak displacement of the pile.

4. Stress State at the Start of Pile Loading

4.1. Stresses Resulting from Pile Installation

The radial effective stresses that act on a pile at the end of equalization depend to a large extent on the stresses induced by pile installation. Installation in lightly overconsolidated clays increases the radial total stresses (σ_r) to values well above those acting in the undisturbed ground (σ_{ho}). In heavily overconsolidated clays, the increases in σ_r are even higher. This is illustrated in Figure 2 by data from the Imperial College experiments in the Bothkennar and London clays. The traces given for the two sites represent the variations in σ_r with depth recorded by individual load cells during pile installation.

A key feature of the results shown on Figure 2 is that the radial total stresses recorded at different locations along the pile shaft decrease with increasing height (h) above the pile tip. The highest stresses were recorded at the leading instrument

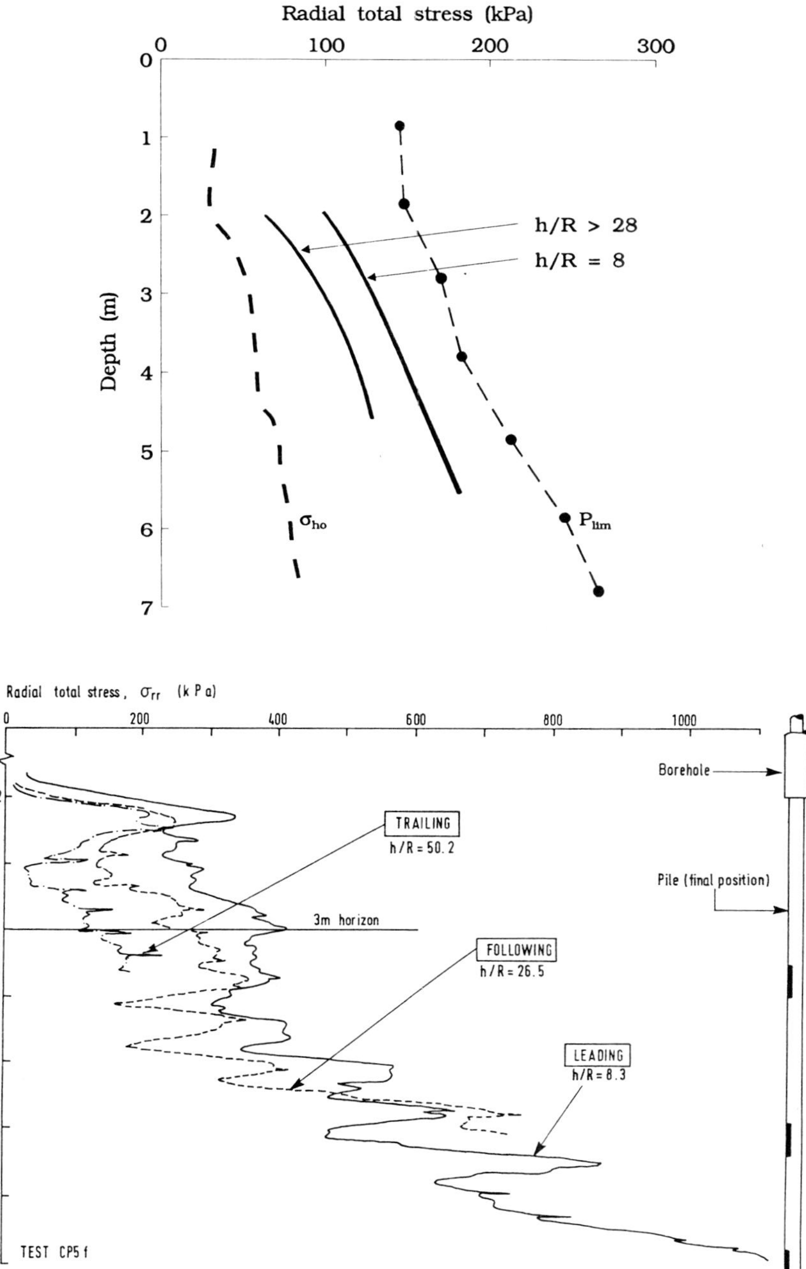

Fig. 2. Variation of radial total stress with depth during pile installation in (a) Bothkennar clay (Lehane and Jardine, 1992b) and (b) London Clay (Bond and Jardine, 1991).

positions ($h/R \approx 8$, where R is the pile's radius) and the lowest stresses at the trailing positions $h/R \approx 50$). As discussed later, this relaxation of stress has an important bearing on the overall shaft capacity of the pile.

4.2. Equalization Period

The changes in stress that occur during equalization depend primarily on whether the clay initially contracts or dilates when sheared.

4.2.1. Contractive (generally low OCR) clays

The general trend in contractive clays is for the radial total stress (σ_r) and the excess pore pressure ($\Delta u = u - u_o$) to decrease with time,[1] and hence for the radial effective stress ($\sigma_r' = \sigma_r - u$) to increase. These trends are illustrated in Figure 3a by data from experiments in the Bothkennar clay. Similar trends have been reported by, amongst others, Azzouz and Lutz (1986) in the Empire Clay and Azzouz and Morrison (1988) in the Boston Blue Clay.

4.2.2. Dilating (generally high OCR) clays

The pattern of behaviour in dilating clays is entirely different, as can be seen from the results obtained in the London Clay (Figure 3b). The radial total stresses show comparatively small changes with time, whereas the pore pressure rise and then fall. The nett effect is for the radial effective stresses to *fall* initially and then to rise. The long term radial effective stresses are comparable with those acting on the pile at the start of equalization. Similar trends have been found in the Gault Clay (Coop and Wroth, 1989) and in the Cowden glacial till.

These observations led Coop and Wroth (1989) and Bond and Jardine (1991) to suggest that piles installed in dilating clays could witness a short-term drop in capacity shortly after installation. The pile's capacity would recover with time but would be unlikely to exceed that recorded at the end of installation.

The experiments at Cowden have confirmed these suggestions. The shaft capacity of a pile tested after 4 days was 70% higher than that of one tested after only 2 hours. Shaft resistances measured at the end of equalization were generally within 20% of those obtained at the end of pile installation.

4.3. Stress State at the End of Equalization

The preceding sections suggest that the radial effective stresses that act at the end of equalization may depend on a number of factors, including the clay's consistency, the height above the pile tip, and whether the soil contracts or dilates when sheared.

[1] u_o is the original (i.e. ambient) pore pressure in the ground

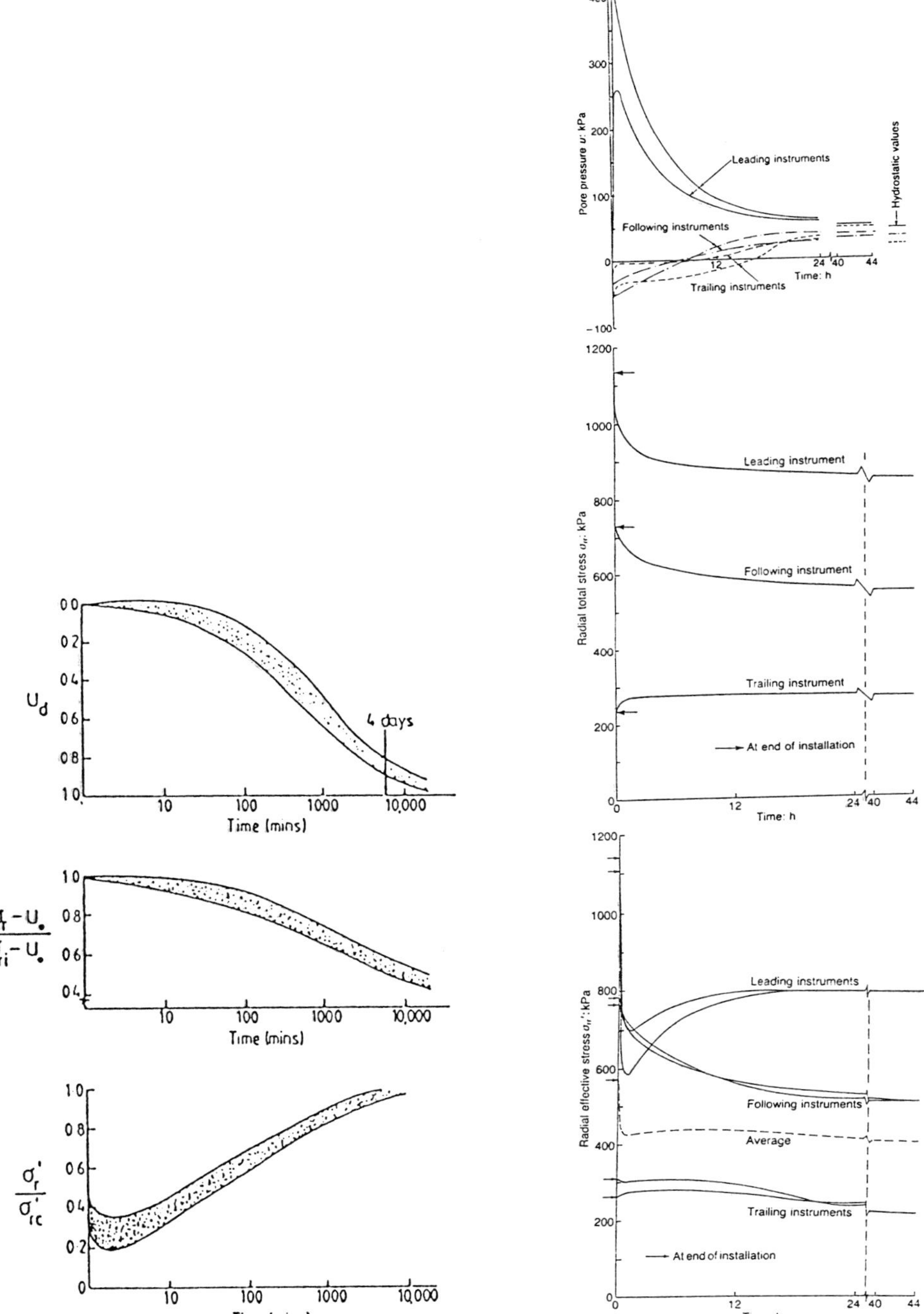

Fig. 3. Variation of earth and pore water pressures during equalization in (a) Bothkennar clay (Lehane and Jardine, 1992b) and (b) London Clay (Bond and Jardine, 1991).

TABLE 2.
Case histories of instrumented pile tests in clay soils.

No	Soil type	Reference	Approx. OCR
1	London Clay (Canons Park)	Bond (1989)	27-40
2	Cowden glacial till	Lehane (1992)	5.5-13
3	Bothkennar soft clay	Lehane (1992)	1.5-1.6
4a	Boston Blue Clay (MIT)	Morrison (1984)	1.25-2.2
4b	Boston Blue Clay (Saugus)	Azzouz & Morrison (1988)	1.2-4
5a	Empire Clay	Azzouz & Lutz (1986)	1.5-1.7
5b	Empire Clay	Bogard & Matlock (1990)	1.5-1.7
6	Haga clay	Karlsrud & Haugen (1985)	4-25
7	Gault Clay	Coop & Wroth (1989)	22-27
8	Silty Somerset clay	Coop (1987)	1.2-1.3
9	Plastic Tokyo clay	Koizumi & Ito (1967)	5-18
10	Rio de Janeiro clay	Soares & Dias (1989)	1.9-2.0
11	Pentre clayey silt	Karlsrud *et al.* (1992)	1.4-1.6
12	Lowestoft till Middle Oxford clay	Karlsrud *et al.* (1992)	10-35 5-10

The latter can be represented approximately by the soil's overconsolidation ratio (OCR).

4.3.1. Variation in K_c with OCR

Reviews of instrumented pile tests in clay soils (Bond, 1989; Lehane, 1992) have revealed that the value of $K = \sigma'_r/\sigma'_{vo}$ at the end of equalization (K_c) is strongly correlated with the overconsolidation ratio (OCR) of the soil. Figure 4 shows data from the 14 most reliable case histories that have been reported to date with closed-ended piles, as listed in Table 2. Individual points are distinguished according to the height above the pile tip (h) at which the measurements were obtained, and the values of h have been normalized by the pile's radius (R).

Although the results are highly scattered, the value of K_c appears to increase with OCR according to a power function of the form:

$$K_c = \Psi \mathrm{OCR}^{\mathrm{b}} \tag{4}$$

where b is a constant (≈ 0.55) and Ψ is a function of h/R.

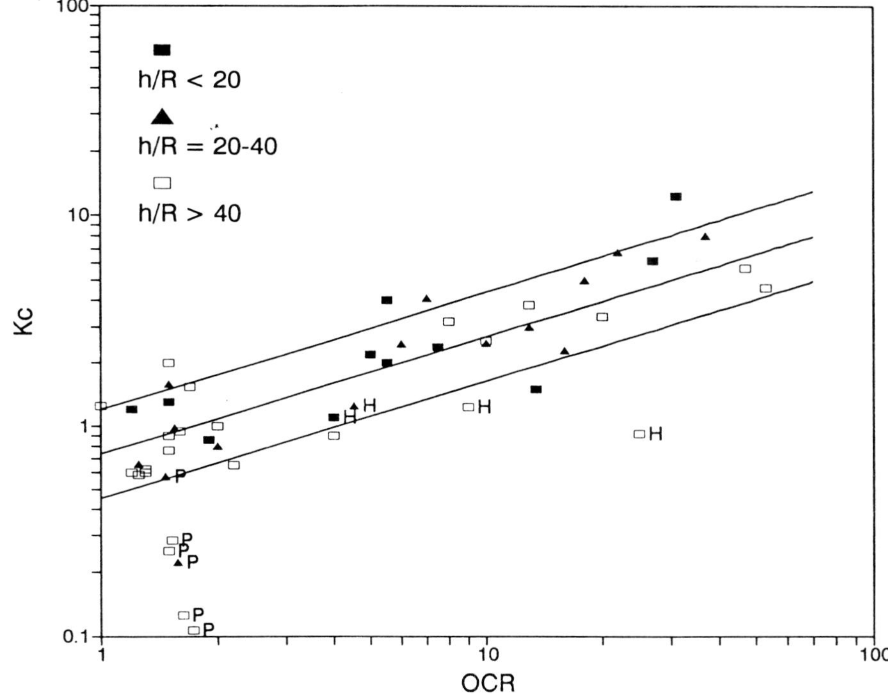

Fig. 4. Variation of K_c with overconsolidation ratio.

Two notable exceptions to this trend are the results obtained in the Haga clay (Karlsrud and Haugen, 1985) and in the clayey silt at Pentre (Karlsrud *et al.*, 1992). These have been marked on Figure 4 with the letters "H" and "P". The Haga clay is a soft, sensitive clay that has derived its apparently high overconsolidation ratio through leaching rather than by the removal of overburden. The clay contracts when sheared and displays many of the characteristics of a lightly overconsolidated clay. In this instance, the apparent OCR is a poor indicator of the soil's likely response to pile installation and equalization.

4.3.2. *Variation in K_c with h/R*

The degree of stress relaxation that takes place above the pile tip during pile installation has a marked effect on the value of K_c.

In the experiments at Canons Park, the installation radial total stresses that were measured at any one instrument position increased steadily with increasing pile penetration. However, the radial total stress that was recorded at any particular soil horizon fell steadily as the pile tip advanced to greater depths. For example, the values of σ_r that were recorded at the 3m horizon decreased from 400kPa

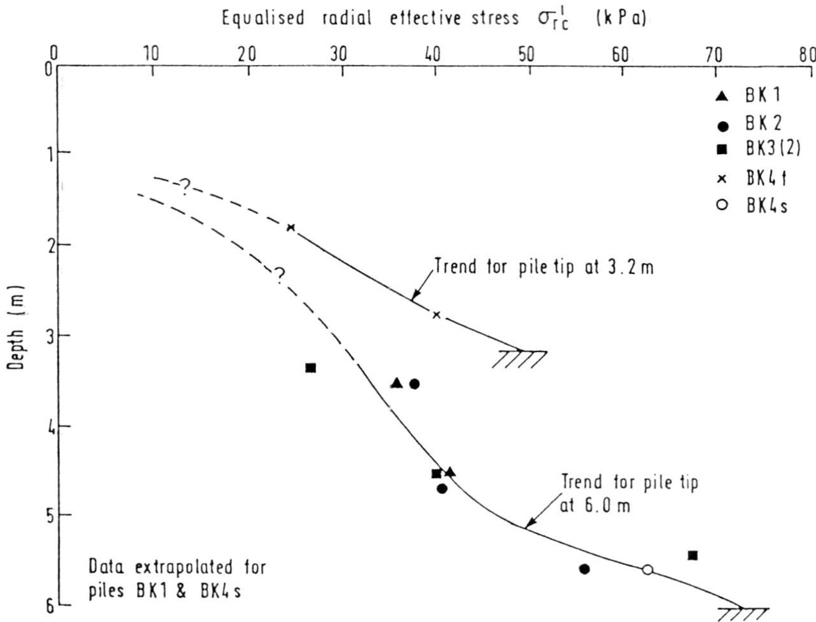

Fig. 5. Variation in radial effective stress with depth at the end of equalization (Bothkennar).

at the leading instrument position ($h/R \approx 8$) to 125kPa at the trailing position ($h/R \approx 50$) – see Figure 2b. Despite the significant variations in stress that take place during equalization, it would not be surprising to discover that the radial effective stresses acting on the pile at the end of equalization also vary with h/R.

4.3.3. *Variation in K_c with h/R – direct evidence*

Unfortunately, there is only limited information with which to test this hypothesis. The clearest proof is obtained by comparing the data recorded in a single soil horizon by piles installed to different depths of penetration. Of the instrumented pile tests listed in Table 2, only those at Labenne, Cowden, and Bothkennar provide this information.

The experiments at these sites were designed to investigate the relationship between the fully equalized radial effective stresses (σ'_{rc}) and the depth of penetration of the pile. At Bothkennar, two piles were installed to a depth of 3.15m and two others to 6m. The radial effective stresses that were recorded at the end of equalization are shown on Figure 5. At the soil depth at which direct comparison is possible (i.e. 3m), the radial effective stresses recorded by the 6m piles were 40% lower than those recorded by the 3.15m piles. The value of σ'_{rc} in any particular horizon clearly depends on the height of that horizon above the pile tip, and hence also on the overall penetration of the pile.

4.3.4. Variation in K_c with h/R – indirect evidence

The variation in K_c with h/R may be inferred indirectly from experiments in which multiple levels of instrumentation are used on a single pile (for example, at Canons Park and Haga), provided the effects of OCR can be accounted for.

Establishing a relationship between K_c and h/R is complicated by the fact that K_c varies with OCR. This can be overcome by plotting values of $\Psi = K_c/\mathrm{OCR}^{\mathrm{b}}$ against h/R, as shown on Figure 6. Data from the Imperial College experiments and those at Haga all follow a similar trend, which is for Ψ to decrease exponentially with h/R. As a first approximation, the lines on Figure 6 can be represented by a function of the form:

$$\Psi(\frac{h}{R}) = \frac{K}{\mathrm{OCR}^{\mathrm{b}}} = e^{-\eta[h/R+C]} \tag{5}$$

where C is a constant and η is a parameter that represents the degree of stress relaxation that occurs behind the pile tip. The value of η depends on soil type, and, for the soils shown on Figure 6, varies between 0.01 and 0.04. The "constant" C is approximately 20 for the Haga, Canons Park, and Cowden clays and approximately 90 for the Bothkennar and Tokyo clays. For the purpose of Figure 6, b has been taken as 0.55.

4.3.5. Variation in K_c with h/R – commentary

The preceding analysis is highly simplified and fails to take account of a number of other factors that affect the magnitude of K_c. These include the soil's critical state angle of friction, its sensitivity, plasticity index, etc.; the original (K_o) stresses in the ground; the degree of cyclic loading imparted to the clay during installation; and drainage conditions around the pile. As shown below, foremost amongst these is the clay's sensitivity.

The dependence of K_c on h/R cannot be determined when piles are equipped with instruments at only one or two locations along the pile shaft. It is important, in future experiments, that measurements are made at a minimum of three locations and preferably more.

4.3.6. Variation in K_c with sensitivity

Lehane (1992) has shown that a large part of the scatter in the K_c data plotted on Figure 4 arises from changes in radial total stress during equalization. There is a much better correlation between the normalized radial total stress developed during pile installation ($H_i = [\sigma_r - u_o]/\sigma'_{vo}$) and OCR than between K_c and OCR. The changes in stress during equalization depend on the initial sensitivity of the clay as illustrated on Figure 7, which shows the variation of the *relaxation coefficient*

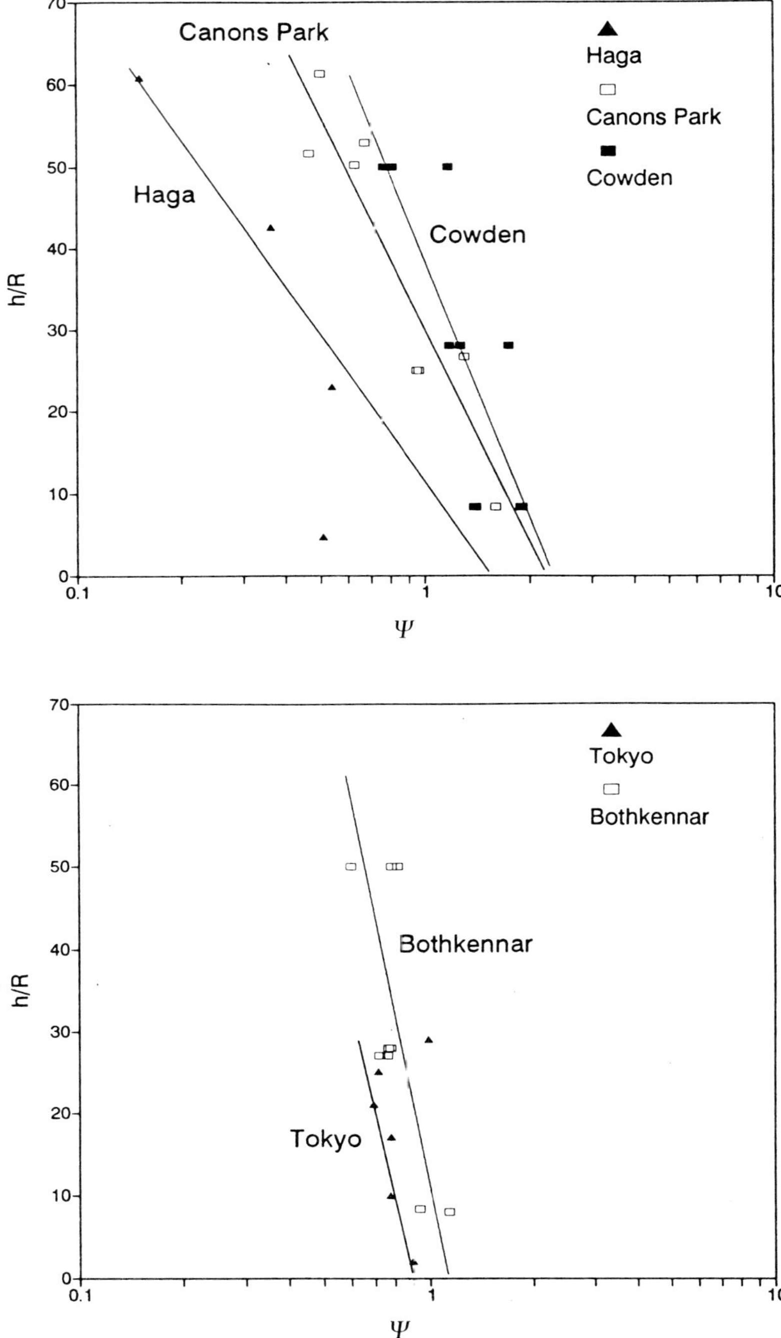

Fig. 6. Variation in Ψ with h/R.

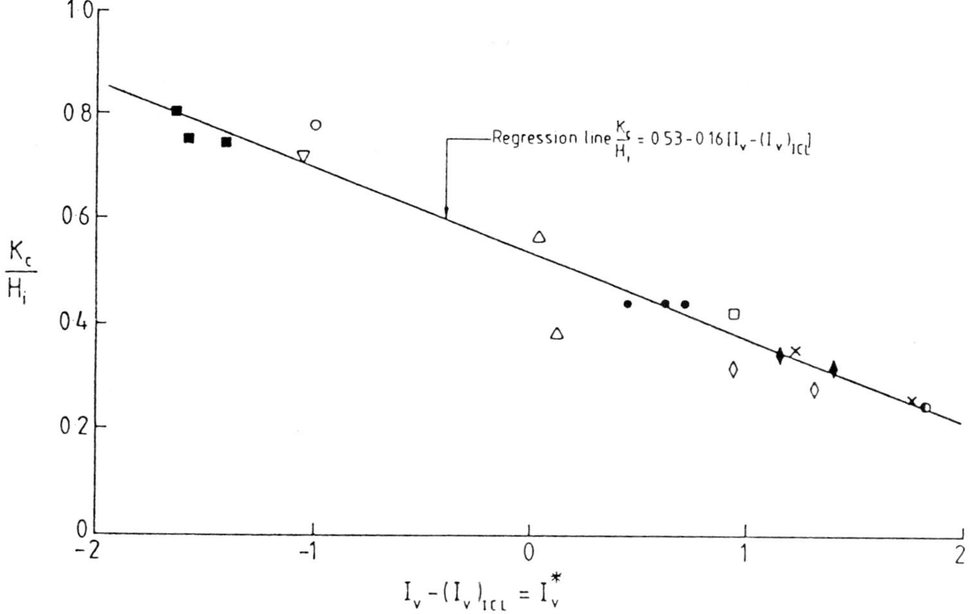

Fig. 7. Variation in the relaxation coefficient (K_c/H_i) with relative void index (I_v^*), after Lehane (1992).

(K_c/H_i) with relative void index (I_v^*). The parameter I_v^* is a measure of the clay's sensitivity and is related to the clay's *in situ* void index[2] (Burland, 1990).

Lehane has proposed an alternative expression for K_c based on the case histories listed in Table 2 and on trends predicted by the Strain Path Method (Baligh, 1985;

[2] $I_v^* = I_v - I_v^{\mathrm{ICL}}$, where I_v is the clay's void index and I_v^{ICL} the value of I_v on the clay's intrinsic compression line.

Whittle *et al.*, 1988):

$$K_c = 4\mathrm{OCR}^{0.4}(\frac{\mathrm{h}}{\mathrm{R}})^{-0.2}[0.5 - 0.2\mathrm{I}_\mathrm{v}^*]. \quad (6)$$

Further instrumented pile tests are required to refine and extend expressions such as (4), (5), and (6).

5. Pile Loading

Various hypotheses have been put forward to explain the behaviour of displacement piles during axial loading to failure. Randolph and Wroth (1981) argued that the soil surrounding the pile fails as a continuum and, because of the boundary conditions, the stress paths followed by an element of soil close to the pile shaft should resemble those obtained in laboratory simple shear tests. The simple shear analogy has been adopted in the work by Whittle *et al.* (1988) and Karlsrud and Nadim (1990).

In contrast to the above, the Imperial College tests suggest that failure occurs through interface sliding, either on a pre-formed residual shear surface (Bond and Jardine, 1991) or when the soil's intact interface resistance is exceeded. Conditions at failure are given by Coulomb's criterion, equation (1). The key parameters that govern the pile's shaft capacity are the pile loading factor (f_L) and the soil's angle of interface friction at failure (δ_f) – see equation (2).

5.1. Stress Paths Followed During Pile Loading

The extensive suite of instruments that are mounted on the Imperial College Instrumented Pile allow the stress paths that are followed during pile loading to be recorded. The data obtained in tests in the London and Bothkennar clays provide particularly interesting illustrations.

Figure 8 summarizes the effective stress paths that were followed by elements of soil adjacent to the pile shaft, during loading tests in London Clay. Both piles were installed by fast-jacking.

The radial effective stresses measured in the compression test fell slightly during pile loading, whereas those measured in the tension test remained largely constant (except at the trailing instrument position). In both cases, the maximum angle of friction that could be mobilized at the pile wall (δ) was close to 14°. The effect of continuing post-peak displacement of the pile was to reduce δ to a minimum of $\approx 8°$ after a few millimetres of axial movement. This implies that long piles installed in London Clay would be particularly susceptible to progressive failure. The critical state angle of friction of the London Clay at Canons Park is approximately $22\frac{1}{2}°$ (Jardine, 1985).

The effective stress paths for the corresponding tests in Bothkennar clay are shown on Figure 9. The radial effective stresses fell slightly in these experiments, both in compression and in tension, and the maximum angle of interface friction that could be mobilized was the same in both cases (between 25 and 30°). The

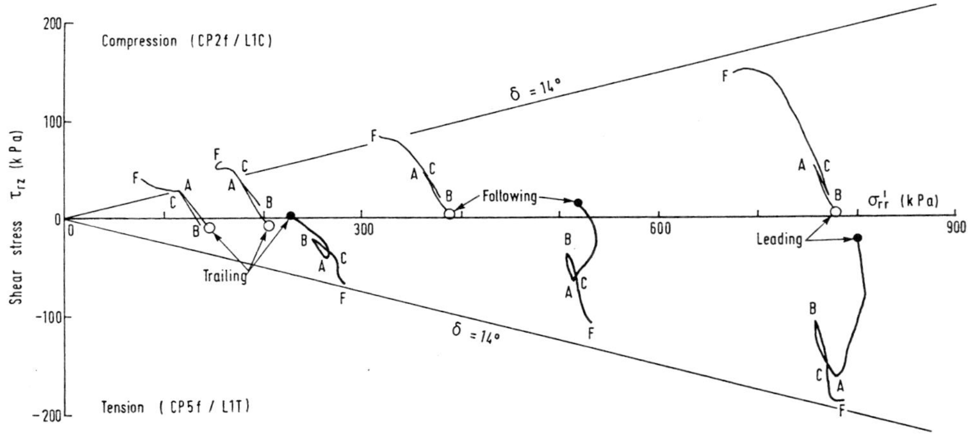

Fig. 8. Effective stress paths followed by elements of soil during pile loading in London Clay (Bond, 1989).

critical state angle of friction of the Bothkennar clay is approximately 36° (Hight *et al.*, 1992).

At Cowden, the maximum angle of interface friction that was mobilized in compression was between 19 and 24° and in tension approximately 24°. This compares with the soil's critical state angle of 26°.

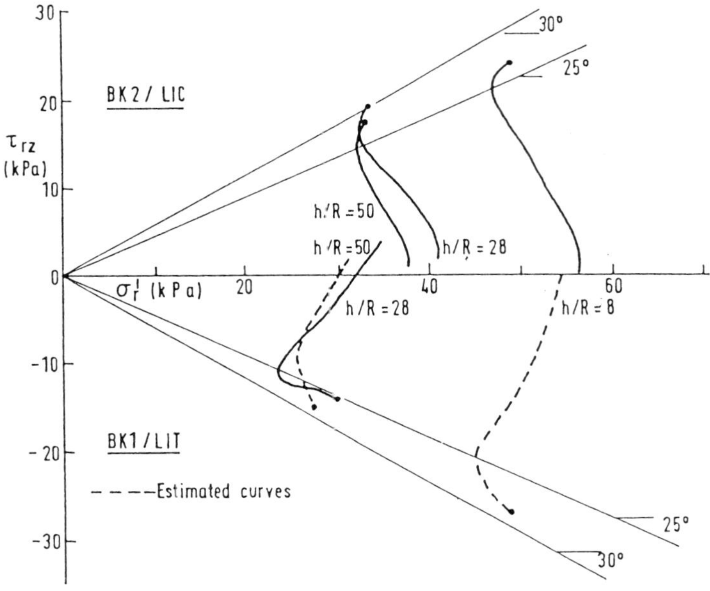

Fig. 9. Effective stress paths followed by elements of soil during pile loading in Bothkennar clay (Lehane and Jardine, 1992b).

TABLE 3.
Values of key parameters during pile loading.

Soil	Pile loading factor ($f_L = \sigma'_{rf}/\sigma'_{rc}$)	Angles of friction		
		δ_f	δ_{lab}	ϕ_{cs}
London Clay	0.90-1.28	11-14°	13-15°	≈22½°
Cowden till	0.8-1.0 (compression) 0.82-0.98 (tension)	19-24° ≈18°	23-28°	≈26°
Bothkennar clay	≈0.85	25-30°	29-33°	≈36°

The stress paths recorded at the three clay sites investigated are different from those obtained in laboratory simple shear tests. Overall, the recorded pile loading factors (f_L) lie in a narrow range, as shown in Table 3. As discussed below, δ values obtained at peak and ultimate conditions match those measured in special ring-shear interface tests.

5.2. Interface Shear Characteristics

A series of special interface ring-shear tests have been performed at Imperial College in parallel with the field studies (Tika *et al.*, 1992; Lehane, 1992). The tests were designed to simulate pile installation and loading, in the following way. First, the soil was sheared against the interface at a *fast* relative velocity of $\approx$ 500mm/min (pile installation). Second, the soil was re-consolidated to a pre-defined vertical stress and left to stand for $\approx$ 24h (equalization). Finally, the soil was sheared again, only this time at a *slow* relative velocity of $<$ 0.01mm/min (drained pile loading). The interfaces that were used in these experiments were of the same material and surface roughness as the instrumented piles.

The results for Canons Park, Cowden, and Bothkennar are listed in Table 3. Excellent agreement was found between the laboratory values of δ and those measured in the field experiments.

Research on a wider range of soil types (Lemos, 1985; Tika, 1989; Lehane and Jardine, 1992c; Ridley and Jardine, 1992) has shown that the variation of the angle of interface friction (δ) with plasticity index (I_p) follows the general trend shown on Figure 10. Separate lines are given for the maximum and minimum values of δ that were measured in the *slow* loading stage of each test. Also drawn on this figure is the trend line for the variation of the critical state angle of friction (ϕ_{cs}) with I_p, as determined in triaxial compression tests, and the results of pile loading tests at the four sites listed in Table 1.

There is clearly a tendency for the highest angles of friction to be developed by lean, low plasticity clays. However, no unique relationship exists between δ and I_p, as the tests on Bothkennar clay demonstrate. This soil has the highest δ value for any of the soils tested, and yet has a very high plasticity. Particle mineralogy, grain shape, and organic content all appear to affect δ and site-specific tests are needed to establish the value of δ for major offshore projects.

6. Summary and Implications for Design

The research described above shows that pile shaft capacity in clays depends on several variables that do not appear in current design methods. The most influential parameters affecting the local values of τ_{rz} appear to be:

- Original vertical effective stress in the ground;
- OCR;
- Clay sensitivity;
- Interface-friction characteristics;
- Pile tip position;
- Time between driving and loading.

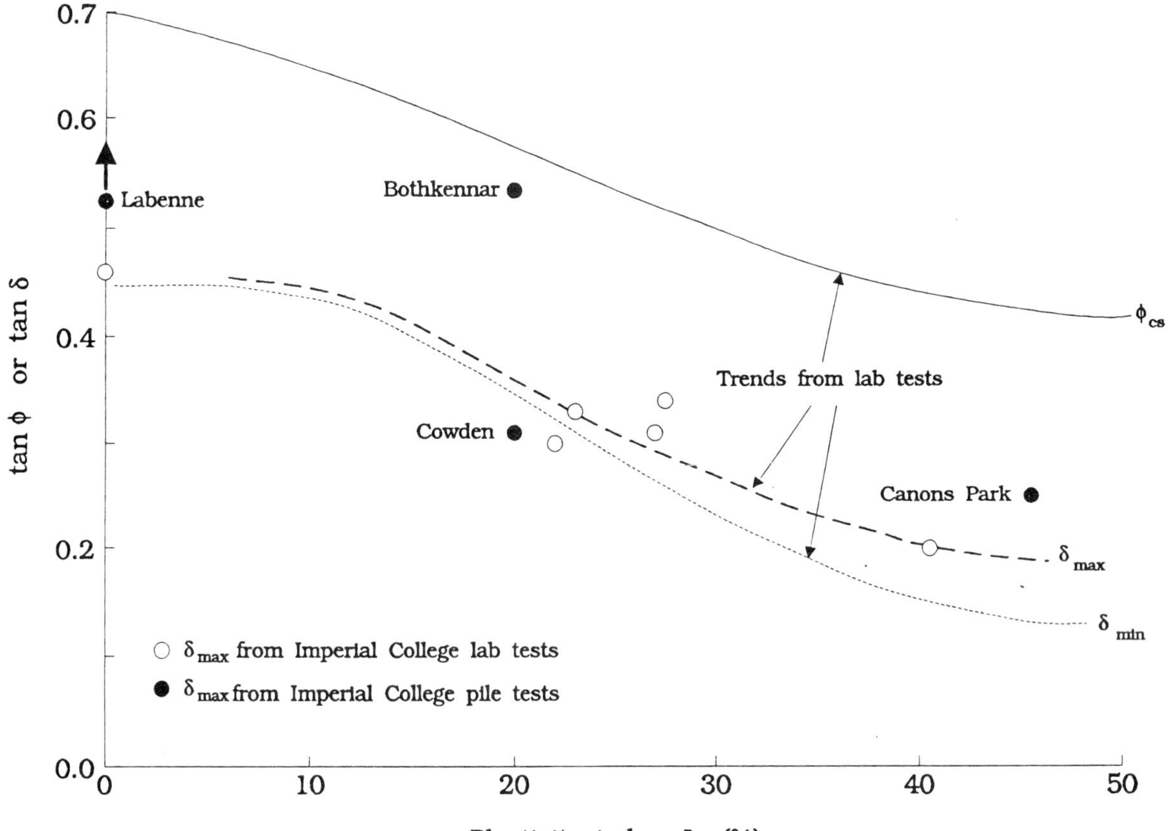

Fig. 10. Variation of interface friction angle with plasticity index (based on data from Lemos, 1985; Tika, 1989; and Lehane and Jardine, 1992c).

Design procedures can be developed that take account of these factors through effective stress analysis. The key steps are:

- To assess K_c from site-specific OCR measurements using an empirical database such as that shown on Figure 4. Alternatively, K_c can be assessed from other validated theories.
- To assess the pile loading factor f_L by similar means.
- To conduct site-specific laboratory tests to determine the soil's angle of interface friction (δ).
- To combine these components in a simple analysis that takes account of progressive failure where this is important. This can be done using spreadsheet, t-z, or finite element programs.

The Authors have applied this method in practice in cases involving North Sea structures (see, for example, Stock *et al.*, 1992).

Some of the implications of this approach are as follows.

6.1. Interface-Friction Characteristics

Shaft capacities can vary greatly from one site to another owing to differences in the soil's angles of interface friction (which have been found to vary from 8 to 30°, see Figure 10). Site-specific tests that model field conditions as closely as possible are needed to select appropriate values of δ for effective stress equations. These tests should reproduce the shearing history, stress level, and interface hardness and roughness of the full-scale pile.

6.2. Pile Tip Position

The Imperial College tests have shown that the peak local shear stress (τ_{rzf}) reduces as the relative depth of the pile tip (h/R) increases. This has many implications, including:

- A tendency for the average shaft resistance to reduce with the pile's length to diameter ratio[3];
- Potentially large differences between the *local* skin friction coefficient (α or β) and its *average* value;
- The pile's overall shaft capacity is sensitive to the way soil properties vary with depth, particularly in layered soils.

[3] This is in addition to reductions in τ_{rzf} owing to progressive failure as δ reduces down the shaft of long, compressible piles.

6.3. Time

Time has different effects on shaft capacity in contractive and dilating clays. Although, in low OCR clays, the radial effective stress acting on the pile increases steadily after driving, in high OCR clays it can reduce dramatically in the short-term to a temporary minimum. This can be very important when it comes to interpreting full-scale tests and has major implications for the safety of offshore structures.

7. Conclusions

The shaft capacity of displacement piles in clay soils is controlled by interface sliding according to an effective stress failure criterion. The stresses in the soil during pile loading depend on a number of factors, of which the most important are: the clay's overconsolidation ratio, sensitivity, and mineralogy; the pile's surface texture, slenderness ratio, and compressibility; the method and rate of pile installation; the time left between pile installation and loading; and (to a lesser extent) the direction of loading. All these factors can be included in an effective stress design approach.

Acknowledgements

The research described above was performed by the Imperial College Soil Mechanics Group in association with a number of projects, several of which were supported by the Science and Engineering Research Council (SERC). Thanks are due to SERC for this funding. The pile tests at Labenne, Cowden and Bothkennar were funded by a consortium of SERC (through MTD Ltd.), Amoco (UK) Ltd., Conoco (UK) Ltd., Exxon, Mobil, Shell (UK), Health and Safety Executive (HSE), Building Research Establishment (BRE) and Laboratoires des Ponts et Chaussées. Their support is acknowledged gratefully.

References

1. Azzouz, A. S. and Lutz, D. G. (1986), 'Shaft behaviour of a model pile in plastic Empire clays', *J. Geotech. Engng.*, Am. Soc. Civ. Engrs., **112**(4), 389–406.
2. Azzouz, A. S. and Morrison, M. J. (1988), 'Field measurements on model pile in two clay deposits', *J. Geotech. Engng.*, Am. Soc. Civ. Engrs., **114**(1), 104–121.
3. Baligh, M. M. (1985), 'Strain path method', *J. Geotech. Engng.*, Am. Soc. Civ. Engrs., **111**(9), 1108–1136.
4. Bogard, J. D. and Matlock, H. (1990), 'In-situ segment model experiments at Empire', Louisiana, *Proc. Offshore Technology Conf*, Houston, **2**, 459–467.
5. Bond, A. J. (1989), 'Behaviour of Displacement Piles in Overconsolidated Clays', PhD thesis, University of London (Imperial College).
6. Bond, A. J. and Jardine, R. J. (1991), 'Effects of installing displacement piles in a high OCR clay', *Géotechnique* **41**(3), 341–363.
7. Bond, A. J., Jardine, R. J., and Dalton, J. C. P. (1991), 'The design and performance of the Imperial College Instrumented Pile', *Geotechnical Testing J.* **14**(4), Dec. 1991, 413–424.
8. Burland, J. B. (1973), 'Shaft friction of piles in clay – A simple fundamental approach', *Ground Engng.* **6**(3), 30–42.

9. Burland, J. B. (1990), 'On the compressibility and shear strength of natural clays', *Géotechnique* **40**(3), 327–378.
10. Chandler, R. J. (1966), 'The shaft friction of piles in cohesive soils in terms of effective stress', *Civ. Engng. Publ. Wks. Rev.*, Jan., 48–51.
11. Coop, M. R. (1987), 'The Axial Capacity of Driven Piles in Clay', DPhil thesis, University of Oxford.
12. Coop, M. R. and Wroth, C. P. (1989), 'Field studies of an instrumented model pile in clay', *Géotechnique* **39**(4), 679–696.
13. Hight, D. W., Bond, A. J., and Legge, J. D. (1992), 'Characterization of the Bothkennar clay: An overview', *Géotechnique* **42**(42), 303–347.
14. Jardine, R. J. (1985), 'Investigations of Pile-Soil Behaviour, with Special Reference to the Foundations of Offshore Structures', PhD thesis, University of London (Imperial College).
15. Karlsrud, K. and Haugen, T. (1985), 'Axial static capacity of steel model piles in overconsolidated clay', *Proc. 11th Int. Conf. Soil Mech. and Fdn. Engng.*, San Francisco, **3**, 140–146.
16. Karlsrud, K. and Nadim, F. (1990), 'Axial capacity of offshore piles in clay', *Proc. 22nd Offshore Technology Conference*, Houston, **1**, 405–416.
17. Karlsrud, K., Borg Hansen, S., Dyvik, R., and Kalsnes, B. (1992), 'NGI's pile tests at Tilbrook and Pentre – Review of testing procedures and results', *Proc. Int. Conf. on Recent Large Scale Fully Instrumented Pile Tests in Clay*, London, Instn. Civ. Engrs./Thomas Telford Ltd. (in press).
18. Koizumi, Y. and Ito, K. (1967), 'Field tests with regard to pile driving and bearing capacity of piled foundations', *Soils and Foundations* **7**(3), 30–53.
19. Lehane, B. M. (1992), *Experimental Investigations of Pile Behaviour Using Instrumented Field Piles*, PhD thesis, University of London (Imperial College).
20. Lehane, B. M. and Jardine, R. J. (1992a), 'The behaviour of displacement piles in glacial till', *Proc. 6th Int. Conf. on Behaviour of Offshore Structures (BOSS '92)*, London (in press).
21. Lehane, B. M. and Jardine, R. J. (1992b), 'The behaviour of a displacement pile in Bothkennar clay', *Proc. Wroth Memorial Conf.*, Oxford (in press).
22. Lehane, B. M. and Jardine, R. J. (1992c), 'Residual strength characteristics of Bothkennar clay', *Géotechnique* **42**(2), 363–368.
23. Lehane, B. M., Jardine, R. J., Bond, A. J., and Frank, R. (1993), 'Mechanisms of shaft friction in sand from instrumented pile tests', *J. Geotech. Engng.*, Am. Soc. Civ. Engrs. (in press).
24. Lemos, L. J. L. (1985), 'The Effect of Rate on Residual Strength of Soil', PhD thesis, University of London (Imperial College).
25. Meyerhof, G. G. (1976), 'Bearing capacity and settlement of pile foundations', *J. Geotech. Engng.*, Am. Soc. Civ. Engrs., **102**(GT3), 197–228.
26. Morrison, M. J. (1984), '*In-situ* Measurements on a Model Pile in Clay', PhD thesis, Massachusetts Institute of Technology.
27. Mullis, C. R. (1992), 'Large diameter pile test project – Overview', *Proc. Int. Conf. on Recent Large Scale Fully Instrumented Pile Tests in Clay*, London, Instn. Civ. Engrs./Thomas Telford Ltd. (in press).
28. Randolph, M. F. and Wroth, C. P. (1981), 'Application of the failure state in undrained simple shear to the shaft capacity of driven piles', *Géotechnique* **31**(1), 143–147.
29. Ridley, A. M. and Jardine, R. J. (1992), 'Internal report on ring-shear tests on North Sea soil samples', Dept. Civ. Engng., Imperial College, London.
30. Soares, M. M. and Dias, C. R. R. (1989), 'Behaviour of an instrumented pile in the Rio de Janeiro clay', *Proc. 12th Int. Conf. Soil Mech. and Fdn. Engng.*, Rio de Janeiro, **1**, 319–322.
31. Stock, P., Jardine, R. J., and McIntosh, W. (1992), 'Foundation monitoring of the Hutton TLP', *Proc. Int. Conf. on Offshore Site Investigation and Foundation Behaviour*, London, Society for Underwater Technology (in press).
32. Tika, T. E. (1989), *The Effect of Fast Shearing on the Residual Strength of Soils*, PhD thesis, University of London (Imperial College).
33. Tika-Vassilikos, T. E., Bond, A. J., and Jardine, R. J. (1992), 'Correlation of the results from field experiments with instrumented model piles in London clay and laboratory ring shear interface tests', *Proc. 6th Int. Conf. on the Behaviour of Offshore Structures (BOSS '92)*, London (in

press).

34. Whittle, A. J., Baligh, M. M., Azzouz, A. S., and Malek, A. M. (1988), 'A model for predicting the performance of TLP piles in clay', *Proc. 5th Int. Conf. on the Behaviour of Offshore Structures (BOSS '88)*, Delft, 97–112.

ASSESSMENT OF AN EFFECTIVE STRESS ANALYSIS FOR PREDICTING THE PERFORMANCE OF DRIVEN PILES IN CLAYS

A. J. WHITTLE
Department of Civil and Environmental Engineering, Massachusetts Institute of Technology, 77 Massachusetts Ave., Cambridge, MA 02139, USA

Abstract. Researchers have advocated systematic analyses, which model changes in effective stresses and soil properties through successive phases in the life of a pile, as a rational method for understanding the factors which control pile performance. Work at MIT has included the development of analytical models which simulate soil disturbance effects associated with pile installation (Strain Path Method), and constitutive models (e.g., MIT-E3) which describe the effective stress-strain behaviour of normally and lightly overconsolidated clays (OCR $\leq$ 4) through successive phases in the life of the pile. This paper summarizes the role of these analyses in predictions of pile shaft behaviour. The results illustrate the effects of soil properties, mode of pile installation and other factors affecting the limiting skin friction which can be mobilized at the pile shaft. Predictive capabilities and limitations of the proposed 'objective analysis' are reviewed based on comparisons with high quality field data measured by the piezo-lateral stress (PLS) cell and by instrumented model pile tests.

1. Introduction

The geotechnical group at MIT has been involved in a sustained research effort to develop more reliable methods for predicting the capacity and performance of friction piles driven in clays. Originally, these efforts were motivated by the uncertainties involved in extrapolating empirical correlations from onshore pile load tests to offshore applications where much larger piles are used and where soil conditions often include deep layers of weak, normally and lightly overconsolidated clays. More recently, the work has focused on the performance of piles supporting Tension Leg Platforms (TLP; Whittle, 1987; Whittle *et al.*, 1988; Malek *et al.*, 1989).

The research makes the fundamental assumption that pile performance should be evaluated using a rational framework in which the changes in stresses and soil properties are described through successive phases in the life of a pile (Esrig *et al.*, 1977; Randolph *et al*, 1979; Baligh and Kavvadas, 1980). For TLP piles this includes: a) the initial, *in-situ* conditions in the ground; b) pile installation; c) soil consolidation or 'set-up'; d) monotonic shearing due to quasi-static tensile forces imposed by mooring of the TLP superstructure; and e) cyclic shearing caused by storm waves. Due to the complexities of these pile-soil interactions, extensive research efforts have been required in four complementary lines of activity:

Volume 28: Offshore Site Investigation and Foundation Behaviour, 607–643, 1993.

1. The formulation of analytical models which are capable of making realistic predictions of pile performance. This work has included the development of: a) the Strain Path Method (Baligh, 1985, 1986a, b) to describe the mechanics of the pile installation process; and b) effective stress soil models (MIT-E1, Kavvadas, 1982; and MIT-E3, Whittle, 1987) which can describe realistically the constitutive behaviour of K_0-consolidated clays which are normally to moderately overconsolidated (OCR $\leq$ 4).

2. *In-situ* measurements on a closed-ended model pile shaft referred to as the Piezo-Lateral Stress Cell (PLS; Morrison, 1984; Azzouz and Lutz, 1986; Azzouz and Morrison, 1988). The PLS cell has the capability of providing simultaneous measurements of the total lateral stress, pore pressures and average skin friction acting on the shaft of a small diameter pile (D = 3.83cm) during installation, consolidation and axial loading.

3. Extensive laboratory testing to support the analytical and field studies, and to develop more comprehensive understanding of complex aspects of clay behaviour. This work has included: a) test programs to characterize *in-situ* soil properties (e.g., Azzouz and Lutz, 1986); b) undrained cyclic direct simple shear testing to simulate pile-soil interaction during TLP storm loading conditions (e.g., Malek *et al.*, 1989); and c) measurement of anisotropic properties in the Directional Shear Cell (DSC) which are used to evaluate the constitutive models (e.g., Whittle *et al.*, 1992).

4. Evaluation of pile shaft predictions was initially accomplished using PLS cell measurements (Whittle and Baligh, 1988; Azzouz *et al.*, 1990). Subsequently, analytical predictions have been compared with field data from instrumented pile tests at a number of sites (e.g., Whittle, 1991b).

This paper describes typical predictions of pile shaft performance using Strain Path analyses in conjunction with the MIT-E3 soil model. The analyses provide objective predictions of soil stresses and pore pressures during installation, consolidation and axial loading , based on specified *in-situ* soil properties and stress conditions. The results illustrate the effects of stress history, mode of pile installation and other factors on the limiting skin friction which can be mobilized at the pile shaft. Predictive capabilities and limitations of the analyses are assessed from comparisons with high quality field data at a number of soft clay sites.

2. The Strain Path Method

Pile driving causes severe disturbances and leads to significant changes in the stresses, pore pressures and properties of the surrounding soils. The analysis of these installation effects represents a highly complex problem due to: a) high gradients of the field variables (displacements, stresses, strains and pore pressures) around the pile; b) large deformations and strains which develop in the soil; c) the complexity of the constitutive behaviour of soils, including non-linear, inelastic, anisotropic, and frictional response; and d) non-linear pile-soil interface characteristics. The Strain Path Method (SPM; Baligh, 1985) assumes that, due to the severe kinematic constraints in deep penetration, the deformations and strains in the surrounding soil are essentially independent of its shearing resistance and can be estimated with reasonable accuracy based only on kinematic considerations and boundary conditions. The application of the Strain Path Method for analyzing piles driven in low permeability clays assumes: a) there is no migration of pore water during penetration and hence, the soil is sheared in an undrained mode; b) pile driving can be modelled as a quasi-static (steady), deep penetration problem (i.e., there is no inherent difference due to pile installation by jacking or driving); and c) the deformations and strains can be estimated from the steady, irrotational flow of an incompressible, inviscid fluid around the pile (Baligh and Levadoux, 1980; Baligh, 1986a; Whittle *et al.*, 1991). By considering two-dimensional deformations of soil elements, the Strain Path analyses provide a more realistic framework for describing the mechanics of deep penetration than one-dimensional, cylindrical cavity expansion methods (CEM; e.g., Kraft, 1982; Randolph *et al.*, 1979). On the other hand, the assumptions of strain controlled behaviour used in the Strain Path Method greatly simplify the penetration problem and avoid the complexity of large deformation finite element analyses (e.g., DeBorst and Vermeer, 1984, Kiousis *et al.*, 1988).

Figure 1 compares SPM solutions of strain paths experienced by individual soil elements for two pile geometries: 1) a closed-ended (or fully plugged) pile of radius, R, with a rounded tip geometry (the 'simple pile'; Baligh, 1985); and 2) an open-ended pile of with aspect ratio, $B/t = 40$ (where $B = 2R$ is the outside diameter and t the wall thickness), which penetrates the soil in an unplugged mode (Chin, 1986). These solutions correspond to the two extreme modes of penetration for large diameter, open-ended pipe piles used in offshore foundations. The strain history at a point is fully described by three independent components of shear strain: E_1, E_2 and E_3 which correspond to triaxial, pressuremeter (cylindrical cavity expansion) and direct simple shear modes, respectively.

For the simple pile geometry, there is a monotonic increase in the E_2 shear component as the pile tip passes the soil element while the components E_1 and E_3 exhibit reversals in direction (which are not included in CEM analyses). The overall magnitude of shear strain in the soil is described by the second invariant of deviatoric strains,

$$E = \frac{1}{\sqrt{2}} \left\{ E_1^2 + E_2^2 + E_3^2 \right\}^{1/2}$$

as shown in Figure 2. At locations around the pile shaft (Figure 2), there is an inner

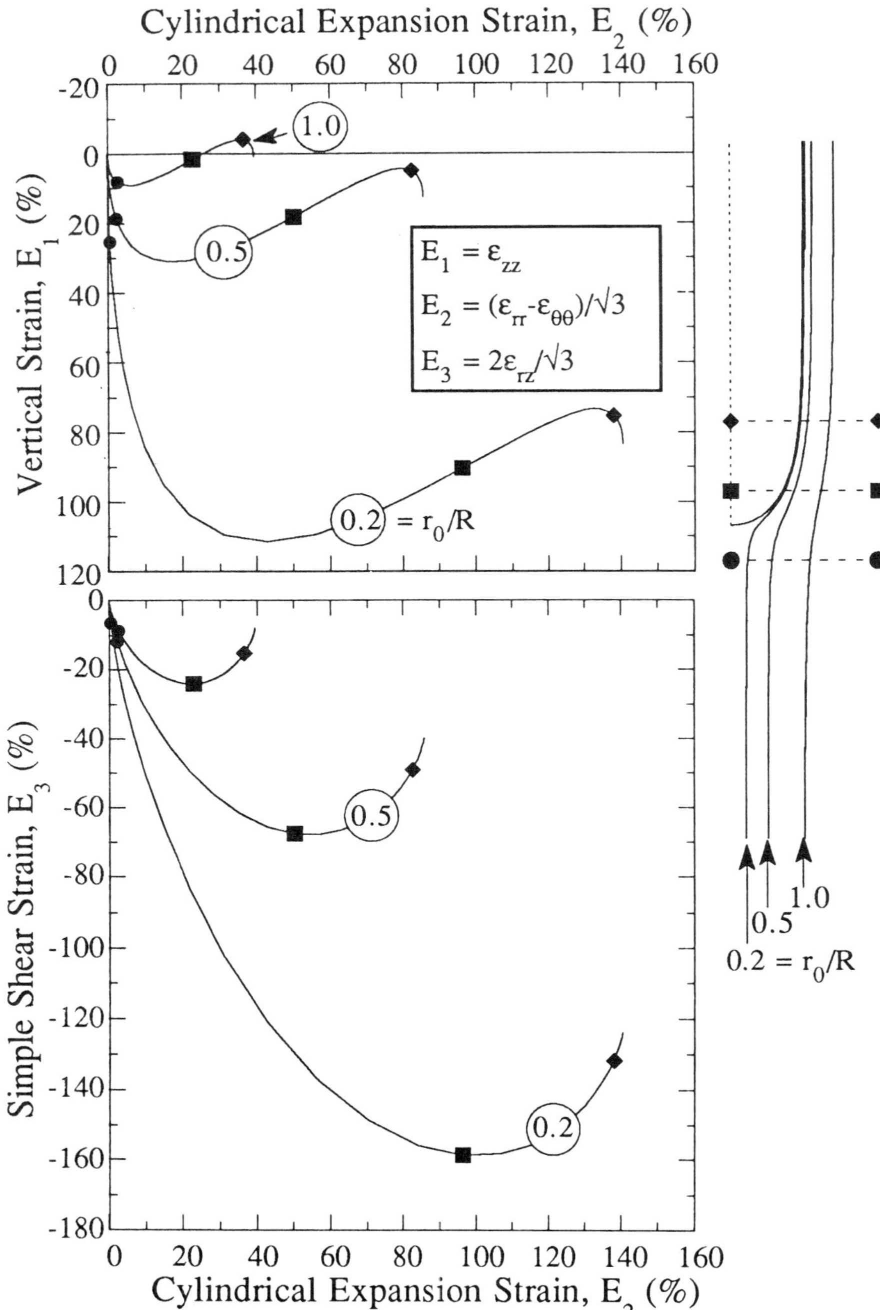

Fig. 1a. Strain paths during simple pile penetration (after Baligh, 1985).

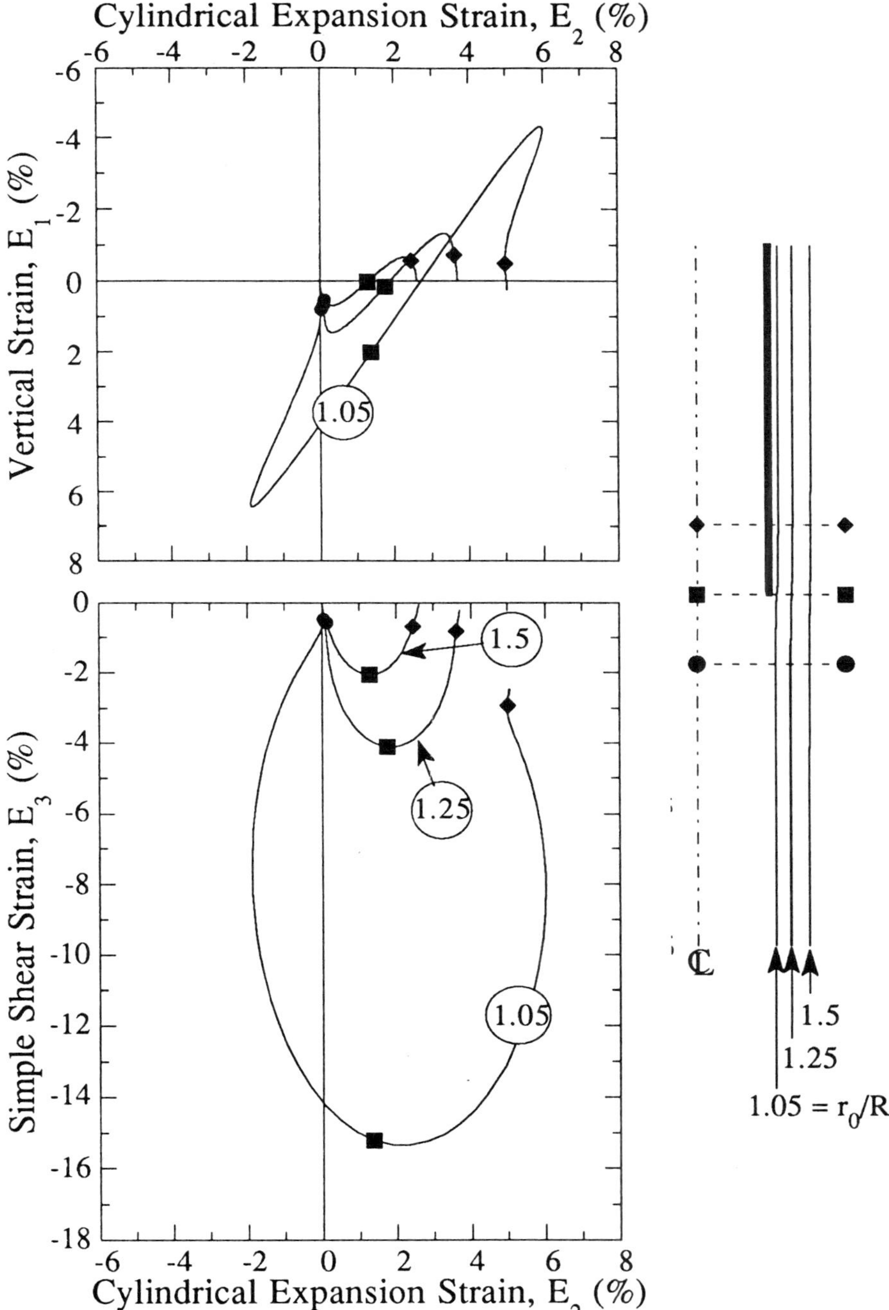

Fig. 1b. Strain paths during unplugged penetration of open-ended pile, $B/t = 40$ (after Chin, 1986).

zone of soil which experiences much larger shear strain levels than can be imposed in conventional laboratory shear tests ($E > 10\%$ at $r/R < 2$) and is characterized also by large net changes in all three strain components (e.g., elements $r_0/R = 0.5$, 0.2; Figure 1a). At radial locations further from the pile shaft, E_1, $E_3 > 0$ (e.g., $r_0/R = 1.0$; Figure 1a) and the final strain state is controlled by the volume of soil displaced by the pile.

The unplugged, open-ended pile causes much less disturbance of the surrounding soil. The zone of high shear strains ($E \geq 10\%$; Figure 2b) is confined to a thin annulus (comparable to the thickness of the pile wall) around the shaft, while far field strain levels are controlled by the volume of soil displaced by the pile. In order to compare the strain levels for the open and closed-ended piles, it is convenient to normalize the radial dimensions by the equivalent radius of a solid section pile, $R_{\text{eq}} \approx \sqrt{Bt}$ (i.e., for $B/t = 40$, $R_{\text{eq}} = 0.316R$) as shown in Figure 2b. The mode of penetration (closed vs. open-ended) also causes important differences in strain paths close to the pile wall, especially in the pressuremeter shear component, E_2.

In this presentation of the Strain Path Method, effective stresses, σ'_{ij}, are computed directly from the strain paths of individual soil elements using a generalized effective stress-strain soil model (see next section). This approach can be contrasted with previous total stress analyses (Levadoux and Baligh, 1980; Baligh, 1986a; Teh and Houlsby, 1991) which compute shear stresses through a deviatoric stress-strain model, and introduce a separate constitutive relationship for shear induced pore pressures. The main advantage of the effective stress analysis is that the same soil model can be used to study stress changes during consolidation and pile loading.

The installation excess pore pressures around the pile shaft are computed from the effective stresses by satisfying conditions of radial equilibrium (Baligh, 1986b). Further predictions of excess pore pressure distributions around the tip of the pile are difficult to achieve due to approximations used in the Strain Path Method. The most reliable estimates of pore pressure distributions are obtained by solving equilibrium conditions in the form of a Poisson equation using finite element methods (Aubeny, 1992; Whittle and Aubeny, 1992).

3. The MIT-E3 Effective Stress Soil Model

The MIT-E3 model (Whittle, 1987, 1990, 1991a) is a generalized effective stress soil model for describing the rate independent behaviour of normally to moderately overconsolidated clays (OCR ≤ 8) which exhibit normalized engineering properties. The model describes a number of important aspects of soil behaviour which have been observed in laboratory tests on K_0-consolidated clays but are not well described by most existing soil models:

1. There is no well defined linear region of soil behaviour, even at small strain levels or immediately following a reversal of loading.

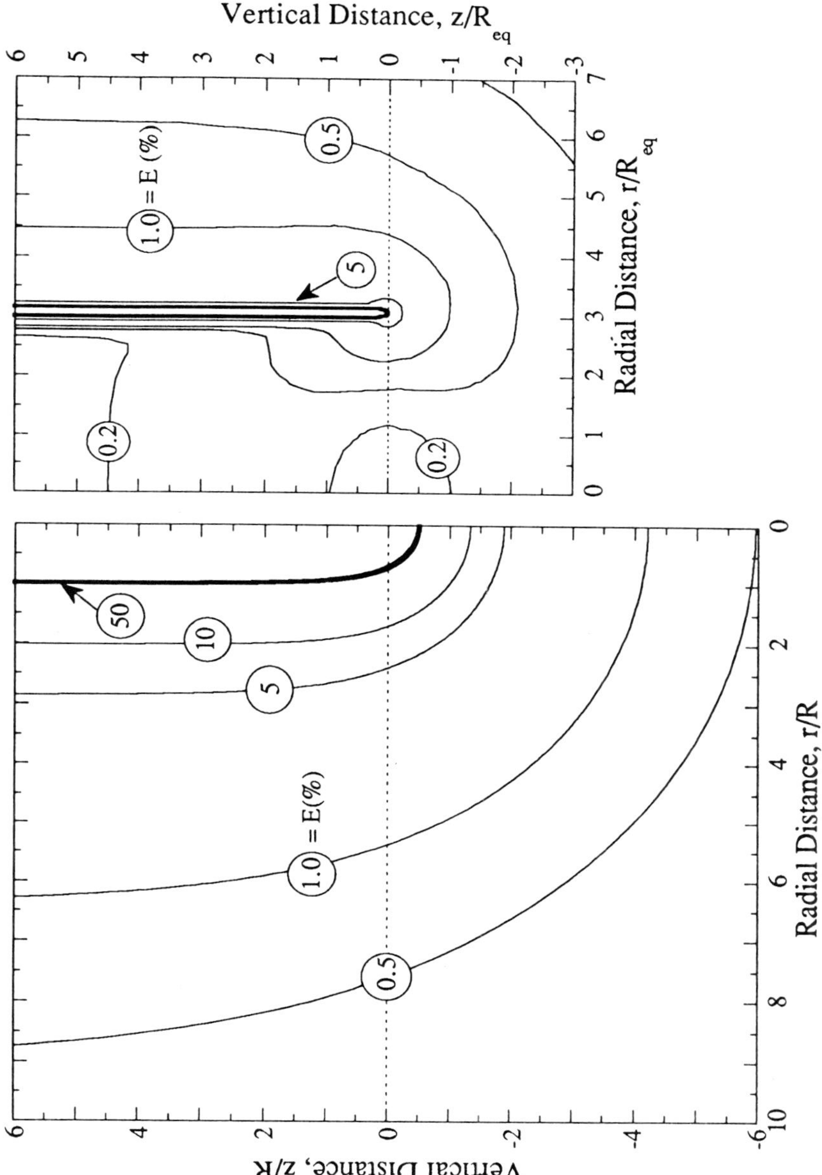

Fig. 2. Shear strains during pile installation. (a) Simple closed-ended pile; (b) Open-ended, unplugged pile, $B/t = 40$.

2. The unload-reload behaviour of clays is characterized by a hysteretic response, but also involves small irrecoverable strains.

3. Clays exhibit anisotropic properties due to their consolidation history and subsequent straining.

4. In some modes of deformation, normally and lightly overconsolidated clays exhibit undrained brittleness.

5. Uniform, undrained cyclic loading of overconsolidated clays causes an accumulation of shear induced pore pressures. Thus coupling of volumetric and shear behaviour is essential to accurate modelling of overconsolidated clays under cyclic loading.

The model formulation comprises three components: 1) an elasto- plastic model for normally consolidated clays, which describes anisotropic properties and strain softening behaviour; 2) equations for the small strain non-linearity and hysteretic stress-strain response in unload-reload cycles; and 3) bounding surface plasticity for irrecoverable, anisotropic and path dependent behaviour of overconsolidated clays. Other observations of clay behaviour, collectively referred to as 'rate effects' (e.g., variation in undrained shear strength with strain rate, undrained creep and secondary compression) are not described by the MIT-E3 model. The model uses 15 input parameters which are determined from standard types of laboratory tests:

1. One dimensional consolidation tests (either incremental oedometer or constant rate of strain consolidation) using a load sequence that includes at least one cycle of unloading-reloading and measurements of lateral effective stresses.

2. Undrained shear tests on K_0-consolidated clay including in triaxial compression (CK_0UC at OCR = 1, 2 and triaxial extension (CK_0UE at OCR = 1. These tests should be performed using SHANSEP consolidation procedures in order to ameliorate the effects of sample disturbance on the measured soil behaviour (Ladd and Foott, 1974).

3. Measurements of elastic shear wave velocity using either resonant column apparatus or field cross-hole tests. Alternatively, reliable measurements of the small strain stiffness can now be obtained from local strain measurements in triaxial tests (e.g., Jardine *et al.*, 1984; Dyvik and Olson, 1989; Clayton *et al.*, 1989; Goto *et al.*, 1991).

The model input parameters have been selected for a number of clays using a standard procedure (Whittle, 1990). Extensive comparisons with measured data in undrained shear tests performed in different modes of shearing and with over-

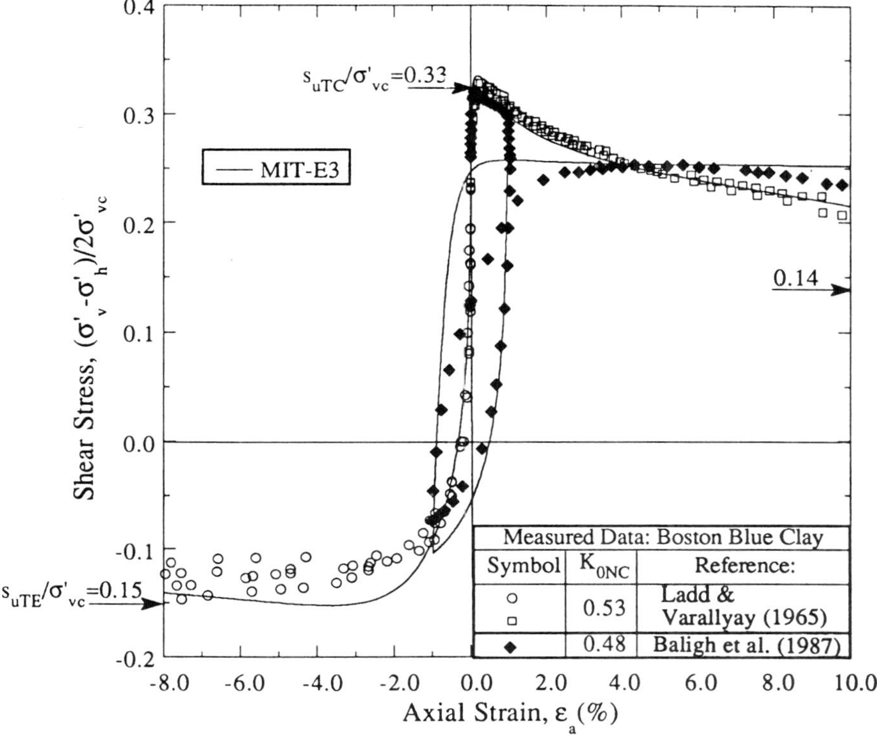

Fig. 3a. Comparison of MIT-E3 predictions and measured data for undrained triaxial shearing of K_0-normally consolidated Boston Blue Clay.

consolidation ratios (OCR) up to 8 have shown that the model a) gives excellent predictions of peak shear resistance and can describe accurately the non-linear stress-strain behaviour, but becomes less reliable for OCR $\geq$ 4. The most comprehensive evaluations have been presented for Boston Blue Clay (BBC), a low plasticity (I_p = 19–23%), illitic, marine clay of moderate sensitivity (s_t = 3–7) whose engineering properties have been studied extensively at MIT (Whittle, 1990, Whittle *et al.*, 1992). Figure 3 compares the computed and measured shear stress strain behaviour for K_0-normally consolidated BBC in the three modes of shearing which occur during pile installation (cf., Figure 1).

The measured data in undrained triaxial compression and extension tests (CK_0UC and CK_0UE) at OCR = 1 illustrate important aspects of the anisotropic behaviour of soft clays: The undrained shear strength in the compression mode ($s_{uTC}/\sigma'_{vc} = 0.33$) is mobilized at very small shear strains ($\varepsilon_{ap} \approx$ 0.3– 0.5%) and there is a significant post-peak reduction in shear resistance. In comparison, the undrained shear strength in triaxial extension is mobilized at relatively large strain levels ($s_{uTE}/\sigma'_{vc} \approx 0.14$ at $\varepsilon_{ap} > 5\%$). There is a large difference in the undrained

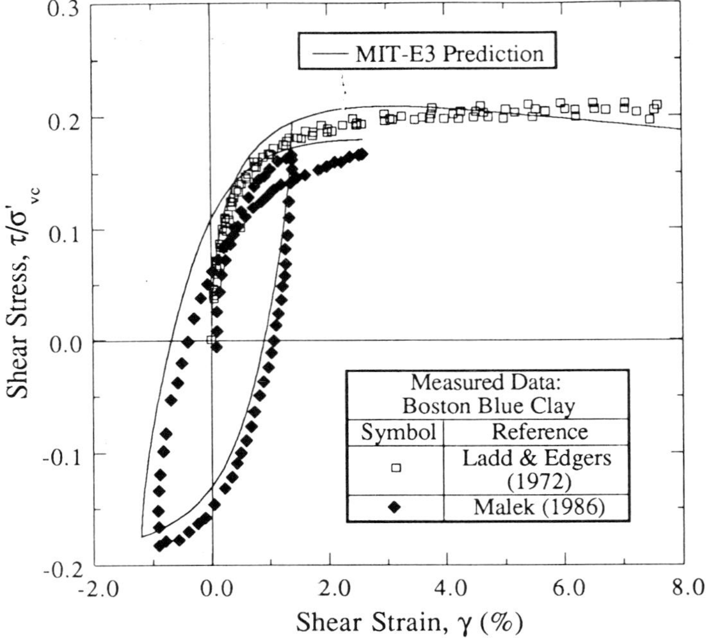

Fig. 3b. Comparison of model predictions and measured data for direct simple shear of K_0-normally consolidated Boston Blue Clay.

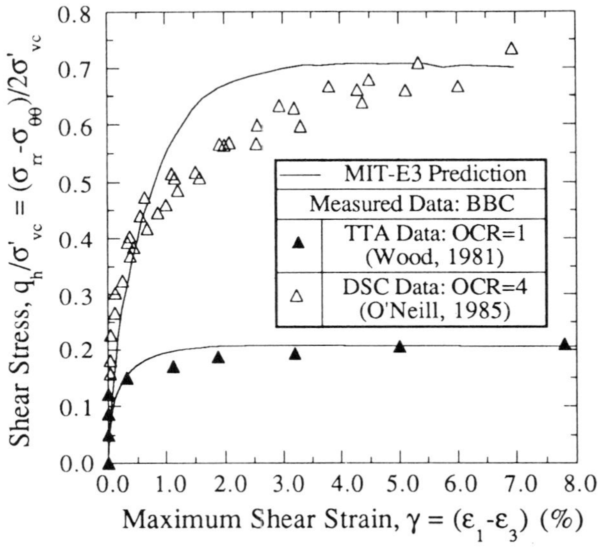

Fig. 3c. Comparison of model predictions and measured data for a pressuremeter mode of shearing of K_0-consolidated BBC.

shear strengths in the two modes of shearing, s_{uTE}/s_{uTC} = 0.42. The MIT-E3 model matches closely the measured undrained shear strengths in both modes of shearing, as well as the axial strain at peak resistance and post-peak strain softening. The measurements in monotonic compression and extension tests are part of the data set used to select input parameters and hence, these comparisons do not constitute an evaluation of model predictions. Model predictions for an undrained triaxial test with a single unload-reload cycle (i.e., two reversals of strain direction) are also shown in Figure 3a. The model describes closely the non-linear and hysteretic nature of the unload-reload process, but tends to overpredict the stiffness during reloading.

Figure 3b compares predictions with measurements in undrained Direct Simple Shear tests (CK_0UDSS) using a Geonor simple shear apparatus. This mode of shearing is also directly relevant to predictions of pile-soil behaviour in axial loading (e.g., Randolph and Wroth, 1981; Azzouz *et al.*, 1990) and is discussed in more detail in section 6. The MIT-E3 model gives very good predictions of the shear stress strain response in tests with monotonic shearing and with reversals of strain direction. The model is in excellent agreement with the measured peak shear resistance ($\tau_{\max}/\sigma'_{vc} = 0.21$) but tends to overestimate the stiffness in reloading.

The pressuremeter shear mode (E_2; Figure 1) is especially important for estimating the effects of pile installation and can be simulated in laboratory element tests using more sophisticated equipment such as the True Triaxial Apparatus (TTA) or Directional Shear Cell (DSC). Unfortunately, there is very little data of this type reported in the literature. Figure 3c compares model predictions with data reported by Wood (1981) using the Cambridge University TTA. The model predictions match the measured peak shear resistance, $s_{uPM}/\sigma'_{v0} = 0.21$, which is mobilized at a shear strain, $\gamma \approx 5\%$. Further comparisons with more comprehensive pressuremeter tests in the DSC at OCR = 4 (O'Neill, 1985; Figure 3c) show similar predictive accuracy for the peak shear resistance, but confirm that the model tends to overestimate the pre-peak shear stiffness.

The shear resistance at large strain levels is important in predicting stress conditions close to the pile shaft during installation. However, it is difficult to obtain reliable large strain measurements in laboratory tests due to non-uniformities in stress conditions, strain localization, etc. The predictions in Figure 3 show that there is a large post-peak reduction in the shear resistance of K_0-normally consolidated BBC in undrained triaxial compression ($s_{ur}/s_u \approx 0.4$), but negligible softening in the pressuremeter shear mode. Strain softening measured in Direct Simple Shear tests may be partly attributed to non-uniform stress conditions in the Geonor apparatus (e.g., DeGroot *et al.*, 1992).

Although it is not possible to duplicate the complex strain paths caused by pile installation using existing laboratory tests, the results in Figure 3 demonstrate the predictive capabilities of MIT-E3 and provide a sound basis for applying the model in conjunction with SPM analyses.

4. Pile Installation

4.1. Predictions of Installation Conditions

Figure 4 presents Strain Path (SPM) and Cavity Expansion (CEM) predictions of installation stresses and pore pressures around the shaft of a pile in K_0-normally consolidated BBC using the MIT-E3 soil model. The individual stress components are normalized by the *in-situ* vertical effective stress, σ'_{v0}, while radial dimensions are related to the equivalent radius of a solid section pile, R_{eq}, in order to unify results for the two limiting modes of pile penetration (plugged and unplugged). The two principal parameters of interest in these analyses are the excess pore pressures, $\Delta u_i/\sigma'_{v0}$, and radial effective stresses, $K_i = \sigma'_{ri}/\sigma_{v0}$ which can be measured at the pile shaft. The results in Figure 4 show the following:

1. For a normally or lightly overconsolidated clay, undrained shearing generates positive shear induced pore pressures and a corresponding net reduction in the mean effective stress, σ'/σ_{v0}, close to the pile shaft. Differences in the magnitude of σ'/σ_{v0} for SPM and CEM analyses ($r/R_{eq} \leq 6$; Figure 4a) reflect how the anisotropic and strain softening properties described by the MIT-E3 model are affected by differences in strain histories.

2. The effects of the analysis used to model installation can be seen most clearly in predictions of the radial effective stress, σ'_r/σ_{v0}, and cavity shear stress, q_h/σ'_{v0} (where $q_h = [\sigma'_r - \sigma'_\theta]/2$ is the maximum shear stress in the horizontal plane). The strain path method predicts very low radial effective stresses ($K_i = 0.08 - 0.10$) and the cavity shear stresses ($q_h/\sigma'_{v0-} \rightarrow 0$) acting at the pile shaft for both modes of penetration. This means that the radial effective stress is similar in magnitude to the mean effective stress (i.e., $\sigma'_r/\sigma'_{v0-} \approx \sigma'/\sigma'_{v0}$) for $r/R_{eq} \leq 15$. In contrast, CEM analyses give higher values of radial effective stress σ'_r/σ'_{v0} and predict that $\sigma'_r/\sigma'_{v0} > \sigma'/\sigma'_{v0}$ over a wide radial zone ($r/R \leq 20$). The cavity shear stress, $q_h/\sigma'_{v0} \approx 0.20$, is approximately constant for $r/R < 7$ and can be deduced from the pressuremeter shear behaviour described in Figure 3c. Strain path predictions of σ'_r/σ'_{v0} are affected significantly by soil properties (including strain softening), while Baligh and Levadoux (1980) show that the geometry of the pile tip has an important effect on the cavity shear stress close to the shaft.

3. The excess pore pressures at the pile shaft are obtained from conditions of radial equilibrium and hence depend on the entire field of effective stresses in the soil. The results in Figure 4a show that undrained pile installation generates large excess pore pressures in the soil which extend to a radial distance $r/R =$ 20–30. Although both CEM and SPM predict a similar accumulation of excess pore pressure in the far field ($3 \leq r/R \leq 30$), there are significant differences in the distribution close to the pile shaft ($r/R \leq 3$). The net result is that the

cavity expansion method predicts excess pore pressures which are typically 20–25% larger than those obtained from corresponding strain path analyses. The characteristic shapes of the pore pressures distributions (CEM and SPM; Figure 4a) have been discussed in detail by Baligh (1986b) and are not affected significantly by the modelling of soil behaviour.

4. The mode of penetration (Figures 4a, 4b) only affects the magnitude and distribution of stresses and pore pressures close to the pile wall, $r/R_{eq} \leq 3$. The strain path predictions of cavity shear stress, radial and mean effective stress components acting at the shaft are very similar for both modes of penetration, while the excess pore pressures are slightly smaller for the unplugged pile. It is interesting to note that the strain path method actually predicts slightly larger shaft pore pressures than the CEM analyses for the unplugged pile.

4.2. Evaluation of Installation Predictions

Simultaneous measurements of shaft pore pressures and lateral earth pressures (radial total stresses) during installation have been obtained at a number of sites using a) instrumented pile shaft elements or probes (PLS cell, Morrison, 1984, t-z and x-probes; Bogard *et al.*, 1985; IMP, Coop and Wroth, 1989), and b) instrumented model piles (Karlsrud and Haugen, 1985; Karlsrud *et al.*, 1992; Bond *et al.*, 1991). Further measurements of installation pore pressures are associated with the development of *in-situ* testing devices such as the piezocone and include both field tests and laboratory experiments in large scale calibration chambers. The reliability of these measurements depends, in large part, on the design of the instrumentation (response time, calibration for thermal changes, etc.) and quality of test procedures (de-airing of porous filters etc.). The method of installation (driving versus jacking, rate of penetration, delay times, etc.) can also affect the measured parameters (e.g., Azzouz and Morrison, 1988) due to factors such as partial drainage which are not considered in the analysis.

Figure 5 compares the predictions of excess pore pressures at the pile shaft with measurements obtained during steady penetration in BBC using the PLS cell (Morrison, 1984; Azzouz and Morrison, 1988) as functions of the stress history (OCR). The measured data are very consistent at low OCR, but exhibit large scatter in the more overconsolidated clay due to the presence of sand seams etc. In general, the strain path predictions underestimate the measured excess pore pressures, particularly at low OCR, while there is better agreement with results from CEM analyses. The figure also includes field measurements from pile and penetrometer tests compiled from five other sites. The results tend to confirm the previous assessment of Baligh and Levadoux (1980) that there is no well defined correlation between the shaft pore pressures and clay type (as described by plasticity index, I_p), stress history (ORC; Figure 5), undrained shear strength or sensitivity (s_t). At low OCR, the excess pore pressures measured at five sites are in the range

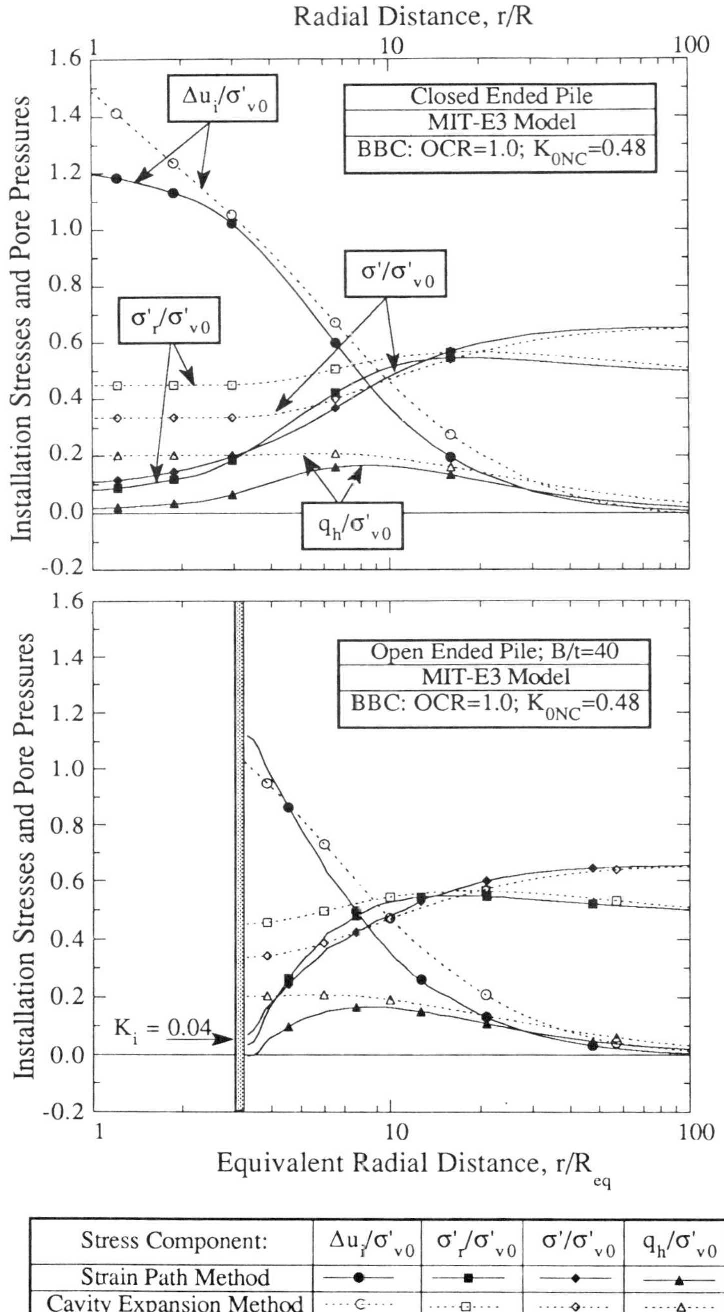

Fig. 4. Strain path predictions of installation stresses in K_0-normally consolidated BBC.

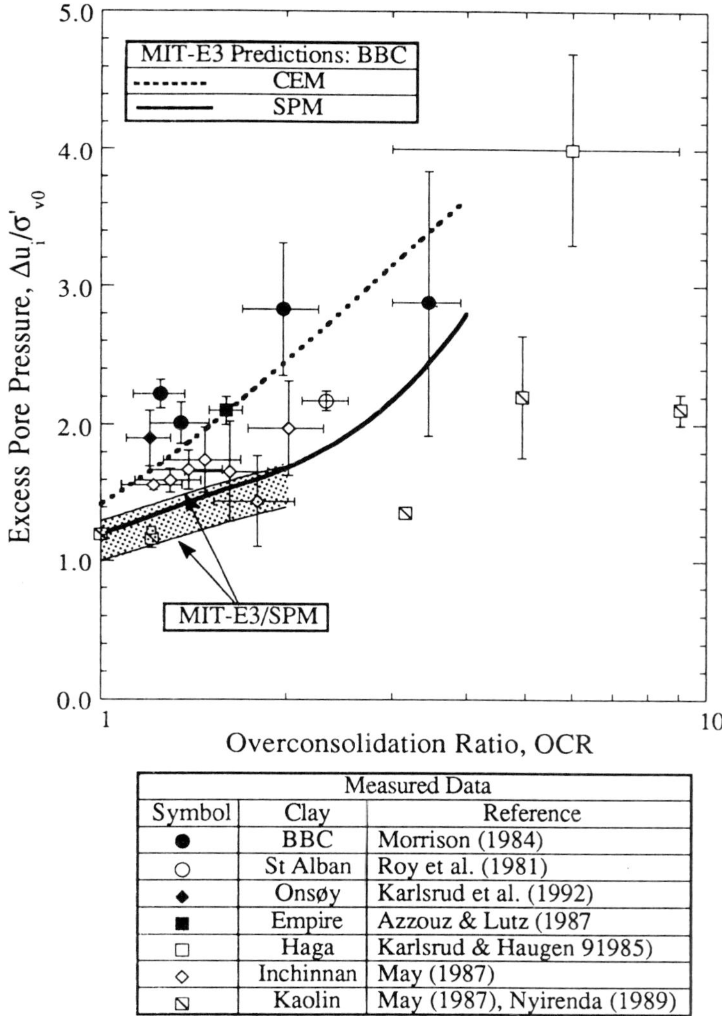

Measured Data		
Symbol	Clay	Reference
●	BBC	Morrison (1984)
○	St Alban	Roy et al. (1981)
◆	Onsøy	Karlsrud et al. (1992)
■	Empire	Azzouz & Lutz (1987
□	Haga	Karlsrud & Haugen 91985)
◇	Inchinnan	May (1987)
⧅	Kaolin	May (1987), Nyirenda (1989)

Fig. 5. Evaluation of installation pore pressures at pile shaft.

$\Delta u_i/\sigma'_{v0} = 2.0 \pm 0.4$. Significantly lower installation pore pressures ($\Delta u_i/\sigma'_{v0}$ = 1.2–1.3; Figure 5) have been reported recently from large-scale, laboratory calibration chamber tests in kaolin (May, 1987; Nyirenda, 1989).

In principle, measurements of the radial distribution of excess pore pressures (using piezometers in the surrounding soil) can provide a more comprehensive evaluation of the strain path predictions. In practice, these measurements are d-ifficult to obtain (especially in the critical region close to the shaft, $1 \leq r/R \leq 5\%$) due to a) interference of the measuring device and the soil deformations, and b) uncertainties in the alignment and position of the piezometers. Figure 6

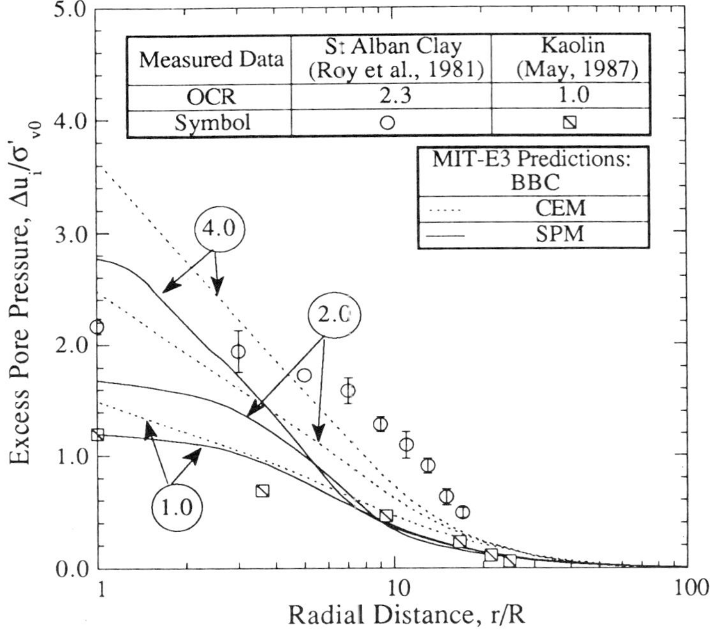

Fig. 6. Distribution of excess pore pressures during installation.

compares MIT-E3 predictions of the excess pore pressure distribution for BBC at OCR = 1.0, 2.0 and 4.0 with 1) field measurements (Roy *et al.*, 1981) around an instrumented pile (R = 11cm) installed in highly sensitive, structured St. Alban clay (liquidity index, $I_L \geq 2$) at OCR ≈ 2.3, and 2) measurements around the shaft of a cone penetrometer installed in K_0-normally consolidated kaolin (R = 1.26cm) within a large calibration chamber (of radius, R_c = 50cm). The SPM predictions at OCR = 1 are in very good agreement with penetration pore pressures measured in kaolin. The predictions at OCR = 2 underestimate both the magnitude and the radial extent of the zone of pore pressure accumulation in the St. Alban clay. These results indicate that the main source of discrepancy between strain path predictions and measured pore pressures are the effective stresses in the far field ($r/R \geq 5$–6, where $\dot{E} \leq 1\%$, Figure 2a), which are affected by small strain properties of the clay and can be addressed through further refinement of the constitutive model. in contrast, CEM analyses do not describe accurately the shape of the pressure distribution and hence, overestimate $\Delta u_i/\sigma'_{v0}$ at the shaft while underestimating pore pressure measured in the far field.

Overall, it is not possible to draw definitive conclusions from the results presented in Figures 5 and 6. Installation pore pressures around the pile shaft are very difficult to evaluate due to the complexity of the analysis and the sensitivity of the predictions to soil non-linearity and inelastic behaviour. Predictions using the MIT-E3 model (with input parameters for BBC) suggest that, although the strain path method underestimates the installation excess pore pressures, it provides a more consistent description of the lateral distribution of effective stresses than corresponding CEM analyses.

Radial effective stresses, K_i, during pile installation are obtained by subtracting the installation pore pressures from measurements of total radial stress at the same location. Figure 5 shows that pile installation in normally and lightly overconsolidated clays generates large excess pore pressures in the surrounding soil. It is therefore apparent that small errors in the measured pore pressures (and/or total stress) can affect significantly the computed radial effective stress. Azzouz and Morrison (1988) report $K_i = 0.05$–0.20 from PLS measurements in the lower Boston Blue Clay (OCR = 1.2±0.1) which are in excellent agreement with strain path predictions (cf., Figure 4). Very small values of K_i are also reported from instrumented pile tests in other sensitive, low plasticity clays (Haga, Karlsrud and Haugen, 1985; Onsøy, Karlsrud *et al.*, 1992). Significantly larger radial effective stresses, $K_i = 0.38$–0.54, were measured in the more plastic, less sensitive Empire clay (Azzouz and Lutz, 1986) and are also in good agreement with strain path predictions using the MIT-E3 model ($K_i = 0.37$–0.45; Whittle and Baligh, 1988).

5. Consolidation

5.1. Non-Linear Analysis of Radial Consolidation

After pile installation, soil consolidation occurs due to the dissipation of excess pore pressures around the pile. For shaft locations far from the pile tip and the mudline, it is assumed that excess pore pressures dissipate radially away from the shaft and there are concomitant changes in the effective stresses due to radial displacements of the soil. Thus the underlying mechanism of 'set-up' (changes in shaft capacity with time after installation) is attributed to changes in effective stresses in the soil at (or close to) the pile-soil interface due to radial consolidation.

Previous studies (Randolph *et al.*, 1979; Baligh and Levadoux, 1980; Baligh and Kavvadas, 1980) have shown that predictions of the change in radial effective stress acting on the pile shaft during consolidation are strongly affected by non-linearity of the soil. For the special case of a linear, isotropic (and elastic) soil, the decrease in excess pore pressure is exactly balanced by an increase in radial effective stress, such that at the end of consolidation, $\sigma'_{rc} = \sigma'_{ri} + \Delta u_i$. Comparison with measured data shows that this leads to a vast overprediction of the set-up around the pile shaft in soft clay deposits. Thus, comprehensive analyses of the coupled non-linear consolidation (i.e., coupling of total stressed and pore pressured, non-linear effective stress-strain and permeability properties of the soil) are required in order to achieve reliable predictions of the set-up process. The analyses described in this section solve the radial consolidation by non-linear finite element methods using with the following assumptions:

1. Initial conditions are described by the radial distributions of effective stresses and pore pressures predicted around the shaft during pile installation using the Strain Path Method.

2. The non-linear effective stress-strain response of soil elements around the pile shaft is represented consistently by the MIT-E3 model (i.e., with the same input parameters used during pile installation). The compressibility of the soil is a function of the radial location of the soil element (due to installation disturbances) and varies with effective stresses during consolidation.

3. Non-linearities associated with changes in soil permeability are not considered in the analysis. This assumption implies that the radial permeability is spatially constant after pile installation and that changes in permeability during consolidation can be neglected. Recent experimental data from model tests on resedimented BBC suggest that permeability can decrease by up to a factor of 3 at locations close to the pile shaft following complete set-up (Ting *et al.*, 1990). However, these changes in permeability are small compared with inherent uncertainties in the measurement of permeability of natural soil deposits and with non-linearities associated with soil stiffness predicted during consolidation.

Figure 7 summarizes predictions of the excess pore pressures, $\Delta u/\sigma'_0$, radial total and effective stresses ($H = (\sigma_r - u_0)/\sigma'_{v)}$, and σ'_r/σ'_{v0}, respectively) acting at the shaft of a pile installed in K_0-normally consolidated BBC. The results are presented using a dimensionless time factor, T, which is defined as

$$T = \frac{\sigma'_0 kt}{\gamma_w R_{eq}^2} \tag{1}$$

where $\sigma'_{v0} = 1/3(1 + 2K_0)\sigma'_{v0}$ is the *in-situ* mean effective stress in the ground, t is the time, R_{eq} is the equivalent radius, k is the (horizontal) coefficient of permeability, and γ_w is the unit weight of water.

The principal parameter of interest in the analysis is the magnitude of the radial effective stress acting on the pile shaft after full dissipation of excess pore pressures $K_c = \sigma'_{rc}/\sigma'_{v0}$. For the closed-ended pile example shown in Figure 7, the analyses using the Strain Path Method and MIT-E3 model predict a final set-up stress ratio, $K_c = 0.37$ (Figure 7a) which is significantly smaller than the initial, *in-situ* earth pressure coefficient ($K_0 = 0.48$), while comparable linear, consolidation analyses (i.e., based on the same installation conditions) would give $K_c = 1.28$ ($= K_i + \Delta u_i/\sigma'_{v0}$). Thus, the predictions of set-up stresses vary by a factor of 3 to 4 depending on the modelling of non-linear soil behaviour. The set-up stresses predicted for the open-ended (unplugged) pile are approximately 25% lower than for the closed-ended pile.

Predictions using cavity expansion analysis of pile installation (CEM) show significantly higher set-up stresses ($K_c = 0.72$; Figure 7a). This result is primarily

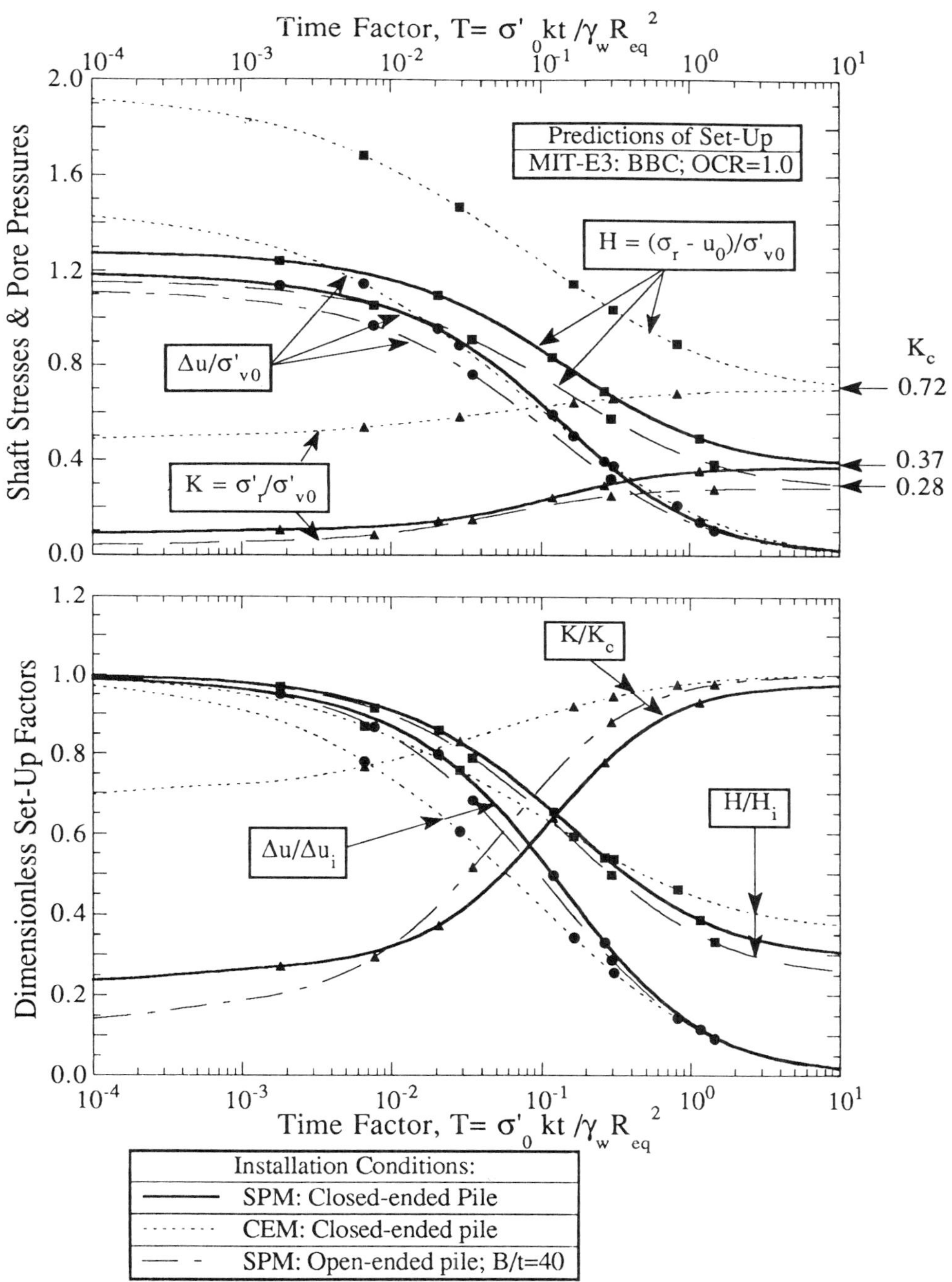

Fig. 7. Typical predictions of consolidation at the pile shaft in K_0-normally consolidated BBC.

due to the predicted initial conditions ($K_i = 0.45$, Figure 4) since the net change in radial effective stress during consolidation is relatively small.

Further insights into factors affecting the set-up predictions can be achieved by introducing the dimensionless stress ratios shown in Figure 7b: The excess pore pressure ratio (degree of consolidation), $U = \Delta u/\Delta u_i$, depends primarily on the normalized radial distribution of installation excess pore pressures ($\Delta u/\Delta u_{sh}$, where Δu_{sh} is the excess pore pressure at the pile shaft; Baligh and Levadoux, 1980). Strain path analyses for both closed and open-ended piles give very similar rates of consolidation (U vs T, Figure 7b), while CEM predictions show more rapid dissipation of pore pressures for $U \geq 0.3$. The set-up effective stress ratio, K/K_c, illustrates most clearly the importance of the installation analysis (SPM vs CEM). The strain path results in Figure 7b show that the largest change in set-up stresses occurs over the time period $0.01 \leq T \leq 1.0$. In contrast, the total stress release ratio, H/H_i, is primarily a function of the soil behaviour. Factors affecting H/H_i include the shear behaviour at large shear strains (i.e, during installation) and the radial compressibility during consolidation. The MIT-E3 predictions for normally consolidated BBC show H_c/H_i = 0.25–0.3 (Figure 7b), while results obtained for other soils and stress histories range from H_c/H_i = 0.2 to 0.6 (Whittle and Baligh, 1988).

5.2. Evaluation of Shaft Stresses during Set-Up

Effective stresses at the pile shaft are currently calculated from measurements of total stresses and pore pressures during consolidation (e.g., Azzouz and Morrison, 198; Karlsrud and Haugen, 1985), although direct measurements of σ'_r have been reported recently in a highly overconsolidated clay (using effective pressure transducers; Karlsrud *et al.*, 1992). In addition to obvious sources of experimental errors, there are two main difficulties in determining accurately the set-up stress ratio, $K_c = \sigma'_{rc}/\sigma'_{v0}$ (Azzouz *et al.*, 1990) 1) incomplete consolidation, and 2) limitations of total lateral stress measurements.

The time delay required to achieve complete consolidation depends on the (equivalent) radius of the pile and the soil properties (permeability and compressibility): For example, Azzouz and Morrison (1988) report consolidation times of 4–8 days using the PLS cell (R = 1.9cm), while 1–2 months of set-up are typical of instrumented model piles with R = 7–10cm (e.g., Karlsrud *et al.*, 1992). In practice, K_c values are frequently quoted from measurements obtained with incomplete consolidation. In these situations, calculations using measurements of σ_r and u tend to underestimate K_c; while those based on σ_r and u_0 (*in-situ* pore pressures) overestimate the effective set-up stress.

Measurements of pore pressures at later stages of consolidation can easily be checked through comparisons with known values of u_0. Thus, most of the uncertainties in σ'_{rc} are due to possible errors in total stress measurements which are commonly caused by: a) significant seating errors (zero reading) of the σ_r cell due to soil arching; b) the cross sensitivity of σ_r measurements to changes of the axial load in the pile; and c) the changes of total stress cell calibration with time after installation (zero shift). Significant improvements in the design of instrumentation (e.g., Bond *et al.*, 1991; Karlsrud *et al.*, 1992) represent an important contribution in reducing these errors.

Figure 8 compares the predictions of the effective stresses at full set-up with measurements obtained by the PLS cell in BBC (Morrison, 1984; Azzouz and Morrison, 1988) as functions of the stress history. The predicted K_c increases significantly with the OCR and range from K_c = 0.37–0.42 at OCR = 1 (for K_{0NC} = 0.48–0.53) to K_c = 1.1–1.30 at OCR = 4 (where K_0 = 0.75–1.0). Although the set-up stress is similar in magnitude to the *in-situ* earth pressure, the average ratio K_c/K_0 ranges from 0.8 at OCR = 1, to 1.4 at OCR = 4. The PLS measurements are generally in very good agreement with the predictions and support the result that K_c increases with OCR, although there is a large scatter in the data for OCR $\geq$ 3. The figure also includes field measurements of K_c at three other sites (Empire, Onsøy and Haga) together with laboratory tests on a miniature pile in kaolin. There is consistent agreement in the data obtained for three clays of moderate sensitivity (s_t = 3–7; BBC, Onsøy and Haga); while higher set-up stresses are measured in the Empire clay and kaolin. Although the predictions of K_c for Empire clay (Figure 8) are 50–60% higher than those described for BBC, they still underpredict the PLS measurement in zone I (cf., Azzouz and Lutz, 1986). No analyses (using SPM and MIT-E3) have yet been performed for the kaolin, however, Azzouz *et al.* (1990) suggest that these data are affected significantly by boundary conditions.

The results in Figure 8 indicate that the combination of strain path installation analyses and non-linear radial consolidation with MIT-E3 are capable of providing good predictions of the effective stresses at the end of set-up. However, more comprehensive predictions are necessary to establish how predictions of K_c are related to engineering properties of the soil.

Figures 9 and 10 present a detailed evaluation of the predictions of set-up behaviour at the pile shaft based on PLS measurements (Morrison, 1984) in the lower deposit of BBC (OCR = 1.3 $\pm$ 0.1). The consolidation predictions are presented at OCR = 1.0 and 1.5 using a modified time factor, $T = \sigma'_p kt/\gamma_w R^2$, where σ'_p is the vertical preconsolidation stress. This time factor unifies predictions of the consolidation rate (U vs T; Figure 9b) for normally and lightly overconsolidated BBC (Aubeny, 1992). Measurements of the total radial stress and excess pore pressure ratios (H/H_i and U, respectively) are then compared with the predictions (Figure 9). An average coefficient of horizontal permeability, $k_h = 8 \times 10^{-7}$cm/sec is selected from laboratory CRS and constant head measurements which range from $3.0 \times 10^{-8} \leq k_h \leq 1.5 \times 10^{-7}$cm/sec in the lower BBC (no significant variation with depth in the deposit). The predictions are in excellent agreement with the measured excess pore pressure ratio throughout consolidation, but tend to underestimate slightly the reduction in total radial stress. Effective stress changes at the pile shaft during consolidation $K(T)$, are calculated from the average measurements

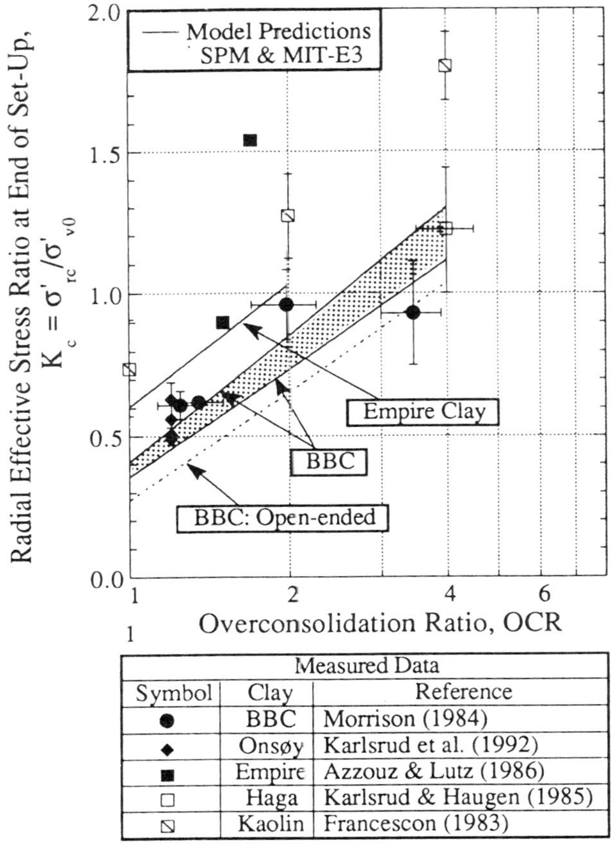

Measured Data		
Symbol	Clay	Reference
●	BBC	Morrison (1984)
◆	Onsøy	Karlsrud et al. (1992)
■	Empire	Azzouz & Lutz (1986)
□	Haga	Karlsrud & Haugen (1985)
⧅	Kaolin	Francescon (1983)

Fig. 8. Evaluation of radial stress at end of set-up.

of total radial stress and excess pore pressure (Figure 10). These data agree with predictions of radial effective stresses for OCR = 1.5 and confirm the capabilities of the analysis for describing pile shaft performance throughout the set-up process.

5.3. Stress Conditions in the Soil after Consolidation

Predictions of the stress state in the soil after consolidation have an important influence on the subsequent prediction and interpretation of pile shaft capacity. Figure 11 presents predictions of the stress distribution around the pile shaft in BBC at OCR = 1 and 2. The results show a net reduction in the mean effective stress (σ'_c/σ'_{v0}) in the soil around the pile (i.e., compared to the *in-situ* conditions). Radial consolidation produces cavity shear stresses (q_{hc}/σ'_{v0}) in the soil extending to a distance $r/R \leq 3$ (cf. Figure 4). At the pile shaft, σ'_{rc} is the major principal effective stress, while $\sigma'_{\theta c}/\sigma'_{rc} \approx K_{0NC}$. Figure 12 compares the volumetric behaviour

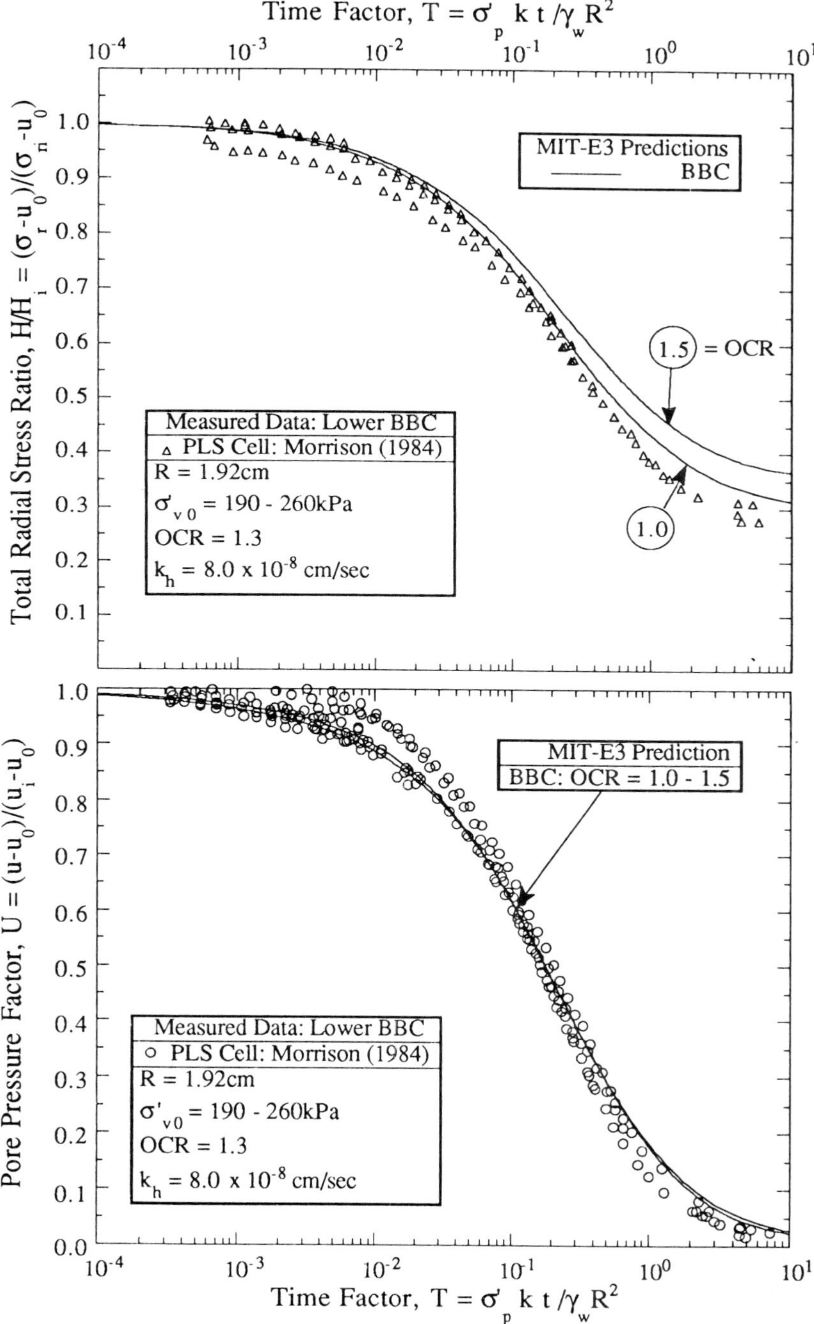

Fig. 9. Comparison of predictions and measurements during consolidation in Boston Blue Clay.

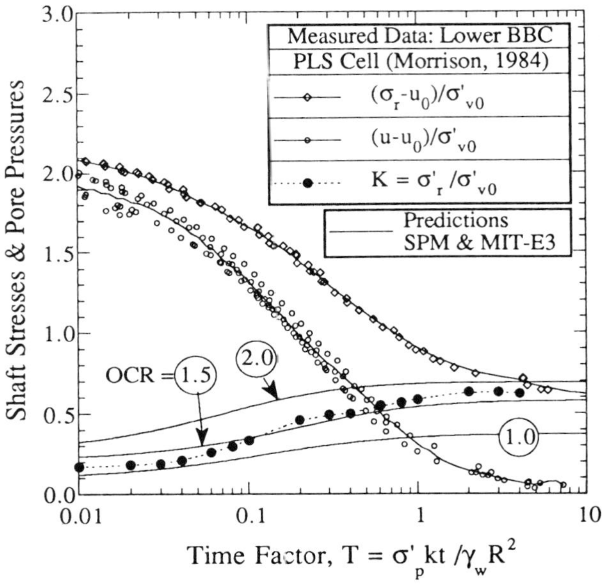

Fig. 10. Evaluation of effective stress set-up in lower BBC.

(i.e., mean effective stress and volumetric strain) of soil elements adjacent to the pile shaft with the K_0-Virgin Consolidation Line (K_0-VCL) predicted for the undisturbed clay (using MIT-E3). Pile installation (paths A-B) generates large shear induced pore pressures (reductions in σ') associated with severe shearing of the soil at constant water content. During set-up (paths B-C) the soil elements do not return towards the K_0-VCL, but instead exhibit a more compressible behaviour such that the final states of stress (C_1, C_2) coalesce on a new re- consolidation line which lies parallel to the compression of the undisturbed clay. This result is qualitatively similar to the observations of the consolidation behaviour of remoulded Haga clay reported by Karlsrud and Haugen (1985).

6. Axial Pile Loading

6.1. Analysis of Shaft Capacity

This section focuses on predictions of the limiting skin friction, f_s, which is mobilized at the pile shaft after full dissipation of the installation excess pore pressures. The analyses are restricted to the case of a long, rigid pile for which the pile tip and the mudline have negligible effects on the shaft resistance. Even after introducing this simplification, there are three main factors which complicate significantly the prediction and interpretation of pile-soil interaction during axial loading:

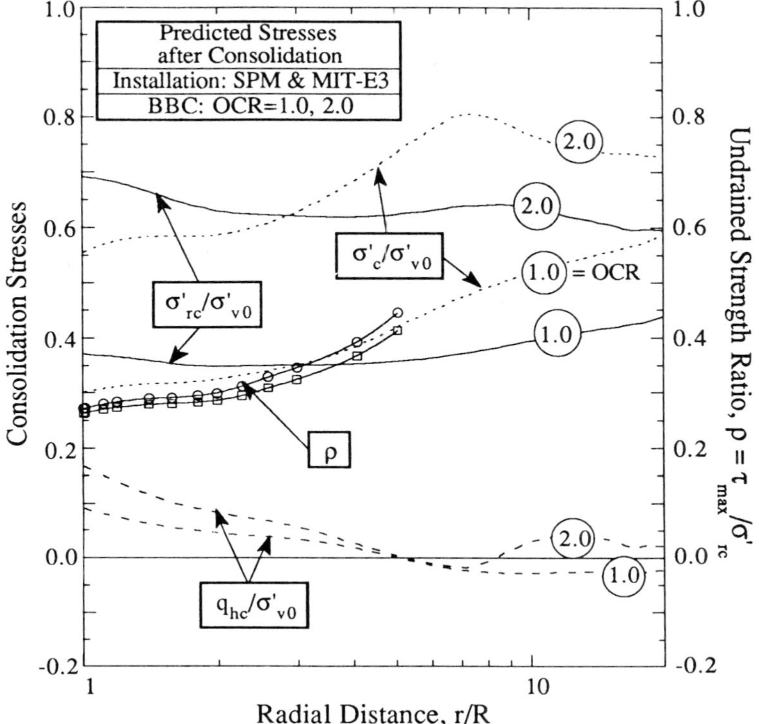

Fig. 11. Predicted stress conditions at the end of consolidation.

1. Although slip surfaces are constrained to form parallel to the pile shaft, the actual slippage may occur either at the pile-soil interface or within the surrounding soil mass. Slippage at the interface is of practical importance when the limiting angle of interface friction, δ', is less than the effective stress obliquity, α'_p, mobilized along vertical planes in the soil at peak shear resistance (i.e., $\alpha'_p = \tau_f/\sigma'_{rf}$).

2. The shear resistance of the soil is controlled by effective stresses and properties around the shaft prior to loading. The previous sections have shown that relatively sophisticated analyses can predict many aspects of the set-up behaviour measured at the shaft for piles installed in normally and lightly overconsolidated clays. However, many aspects of the predictions cannot be evaluated from field measurements and hence, represent a source of uncertainty in the subsequent comparison with measured capacity.

3. Drainage conditions during pile loading are not well defined. For most conventional test procedures, the pile is loaded to failure over a period of several

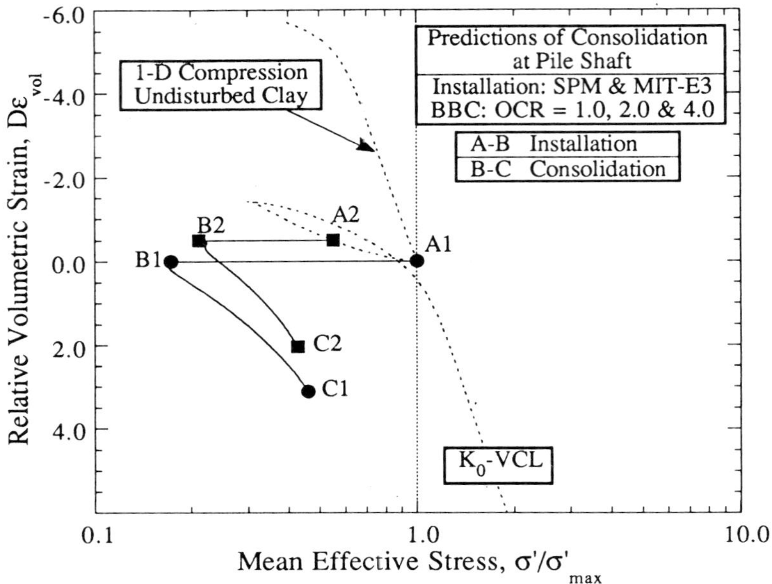

Fig. 12. Predicted volumetric behaviour of soil elements adjacent to the pile shaft.

hours, such that the surrounding soil is sheared under conditions of partial (or complete) drainage which must be interpreted using effective stress methods.

For piles installed in lightly overconsolidated clays, the critical conditions (i.e., minimum shaft capacity) are likely to occur when the pile is loaded rapidly with no radial migration of pore water, and hence undrained shearing of the clay. in this situation, the mode of deformation corresponds to the shearing of concentric cylinders around the pile shaft (i.e., γ_{rz} is the only non-zero strain component), while vertical equilibrium controls the radial distribution of shear stresses ($\tau_0 R = \tau r$, where τ_0 is the interface shear stress). Figure 13a shows MIT-E3 predictions of the normalized effective (σ'_r/σ'_{rc}, τ/σ'_{rc}) and total stress paths ($[\sigma_r - u_0]/\sigma'_{rc}$, τ/σ'_{rc}) acting in the vertical plane for undrained shearing of soil elements adjacent to the pile shaft in Boston Blue Clay. The results show the following:

1. The initial stress history of the soil does not affect the normalized effective stress paths predicted at the pile shaft after full re-equilibration of the installation excess pore pressures for OCR $\leq$ 4.

2. The peak shear resistance of the soil elements can be described by an undrained strength ratio, $\rho = \tau_f/\sigma'_{rc} \approx 0.26$ (Whittle *et al.*, 1988; Azzouz *et al.*, 1990) which is mobilized at an effective stress obliquity, $\alpha'_p \approx 24°$ (which is signifi-

cantly lower than the reference critical state friction angle measured in triaxial compression tests, $\phi'_{TC} = 33.4°$.

3. There is a reduction in the radial effective stress, $\Delta\sigma'_{rf}/\sigma'_{rc} \approx 0.4$, together with positive excess pore pressures, $\Delta u_f/\sigma'_{rc} \approx 0.4$–$0.5$, and a small net increase in the total stress ($[\sigma_r - u_0]_f/\sigma'_{rc} \approx 1.05$, Figure 13a) for shearing up to peak shaft resistance. The zone of excess pore pressures extends laterally to a distance, $r/R \leq 5$ (Figure 13b).

The analyses can be extended to consider the radial distribution of undrained shear resistance $\rho(r/R)$ by computing the response of individual soil elements to the same mode of shearing. The predictions in Figure 11 show that a) the shaft capacity is limited by the undrained shear strength of soil elements adjacent to the pile, and b) the shear resistance, $\rho(r/R)$, is proportional to the mean effective stress predicted at the end of consolidation (σ'_c/σ'_{rc}) for $r/R \leq 2$.

The mode of shearing assumed for soil elements adjacent to the pile shaft is identical to that imposed in laboratory constant volume (undrained) Direct Simple Shear (DSS) tests. Thus, it is useful to compare model predictions of soil behaviour for the remoulded soil around the pile shaft with that of the undisturbed K_0-consolidated clay in undrained Direct Simple Shear tests. Figure 13a shows MIT-E3 predictions of the normalized effective stress paths (σ'_v/σ'_{vc}, τ/σ'_{vc}) for K_0-consolidated BBC at OCR = 1.0, 1.2 and 1.5. The undrained strength ratio ($s_{u\mathrm{DSS}}/\sigma'_{vc} = 0.25$) and effective stress path of the undisturbed clay at OCR = 1.2 are in close agreement with the normalized behaviour predicted at the pile shaft, although there is a difference ($\Psi = 20°$ versus $\alpha'_p = 24°$) in the stress obliquity mobilized at peak shear resistance. Differences in the normalized response at the pile shaft and the K_0-normally consolidated clay (OCR = 1.0) can be attributed mainly to the predicted ratios of effective stress components in the soil at the end of consolidation (i.e., $\sigma'_{rc} : \sigma'_{zc} : \sigma'_{\theta c}$ versus $\sigma'_h = K_{0NC}\sigma'_v$) (Whittle, 1987).

Azzouz *et al.* (1990) have proposed that the undrained strength ratio, ρ, can be used to provide realistic estimates of the limiting value of f_s for the design of friction piles in lightly overconsolidated clays (OCR $\leq$ 4). The limiting skin friction is written as:

$$f_s = K_c \rho \sigma'_{v0} \tag{2}$$

where σ'_{v0} is the *in-situ* effective overburden stress.

This method re-expresses the well known β parameter ($\beta = f_s/\sigma'_{v0}$; Chandler, 1968; Burland, 1973) as the product of the lateral earth pressure coefficient, K_c, and the undrained shear strength ratio, ρ, of the soil adjacent to the pile shaft. Predictions for normally and lightly overconsolidated clays show that K_c is affected by the stress history of the clay (OCR) and by its sensitivity, while ρ can be estimated from the strength ratio measured in undrained direct simple shear tests on the undisturbed clay at OCR $\approx$ 1.2.

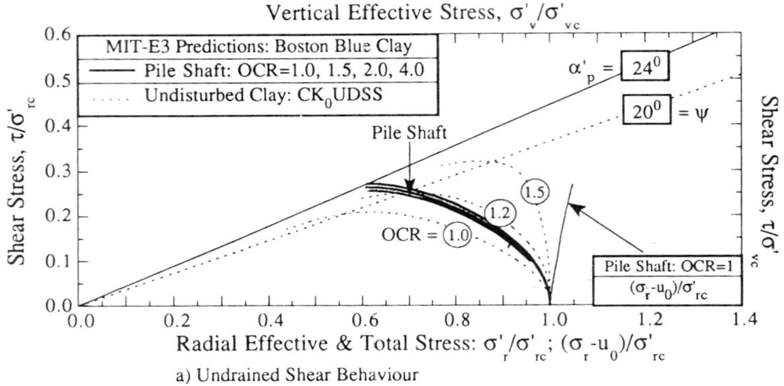

a) Undrained Shear Behaviour

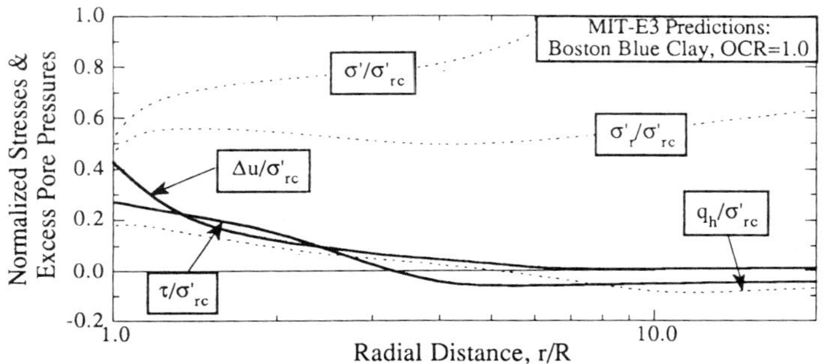

b) Radial Distribution of Stresses and Excess Pore Pressures at Peak Shear Resistance

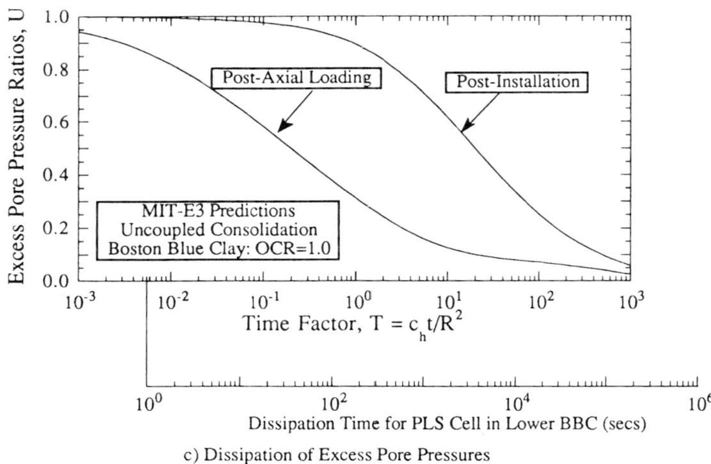

c) Dissipation of Excess Pore Pressures

Fig. 13. MIT-E3 predictions of soil behaviour for elements at the pile shaft during axial loading.

In principle, lower values of the limiting skin friction can occur if the interface friction angle, δ', is smaller than the effective stress obliquity, α'_r, mobilized at peak shear resistance in the soil. Measurements in the ring shear apparatus show that δ' depends on numerous factors including soil mineralogy and fabric, consolidation and shear history, rate of shearing, interface roughness and hardness (Lemos, 1986). Comparisons between MIT-E3 predictions of α'_p for a number of clays (in the range, $\alpha'_p = 17°–24°$) and recent correlations for δ' (Jardine and Christoulas, 1991) show that $[\alpha'_p - \delta'] \leq 5°$, and suggest that interface slippage can reduce the calculated undrained shear resistance, ρ, by up to 25%.

The model predictions also provide a basis for evaluating drainage conditions during pile loading. Figure 13b shows the radial distribution of stresses and excess pore pressures at conditions corresponding to peak undrained shear resistance in the soil at the pile shaft (for BBC at OCR = 1). Subsequent dissipation of the excess pore pressures provides guidance on the characteristic time required to achieve undrained loading of the pile. Figure 13c compares predictions of the uncoupled dissipation of excess pore pressures generated during installation and axial loading. For the PLS cell (R = 1.92cm) installed in the lower deposit of BBC ($c_h \approx 0.02\text{cm}^2/\text{sec}$; Baligh and Levadoux, 1980), the prediction show $t_{50} \approx 40$secs, while undrained conditions are only achieved when the shaft is loaded to failure within 1sec. In this situation, the soil around the pile is sheared at an average strain rate which is significantly higher than that imposed in conventional laboratory CK_0UDSS tests ($\dot{\gamma}$ = 5%/hr). Data from CK_0UDSS tests with $t_f \approx$ 1–10secs (e.g., Lacasse, 1979) show a 15–25% increase in the undrained strength ratio and develop much smaller shear induced pore pressures compared to tests performed at conventional strain rates. These measurements suggest that predictions using the rate independent MIT-E3 model (Figure 13a) will tend to underestimate the shaft resistance and overestimate the excess pore pressures for undrained loading.

The one-dimensional (radial) predictions described in this section implicitly assume that the pile shaft performance is similar for loading in either axial compression or tension. In practice, the local shaft resistance can be affected by factors such as proximity to the pile tip (e.g., Lehane and Jardine, 1992) or residual axial loads in the pile (e.g., Whittle, 1991b). More comprehensive two-dimensional analyses of set-up and loading are required in order to evaluate these effects. However, some insight can be obtained from simple model predictions of material behaviour in laboratory element tests. For example, residual axial loads in the pile generate shear stresses during set-up at the pile-soil interface (i.e., $\rho_c = \tau_c/\sigma'_{rc}$). These effects can be simulated in laboratory CK_0UDSS tests by consolidating the soil under an applied shear stress, τ_c/σ'_{vc}. Figure 14 compares MIT-E3 predictions with the measured effective stress paths and shear stress-strain response (DeGroot, 1989) for CK_0UDSS tests on normally consolidated BBC for consolidation shear stress ratios $-0.2 \leq tau_c/\sigma'_{vc} \leq 0.2$. The results show the following:

1. Specimens consolidated with shear stresses applied in the same direction as the subsequent undrained loading ($\tau_c/\sigma'_{vc} > 0$) exhibit higher undrained shear strengths, smaller shear induced pore pressures, lower effective stress obliquity and stiffer stress-strain response (at $\tau_c/\sigma'_{vc} = 0.2$: $s_{u\text{DSS}}/\sigma'_{vc} = 0.29$, $\Delta u_s/\sigma'_{vc} = 0.07$, $\alpha'_p = 17°$, and $\gamma_p = 0.7\%$) than standard tests performed at $\tau_c/\sigma'_{vc} = 0$. In contrast, the consolidation shear stress has little effect on the undrained strength for $\tau_c/\sigma'_{vc} < 0$, but affects significantly the stress- strain response and the development of shear induced pore pressures.

2. The MIT-E3 model is in very good agreement with the measurements of pre-peak stress-strain behaviour and effective stress paths in these tests. However, the model tends to overpredict the undrained shear strength for $\tau_c/\sigma'_{vc} > 0$ and does not describe accurately the post-peak strain softening measured in these tests.

Overall, the results in Figure 14 show important aspects of soil behaviour which can be related directly to the effects of residual axial loads on pile shaft performance.

The ρ-method of estimating capacity (Azzouz *et al.*, 1990) assumes that there is full dissipation of installation excess pore pressures prior to pile loading. However, for large diameter offshore piles, loading is often carried out under conditions of incomplete set-up. These situations can also be readily analyzed using the proposed framework. Figure 15 illustrates predictions of the normalized shaft capacity β/β_∞ (where $\beta_\infty = K_c\rho$) and pore pressure ratio, U, as functions of the dimensionless time factor, T. The results show that the shaft capacity increases by a factor of 2–3 during the set-up process, with most of the strength gain occurring over the time frame, $0.01 \leq T \leq 1.0$.

6.2. Evaluation of Shaft Capacity

The predictions of the undrained strength ratio, ρ, can be evaluated by dividing the measured values of β from rapid axial load tests (which ensure undrained shearing of the clay) and the lateral effective stress ratio measured at the end of set-up, K_c. Measurements of β can be obtained with a high degree of reliability from measurements of the distribution of axial load along the axis of the pile. In contrast, concurrent measurements of the local pore pressures, shear and normal lateral stresses are required in order to interpret the effective stress paths of soil elements at the pile shaft. Reliable data of this type have only recently been reported from high quality instrumented pile tests in a soft clay (Lehane and Jardine, 1992).

Figure 16a compares the MIT-E3 predictions and measurements of ρ from axial load tests obtained using the PLS cell and instrumented model piles. The PLS data in BBC and Empire clay (R = 1.92cm, t_f = 10–40secs) (Morrison, 1984; Azzouz and Lutz, 1986) are obtained from axial load cell measurements and provide the

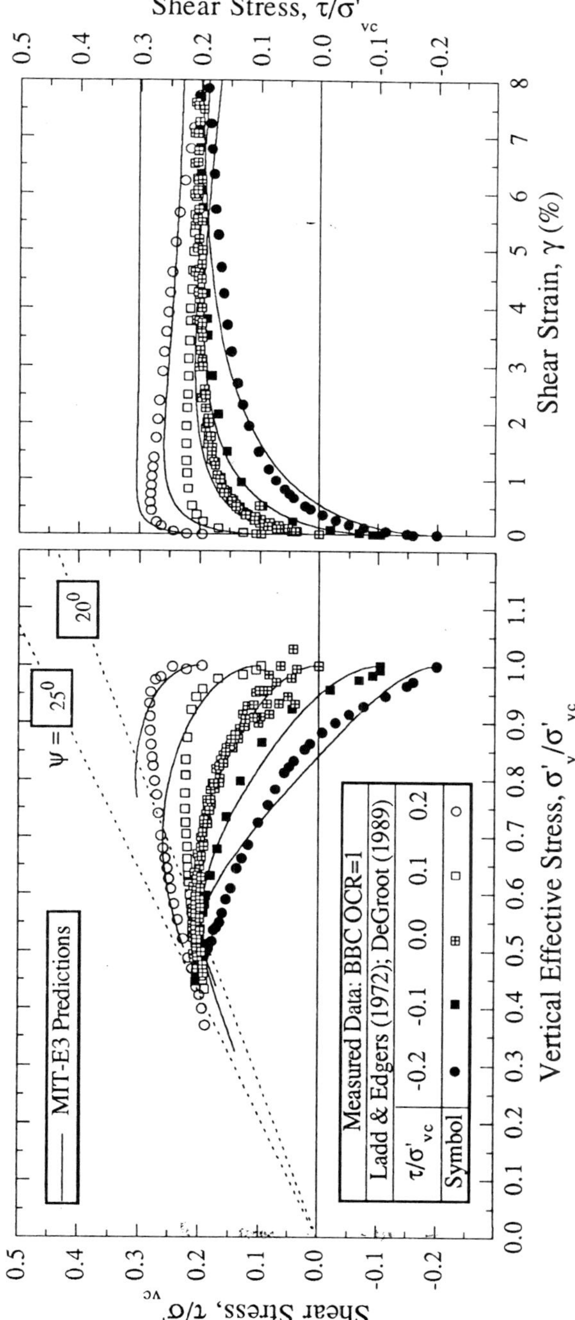

Fig. 14. Effect of consolidation shear stress in undrained direct simple shear tests on BBC.

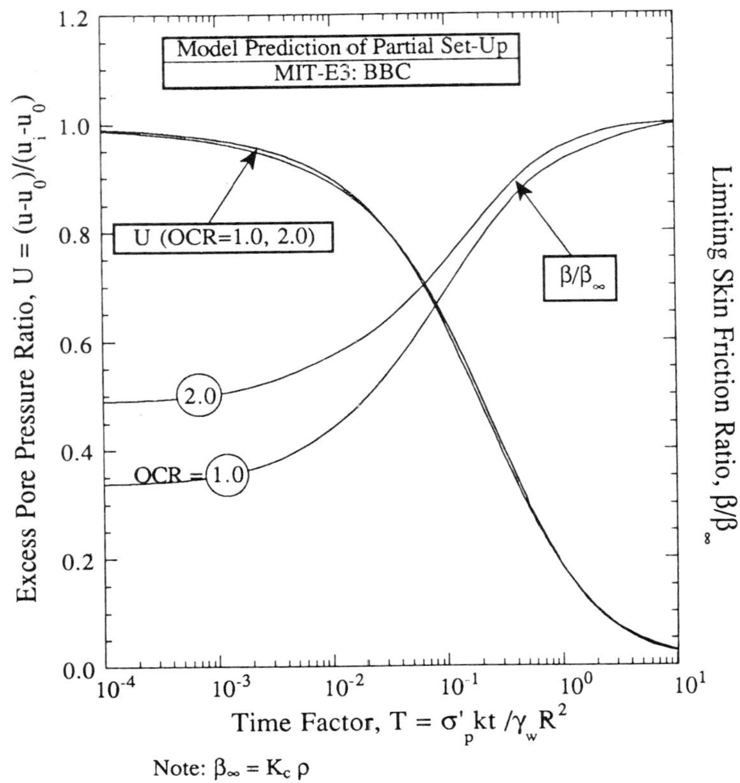

Fig. 15. Prediction of shaft capacity for partial set-up in BBC.

average shaft resistance acting between the pile tip and the elevation of the PLS cell. In contrast, six levels of strain gauges are used to estimate the distribution of shaft friction along the length of the Haga piles (L = 6m, R = 7.6cm, t_f = 20mins) (Karlsrud and Haugen, 1985). The results show very good agreement between predictions and measurements of ρ from PLS tests, but underestimate significantly the skin friction ratio reported in the Haga tests. This result can be explained, in large part, by residual loads in the piles which generate significant consolidation shear stresses in the upper half of the pile. Figure 16b compares model predictions of the undrained shear resistance, ρ, as a function of the consolidation shear stress, ρ_c. The results show good agreement with data at depths, z = 2 and 3m, but still underpredict the behaviour at z = 4m probably due to factors such as the proximity of the pile tip (Whittle, 1991b).

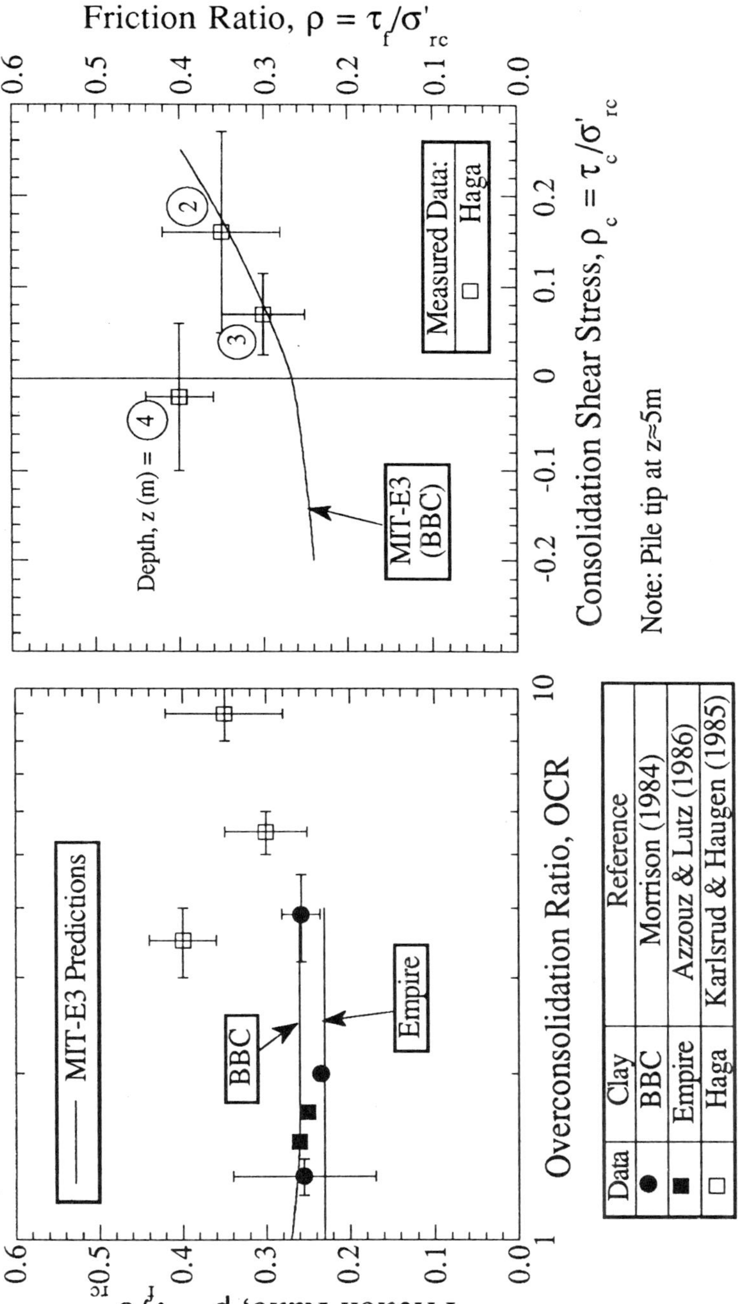

Fig. 16. Evaluation of the undrained shear strength ratio, ρ, of soil adjacent to the pile shaft.

7. Summary and Conclusions

This paper has described the application of a systematic analysis for predicting the performance of driven piles in lightly overconsolidated clays (OCR ≤ 4). The predictions are evaluated through comparisons with field measurements from the Piezo-Lateral Stress (PLS) cell and instrumented model piles. The key components of the analysis are: a) the Strain Path Method, which models the effects of the severe soil disturbances caused by pile installation; and b) MIT-E3, a generalized effective stress soil model with well documented capabilities for describing the non-linear and anisotropic behaviour of K_0-consolidated clays with normalized, rate independent properties. The results focus on conditions during pile installation, stress and pore pressure changes during subsequent consolidation, and the shaft capacity during axial loading.

The Strain Path Method (SPM) provides a realistic model of the mechanics of undrained deep penetration in clays. Predictions using the SPM with the MIT-E3 model give good agreement with the radial effective stresses measured at the pile shaft during steady penetration, but tend to underestimate the installation excess pore pressures. This latter result probably reflects limitations of the constitutive model for describing the behaviour of structured natural clays.

Radial consolidation around the pile shaft is solved by a coupled, non-linear finite element method with initial conditions from the installation phase and using the MIT-E3 model to describe the effective stress-strain properties of the soil. The consolidation process can be characterized by the pore pressure ratio, $U = \Delta u/\Delta u_i$, the total stress release, H/H_i (where $H = \sigma_r - u_0$), and the set-up effective stress, $K = \sigma'_r/\sigma'_{v0}$, as functions of a dimensionless factor, T (Figure 7). The analyses show that the radial distribution of installation excess pore pressures control U, while H/H_i is primarily a function of the non-linear radial compression behaviour of the soil. The paper presents detailed comparisons between predictions and PLS measurements in Boston Blue Clay using average values for the horizontal coefficient of permeability from laboratory tests. The predictions consistently describe the changes in the effective stresses, the total stress release, H/H_i, and the pore pressure ratio, U, throughout consolidation. The predictions are in good agreement with measurements of the radial effective stress at the end of consolidation ($K_c = \sigma'_{rc}/\sigma'_{v0}$) as reported from a number of field sites (Figure 8). The parameter K_c is affected by the *in-situ* overconsolidation ratio and sensitivity of the clay.

Predictions of pile shaft performance are presented for rapid axial loading of a long, rigid pile after full dissipation of installation excess pore pressures. The analyses assume that minimum values of the limiting skin friction, f_s, can be estimated from the undrained shear resistance in the soil adjacent to the pile shaft. Results using the MIT-E3 model show that the normalized shear resistance of the soil, ρ ($= f_x/\sigma'_{rc}$; where σ'_{rc} is the radial effective stress at full set-up) is independent of the initial OCR, and can be equated approximately with the undrained strength ratio ($s_{u\text{DSS}}/\sigma'_{vc}$) of the undisturbed clay measured in an undrained direct simple shear test at OCR $\approx$ 1.2 (Figure 13a). The paper discusses other factors affecting the shaft resistance of the pile including partial set-up, drainage conditions in the soil, loading rate and residual loads in the pile. Predictions of the undrained strength ratio, ρ, are in good agreement with field measurements obtained using the PLS cell measured at two sites (Figure 16).

Acknowledgements

The author would like to thank Prof. M. M. Baligh who initiated and guided the MIT research program on piles in clays. The developments described in this paper were supported by the MIT Sea Grant College program through grant NA86AA-D-SG089 and by the Henry L. Doherty Professorship in Ocean Engineering. Additional support was provided by a consortium of oil companies including Amoco Production Company, Chevron Oil Field Research, Exxon Production Research, Mobil Research and Development, and Shell Development Company.

References

1. Aubeny, C. P. (1992), 'Rational Interpretation of *in-situ* Tests in Cohesive Soils', PhD Thesis, Dept. of Civil Engineering, MIT, Cambridge, MA.
2. Azzouz, A. S. and Lutz, D. G. (1986), 'Shaft behaviour of a model pile in plastic Empire clays', *ASCE Journal of Geotechnical Engineering* **112**(4), 389–406.
3. Azzouz, A. S. and Morrison, M. J. (1988), 'Field measurements on model pile in two clay deposits', *ASCE Journal of Geotechnical Engineering* **114**(1), 104–121.
4. Azzouz, A. S., Baligh, M. M., and Whittle, A. J. (1990), 'Shaft resistance of friction piles in clay', *ASCE Journal of Geotechnical Engineering* **116**(2), 205–221.
5. Baligh, M. M. (1985), 'Strain path method', *ASCE Journal of Geotechnical Engineering* **111**(9), 1108–1136.
6. Baligh, M. M. (1986a), 'Undrained deep penetration: I Shear stresses', *Géotechnique* **36**(4), 471–485.
7. Baligh, M. M. (1986b), 'Undrained deep penetration: II Pore pressures', *Géotechnique* **36**(4), 487–501.
8. Baligh, M. M. and Kavvadas, M. (1980), 'Axial Static Capacity of Offshore Friction Piles in Clays', Research Report R80-32, Dept. of Civil Engineering, MIT, Cambridge, MA.
9. Baligh, M. M. and Levadoux, J.-N. (1980), 'Pore Pressure Dissipation After Cone Penetration', Research Report R80-11, Dept. of Civil Engineering, MIT, Cambridge, MA.
10. Baligh, M. M., Azzouz, A. S., and Chin, C. T. (1987), 'Disturbances due to 'ideal' tube sampling', *ASCE Journal of Geotechnical Engineering* **113**(7), 739–757.
11. Bond, A. J., Jardine, R. J., and Dalton, C. P. (1992), 'Design and performance of the Imperial College instrumented pile', *ASTM Geotechnical Testing Journal* **14**(4), 413–425.
12. Boagard, D., Matlock, H., Audibert, J. M. E., and Bamford, S. R. (1985), 'Three years' experience with model pile segment tool tests', *Proc. 17th Offshore Tech. Conf.*, Houston, Paper 4848.
13. Burland, J. B. (1973), 'Shaft friction of piles in clay – A simple fundamental approach', *Ground Engineering* **6**(3), 30–42.
14. Chandler, R. J. (1968), 'The shaft friction of piles in cohesive soils in terms of effective stresses', *Civ. Engrg. & Public Works Review*, Jan., 48–51.
15. Chin, C. T. (1986), 'Open-Ended Pile Penetration in Saturated Clays', PhD Thesis, Dept. of Civil Engineering, MIT, Cambridge, MA.
16. Clayton, C. R. I., Khatrush, S. A., Bica, B. V. D., and Siddique, A. (1989), 'The use of Hall effect semi-conductors in geotechnical instrumentation', *ASTM Geotechnical Testing Journal* **12**(1), 69–76.

17. Coop, M. R. and Wroth, C. P. (1989), 'Field studies of an instrumented model pile in clay', *Géotechnique* **39**(4), 679–696.
18. DeBorst, R. and Vermeer, P. A. (1984), 'Possibilities and limitations of finite element for limit analysis', *Géotechnique* **32**(2), 199-210.
19. DeGroot, D. J. (1989), 'The Multi-Directional Simple Shear Apparatus with Application to the Design of Offshore Arctic Structures', PhD Thesis, Dept. of Civil Engineering, MIT, Cambridge, MA.
20. DeGroot, D. J., Ladd, C. C., and Germaine, J. T. (1992a), 'Direct Simple Shear Testing of Cohesive Soils', Research Report R92-18, Dept. of Civil Engineering, MIT, Cambridge, MA.
21. Dyvik, R. and Olsen, T. S. (1989), 'G_{max} measured in oedometer and DSS tests using bender elements', *Proc. 12th Intl. Conf. on Soil Mechs. and Found. Engrg.*, Rio de Janeiro, Vol. 1, pp. 39–42.
22. Esrig, M. I., Kirby, R. C., Bea, R. G., and Murphy, B. S. (1977), 'Initial development of a general effective stress method for the prediction of axial capacity of driven piles in clay', *Proc. 9th Offshore Tech. Conf.*, Houston, 3, pp. 495–506.
23. Francesçon, M. (1983), 'Model Pile Tests in Clay – Stresses and Displacements due to Installation and Pile Loading', PhD Thesis, University of Cambridge.
24. Goto, S., Tatsuoka, F., Shibuya, S., Kim, Y. S., and Sato, T. (1991), 'A simple gauge for small strain measurements in the laboratory', *Soils and Foundations* **31**(1), 169–180.
25. Jardine, R. J., Symes, M. J., and Burland, J. B. (1984), 'The measurement of soil stiffness in the triaxial apparatus', *Géotechnique* **34**(3), 323–34.
26. Jardine, R. J. and Christoulas, S. (1991), 'Recent developments in defining and measuring static piling diameters', *Proc. Intl. Conf. on Deep Foundations*, Paris, Press Nationale de l'École Nationale des Ponts et Chausées, pp. 713–745.
27. Karlsrud, K. and Haugen, T. (1985), 'Axial static capacity of steel model piles in overconsolidated clay', *Proc. 11th Intl. Conf. on Soil Mechs. and Found. Engrg.*, San Francisco, pp. 1401–1406.
28. Karlsrud, K. and Nadim, F. (1990), 'Axial capacity of offshore piles in clay', *Proc. 22nd Offshore Tech. Conf.*, Houston, Paper 4883.
29. Karlsrud, K., Nadim, F., and Haugen, T. (1986), 'Piles in clay under cyclic axial loading, field tests and computational modelling', *Proc. 3rd Intl. Conf. on Numerical Methods in Offshore Piling*, Nantes, France, pp. 166-190.
30. Karlsrud, K., Borg Hansen, S., Dyvik, R., and Kalsnes, B. (1992), 'NGI's pile tests at Tilbrook and Pentre – Review of testing procedures and results', *Proc. ICE Conf. on Recent Large Scale Fully Instrumented Pile Tests in Clay*, London, July.
31. Karlsrud, K., Kalsnes, B., and Nowacki, F. (1992), 'Response of piles in soft clay and silt deposits to static and cyclic loading based on recent instrumented pile load tests', *Proc. SUT Conf. on Offshore Site Investigation and Foundation Behaviour*, London, September.
32. Kavvadas, M. (1982), 'Non-Linear Consolidation around Driven Piles in Clays', ScD Thesis, Dept. of Civil Engineering, MIT, Cambridge, MA.
33. Kiousis, P. D., Voyiadis, G. Z., and Tumay, M. T. (1988), 'A large strain theory and its application in the analysis of the cone penetration mechanism', *Intl. J. for Num. & Anal. Meth. in Geomechanics* **12**(1), 45–60.
34. Kraft, L. M. (1982), 'Effective stress capacity model for piles in clay', *ASCE Journal of Geotechnical Engineering* **108**(11), 1387–1404.
35. Kraft, L. M., Focht, J. A., and Amerasinghe, S. F. (1981), 'Friction capacity of piles driven into clay', *ASCE Journal of Geotechnical Engineering* **107**(11), 1521–1541.
36. Lacasse, S. (1979), 'Effect of Load Duration on Undrained Behaviour of Clay and Sand', NGI Internal Report 40007-1.
37. Ladd, C. C. and Edgers, L. (1972), 'Consolidated-Undrained Direct-Simple Shear Tests on Saturated Clays', Research Report R72-82, Dept. of Civil Engineering, MIT, Cambridge, MA.
38. Ladd, C. C. and Foott, R. (1974), 'New design procedure for stability of soft clays', *ASCE Journal of Geotechnical Engineering* **100**(7), 763–786.
39. Lehane, B. and Jardine, R. (1992), 'The behaviour of a displacement pile in Bothkennar clay', *Proc. Wroth Memorial Symposium*, Oxford.

40. Lemos, L. J. (1985), 'The Effects of Rate on the Residual Strength of Soil', PhD Thesis, Imperial College, London University.
41. Levadoux, J.-N. and Baligh, M. M. (1980), 'Pore Pressures in Clays due to Cone Penetration', Research Report R80-15, Dept. of Civil Engineering, MIT, Cambridge, MA.
42. Malek, A. M., Azzouz, A. S., Baligh, M. M., and Germaine, J. T. (1989), 'Behaviour of foundation clays supporting compliant offshore structures', *ASCE Journal of Geotechnical Engineering* **115**(5), 615–636.
43. May, R. E. (1987), 'A Study of the Piezocone Penetrometer in Normally Consolidated Clay', PhD Thesis, Dept. of Engineering Science, University of Oxford.
44. Morrison, M. J. (1984), '*In Situ* Measurements on a Model Pile in Clay', PhD Thesis, Dept. of Civil Engineering, MIT, Cambridge, MA.
45. Nyirenda, z. M. (1989), 'The Piezocone in Lightly Overconsolidated Clay', PhD Thesis, Dpt. of Engineering Science, University of Oxford.
46. O'Neill, D. A. (1985), 'Undrained Strength Anisotropy of an Overconsolidated Thixotropic Clay', SM Thesis, MIT, Cambridge, MA.
47. Randolph, M. F., Carter, J. P., and Wroth, C. P. (1979), 'Driven piles in clay: Effects of installation and subsequent consolidation', *Géotechnique* **29**(4), 361–393.
48. Randolph, M. F. and Wroth, C. P. (1981), 'Application of the failure state in undrained simple shear to the shaft capacity of driven piles', *Géotechnique* **31**(1), 143–157.
49. Roscoe, K. H. and Burland, J. B. (1968), 'On the generalized behaviour of wet clays', *Engineering Plasticity*, J. Heymann and F. A. Leckie (eds.), Cambridge University Press, pp. 535–609.
50. Roy, M., Blanchet, R., Tavenas, F., and LaRochelle, P. (1981), 'Behaviour of a sensitive clay during pile driving', *Canadian Geotechnical Journal* **18**(1), 67–85.
51. Teh, C.-I. and Houlsby, G. T. (1991), 'An analytical study of the cone penetration test in clay', *Géotechnique* **41**(1), 17– 35.
52. Ting, N.-H., Onoue, A., Germaine, J. T., Whitman, R. V., and Ladd, C. C. (1990), 'Effects of Disturbance on Soil Consolidation with Vertical Drains', Research Report R90-11, Dept. of Civil Engineering, MIT, Cambridge, MA.
53. Whittle, A. J. (1987), 'A Constitutive Model for Overconsolidated Clays with Application to the Cyclic Loading of Friction Piles', ScD Thesis, Dept. of Civil Engineering, MIT, Cambridge, MA.
54. Whittle, A. J. (1990), 'A Constitutive Model for Overconsolidated Clays', MIT Sea Grant Report, MITSG90-15.
55. Whittle, A. J. (1991A), 'Evaluation of a constitutive model for overconsolidated clays', *Géotechnique*, to appear.
56. Whittle, A. J. (1991b), 'Interpretation of pile load tests at the Haga site', *Proc. ASME Conf. on Offshore Mechs. and Arctic Engrg. (OMAE '91)*, Stavanger, Vol. 4, pp. 267–275.
57. Whittle, A. J., Baligh, M. M., Azzouz, A. S., Malek, A. M. (1988), 'A model for predicting the performance of TLP piles in clay', *Proc. 5th Intl. Conf. on Behaviour of Offshore Strctures (BOSS '88)*, Trondheim, Vol. 1, pp. 97–112.
58. Whittle, A. J. and Baligh, M. M. (1988), 'The Behaviour of Piles Supporting Tension Leg Platforms. Results of Phase III', Report submitted to Join Oil Industry, Dept. of Civil Engineering, MIT, Cambridge, MA.
59. Whittle, A. J., Aubeny, C. P., Rafalovich, A., Ladd, C. C., and Baligh, M. M. (1991), 'Interpretation of *In-Situ* Tests in Cohesive Soils using Rational Methods', Research Report R91-01, Dept. of Civil Engineering, MIT, Cambridge, MA.
60. Whittle, A. J., DeGroot, D. J., Ladd, C. C., and Seah, T. H. (1992), 'Model prediction of the anisotropic properties of Boston Blue Clay', *ASCE Journal of Geotechnical Engineering*, to appear.
61. Wood, D. M. (1981), 'True triaxial tests on Boston Blue Clay', *Proc. 11th Intl. Conf. on Soil Mechs. and Found. Engrg.*, Stockholm, pp. 825–830.

SHAFT FRICTION OF PILES IN CARBONATE SOILS

M. R. COOP
GERC, City University, Northampton Square, London EC1V 0HB, UK

and

J. D. MCAULEY
BP Engineering, 1 Harefield Rd., Uxbridge, Middlesex UB8 1PD, UK

Abstract. The paper describes the results of an investigation into the behaviour of friction piles in carbonate soils prompted by problems at the North Rankin site, and following on from work by Coop (1990) on the fundamental mechanics of these soils. Firstly a series of shear box tests was conducted which was aimed at examining the behaviour of the soil at the pile interface. These were followed by a short series of triaxial tests which confirmed that the mechanics of the soils were qualitatively similar to those tested by Coop. The *in situ* states of the Rankin soils were then examined within a critical state framework, correlating the soil state at any particular depth with published pile test data for the site. It is suggested that the flexibility of the pile may be an important influence on the capacity of steel piles. This may be the result of progressive failure, as the steel interface shear box tests revealed brittle strain softening behaviour.

1. Introduction

In 1982 the North Rankin platform was installed in 125 m water depth off North West Australia. The design of the piles had been based on shaft friction measurements made using a small down-hole model pile called a steel friction tool (SFT) during the site investigation. The unexpected ease with which the 120 m long piles were driven indicated that they might not develop the required shaft friction and a programme of foundation strengthening was implemented at a total cost of Aus$380m (King and Lodge, 1988). It was this experience which has prompted much recent research into the behaviour of carbonate soils. Semple (1988) and Coop and Lee (1992) have shown that the mechanics of carbonate soils are fundamentally similar to other more commonly encountered granular soils at similar densities, and that this behaviour could be conveniently described within a critical state framework. Isotropic and one-dimensional loading of triaxial samples defined unique, normal compression lines in $v : ln p'$ space, where v is the specific volume $[= 1 + e]$ and p' is the mean normal effective stress $[= (\sigma'_a + 2\sigma'_r)/3]$. These were independent of the soil's initial density and as for clay soils were found to be straight and parallel. Typical data are shown in Figure 1. Coop and Lee (1992) showed that the volumetric compression associated with these normal compres-

Volume 28: Offshore Site Investigation and Foundation Behaviour, 645–659, 1993.

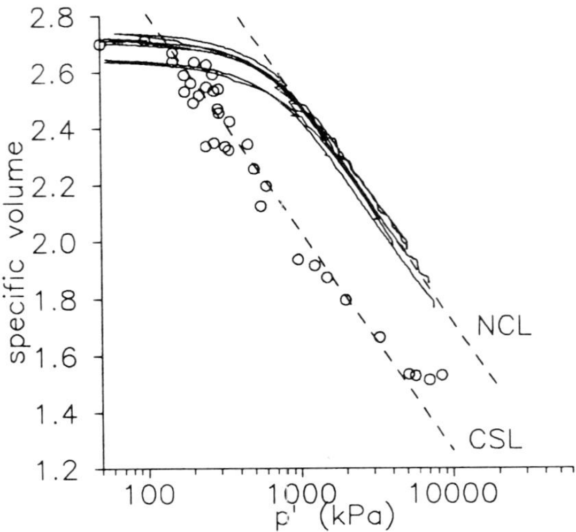

Fig. 1. Isotropic compression and critical states of Dogs Bay carbonate sand.

sion lines resulted from particle breakage, and that the unusual feature of biogenic carbonate sands such as are encountered at Rankin is that the angular nature of the shell fragments which comprise these soils results in very high voids ratios. Compression of these soils therefore results in yield, corresponding to the onset of particle breakage as the soil reaches the normal compression line at relatively low stresses. This typically may be in the region of 100-1000 kPa depending on the precise soil type and its initial density, so that these soils can be extremely compressible in the engineering stress range. Coop (1990) also showed that despite the continued particle breakage which occurred during shearing, triaxial tests would eventually reach critical or constant volume states which defined a unique critical state line which was independent of the loading path and for the soil tested was straight and parallel to the normal compression line in $v : lnp'$ space, as again shown in Figure 1.

2. Shear Box Tests

A series of monotonic loading shear box tests has been conducted on Dogs Bay sand, a biogenic carbonate sand chosen by Evans (1987) for its similarity to those at Rankin. The principal aim of these tests was to examine the interface shearing characteristics of a carbonate sand against both steel and grout, investigating also the influence of the soil stiffness on the stress:strain behaviour at the pile interface. This was achieved by making the change in vertical stress during shearing proportional to the vertical displacement of the shear box platen by means of a constant normal stiffness (CNS) as used by Johnston *et al* (1988). The tests were conducted on 100 mm diameter circular samples in a computer controlled shear box developed at City University, the interface being placed in the lower half of the box and a soil sample about 16 mm thick in the upper.

Figures 2 and 3 show CNS test data for grout and steel interface respectively. In each case the shear displacement (δh) has been normalised by the initial thickness of the sample (Ho). The shearing behaviour of a soil against an interface will clearly depend on the roughness of the interface but Kishida and Uesugi (1987) found that the roughness should be normalised with respect to the particle size:

$$R_n \approx \frac{R_{\text{max}}}{D_{50}}$$

where R_n is the normalised roughness, R_{max} the maximum asperity size over a horizontal distance equal to the mean particle size (D_{50}) of the soil. For offshore sites of carbonate soils such as Rankin, typically a large variety of particle sizes are found over the depth range of the piles from clays to coarse sands. The pile load test data discussed in later sections also covers a wide range of pile types with a variety of surface roughnesses. No attempt was therefore made to model the precise roughness of a particular pile. Two of the interfaces were instead made as smooth as easily achievable to give the lowest likely values of R_n (0.044 for the grout and 0.005 for the steel) and a conservative assessment of the interface load:deflection behaviour. One test was also carried out on a "rough" steel interface which had a value of R_n of 0.11 which is more typical of a prototype pile.

The tests were each conducted on samples of Dogs Bay sand which were initially loose but which had been compressed to a vertical stress of 1000 kPa at which the state of the soil lay on the K_o normal compression line. This is a higher stress than those *in situ* at Rankin for the depths of pile tests. However it will be shown later that the Rankin soils reach a normal compression line at lower stresses than Dogs Bay sand, so that below about 30 m depth, the soil is normally compressed and qualitatively similar behaviour might be expected to that seen in these shear box tests. Other tests were also carried out at 100 kPa and gave similar data.

For the grout interface, only two tests have been conducted, representing the extremes of possible behaviour with a traditional constant vertical stress test (CNS = 0) and a test in which the sample height has been held constant (CNS = ∞). In the latter the sample volume is effectively held constant, and the test is therefore equivalent to undrained loading. The stress path is similar to those obtained by Coop (1990) for undrained triaxial tests on normally compressed samples of this sand with a large and rapid reduction in the normal stress. An interface friction angle ($\delta' = \tan^{-1} \tau'/\sigma'_v$) of 42° can be identified from this test, slightly above the critical state friction angle (ϕ'_{cs}) of 39° measured in the triaxial, the higher value perhaps resulting from plane strain loading. The constant vertical stress test reaches much higher shear stresses than that with constant height, but the data show that large displacements would be required to bring the soil to an ultimate state under this loading condition. Although the tests shown in Figure 2 were carried

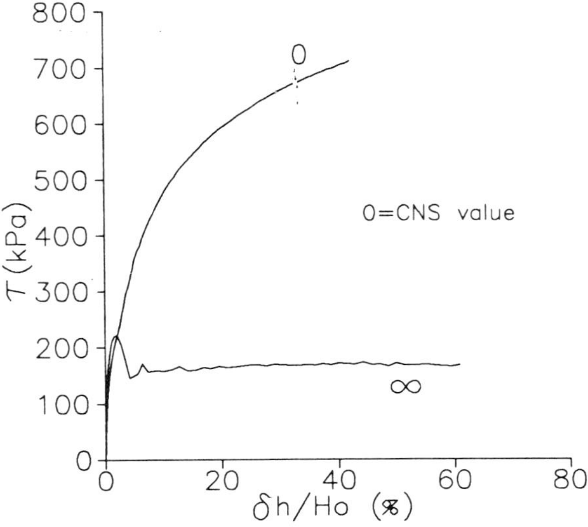

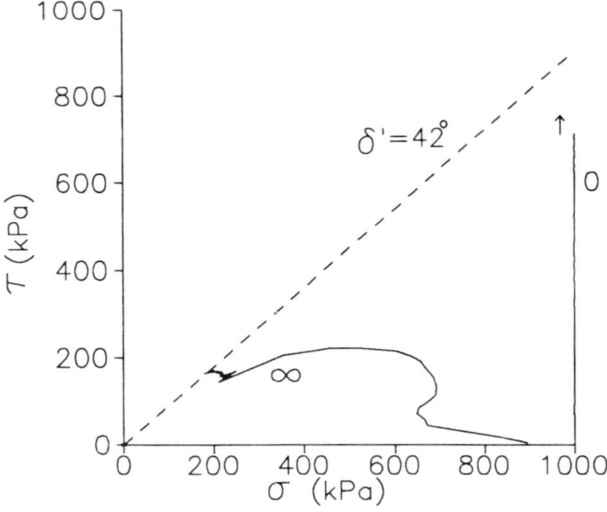

Fig. 2. Grout interface shear box tests.

out on soil compressed against a smooth grout interface which had already set, other tests showed that any penetration of the grout into the soil had a relatively minor influence on the shearing behaviour, indicating that the roughness of the grout interface may be unimportant.

The tests for the steel interface identify a very different pattern of behaviour.

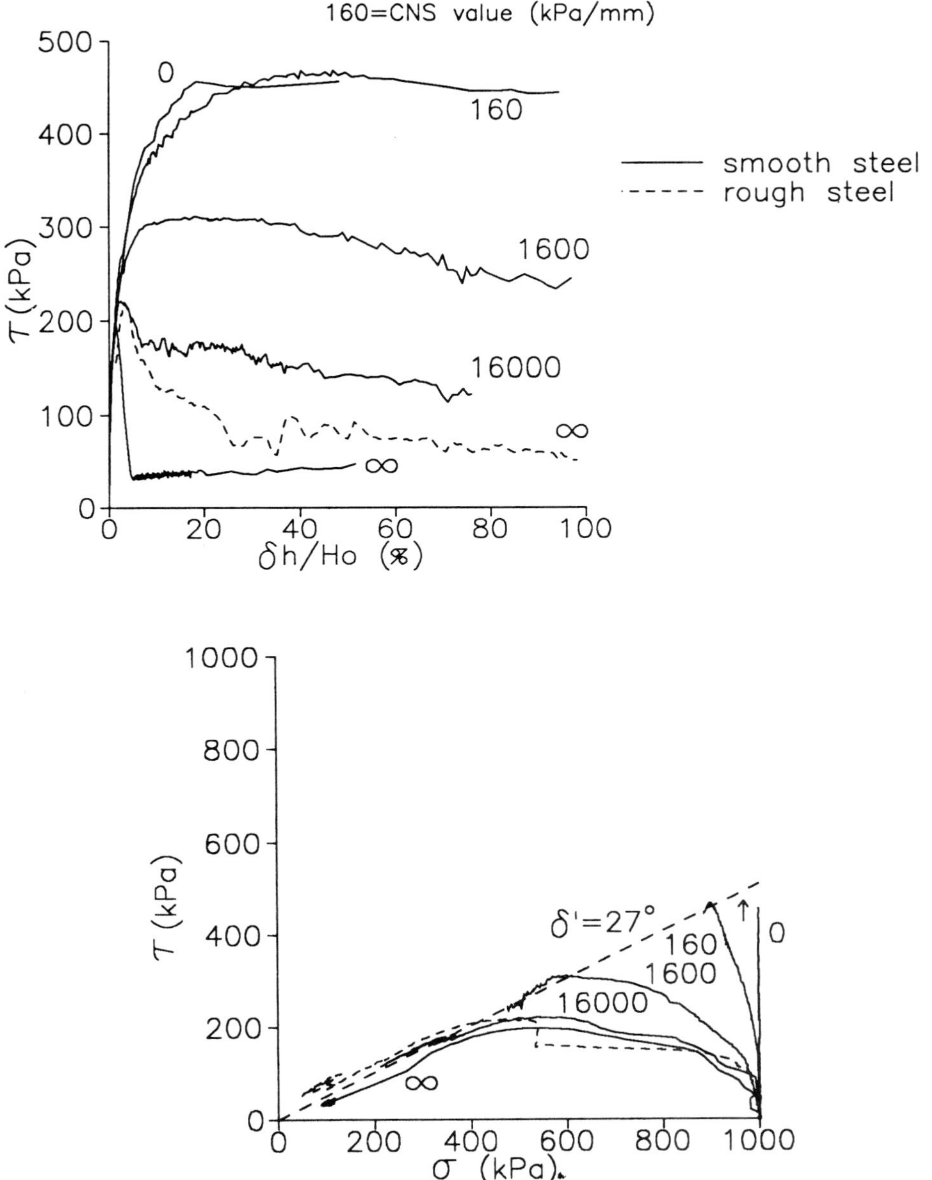

Fig. 3. Steel interface shear box tests.

The failure envelope defined by the stress paths give a much lower interface friction angle of 27° for the smooth steel and 31° for the rough. Perhaps of greater importance as will be discussed later, is the strain softening behaviour at higher CNS values. This is seen most dramatically for the constant height test on the smooth steel interface. For the rough interface the data indicate a stick-slip behaviour and a similar degree of strain softening but over a larger displacement. Strain softening of this type is not seen for the grout interface even though its normalised roughness is lower than that of the rough steel. The strain softening therefore appears to result not from the smoothness of the interface but its hardness.

3. Mechanical Behaviour of Rankin Soils

Figure 4 shows triaxial test data from tests conducted at City University on samples of calcarenite from Rankin which had a medium degree of cementing and also two samples of uncemented soils from the neighbouring Goodwyn field. Each of the samples was first isotropically compressed and the data show little consistency, because as Coop and Atkinson (1993) have shown, the isotropic compression behaviour of carbonate soils is strongly dependent on both the grading of the soil and the presence of cementing. In contrast, the critical states from shearing stages are much closer, agreeing with Coop and Atkinson's observations that the grading of soil has a much smaller influence on the position of the critical state line and that whether a soil of a particular grading was cemented or uncemented had no effect on its location. A tentative critical state line has therefore been sketched through these data, which is curved even in the $lnv : lnp'$ axes chosen. In Figure 5 this line is compared with a compilation of end of test states from commercial triaxial test data. Although the data scatter is poor, the critical state line identified from the tests conducted at City University passes through the centre of these data. In addition to site investigation data from the Rankin NRA site, which was the site chosen from the platform, data are also shown for other sites within the Rankin field (NRAA and NRB), and also for Goodwyn. Within the scatter of data, there appear to be no clear differences between the various sites, nor could any consistent influence of soil grading or the degree of cementing be detected on the location of the critical states.

4. In Situ States of Rankin Soils

In situ states for the soils at the various Rankin and Goodwyn sites have been estimated from the site investigation data and are shown in Figure 6. For each site, mean *in situ* water contents were calculated from all the available data within depth intervals of around 10 m. These have been used to calculate the specific volumes (v) which are then plotted against values of p' estimated from the effective unit weight of the soil (8 kN/m^3) and a value of K_o of 0.43 estimated from K_o triaxial tests conducted for the site investigation.

The data are similar to the isotropic compression data shown in Figures 1 and 5 with samples at shallow depths which are at low stresses lying to the left (dry) side of the critical state line. The soils are considered in this analysis are predominantly sands and silts which increase in age from "recent" at the surface to

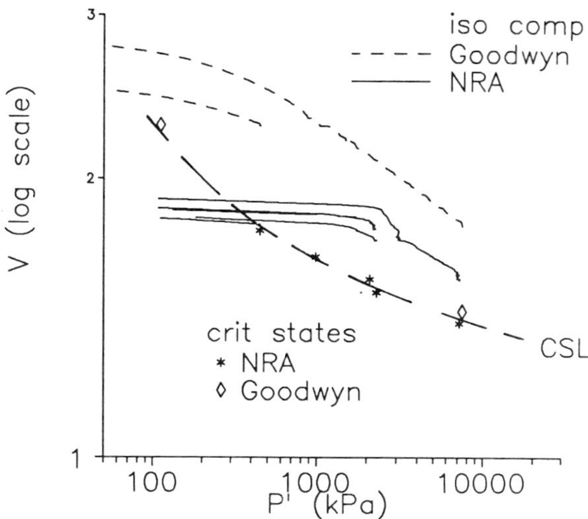

Fig. 4. Isotropic compression and critical states of Rankin and Goodwyn soils.

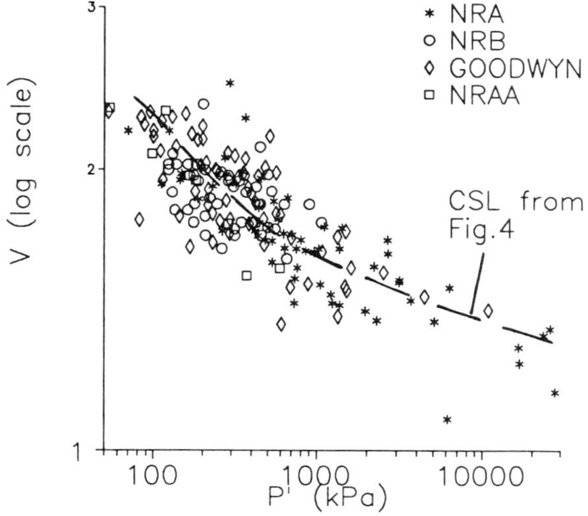

Fig. 5. Critical states of Rankin and Goodwyn soils.

Early Pleistocene below 110 m. The surface soils are currently under the maximum stresses they have experienced and their current specific volume is determined by the density of the packing of the soil particles as they are deposited. They have not reached a state dry of critical by overconsolidation as would be typical of a clay soil.

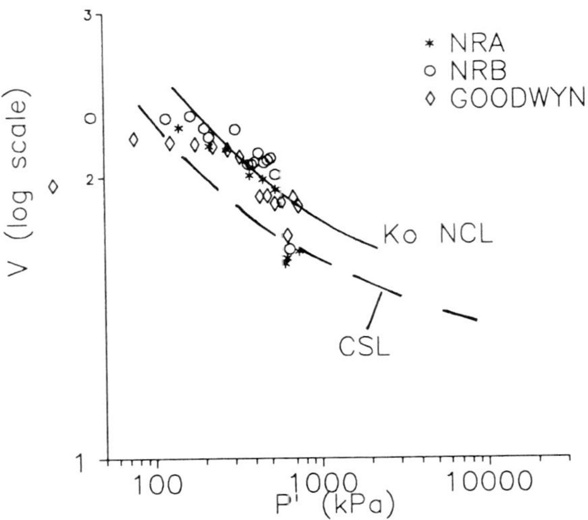

Fig. 6. *In situ* states at Rankin.

As the depth and stress increases, so the soil states move to the right (wet) side of critical, but initially there is little change in specific volume. Only at a stress of around 200-300 kPa, corresponding to a depth of 40-60 m does the soil yield. Below this depth the soil states appear to define a K_o compression line which has been drawn parallel to the critical state line. Again the data are similar for each of the sites considered. Most of the soils and soft calcarenites found at the Rankin and Goodwyn sites therefore exist on the wet side of critical, many on or near a K_o compression line. However, at 110 m depth there is a sudden reduction in the specific volume of the soils. This corresponds to a change at this depth from uncemented and lightly cemented soils to better cemented calcarenites. Apthorpe *et al* (1988) have identified that the latter were formed during an Early Pleistocene fall in sea level so that deposition occurred in shallower seas. For carbonate soils the cementing process is largely coincident with deposition and is more pronounced in shallower, warmer seas. The well cemented calcarenites formed during this period were also exposed to form an eroded land surface, before being re-submerged and buried beneath subsequent carbonate deposition. The jump in specific volume which can be seen in Figure 6 therefore represents a geological non-conformity, although the erosion of this surface was limited, so that the soils below 110 m are currently under the maximum stress they have experienced. Their state has moved away from the normal compression line probably not as a result of overconsolidation, but as Coop and Atkinson (1993) have observed, because of the increase in density of a soil when it is cemented as a result of in filling of the pore spaces.

5. Analysis of Pile Shaft Friction

From their cavity expansion analyses, Randolph *et al* (1979) determined that a key influence on the shaft friction of piles in clay soils would be its overconsolidation ratio (OCR). As we have seen that carbonate sands like clays may be described within a critical state framework, it follows that OCR should have a similar effect for these soils. However, as was shown in the previous section, the soils at Rankin are currently under their maximum stresses and so OCR loses any meaning. It is suggested that what will control the pile capacity is the *in situ* state of the soil relative to the critical state and normal compression lines. At any given initial stress soils on the dry side of critical should dilate during shearing giving higher shaft friction than those on the wet side which should compress.

The state of the soil may be quantified in a variety of ways, such as OCR for clays or Been and Jefferies' (1985) state parameter for sands. Here the mean normal effective stress *in situ* (p'_o) has been normalised with respect to the value of p' on the critical state line at the *in situ* water content (p'_{cs}). This has been chosen in preference to a Hvorslev equivalent pressure (p'_e) because as has been seen, a unique normal compression line cannot be defined for these soils. Although on Figure 5 a single critical state line has been drawn, it is likely that the data scatter conceals considerable variation in the location of this line with site and soil type. In the choice of p'_{cs} representative for each of the pile test analysed, only the critical state data from the samples closest to the pile test site and at similar depths have been considered.

The pile shaft friction may be normalised either with respect to the undrained shear strength of a soil (S_u) to give an "α" factor or the *in situ* vertical effective stress to give a "β" factor. Randolph *et al* concluded that for clays both would be influenced by OCR and that if that influence was accounted for, either could be used. For these largely granular soils undrained shear strengths are not easily defined, and β has therefore been chosen. Figure 7 shows a large amount of data normalised in this way for both model piles or test pile sections at the Rankin site together with one data point representing an estimate of the installed pile capacity based on wave equation analyses of the first blows of redriving by Dolwin *et al* (1988). Also included are some data from pile tests conducted at an onshore test-bed site (Fahey *et al*, 1992) chosen for the similarity of the calcarenites to those below 110 m at the Rankin and Goodwyn sites.

The data on Figure 7 cover a large range of pile types and sizes, both driven and bored, tested at various loading rates in a wide variety of soils from strong calcarenites to completely uncemented soils, and this is clearly responsible for much of the data scatter. There is, however, a clear trend of increasing β as p'_{cs}/p'_o increases from a minimum of around 0.3 to 0.4 for soils in normally compressed states to values greater than unity for soils on the dry side of critical.

For cemented soils this analysis can only be applied if the soil around the pile behaves plastically. The behaviour of a pile driven into a cemented soil at shallow

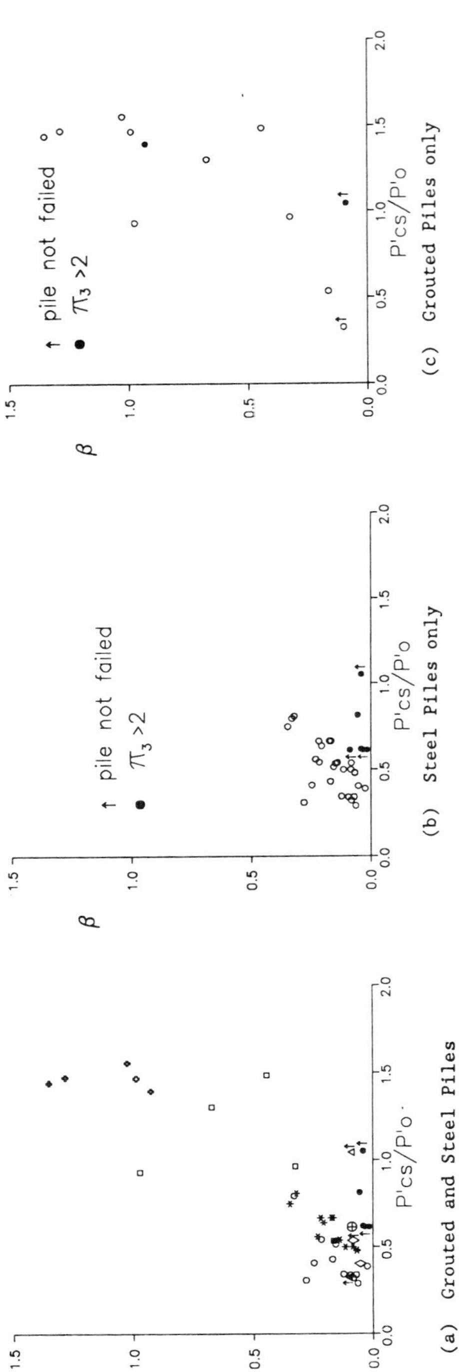

Fig. 7a. Pile test data.

Symbol	Reference	Pile Type/Location	Soil Type
△	Renfrey et al (1988)	grouted sections (1974) NRAA	1 in uncemented soil 1 in calcarenite
◇	Renfrey et al (1988)	driven sections (1974) NRAA	uncemented/lightly cemented soil
□	Williams & van der Zwaag (1988)	grouted sections	calcarenite
■	Renfrey et al (1988)	grouted section (1978) NRA	uncemented/lightly cemented soil
✳ ○	Khorshid et al (1988)	SFT (1978) NRA SFT (1981) NRB	uncemented/lightly cemented soil
●	Ripley et al (1988)	conductor load tests (1983-84) NRA	uncemented/lightly cemented soil
⊕	Dolwin et al (1988)	redrive analysis of foundation piles (1982) NRA	uncemented/lightly cemented soil
♧	Fahey et al (1992)	grouted sections overland corner	calcarenite

Key to Figure 7

Fig. 7b. Pile test data.

depths (high p'_{cs}/p'_o) is a complex and it may suffer among other things from a post-holing effect, giving unusually low β values. In contrast a drilled and grouted pile in the same soil would mobilise the cohesive component of the soil strength arising from the cement, giving very high values of β. Coop and Atkinson (1993) showed that for the range of p'_{cs}/p'_o values given on Figure 6(a) the strength of the cemented soils at Rankin would be predominantly frictional.

Figures 7(b) and 7(c) show the same database as Figure 7(a) but sub-divided into driven and grouted piles respectively. The driven piles tests were predominantly in uncemented or lightly cemented soils with values of p'_{cs}/p'_o below 0.8, whereas the grouted piles were usually tested in better cemented calcarenites with higher p'_{cs}/p'_o values. Apart from the steel piles with high Π_3 values which will be discussed in the following section, where the two sets of data overlap, they suggest that the difference in capacities between grouted and steel piles is not very large. This is however a conclusion which needs to be verified by conducting tests on steel piles in well cemented soils.

6. Influence of Pile Flexibility

For the steel piles the data scatter is accentuated by a number of particularly low β values which on Figure 7(b) have been identified as being associated with high pile flexibility. Murff (1980) defined a pile flexibility parameter as follows

$$\Pi_3 = \frac{\Pi D L^2 \tau_p}{(AE)_p Z_c}$$

where D is the pile diameter, L its length and $(AE)_p$ its cross-sectional rigidity. The peak shaft friction, τ_p is mobilised at a critical displacement Z_c. He found that flexible piles in soils with a strain-softening shaft friction behaviour would have a reduced capacity because of progressive failure. Figure 2 showed that it is unlikely that the interface shearing behaviour of a carbonate sand against a grout interface would be strongly strain-softening and for the grouted piles in Figure 7(c) the only one brought to failure which had a high Π_3 value does not have a reduced β. In contrast, it was shown that a smooth steel interface might give rise to dramatically strain-softening behaviour for undrained loading or for high CNS values, and this may account for the low β values for the highly flexible piles. Many of the steel piles identified as having high Π_3 values were driven conductors which were loaded in tension, and the reduction in pile diameter during loading would further reduce the normal stress on the pile so accentuating the strain softening behaviour of the interface.

Figure 8 shows a comparison between the average values of β measured for each type of steel pile and their flexibilities considering a single value of p'_{cs}/p'_o of 0.62 and extrapolating the test data at other *in situ* states to that value where necessary. The reduction in shaft friction with increasing Π_3 is evident, and this may at least in part be accounted for by progressive failure. Randolph (1988) has

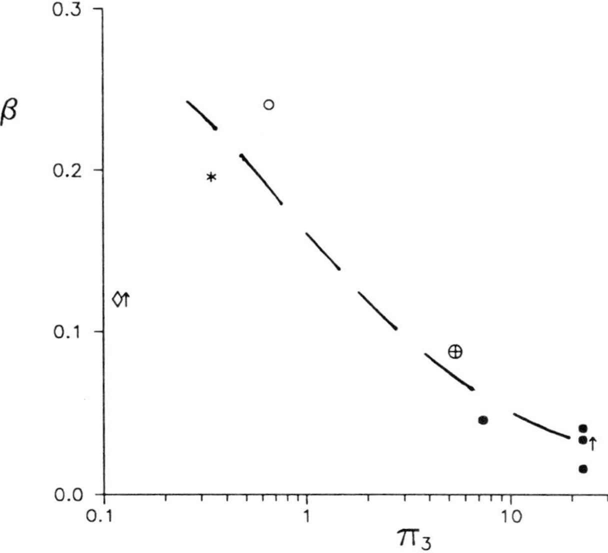

Fig. 8. Influence of π_3 on β ($p'_{cs}/p'_o = 0.62$).

however identified several other mechanisms of pile behaviour which may also contribute to this observation, or at least add to the data scatter on Figure 7. In particular he showed that pile diameter should affect the shaft friction capacity of all types of pile, and that for driven piles a high lateral flexibility would result in whip during driving, so reducing the normal stress on the pile and its capacity.

The ease with which piles are driven into carbonate soils may be attributed to their *in situ* states which are on the wet side of critical as seen from Figures 4 and 6. The constant height interface shear test (Figure 3) indicated that under undrained conditions for soils in a normally compressed state large pore pressure would be generated by shearing on the interface resulting in a very low effective normal stress and shaft friction. Subsequent dissipation of the pore pressures would give an increased capacity. However, as Randolph observed, the degradation of normal stress that occurred during drained shearing would not be recovered and the overall capacity would be reduced. Pile tests in dry soils or slow drained installations in saturated soils therefore result in low shaft frictional capacity and breaks in installation such as redriving should be avoided. Dolwin *et al* showed that on redriving the shaft friction degraded from around 43 kPa initially to around 10 kPa after several hundred blows. The database shown in Figure 7 is unfortunately insufficient to examine the importance of either this or the other two factors identified by Randolph.

7. Conclusions

In common with other carbonate soils the behaviour of those found at Rankin may be simply described within a critical state framework. For clay soils a natural consequence of the critical state model is that the state of the soil as quantified by overconsolidation ratio should be a controlling factor for the soil's behaviour as well as that of any structure built upon or within it. The work described in this paper has therefore examined if the *in situ* soil state influences the capacity of piles in carbonate sands. Analysis of published data suggests that the combination of stress level and specific volume of the soil may determine the behaviour of piles with distinct differences between the behaviour of soils on the wet and dry sides of the critical state line. Values of the *in situ* p'_{cs}/p'_o may be simply estimated from *in situ* water contents and a critical state line identified from a short series of triaxial tests. This factor is worth considering in the interpretation of either laboratory model or field pile tests, and in any extrapolation to prototype design.

The large discrepancies between driven and grouted pile performance at Rankin may be partly attributed to the fact that the driven piles have largely been tested in soils on the wet side of critical but the grouted in soils or soft rocks on the dry side. In addition, particularly low capacities are observed for the most flexible steel piles and shear box tests have shown that this may be due to pronounced strain softening on the steel interface resulting in progressive failure. Otherwise the pile test data indicate that the inherent difference in the capacities of steel and grouted piles in these soils may not be large.

Acknowledgements

This work was carried out partly within a project at City University sponsored by BP International. The authors are grateful to the project Steering Committee, in particular Mr. M. Sweeney, Professor J. H. Atkinson and Professor G. T. Houlsby for their comments on the work, and to Mr. J. Rigden of BP who provided the initial impetus for this research.

Thanks are also due to Woodside Offshore Pty Ltd and the North-West Joint Venture participants for the supply of samples and for the permission to publish certain of the test data from the Rankin and Goodwyn sites. The participants are: Woodside Petroleum Ltd (through subsidiaries), the Shell Development (Australia) Pty Ltd, BP Developments Australia Ltd, California Asiatic Oil Company and Japan Australia LNG (MIMI) Pty Ltd.

The views expressed in this paper are those of the authors.

References

1. Apthorpe, M., Garstone, J., and Turner, G. J. (1988), 'Depositional setting and regional geology of North Rankin "A" foundation sediments', *Proc. Int. Conf. Calcareous Sediments*, Perth 2, pp. 357–366.
2. Been, K. and Jefferies, M. G. (1985), 'A state parameter for sands', *Géotechnique* **35**(2), 99–112.

3. Coop, M. R. (1990), 'The mechanics of uncemented carbonate sands', *Géotechnique* **40**(4), 607–626.
4. Coop, M. R. and Lee, I. K. (1992), 'The behaviour of granular soils at elevated stresses', *Proc. Wroth Memorial Symposium*, Oxford, pp. 101–112.
5. Coop, M. R. and Atkinson, J. H. (1993), 'The mechanics of cemented carbonate sands', *Géotechnique*, to be published.
6. Dolwin, J., Khorshid, M. S., and van Goudoever, P. (1988), 'Evaluation of driven pile capacity – Methods and results', *Proc. Int. Conf. Calcareous Sediments*, Perth 2, pp. 493–502.
7. Evans, K. M. (1987), 'A Model Study of the End Bearing Capacity of Piles in Layered Carbonate Soils', D.Phil. thesis, Oxford University.
8. Fahey, M., Jewell, R. J., Khorshid, M, and Randolph, M. F. (1992), 'Parameter selection for pile design in calcareous sediments', *Proc. Wroth Memorial Symposium*, Oxford, pp. 169–180.
9. Johnston, I. W., Carter, J. P., Novello, E. A., and Ooi, L. H. (1988), 'Constant normal stiffness direct shear testing of calcarenite', *Proc. Int. Conf. Calcareous Sediments*, Perth 2, pp. 541–554.
10. Khorshid, M. S., Haggerty, B. C., and Male, R. (1988), 'Development of geotechnical aspects of the investigation programme', *Proc. Int. Conf. Calcareous Sediments*, Perth 2, pp. 377–386.
11. King, R. and Lodge, M. (1988), 'North West shelf development – The foundation engineering challenge', *Proc. Int. Conf. Calcareous Sediments*, Perth 2, pp. 333–342.
12. Kishida, H. and Uesugi, M. (1987), 'Tests of the interface between sand and steel in the simple shear apparatus', *Géotechnique* **37**(1), 45–52.
13. Murff, J. D. (1980), 'Pile capacity in a softening soil', *Int. Journal Num. and Anal. Methods in Geomech.* **4**, 185–189.
14. Randolph, M. F. (1988), 'The axial capacity of deep foundations in calcareous soil', *Proc. Int. Conf. Calcareous Sediments*, Perth 2, pp. 837–860.
15. Randolph, M. F., Carter, J. P., and Wroth, G. P. (1979), 'Driven piles in clay – The effects of installation and subsequent consolidation', *Géotechnique* **29**(4), 361–393.
16. Renfrey, G. E., Waterton, C. A., and van Goudoever, P. (1988), 'Geotechnical data used for the design of the North Rankin "A" platform foundation', *Proc. Int. Conf. Calcareous Sediments*, Perth 2, pp. 343–356.
17. Ripley, I., Keulers, A. J. C., and Creed, S. G. (1988), 'Conductor load tests', *Proc. Int. Conf. Calcareous Sediments*, Perth 2, pp. 429–438.
18. Semple, R. M. (1988), 'The mechanical properties of carbonate soils', *Proc. Int. Conf. Calcareous Sediments*, Perth 2, pp. 807–836.
19. Williams, A. F. and van der Zwaag, G. L. (1988), 'Analysis and evaluation of grouted section tests', *Proc. Int. Conf. Calcareous Sediments*, Perth 2, pp. 493–502.

Discussion

Question from R. Hobbs, Lloyd's Register: Figure 8 (or a similar plot with length rather than π_3 as the abscissae) clearly demonstrates the importance of exercising caution when extrapolating the results of small scale pile tests to field conditions, particularly for driven piles in carbonate soils.

Some time after the North Rankin A experience, Lloyd's Register reviewed a pile design in which high unit frictions were proposed, on the basis of short-section pile tests, for a soil very similar to that at Rankin. Fortunately we were given permission to release the Rankin data (not then in the public domain) and, as a result, were able to convince the client that more modest frictions should be used and pile penetration into more competent underlying soils increased. Freefall of the piles from a thin caprock to 70m penetration appears to have justified the downgrading.

FRICTION COEFFICIENTS FOR PILES IN SANDS AND SILTS

R. J. JARDINE
Senior Lecturer, Imperial College, London

B. M. LEHANE
Senior Engineer, Arup Geotechnics, London (formerly Research Assistant Imperial College)

and

S. J. EVERTON
Sir Alexander Gibb & Partners (formerly Post-graduate student Imperial College)

Abstract. The results from a series of direct shear interface tests on a range of cohesionless soils are presented. The tests used steel interfaces with properties comparable to those of industrial piles and investigated the influence on the shearing resistance of relative density, mean particle size and stress level. These results are compared (and seen to be in good agreement) with measurements obtained using a heavily instrumented displacement pile installed in a medium dense sand. Some of the more important factors affecting the friction coefficients developed by piles in cohesionless soils are identified and their implications for design are discussed.

1. Introduction

Recent reviews describe how the shaft capacity of piles in sand is usually assessed by one of three routes; Briaud and Tucker (1988), Poulos (1989) or Jardine and Christoulas (1991). One popular approach is to assume a direct relationship between the local skin friction (τ_f) and an *in-situ* test parameter such as SPT N, CPT q_c of pressuremeter p_{lim}. Alternatively, τ_f is assumed to be a simple multiple (β) of the initial free-field vertical effective stress (σ'_{v0}). The third route is to assume a Coulomb failure criterion and to assess the normal effective stresses and frictional coefficients acting at failure separately:

$$\tau_f = \sigma'_{rf} \tan \delta'. \tag{1}$$

Although the third route is the most attractive analytically, it is not clear how σ'_{rf} should be evaluated and δ' is rarely measured in appropriate laboratory tests. As a result, approximate rules of thumb (such as taken $\delta' = 0.8\phi'$, after Potyondy, 1961) are adopted for design calculations.

The API RP2A recommendations for offshore piles in sand suggest that Equation (1) should be evaluated assuming that $\sigma'_{rf}/\sigma'_{v0}$, or K_f, is equal to 0.8 and 1.0 for open and closed-ended piles respectively and that δ' varies with soil type and relative density (D_r); a graphical interpretation of the implied variations of

Volume 28: Offshore Site Investigation and Foundation Behaviour, 661–677, 1993.

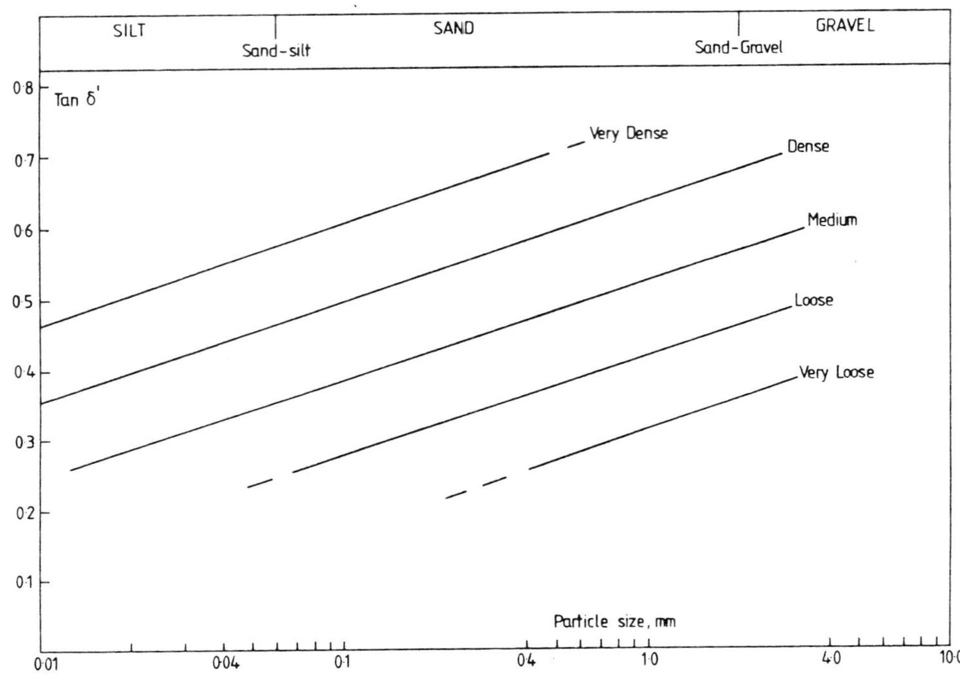

Fig. 1. Graphical interpretation of API recommendations: δ' values in granular soils.

δ' with D_r and mean grain size (D_{50}) is given in Figure 1. Upper limits to τ are also specified, but we should note that the API recommendations were derived by back-analyzing tests on un-instrumented piles and the individual parameters may be more notional than realistic.

For advances to be made, it is necessary to establish real values for σ'_{rf} and δ'. This paper explores the factors affecting the values of δ' using data from instrumented pile tests and laboratory direct shear interface tests. Factors affecting σ'_{rf} are also discussed.

2. Previous Studies of Soil-Interface Shearing Resistance

Potyondy (1961) presented some of the best known work on soil-interface shearing resistance. His study involved direct shear tests on five soils (ranging from sand to clay) using concrete, steel and timber interfaces. Broadly, he showed that the peak resistance (δ'_{peak}) depended on soil grading, density, stress level, interface material and surface roughness. For very rough surfaces, δ'_{peak} approached the ϕ'_{peak} value seen in equivalent soil-to-soil shear tests, but, in most cases, lower δ' values were obtained. Tests on silt gave higher δ' values than equivalent experiments on a medium sized sand, indicating the opposite trend to that projected in Figure 1.

The trend for δ' to increase as the grain size falls breaks down if the soil contains a significant proportion of platy clay particles. Comprehensive research using the ring-shear apparatus (e.g. see Lupini *et al*, 1981, Lemos, 1986, Tika-Vassilikos, 1991) has since shown that the interface shear behaviour of such clays can be dominated by the development of thin zones of oriented (residual) fabric which can show very low δ' values.

Leaving aside the question of residual clay fabric, Potyondy's results for sands and silts cannot easily be generalised as the roughness and hardness of his interfaces were not quantified, just three granular soils were tested and only the peak strength data are reported. However, more recently, several research projects on the interface shearing resistance of sans have been reported, including those by Yoshimi and Kishida (981), Lemos (1986), Boulon and Foray (1986), Kishida and Uesugi (1987), Everton (1991) and Lehane (1992). Most of these studies will be referred to later, but we will consider first the results obtained by Kishida and his co-workers using a specially designed simple shear apparatus.

Kishida's apparatus made it possible to distinguish the point at which slippage starts between the soil and interface in direct shear. It turns out that, with dense sands, this yielding condition coincides with the attainment of δ'_{peak}, and therefore with the peak in measured shear resistance. For samples denser than their critical density, the first slippage also coincides with the start of measurable dilation; the interface resistance falls as shearing continues until, after a few millimetres of relative displacement, an ultimate condition is reached where dilation ceases and the resistance remains constant. With loose sands, slippage appears to start significantly before the maximum δ' value is achieved.

Kishida and Uesugi (1987) report that the soil parameters which affect δ'_{peak} in sand-to-steel tests include the shape, size (D_{50}) and hardness of the sand grains[1]. They showed that the steel interface's roughness was an important factor, but considered that normal effective stress had little overall influence. Uesugi and Kishida (1986) propose straight line relationships for each sand between $\tan \delta'_{\text{peak}}$ and roughness as shown in Figure 2. This relationship is truncated by an upper limit, where $\delta' = \phi'$. When their data were re-plotted using a normalised Roughness Index (R_{max}/D_{50}) with account being taken of the angularity of the particles, a general bi-linear relationship was found to hold for all sands composed of the same minerals.

3. Instrumented Field Pile Tests in Labenne Sand

Field research with instrumented piles is required to assess the relevance of laboratory interface tests to the design of offshore foundations.

[1] Kishida and Uesugi defined roughness as the maximum asperity measured over a gauge length equal to D_{50}.

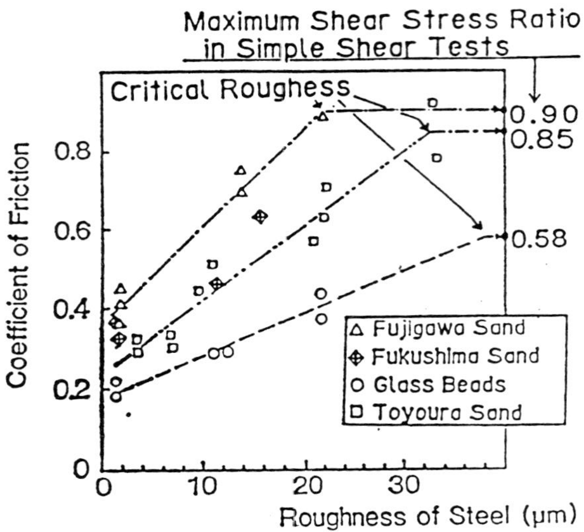

Fig. 2. Effects of steel interface roughness and sand type on $\tan\delta'$ (coefficient of friction) after Uesugi and Kishida (1986a).

A programme of pile tests was performed in 1989 at the Laboratoires des Ponts et Chausséees sand research site at Labenne, S.W. France; full details of these experiments are given by Lehane (1992), Jardine *et al* (1992) and Lehane *et al* (1993). The tests used the Imperial College instrumented Pile (ICP)[2], described by Bond *et al* (1992), which allowed the local values of σ_r' and τ_{rz} developed at a series of points on the pile shaft to be monitored with unprecedented accuracy. Data were obtained throughout the processes of installation, equalisation and load testing.

One of the key results obtained was that the equalised radial effective stress (σ_{rc}') was not a constant multiple of σ_{v0}', but was strongly and systematically dependent on the relative density of the sand (which varied with depth at Labenne) and also on the relative position of the pile tip. Lehane (1992) explores the many practical implications of this finding. We will return to this point later, but the principal point of interest here is the way in which σ_r' varied with τ_{rz} during pile loading.

Figure 3 presents a typical plot from one of the Labenne ICP (compression) pile tests, using data from an instrument cluster located in medium dense quartz sand. The load cells show that shaft loading causes relatively small reductions in σ_r' at first, but that the response becomes more 'dilatant' as τ_{rz}/σ_r' approaches a limiting ratio. The stress path then deviates sharply to the right, mobilising its peak δ' angle *before* the maximum shaft resistance is attained; the ultimate δ' is $\approx 3°$ less than δ'_{peak}. Further pile data from Labenne showed that the ultimate δ' value (i) did not

[2] The ICP is shot blasted before use to have the typical surface roughness of an offshore pile, i.e. $\approx 8\mu$m Centre-Line Average (CLA).

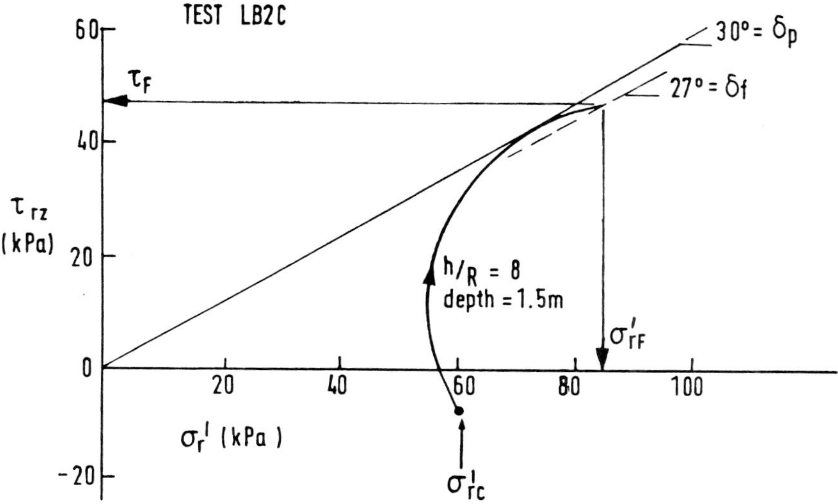

Fig. 3. Variations of σ_r' with τ during pile loading at Labenne, after Lehane *et al* (1993).

vary with the initial undisturbed sand density, (ii) was the same for wet and dry sand and (iii) was independent of the pile displacement rate (Lehane, 1992).

Uesugi and Kishida's laboratory studies suggest that the increases in σ_r' seen in the Labenne tests were associated with sand grains being displaced radially outwards (by an amount Δr) as local slippage started at the pile shaft. The ultimate failure (or critical state) was reached when the grains became 'unlocked' from the rough steel and large scale relative displacements become possible without any further dilation or change in shear stress. The operational δ' value is the critical state angle (δ'_{cs}) and not the peak angle – which has been the focus of most previous research.

Unlike the conventional shear box test, dilation at the pile-soil interface is resisted by the stiffness of the surrounding ground, causing σ_r' to increase once slippage starts. As a result peak obliquity (δ'_{peak}) and maximum shaft resistance (τ_f) do not coincide. Following the analysis given by Boulon and Foray (1986), Johnston *et al* (1987) and others it is surmised that the change in σ_r' can be related to the pile radius (r_0), the average radial movement of the soil grains (Δr) and the secant 'pressuremeter' shear modulus (G_p) (which falls as $\Delta r/r_0$ increases) by the expression:

$$\Delta\sigma_r' = 2G_p\frac{\Delta r}{r_0} \tag{2}$$

The radial movement depends on roughness, D_{50}, stress level and D_r, but is independent of pile radius per se (Boulon and Nova, 1990). We can therefore expect $\Delta\sigma_r'$ to vary inversely with pile radius and infer (from the Labenne experiments) that dilation effects will dominate the capacity of small model piles, but make no significant contribution to the shaft resistance of large offshore piles. Laboratory model pile tests reported by Lebègue (1964), Hettler (1982) and others are collated by Lehane (1992) and are shown to be consistent with this trend.

The controlled normal stiffness interface shear tests advocated by Boulon and his co-workers provide one way of investigating by interactions between dilation, G_p, pile radius and shaft resistance.

4. Laboratory Studies on Labenne Sand

Laboratory studies were performed in parallel with the Labenne field pile research. These included index tests, triaxial stress path experiments, shear box tests and soil-interface tests using the shear box and ring-shear apparatus (Lehane, 1992). Labenne sand has the uniform and consistent particle size distribution illustrated in Figure 4a (soil 10); the principal results from soil-soil and soil- interface direct shear tests are presented in Figures 5, 6 and 7. The tests were conducted over a relatively narrow range of vertical effective stress (70±45 kPa).

Figure 5 shows the marked influence of the sand's initial voids ratio (and D_r) on its peak ϕ' angle, with the densest samples giving angles 10° higher than the loosest. However, in all cases ϕ' tended towards similar ultimate (or critical state) values ($\approx 33°$) as dilation ceased and the shear stresses stabilised. Figure 6 shows the individual tests data from one of the three suites of interface tests. A range of initial void ratios and normal stresses were investigated for a steel interface which had a similar roughness to that of the Labenne pile. The main features seen were:

- Dense samples showed reductions in shear stress (τ) after achieving their peak values at relatively small displacements, but loose samples showed no clear peak in resistance.

- The maximum rate of dilation for dense samples occurred near the point at which τ was greatest. The loose samples showed either a net contraction, or no overall dilation.

- All samples tended towards approximately constant volume conditions after relative displacements of 2 to 4mm.

Figure 7 combines the data from Figure 6 with those from other tests on Labenne sand involving (i) a slightly smoother steel interface and (ii) a Teflon interface which was both smoother and softer. The picture that emerges is similar to that from the sand-sand tests: for each particular interface, the initial relative density affects the amount of dilation and δ'_{peak}, but not δ'_{cs}. For the two slightly different steel interface roughnesses considered the δ'_{cs} values were comparable; peak angles were more sensitive to the changes in R. Substantially lower angles were mobilised against

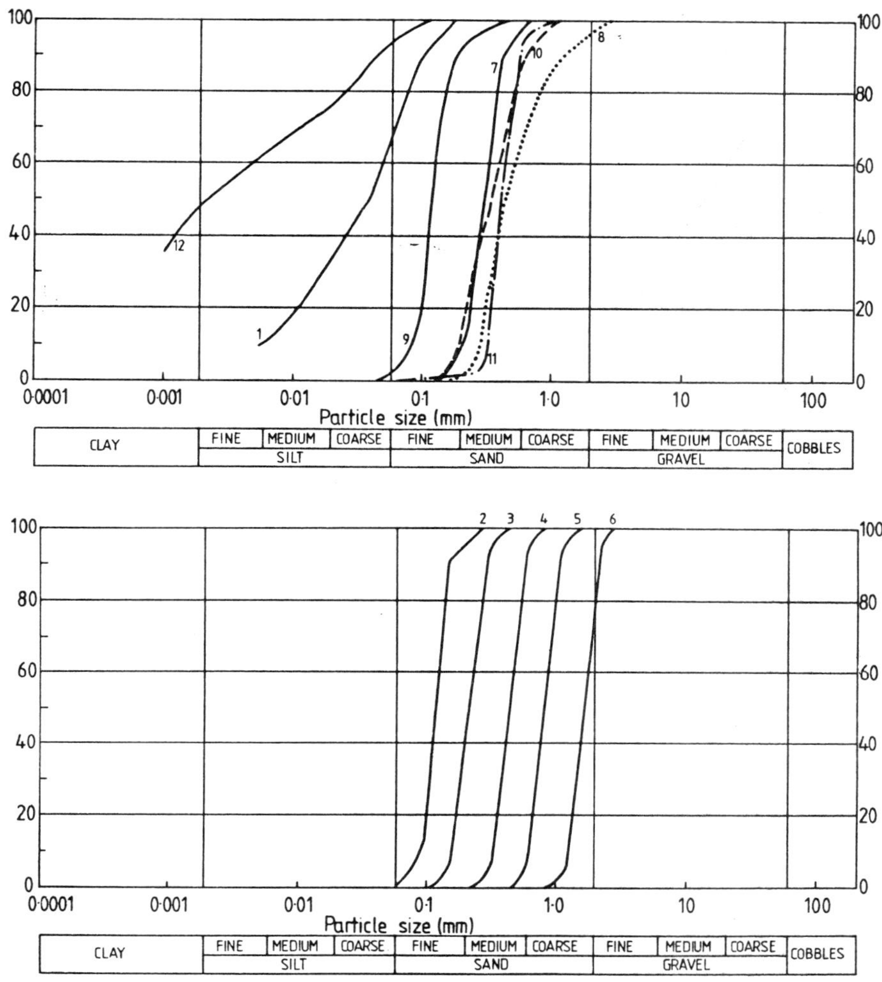

Fig. 4. Particle size distribution curves for soils in parametric study; (a) ungraded soils, (b) graded Cretaceous sand.

Teflon interfaces.

As the direct shear box has imperfect boundary conditions, check tests were run (against an 8μm steel interface) in the ring-shear apparatus described by Bishop *et al* (1971); δ'_{cs} was found to be within 1° of the average equivalent shear box value. Stress-path triaxial tests gave the same ϕ'_{cs} as the soil-soil shear box tests. However, the most interesting result was that the laboratory δ' values were virtually identical to those measured in the field instrumented pile tests (see Figure 3). The same agreement between laboratory and field was found in a later programme of instrumented pile tests in the Bothkennar clay-silt (Lehane and Jardine, 1992a).

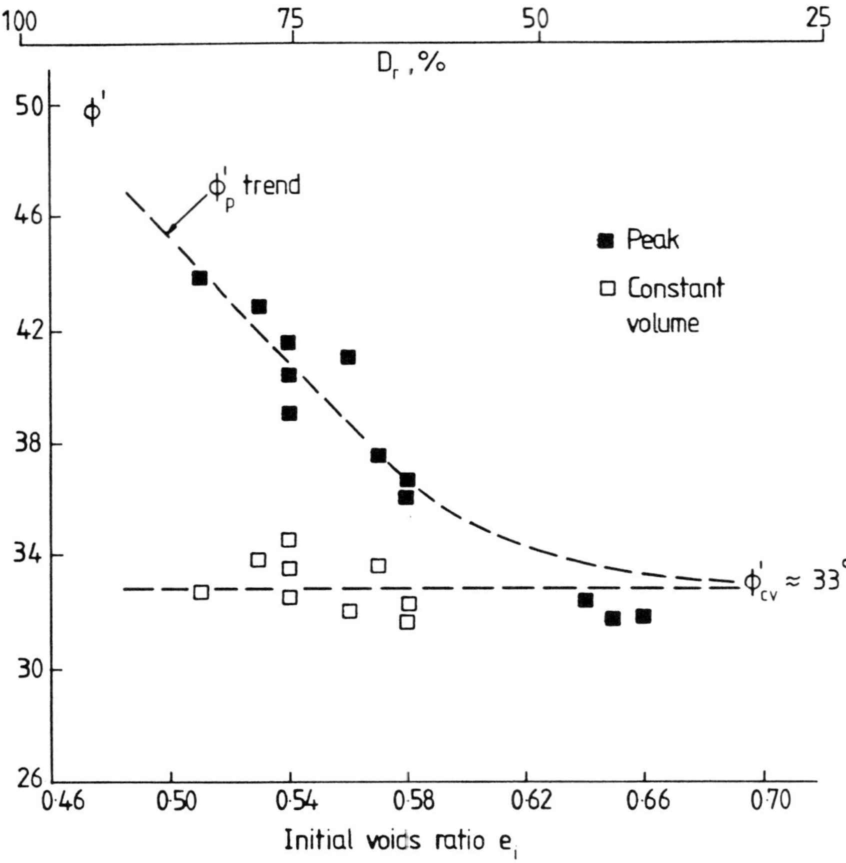

Fig. 5. Variation of ϕ' with D_r for Labenne sand (σ'_n = 75 =/- 45 kPa).

5. Parametric Study of Sand-Interface Shear

A parametric laboratory study was performed by Everton (1991) to supplement the Labenne research. This involved direct shear tests on eight other granular soils in which the initial relative density, interface roughness and normal stress level were varied. Control soil- soil tests were also performed. Five of Everton's sands were fractions graded from the same deposit of Cretaceous Greensand[3]. The remaining three soils were: an industrial rock flour silica silt (HPF4), Ham River sand and a calcareous sand from the Middle East (Al Shattie).

Everton's data are collated here with: the Labenne data, results for Leighton Buzzard sand (given by Lemos, 1986), ring-shear interface data for a North Sea glaciomarine sand (Ridley and Jardine, 1992) and the Bothkennar clay-silt ring

[3] A predominantly silica sand produced by the David Ball Company of Cambridge, UK.

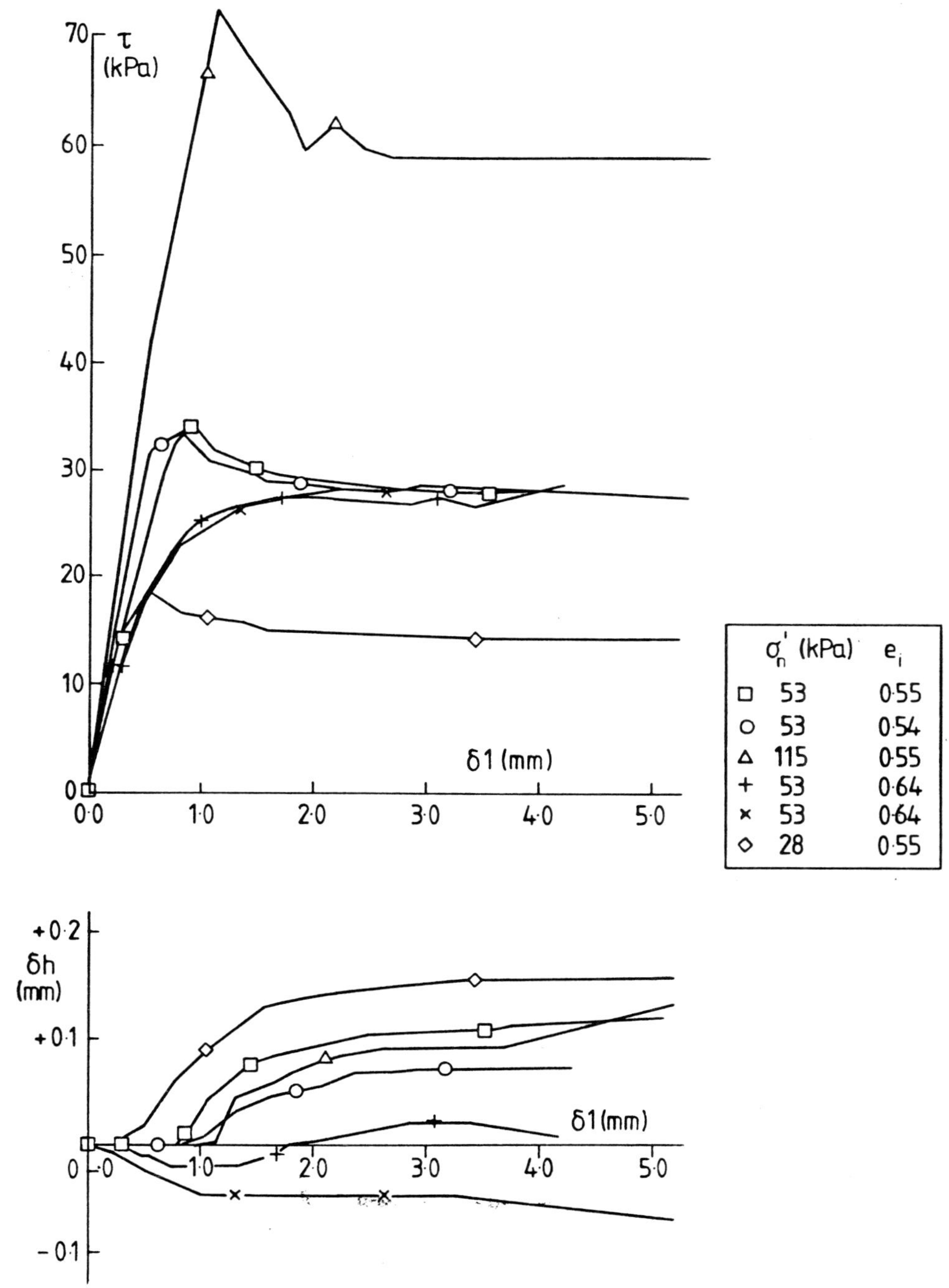

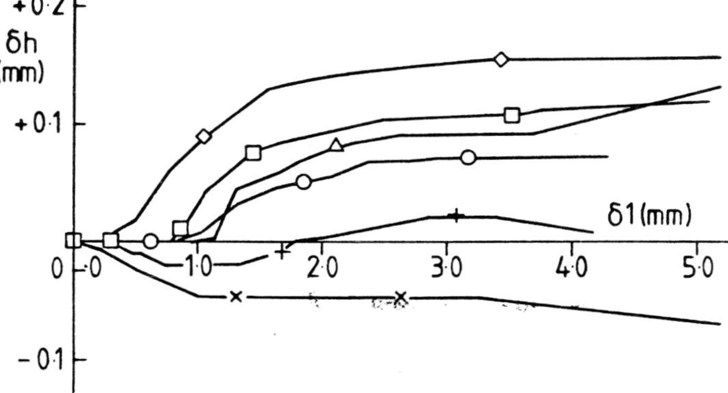

Fig. 6. Interface shear tests on Labenne sand (R = 9.5 μm, σ_n' = 75 =/- 45 kPa).

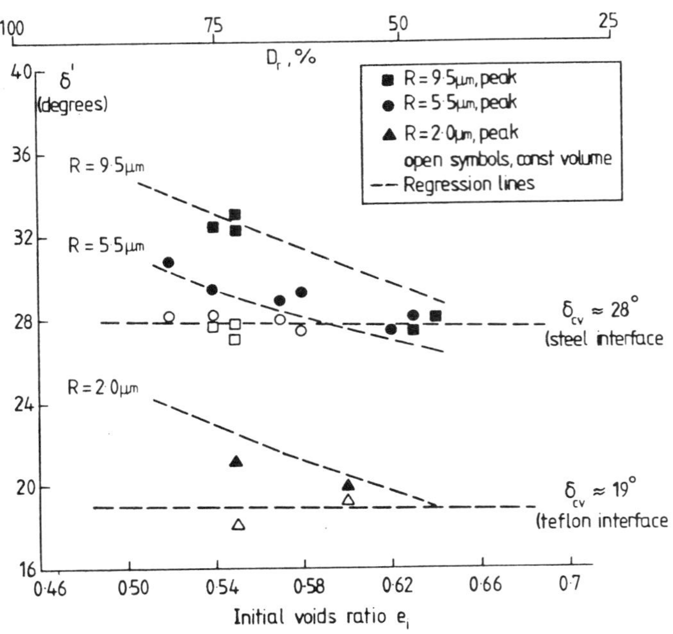

Fig. 7. Variation of δ' with D_r for Labenne sand (σ'_n = 75 =/- 45 kPa).

TABLE 1.
Soils considered in the study.

Code	Soil	D_{50}mm
1.	HPF4 – Crushed industrial rock flour; angular	0.04
2.	100/170 graded sand; sub-rounded	0.10
3.	52/100 graded sand; sub-rounded	0.22
4.	25/52 graded sand; sub-rounded	0.50
5.	14/25 graded sand; sub-angular	0.85
6.	7/14 graded sand; sub-angular	1.50
7.	Ham River Sand; sub-rounded to sub-angular	0.32
8.	Al-Shattie calcareous sand; rounded to angular	0.44
9.	North Sea glaciomarine sand	0.12
10.	Labenne dune sand; sub-rounded to sub-angular	0.32
11.	Leighton Buzzard sand	0.45
12.	Bothkennar clay-silt	0.025

shear tests described by Lehane and Jardine (1992b). The twelve soils are identified in Table 1, the grading curves are given in Figure 4 and Figure 8 presents the minimum and maximum void ratios, as determined by British Standard procedures. The limiting void ratios appear to be controlled principally by the grain size,

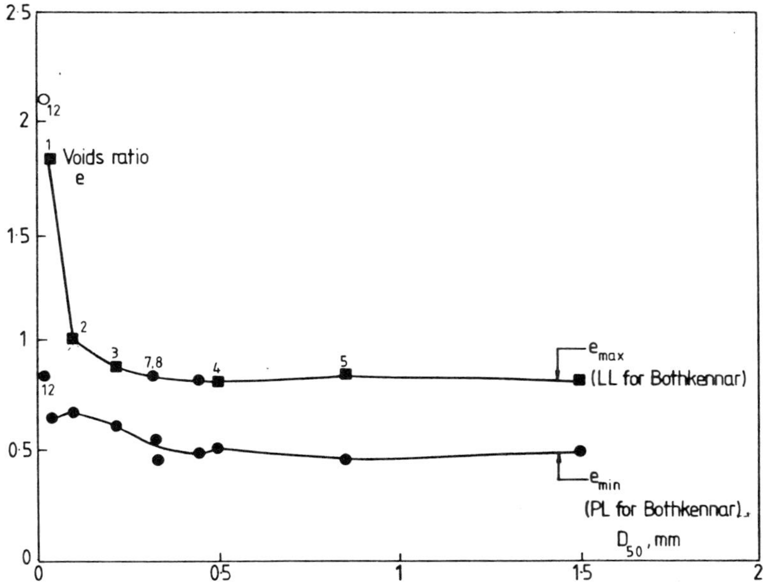

Fig. 8. Variations of $e_{\max}$ and $e_{\min}$ for soils in parametric study (note the void ratios plotted for Bothkennar clay-silt are those at the Atterberg limits).

although grading and particle shape are also likely to be influential. Note that the silt and clay-silt can exist in much looser states than the sands.

Everton's shear tests confirmed many of the trends identified for δ'_{peak} by Uesugi and Kishida (1986): the measured angles increased linearly with normalised surface roughness (R/D_{50}) and were upper-bounded by the soil to soil ϕ'_{peak} values. Also, as with Labenne sand, δ'_{peak} increased with relative density. Contrary to the findings of Kishida and his co-workers, δ' was, in some cases, strongly affected by normal stress level (σ'_n). Because δ'_{peak} is affected by these three factors, as well as particle shape and hardness, the δ'_{peak} measurements spanned the relatively wide range of 21.1° to 36.8°.

To clarify the picture we focus first on tests performed at moderate stresses against steel interfaces with $R \approx 10\mu$m. Figure 9 plots δ'_{peak} against normalised roughness for tests with $\sigma'_n = 100$ kPa; the mean values of δ'_{cs} are also shown. Figure 10 summarises the general trends for the total vertical displacement (Δh) which each sample had accumulated by the time it had achieved its ultimate condition. The main features to note are:

- The mean δ'_{cs} shown by each soil is independent of the initial relative density.

- δ'_{cs} reduces in inverse proportion to D_{50} and has an upper limit of ϕ'_{cs}. A linear

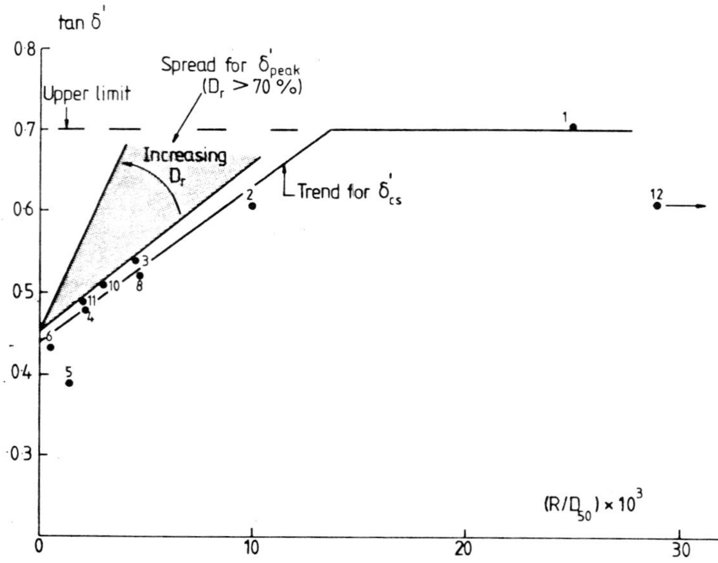

Fig. 9. Summary of δ' values from parametric study, $\sigma'_n \approx 100$ kPa.

relationship is obtained when tan δ'_{cs} is plotted against normalised roughness (R/D_{50}).

- Sand samples tend to contract after 'yield' if D_r is smaller than $\approx 60\%$ and dilate if D_r exceeds $\approx 70\%$. All the silt and silt-clay samples contract, reflecting their smaller particle size and greater initial void ratios.

- In keeping with the dilation characteristics, sand samples with $D_r > 70\%$ mobilise δ'_{peak} angles that exceed δ'_{cs} by a margin that increases with D_r. None of the silt samples showed a peak in τ, but their δ'_{cs} values were the highest seen in the parametric study.

The mean ϕ'_{cs} values and the ranges for δ'_{cs} (as measured in 53 tests, with 50 kPa $< \sigma'_n <$ 110 kPa) are plotted against D_{50} on Figure 11. ϕ'_{cs} varies only slightly from a mean value of 34.5°. For D_{50} less than 0.1mm, δ'_{cs} is only slightly smaller than ϕ'_s. However, the margin between the two angles increases systematically with grain size and amounts to $\approx 15°$ for coarse sand.

Limited studies of the effects of normal and effective stress level were made for three of the soils. The most extensive was the Leighton Buzzard sand shear box series (Lemos 1986), which showed that increasing the stress level from 100 to 200 kPa was sufficient to suppress dilation in even very dense samples. The resulting envelope for δ'_{cs} is shown in Figure 12 along with the shear box results

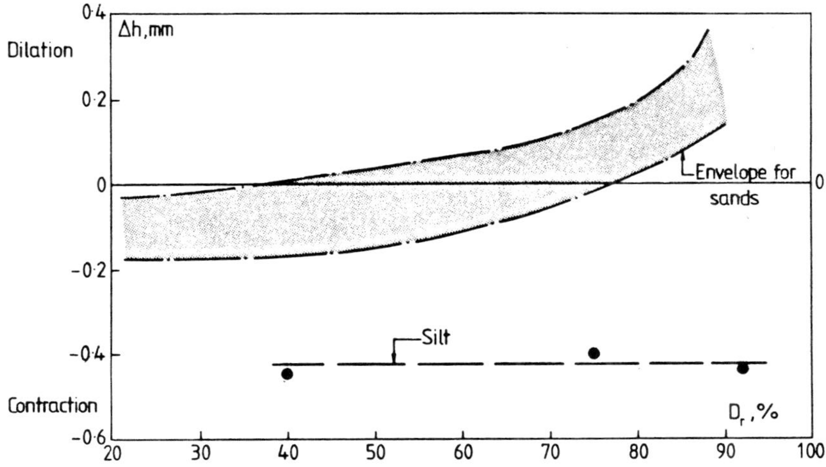

Fig. 10. Variations of Δh with D_r for soils in parametric study, $\sigma'_n \approx 100$ kPa.

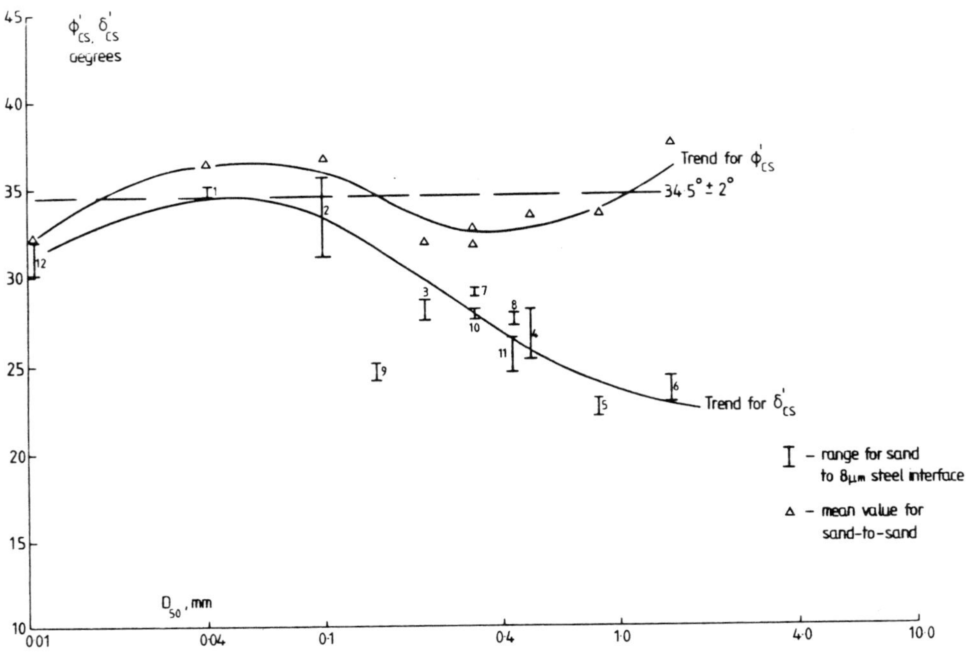

Fig. 11. Effect of particle size on ϕ'_{cs} and δ'_{cs} for soils in parametric study ($50 < \sigma'_n < 110$ kPa).

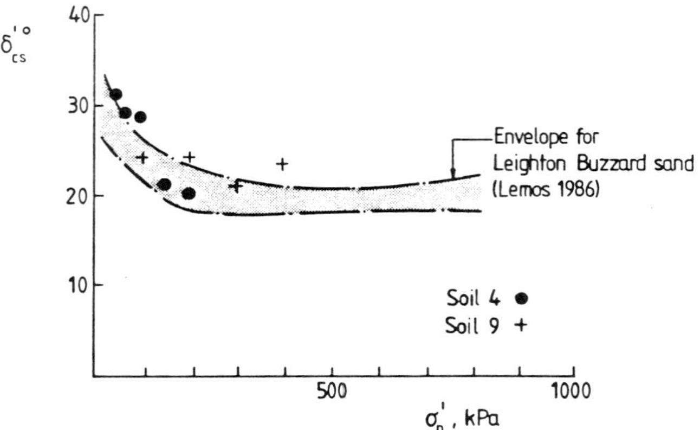

Fig. 12. Effect of stress level on δ'_{cs} for three sands.

for soil 4 and the right shear results for soil 9. A strong σ'_n influence is evident for Leighton Buzzard sand and soil 4 when tested at stresses below a threshold of 150 to 200 kPa, but the ring-shear experiments (on soil 9) show a much less pronounced dependence on σ'_n. Clearly, more research is required into this aspect of interface shear behaviour.

6. Pile Design in Sands and Silts

The preceding data have the following implications for pile design:

1. The Labenne and Bothkennar instrumented pile tests have shown that interface shear tests may be used to assess the operational values of δ'. At Labenne, dilation made a significant contribution to the shaft capacity and δ'_{cs} controlled local failure. The effects of dilation on the radial effective stresses will be smaller for large offshore piles and local failure may be associated with δ'_{peak} rather than δ'_{cs}. However, such piles will fail progressively because of their higher compressibility and the operational δ' value for assessing the overall shaft capacity will tend towards δ'_{cs}.

2. API RP2A suggests notional δ' values that increase with grain size and increase with relative density (Figure 1). However, measured δ'_{cs} values for 12 different granular soils were independent of D_r and tended to fall as D_{50} increased (Figure 11).

3. The important finding that the operational δ' is independent of D_r does not mean that the *overall shaft* capacities of piles in sand are independent of relative density. Whilst the API code suggests that $\sigma'_{rf}/\sigma'_{v0} = K_f$ is not affected by $_r$, the Labenne tests, and other data reviewed by Lehane (1992), show that

the ratio $K_c = \sigma'_{rc}/\sigma'_{v0}$, which quantifies the radial effective stresses set up by installing displacement piles, increases steeply with the initial undisturbed relative density (and also declines significantly with distance from the pile tip (see Lehane *et al*, 1993).

The API recommendations appear to allow for relative density by weighting the wrong parameter. However, if the methods are to be modified it is vital that alternative ways are found of predicting K_c accurately. Lehane (1992) proposes two ways of achieving this aim. If such approaches are accepted, the true operational values of δ' may be determined in relatively simple direct shear tests, provided that the care is taken to ensure that the interface materials, roughnesses and stress levels are compatible with those experienced in the field. Curves such as those given in Figure 11 might provide some guidance, but cannot be expected to be reliable when applied to soils which have less uniform gradings, different particle shapes (or minerals) or when the piles have different surface finishes or experience different stress levels. Site specific tests are therefore likely to be worthwhile in all cases.

7. Conclusions

Five conclusions follow directly from the experiments reported in this paper.

1. The controlling frictional parameter for piles in sands and silts is the critical state interface angle δ'_{cs}.

2. δ'_{cs} does not depend on relative density and, for a given interface roughness, reduces sharply as D_{50} increases.

3. For uniform soils a linear relationship exists between $\tan \delta'_{cs}$ and normalised interface roughness.

4. Direct interface shear tests are relatively simple to perform and should be incorporated into many more offshore site investigations.

5. Ring-shear tests in which the normal stiffness is controlled to follow Equation (2) may provide the best way of assessing the potential effects of dilation on drained pile loading.

Acknowledgements

The tests described above were performed in the Imperial College Soils laboratory in association with a number of projects, several of which were supported by the Science and Engineering Research Council (SERC). Thanks are due to SERC for this funding; the contributions made by Dr. L. Lemos and other present and former colleagues are acknowledged gratefully. The work at Labenne was supported by a consortium of SERC (through MTD Ltd.), Amoco (UK) Ltd., Conoco (UK) Ltd., Exxon, Mobil, Shell (UK), Health and Safety Executive (HSE), Building Research Establishment (BRE) and Laboratoires des Ponts et Chaussées.

References

1. Bond, A. J., Jardine, R. J., and Lehane, B. M. (1993), 'Factors affecting pile capacity in clay soils', *Proc. Int. Conf. on Offshore Site Investigations and Foundation Behaviour*, SUT, London, September 1992.
2. Boulon, M. and Nova, R. (1990), 'Modelling of soil-structure interface behaviour. A comparison between elastoplastic and rate type laws', *Computers and Geomechanics* **9**, 21–46.
3. Boulon, M. and Foray, P. (1986), 'Physical and numerical simulation of lateral shaft frictions along offshore piles in sand', *Proc. 3rd Int. Conf. on Numerical Methods in Offshore Piling*, Nantes, Editions Technip, pp. 127–147.
4. Briaud, J. L. and Tucker, L. M. (1988), 'Measured and predicted axial response of 98 piles', *J. Geotech. Eng. Div., ASCE* **114**(9), 984–1002.
5. Everton, S. J. (1991), 'Experimental Study of Frictional Shearing Resistance between Non-Cohesive Soils and Construction Materials', MSc Dissertation, Univ. of London (Imperial College).
6. Hettler, A. (1982), 'Approximation formulae for piles under tension', *Proc. IUTAM Conf. on Deformation and Failure of Granular Materials*, Delft, pp. 603–608.
7. Jardine, R. J. and Christoulas, S. (1991), 'Recent developments in defining and measuring static piling parameters', Gen. Report., *Int. Conf. on Deep Foundations*, Paris, Presse de l'École de Ponts et Chaussées, pp. 713–746.
8. Jardine, R. J., Bond, A. J., and Lehane, B. M. (1992), 'Field experiments with instrumented piles in clays and sands', *Piling: European Practice and Worldwide Trends*, ICE, London, pp. 59, 66.
9. Johnston, I. W., Lam, T. S. K., and Williams, A. F. (1987), 'Constant normal stiffness direct shear testing for socketed pile design in weak rock', *Géotechnique* **37**(1), 83–89.
10. Kishida, H. and Uesugi, M. (1987), 'Tests of the interface between sand and steel in the simple shear apparatus', *Géotechnique* **37**(1), 45–52.
11. Lebègue, M. Y. (1964), 'Étude expérimental des efforts d'arrachage et de frottement négatif sur les pieux en milieu pulvérant', *Annales de l'Institut du Bâtiment et de Travaux Publics, Série Sols et Fondations* **17**, 199–200.
12. Lehane, B. M. (1992), 'Experimental Investigations of Pile Behaviour Using Instrumented Piles', PhD Thesis, Univ. of London (Imperial College).
13. Lehane, B. M., Jardine, R. J., Bond, A. J., and Frank, R. (1993), 'Mechanisms of shaft friction in sands from instrumented pile tests', *J. Geotech. Eng., ASCE*, Jan. 1993.
14. Lehane, B. M. and Jardine, R. J. (1992a), 'On the residual strength of Bothkennar clay', *Géotechnique* **42**(2), 363–368.
15. Lehane, B. M. and Jardine, R. J. (1992b), 'The behaviour of a displacement pile in Bothkennar clay', *Proc. Wroth Memorial Symposium on Predictive Soil Mechanics*, Oxford (in press).
16. Lemos, L. J. (1986), 'The Effects of Rate on the Residual Strength of Soil', PhD Thesis, Univ. of London (Imperial College).
17. Lupini, J. F., Skinner, A. E., and Vaughan, P. R. (1981), 'The drained residual strength of cohesive soils', *Géotechnique* **31**(2), 181–213.
18. Potyondy, J. G. (1961), 'Skin friction between various soils and construction materials', *Géotechnique* bf 11(4), 229–353.
19. Poulos, H. G. (1989), 'Pile Behaviour – Theory and Application', Rankine Lecture, *Géotechnique* **34**(2), 365–415.

20. Ridley, A. M. and Jardine, R. J. (1992), 'Ring-Shear Interface Tests on North Sea Clays and Sand', Internal Report, Dept. of Civil Eng., Imperial College.
21. Tika-Vassilikos, T. (1991), 'Clay-on-steel ring shear tests and their implications for displacement piles', *Geotech. Testing J., ASTM* **14**(4), 457–463.
22. Uesugi, M. and Kishida, H. (1986a), 'Influential factors of friction between steel and dry sands', *Soils and Foundations* **26**(2), 33–46.
23. Uesugi, M. and Kishida, H. (1986b), 'Frictional resistance at yield between dry sand and mild steel', *Soils and Foundations* **26**(4), 139–149.
24. Yoshima, Y. and Kishida, H. (1981), 'A ring torsion machine for evaluating friction between soil and material surfaces', *Geotech. Testing J., ASTM* **4**(4), 145- -152.

SESSION 8

DESIGN CRITERIA

OFFSHORE EXPERIENCE WITH LATERALLY LOADED PILES

R. MARTIN
Anchortech Ltd, Stena House, Aberdeen AB32 6TQ

and

E. BURLEY
Queen Mary & Westfield College, Mile End Road, London E1 4NS

Abstract. The soil reaction for laterally loaded piles is generally determined by the methods described in API Recommended Practice 2A. The number of offshore platforms currently installed worldwide is testimony to the success of these methods. These design methods are based, for reasons of cost, on small diameter piles. Large diameter piles are rarely instrumented for lateral capacity, and such data as are available from jacket piles are subject to axial as well as lateral loading. Pile dead- man anchors used for pipeline initiation are subject to lateral loads – depending on pipe size – in the range 10–100 tonnes, and constitute a data source for confirmation of design data at high loading. This is particularly true for short piles where the pile toe is allowed to move so that any permanent set will be a function of soil plasticity rather than pile elasticity. This Paper presents a comparison between theoretical and observed pile performance under these conditions.

1. Introduction

The load-displacement behaviour of laterally loaded piles is commonly predicted using the methods described in American Petroleum Institute Recommended Practice 2a [1]. These consist of modelling the soil as a series of non-linear springs along the pile's length, the spring response taking the form of P-Y curves of soil resistance P against soil deflection Y (Figure 1).

The geometry of the P-Y curves is developed by different methods, depending on whether the soil is soft clay, stiff clay, or sand. The curve is also affected by soil strength parameters, depth below mudline, and by the diameter of the pile.

The effectiveness of these methods is demonstrated by the number of offshore platforms safely operating worldwide, although bearing capacity rather than lateral restraint is generally the overriding criterion for the design of jacket piles. This results in a pile that is longer than would be required for lateral capacity, along with the lateral load-displacement behaviour being more a function of the elastic properties of the pile embedded in the undisturbed lower soil strata than of the P-Y curves themselves.

The problem of monitoring laterally loaded piles at large scale is not easily solved. The behaviour of jacket piles under lateral loading is difficult to isolate

Volume 28: Offshore Site Investigation and Foundation Behaviour, 681–690, 1993.

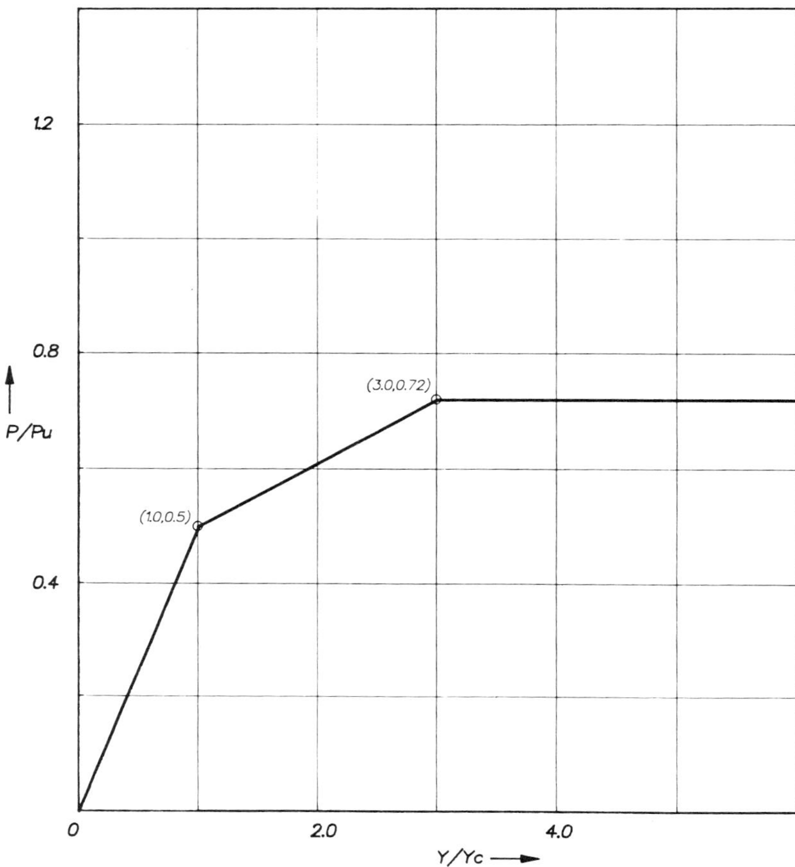

Fig. 1. General form of P-Y curve for soft clay.

from the effects of axial loading, not to mention the difficulties of measuring the applied lateral load (which tends to be due to environmental loading and of short duration), the percentage of this loading actually applied to the piles, and how to measure the displacement within the soil.

It is theoretically possible to observe lateral pile response in isolation by monitoring the effect of lateral loading from mooring chains on anchor piles. In practice this is not so simple to achieve as the loading padeye on the pile is generally some distance below the mudline, so the load experienced by the pile will contain some uplift due to the "inverse catenary" of the chain within the ground. Moreover the head of the pile is frequently driven below the mudline to avoid a potential snagging point on the seabed, so while the applied load in the anchor line may be known with some accuracy, the movement of the pile head cannot be seen. This problem is exacerbated by the anchor pile being hundreds of metres away from the point of

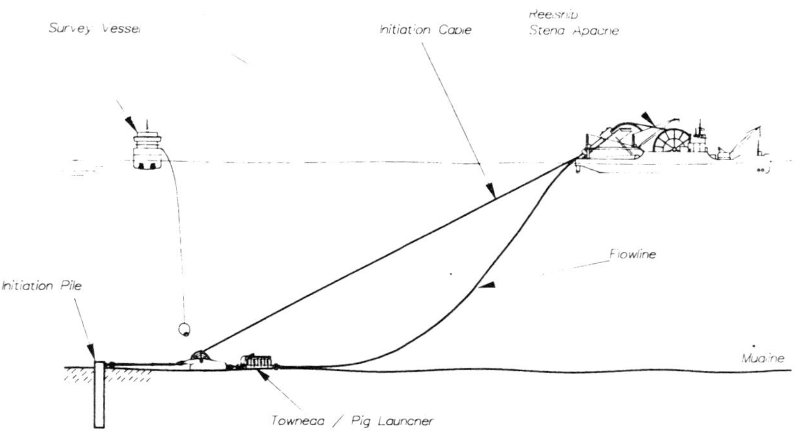

Fig. 2. General arrangement of piled flowline lay initiation.

load application on the vessel, without any local reference datum.

It is therefore not surprising that the most accurate experimental data results from scale model tests in controlled conditions, while little opportunity for confirming predictions at full scale exists. Large scale experiments can be performed on dry land, but the results then have to be adjusted for underwater effects unless the water table is very close to the surface. Moreover certain types of soil – such as very soft unconsolidated marine clays – simply do not exist in a dry environment.

Very soft clays occur in many offshore areas worldwide, notably in much of the central North Sea basin, and constitute difficult anchoring and foundation conditions. The need for more economic methods of foundation in these conditions is therefore combined with logistical difficulty in performing any large-scale tests upon which some improvement could be based.

This Paper describes the design for a laterally-loaded pipeline initiation pile and observations during pipelay operations offshore as part of Amerada Hess Limited's Scott Project.

2. Description of Project

The pipelay operations were performed from the reelship "Stena Apache" (Figure 2), as a part of the installation of infield flowlines for Amerada Hess Limited's Scott Field Development. These flowlines were 8" and 10" dia steel pipe, the ends of which were required to be laid precisely within a target box 3m × 2m on the seabed. This required accuracy and the very soft clay seabed were among the reasons for selecting the pile method of pipeline startup.

Anchortech Piles were used for the flowline initiation. These were short wide piles with a triangular "A-shaped' section (Figure 3). They were designed to restrain

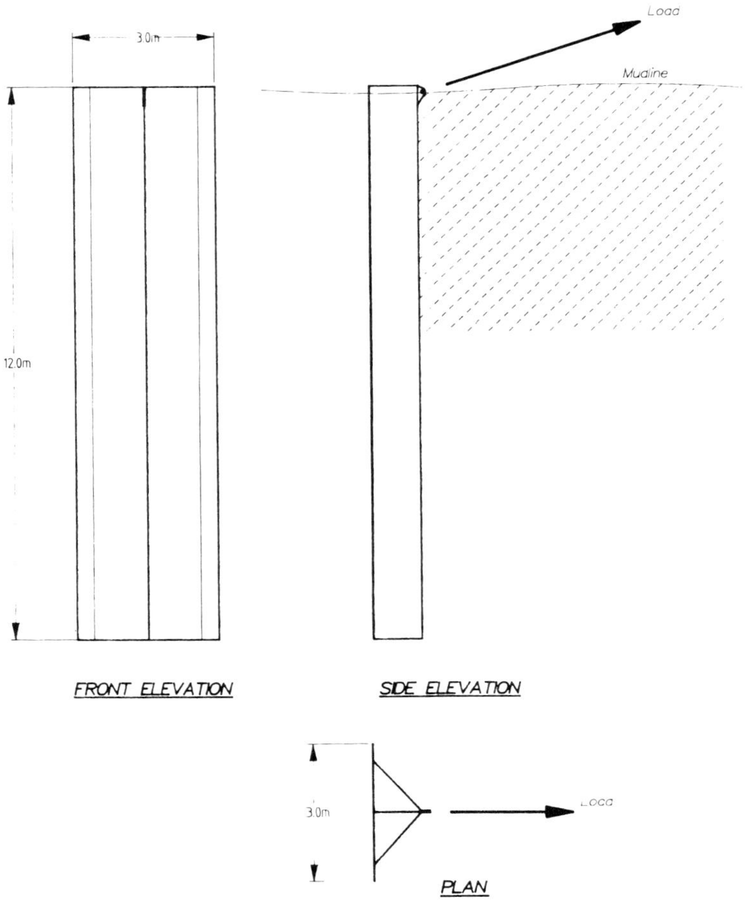

Fig. 3. General arrangement of Anchortech flowline initiation pile.

lateral loads of 210kN for the 8" flowlines, and 289kN for the 10" flowlines. The loading padeye on the pile was situated at the mudline. A 20% uplift component (i.e. 42kN and 58kN respectively) was also to be restrained, but this would be accommodated by the self-weight of the pile itself. The piles were designed to a common cross-section 3m wide, with penetration below mudline of 10m for the 8" flowlines, and 12m for the 10" flowlines.

2.1. Pile Design

The piles were designed using P-Y curves generated by the API-RP2A method:

$$\begin{array}{lcccc} P/P_u: & 0 & 0.5 & 0.72 & 0.72X/X_R \\ Y/Y_c: & 0 & 1.0 & 3.0 & 15.0 \end{array}$$

where

$$P_u = 3c + \mu X + JcX/D, \quad \leq 9c.$$
$$Y_c = 2.5E_cD$$

where

Undrained shear strength	c	=	varies with depth X as Figure 4
Submerged density	μ	=	$7kN/m^3$
Dimensionless constant	J	=	0.5
Pile "diameter" (width)	D	=	$3.0m$
50% strain	E_c	=	0.02.

Note that the early part of the P-Y curve defined above is a linearised form of the equation:

$$(2P/P_u)^3 = Y/Y_c.$$

Since the first point on the API linearised P-Y curve occurs at 150mm displacement for a 3m wide pile, additional points were included on the curve at lower displacement (Figure 4). This has the effect of "stiffening" the pile response at small deflections to fit the P-Y cubic equation more closely. These additional points were as follows:

P/P_u :	0.17	0.34
Y/Y_c :	0.039	0.314

The piles were designed to a "design profile" of shear strength with depths as illustrated in Figure 5. The 12m long pile was also back- analysed against the upper bound shear strength profile (Figure 5); this profile is notable for its similarity to the design profile in the top 8m below mudline, indicating very uniform conditions in this region.

Load-deflection curves for the pile padeye produced by these analyses are illustrated in Figure 6. Note that these curves are stiffer (i.e. deflections are smaller) than would have been obtained by using the unmodified API method.

These predicted head displacements for the 12m pile correspond to rotational movement about a point 9m–10m below the seabed as shown in Figure 7. This also illustrates the potential for void formation around the pile.

2.2. Observations Offshore

The piles were driven to full 10m/12m penetration using an ICE 1412 vibratory hammer. In the very soft soil conditions, the piles penetrated the seabed by up to 8m under self-weight with the hammer, so very little driving was required to achieve full penetration. The piles were positioned by diver on a pre-positioned target to

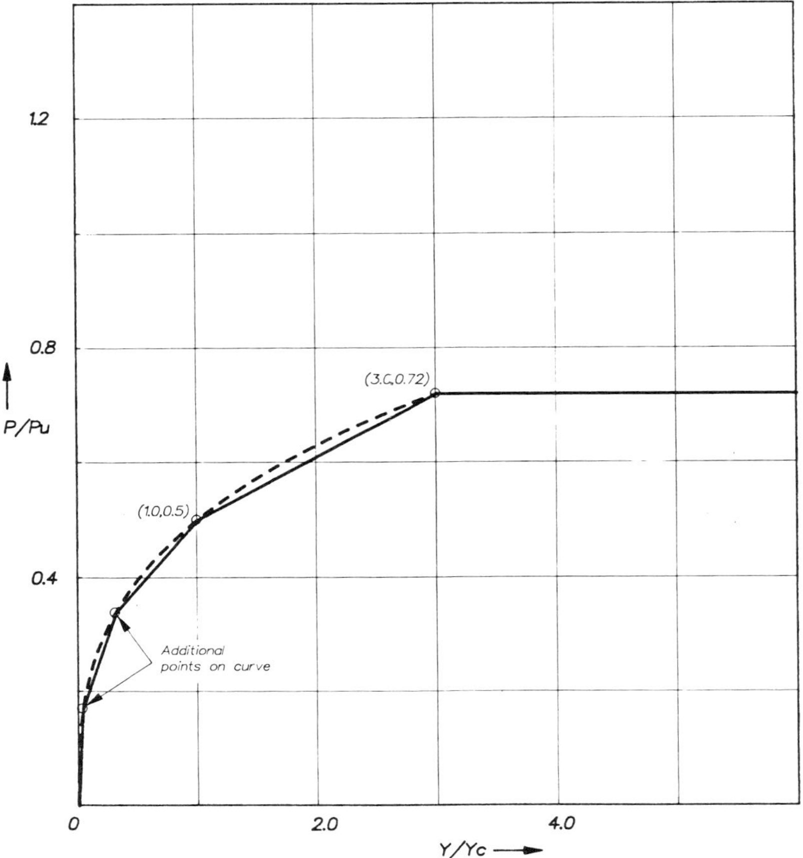

Fig. 4. Additional points on P-Y curve for soft clay.

ensure accuracy of the flowlines within the target box. No pile guide frame was used.

On completion of piling, the piles were connected to the flowline initiation rigging and surveyed. This rigging was then picked up by the reelship "Stena Apache" which performed a test load on the pile before commencing pipelay operations.

The piles performed better than predicted by the design throughout the pipelay operations, with no significant movement being observed, and the flowline initiation heads being installed within the target boxes. On one occasion in particular, a 12m pile was proof-loaded at twice the design load, with a force of 600kN being repeatedly applied by the vessel over a 1/2-hour period. This loading translated into a 550kN lateral load being applied at the head of the pile. This pile was re-surveyed before commencing pipelay operations, and no observable movement

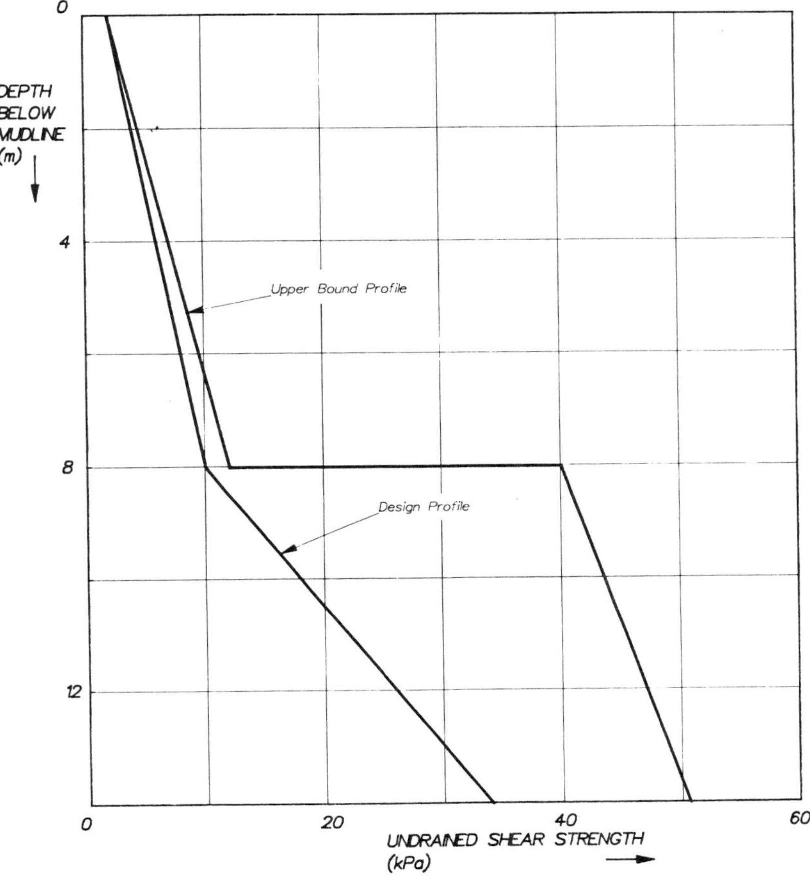

Fig. 5. Variation of undrained shear strength with depth.

was found. The resolution of the ROV equipment was such that the pile movement could not have been more than 0.1m, and was probably significantly less.

On completion of pipelay operations, the initiation rigging was disconnected by diver, and the piles were recovered with a direct vertical pull of over 500kN.

2.3. Results

The observed lateral deflections at the pile head were much smaller than predicted by the design. The design analyses predict about 0.2m movement under the design load, and about 1.0m peak movement, as shown in Figure 6; the effect of this on a 12m pile is illustrated in Figure 7. It can be seen that the pile body is relatively stiff, so that the entire pile rotates in the soil with little potential for recovery due to pile elasticity and the pile toe being fixed. It is therefore clear that the residual

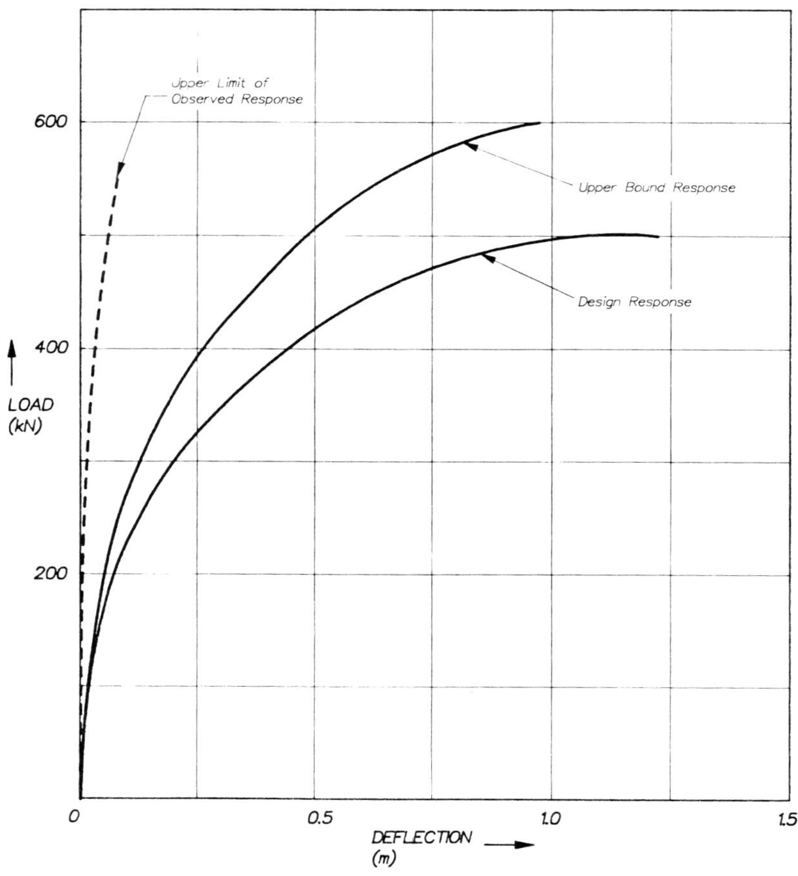

Fig. 6. Pile load/deflection behaviour – Design and Observed

pile displacement on removal of the load would be similar to that of the pile under load, which could not be observed for operational reasons.

The seabed characteristics were such that, although soft, surface deformations were slow to backfill and could be observed days afterwards. This was particularly noticeable after trenching operations in the same area. It is therefore unlikely that the piles would have moved significantly and backfilled so that such movement could not be detected. Furthermore the vertical extraction loading was consistent with a calculated vertical capacity assuming no loss of external adhesion due to lateral movement.

The upper limit of pile deflection as observed – 0.1 m – is shown with the design load-deflection curves in Figure 6. This represents about 10% of the predicted deflection. The following factors may be considered in relation to this discrepancy:

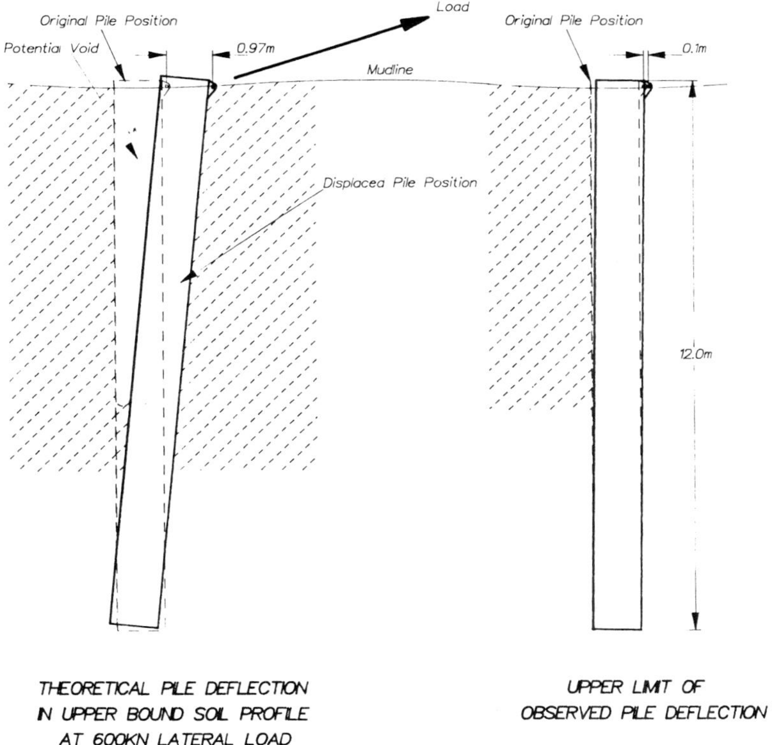

Fig. 7. Initiation pile deflections – Theoretical and Observed.

- Design soil shear strength too low. This is considered unlikely due to the extent of soil testing in the field, and the close agreement between design and upper bound profiles shown in Figure 5. Increased shear strength below 8m has little effect.

- Design soil strain parameter – 50% strain E_c – too high. The design value of 0.02 is less than often recommended for very soft clays, but with the benefit of hindsight a value between 0.01 and 0.02 might have been more appropriate. A value of 0.01 throughout – which would have halved predicted deflections – is not justified by the soil data.

- The use of the unmodified API method without additional points to "stiffen" the P-Y curve would result in still larger predicted deflections.

- The use of cyclic loading P-Y parameters – as applied for oscillating loads on jacket piles – may be too conservative for pulsating loads in a single direction as applied to the flowline initiation pile. However by the API method this only increases holding power at displacements greater than 0.45m for the 3m pile, much larger than anything observed in this context.

- Pile diameter D taken as pile width for the triangular pile. This simple assumption seems reasonable except that no account is taken of shape and the formation of a soil wedge in front of the pile. Since the entire face of the triangular pile is at more than 45° to the direction of load (cf. 71% for a tubular pile) and particularly the extremities are perpendicular to the load, this may be too simplistic. Related to this is the anomaly with the API method that at small loads a wide pile moves further than a narrow pile of the same length.

- Work carried out by Abbassian [2], investigating the behaviour of piles with enlarged heads subjected to lateral loading, had uncovered a similar discrepancy between theoretical and measured values of deflection, for piles in sand. This work suggested that in a P-Y analysis, the relationship between soil deflection and pile diameter should be $Y \propto D^{0.2}$ rather than $Y \alpha D$, to give a closer fit of results over the range of sizes detailed in the literature. The results observed offshore also strongly suggest that deflection is not directly proportional to diameter, but perhaps to some similar lower power of diameter.

3. Conclusions

The observations from flowline initiation piles installed in very soft clay and subjected to lateral loading offshore indicate that the predictions based on the method described in API RP2A are very conservative, with observed deflections 10% of those predicted. Ultimate capacity is probably also conservative, although not necessarily by the same order.

While the tests on which these observations have been based are by no means exhaustive, they do indicate a scope for further offshore testing of large piles under lateral loading, subject always to the logistical difficulties involved.

Acknowledgement

The authors wish to thank Amerada Hess Limited for making this publication possible.

References

1. American Petroleum Institute (1991), 'Recommended Practice for Planning, Designing and Constructing Fixed Offshore Platforms', Recommended Practice 2A, 19th edition, Washington, DC 20005.
2. Abbassian, F. (1988), 'Behaviour of Simple and Compound Piles in Saturated Sand under Horizontal and Oblique Pull', PhD Thesis, University of London.

ANALYSIS OF LONG TERM JACK-UP RIG FOUNDATION PERFORMANCE

DEREK W. F. SENNER
Derek W. F. Senner Ltd., 69 High Street, Bideford, Devon EX39 2AA, U.K.

Abstract. Jack-up rigs are now being considered for use in deeper waters and more hostile environments, and for long term functions in hydrocarbon development projects. Much work has been undertaken over the past few years to better understand the performance of jack-up rig foundations. This paper discusses some of the important aspects considered when evaluating long-term jack-up rig stability.

1. Introduction

Prediction of jack-up rig footing behaviour is an interesting challenge for geotechnical engineers. Initially, footings are proof loaded by preloading, for which a prediction of load-penetration is required. During the jack-up rig's time on location, footings need to carry combinations of vertical and horizontal loading with tolerable displacements. This can generally be assured if preloading exceeds the design loading effect. However, in deep, hostile waters it may not be possible to prove the foundations. In such cases foundation stability analysis is required to assess foundation adequacy.

Still the most sensitive aspect of foundation behaviour is footing penetration in layered soils, with punch-through remaining a major hazard. Procedures for assessing punch-through are presented and may be used to evaluate footing performance during preloading.

Jack-up rig foundation fixity can have a significant impact on predicted structural performance. The evaluation of foundation fixity is considered in the light of recent published research work by model testing, and structure monitoring is discussed. An approach to stability analysis for footings in uniform soils developed by Schotman and Efthymiou (1989) is described. In this partial factor approach, foundation adequacy is achieved if the loading is within the vertical-horizontal load failure envelope. For loading beyond this yield surface, footing displacement will occur. In complex soil conditions where large displacements could occur foundation performance should be considered using a permissible stress approach.

In the concluding sections of the paper other important design considerations are addressed. These include scour effects, the influence of jack-up rig footings on adjacent pile foundations and cyclic loading. Some recommendations are given for future research into footing performance.

Volume 28: Offshore Site Investigation and Foundation Behaviour, 691–716, 1993.

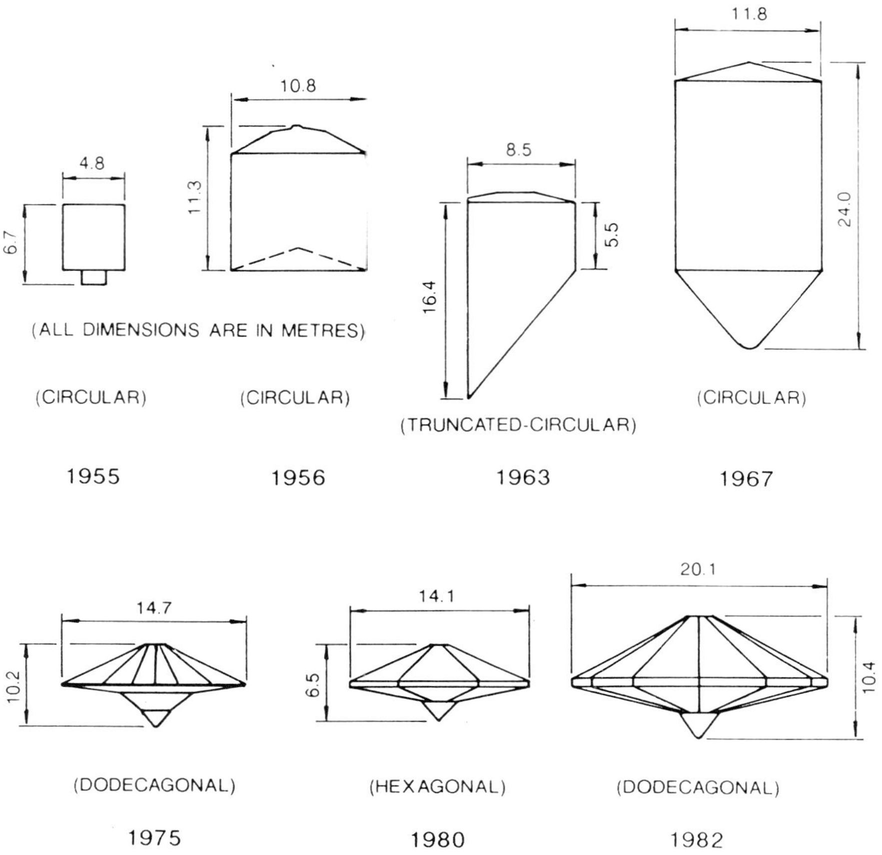

Fig. 1. Development of jack-up rig footings (spud cans).

2. Background

2.1. The Development of Jack-Up Rigs

Seabed-supported rigs have been used for hydrocarbon development projects since about 1949. In 1951 footings, termed "spud cans", were put on to the leg of Offshore Company's Rig 51 to reduce leg penetration. Initially jack-up rigs had 8 to 12 legs. However, in 1955 R. G. Le Tourneau Company designed and built the three-leg rig "Scorpion" for Zapata Off-shore. This was a significant development, and now most jack-up rigs used offshore are the three-leg variety. Many shapes and sizes of spud cans have been used to meet specific performance needs. Bearing areas have tended to increase with loads to keep bearing pressures similar. Figure 1 shows the development of spud cans. Now most jack-up rigs use the near circular pointed ("flying saucer") footings shown at the bottom of the figure.

2.2. Footing Loads

Jack-up rig footings are subjected to both gravity and environmental loads. Typically about 80% of the gravity load is from the weight of the rig and its permanent equipment, termed "Operational Light Ship Load'. The remaining gravity load, termed "Variable" is from movable supplies such as drill pipe, fluids in holding tanks etc. Gravity loads for older rigs may be as much as 10% different to those assumed due to new equipment, modifications etc. The three components of environmental loads are wave, wind and current, and these typically produce about 55 to 65%, 25 to 25% and 10% of the lateral load, respectively. During a storm the overturning moment due to environmental forces may increase the vertical footing load by more than 50% of the gravity load, and the load inclination may approach one-third.

2.3. Preloading

When a jack-up rig moves onto location the hull is lifted out of the water and additional loading, termed preload, is applied to the footings. Typically ballast preloading is undertaken whereby seawater is pumped into holding tanks. For jack-up rigs with four legs or more, mechanical preloading of one or a pair of legs is achieved by using the rig as a deadweight reaction.

The purpose of preloading is to proof load the footings to greater than the expected maximum loading effect. This can generally be achieved in shallow, relatively calm environments. However in deep water, hostile environments the maximum preload available may be insufficient to prove the foundation for the combined vertical and horizontal extreme storm loading. Therefore, a stability analysis needs to be performed to assess the consequences of footing loads exceeding the preload, as discussed later in the paper.

2.4. Accident Record

McClelland, Young and Remmes (1981) reported that up to about 2.5% of the jack-up fleet experienced major accidents each year during the period 1970 to 1980. Leg accidents, which includes foundation failures, were the most significant proportion. In a comprehensive review of jack- up "mishaps" from 1979 to 1988, Sharples, Bennett and Trickey (1989) report that foundation related problems are the most frequent single category. The proportion of foundation problem types are given in Table 1 below.

2.5. Leg Design Optimisation

Examination of jack-up rig accident data by Sharples *et al* (1989) indicates that there may be risk trade-off in the design. Slender legs may reduce wave loading, however, the structural resistance to rapid leg penetration may also be reduced.

TABLE 1.
Causes of foundation problems. (Data from Sharples, Bennett and Trickey, 1989).

CAUSE OF FOUNDATION PROBLEM	PROPORTION
Punch Through	70%
Severe Storm Loading	16%
Scour Around Footings	5%
Interaction with Previous Footprints	3%
Others	6%

There appears to be a trend of increasing the ratio of hull-weight to leg-strength for more modern rigs, resulting in a reduced ability to withstand footing movements. With the move towards more optimisation in design, it is imperative that detailed assessments of foundation behaviour be undertaken, and the impact on structural performance determined.

2.6. Accident Avoidance

Foundation related jack-up rig problems are avoidable. It is quite frankly amazing that after two decades with major accidents occurring annually, jack-up rigs are still moved to location where little or no soil data are available and where there has been no input from experienced geotechnical engineers.

McClelland, Young and Remmes (1981) and Young, Remmes and Meyer (1984) provide comprehensive descriptions of foundation problems, and discuss ways by which such problems can be avoided. Rappoport and Alford (1989) provide some useful ideas for preloading procedures at difficult sites, including the way tidal variation can be used to advantage.

A detailed discussion of jack-up rig foundation problems is beyond the scope of this paper. However, as shown in Section 2.4, the major causes of foundation problems are known, and it is essential that the risk of these problems occurring is thoroughly evaluated prior to a jack-up rig being sited.

3. Footing Penetration Analysis

3.1. Purpose of Analysis

Analyses should be undertaken of the footing behaviour during preloading for two main purposes: (i) to assess footing preparation at full preload, and (ii) to assess the likelihood of a sudden penetration of one or more of the footings, termed "punch-through". Such predictions enable the rig operator to evaluate both rig suitability for a particular site, and appropriate installation procedures.

Such analyses are intended to provide a realistic estimate of footing performance, not a conservative assessment of bearing capacity.

3.2. Uncertainties

There are uncertainties in predicting footing penetration, including: (a) soil variability, (b) inaccuracies in applied loading, (c) rate of footing penetration, and (d) application of simple theories to complex footing configurations and soil conditions. It is therefore important to use actual rig performance data, rather than to rely solely on theoretical procedures. Performance data should be used to calibrate theory.

3.3. Footing Penetration in Clay

Gemeinhardt and Focht (1970) present results of footing penetrations at 120 locations in the Gulf of Mexico. This database was reviewed and extended in the mid 1980's. The fit between observed and predicted penetrations is good considering the range of footing types and penetrations.

In clays the bearing capacity of a footing, Q at any depth can be assessed using the method proposed by Skempton (1951), viz:

$$Q = S_u N_c A + \gamma' V \tag{1}$$

where:

S_u = undrained shear strength
N_c = bearing capacity factor
A = bearing area
γ' = submerged unit weight of soil
V = soil volume displaced by footing.

If the shear strength varies by $\pm$ 50% or more of the average value over two-thirds of the diameter beneath the foundation level, Equation (1) is not valid. For such conditions the bearing capacity of a layered system should be used.

For a circular (or approximately circular) footing, the bearing capacity factor, N_c, is given by:

$$N_c = 6[1 + 0.2D/B] \geq 9 \tag{2}$$

where;

D = depth to widest part of footing in contact with soil
B = equivalent diameter of footing.

3.3.1. *Undrained shear strength*

Many factors can affect measured undrained shear strength, including: sampling method, sample size, test type, stress history, soil structure and sensitivity. Sampling

TABLE 2.
Undrained shear strength.

Sample Type	Sample Size	Soil Consistency	Strength Test Type
Driven	50mm	Soft/Firm	Mini Vane (1)
Pushed	75mm	Firm/Stiff	UU Triaxial (2)
Driven	50mm	Firm/V. Stiff	UU Triaxial (3)
Pushed	75mm	Stiff/V. Stiff	UU Triaxial (2)

Notes: (1) mean to upper bound
(2) mean
(3) mean to upper bound

method can have a particularly marked effect on undrained shear strength in weak normally or underconsolidated clays (Quiros, Young, Pelletier and Chang, 1983), the soils in which deep footing penetrations can occur. In stiff clays, where footing penetration is likely to be limited, although sampling method can affect soil strength (Semple and Rigden, 1983), the magnitude of strength differences will probably be insignificant for penetration analysis.

Table 2 gives broad recommendations for the type of shear strength to be used to be consistent with the empirical bearing capacity procedure described herein. Where possible, performance data from other jack-up rigs in similar soil should be used to calibrate the analysis. Care should also be taken when dealing with sensitive or gaseous soils.

3.3.2. Pointed footings

For pointed footings, field observations have shown that the procedure given by Equations (1) and (2) give reasonable predictions if a modification for apparent foundation depth, D', is made. As shown on Figure 2, D' is given by:

$$D' = D + V_p/A \tag{3}$$

where:

$V_p =$ volume of footing below D.

Osborne, Trickey, Houlsby and James (1991) present theoretically calculated bearing capacity factors for conical footings on clay (Houlsby, 1982). The bearing capacity factor is a function of cone angle β and a surface adhesion (roughness) factor α. Osborne *et al* (1991) indicate that the estimated value of α ranges from 0.5 to 1, and tends to 1 for weaker clays, where footing penetration is likely to occur. Craig and Higham (1985) report that cone angle has no appreciable effect on bearing capacity in clay.

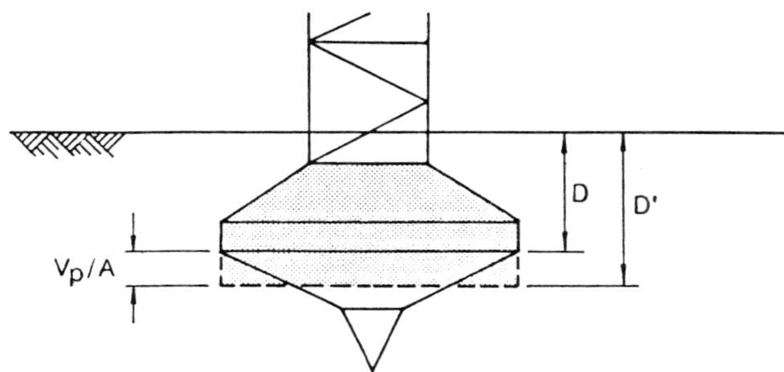

Fig. 2. Apparent foundation depth.

Figure 3 shows a comparison between the theoretical bearing capacity factors (Osborne *et al*, 1991) and equivalent factors obtained using the apparent foundation depth approach. From simple geometry the increase in depth from the actual to the apparent foundation ($d = D' - D$) is given by:

$$d = B\frac{1}{6\tan(\beta/2)}. \tag{4}$$

Figure 3, for a surface footing, shows the value of bearing capacity factor obtained when d is substituted into Equation (2).

For most footings the cone angle β is greater than 120° and for typical jack-up rig footing bearing stresses, significant penetration is only likely to occur for soft to firm clays with undrained shear strength less than about 50 kPa. For these practical ranges the theoretical and apparent foundation depth approach give similar results. The apparent foundation depth approach clearly has the advantages that: (i) it is based on field observations, (ii) complex pointed shapes can be readily handled, (iii) surface roughness does not have to be estimated since this is implicitly in the database for steel spud cans, and (iv) it is easy to apply.

3.3.3. *Infilling above footing*

For deeply penetrating footings where the spud can is totally below the seafloor, whether or not infilling occurs above the spud can may significantly affect the bearing capacity. An assumption in the database for footing penetration performance is that only the spud can volume is displaced during preloading. There are uncertainties regarding the mechanism of infilling, but there are no reported cases of foundation failure or punch-through due to infilling. Deep penetrations will only occur in weak soils where it is likely that soils will flow over the top of a footing during jacking.

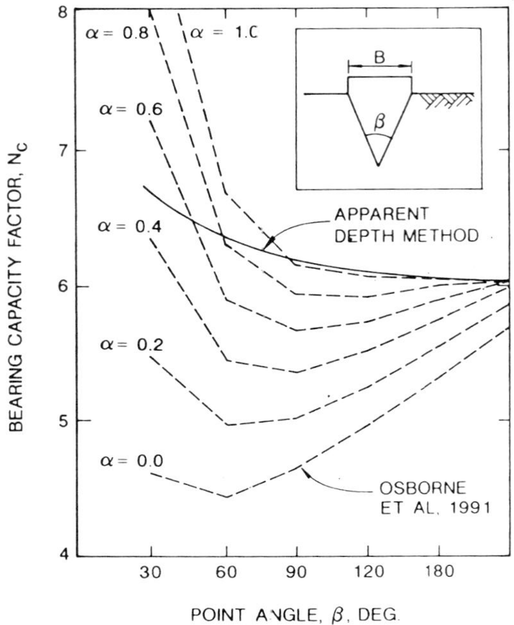

Fig. 3. Bearing capacity factor using apparent depth method.

3.3.4. Bearing capacity of weak clays

The bearing capacity of weak clays is important not only for assessing footing penetration during preloading but also for evaluating the likelihood of punch-through. The procedure given above for determining bearing capacity although empirical does include several important factors including: (a) footing depth, (b) footing shape and (c) the likelihood of soil infilling above deeply penetrating footings.

3.4. Footing Penetration in Sand

For an approximately circular footing resting on sand the gross bearing capacity, Q, is given by:

$$Q = A[0.3\gamma_1' B N_\gamma + \gamma_2' D(N_q - 1)] + \gamma_2' V \quad (5)$$

where:

γ_1' = submerged unit weight of soil to $B/2$ below foundation
γ_2' = submerged unit weight of soil above foundation
N_q, N_γ = bearing capacity factors.

The bearing capacity factors N_q and N_γ are functions of the soils angle of internal friction, ϕ, as shown in Table 3.

TABLE 3.
Bearing capacity factors (after Vesic, 1975).

ϕ	N_q	N_γ
20	6.4	5.4
25	10.7	10.9
30	18.4	22.4
35	33.3	48.0
40	64.2	109.0

TABLE 4.
Friction angles for bearing capacity analyses.

Analysis	Friction Angle
Minimum Footing Penetration	ϕ
Likely Footing Penetration	$\phi - 5°$
Bearing Capacity of Sand Layer	$\phi - 5°$

For silica sands with $\phi > 25$ to 20° most footings will not penetrate beyond the widest cross-section during preloading. A lack of penetration may be significant for long term stability, as discussed later.

For typical "flying saucer" spud cans the bearing area increases rapidly for little additional penetration before the widest part of the can contacts the seafloor. Slight inaccuracies in the measured penetration can lead to appreciable differences in inferred friction angle. It is to be expected that for typical footings of 10 to 15m diameter, bearing capacity will be less than predicted from theory because of scale effects. There is some evidence to support this expectation.

The angle of internal friction, ϕ, is the value obtained from drained triaxial tests performed at the appropriate relative density and stress level. To reflect uncertainties with respect to bearing capacity, Table 4 gives recommendations for friction angles.

3.5. Punch-Through

Where a footing rests on a strong soil of limited thickness overlying a relatively weaker soil there is the possibility that during preloading or storm loading the underlying soil will become overstressed and the footing will penetrate completely through the strong soil. This type of footing penetration is termed punch-through. Because of soil variability and differences in leg loads, punch-through will typically only affect one leg, leading to rig tilting and possibly damage to the punching leg.

Where the potential for punch-through exists during preloading the hull is generally kept close to the water. A 1.5m air gap is widely used so that should a

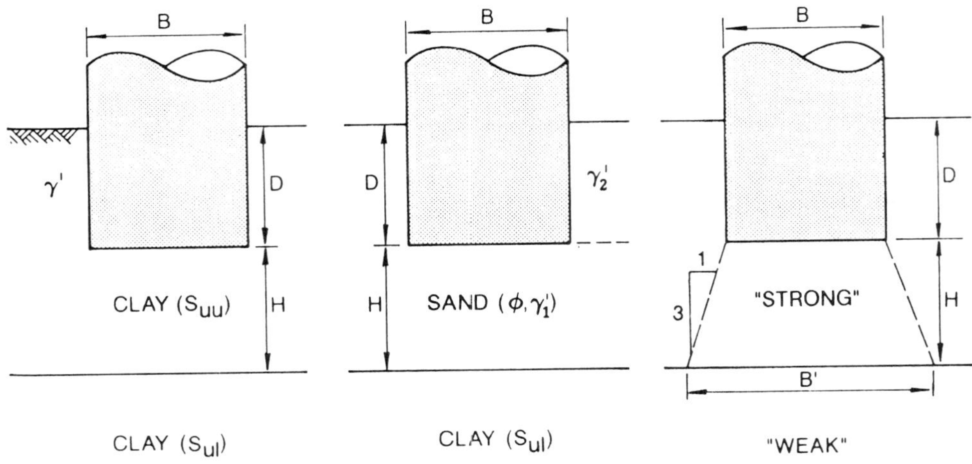

Fig. 4. Bearing capacity in layered soils.

leg suddenly plunge, the buoyancy of the hull will reduce the punching leg load.

Punch-through is the main foundation concern for jack-up rigs, as indicated by the accident data discussed earlier. The remainder of this section deals with basic methods for assessing punch-through potential. Typically, punch-through is associated with layered soil conditions. Such layering can be complex and highly variable, and a combination of methods will be required to assess likely footing performance. Considerable judgement will be required when applying the following procedures to layered soil conditions.

3.5.1. Strong clay overlying weak clay

Brown and Meyerhof (1969) published results from model tests on circular footings bearing on layered clays. They proposed a modified bearing capacity factor, N_m, to be applied to the upper clay to represent the effect of layering. The value of N_m is a function of both the ratio of footing width to upper layer thickness (B/H) and the ratio of upper to lower layer undrained shear strengths, S_{uu}/S_{u1}. Figure 4a represents the layered condition considered.

The gross bearing capacity for a footing resting at the surface on a strong upper clay overlying a weak lower clay is:

$$Q = S_{uu} N_m A + \gamma' V. \tag{6}$$

Although Brown and Meyerhof (1969) do not include for footing at depth in their work, experience indicates that the bearing capacity factor can be further modified, N_{md}, to account for depth using:

$$N_{md} = N_m(1 + 0.2\frac{D}{B}) \geq 9. \tag{7}$$

3.5.2. Sand overlying clay

Hanna (1978) undertook model tests of strip footings on dense sand above soft clay. The results from these tests were compared with a theoretical analysis of a truncated cone of sand punching into the clay (Hanna and Meyerhof, 1980). In their theoretical analysis the influence of the weak clay is represented by the earth pressure coefficient of punching shear, K_s, that acts on the failure planes around the punching sand plug. K_s reduces with undrained shear strength of the underlying clay.

For a footing resting on the sand, as shown in Figure 4b, the gross bearing capacity is:

$$Q = A[6S_{u1} + \frac{2H}{B} K_s \tan\phi(H\gamma_1' + 2\gamma_2' DH0] + \gamma'/V. \tag{8}$$

The coefficient K_s is a function of the bearing capacity ratio of the sand and clay, the function being reasonably linear for a given ϕ value. A reasonable approximation for K_s may be obtained from:

$$K_s \tan\phi = \lambda S_{u1}/B \tag{9}$$

where:

$$\lambda = 0.4 \quad (\text{for } S_{u1}/B = 2.5 \text{ KN/m})$$
$$\lambda = 0.3 \quad (\text{for } S_{u1}/B = 10\text{KN/m}).$$

The coefficient λ varies approximately linearly in the range considered. When using this procedure the following limits are recommended:

i) $\phi \not> 35°$

ii) $S_{u1}/B \geq 2.5$KN/m.

3.5.3. Strong overlying weak load spread

The bearing capacity of a strong soil overlying a weak soil can be considered by a simple load spread analysis. Young and Focht (1981) recommend the use of a "3-on-1" load spread, as shown on Figure 4.

The apparent footing area A' on the weak soil is given by:

$$A' = A[1 + \frac{2H}{3B}]^2 \tag{10}$$

The gross bearing capacity on the apparent footing, Q, should not exceed the bearing capacity of the actual footing in the strong soil, and Q' is given by:

$$Q' = 6S_{u1}A'[1 + 0.2\frac{D'}{B'}]. \tag{11}$$

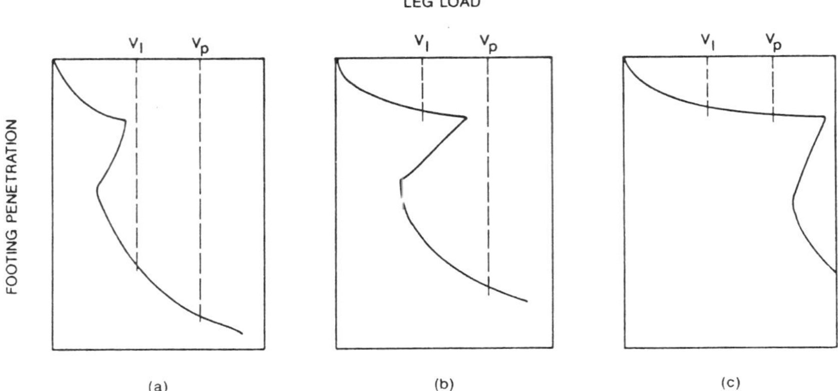

Fig. 5. Punch-through footing load-penetration curves.

3.5.4. Thin weak layers

Where weak layers are relatively thin compared to the footing diameter, their bearing capacity will exceed that of a thick layer of the same strength. The reason for this is that the failure mechanism is one of lateral squeezing. Results from Meyerhof and Chaplin (1953) and Brown and Meyerhof (1969) can be used to evaluate the strength of an apparent thick stratum with bearing capacity equivalent to that of the thin layer. This apparent strength is then used in bearing capacity analysis.

3.5.5. Punch-through potential

To assess punch-through potential, footing bearing capacity is evaluated at increasing penetrations and a curve of leg load versus tip penetration developed. Leg loads during rig installation are compared to bearing capacity to obtain the likely penetration at the important stages, namely: light ship plus variable load, V_1, (as the hull lifts clear of the water), and full preload, V_p.

Figure 5 shows leg load-penetration curves, with a reduction in the leg load that can be supported for a range of penetrations, as is characteristic of a punch-through. In Figure 5a the punch-through should occur before the hull is clear of the water, causing little concern. However, if the punch-through is likely to occur during the preloading phase, as illustrated in Figure 5b, then the suitability of the rig for the site and the installation procedures need careful reflection. Figure 5c represents the case where punch-through is unlikely to occur at full preload, and with an adequate safety margin, installation should be safe. However, for this latter case, if the preload is less than the design storm loading effect, long term stability will require thorough assessment.

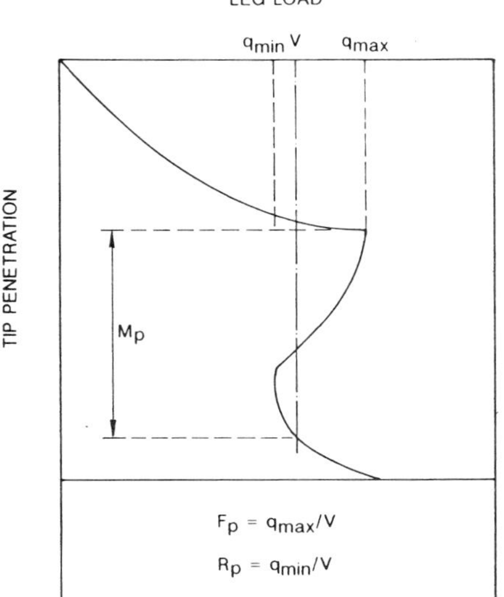

Fig. 6. Punch-through characteristics.

3.5.6. Punch-through characteristics

There are three characteristics that can be used to evaluate the likelihood and consequences of a punch-through, namely: (i) factor of safety, F_p, (ii) load to minimum capacity ratio, R_p, and (iii) penetration magnitude, M_p. Figure 6 defines each of these terms. The factor of safety and load capacity ratio are used to determine the likelihood of punch-through, and the magnitude provides a measure of the penetration that will occur during a punch-through. Some rigs are better able to tolerate rapid leg penetration than others, and if the magnitude is acceptable then jacking could proceed even though punch-through is predicted during preloading.

3.5.7. Punch-through safety margin

Back analysis of footing performance has been used to establish safety margins with respect to punch-through. This footing performance experience has been used to develop the chart shown on Figure 7, which can be used to assess the suitability of a rig at a site. The information from Figure 7 can be summarised as punch-through is unlikely to occur if:

1) $F_p > 1.5$;

2) $F_p > 1.35$ and $R_p > 1.2$ to 1.3.

The range of load to capacity ratio, R_p, depends on the method used to assess bearing capacity.

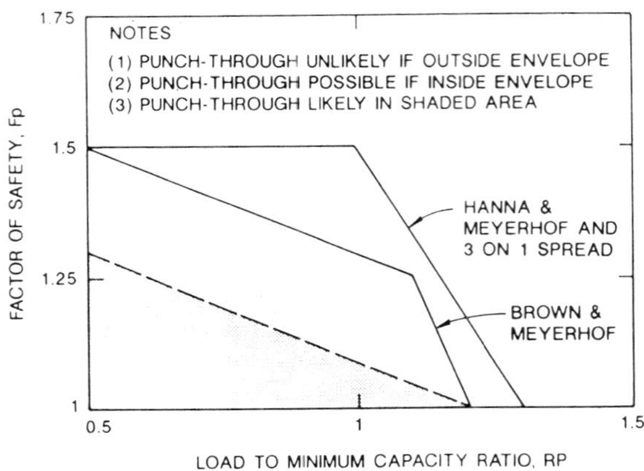

Fig. 7. Assessment of punch-through potential.

3.5.8. Footing hang-up

At a location where punch-through is likely to occur at less than full preload, one or more legs may not penetrate into the lower, weaker soil, but "hang-up" in the overlying stronger soil. It may be possible to increase leg loads or force penetration of the hanging leg. If not, additional geotechnical assessment will be required, which may include soil borings at the shallow legs, before long term stability can be established.

3.6. Calcareous Soils

Calcareous sands may exhibit relatively high friction angles in laboratory tests. If these friction angles are used, footing penetration may be significantly underestimated (Dutt and Ingram, 1988). The reason for this is either the soil's high compressibility or the high voids ratio and consequent collapsible structure. Where a calcareous soil is believed to be compressible or open-structured, the friction angle used in footing penetration analysis should be 5° to 10° less than the value measured in the laboratory. Long term stability under transient loading will need careful examination.

4. Foundation Performance Assessment

4.1. Requirement

The foregoing sections have concentrated on footing performance during preloading. For shallow water sites the available preload can be well in excess of design loads and the jack-up rig will operate comfortably within its structural limits. However, in deeper waters detailed analysis is required to evaluate structure adequacy. Part of this analysis will be an assessment of foundation performance.

For many jack-up rigs, foundation performance is considered under combined vertical and horizontal loading with the footings assumed to be pinned, i.e. no rotational restraint. However, as discussed below, if footings can carry moments there are potential benefits in terms of structure performance. These benefits become more important in deeper, more exposed waters.

A significant difference between jack-up rig footings and other foundations is that it may not be possible with the available preload to develop an "adequate" safety factor with respect to bearing capacity. However, this does not mean that the jack-up rig is unsuitable. As will be discussed below, overloading of the footing will result in settlement, until a factor of safety of unity is achieved with respect to the particular load combination. If the resulting displacement is tolerable then the situation is satisfactory. Rig levelling is therefore likely to be required from time to time, particularly in deeper waters where preloading is relatively limited in comparison to design loads.

4.2. Foundation Fixity

The water depth in which a jack-up rig can operate may be limited by the foundation rotational restraint. This is clearly demonstrated by Murff *et al* (1991) who investigated the global behaviour of a jack-up rig using a simplified model. Their results indicate that the consequence of considering the footing being fully restrained (fixed) rather than pinned is to halve the moment at the deck-leg connection and reduce the lateral (sway) stiffness, k_l, by a factor of four. This latter factor can significantly affect the rig's response because of dynamic amplification. The natural sway period, T, is approximately proportional to $(k_l)^{-0.5}$. In deeper waters where the rigs natural period will increase, an underestimate of foundation fixity will push the computed period towards that of the waves, and therefore the dynamic response could become an important consideration.

4.3. Evaluation of Foundation Fixity

The significance of foundation fixity on structural performance of jack-up rigs in deeper waters has led to much research in recent years. This research has taken the form of centrifuge testing and actual structure monitoring, inter alia.

4.3.1. Model tests

An on-going series of model tests at the Geotechnical Centrifuge Testing Centre of Cambridge University has been set up to investigate jack-up rig footing fixity on sand. Initial results have been published by Murff *et al* (1991), Murff *et al* (1992) and Tan (1990). Specifically, the tests have examined footing performance under combined vertical, horizontal and moment loading. The aim of the test programme is to develop a foundation model that can be used in detailed structure analysis.

4.3.2. *Structure monitoring*

Indirect measurements of foundation fixity have been conducted on several jack-up rigs (Brekke *et al* (1989); Brekke *et al* (1990); Hambly, Imm and Stahl, 1990).

One rig was located in 65.5m of water in the Gulf of Mexico and founded on soft clay. The 12m diameter footings penetrated to 15.5m where a 4.6m sand stratum was encountered. Results indicate that: a) although foundation fixity reduces with increasing load due to soil non-linearity, there is considerably more fixity than for a pinned footing, and b) the soils above the spud can provide lateral resistance to leg rotation (Brekke *et al*, 1989). Lateral resistance on leg members can be evaluated using P-Y data (Matlock, 1970) where soil flows back over the footing, taking due regard of soil remoulding.

In a second project a comprehensive structural measurement programme was carried out on a jack-up rig operating at a sand site in 70m water depth in the southern North Sea (Brekke *et al*, 1990). During the monitoring period wave heights up to about 6m were seen. Observations from the programme include: a) even on a sand site with shallow footing penetration significant fixity occurs, and b) the effect on natural period of the jack-up due to foundation fixity could be modelled. The authors give a simple method to evaluate rotational stiffness, whilst warning about the sensitivity of stiffness predictions to the assumed contact area. However, there is still considerable uncertainty regarding the degree of fixity during extreme storms, and the influence of storm induced scour.

4.4. Combined Loading

Bearing capacity of a footing is reduced when horizontal and moment loading are applied. For the pinned case, where a footing is assumed unable to carry moments applied at the bottom of the leg, bearing capacity is assessed for inclined loading (H/V). Figure 8 shows typical bearing capacity envelopes for footings on cohesive and cohesionless soils subjected to inclined loading. As indicated on Figure 9 the size of these envelopes is reduced if moment loading is also applied.

4.4.1. *Yield surface*

Centrifuge model testing work (Murff *et al*, 1991) shows clearly the effect on bearing capacity of combined V, H and M loading, in what the authors describe as "a remarkable demonstration of the concept of a yield locus". For loading within the yield locus, displacements are reasonably elastic, although sub-yield plastic deformations occur.

Where bearing capacity increases with depth the yield surface can be extended by additional footing penetration. In soil models tentatively proposed, penetration is taken as the work hardening parameter. However, such models are inappropriate where bearing capacity reduces with depth.

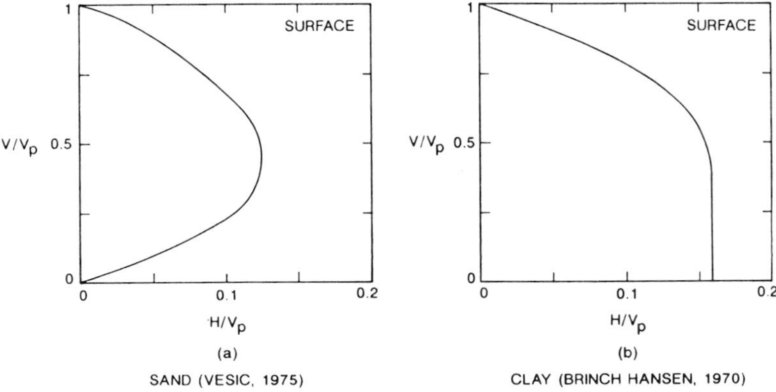

Fig. 8. Bearing capacity envelope for combined loading (pinned footing).

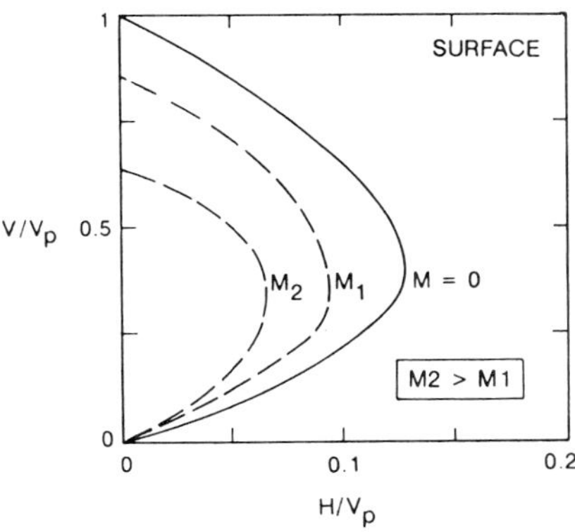

Fig. 9. Bearing capacity envelope for combined loading (fixed footing).

4.4.2. Bearing capacity

It appears from centrifuge testing that the model yield surface approximates the theoretical value at low load levels, but at high load levels the theoretical bearing capacity is conservative. This is to be expected since empirical bearing capacity factors have been developed for application to foundation design. Until actual structure performance data are available it is recommended that bearing capacity be evaluated using Brinch Hansen (1970) for clays and Vesic (1975) for sands. Definitions of the bearing capacity factors for load inclination and eccentricity are given in the source references.

4.5. Foundation Stiffness

For loading within the yield surface the footing may be considered as a rigid circular plate on an elastic half space. The Boussinesq elastic stiffness values, K, are:

$$\text{Vertical:} \quad K_v = \frac{2BG}{(1-\nu)}$$

$$\text{Horizontal:} \quad K_h = \frac{16(1-\nu)BG}{(7-8\nu)}$$

$$\text{Rotation:} \quad K_\theta = \frac{B^3G}{3(1-\nu)}$$

where:

G = shear modulus

S = Poisson's ratio (0.5 undrained, 0.2 drained).

The determination of an appropriate shear modulus is difficult. James (1987) found that the modulus for assessing K_θ on sands is about one third the vertical value. Centrifuge tests on sands show that shear modulus for rotation and horizontal displacements are about 20 to 40% and 5 to 15% the vertical value, respectively. For sands, the shear modulus is reasonably linear whereas for clays it is highly load level dependent.

Shear modulus should be evaluated for specific site conditions. Where data is lacking, it is recommended that Hardin and Drnevich (1972) is used to evaluate G in sands. For clays results from laboratory and *in situ* tests should be used, however from extreme storm loading G should not exceed about $40S_u$.

4.6. Degree of Fixity

If a footing is fully fixed it will carry moment and thus have reduced bearing capacity, the vertical load will be reduced. Therefore, footing performance analysis needs to account for the effect of fixity on loading. Where a footing is highly loaded the reserve foundation strength available to provide rotational restraint is limited. Conversely, lightly loaded footings may have relatively high fixity.

An iterative analysis can be undertaken with the degree of fixity modified until it is compatible with the applied footing loads. As a first step it would be reasonable to assume the footing is 50% fixed (mid-way between pinned and fully fixed). If loading is within the allowable envelope the degree of fixity can be increased, and vice versa. In this way the benefits of fixity on rig structural performance can be used, and the effect on foundation capacity due to moment loading evaluated.

4.7. Three-Level Stability Assessment Procedure

Schotman and Efthymiou (1989) have proposed a three-level foundation stability assessment procedure. The assessments at each level become progressively more refined, but once stability requirements are satisfied there is no need to check at a higher level. Partial load and material factors are applied throughout, as defined in the source reference.

Under storm conditions loading will vary between footings. Storm induced moments will lead to down wind, "leeward", footings carrying high vertical loads, whereas up wind, "windward", footings will have smaller vertical loads but consequent higher load inclination. The assumption of leeward and windward legs bound the range of footing performance.

4.7.1. Level I: Preload and sliding checks

Under environmental design conditions the leeward leg load will be mainly vertical; similar to the preload. For a factored horizontal load of less than 0.1 times preload, stability requirements will be met if the factored preload exceeds the factored vertical loads. In contrast, the windward leg will be subject to relatively high load eccentricity and therefore the stability check should be made by comparing factored horizontal load and factored sliding resistance.

If either the preload check or sliding check does not satisfy the foundation stability requirement, then proceed to Level II.

4.7.2. Level II: Capacity check

In this check, the factored footing loads for all legs should remain within the bearing capacity envelope. Care should be taken to ensure that load factors are applied correctly, depending on whether the load forces or resists foundation failure. Figure 10 illustrates the check for both a windward and leeward leg.

4.7.3. Level III: Displacement check

If a footing is loaded beyond the allowable capacity envelope, then displacements beyond the assumed elastic range are likely. As long as the structure can take additional footing displacements, stability requirements will be satisfied if: a) the additional footing penetration does not lead to a reduction in bearing capacity, and b) the rotation of the jack-up rig does not exceed 2 degrees.

In deeper waters the 2 degree tolerance on rotation may be too large. Hambly (1985) presents a method to estimate leaning instability. This method is conservative, and therefore if potential problems are indicated, more detailed analysis should be undertaken.

Schotman (1989) presents a non-linear load-displacement model to evaluate the effects of footing displacements on jack-up rig stability. The model, which has been calibrated by finite element analysis, may be used to investigate load transfer between footings, and thus to refine the prediction of structure stresses.

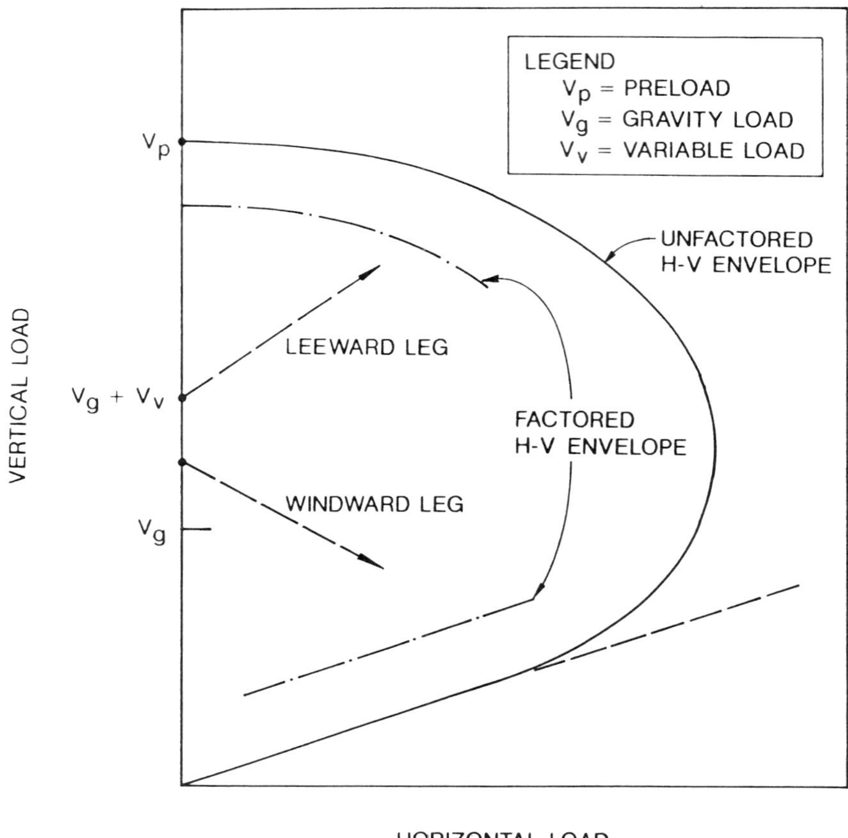

Fig. 10. Allowable combined load on jack-up footing.

4.7.4. *Limitations*

The above stability assessment procedure is suitable for well conditioned foundations. However, where either non-uniform soils or complex loading conditions pertain, a straight application of the procedure could be inappropriate. Rig performance monitoring will be required to assess the applicability of any foundation load-displacement model used in stability analyses.

4.8. Foundation Displacements

4.8.1. *Short term*

Within the permissible load envelope, displacements are considered to be elastic. For loading beyond this envelope, foundation settlement is obtained from the predicted load-penetration behaviour. The vertical load used to assess settlement is the pure vertical load on the combined load envelope. Where the load vector exceeds the allowable load envelope because of high load inclination this approach is inappropriate since the displacement will be largely horizontal. For such cases, foundation adequacy should be assessed by a permissible stress approach.

4.8.2. Long term

Long term settlement is assessed as for gravity structures, with due recognition of long term transient loading (Andersen, 1976; Andersen and Lauritzsen, 1987; Andersen, Kleven and Heien, 1988). An advantage with jack-up rigs is that settlement can be monitored and compared with predictions, and the jack-up can be re-levelled from time to time.

4.8.3. Layered soil conditions

In layered soils where punch-through is a consideration, settlement predictions should be made using a range of techniques including finite element modelling. In such soils frequent monitoring is essential in order that accelerations in settlement may be detected and evaluated.

5. Other Considerations

There are other important considerations to be made when assessing the suitability and performance of a jack-up rig at a site. These are briefly discussed in the following paragraphs. Other considerations such as seafloor instability due to wave or seismic loading and relocation of a jack-up site will need to be specifically addressed as the need arises.

5.1. Scour

Scour around spud cans bearing on cohesionless soils usually leads to additional footing penetration. Sweeney, Webb and Wilkinson (1988) report that in the southern North Sea scour induced settlements in excess of 1m. For several rigs offshore China (in similar water depths) scour has led to additional penetrations of up to 5m, with a maximum rate of about 0.8m/day. The depth of scour is difficult to assess, but is dependent upon many factors, including: footing shape, bearing pressure, soil properties, seabed wave loading and seabed currents. Scour has been believed to accelerate during storms but can also be significant in ambient conditions. Observations offshore China show that scour had reached a stable depth of about 5m in three months, before any storms occurred at the site (Sweeney *et al*, 1988). More, detailed observation is needed in order to obtain a better understanding of the complex, interactive factors affecting scour.

The consequences of scour could be particularly important in deeper waters. Scour can lead to load eccentricity, and Sweeney *et al* (1988) cite one case where this has led to spud can damage. Watson (1973) reports incidents where scour settlements have led to overstressed jack houses and legs. If scour increases footing penetration closer to an underlying weak layer there is a potential for punch-through. Rig displacements caused by scour may lead to problems where a rig is operating in work-over mode adjacent to a fixed platform.

Field observations show that more angular spud can shapes lead to increased scour. In vigorous scour environments scour stability may only be achieved when the footing penetrates fully below the original seabed. Many techniques are available to reduce the effects of scour. These include both scour prevention and methods of increasing footing penetration. The placement of sand bags, rock, slag or oyster shells have all been used successfully, as has artificial seaweed. For long term scour prevention dumped materials should be designed with an efficient filter (Sweeney *et al*, 1988). Air lift-water jet systems have been used to increase penetration, but for larger footings this additional penetration is unlikely to fully penetrate the can as required in order to stabilise against scour.

In deeper waters where footing fixity is assumed in a structural analysis, scour will reduce fixity and hence increase leg-deck connection bending moments. Also, rig displacements above water will be amplified. Therefore scour prevention measures should be implemented at the time of the rig moving onto location. If significant scour occurs, footings should be preloaded again when the scouring has stabilised.

5.2. Jack-Up Rig Adjacent to Pile-Supported Structures

Jack-up rigs are being increasingly used in "work over" mode adjacent to jacket structures. In this situation the maximum spacing that can be left between the jack-up rig and structure at deck level is limited. Jacket structures typically have sloping legs, consequently, the spacing between spud cans and piles can reduce as water depth and footing penetration increase. To reduce interaction effects the edge-to-edge spacing between a spud can and a pile should be kept as large as possible, with a target of at least one spud can diameter, particularly in weak soils.

5.2.1. Shallow footing penetration

At stiff clay or sand sites footing penetration under full preload is normally shallow, with the widest cross-section close to seabed. The effect of footing loading on adjacent piles is only likely to be important if the pile stresses are close to permissible prior to jack-up installation. Lyons and Wilson (1986) and Chow (1987) both discuss the evaluation of small soil displacement loading on piles. At sand sites, scour around a spud can could be of greater importance to lateral behaviour of adjacent piles.

5.2.2. *Deep footing penetration*

In soft clays footing penetration causes the displacement and remoulding of large soil volumes. Such soil movement and strength reduction can both load existing piles and reduce the piles ability to withstand lateral forces.

Siciliano *et al* (1990) present results from centrifuge tests of a footing penetrating into kaolin with a linearly increasing undrained shear strength. Non-linear finite element analyses significantly overpredicted displacements, whereas a linearly elastic soil model underpredicted. The authors developed an empirical model for prediction of displacements, but it is limited to similar conditions to those tested. However, the test results show that large soil lateral displacements occur within a zone about one diameter from the footing centre. At the edge of this zone displacement is small relative to footing (about 0.01 times footing diameter), but is large for typical piles at upto about 0.1 times pile diameter. This zone of influence and order of displacement may represent a lower bound.

Loading on piles due to soil movements is considered for structures designed to operate in possible mudslide areas (Kraft and Ploessel, 1986). Such loading can be severe, probably more severe than for the footing installation problem (Mirza *et al*, 1988), and represent an upper bound solution.

The effect of jack-up rig should be addressed at the jacket structure design stage, and not considered in isolation. Pile performance monitoring would provide valuable information for developing efficient design methods for this highly complex problem.

5.3. Cyclic Loading

The effect of cyclic wave loading is likely to be increased settlements, which is more important for jack-ups used as fixed production structures than for their more usual function. If settlements due to cyclic loads were large there would be far more discussion in the literature than appears to date. Large penetrations due to sand liquefaction have not been observed.

In deeper waters cyclic shear stress ratios will exceed those in shallower waters. This effect can be addressed using techniques developed for gravity structure design (Andersen, 1976; Andersen, Kleven and Heien, 1988; Andersen and Lauritzsen, 1988).

6. Concluding Comments

1. The largest single cause of jack-up rig mishaps are foundation related problems, of which about 70% are due to punch-through during preloading. Punch-through can be avoided with proper geotechnical assessments prior to rig installation.

2. The likelihood and effect of punch-through can be assessed using three parameters, namely: (i) factor of safety, F_p, (ii) load to minimum capacity ratio, R_p, and (iii) penetration magnitude, M_p. More case histories of footing performance in layered soils, including raw laboratory and *in situ* test data, are required to better calibrate analytical bearing capacity procedures.

3. Foundation fixity can have a significant impact on leg-deck connection bending moments and lateral sway stiffness (and hence natural period) for a given jack-up rig. This is particularly important in deep waters. Actual structure performance monitoring and centrifuge model testing have shown foundation fixity occurs even at shallow footing penetrations. Caution should be exercised where assessing the degree of fixity under extreme loading conditions.

4. In deep waters it may not be possible to preload footings to sufficiently beyond the design load effect. A partial factor stability assessment procedure (Schotman and Efthymiou, 1989) can be used in favourable soil and loading conditions. However, in complex soil conditions or where horizontal loading dominates, a permissible stress approach is recommended to evaluate actual footing performance.

5. Scour around jack-up footings in sand can cause significant additional penetration. Scouring may not stabilise until the footing is fully below the original seabed level. Where scouring occurs the rig will require relevelling from time to time, and consideration should be given to preloading again after scouring has stabilised. The potential for scour should be assessed prior to installing a jack-up rig. Scour penetration techniques are available.

6. Where a jack-up rig is located adjacent to an existing jacket structure its effect on pile performance should be addressed at the jacket design stage. The pile to footing edge spacing should be kept as large as possible, with a minimum target of about one footing diameter. Scour around a jack-up rig footing could affect pile performance.

References

1. Andersen, K. H. (1976), 'Behaviour of clay subject to undrained cyclic loading', *Proceedings International Conference on the Behaviour of Off-shore Structures*, Trondheim, Vol. 1, pp. 392–403.
2. Andersen, K. H., Kleven, A., and Heien, D. (1988), 'Cyclic soil data for design of gravity structures', *Journal of Geotechnical Engineering, ASCE* **114**(5), 517–539.
3. Andersen, K. H. and Lauritzsen, R. (1988), 'Bearing capacity for foundations with cyclic loads', *Journal of Geotechnical Engineering, ASCE* **114**(5), 540–555.
4. Brekke, J. N., Murff, J. D., Campbell, R. B., and Lamb, W. C. (1989), 'Calibration of jackup leg foundation model using full-scale structural measurement', *Proceedings 21st Offshore Technology Conference*, Houston, Vol. 4, pp. 49–58.

5. Brekke, J. N., Campbell, R. B., Lamb, W. C., and Murff, J. D. (1990), 'Calibration of a jackup structural analysis procedure using field measurements from a North Sea jackup', *Proceedings 22nd Offshore Technology Conference*, Houston, Vol. 4, pp. 357–368.
6. Brown, J. D. and Meyerhof (1969), 'Experimental study of bearing capacity in layered clays', *Proceedings 7th International Conference of Soil Mechanics and Foundation Engineering*, Mexico City, Vol. 2, pp. 45–51.
7. Chow, Y. K. (1987), 'Interaction between jack-up rig foundations and offshore platform piles', *International Journal for Numerical and Analytical Methods in Geomechanics*, **11**, 325–344.
8. Craig, W. H. and Higham, M. D. (1985), 'The application of centrifugal modelling to the design of jack-up rig foundations', *Proceedings SUT International Conference, Advances in Underwater Technology and Offshore Engineering*, Vol. 3, Offshore Site Investigation, pp. 293–310.
9. Dutt, R. N. and Ingram, W. R. (1988), 'Bearing capacity of jack-up footings in carbonate granular sediments', *Proceedings International Conference on Calcareous Sediments*, Perth, pp. 291–296.
10. Gemeinhardt, J. P. and Focht, J. A., Jr. (1970), 'Theoretical and observed performance of mobile rig footings on clay', *Proceedings Second Offshore Technology Conference*, Vol. 1, pp. 549–558.
11. Hambely, E. C. (1985), 'Punch-through instability of jack-up on seabed', *Journal of Geotechnical Engineering, ASCE* **111**(4), 545–550.
12. Hambly, E. C., Imm, G. R., and Stahl, B. (19990), 'Jack-up performance and foundation fixity under developing storm conditions', *Proceedings 22nd Offshore Technology Conference*, Vol. 4, pp. 369–380.
13. Hanna, A. M. (1978), 'Bearing capacity of footings under vertical and inclined loads on layered soils', Ph.D. Thesis, Nova Scotia Technical College, Halifax, N. S.
14. Hanna, A. M. and Meyerhof, G. G. (1980), 'Design charts for ultimate bearing capacity of foundations and sand overlying soft clay', *Canadian Geotechnical Journal* **17**(2), 300–303.
15. Hardin, B. O. and Drnevich, V. P. (1972), 'Shear modulus and damping in soils: Design equations and curves', *Journal of the Soil Mechanics and Foundation Division, ASCE*, **98**(SM7), 667–692.
16. Hansen, J. B. (1970), 'A Revised and Extended Formula for Bearing Capacity', Bulletin No. 28, Danish Geotechnical Institute, Copenhagen, pp. 5–11.
17. Houlsby, G. T. and Wroth, C. P. (1982), 'Direct solution of plasticity problems in soils by the method of characteristics', *Proceedings Fourth International Conference on Numerical Methods in Geomechanics*, Edmonton, Vol. 3.
18. Kraft, L. M., Jr. and Ploessel, M. R. (1986), 'Stability of submarine slopes', *Planning and Design of Fixed Offshore Platforms*, Van Nostrand Reinhold, pp. 440–516.
19. James, R. G. (1987), *Foundation Fixity of Jack-up Units*, Cambridge University Research, Appendix B to Noble Denton Report to Joint Industry Sponsors.
20. Lyons, R. H. and Willson, S. (1986), 'Effects of spud cans on adjacent pile foundations', *The Jack-Up Drilling Platform*, F. L. Boswell (ed.), pp. 8–22.
21. McClelland, B., Young, A. G., and Remmes, B. D. (1981), 'Avoiding jack-up rig foundation failures', *Symposium on Geotechnical Aspects of Offshore and Nearshore Structures*, Bangkok.
22. Meyerhof, G. G. and Chaplin, T. K. (1953), 'The compression and bearing capacity of cohesive layers', *British Journal of Applied Physics* **4**(1), 20–26.
23. Mirza, U. A., Sweeney, M., and Dean, A. R. (1988), 'Potential effects of jackup spud can penetration on jacket piles', *Proceedings 20th Offshore Technology Conference*, Houston, Vol. 3, pp. 147–157.
24. Murff, J. D., Hamilton, J. M., Dean, E. T. R., James, R. G., Kusakabe, O., and Schofield, A. N. (1991), 'Centrifuge testing of foundation behaviour using full jackup rig models', *Proceedings 23rd Offshore Technology Conference*, Houston, Vol. 1, pp. 165–178.
25. Murff, J. D., Prins, M. D., Dean, E. T. R., James, R. G., and Schofield, A. N. (1992), 'Jackup rig foundation modelling], *Proceedings 24th Offshore Technology Conference*, Houston, Vol. 1, pp. 35–46.
26. Osborne, J. J., Trickey, J. C., Houlsby, G. T., and James, R. G. (1991), 'Findings from a joint industry study on foundation fixity of jackup units', *Proceedings 23rd Offshore Technology Conference*, Houston, Vol. 2, pp. 517–533.

27. Quiros, G. W., Young, A. G., Pelletier, J. H., and Chan, J. H.-C. (1983), 'Shear strength interpretation for Gulf of Mexico clays', *Proceedings of the Conference on Geotechnical Practice in Offshore Engineering*, Austin, April, pp. 144–165.
28. Rappoport, V. and Alford, J. (1989), 'Preloading of independent leg units at locations with difficult seabed conditions', *Second International Conference on the Jack-Up Drilling Platform: Design, Construction and Operation*, City University, London, pp. 271–282.
29. Schotman, G. J. M. (1989), 'The effect of displacements on the stability of jackup spud-can foundations', *Proceedings 21st Offshore Technology Conference*, Houston, Vol. 2, pp. 515–524.
30. Schotman, G. J. M. and Efthymiou (1989), 'Stability of jack-up spud-can foundations', *Second International Conference on the Jack-Up Drilling Platform: Design, Construction and Operation*, City University, London, pp. 245–269.
31. Semple, R. M. and Rigden, W. J. (1983), 'Site investigation for magnus', *Proceedings 15th Offshore Technology Conference*, Houston, Vol. 1, pp. 205–216.
32. Sharples, B. P. M., Bennett, W. T., and Trickey, J. C. (1989), 'Risk analysis of jackup rigs, *Second International Conference on the Jack-Up Drilling Platform: Design, Construction and Operation*, City University, London, pp. 101–123.
33. Siciliano, R. J., Hamilton, J. M., Murff, J. D., and Phillips, R. (1990), 'Effect of jackup spud cans on piles', *Proceedings 22nd Offshore Technology Conference*, Houston, Vol. 4, pp. 381–390.
34. Skempton, A. W. (1951), 'Bearing Capacity of Clays', Building Research Congress, London, Division 1, pp. 180–189.
35. Sweeney, M., Webb, R. M., and Wilkinson, R. H. (1988), 'Scour around jackup rig footings', *Proceedings 20th Offshore Technology Conference*, Houston, Vol. 3, pp. 171–180.
36. Tan, F. S. G. (1990), 'Centrifuge and Theoretical Modelling of Conical Footing on Sand', Ph.D. Thesis, Cambridge University.
37. Vesic, A. S. (1975), 'Bearing capacity of shallow foundations', *Foundation Engineering Handbook*, Van Nostrand, pp. 121–147.
38. Watson, T. (1973), 'Scour in the North Sea', *2nd Annual European Meeting of SPE and AIME*, London, Paper No. SPE 4324, 12pp.
39. Young, A. G. and Focht, J. A., Jr. (1981) 'Subsurface hazards affect mobile jack-up rig operations', *Soundings* **3**(2), McClelland Engineers Inc., Houston, 4–9.
40. Young, A. G., Remmes, B. D., and Meyer, B. J. (1984), 'Foundation performance of offshore jack-up rig drilling rigs', *ASCE Journal of Geotechnical Engineering*, **110**(7), 841–859.

NOVEL FOUNDATION CONCEPT FOR A JACKET FINDING ITS PLACE

T. I. TJELTA and G. HAALAND
STATOIL, P.O. Box, 4001 Stavanger, Norway

Abstract. A major departure from traditional foundation practice for steel jackets has been developed based on the experience with skirts for huge concrete gravity platforms. The method, using plate-skirt foundations or perhaps better described as "skirted mudmats" or skirtpiles, will provide bearing capacity for compressive loads in the same way as gravity structures. Overturning uplift forces will be resisted by suction. This is a new concept in sand, and has been verified by offshore field tests and onshore laboratory tests. This is a description of the philosophy and efforts behind the development. The principles will be applied for the Europipe riser platform at Norwegian block 16/11.

1. What Is New?

This paper presents a new foundation concept for jackets, see Figure 1. In reality it is not a completely new foundation. It is the evolution of the raft foundation combined with piles. To fully appreciate the particularities of this foundation, it is necessary to look back on the use of gravity foundations in the North Sea.

The history of gravity foundations in offshore engineering starts with the Ekofisk storage tank in 1972. Since then several gravity foundations have been placed in the North Sea. The common principle is that gravity provides sufficient stability against the overturning loads from environmental loads. Horizontal sliding, which is sometimes the governing foundation failure mode, is taken care of by sufficient gravity load and often assisted by skirts penetrating into the seabed. Skirts are also usually an important measure against scour.

In the eighties, skirts were actively used to transfer loads through soft top soils down to more competent soil layers, e.g. the Gullfaks C structure reference [1]. In that respect, the skirts were now looking like giant piles, with diameter in the range of 20–30 meters and lengths of the same magnitude. These skirts resemble the piles in many ways. The Gullfaks C skirts carry most of the platform deadweight as friction on the skirt walls. The intention of the skirts was to transfer environmental forces down to stronger soil layers, which is also often the case with a pile foundation. Hence the name "skirtpile" is adopted for this kind of foundation.

However, an important difference is still visible. The skirtpile, with its giant diameter, finds the main capacity contribution from bearing capacity of the cross sectional area. This is very visible for the Snorre TLP foundations, reference [2], which is another application of the very efficient principle of skirtpiles. Both

Volume 28: Offshore Site Investigation and Foundation Behaviour, 717–728, 1993.

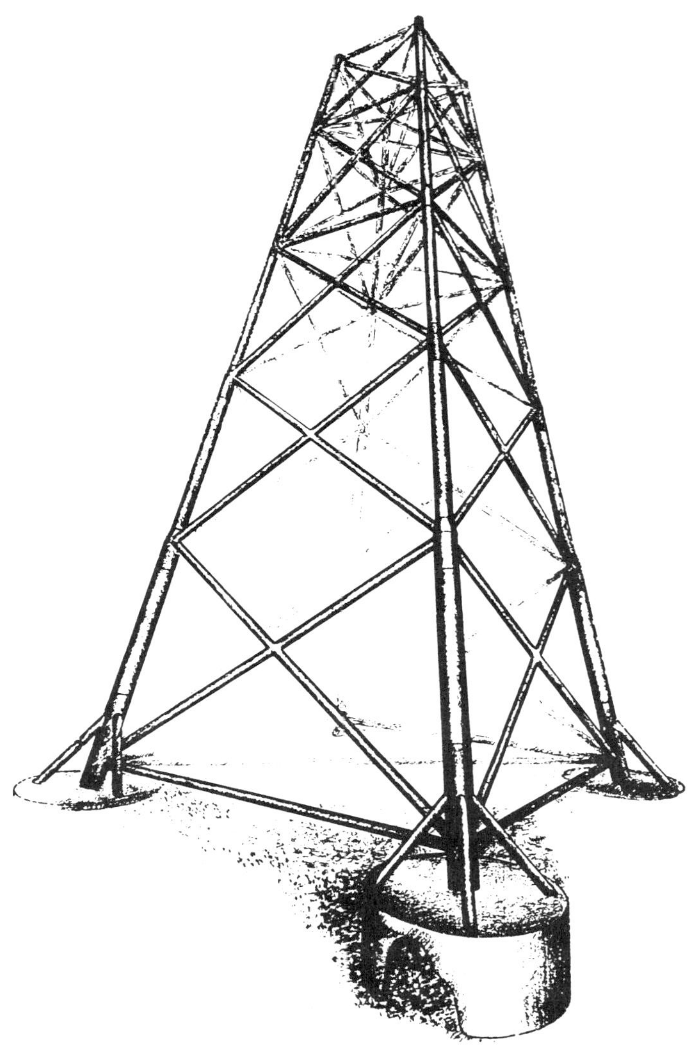

Fig. 1. Artist view of jacket with plate-skirt foundations.

compression and tension capacity of such foundations in clay, which are almost identical, are dominated by the bearing capacity contribution whilst the friction capacity in the unconsolidated state (i.e. after installation when high porewater pressure exists), is almost negligible. The fact that compression and tension capacity is almost identical is another difference between piles and skirtpiles. This difference, however, is believed to be more out of tradition than of an actual conceptual and rational reasoning. There is no reason why the end bearing of piles should not be equal for compression and tension loading in clay, and be included in the capacity predictions.

The first time tension capacity or uplift capacity of skirtpiles were considered, was for the Gullfaks C structure. Although only to a very modest degree, the tensile stresses at the heel of the platform from an elastic distribution of overturning moment were conceptually taken care of by a reversed type of bearing capacity consideration. The next time real tension capacity were utilized was the Snorre foundations. However, this was in clay.

The status with respect to gravity and TLP foundations can then be summarized:

- Standard gravity foundations have developed over the last 20 years to include skirts which in addition to providing scour protection and sliding resistance also transfer loads down to deeper and often stronger strata.

- Skirts have dramatically changed the concept of a gravity foundation in clay to being an universal foundation that can take not only compressive loads, but also tensile loads. Conceptually these skirtpile foundations are different from a pile foundation in that reversed bearing capacity rather than friction, is the main source for tension capacity.

- So far, only undrained tension capacity in clay have been considered. The next logical step would then be to look at drained and partly drained conditions in noncohesive soils.

2. Is Tension or Uplift Capacity in Sand Possible?

Why has not tension capacity of foundations on sand been much considered in the past? As mentioned above, gravity has often been the principle for foundation of concrete structures, and for jackets the limited tension capacity required for the often huge and heavy structures have been noncontroversial for the pile groups in question. It also adds to the fact that for long offshore piles, a significant portion of the pile is often in clay. Consequently, no great need for tension capacity in sand has been present.

The new concept of skirtpiles has changed this need. The skirtpile principle applied on jackets, one per leg, has raised the question of a possible contribution from suction capacity. Applied on jackets, this principle has evolved from two directions. The first being the scaling down of skirtpiles from gravity structures. The second one from the need to improve unpiled stability of jackets by providing mudmats with skirts, first looked into for the Veslefrikk jacket, and since then studied for a number of jackets, reference [3]. In this application the principle has been named “skirted mudmats”, “bucket foundation” and recently “plate-skirt foundation”, the latter probably the most descriptive name.

When looking into the uplift capacity of a plate-skirt foundation on sand, one immediately has to decide whether the nature of the actual problem is drained or undrained. And if the environmental loads on a jacket can be judged to give undrained or partly drained behaviour of the actual soil, still an important question remains: What maximum gradient can be allowed in order to avoid erosion, local failure and loss of capacity? As can be seen in reference [3], plate-skirt foundations on sand with the inherent limitations that lay in a conservative application of traditional theory for static waterflow in permeable soils yield no large cost savings over pile foundations. The need for additional ballast to limit the uplift forces on the foundation is the main reason for this conclusion.

Also a strong limitation in the application of plate-skirt foundations on sand is that skirt depth is an important parameter both to provide scour protection and to provide suction capacity. But again, the limitations in current empirical correlations do not support any significant skirt penetration in dense sand, and suction as additional driving force to penetrate skirts is questionable when only limited initial penetration has been achieved from deadweight. Again, an obstacle provided by current knowledge that prevents plate-skirt foundations to be an efficient alternative to pile foundations for jackets.

And that is really what this is all about: How is it possible to provide an alternative to pile foundations for jackets? And it all boils down to two simple questions: Can uplift capacity of skirtpile type of foundations exceed the very limited drained or nearly drained capacity? And is it possible to penetrate skirts by deadweight and suction to any required depth?

To answer these questions, it might be appropriate to look at experience that many marine practitioners have. It is not unusual that diving companies and offshore contractors experience that equipment and structures placed on seabed get stuck when trying to retrieve them. This is not only the case for clay conditions. It is also often experienced when the seabed consists of sand. Two examples will be mentioned here.

The first relates to a jacket installation in 1980 at the Valhall field south of Ekofisk. The soil condition is a top dense to very dense sand layer of several meters thickness. When the jacket was placed on the seabed, it hit with one leg first and penetrated slightly more on this corner than the others, resulting in a jacket being out of level and failing to meet the installation criteria. Placing piles in the other corners did not change this fact, and attempts were made to lift the low corner with the heavy lift vessel. The jacket submerged weight in this phase was only some 300 tons since buoyancy tanks were still attached. After several attempts with average hook load in excess of 300 tons applied at the low corner of the jacket, no movement could be observed. The jacket had no mudmats, only an approximately 2 meter diameter packer box was in touch with the seabottom at each corner in addition to the stabbed main piles inside the legs. With continued attempts to lift the jacket, swell caused the crane boom to move and peak loads exceeding 600 tons and a maximum load of 1200 tons were observed by the crane operator. Swell period was approximately 20 seconds. Only after many attempts was it possible to achieve a level jacket. Although these are dynamical loads influenced by the inertia of the total jacket load, they do indicate that a significant short term suction capacity was present in the sand.

The second example is from an initial soil investigation at the Gullfaks C location in November 1983. Because of the soft soil conditions it was considered beneficial to have one meter deep skirts on the seabed frame used for control and re-entry of the drill string. The frame was equipped with skirts around the perimeter, i.e. 3×3 meters square, and the submerged weight of the frame was approximately 8 tons. Soil condition is 1–1.5 meter loose sand over normally consolidated clay. When bad weather made it necessary to abort the borehole and retrieve the seabed frame from the seabed, it was found to be stuck. With the constant tension winch pulling at its maximum capacity of 40 tons for more than one hour it was finally possible to retrieve the frame. Again this was an example of suction capacity in sand beyond what would be expected. It is also interesting to note that this capacity was not of a short term character.

These are only two examples out of several that indicate that suction capacity in sand is present. Based on these observations, small scale testing took place in the years from 1989 to 1991, some of them with a tin can and a small fishing scale on the beach. Others were more scientific and took place in the geotechnical laboratory with assistance from professor Lars Grande and his students in Trondheim. All tests indicated that suction capacity of plate-skirt foundations in sand was a real phenomenon, but difficult to investigate in small scale testing since tests behaved partly drained due to short drainage paths. Only in silt was it possible to perform undrained tests, and these tests showed capacities outside the range that could be predicted.

The conclusion from all work performed, both theoretical and experimental pointed in the direction that undrained suction capacity in sand is very dependent on permeability, or more precisely coefficient of consolidation, C_v. The next logical step to take from here would be to initiate a comprehensive research programme to document these initial findings and to develop the necessary models that could predict design capacities. But as already mentioned, the problem of performing realistic modeltests in sand was not only the traditional one of getting realistic effective stresses, but also the problem of getting realistic drainage conditions in relation to load periods. This is very fundamental as drained, partly drained or undrained behaviour is difficult to predict and even more difficult to agree on. Consequently, only large scale tests were believed to be able to provide convincing arguments and test results that could support necessary model development. Further, only field tests could provide realistic penetration results.

3. Field Tests

The benefits of performing field tests soon were attached to two primary needs which both concerned the feasibility of an economical plate-skirt foundation, i.e. the penetration and the suction capacity in sand. But also the effect of cyclic loading

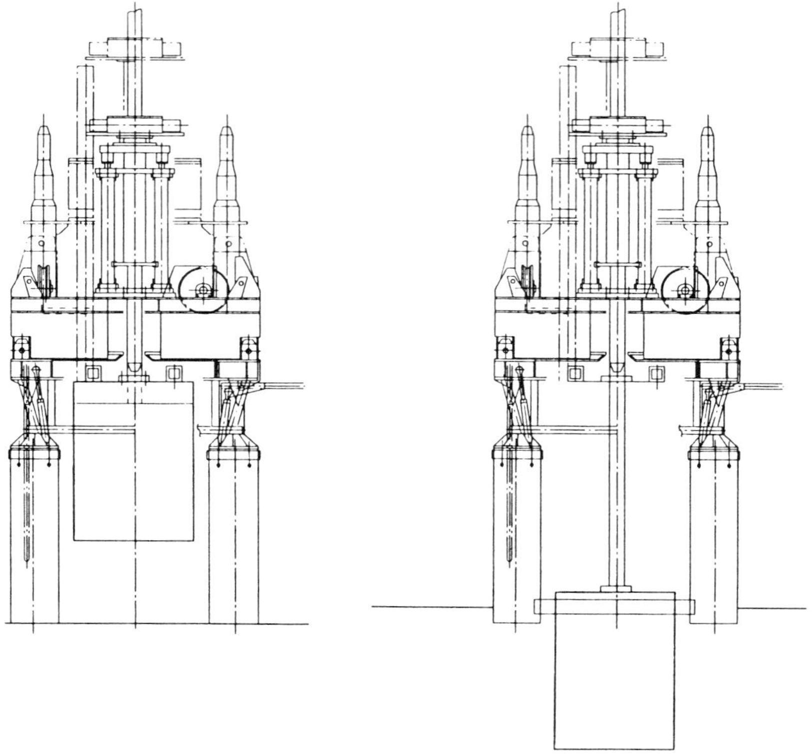

Fig. 2. Field test rig.

on such a foundation was a topic of great interest and the decision to include the ability to perform cyclic loading complicated the loading system.

The problem of skirt penetration was causing much concern prior to the field tests, and the limited initial penetration calculated was not felt to be sufficient to provide a good seal for continued suction penetration, see Figure 3. Predicted penetration resistance, dense sand.

The test structure that was arrived at for performing the field tests was restricted in dimensions to 3 $\times$ 3 meters cross sectional area to fit a standard moonpool of the soil investigation ships. The caisson simulating the plate-skirt foundation model was 1.5 meters diameter and 1.7 meters high, see Figure 2. This caisson was fitted within four legs and could be penetrated to its full depth with suction alone or in combination with deadweight. Normally, the caisson was penetrated by weight to an initial penetration of up to 0.9 meters followed by suction penetration to the full depth.

Tests were carried out at three different sites, one at block 16/11 and two at the Sleipner field. Soil conditions at the sites are dominated by sand. At Sleipner T and block 16/11 the sand is very dense and at Sleipner B silty and clayey and of

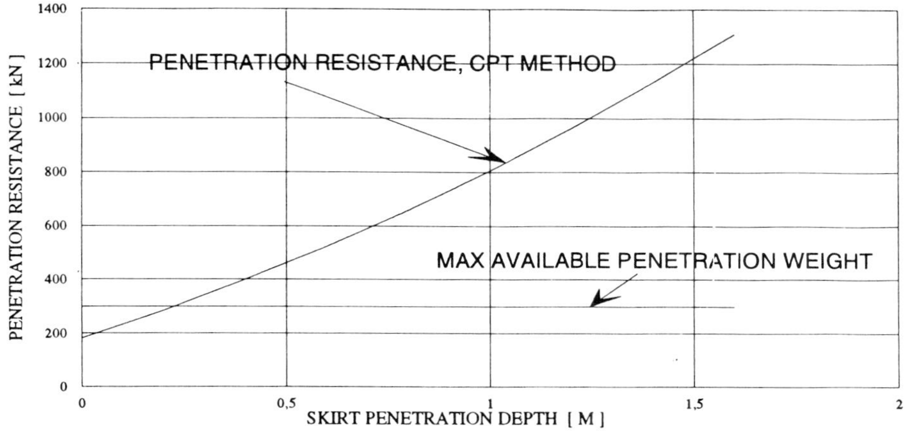

Fig. 3. Predicted penetration resistance, dense sand.

medium density. A typical test programme at one location consisted of:

1. Penetration by weight
2. Permeability test
3. Static pullout test
4. Cyclic test programme to failure
5. Cyclic programme, long term (1 hr.) at constant load, 50–60% of static capacity
6. Static pullout test
7. Drained pullout test.

A total of 15 test locations at the three sites were carried out and the various test parameters were systematically varied. Examples of test results are shown in Figure 4 (penetration results) and Figure 5 (cyclic test results). A comparison between Figure 3 and Figure 4 is quite interesting from a geotechnical point of view and clearly shows the limitations of using empirical relations, but also shows the benefits of performing site specific field tests. In short, the problem of achieving sufficient skirt penetration is almost eliminated.

The cyclic test programme showed in Figure 5 gives a clear indication that cyclic loads compact the sand. It is further observed that even when average load in the cyclic load programme becomes tensile after 1400 seconds, the caisson still takes further load increase and many cycles before failure starts to develop.

The effect of cyclic loading on the static pullout capacity is shown in Figure 6 (static pullout test results from test location T4 and T5). Pull 1 is static pullout capacity after initial penetration; Pull 2 is after first cyclic programme; Pull 3 after second cyclic programme, etc. The cyclic programme leads to an increased pullout capacity. Test T5 is giving an even higher capacity, and is showing the effect of a

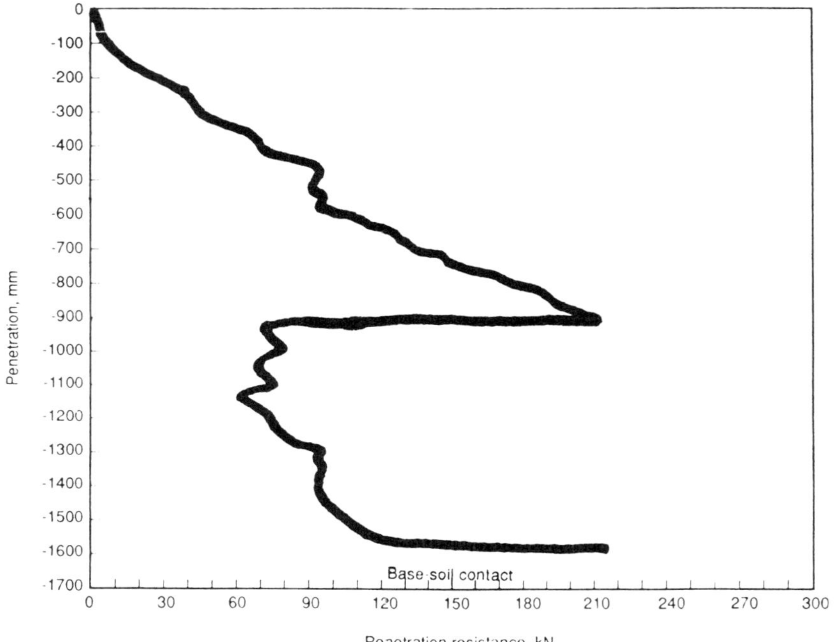

Fig. 4. Penetration results.

different penetration technique in addition to a comprehensive cyclic programme.

The results from the field tests were very encouraging. But could we now relax and conclude that the plate-skirt foundations were feasible?

4. Scepticism and Further Work

Following the field tests there were still many questions left to be answered. Offshore only the vertical pullout capacity had been tested. Questions left to be answered were: The effect of combined loads, would horizontal loads and moments open a crack at the rear of the skirt and how could amount of drainage be scaled to prototype dimensions.

These questions were addressed in two onshore laboratory test setups. One at the Geotechnical Institute, University of Trondheim/Sintef, and the other at NGI in Oslo. At Sintef a 0.8 m diameter plate with 0.6 m deep skirts was tested in a 4 m by 4 m bin filled with saturated sand. The Sintef tests aimed at investigating the effects of combined loads and through back calculations of both field tests and laboratory tests calibrate theoretical models for capacity calculations. Good correlation between theoretical models and results from model tests offshore and onshore was achieved.

The NGI tests aim at a fundamental understanding of failure mechanisms and

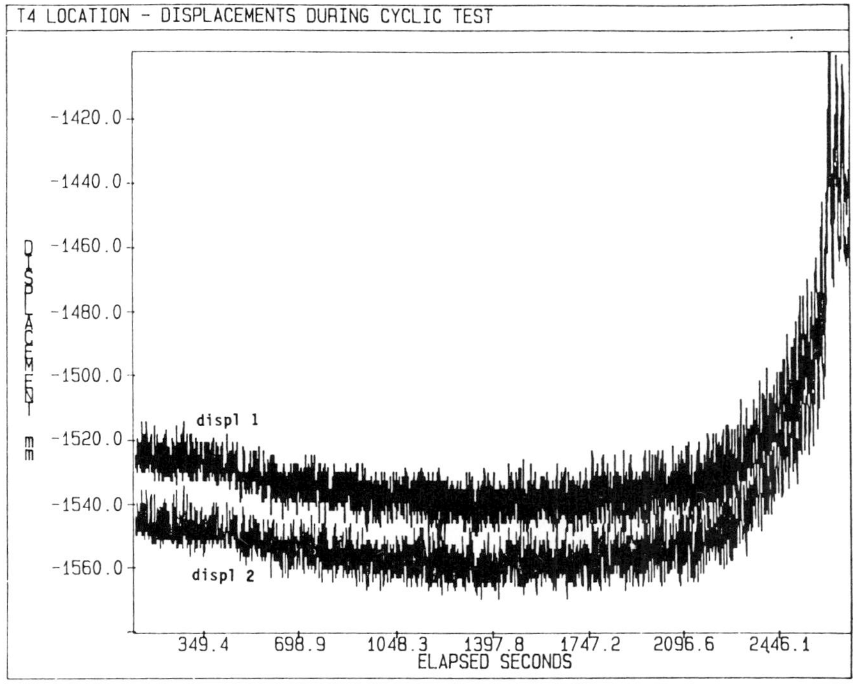

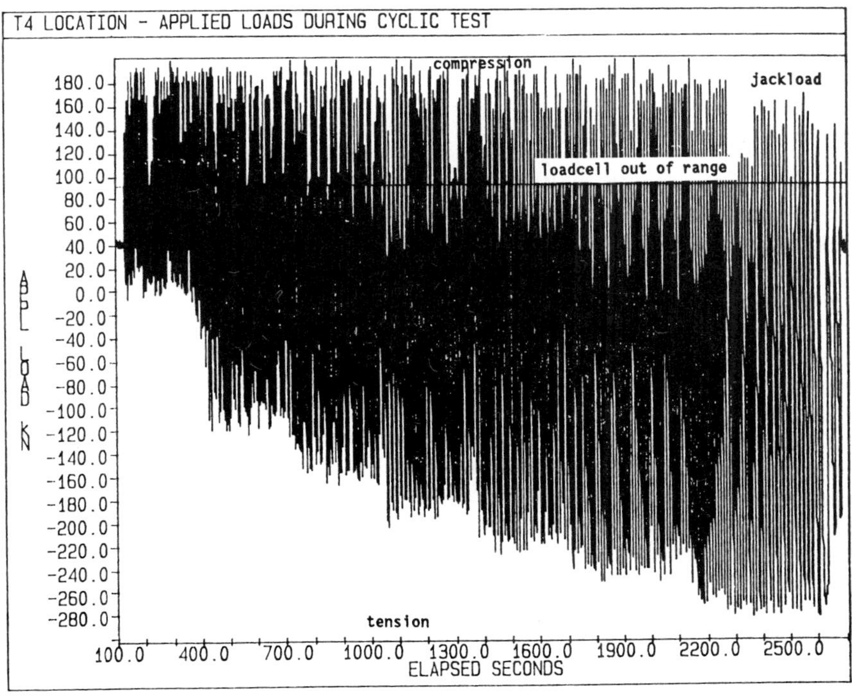

Fig. 5. Cyclic test results.

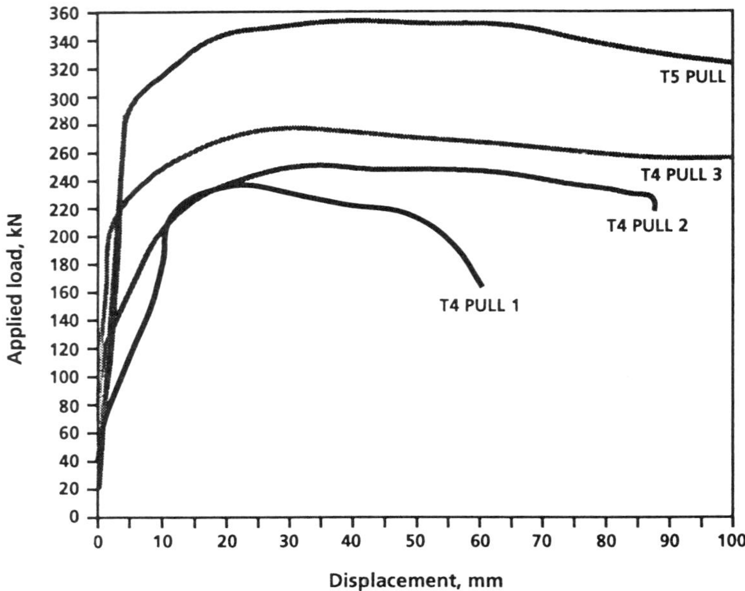

Fig. 6. Static pullout tests.

how cyclic pore pressures and gradients develop during loading to failure. In these test series the plate-skirt foundation was placed within a container making it possible to establish soil and porewater pressures similar to those existing at the tip of a 5 m deep skirt at 70 m of water. This programme is still going, but very interesting results are emerging from the tests. It is seen that cyclic gradients by far exceed the limitations that apply to stationary conditions.

All this work together with special features tested offshore, e.g. the test programme with a 15 mm diameter hole in the caisson to see the effect of a non-sealed compartment, was aimed at satisfying all the scepticism that can realistically be mobilised.

5. The Europipe 16/11-E Riser Platform

The plate-skirt foundation principle will be used for the Europipe riser platform. This is a four-legged platform with an on-bottom weight of some 4000 to 6000 tons and base dimensions of 40 m by 40 m. Instead of the conventional mudmats and piles one 12 m diameter plate with 6 m deep skirt makes up the foundation for each leg. Each foundation carries a deadweight of 2500 tons and a maximum wave load of ± 4000 tons. The deadweight includes up to 1000 tons ballast on each foundation. The ballast is placed on top of the foundation plate inside walls made by extensions of the skirts above mudline as shown in Figure 7.

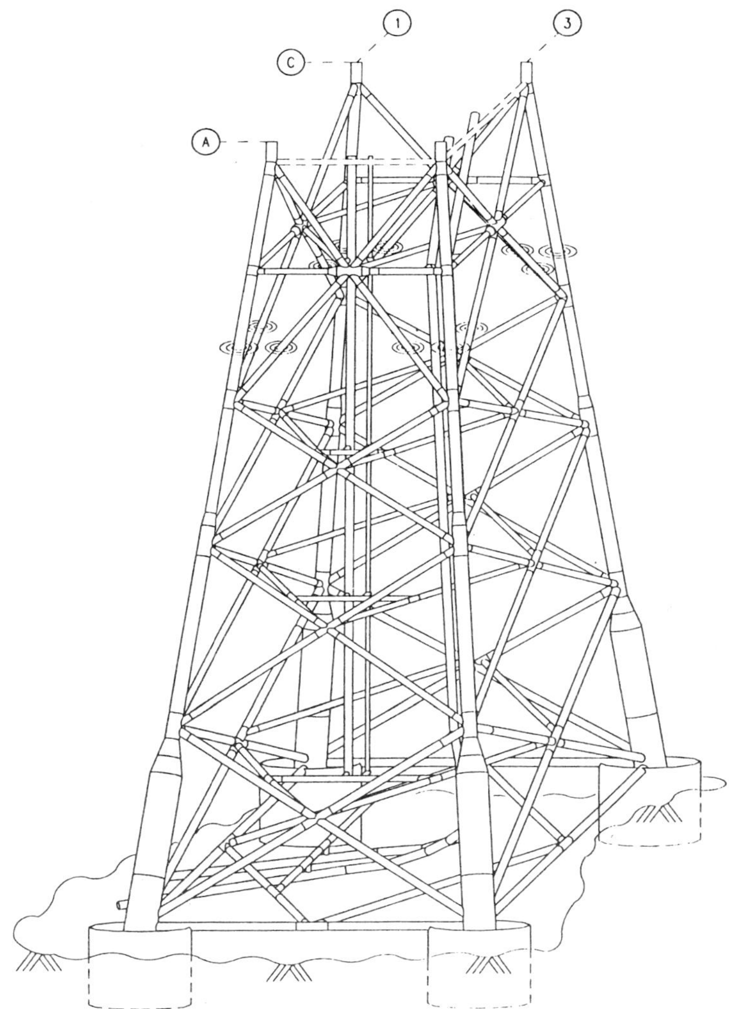

Fig. 7. Europipe riser platform.

The soil conditions at the 16/11-E location consist of a very dense fine sand in the upper 25 m. The tip resistance measured from cone penetration tests exceed 50 MPa from 2 m downwards. At 70 m water depth, the seabottom is very flat.

Installation of the platform is planned to be performed by a heavy lift vessel placing the jacket on the seabottom and let it settle under selfweight. During this phase of installation, hatches on top of the plates will be open to accommodate free flow of water out of the skirts. The selfweight penetration is estimated to be approximately 1 m based on results from the site specific field tests. Penetration down to the 6 m target penetration is accomplished by applying underpressure within the foundations by pumping out water. The maximum necessary underpressure is estimated to some 100 kPa. After penetration the void under the plates is filled with grout to ensure full contact between plate and soil.

The bearing capacity of the foundations is very large for drained compression loads which leads to large margins to carry the platform weight. The capacity to withstand the wave loading is assumed to be developed at close to undrained conditions. Only the largest waves induce uplift forces at one leg. The uplift conditions only last for a few seconds for the largest waves in the design storm.

Due to the novelty of the foundation only a part of the calculated uplift capacity of the skirt/plates is utilised. This uplift capacity is not much more than the sum of friction on skirts and weight of the soil plug inside the skirts. During the time until platform installation in May 1994, further work will be carried out in order to minimize or even eliminate the current amount of ballast.

The main advantage for the Europipe project in using the plate-skirt foundations is cost saving, both from savings in total steel weight, and shorter installation time.

Acknowledgement

The authors acknowledge Statoil for the opportunity to work with such challenging projects. Also colleagues in Statoil, Asle Eide, Eiliv Skomedal and Morten Bærheim, as well as outside Statoil, in particular Lars Grande (NTH), Per Sparrevik (NGI) and Hermann Zuidberg (Fugro-McClelland) are acknowledged together with the remaining offshore field-test team of NGI, Fugro-McClelland and Coe Metcalf.

References

1. Tjelta, T. I., Aas, P. M., Hermestad, J., and Andenæs, E. (1990) 'The skirt piled Gullfaks C platform installation', *Proceedings of 22nd Anual OTC*, May 1990, Vol. 4, 453–462.
2. Christophersen, H. P. (1993), 'The non piled foundation systems of the Snorre field', *Proc. Offshore Site Investigation and Foundation Behaviour*, SUT, London, September 1992.
3. Baerheim, M. and Tjelta, T. I. (1992), 'Skirt-plate foundations for offshore jackets', ISOPE June 1992.

A REVIEW OF THE DESIGN AND CERTIFICATION OF OFFSHORE PILES, WITH REFERENCE TO RECENT AXIAL PILE LOAD TESTS

R. HOBBS
Lloyd's Register Offshore Division, 71 Fenchurch Street, London, EC3M 4BS

1. Introduction

Design of offshore piling should commence with a carefully planned, integrated site investigation including geological appraisal of the immediate location, a shallow geophysical survey and an adequate number of detailed geotechnical boreholes. Brief guidelines for such investigations are presented. Certification includes independent interpretation by the certifying authority of the data resulting from such site investigation, to derive appropriate design parameters for use in its foundation and structural analyses.

Recent axial load tests of driven piles in normally consolidated and overconsolidated clays have enabled the validity of design procedures used by the offshore industry to be reviewed. This paper summarises such a review. Similar tests in sands are still outstanding and the design procedures proposed in recent editions of the American Petroleum Institute's RP2A code are discussed. Introduction of a "Load and Resistance Factor Design" version of this document and its possible adoption as a Eurocode are considered.

Areas of uncertainty in driven pile axial capacity design are also addressed, for example, the build-up and drop-off of pile end bearing in layered soils.

2. Certification

2.1. Legal Background

Under the Mineral Workings (Offshore Installations) Act of 1971, the regulations governing the design and certification of fixed offshore installations in UK waters are given in Statutory Instrument 289 (1974). Lloyd's Register (LR) has acted as a Certifying Authority under these regulations since their introduction.

Volume 28: Offshore Site Investigation and Foundation Behaviour, 729–750, 1993.

SI 289 is supported by Guidance on Design and Construction, issued and periodically updated until recently by the UK Department of Energy (1990). The responsibility now rests with the Health and Safety Executive (HSE). This document and an accompanying Background document (Semple, 1986) include recommendations on site investigation and design of foundations for offshore structures, but are not prescriptive.

The most widely used document for offshore structure design is the American Petroleum Institute's Recommended Practice for Planning, Designing and Constructing Fixed Offshore Platforms, RP2A (API, 1991). This, however, has no legal force and, in a number of respects, its recommendations differ from established UK practice for foundation design, as discussed herein.

In addition to the UK, a number of other countries also have a legal framework for certification or verification. However, this paper primarily reflects UK practice.

2.2. Role of Certifying Authority

The role of the certifying authority is to ensure compliance with legal requirements or to provide a third party check. In practice this process includes, but is not limited to:

a) review of data and assumptions on environmental and soil conditions;
b) independent design analysis of primary structure and foundations;
c) review of topsides design;
d) review of specifications for materials, welding procedures, non-destructive testing (ndt), fabrication and installation;
e) attendance during steel manufacture and review of material tests;
f) qualification of welders;
g) attendance during fabrication and review and approval of ndt reports;
h) monitoring of loadout and tow;
i) attendance during installation and hook-up;
j) periodical survey.

In respect of piled foundations, it is LR's practice to perform a totally independent assessment of soil parameters based on a review of all soils data and, where necessary, examination of samples. Calculations are performed of pile capacity and pile response, the latter being used in structural analyses for both static and fatigue loading conditions. The loads generated in these independent structural analyses are then used to calculate pile axial capacity factors of safety and to check pile stresses against allowables.

Checks are made to ensure that predicted stresses in the piles during driving are tolerable and will not result in significant loss of fatigue life. The detailed design of structural appurtenances is also reviewed to ensure that these are robust enough to withstand any vibrations which may be induced by pile driving (Hobbs and Waller, 1992).

The independent analysis carried out by experienced engineers is seen by the

Author as one of the strengths of the UK certification scheme.

In the majority of cases specialist engineers from LR attend installation to confirm that piles achieve acceptable penetrations. Pile driving is monitored and independent records are taken. They also monitor and approve grouting, swaging and welding operations, confirm that as-installed structural elevations relative to LAT allow the required minimum airgap to be achieved and review any structural damage which may occur during installation.

While not a requirement of certification, specialist foundation engineers from LR also frequently review proposals for and attend during site investigation to ensure that sufficient and acceptable data are obtained, thus avoiding possible additional expense in further investigation or undue conservatism in design assumptions.

This paper gives brief guidance on an acceptable basis for site investigation and design of driven piles for certification purposes, with particular reference to axial capacity. These supplements recommendations given in Lloyd's Register's (1989) rule for the classification of fixed offshore structures and in the HSE Guidance (DEn, 1990, Semple, 1986). Special considerations for piles permanently in tension are not addressed herein.

3. Site Investigation

3.1. Integrated Approach

A properly planned offshore site investigation will usually start with a desk study of existing regional geological, shallow geophysical and borehole data. Some information may also be gleaned from exploration geophysical and drilling and British Geological Survey data may be available for UK locations. Lloyd's Register has a comprehensive library of site investigation reports, with worldwide coverage including the majority of existing fixed installations in the UK and Dutch sectors of the North Sea.

It is preferable to perform site-specific shallow geophysical surveys prior to planning borehole locations, since the presence of buried geological features such as channels or faults, or of shallow gas, may dictate a re-evaluation of the final platform location or at least a review of the number of borings. Initial "ground truthing" of geophysical surveys can be achieved by reference to any existing nearby boreholes. In some cases exploratory boreholes may be justified.

High resolution seismic reflection surveys should aim to identify soil stratification to beyond the anticipated penetration of piles to an accuracy of less than a metre. Spacing of survey lines should be such that important features are not missed. Survey positioning should be accurate to within about 10m and it is desirable that, if possible, the same system be used as for borehole and jacket positioning, where tighter tolerances can be applied. Survey lines should preferably incorporate borehole locations.

3.2. Number of Borings

The number of borings to be drilled depends upon the lateral extent of the jacket at seabed and batter of the piles, the complexity of the geology in terms of uniformity of strata horizons and the amount of soils data already available for the immediate location.

For a small jacket, in the southern North Sea for example, it may be sufficient to drill two deep boreholes if stratification and soil strength are expected to be uniform. In some cases, where there are existing boreholes and previously installed jackets nearby, for which detailed pile driving records are available, this may be reduced to just one.

For large jackets, in the northern North Sea for example, four or more deep borings may be required, again depending upon expected variability over the site.

Where mudmat stability or pile lateral behaviour are critical, consideration may be given to additional shallow boreholes or seabed cone penetrometer tests.

Boreholes should be positioned to cover the areal extent of proposed pile locations.

3.3. Sampling and In Situ Testing

In general, boreholes should include a mixture of push samples and *in situ* piezocone penetrometer tests. For clay strata it is recommended that a sequence of two, thin-wall, push samples of up to 1m length followed by a 1.5 to 3m stroke piezocone test be adopted, while for sands a sequence of alternate samples and up to 3m stroke piezocones should be adopted, with additional (disturbed) samples taken within the cone stroke where there are indications of inclusions or variations.

For small structures in predominantly sand sites, expected to be uniform, an alternative is to have one sampled boring and one piezocone boring.

With the aim of achieving near-continuous coverage, spacing of samples and cone tests should be such that no gap greater than about one metre occurs between the bottom of one sample or cone penetrometer test (cpt) stroke and the start of the next, within depths of interest to pile design. This spacing may be reduced in potential end bearing zones, especially where there is evidence of clayey inclusions in sand. It is usual to extrude all samples offshore.

Sample lengths in clay should be sufficient to enable unconsolidated undrained (UU) triaxial tests on full diameter (generally 70–75mm) samples to be carried out. Further attempts should be made if insufficient recovery is achieved, to allow a complete profile of UU shear strengths to be obtained. As much of this testing as possible should be done offshore to minimise the effects of disturbance. Additional, qualitative, indications of consistency may be obtained by use of pocket penetrometer and hand vane. For very soft clays, miniature vanes may be employed. Suitable clay samples which remain should be sealed in wax and carefully packaged for transport to the onshore laboratory.

Where recovery from 70mm diameter push sampling becomes insufficient, for example in very hard clays and very dense sands, 50mm hammer samples may be used. Special consideration should be given to drilling in soft rocks and, in some cases, coring may be required.

All sampled material, whether disturbed or undisturbed, should be retained to enable subsequent inspection. Comprehensive sample descriptions should be made and reported.

Boreholes should extend to at least 5 pile diameters or 10m beyond the calculated required pile penetration (or likely depth of influence of pile groups) based on a conservative assessment of axial pile load and soil conditions. In this respect it is important that the site investigation has the flexibility to allow modification of proposed borehole depths (and indeed number) on the basis of conditions encountered.

3.4. Laboratory Testing

Much of the required testing will be done offshore, but further tests should be carried out in the onshore laboratory to enable a full picture to emerge of, for example, unit weights, natural water contents, particle size distributions, Atterberg limits and UU triaxial strengths. Additional tests may include consolidated undrained triaxial and oedometer tests on clays and consolidated drained triaxial tests on sand compacted to estimated *in situ* densities. Where it may aid understanding of the soils, geological and mineralogical testing should be carried out and colour photographs taken.

4. Basis of Design Soil Profile

4.1. General

Pile design is usually based on the "weakest" boring at the location. This, of course, does not necessarily reflect the worst soil conditions that may be encountered by any pile. Account should also be taken of variations in stratigraphy interpreted from geophysical data, particularly where there is reliance in the pile design on a given penetration into a potential end bearing stratum, or where such a stratum is of limited thickness. Where there are insufficient data, for whatever reason, conservative assumptions should be made. Selection of parameters is, to some extent, subjective and uncertainty in parameters may, in some cases, be similar to that associated with the method of calculation of axial capacity.

4.2. Clay Strata

For clay strata the reference strength for axial pile capacity calculation is that from unconsolidated undrained triaxial tests on full diameter push samples. Cone penetrometer data between samples indicate qualitative trends in undrained shear strength, S_u, and quantitative interpretations can only be made on the basis of layer by layer correlations. In the absence of such correlation, $S_u = q_c/20$ may be locally assumed (where q_c is cone point resistance) providing this is consistent with other data. Available UU data should be checked against geological setting and empirical index property correlations. The design shear strength profile should then be essentially a trend line through the UU data with any bias being towards lower results where there are anomalies which cannot be convincingly explained. Unit frictions calculated from design S_u values should be checked against cone sleeve frictions and anomalies (i.e. lower cone frictions) investigated.

4.3. Sand Strata

For sand strata, design parameters are assigned on the basis of piezocone data, grading curves and sample descriptions. *In situ* relative density is best interpreted from cone resistance, but selected parameters should also take account of fines content and the presence of any inclusions.

Parameters required are angle of friction, ϕ', and limiting values of unit friction and unit end bearing. Where no cone data are available lower maximum values of limiting unit friction and unit end bearing are assigned. Selection of ϕ' and limiting values are discussed further below. As for clays, calculated unit friction should be checked against cone sleeve friction.

4.4. Interbedded Soils

Where sand and clay are closely interbedded, such that it is difficult to identify distinct strata, it is prudent to assume that the soil's behaviour is governed by whichever material gives the lower friction and end bearing respectively.

5. Pile Friction in Clay

5.1. Established Practice

In the North Sea it has been the practice to adopt criteria given in API RP2A for calculating the unit friction, f, on driven piles in cohesive soils. Specifically, "API method 2" (API 1986 para. 2.6.4b2 and API, 1991, Commentary) in which f varies with S_u alone, has been used ($f = \alpha.Su$). For S_u less than 24 kPa α is taken as 1.0, while for S_u greater than 72 kPa α is taken as 0.5, with a linear variation between these values.

To account for the effects of pile whip during driving and lateral pile movements due to environmental loading, it has been Lloyd's Register's practice to modify the above procedure to limit unit friction near seabed to that sustainable by passive pressure acting on the pile wall (Hobbs, 1992). Where shear keys are provided on the pile wall to improve grout bond between pile and sleeve, this limit may also

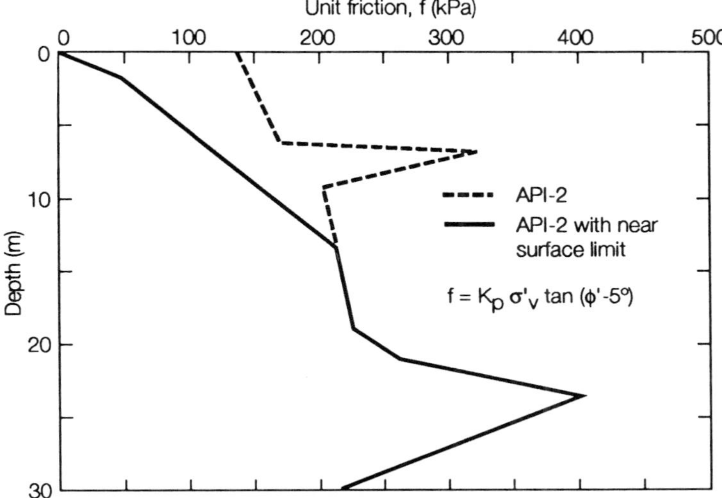

Fig. 1. Effect of near surface limit on unit friction calculated using API method 2 in clay

cover the effect of driving these protrusions into the soil. The result of this limit for the Tilbrook Grange pile load tests, discussed below, is shown in Figure 1.

5.2. Recent Developments

While the above method is an empirical procedure, based on the results of a number of pile load tests, it has been recognised that it takes no direct account of the soil's stress history, which is considered to be one of the factors which should control unit friction. To redress this concern, API (1987) introduced into the main text of RP2A new procedures for calculating unit friction in clays, based on a review of pile load test data (Olson and Dennis, 1982) by Randolph and Murphy (1985). In these procedures, the ratio of undrained shear strength to effective overburden pressure, rather than S_u alone, is used to define α. Thus:

$$\alpha = 0.5(S_u/\sigma'_v)^{-0.5} \quad \text{for} \quad S_u/\sigma'_v \leq 1 \tag{1}$$

$$\alpha = 0.5(S_u/\sigma'_v)^{-0.25} \quad \text{for} \quad S_u/\sigma'_v > 1 \tag{2}$$

(where σ'_v is the effective overburden pressure), with the proviso that $\alpha \leq 1.0$. In the original paper by Randolph and Murphy the factor 0.5 in the above equations is replaced by the ratio $(S_u/\sigma'_v)^{0.5}$ for a normally consolidated soil.

While the original paper suggests that an appropriate "length effect" modification be used to account for reduction in the degree of mobilisation of skin friction as the pile length increases, no clear recommendation is given in RP2A (API, 1991) to this effect, although the topic is discussed. In practice, it may be introduced by use of an overall "length factor" or by assigning "falling branch" t–z axial response curves to a pile-soil model. However, it is not clear how much of this effect is related to the amount of pile driven past an element of soil, how much due to lateral movements during driving and how much due to progressive failure.

5.3. Recent Pile Load Tests

Because the pile load test database on which the API criteria are based is limited, there being no tests in overconsolidated clays with undrained shear strengths in excess of 400 kPa and no tests on long piles in stiff normally consolidated clays, the UK Department of Energy instigated a study in 1982 which lead to the so-called "Large Diameter Pile Tests". Details of these tests are summarised in Hobbs (1992) and in other papers to the 1992 International Conference on Large Scale Fully Instrumented Pile Tests in Clay, held in London.

Two sites were selected as having soils representative of North Sea conditions. Site investigation, laboratory testing and pile installation techniques, together with pile stiffness, were also selected to match as closely as possible those applicable offshore.

At the essentially normally consolidated site at Pentre, a 762mm diameter pile was driven 40m below a 15m sleeve into silty clay with a mean shear strength of 89 kPa. The back-figured average α value of 0.61 from the initial constant rate of penetration compression test compares with a predicted value of 0.54 for "API method 2" and 0.95 for the API RP2A main text procedure without "length effect" correction. Calculated and measured distributions of friction are given in Figure 2. It is concluded that, without some modification, the API main text procedure is unconservative for sites like Pentre.

In order to investigate appropriate "length effect" corrections, the Author analysed the Pentre test using the peak API main text unit frictions and a "falling branch" t–z curve approach (Figure 3) to model progressive failure down the pile with increasing load. In this limited study, it was found to be impossible to match the measured post-peak behaviour of the pile head, and the behaviour up to peak was only matched if a very low residual unit friction was assumed. A more satisfactory approach would appear to be use of an overall factor on calculated friction. The factor suggested by Semple and Rigden (1984), based on embedded length divided by pile diameter is insufficient, reducing the calculated α only from 0.95 to 0.91, while that of Flaate and Selnes (1977), based solely on length, gives a reasonable prediction ($\alpha = 0.57$).

An alternative explanation put forward for the high friction calculated by the API main text procedure for Pentre is that the method is inappropriate for low plasticity clays. However, it is noted that API method 2, derived from exactly the same database, predicts the Pentre result to within 11 per cent and that no clear correlation with soil plasticity emerges from a review of that database.

At the overconsolidated site at Tilbrook Grange, a 762mm diameter pile was

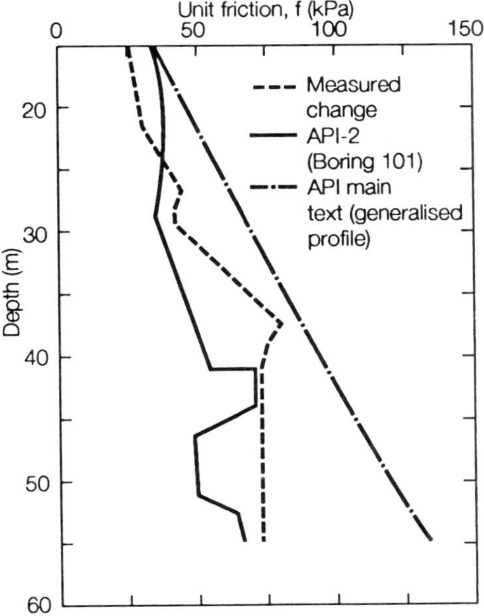

Fig. 2. Calculated and measured friction distribution for pile test at Pentre.

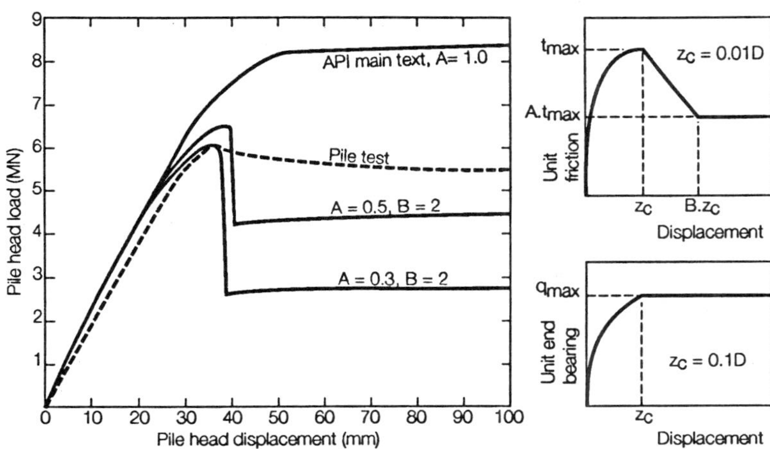

Fig. 3. Effect of "falling branch" t–z analysis on predicted capacity of Pentre pile using API main text method.

driven 30m below ground level in clays with an overall mean shear strength of 478 kPa. For this site, the backfigured average α value of 0.43 for the initial compression test compares with a predicted value also of 0.43 for API method 2

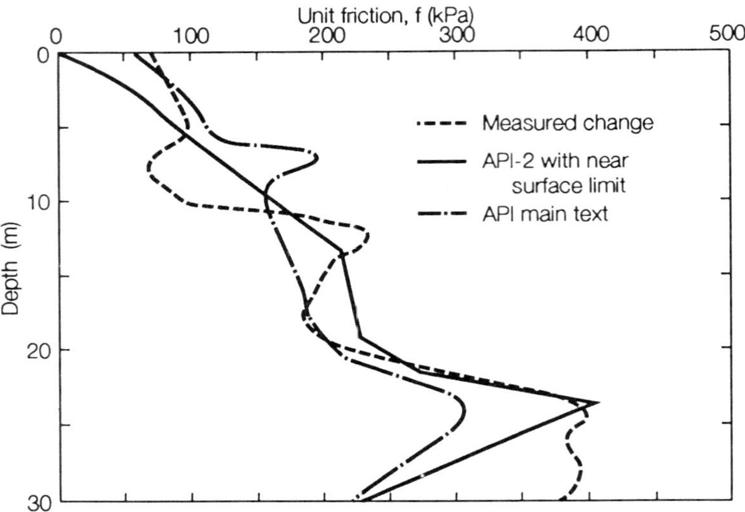

Fig. 4. Calculated and measured friction distribution for compression pile test at Tilbrook Grange.

with near surface limit, 0.5 for this method without limit and 0.40 for the API main text procedure (without "length effect" modification). The friction distribution down the pile is modelled reasonably well by both API method 2 with limit and the main text procedure (Figure 4) while the unlimited method 2 overestimates friction in the upper 10m or so.

At Tilbrook Grange, a tension test was also carried out (Clarke *et al*, 1992) on an adjacent 762mm diameter pile driven 31m below ground level (29.45m embedment). This gave a different friction distribution on the upper part of the pile (Figure 5) although the average α (0.47), based on an overall mean S_u of 487 kPa is again similar to that predicted using API method 2 with limit ($\alpha = 0.44$). The API main text procedure gave an α of 0.40. This additional test showed little gain in frictional capacity compared to the compression test, despite a much longer period to allow soil set-up after driving.

Recent load tests, on much smaller, shorter piles, have been carried out by NGI at Pentre, Tilbrook, Lierstranda and Onsøy (e.g. Karlsrud *et al*, 1992, 1993) and by Imperial College at Canons Park (e.g. Bond and Jardine, 1990, 1991).

The NGI tests at Pentre and Tilbrook gave results broadly consistent with the larger diameter pile tests at those sites. Those at Onsøy and particularly at Lierstranda gave α values lower than API method 2, as did the previously published tests at Haga (Karlsrud and Haugen, 1985). Caution should, however, be exercised in interpreting these results, since the shear strength profiles were based on tests other than UU triaxials and due to the relatively high sensitivity of these Norwegian soils, in addition to any scale effects which may exist.

The small scale Canons Park tests gave peak α values consistent with or above

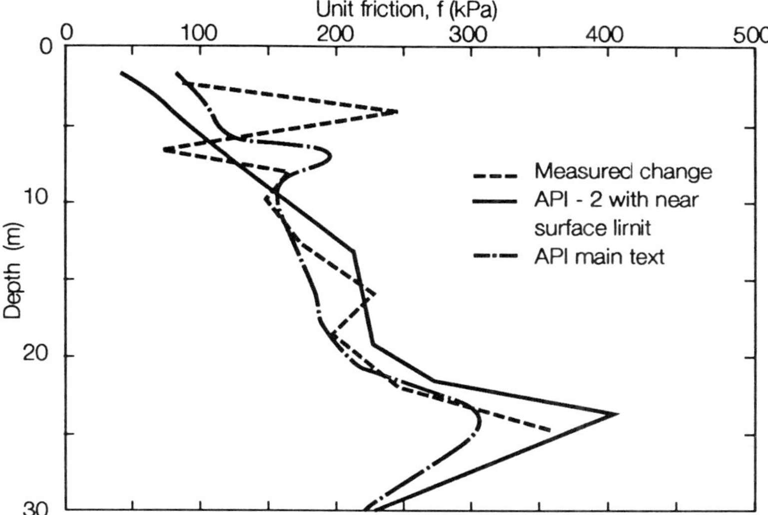

Fig. 5. Calculated and measured friction distribution for tension pile test at Tilbrook Grange.

API method 2 values, but indicated rapid post-peak strain softening.

While the recently published pile tests in clay appear to confirm the correlation between lateral effective stress and friction on a pile wall, a coherent theory to accurately predict the lateral stress (and the interface friction angle) is still awaited. Until the fundamental physics of the problem are fully understood, empirically derived, practical approaches have to be adopted. In the context of certification, these procedures should also preferably be codified.

5.4. Recommended Procedure

Following a review of the Pentre and Tilbrook Grange compression tests, it was concluded (Hobbs, 1992) that:

(a) For normally consolidated clays, unit friction should be calculated using "API method 2". The API (1991) main text method should only be used with an appropriate "length effect" correction;

(b) For overconsolidated clays, unit friction should be calculated using "API method 2" with a limit of

$$f = K_p \sigma'_v \tan(\phi' - 5^\circ) \tag{3}$$

where

$$K_p = \tan^2(45^\circ + \phi'/2). \tag{4}$$

The API (1991) main text method is recommended as an additional check;

(c) The above methods should be employed incrementally with depth.

Review of the additional pull out test on a 762mm diameter pile at Tilbrook Grange and of the smaller scale NGI and Canons Park tests have not altered these recommendations, although the reported low frictions at Lierstranda warrant further investigation.

Notwithstanding their limitations from a theoretical viewpoint, these recommendations essentially suggest no change in established North Sea practice.

6. Pile End Bearing in Clay

Pile unit end bearing in cohesive soils is conventionally taken as $9S_u$.

For the 762mm piles at Pentre the factor varied from 13.7–14.4 at peak load to 19.5–20.4 at large pile head displacement (depending upon assumed S_u profile), while at Tilbrook Grange it varied from 7.1 at peak to 10.2 at large displacement.

Since end bearing in clay usually represents only a small proportion of total capacity and in view of other data supporting the value of 9, no change is recommended in this factor.

7. Pile Friction in Sand

7.1. Established Practice

In principle, North Sea practice for calculating unit friction for driven piles in (non-carbonate) cohesionless soils is based on procedures given in the 13th edition of RP2A (API, 1982). Unit friction at any given point on a pile is defined as

$$f = K\sigma_v' \tan\delta \tag{5}$$

where K = earth pressure coefficient, σ_v' = effective overburden pressure, with due account of local and global scour and $\delta = \phi' - 5°$.

The coefficient K is taken as 0.7 for compression and 0.5 for tension, recognising experimental and theoretical evidence for lower unit friction in tension. API (1982) gives a range of values for K for compression, but the value of 0.7 is adopted in practice.

While API (1982) gives guidance as to the value of ϕ' to use for medium dense to dense soils, based on grading along (Table 1), in practice, considerable weight is given to cone penetrometer data as well as sample descriptions in the selection of parameters.

Limiting values are applied to unit friction to reflect experimental evidence that average values do not continue to rise in accordance with equation (5) (e.g. Vesic, 1970, Lehane *et al*, 1993). Specific limits are not given in API (1982) but, historically, values similar to those given by McClelland (1974) were applied

TABLE 1.
API RP2A 13th Edition Recommended Pile Design Parameters for Cohesionless Soils with Limiting Values after Mc Clelland (1974)

Soil description	ϕ (degrees)	δ (degrees)	Limiting unit friction (kPa)	N_q	Limiting unit end bearing (MPa)
Silt	20	15	47.8	8	1.9
Sandy silt	25	20	67.0	12	2.9
Silty sand	30	25	81.3	20	4.8
Clean sand	35	30	95.7	40	9.6

(Table 1). Again, in practice, selected limiting values take account of cpt results as well as sample descriptions.

7.2. Recent Developments

Following a comprehensive review of available axial pile load test data for API by Olson and Dennis (1982), the RP2A criteria for calculating pile capacity of driven piles in sand were revised (API, 1985).

As well as giving more detailed guidance on parameter selection, which reflects both grain size and density, specific limiting values were defined (Table 2) ranging from 48 to 115 kPa. A new maximum δ value of 35° (previously 30°) was introduced.

A major change was in the values specified for the earth pressure coefficient K. For the open-ended pipe piles almost universally used offshore, this parameter was set at 0.8 for both compression and tension. This change, in particular, provoked considerable debate and further review of the available pile test data ensued (Lings (1985), Briaud *et al* (1987), Olson (1988), Olson and Al-Shafei (1988), Toolan and Ims (1988), Olson (1990), Toolan *et al* (1990), Kraft (1991a, b) and Hossain and Briaud (1991), for example).

While these studies led to different interpretations of the validity of the new proposals, they highlighted, inter alia:

(a) the limited number of tests with reliable data for assessment of the criteria;

(b) lack of similarity of the tests, in terms of pile type (closed rather than open-ended), dimensions, penetration and soil density, to North Sea piling;

(c) considerable scatter in the results;

TABLE 2.
API RP2A 19th Edition Recommended Pile Design Parameters for Cohesionless Soils

Density	Soil description	δ (degrees)	Limiting unit friction (kPa)	N_q	Limiting unit end bearing (MPa)
Very loose Loose Medium	Sand Sand-silt Silt	15	47.8	8	1.9
Loose Medium Dense	Sand Sand-silt Silt	20	67.0	12	2.9
Medium Dense	Sand Sand-silt	25	81.3	20	4.8
Dense Very dense	Sand Sand-silt	30	95.7	40	9.6
Dense Very dense	Gravel Sand	35*	114.8	50*	12.0

* not recommended by Author

(d) the likelihood that unit friction in tension may indeed be lower than in compression and

(e) the suggestion that for longer piles the method may be unconservative.

Toolan and Ims (1988) included pull-out tests on conductors driven to 30 and 40m in the southern North Sea in their review and concluded that the API (1985) criteria led to overprediction of 40 to 50 per cent. Further, the large scatter in the data suggested that higher factors of safety should be used with this method.

The 19th edition of RP2A (API, 1991) acknowledged some of these reservations but only suggested that "in unfamiliar situations the designer may want to account for this uncertainty through a selection of conservative design parameters and/or safety factor".

In the North Sea, the 1985 criteria have not been adopted and the K values associated with the 13th edition (1984) RP2A have been retained, as recommended by the UK Department of Energy (Semple, 1986), Toolan and Ims (1988) and Lloyd's Register (1989).

Alternative approaches have been proposed by e.g. Toolan *et al* (1990) and Kraft (1991b), the former recognising the apparent reduction in local unit friction at any point as the pile penetration increases.

Small scale pile tests (e.g. Brucy *et al*, 1991, Lehane *et al*, 1993) and chamber tests (e.g. Foray *et al*, 1991) may help to improve understanding of the basic physical processes involved in axial capacity. However, until tests are carried out on piles more representative of those offshore in scale and stress state, installed using similar equipment and in similar soil conditions, and with site investigation techniques adopted like those in the North Sea, it is not considered to be justified to change from established practice, where this would lead to less conservative pile design. Such tests should be performed on well-instrumented piles, to investigate the effects of initial residual stresses and at various embedments, to investigate length effects. The greatest embedded length should approach that typical offshore. Both compression and tension tests should be carried out.

The above discussion is limited to soils which are derived predominantly from silica. Soils with a high carbonate content pose special problems, as discussed in papers to the 1988 International Conference on Calcareous Sediments, held in Perth and by Hobbs and Price (1989).

7.3. Recommended Procedure

As discussed above, K values of 0.7 for compression and 0.5 for tension are recommended. Parameters including limiting values should be selected on the basis of cone penetrometer data and sample descriptions, and be broadly in line with the values given in Tables 1 and 2, save that δ values in excess of 30° are not recommended.

It should not be assumed that this practice is consistently more conservative than use of the 1991 RP2A recommendations, at least for compressive loading, since pile penetrations are typically 60m in the southern North Sea while limiting values are reached at about 30m. This means that selected limiting values predominate the calculated friction capacity. Cone penetrometer data may well enable higher limiting frictions to be used than soil descriptions alone would justify and higher unit end bearing than recommended by RP2A may also be justified.

A unit friction of 120 kPa, which exceeds the highest value in RP2A is used for very dense, clean sands. Where cone data are not available, however, unit friction may be limited to 100 kPa. Aids to interpretation of relative density are given, for example, by Lunne and Christoffersen (1983) and Baldi *et al* (1986).

"Dirty" sands which have some clay content, particularly where this occurs as "occasional" inclusions or thin layers, present particular problems for parameter selection. While each case should be viewed on its merits, in general it is considered that conservative assumptions regarding unit friction should be made, as discussed later for end bearing.

The above procedure is regarded by the Author as a total package. Individual parameters K, σ'_v, δ and limits may not relate directly to theoretical or actual conditions at any specific point on the pile shaft and the distribution of friction along the pile may be erroneous.

8. Pile End Bearing in Sand

Much of the above discussion also applies to pile end bearing. However, there is less controversy in practice, with unit end bearing generally being based on some proportion of cone tip resistance with allowance for soil description. Limiting values usually apply at typical design penetrations.

Up to the limiting value, the unit end bearing is given by

$$q = N_q.\sigma'_v \tag{6}$$

where N_q is a function of ϕ' (Tables 1, 2).

For clean, dense sands, with cone resistance consistently in excess of 60 MPa, a limiting unit end bearing of 15 MPa may be used. This exceeds the maximum RP2A limit of 12 MPa (Table 2). As a general guide, for very dense sand unit end bearing of $q_c/4$ may be used while for loose or clayey materials a higher proportion of cone resistance may be adopted, approaching that for clay ($q_c/2.25$).

Where no cone data are available, limiting values should generally not exceed 10 MPa.

9. Special Considerations – End Bearing

9.1. Plugging

It is conventional to consider piles to be plugged (that is, behave as closed-ended) during quasi-static storm loading, if the sum of the internal friction (assumed to have the same unit values as external friction) and annulus end bearing (assumed to have the same unit value as for the full plug area) exceeds the fully plugged end bearing capacity. There is experimental evidence, on the small scale at least, that plugging may require a shorter length of soil (e.g. Brucy *et al*, 1991, Murff *et al*, 1992). In many practical cases, however, these assumptions have been found to have marginal impact on pile design penetrations, providing that the soil inside the pile remains close to seabed level.

Driving shoes, in which the internal diameter of the pile is reduced at the tip, are not used in current North Sea pile designs. Where they are employed, a reduction in internal friction, and thus possibly in available end bearing capacity, is assumed.

9.2. Build-up and Drop-off

At the interface between relatively weak and relatively strong end bearing strata, for example between clay and very dense sand, it is assumed that there is a transition zone before the full end bearing potential of the more competent stratum is realised (or lost).

There are conflicting data and recommendations in the literature regarding this transition. For example, Meyerhof (1976) suggests linear variation over 10 pile diameters. Semple (1980) suggests that, since at depths associated with offshore piling, the dilatancy of dense sands beneath a pile tip is suppressed by the high overburden pressure, this transition may take place over 3 pile diameters (3D).

It has been LR's practice to assume a 3D linear build-up and drop-off of end bearing and this is consistent with the recommendations of API (1991) that "full end bearing may be taken where the pile penetrates two or three diameters or more" into the stronger stratum and the tip is "approximately three diameters" above the bottom of the stratum "to preclude punch through".

9.3. Practical Considerations

Clearly the true build-up of end bearing capacity is likely to be more complicated than such a simple model implies. However, it would be imprudent to design a pile penetration on this part of a capacity curve, be it based on this or any other assumption, because of the sensitivity of the calculated capacity to small changes in penetration and uncertainty regarding the exact horizon of the stratum change. The assumed depth is affected, for example, by uncertainty in the depth accuracy of samples and cone tests, in interpolation of stratum changes between these, in variations across the site, in changes in seabed elevation (scour or accretion) between when the boreholes were drilled and the jacket installed and by differences between nominal and actual pile batter.

The same concerns regarding the exact location of underlying weaker strata lead to the following general recommendations, to guard against punch through failure:

(a) found piles at least 5 diameters above a significantly weaker stratum or

(b) ensure that the factor of safety in the underlying stratum is sufficiently high.

It follows that piles should not be designed to rely upon the peak capacity of a "blip" in the capacity curve.

This can, however, have economic consequences. Figure 6 shows the axial capacity curve for a 1.219m pile for a southern North Sea site. An 8m (6.6D) zone at 43.75m provides a calculated capacity rising from 13.4MN to 25.5MN within only 3.4m, and 26.3MN within 4.5m. An ultimate pile capacity of, say, 25MN is just achieved at about 47m and, unless advantage is taken of this, the pile would have to penetrate to 76m to achieve the required capacity. While the additional driving time may not be long with modern hammers, if another add-on has to be welded to piles installed through the jacket legs, more substantial delays may ensue. An alternative is to increase the pile diameter. Again, if the piles are to be installed through the jacket legs, the consequences of increasing pile diameter are significant because of the knock-on effects on leg diameter and thus wave loads. If the piles are to be installed through vertical sleeves, the impact is less significant, both for increased length and diameter.

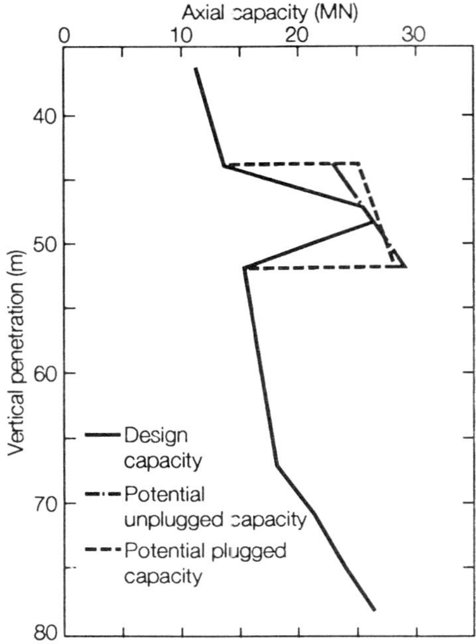

Fig. 6. Example of effect of end bearing stratum of limited extent on calculated axial pile capacity.

9.4. Clayey Inclusions in Dense Sand

Where piles are to be founded in interbedded sand and clay layers, end bearing should be based on the weaker soil, i.e. clay.

Where clayey inclusions appear to be occasional, from the evidence of available cone penetrometer data and examination of samples, great care has to be taken in assigning unit end bearing. While it may be argued that rare clayey inclusions, small in comparison to pile diameter, will have little effect on plugged end bearing capacity, the variability of soil due to its depositional history and the incomplete coverage of stratigraphy by *in situ* testing or sampling, even in the best site investigations, may mean that, at the pile location, the inclusions are more frequent. Experience from actual pile installations suggests that "occasional" clayey inclusions should not be discounted and that "broadbrush" parameters should be avoided.

10. Eurocode/API RP2A LRFD

A draft "Load and Resistance Factor Design" (LRFD) version of RP2A, based on limit state principles, has been published for industry review (API, 1989b).

It is understood (Offshore Engineer, 1991) that, initially, a revised version of this document may be adopted by ISO as an interim code, with appendices for various geographical areas reflecting specific regional aspects and regulations. At a later date, the ISO code may be adopted as the Eurocode for fixed offshore structures.

The 1989 draft is based on the 18th (Working Strength Design, WSD) edition of RP2A (API, 1989a) with load and resistance factors selected with the stated aim of achieving the same average reliability or risk.

Differences between North Sea practice and the recommendations of RP2A regarding calculation of axial pile capacity have been discussed in previous sections. Any differences which persist when the LRFD code is finalised may be reflected in the regional appendix. However it is noted that RP2A, like other codes, does not preclude the use of alternative methods.

Comparison between the LRFD and WSD approaches is complicated by the different factors applied to the various components forming the overall gravity load, the different proportions of gravity and environmental load from platform to platform, and the fact that load distribution to individual piles will depend upon foundation nonlinearity. These factors may result in either more or less conservative design by LRFD compared to WSD.

Because the required structural analysis includes an (unfactored) nonlinear foundation model, and loads are factored before the analysis is carried out, the effects of the foundation nonlinearities will be increased. Apart from any effects on axial capacity design, this will change the distribution of axial loads and bending moments down the pile. Checks should be made to ensure that this does not lead to less conservative design of pile strength.

A preliminary review of the first draft (API, 1989b) by Lloyd's Register concluded that for axial pile design, compared to the WSD approach, the LRFD procedure would generally result in:

a) less conservative design of piles for still water and operating conditions;

b) less conservative design of piles for shallow water production platforms, when gravity load predominates;

c) more conservative design of piles for deep water platforms, when the environmental load component predominates.

It should be noted that the API RP2A LRFD document referred to herein is in draft form only and it is understood that partial factors are under review, particularly those for still water and operating conditions. It is likely that alternative partial factors will be included in the regional appendices.

11. Installation

Specialist Lloyd's Register engineers attend pile installation offshore to ensure that adequate pile penetration is achieved, this being judged on the basis of actually encountered stratification (as interpreted from driving behaviour), individual pile or pile group loads from independent structural analysis and independently derived axial capacity curves.

Experience with properly selected, established hydraulic hammers indicates that pile "refusal", for current pile designs, is unlikely. However, occasionally "refusal", at a depth less than design penetration, does occur – perhaps with an undersized steam hammer – and more capable hammers are not available on site or may overstress the pile.

In these situations the first step is to review factors of safety for the as-installed piles taking account of the factors above. In many cases, review of design criteria may reveal that conservative assumptions have been made about possible deck loadings, future wells and risers or, indeed, about extreme storm conditions. Since remedial action involving soil removal may be detrimental to pile capacity then, if the calculated factor of safety of the foundation as a whole is close to that required, consideration would be given to accepting the consequences of the reduced factors of safety.

A more common occurrence is blowcounts less than predicted. If this gives cause for concern, a review similar to that for refusal can be performed and re-drive tests can be carried out, after a suitable delay, to give an indication of "set up" in cohesive soils.

While pile or hammer instrumentation may aid the interpretation of soil stratification or of set-up effects, by identifying blowcount changes due solely to changes in hammer energy, it is not considered that the state of the art is such that the link between dynamic and static capacity for offshore pipe piles has been clearly established. The mechanism involved in pile driving (basically unplugged) differs from that during environmental loading (usually plugged) and rates and duration of loading also differ. Experience indicates that non-unique "static resistance at time of driving" may be back-analysed. Unlike onshore practice, site-specific correlation between pile tests and driving resistance is not available to the offshore geotechnical engineer, nor are time and cost restraints likely to allow this.

12. Conclusions

A basis for site investigation and design for the axial capacity of offshore piling has been presented which should satisfy the requirements of certification.

A review has been made of recent research into the axial capacity of driven piles, including load tests on piles in clays selected to be representative of North Sea conditions. It is concluded that, in the main, previously established design procedures should be retained until reliable, practical methods are developed which capture the fundamental physics of axial behaviour of driven pipe piles and, in the case of sand sites, until representative pile tests have also been carried out

Guidance is given to the selection of parameters for interbedded soils and for founding piles on relatively thin end bearing strata. In the latter case practical, rather than theoretical, considerations may govern.

Acknowledgements

The Author wishes to thank Lloyd's Register for permission to publish this paper. However, opinions expressed are those of the Author and do not necessarily reflect Lloyd's Register policy. Thanks are also due to colleagues in Offshore Division for comments on the draft manuscript.

References

1. API (1982, 1985, 1986, 1987, 1989a, 1991), 'RP2A: Recommended Practice for Planning, Designing and Constructing Fixed Offshore Platforms', 13th, 15th, 16th, 17th, 18th, 19th editions.
2. API (1989b), 'RP2A-LRFD: Draft Recommended Practice for Planning, Designing and Constructing Fixed Offshore Platforms – Load and Resistance Factor Design', 1st edition.
3. Baldi, G., Bellotti, R., Ghionna, V. N., Jamiolkowski, M., and Lancelotta, R. (1986), 'Interpretation of CPT's and CPTU's, Part 2: Drained penetration in sands', *Proc. 4th Intl. Geotech. Seminar on Field Instrumentation and In Situ Measurements*, Singapore.
4. Bond, A. J. and Jardine, R. J. (1990), 'Research on the Behaviour of Displacement Piles in an Overconsolidated Clay', UK Dept. of Energy Report OTH 89 296, HMSO.
5. Bond, A. J. and Jardine, R. J. (1991), 'Effects of installing displacement piles in a high OCR clay', *Géotechnique* **41**(3), 341–363.
6. Briaud, J.-L., Anderson, J., and Perdomo, D. (1987), 'Evaluation of API method using 98 vertical pile load tests', *Proc. 19th Offshore Technology Conf.*, Houston, Vol. 1, Paper OTC 5411, pp. 447–451.
7. Brucy, F., Meunier, J., and Nauroy, J. F. (1991), 'Behaviour of pile plug in sandy soil during and after driving', *Proc. 23rd Offshore Technology Conf.*, Houston, Paper OTC 6514, pp. 145–154.
8. Clarke, J., Long, M. M., and Hamilton, J. (1992), 'The axial tension test of an instrumented pile in over-consolidated clay at Tilbrook Grange', *Proc. Conf. on Recent Large Scale Fully Instrumented Pile Tests in Clay*, ICE, London.
9. Department of Energy (1990), 'Offshore Installations: Guidance on Design, Construction and Certification', 4th edition, HMSO.
10. Flaate, K. and Selnes, P. (1977), 'Side friction of piles in clay', *Proc. 9th Intl. Conf. Soil. Mech. Fndn. Eng.*, Tokyo, Vol. 1, pp. 517–522.
11. Foray, P., Labanieh, S., Mokrani, L., and Colliat-Dangus, J. L. (1991), 'Étude de la capacité portante des pieux dans les sables à partir d'essais en chambre d'étalonnage', *Proc. Colloque Intl. des Fondations Profondes*, Presses Ponts et Chaussées, Paris, pp. 169–176.
12. Hobbs, R. (1992), 'The impact of axial pile load tests at Pentre and Tilbrook on the design and certification of offshore piles in clay', *Proc. Conf. on Recent Large Scale Fully Instrumented Pile Tests in Clay*, ICE, London.
13. Hobbs, R. and Price, J. L. (1989), 'Certification of offshore structures founded on carbonate soils', *Proc. 8th Intl. Conf. on Offshore Mechanics and Arctic Eng.*, The Hague, Vol. 1, pp. 541–548.
14. Hobbs, R. and Waller, J. (1992), 'Pile Driving Vibrations Database', Final Report to Health and Safety Executive, Project P2393, Report OTO-92-008.
15. Hossain, M. K. and Briaud, J.-L. (1991), 'Critical assessment of existing data for pipe piles in sand subjected to monotonic axial loading', Dept. of Civil Engineering, Texas A and M University.
16. Karlsrud, K., Hansen, S. B., Dyvik, R., and Kalsnes, B. (1992) 'NGI's pile tests at Tilbrook and Pentre - Review of testing procedures and results', *Proc. Conf. on Recent Large Scale Fully Instrumented Pile Tests in Clay*, ICE, London.
17. Karlsrud, K. and Haugen, T. (1985), 'Axial static capacity of steel model piles in overconsolidated clay', *Proc. 11th Intl. Conf. Soil Mech. Fndn. Eng.*, San Francisco.

18. Karlsrud, K., Kalsnes, B., and Nowacki, F. (1993), 'Response of piles in soft clay and silt deposits to static and cyclic loading based on recent instrumented pile load tests', *Proc. Intl. Conf. on Offshore Site Investigation and Foundation Behaviour*, SUT, London.
19. Kraft, L. M. (1991a), 'Performance of axially loaded pipe piles in sand', *J. Geotech. Eng., ASCE*, **117**(GT2), 272-296.
20. Kraft, L. M. (1991b), 'Computing axial pile capacity in sands for offshore conditions', *Marine Geotechnology* **19**(1), 61–92.
21. Lehane, B. M., Jardine, R. J., Bond, A. J., and Frank, R. (1993), 'Mechanisms of shaft friction in sand from instrumented pile tests', *J. Geotech. Eng., ASCE* **119**(GT1), 19–35.
22. Lings, M. (1985), 'The Skin Friction of Driven Piles in Sand', MSc. dissertation, Univ. London.
23. Lloyd's Register (1989), 'Rules and Regulations for the Classification of Fixed Offshore Installations', Part 3, Chapter 2.
24. Lunne, T. and Christoffersen, H. P. (1983), 'Interpretation of cone penetrometer data for offshore sands', *Proc. 15th Offshore Technology Conf.*, Houston, Vol. 1, Paper OTC 4464, pp. 181–192.
25. Meyerhof, G. G. (1976), 'Bearing capacity and settlement of pile foundations, 11th Terzaghi Lecture', *J. Soil Mech. Fndn. Eng., ASCE* **102**(GT3), 197–228.
26. McClelland, B. (1974), 'Design of deep penetration piles for ocean structures', *J. Geotech. Eng., ASCE* **100**(GT7), 709–747.
27. Murff, J. D., Raines, R. D., and Randolph, M. F. (1990), 'Soil plug behavior of piles in sand', *Proc. 22nd Offshore Technology Conf.*, Houston, Vol. 4, Paper OTC 6421, pp. 25-32.
28. Offshore Engineer (1991), 'Industry keeps its hands on offshore Eurocode', Offshore Engineer, February, p. 16.
29. Olson, R. E. (1988), 'Comparison of Measured Axial Load Capacities of Steel Pipe Piles in Sand with Capacities Calculated Using the 1986 API Recommended Practice (RP2A)', Final Report to API, Project PRAC 86-29A, Univ. Texas at Austin.
30. Olson, R. E. (1990), 'Axial load capacity of steel pipe piles in sand', *Proc. 22nd Offshore Technology Conf.*, Houston, Vol. 4, Paper OTC 6419, pp. 17–24.
31. Olson, R. E. and Al-Shafei, K. S. (1988), 'Axial load capacities of steel pipe piles in sand', *Proc. 2nd Intl. Conf. on Case Histories in Geotech. Eng.*, Univ. Missouri, Rolla, Vol. 3, pp. 1731–1738.
32. Olson, R. E. and Dennis, N. D. (1982), 'Review and Compilation of Pile Test Results, Axial Pile Capacity', Final Report to API on project PRAC 81-29, Univ. Texas at Austin.
33. Randolph, M. F. and Murphy, B. S. (1985), 'Shaft capacity of driven piles in clay', *Proc. 17th Offshore Technology Conf.*, Houston, Vol. 1, pp. 371–378.
34. Semple, R. M. (1980), 'Discussion', *Recent Developments in the Design and Construction of Piles*, ICE, London, Addenda and Corrigenda, p. 5.
35. Semple, R. M. (1986), 'Background to Guidance on Foundations and Site Investigations for Offshore Structures', Report of Department of Energy Guidance Notes Revision Working Group, HMSO.
36. Semple, R. M. and Rigden, W. J. (1984), 'Shaft capacity of driven pipe piles in clay', *Proc. Symp. on Analysis and Design of Pile Foundations, ASCE*, San Francisco.
37. Statutory Instrument 289 (1974), 'The Offshore Installations (Construction and Survey) Regulations', HMSO.
38. Toolan, F. E. and Ims, B. W. (1988), 'Impact of recent changes in the API recommended practice for offshore piles in sand and clays', *Underwater Technology* **14**(1), 9–13, 29.
39. Toolan, F. E., Lings, M. L., and Mirza, U. A. (1990), 'An appraisal of API RP2A recommendations for determining skin friction of piles in sand', *Proc. 22nd Offshore Technology Conf.*, Houston, Vol. 4, Paper OTC 6422, pp. 33–42.
40. Vesic, A. S. (1970), 'Tests on instrumented piles, Ogeechee River site', *J. Soil Mech. Fndn. Eng., ASCE* **96**(SM2), 561–584.

THE EVOLUTION OF OFFSHORE PILE DESIGN CODES AND FUTURE DEVELOPMENTS

F. E. TOOLAN and M. R. HORSNELL
Fugro-McClelland Limited, 18 Frogmore Road, Hemel Hempstead, Hertfordshire HP3 9RT

Abstract. This paper reviews the evolution of offshore codes, and in particular API RP2A, in relation to axial pile capacity design procedures and comments on the need for, and scope of, future changes. It also discusses progress towards a new international code. The conclusion is that the API system has shown itself responsive of industry needs and that any new system should try to emulate this. With regard to design practice, the paper concludes that this has improved significantly over the years for piles in clay, but much less so for piles in sand.

1. Introduction

The offshore industry and various regulatory bodies have recently agreed to commence work on an international code for the design and construction of offshore structures. Such a document will include sections on pile design and some governments may make compliance mandatory in their economic zones; elsewhere it will have a strong influence on design practice. Thus the drafting process is something which foundation engineers should follow closely and positively influence when necessary.

Those writing this new document will not start with a blank sheet of paper, but will modify sections of existing codes into an internationally acceptable format. In this paper the Authors have had to use the term "code" rather loosely, and for the sake of simplicity, to cover various rules, regulations, recommended practices and guidance notes, as well as codes themselves.

With regard to pile design procedures there are two types of code in current use: prescriptive and performance related. Prescriptive codes recommend the use of certain computational procedures whereas performance related codes provide "motherhood" statements as to what should be considered in the design process. All prescriptive and most performance related codes also specify minimum factors of safety or material and load factors as the case may be. Prescriptive codes require revision whenever new test results or field experiences show a design formula needs amendment. Performance related codes only need to be revised when the philosophy of design changes. On the face of it performance related codes have all the advantages: they give the designer flexibility, are easier and cheaper for the codifying authority to administer and rarely require revision.

Volume 28: Offshore Site Investigation and Foundation Behaviour, 751–772, 1993.

In practice the special demands of the offshore industry frequently result in such codes either having Appendices with suggested design procedures or an extensive bibliography directing the user to approved techniques. Complications may arise with the performance criteria specifying factors of safety which are not compatible with the reliability of "freedom of choice" design procedures. Finally, if an industry accepted prescriptive code and a performance code co-exist, the designer is likely to check his "freedom of choice" result against the formulae in the prescriptive code. Thus the prescriptive codes become the benchmarks for all the others.

For offshore codes the benchmark pile design procedures are contained in the Recommended Practice for Planning, Designing and Constructing Fixed Offshore Platforms published by the American Petroleum Institute (API), i.e. API Recommended Practice 2A (RP2A). These have evolved over the years since the first edition of RP2A was published in October 1969 [1]. Changes to the pile design recommendations have regularly been made due to the availability of new pile test data from around the world, field observations of the behaviour of offshore piles, different installation techniques and changes to offshore practices in general. In addition, local experiences have given rise to variations on the API Recommendations and such variations have become established within the application of design procedures without necessarily appearing within RP2A or any other similar publication. In this context RP2A states unequivocally that use of its procedures does not obviate the need for applying sound judgement nor should engineers be inhibited from using other design practices. The evolution of the "Foundation Section" of API RP2A is shown in Table 1.

This paper reviews the evolution of offshore codes, and in particular API RP2A, in relation to pile design procedures and the need for, and scope of, future changes. Due to space limitations the Authors have focused only on the determination of axial pile capacity. This paper should prove useful to those involved with the new international code both as drafters and subsequently as users. In addition, for engineers currently involved in recertification or safety case preparation for existing structures, it may provide clarification as to why the original foundation design does not comply with current practice.

2. Design Practice Prior to API RP2A

The first paper specifically addressing the design of piles for offshore platforms was presented at a conference in 1967 and subsequently published in 1969 by McClelland *et al.* [2]. It is understood that the design procedures described therein were developed in 1963. The paper set out the basic formulae for determining the static capacity of piles which were carried over to API RP2A and remain current today (and where definition of terms may be found), i.e.

TABLE 1.
Evolution of API RP2A foundation design section.

EDITION	DATE	NO. OF PAGES	COMMENT
1st.	Oct-69	1.5	Sub-heading "Allowable Soil Stress"
2nd.	Jan-71	2	
3rd.	Jan-72	2	
4th.	Oct-72	3	Sub-heading "Foundation Design"
5th.	Jan-74	4	
6th.	Jan-75	7.5	
7th.	Jan-76	7.5	
8th.	Apr-77	7.5	
9th.	Nov-77	7.5	
10th.	Mar-79	11.5	Inclusion of Commentary on Foundations
11th.	Jan-80	16	
12th.	Jan-81	16	
13th.	Jan-82	16	
14th.	Jul-84	16.5	
15th.	Oct-84	16.5	
16th.	Apr-86	16.5	
17th.	Apr-87	17	Inclusion of Commentary on Pile Capacity
18th.	Sep-89	21	Inclusion of Commentary on Pile Capacity for Axial Cyclic Loads
19th.	Aug-91	23	Inclusion of commentary on Carbonate Soils

$$
\begin{aligned}
Q = Q_s + Q_p &= fA_s + qAp \\
\text{for clay } f &= \alpha c \qquad \text{and} \qquad q = N_c c \\
\text{for sand } f &= Kp'_o \tan\delta \qquad \text{and} \qquad q = p_o N_q
\end{aligned}
$$

McClelland *et al.* noted the importance of stress history in determining α and the considerable variation in the computed capacities of piles in sand from different published values of N_q and K, all from respected sources. The question of limiting values for friction in both sand and clay was discussed as was limiting end bearing in sand. The recommended values were:

for normally consolidated clay	f	$=$	c
for overconsolidated clay	f	$=$	48 kPa,
	or f	$=$	$0.25p_o$
	whichever is greater		
	but f	$\not>$	c
for all clays	q	$=$	$9c$
for medium dense clean sand	δ	$=$	$30°$
in compression	K	$=$	0.7
in tension	K	$=$	0.5
	f_{max}	$=$	96 kPa
	N_q	$=$	41
	q_{max}	$=$	9.6 MPa

Factors of safety for use in conjunction with these parameters are not explicitly quoted, but the impression given is that they were low and carried a real risk of failure in a hurricane when platforms would be evacuated and there would be no danger to life. McClelland *et al.* point out the urgent need for more pile load tests to calibrate design procedures. The subsequent changes in values of design parameters in API RP2A are traced below, separately for piles in clay and sand.

3. Pile Capacity in Clay

3.1. API RP2A – 1st to 5th Editions

When first published in October 1969, API RP2A reflected design practice for the US offshore industry at that time. The majority of offshore developments were concentrated in the Gulf of Mexico along the Texas and Louisiana coastlines where pile design was governed by substantial deposits of normally consolidated clays, with occasional inter-bedded sand strata. Recommended practice for pile design in clay was that friction should not exceed the undrained shear strength of the clay as measured by unconfined compression tests or miniature vane tests. For penetrations of less than 30.5 m friction should not exceed 48 kPa. For penetrations greater than 30.5 m:

$$f \not> 0.33p_o.$$

It should be noted that there was no explicit comment differentiating between normally and overconsolidated clays and that there was an increase over McClelland *et al.* [2] recommendations in the possible relationship between friction and

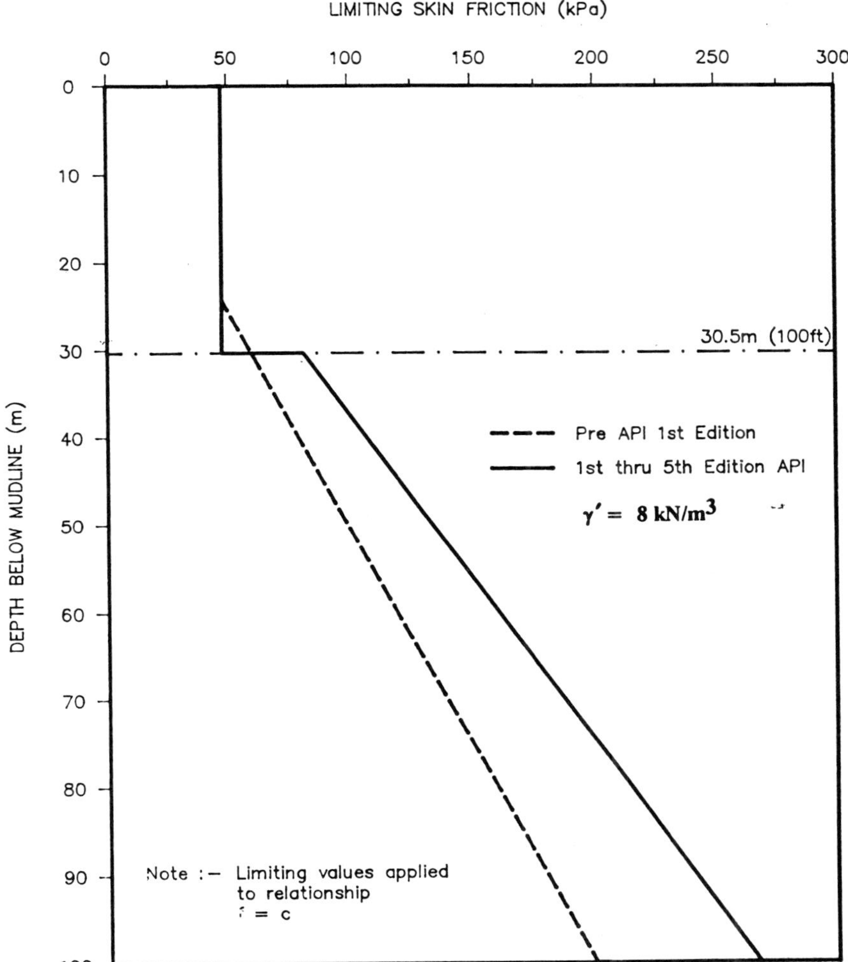

Fig. 1. Comparison of pre API and API 1st – 5th design criteria for clay.

overburden pressure. A comparison of limit values is presented in Figure 1. End bearing was as previously recommended and has remained so ever since. Factors of safety for use in conjunction with the procedures were provided and they were:

for the design environmental condition 1.5
for the operating environmental condition 2.0.

And these have also remained unchanged to the present day.

Dissent began almost immediately, due to new pile test data from the North Sea being published by Fox *et al.* [3] and the API recommendations were probably followed more in breach than compliance. A new pile design method, the Lambda method, was published in 1972 by Vijayvergiya and Focht [4] who reanalysed a total of 42 tests which became the basis of the API data base. This method was almost universally used for Gulf of Mexico platforms and may be summarised as follows:

$$\begin{aligned} Q_s &= \lambda(\sigma_m + 2c_m)A_s \\ \lambda &= \text{factor which reduced with pile penetration} \\ \sigma_m &= \text{mean effective vertical stress on the pile} \\ c_m &= \text{mean undrained shear strength along the pile shaft} \end{aligned}$$

The Lambda method was used in the North Sea but its application to highly stratified soils was difficult and attempts to develop an incremental approach gave rise to serious intellectual difficulties. For example, if a hard clay layer overlay a stiff layer the pile capacity could be much greater than for a stiff layer overlying a hard layer, all other factors being equal. Not only was this difficult to explain, but it was contrary to other theories and observations of driving behaviour.

Most platforms in the North Sea were designed using a variety of methods of computing skin friction and, in particular for overconsolidated clays:

$$f = 0.5c$$

with c being determined from undrained triaxial compression tests and correlations with *in situ* cone penetration tests (CPT's). A range of acceptable pile penetrations was calculated with driving observations being used to determine final penetration and/or capacity.

3.2. API RP2A – 6TH EDITION

The first major change to pile capacity codes for clays appeared in the sixth edition of RP2A published in January 1975 and reflects the increased requirement for a design code for piles in soils other than those encountered in the Gulf of Mexico. An attempt was also made to bring API recommendations into line with the practice at that time in both the Gulf of Mexico and the North Sea. This was done by making α equal to 1 for shear strengths up to 24 kPa, decreasing linearly to 0.5 at a strength of 72 kPa and remaining constant at 0.5 for all strengths in excess of 75 kPa. This criterion, together with the API database, is presented on Figure 2.

The influence of this change in recommendation on predicted pile capacity is shown in Figure 3 for a typical North Sea type soil profile and Figure 4 for Gulf of Mexico soil conditions.

It must be remembered that the Lambda method, which was in use in the Gulf of Mexico at that time, gave results for typical piles only 10 to 20 percent higher than this new approach.

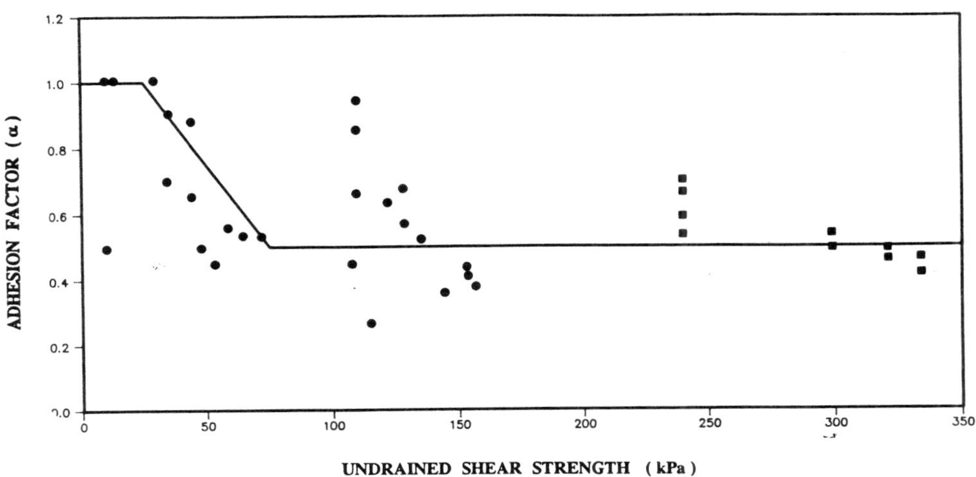

Fig. 2. Comparison of API database and API 6th edition design criteria for clay.

3.3. API RP2A – 7th to 16th Edition

The recommended criteria of the 6th edition took no account of the stress history of the soil. Thus, for a stiff normally consolidated clay, as encountered at depth in the Gulf of Mexico, the 6th edition recommended an adhesion factor of 0.5 compared with a pre 6th edition value of 1.0.

Questions needed to be asked about the safety of structures installed prior to 1972 when the Lambda method came into use. A study by McClelland and Cox [5] published in 1976 into the actual performance of piled offshore structures showed that although piles had been overloaded during hurricanes, none had actually failed due to lack of soil adhesion. In fact, there was evidence that Gulf of Mexico design practice prior to 1972 was actually overconservative during transient loading, although not for sustained loading. In addition, results from new pile load tests from the Gulf of Mexico became available [6] although these were not published until 1979. Thus without delay API revised the recommendations in the 7th edition, published in January 1976, to incorporate both methods, i.e. Method 1 (pre 6th edition) for "highly plastic Gulf of Mexico type soils" and Method 2 (6th edition) for "other soil types".

In reintroducing the pre 6th edition method for highly plastic clays, a few minor changes were made. The 30.5 m definition was removed such that for "under" or "normally" consolidated clays, the unit friction was equated to the undrained shear strength of the clay. For "overconsolidated" clays, unit friction was equated to an

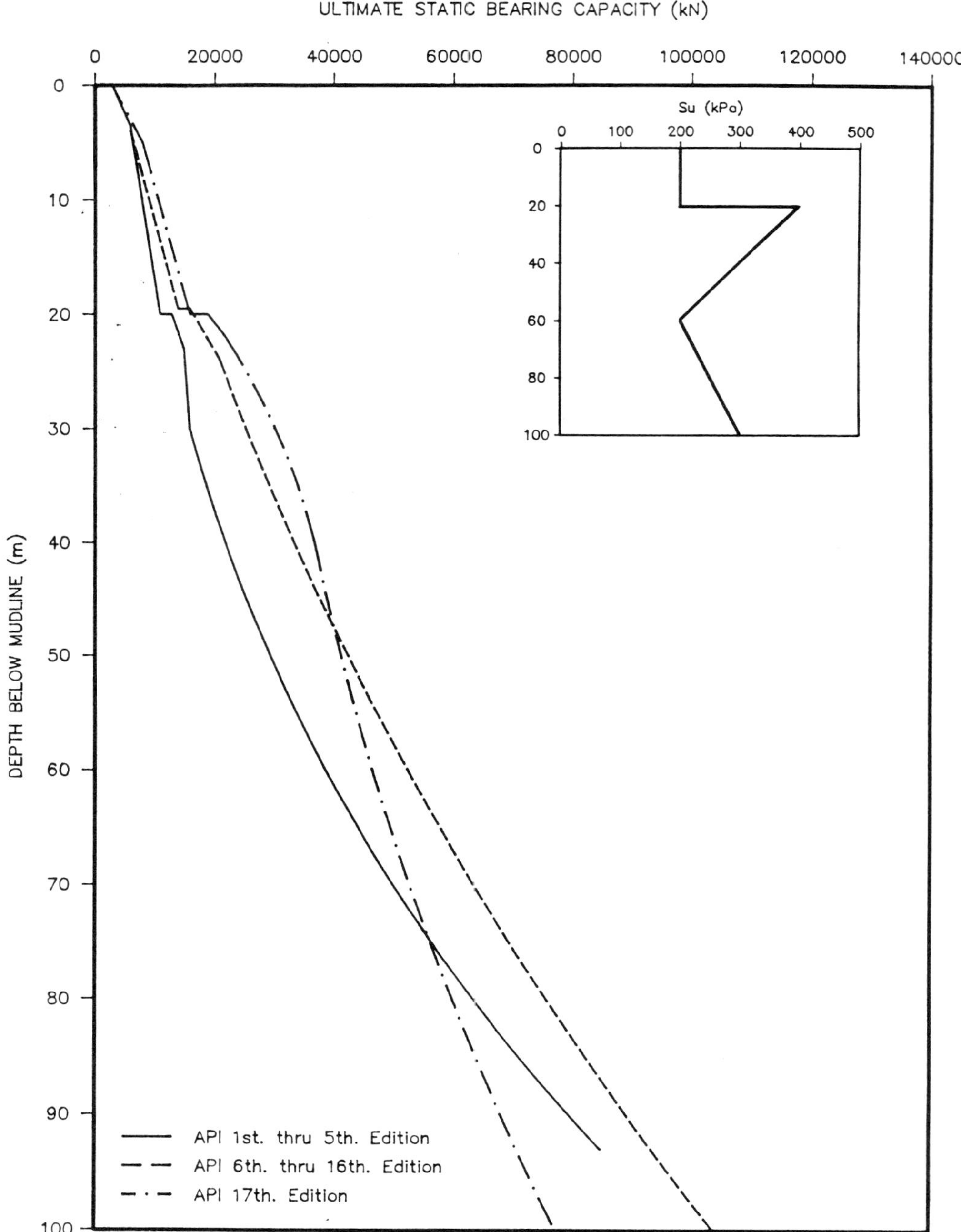

Fig. 3. Comparison of API pile design criteria for clay – stiff clay profile.

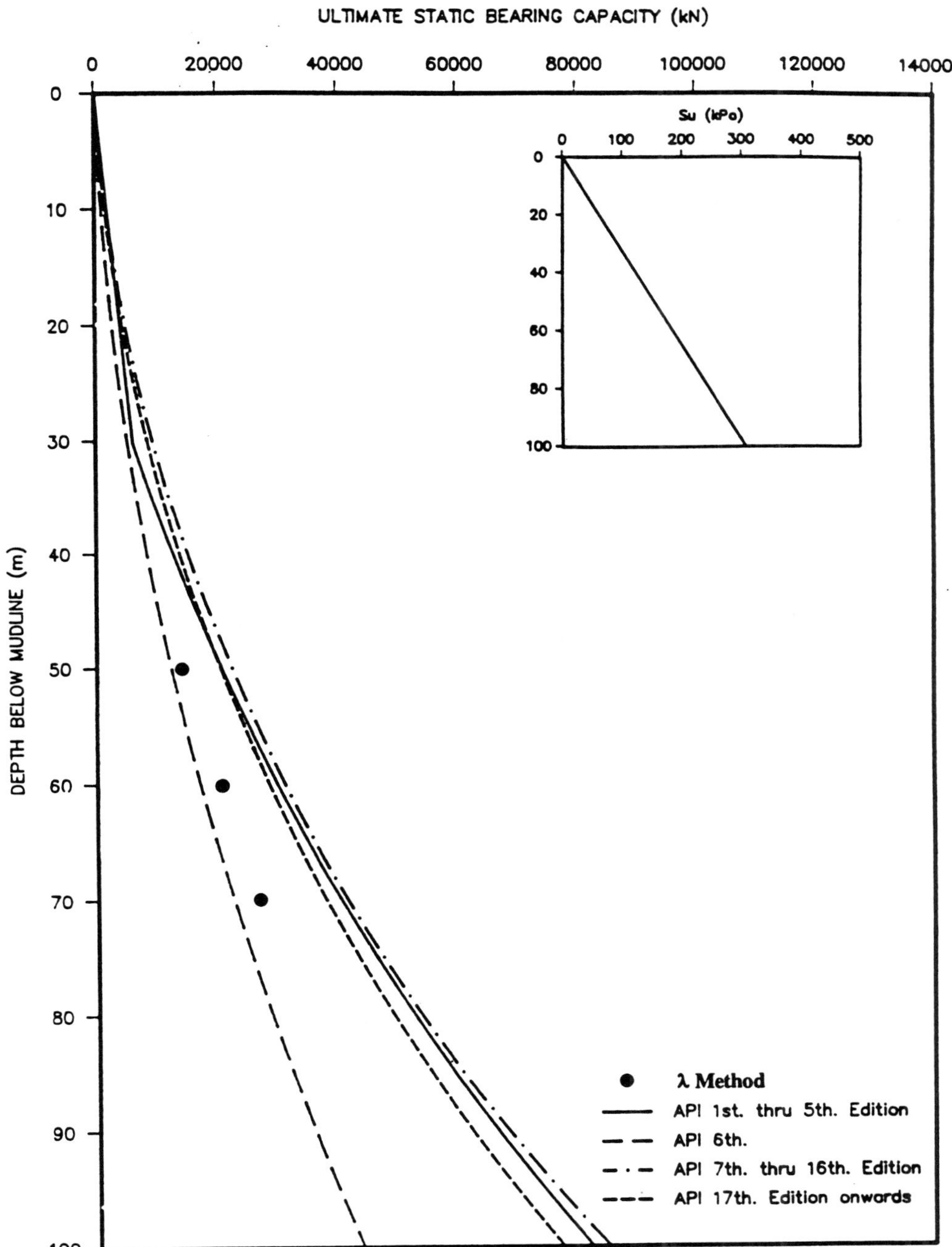

Fig. 4. Comparison of API pile design criteria for clay – normally consolidated clay.

equivalent normally consolidated shear strength, or a value of 48 kPa, whichever was the greater. As can be seen from Figure 4, the removal of the 30.5 m limit resulted in a slight increase in overall capacity.

The use of the Lambda method declined, but engineers dealing with long piles started taking account of length effects, particularly when using Method 1.

3.4. API RP2A – 17TH EDITION ONWARDS

The next significant change to the design code for clays appeared in the 17th edition published in April 1987. The division of soils into two types, based apparently on plasticity and geography, was not considered satisfactory and much research effort went into developing a design procedure applicable to all types of clay. Of concern in North Sea soils was the level of frictional capacity that could be derived from the heavily overconsolidated soils frequently encountered at shallow depths. API Method 2 suggested a value of one-half of the shear strength whereas theoretical considerations suggested that due to the level of overconsolidation the soil was at, or close to, a state of passive failure and hence unable to sustain high shear stresses at the pile-soil surface. Several studies were undertaken to establish a relationship between adhesion factor and *in situ* stress conditions, the latter commonly being referenced by the shear strength ratio c/p'_o (see Kraft *et al.*, [7], Semple and Rigden, [8], and Randolph and Murphy, [9]). As a result, the 17th edition of RP2A recommended that α be taken as equal to one-half of the shear strength ratio raised to the power of either -0.5, for shear strength ratios of 1 or less, or -0.25 for ratios greater than 1. A comparison of these recommendations with API Method 2 is presented in Figure 5. As can be seen, for hard clays at shallow depth the adhesion factor drops below 0.5 and approaches a value of approximately 0.3. For normally consolidated clays a value approaching 1.0 is computed. The maximum permitted value of α is 1.0. A comparison of the recommendations with the API database is presented in Figure 6.

The need for judgement in using this approach is stressed in particular for long piles. A commentary is now provided which recommends that c should be determined from unconsolidated undrained triaxial compression tests on specimens taken by thin walled push samplers. The commentary also includes Methods 1 and 2 from the previous editions and much more information on the length effect.

The 18th edition introduced a separate and important commentary on pile capacity for cyclic loading.

3.5. VARIATIONS TO API RP2A

As discussed above, there has always been general concern regarding the application of predicted high friction values in very stiff to hard clays encountered at shallow penetrations based upon the variable α approach first proposed in the 6th edition of API. Certainly applications of the code published in the 17th edition

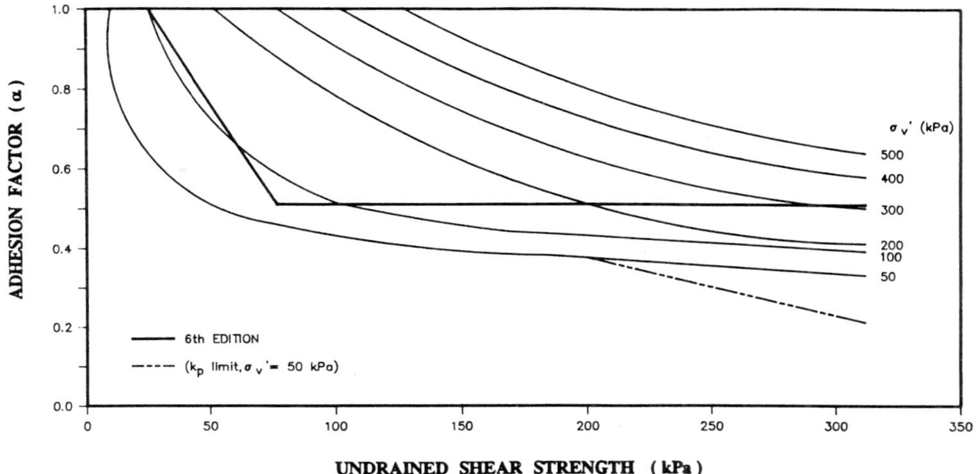

Fig. 5. Comparison of "old API" and "new API" pile design criteria for clay.

will result in lower friction values for such soils, however in addition to this some designers advocate the use of a K_p limit, i.e. a limit based upon the relationship:

$$f_s = K_p p'_o \tan\delta.$$

This relationship is plotted on Figure 5 for a depth of approximately 5 m below mudline. Unlike sands (discussed later in this paper) no design code has ever advocated the adoption of a maximum numerical value of shaft friction. Again, however, within the industry "unofficial" limits have been proposed on occasions, the most common being 250 kPa, a somewhat artificial value based upon the limit of available pile load test data. Recent results from the Large Diameter Pile Load Test (LDPT) sponsored by BP [10] indicate that friction values well in excess of this value can be mobilised (see Figure 7).

4. Pile Capacity in Sand

4.1. API RP2A – 1st and 2nd Editions

For pile design in sands, the first edition of API RP2A reflected current practice at that time in the Gulf of Mexico. Values of d, $f_{\max}$, N_q and $q_{\max}$ were tabulated for various sand and silt contents with the proviso that they were only applicable for medium dense to dense materials. K was to be taken as 0.7 for compressive loads and 0.5 for tension. The values were:

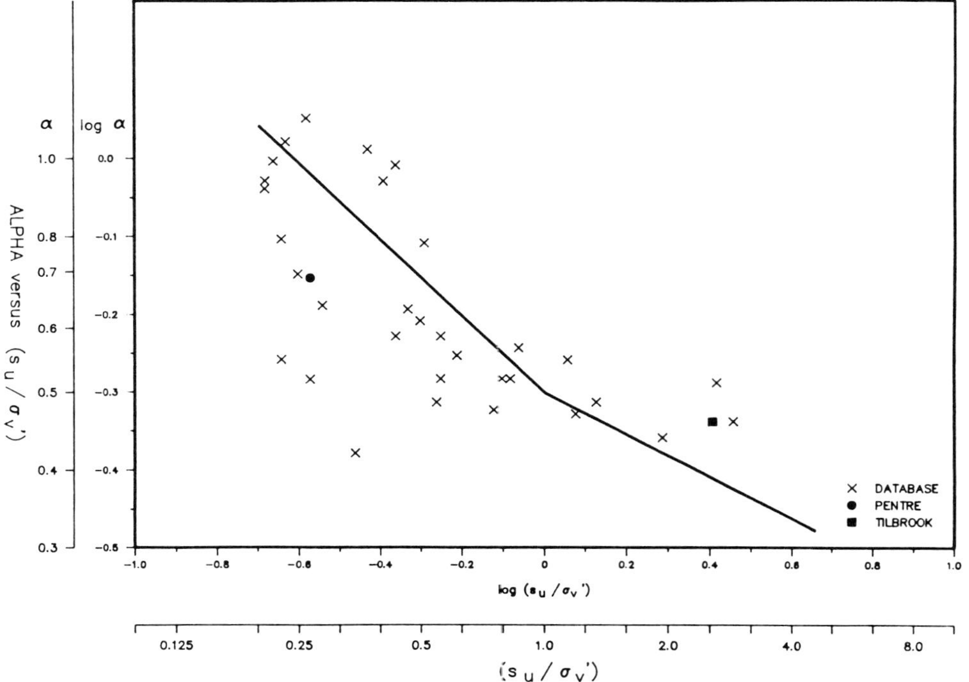

Fig. 6. Comparison of API 17th edition design criteria for clay with API database.

Soil Type	δ (°)	f_{max} (kPa)	N_q	q_{max} (MPa)
Clean sand	30	96	40	9.6
Silty sand	25	82	20	4.8
Sandy silt	20	67	12	2.9
Silt	15	48	8	1.9

The above are almost identical to the recommendations given by McClelland *et al.* [2].

4.2. API RP2A – 3rd Edition to 14th Edition

From 1972 onwards the same table of values was provided except that the recommended limit values were withdrawn and replaced by a comment that they should

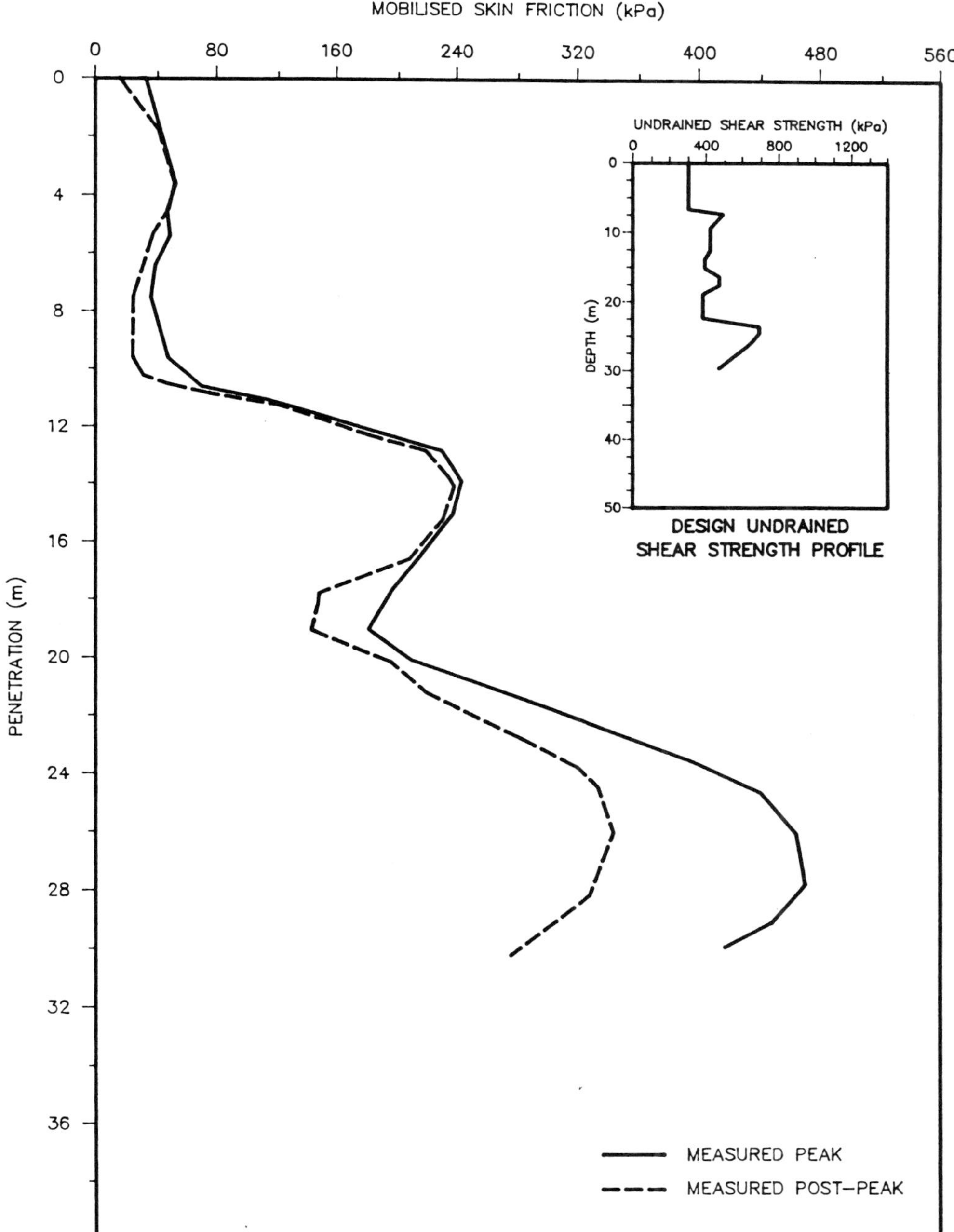

Fig. 7. Mobilised skin friction BP LDPT – Tilbrook.

be determined for local conditions. Also the recommendation that K should be 0.7 for compressive loads was changed to K being in the range 0.5 to 1.0 for compressive loads. The recommendation for tension remained unchanged.

The background to these changes is not clear, but may be associated with results presented in 1972 by McClelland at his Tergzaghi lecture [11], but not published until 1974. This disclosed the results of some new pile tests in silica sand and discussed problems associated with designing piles in carbonate sands.

In the 6th edition, January 1975, the problem of designing in carbonate soils was explicitly noted, but no quantitative guidance was provided.

A statement that adequate penetration into a sand layer had to be obtained to guarantee recommended values of N_q was included in the 9th edition of November 1977.

4.3. API RP2A – 15th Edition

Up until the publication of the 15th edition of RP2A in October 1984, the industry in general adopted the design criteria set out in the early editions, keeping to values of 0.7 and 0.5 for K for compression and tension respectively. Higher limit values of 15 MPa and 120 kPa were adopted for unit end bearing and shaft friction in soils where CPT data could be used to justify them. Based upon the trends within the industry and the result of studies performed by Denis and Olson [12] API RP2A introduced a new range of design characteristics and limit values, similar to those that had been removed in the 5th edition, but covering a wider range of soil types. These are presented in Table 2.

In addition to reintroducing limit values and increasing the range of soil types, the 15th edition also recommended that K in both tension and compression could be taken as 0.8. This obviously had a major impact on pile design, as up to the depth at which limit values were reached, unit frictions in tension and compression could be increased by 60% and 14% respectively, see Figure 8. Not all engineers adopted the new recommendations, and this led to some technical disputes. In the North Sea, designers continued to adopt the old criteria for coefficients of lateral earth pressure based upon the work of Toolan and Ims [13] who indicated that for piles in North Sea sands the revised API criteria could be non-conservative (see Figure 9).

To resolve the issue, various research programmes were started which related to adding to databases, calibrating existing design procedures to them and generating alternative methods. The first results of this work were included in the 19th edition in 1991 as a caution to users as to the reliability of existing procedures.

The 19th edition also included a commentary on carbonate soils.

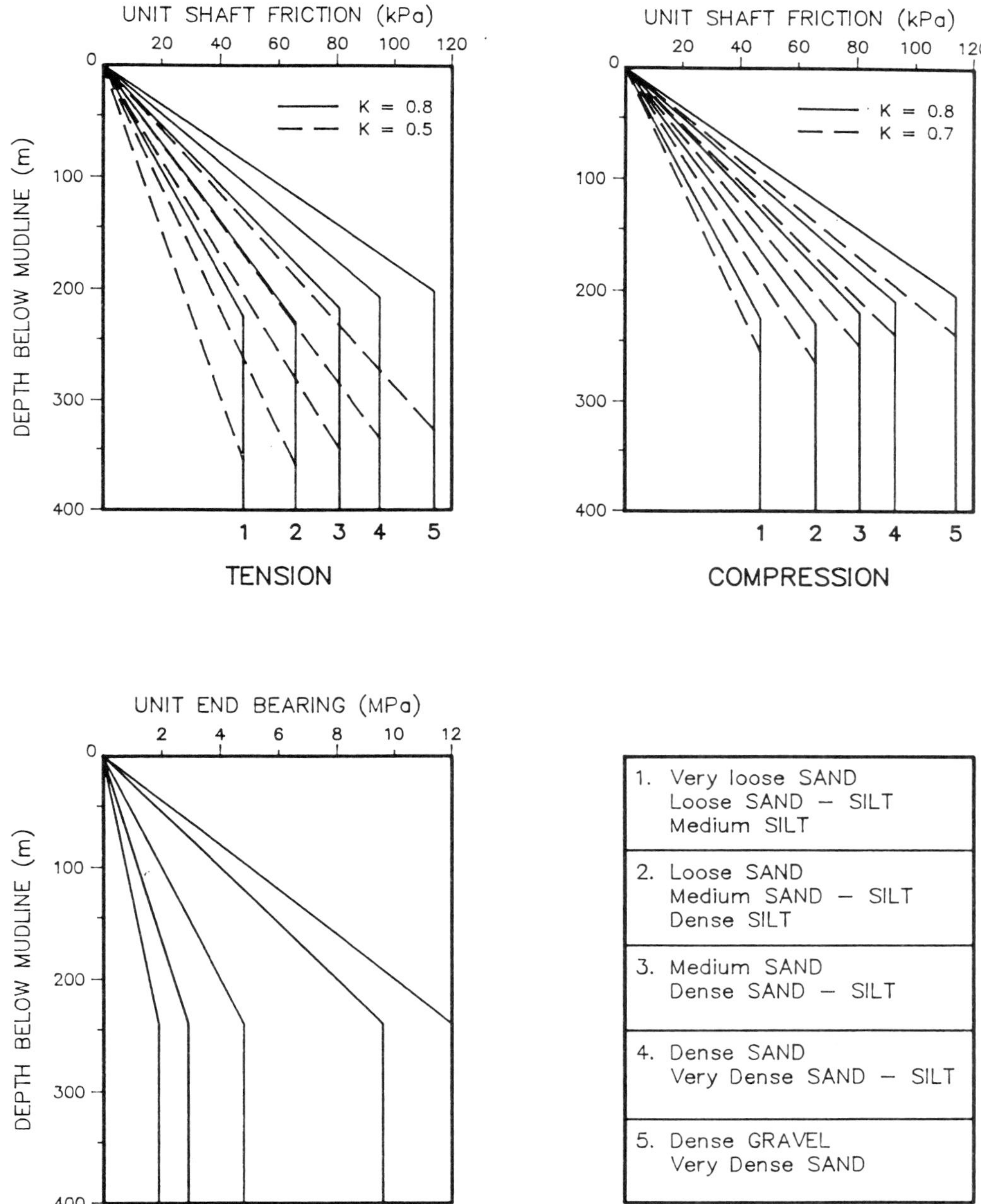

Fig. 8. API pile design criteria in sand comparison of limit values.

TABLE 2.

Density	Soil Type	δ (°)	f_{max} (kPa)	N_q	q_{max} (MPa)
Very Loose Loose Medium	Sand Sand-Silt Silt	15	48	8	1.9
Loose Medium Dense	Sand Sand-Silt Silt	20	67	12	2.9
Medium Dense	Sand Sand-Silt	25	81	20	4.8
Dense Very Dense	Sand Sand- Silt	30	96	40	9.6
Dense Very Dense	Gravel Sand	35	115	50	12.0

5. Future Trends in Codes

Norwegian practice offshore and general European practice onshore is to use codes that are based on factored loads and material properties rather than using overall simple factors of safety.

In December 1989, API published a draft edition of their Load and Resistance Factor Design (LRFD) version of API RP2A [14] as opposed to the standard Working Stress Design (WSD) version. It is the LRFD version which it is intended to incorporate into the international code. Since the factor proposed for environmental loading is 1.35 and the resistance factor for pile capacity is 0.8, the "factor of safety" against extreme conditions is effectively 1.69, ignoring gravity loads. This compares with the existing requirement under WSD of 1.5, and hence piles may be longer. Figure 10 shows a plot of the ratio of pile length computed by LRFD compared to the existing WSD requirement for various combinations of environmental to gravity loading in uniform sand and normally consolidated clay. The increase will tend to be greater for piles in sand rather than in clay because of the limiting values for friction and end bearing applied in sand. The situation is complicated by the different treatment of tension loading in the two versions of the codes and

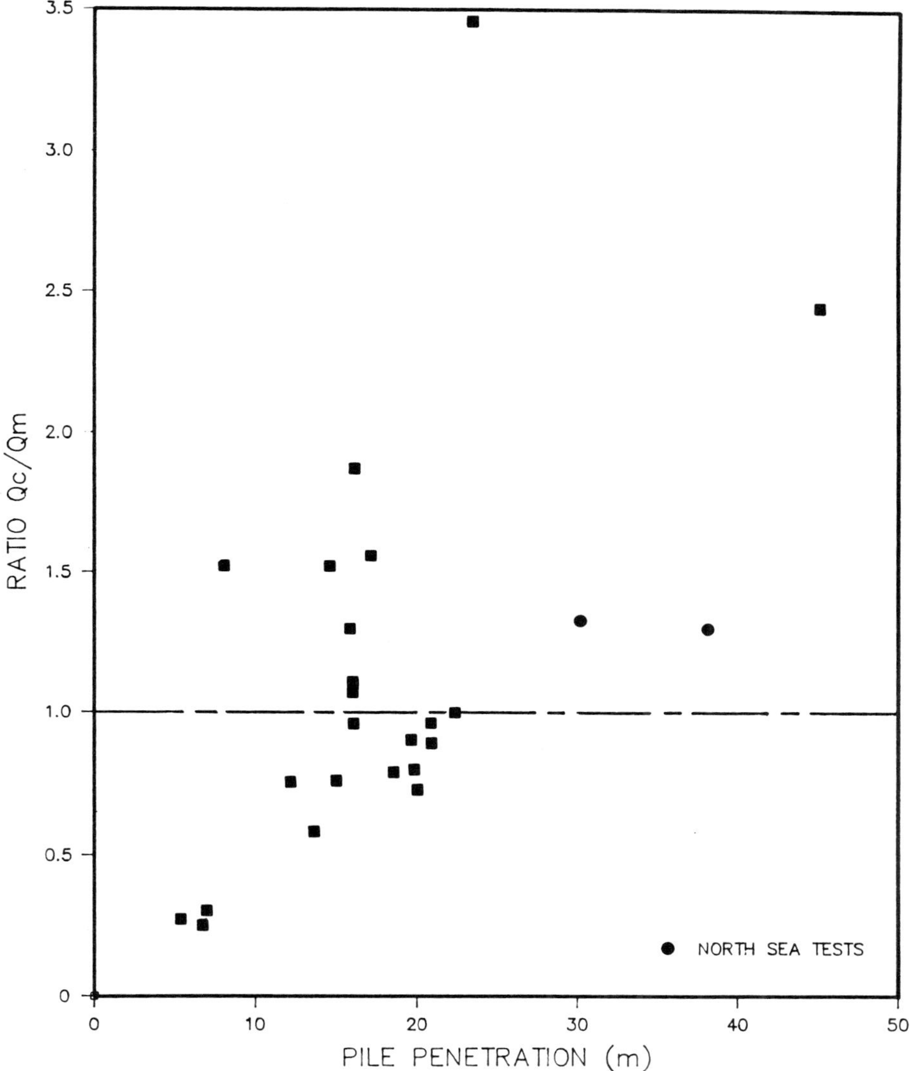

Fig. 9. Comparison of measured and calculated pile capacity based upon API 15th edition.

means that in uniform soils piles in sand could be up to 25% longer under LRFD than for WSD. For piles in variable strata the increases in length could be very much greater or less, depending on whether the soil below the tip of the pile is decreasing or increasing in strength. Similar but less divergent results would be obtained by following Norwegian practice.

One problem with the LRFD approach is that it tends to give lower "factors of safety" against gravity loads, because they are more certain, than against en-

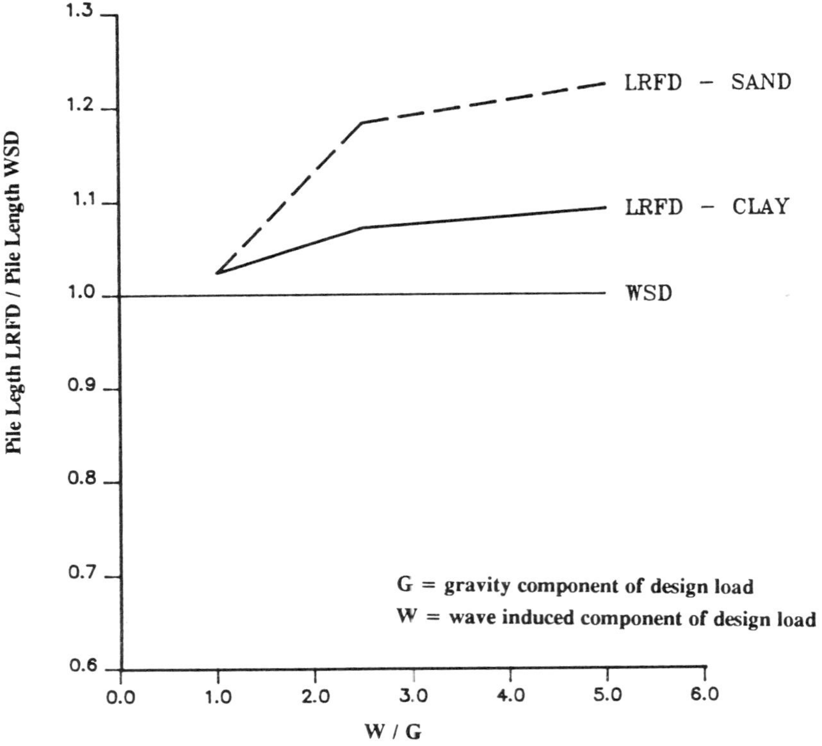

Fig. 10. Ratio of required pile length from LRFD to WSD.

vironmental loads. This appears logical to most engineers but flies in the face of experience for those charged with designing piles. This is because there is strong evidence that pile capacity under a load that increases and then decreases rapidly, such as that due to waves, is significantly higher than under a load that is built up and hen held constant, such as platform weight. This is called the rate effect and played a part in the reasoning in selecting factors of safety for piles of 2.0 for operational conditions and 1.5 for extreme environmental conditions in the WSD code [5].

In the LRFD code, for extreme conditions, different load factors are applied to the constant gravity load, variable gravity load and environmental load. However, a single resistance factor is applied to the pile capacity. There is probably scope for having a resistance factor which varies with the ratio of wave to gravity loads and with the degree of cyclic loading. This would have the advantage of providing foundations with a consistent level of reliability and provide vital cost savings for structures with relatively high environmental loads, such as lightweight jackets and deepwater platforms. It should be noted that for the case of gravity loads alone, a lower resistance factor is being recommended for pile design, and the proposal above is therefore merely a positive extension of the same philosophy.

The question as to whether the LRFD approach is an improvement on WSD is frequently asked. In the Authors' opinion the LRFD approach has the potential, so far untapped, of taking risk and reliability into account and in the longer term, with fine tuning, should result in lower cost for a given specified level of safety. In addition, it is more conductive to addressing questions such as whether the same level of reliability should be required for the foundation of a small unmanned platform as for a large drilling, production and accommodation platform.

The ongoing research into the reliability of pile design methods is likely to arouse interest once more into the use of both static design techniques and dynamic testing to determine a most probably capacity.

6. Future Trends in Pile Design in Clay

In the mid-1980's several large scale pile tests were performed in clay soils. The results have now been released into the public domain [10]. Many of the lessons learnt from these tests have already found their way into industry practice and subsequently it can be expected that they will be incorporated into codes. The first of these is likely to be new T–Z curves to replace those recommended by Coyle and Reese [15].

There is considerable interest in developing reliable design techniques for predicting the variation of pile capacity in the very short term and very long term. This is because (a) changes in installation practice mean that piles may have to carry maximum gravity load and possibly a significant environmental load a few days after driving and (b) old structures may have to be subject to safety case assessments for loadings developed by current design techniques which could be higher than originally envisaged.

7. Future Trends in Pile Design in Sand

API RP2A recommends that for open ended piles driven into a sand layer the end bearing be computed as the lesser of: the friction of the soil plug against the inside wall of the pile, or the end bearing of a closed ended pile of the same outer diameter. Frequently the closed ended pile capacity may be as much as twice that based on internal friction, see Figure 11. Many organisations have conducted tests to determine what length of plug of sand is required inside a pile to mobilise maximum end bear. The conclusion each time is that the required length is substantially shorter than would be conventionally calculated from "inside friction". Unfortunately, most of this research is confidential and has not yet been collected into a coherent package which would facilitate widespread acceptance of a new and more cost-effective design procedure. However, the Authors are certain this will come about with time.

Another problem is related to defining pile failure. The API database uses

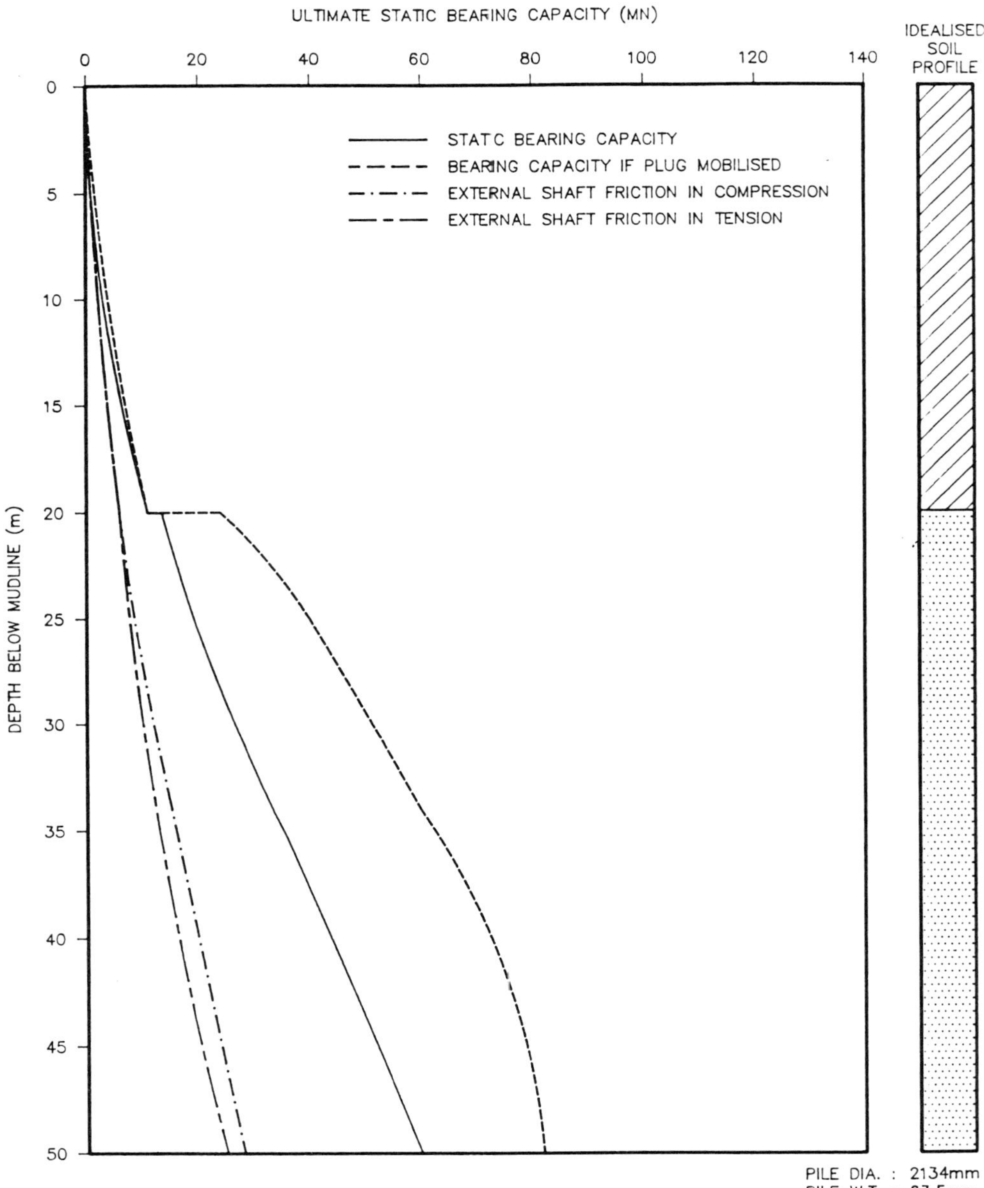

Fig. 11. Example of the influence of plugging on the ultimate static bearing capacity of piles end-bearing in sand.

Davisson's [16] criterion, which is deflection based, and for large diameter piles end bearing in sand such 'failure" will occur long before peak load is reached. This may be an unnecessarily conservative approach for platforms if they can tolerate deflection under transient loading. This could be relatively easily tackled in a revision to design procedures. Other uncertainties which may be resolved, related to the end bearing of piles in sand, are the application of limiting values and the rate of build up or decrease of end bearing as a pile enters or leaves a sand stratum.

It is possible that design procedures will become more explicit and use a quantitative rather than qualitative assessment of the relative density and elastic modulus of the sand. New site investigation techniques such as the cone pressuremeter, a re-evaluation of the existing pile test database and some additional pile tests should make this possible. The benefits of this would be more reliable and consistent designs and a reduction in cost for piles in dense and very dense sands. There would also be fewer surprises, which cost money, in loose and very silty sands.

8. Conclusions

The history of offshore pile design practice shows the willingness of the industry to respond to the needs for change – sometimes quickly, other times more slowly. The forthcoming international code should be treated as another opportunity for improvement. The codifying body must ensure that, in the future, it can issue revisions whenever circumstances demand, just as API has done in the past.

In comparing the situation now with that faced by McClelland *et al.* [2] in the 1960's, it can be concluded that for axial capacity in clay design reliability has greatly improved, but that unfortunately much less progress in design has been made for granular materials.

References

1. American Petroleum Institute, 'Recommended Practice for Planning, Designing and Constructing Fixed Offshore Platforms', API RP2A, 1st edition, October 1969.
2. McClelland, B., Focht, J. A., and Emrich, W. J., 'Problems in design and Installation of offshore piles', *Journal SMFE, ASCE* **95, SM6**, November 1969.
3. Fox, D. A., Parker, G. F., and Sutton, V. J. R., 'Pile driving in the North Sea boulder clays', *Proceedings 2nd Offshore Technology Conference*, Houston, 1970.
4. Vijayvergiya, V. N. and Focht, J. A., 'A new way to predict capacity of piles in clay', *Proceedings 4th Offshore Technology Conference*, Houston, 1972.
5. McClelland, B. and Cox, W. R., 'Performance of pile foundations for fixed offshore structures', *Proceedings of 1st International Conference on Behaviour of Offshore Structures*, Trondheim, 1976.
6. Cox, W. R., Kraft, L. M., and Verner, E. A., 'Axial load tests on 14 inch pipe in clay', *Proceedings 11th Offshore Technology Conference*, Houston, 1979.
7. Kraft, L. M., Focht, J. A., and Amerasinghe, F. S., 'Friction capacity of piles driven into clay', *Journal of the Geotechnical Division, ASCE*, November 1981.
8. Semple, R. M. and Rigden, W. J., 'Shaft capacity of driven pipe piles in clay', *ASCE Symposium on Analysis and Design of Pile Foundations*, October 1984.

9. Randolph, M. F. and Murphy, B. S., 'Shaft capacity of driven piles in clay', *Proceedings 17th Offshore Technology Conference*, Houston, 1985.
10. Proceedings, *Recent Large Scale Fully Instrumented Pile Tests in Clay*, Institution of Civil Engineers, London, 1992.
11. McClelland, B., 'Design of deep penetration piles for ocean structures', *Journal of the Geotechnical Engineering Division, ASCE*, July 1974.
12. Denis, N. D. and Olson, R. E., 'Axial capacity of steel pipe piles in sand', *Proceedings of Conference on Geotechnical Practice in Offshore Engineering*, ASCE, 1983.
13. Toolan, F. E. and Ims, B. W., 'Impact of recent changes to API recommended practice for offshore piles in sand and clays', *Journal of Underwater Technology* **14:1**, 1988.
14. American Petroleum Institute, 'Draft Recommended Practice for Planning, Designing and Constructing Fixed Offshore Platforms – Load and Resistance Factor Design', API RP2A-LRFD, 1st edition, December 1989.
15. Coyle, H. M. and Reese, L. C., 'Load transfer for axially loaded piles in clay', *Journal of the Geotechnical Engineering Division, ASCE* **SM2**, 1966.
16. Davisson, M. T., 'High capacity piles', Soil Mech. Div III Section, ASCE, Chicago, 1973.

Discussion

R. Hobbs, Lloyd's Register:

I have a few points of clarification:

1. In Figure 2 of the paper, the data plotted at undrained shear strengths of about 240 to 340 kPa refer to tests carried out on conductors at BP's West Sole C platform. Subsequent site investigation at the site, using improved techniques, resulted in higher S_u's being assigned to the clays and thus lower adhesion factors (Clarke *et al*, 1985);

2. The Empire pile load tests reported in reference (6), which showed high α values compared to S_u, were on relative short (12–15 m) embedded lengths of pile installed through sleeves from depths of 35 to 79 m. They therefore did not include "length effects";

3. In my organisation, the "K_p limit" for near surface clayey soils is used only with the API RP2A "method 2" and not with the 17th to 19th edition "main text" procedure as in the paper.

References

1. Clarke, J., Rigden, W. J., and Senner, D. W. F. (1985), 'Reinterpretation of the West Sole platform 'WC' pile load tests', *Géotechnique* **35**(4), 393–412.

CLOSING ADDRESS

J. B. BURLAND
Imperial College of Science, Technology and Medicine, South Kensington, London SW7, U.K.

The papers to this Conference cover a very wide range of topics and experience. It is clear that a lot of careful thought has been given to the choice of topics. To pick up a major theme of the Conference – the programme and content has been *carefully integrated* to give a coherent and exceptionally complete coverage of the whole field of offshore site investigation and foundation design, installation and performance. In view of this, to attempt to summarise the Conference would be an impossible task and would only diminish its value. All I can offer are a few very subjective and personal reflections.

The keynote address by Malcolm Birkinshaw is very important. He has clearly studied the proceedings of the 1979 and 1985 SUT Conference on Offshore Site Investigation and that in itself is significant and to be welcomed. He stressed the importance of communication, particularly in relation to risk. Communication requires a common language and it is necessary for us to be very precise in the terminology we use when discussing safety and risk. On the matter of Safety Assessment I believe that the HSE are taking a mature approach in seeking statements of those risks that have been identified. He offered to cooperate with the SUT Offshore Site Investigation and Geotechnics Committee. I believe that the Committee is well placed to consider uncertainties and risks associated with the investigation, design, construction, deployment and operation of offshore foundations. I hope that the Committee will accept the HSE offer as it has an important role to play in representing objectively all sectors of the industry.

Since the 1979 Conference there have been significant advances in a number of areas and a few of them are listed below:

1. In the field of geophysics, tremendous advances have been made in equipment, data acquisition and processing. There are a number of important papers illustrating these developments and some useful and stimulating discussions took place during the Conference.

2. *In situ* and laboratory testing techniques have improved immensely as is apparent from papers covering these topics.

3. The development of robust and precise geotechnical instruments have resulted

Volume 28: Offshore Site Investigation and Foundation Behaviour, 773–776, 1993.

in the measurement of the performance of the foundations of a number of offshore structures and are very important research on large fully instrumented piles.

4. There are not many papers on pure geology put it is evident that our understanding of geological processes, particularly during the Quaternary, has developed significantly.

5. There have also been very significant developments in our understanding of the properties of natural soils – particularly their stiffness characteristics at small and large strains. This understanding is crucial to the interpretation of *in situ* tests and the prediction of foundation performance.

In the papers describing specific site investigations there is clear recognition and acknowledgement of the importance of understanding geological processes and regional geology in planning an offshore facility. I hope that these papers will be used to convince clients of the value of a *planned, integrated* approach to site investigation using data from a variety of sources. This was certainly one of the objectives that Dennis Ardus and I had in setting up the SUT Site Investigation Committee in the mid seventies.

As a laymen in geophysics and geology, I particularly enjoyed the paper by Stoker *et al* on seismic facies analysis because, while acknowledging the value of high quality data, he also gives a timely reminder of how misleading it can be to read too much into the seismic records unless you have good geological control.

We all enjoyed Kerry Campbell's presentation, particularly the video illustrating the complex processes operating on a vast scale on the continental slopes of the Gulf of Mexico.

I wish to turn now to the papers dealing with the application of geophysical measurements as an aid to determining geotechnical properties. This is a very desirable goal and one that the SUT Site Investigation Committee encouraged strongly – certainly in the days that I was a member of it.

However there is a need for caution and a real danger of attempting too much. I see the shear wave velocity as a particularly important parameter. It is controlled largely by the shear modulus G_{max} which is a property of the soil skeleton at small strain, i.e. it is an effective stress parameter. Moreover, evidence from laboratory and field tests shows that the static small strain shear modulus is not significantly less than the dynamic shear modulus. The use of 'p' wave velocity is much more difficult as it is not directly linked with the stiffness of the soil skeleton being dependent on the pore fluid as well.

In my opinion we should distinguish clearly between those properties which *directly control* the acoustic response and those that can be *empirically correlated* with it. For example G_{max} (together with density) *directly controls* shear wave velocity. In this sense it is a 'prime' material parameter and it has a very clear meaning in any idealised elastic or visco-elastic model. But $G_{\max}$ depends on a large number of factors such as the effective stress, the properties of the grains, the micro- and macro-structure etc. Thus shear wave velocity can be expected to correlate with any of these in ways which will vary from sediment to sediment. These factors *indirectly* control acoustic response.

Thus we need to distinguish clearly and rigorously between parameters which *directly control* an acoustic property and factors which *correlate* with an acoustic property. If we are careful to do this then I believe that we can make real progress.

I note that in the GEOSIS study, correlations will be made with cone penetration data and with cores. It is important to appreciate that the interpretation of the cone in terms of basic geotechnical parameters is itself an empirical process. It is most important to tie back into the best possible testing of high quality samples using the most up-to- date methods of testing. I also welcome the initiatives to set up databases of geotechnical tests but the correlations should be carried out with care, knowledge and rigour.

Much of the final day was devoted to the design and performance of steel driven piles. The many papers on this topic constitute a useful summary of much recent field research on this topic. The discussions were detailed and sophisticated and may well have left some of the practitioners a little confused. At the risk of gross oversimplification I will summarise the key differences between the traditional API approach to the design of piles in clay and the approach via effective stress.

In the API approach the shaft friction τ_{sf} is related empirically to the undrained strength s_u. The successful application of a purely empirical approach requires that *all* the conditions of the empirical correlation be met. The difficulties with the API method stem from the fact that s_u is only indirectly related to τ_{sf}. Moreover s_u is not unique for a given soil but depends significantly on the method of test. There is therefore uncertainty about the appropriateness of the API database for many soils and there are very real difficulties in determining appropriate values of s_u.

Turning now to the effective stress approach, research has shown that τ_{sf} is governed by the Coulomb equation

$$\tau_{sf} = \sigma'_{rf} \cdot \tan \delta'$$

where σ'_{rf} is the radial effective stress normal to the shaft at failure and δ' is the angle of interface friction. Bond, Jardine and Lehane have expressed this equation as

$$\tau_{sf} = f_L \cdot K_c \cdot \sigma'_{vo} \cdot \tan \delta'$$

where f_L is a factor expressing the change of σ'_r during loading and K_c is the ratio between the radial and initial vertical effective stresses after installation and prior to loading.

The key points to note about the effective stress approach are:

1. τ_{sf} is expressed in terms of the basic geotechnical parameters controlling it.

2. It is clear that some empiricism will always be necessary in evaluating σ'_{rf} because of the complexity of the physical processes involved.

3. Empiricism applied to the fundamental controlling parameters is intrinsically more satisfactory than the traditional API approach – especially when dealing with unusual ground conditions.

We may be still a little way from routinely applying effective stress methods to pile design. However, the effective stress approach, coupled with field measurements on instrumented piles, has given us a much clearer idea of the *mechanisms of behaviour* during installation and loading. It is an appreciation of these factors which leads to good engineering.

The Conference contains some important papers describing the instrumentation and observations from full-scale structures. The innovative and economic foundation solutions that are being developed by the Norwegians are very impressive. I believe that many of these developments stem from the understanding and confidence the designers have gained from the extensive instrumentation of their structures.

We in the UK are just now beginning to see the benefits of instrumentation of piled foundations. In general, performance seems to be better than predicted. Whether we can achieve significant economies as a result of these measurements is too early to say. It is clear that an integrated assessment is needed involving many other disciplines – oceanographers, hydraulics engineers, structural engineers, etc. I trust that the Conference will help to convince clients and regulatory authorities of the benefits of instrumentation of offshore foundations.

Following the excellent presentation of innovative foundation systems by Hans Peter Christoffersen and Tor Inge Tjelta we had an interesting discussion on the application of recent research. The research helps to clarify the mechanisms of behaviour and this is very important to innovation. I believe that we should focus on the improvement of foundation systems as well as on refining the design of traditional systems.

In conclusion, this Conference has been quite exceptional both in its integrated coverage of offshore site investigation and foundation performance and also in the quality of the papers. I hope that the proceedings will come to be regarded as required reading by all those engaged in offshore geotechnical work.